The Dynamics of Heat

Springer
New York
Berlin
Heidelberg
Barcelona
Budapest
Hong Kong
London
Milan
Paris
Santa Clara
Singapore
Tokyo

Hans U. Fuchs

The Dynamics of Heat

With 229 Illustrations

Springer

Hans U. Fuchs
Department of Physics
Technikum Winterthur
CH-8401 Winterthur
Switzerland

Library of Congress Cataloging-in-Publication Data
Fuchs, Hans U.
The dynamics of heat / Hans U. Fuchs.
p. cm.
Includes bibliographical references and index.
ISBN 0-387-94603-9 (hardcover : alk. paper)
1. Thermodynamics I. Title.
QC311.F83 1996
536′.7—dc20 95-44883

Printed on acid-free paper.

Production managed by Frank Ganz; manufacturing supervised by Jeffrey Taub.
Photocomposed copy prepared from the author's files.
Printed and bound by R.R. Donnelley and Sons, Harrisonburg, VA.
Printed in the United States of America.

9 8 7 6 5 4 3 2

ISBN 0-387-94603-9 Springer-Verlag New York Berlin Heidelberg SPIN 10549064

To Robin and Kristin

Preface

The last few decades have seen the development of a general approach to thermodynamic theory. Continuum thermodynamics has demonstrated to us how we can build a theory of the dynamics of heat rather than of statics. In this book I would like to transfer what I have learned about the general theory to an introductory level and to applications in the sciences and engineering.

Two elements combine to make this presentation of thermodynamics distinct. First of all, taking as the foundation the fundamental ideas that have been developed in continuum thermodynamics allows one to combine the classical theory of thermodynamics and the theory of heat transfer into a single edifice. Second, didactic tools have been built that make it not just simple, but rather natural and inevitable to use entropy as the thermal quantity with which to start the exposition. The outcome is a course that is both fundamental and geared toward applications in engineering and the sciences.

In continuum physics an intuitive and unified view of physical processes has evolved: That it is the flow and the balance of certain physical quantities such as mass, momentum, and entropy which govern all interactions. The fundamental laws of balance must be accompanied by proper constitutive relations for the fluxes and other variables. Together, these laws make it possible to describe continuous processes occurring in space and time. The image developed here lends itself to a presentation of introductory material simple enough for the beginner while providing the foundations upon which advanced courses may be built in a straightforward manner. Entropy is understood as the everyday concept of heat, a concept that can be turned into a physical quantity comparable to electrical charge or momentum. With the recognition that heat (entropy) can be created, the law of balance of heat, i.e., the most general form of the second law of thermodynamics, is at the fingertips of the student.

The book contains two lines of development which you can either combine (by reading the chapters in the sequence presented) or read separately. In addition to the four chapters which represent the main line, you will find a Prologue, an Interlude, and an Epilogue which discuss some subjects at a somewhat higher level.

The four chapters that form the main body of the text grew out of my experience in teaching thermodynamics as a part of introductory physics, but represent an extension both in content and level of what I commonly include in those courses. The extension mostly concerns subject matter treated in courses on engineering thermodynamics and heat transfer and applications to solar energy engineering. Still, the chapters maintain the style of a first introduction to the subject. Previous knowledge of thermal physics is not required, but you should be familiar with basic electricity, mechanics, and chemistry, as they are taught in introductory college courses. With the exception of one or two subjects, only a modest amount of calculus is used. Chapter 1 provides an introduction to basic quantities, concepts, and laws. Entropy is introduced as the quantity which is responsible for making bodies warm or for letting ice melt, and the law of balance of entropy is formulated directly on the basis of ideas taken from everyday images of heat. The relation between currents of heat (entropy) and currents of energy is motivated along the lines of Carnot's theory of heat engines, yielding a law which makes the development of thermodynamics rather simple. (The relation is proved later on the basis of some alternative assumptions in the Interlude.) Then, some simple applications which do not rely too heavily upon particular constitutive relations are developed. First among them is a treatment of irreversibility and the loss of power in thermal engines, a subject which teaches us about the importance of the rule of minimal production of heat. Chapters 2, 3, and 4 furnish introductions to constitutive theories. The first of these deals with uniform bodies, which respond to heating by changing mechanical or other variables. A simple version of the constitutive theory of the ideal gas is developed, which leads to a theory of the thermodynamics of ideal fluids. In addition, blackbody radiation and magnetic bodies are treated. A short exposition of the concepts of thermostatics exposes the reader to the difference between dynamics and statics in the field of thermal physics. Chapter 3 deals with theories of heat transfer excluding convection. The general form of the equation of balance of entropy for bodies and control systems is given and applied to various cases. Production rates of heat in conduction and radiation are calculated and applied, among others, to the computation of the maximum power of solar thermal engines. In this chapter, continuous processes are treated for the first time in the context of one-dimensional conduction of heat. The radiation field and the issue of the entropy of radiation are discussed extensively, and a section on solar radiation concludes this Chapter. Chapter 4 extends the theory of heat to processes involving the change and the transport of substances. Subjects such as chemical reactions, phase changes, and convection, and applications to power engineering and to heat exchangers form the body of this Chapter. All of these Chapters include a large number of solved examples in the text.

The second track of the book treats thermodynamics in a more advanced and formal manner. The Prologue provides a brief view of a unified approach to classical physics. Except for the first section, which you definitely should read before starting with Chapters 1 - 4, the Prologue presents several subjects of physics at a relatively quick pace, demonstrating the unified approach to dynamical processes which forms the backbone of the entire book. (The concepts are introduced at a more leisurely pace in the main chapters on thermodynamics.) If you wish, you can then try to read the Interlude which introduces the subject of the thermodynamics of uniform fluids on the basis of the caloric theory of heat. This Chapter repeats the subject of part of Chapter 1 and most of Chapter 2 at a higher mathematical level. In contrast to those chapters, the Interlude also provides a first proof of the relation between currents of entropy and of energy, which shows that the ideal gas temperature can be taken as the thermal potential. Finally, the Epilogue takes the first simple steps into the field of continuum thermodynamics, exposing you to the ideas behind the more advanced subjects which have been the focus of development over the last few decades.

If I seem to succeed in introducing you to an exciting new view of a classical subject, the individuals actually responsible for this achievement are the researchers who have developed this field. Carnot, who gave us an image of how heat works in engines, a view which I have taken as the starting point of my exposition. Gibbs demonstrated how to deal with chemical change and heat. Planck's theory of heat radiation still is one of the clearest expositions of the thermodynamics of radiation. Also, there are the researchers who have built continuum thermodynamics, mainly since the 1960s and who have contributed so much toward clarifying the foundations of the dynamics of heat. They deserve our respect for one of the most fascinating intellectual endeavors.

When it comes to applications we nowadays can turn to computational tools which can make life so much easier. Two such tools which I have used deserve to be mentioned—the system dynamics program Stella (High Performance Systems, Inc., Hanover, New Hampshire), and the program EES (Engineering Equation Solver; Klein, 1991) which provides for extensive thermophysical functions in addition to a solver for nonlinear equations and initial value problems. Also, in the fields of engineering applications, including solar engineering, I have been inspired by such excellent textbooks as those of Bejan (1988), Moran and Shapiro (1992), Rabl (1985), and Duffie and Beckman (1991).

I am grateful to all my friends, colleagues, and teachers who, through their encouragement and support, have contributed toward the writing of this book. Robert Resnick and Roland Lichtenstein of RPI gave me the courage to take up the project. Walter Cohen, Werner Maurer, and Martin Simon read the book and gave me valuable feedback. Heinz Juzi, Heinz Winzeler, and Klaus Wüthrich helped me with discussions of applications, and many more colleagues gave me kind words of encouragement. Most important, however, has been Werner Maurer's friendship and professional companionship in this endeavor. He and I developed the system dynamics approach to the teaching of physics which you will find in this book.

I would like to acknowledge generous grants made available by the Federal Government of Switzerland and my school, which allowed for the development of labs and courses dealing with renewable energy sources, and I would like to thank my thesis students whose work in solar energy engineering has led to many interesting applications included here.

Finally, let me express my gratitude toward all those at Springer-Verlag, who have made the production of the book possible. Thomas von Foerster, Frank Ganz, and Margaret Marynowski turned the manuscript of an amateur madly hacking away on a Macintosh into a professional product. They were very supportive and encouraging, always with an open mind for my wishes.

This has been a long journey. My wife and my daughter have gone through it with me all the way. I would like to thank them for their love and their patience. When my daughter was very little, she asked me if I would dedicate this book to her. I hope it has been worth waiting for.

Honolulu, 1995 *Hans Fuchs*

Table of Contents

PROLOGUE **A Unified View of Physical Processes 1**

P.1 The Flow of Water and Charge 2
P.2 Transport Processes and Laws of Balance 12
P.3 The Properties of Bodies 20
P.4 Energy and Physical Processes 28
P.5 Continuum Physics, System Dynamics, and the Teaching of Physics 38
Questions and Problems 42

CHAPTER 1 **Hotness, Heat, and Energy 51**

1.1 Thermal Phenomena, Concepts, and Images 51
1.2 Temperature and Thermometry 66
1.3 Some Simple Cases of Heating 73
1.4 Engines, Thermal Power, and the Exchange of Heat 82
1.5 The Production of Heat 101
1.6 The Balance of Entropy 109
1.7 Dissipation and the Production of Entropy 119
Questions and Problems 144

CHAPTER 2 **The Response of Uniform Bodies to Heating 153**

2.1 The Model of Uniform Processes 153
2.2 The Heating of Solids and Liquids: Entropy Capacity 157
2.3 The Heating of the Ideal Gas 170
2.4 Adiabatic Processes and the Entropy Capacities of the Ideal Gas 181
2.5 Some Applications of the Thermomechanics of the Ideal Gas 197
2.6 Black Body Radiation as a Simple Fluid 207
2.7 The Coupling of Magnetic and Thermal Processes 215
2.8 The General Law of Balance of Energy 220
2.9 Thermostatics: Equilibrium and Changes of State 229
Questions and Problems 237

INTERLUDE Heat Engines and the Caloric Theory of Heat 243

I.1 Thermal Equations of State 244
I.2 A Theory of Heat for Ideal Fluids 250
I.3 Interaction of Heat and Motion: Carnot's Axiom 267
I.4 Internal Energy and Thermodynamic Potentials 285
I.5 Caloric and Mechanical Theories of Heat 289
Questions and Problems 296

CHAPTER 3 The Transport of Heat 301

3.1 Transport Processes and the Balance of Entropy 302
3.2 Some Simple Applications of the Flow of Heat 312
3.3 Heat Transfer and Entropy Production 343
3.4 The Balance of Entropy and Energy in Conduction 357
3.5 Radiative Transport of Heat 378
3.6 Solar Radiation 415
Questions and Problems 439

CHAPTER 4 Heat and the Transformation and Transport of Substances 447

4.1 The Concept of Amount of Substance 448
4.2 Chemical Reactions and the Chemical Potential 461
4.3 Phase Changes, Solutions, and Mixtures of Fluids 484
4.4 Flow Processes and the Chemical Potential 507
4.5 Vapor Power and Refrigeration Cycles 536
4.6 Applications of Convective Heat Transfer 554
Questions and Problems 583

EPILOGUE Steps Toward Continuum Thermodynamics 591

E.1 Thermodynamics of Uniform Fluids 592
E.2 Equations of Balance for Continuous Processes 611
E.3 The Energy Principle 624
E.4 Thermodynamics of Viscous Heat-Conducting Fluids 629

E.5 Inductive Thermal Behavior: Extended Irreversible Thermodynamics 648
E.6 The Lessons of Continuum Thermodynamics 654
Questions and Problems 655

APPENDIX **Tables, Symbols, Glossary, and References 657**

A.1 Tables of Thermodynamic Properties 657
A.2 Symbols used in the text 675
A.3 Glossary 687
A.4 References 691

Subject Index 697

PROLOGUE

A Unified View of Physical Processes

πavτα ρει

Heraclides (550-480)

Everything flows. Water and air flow on the surface of the Earth, where they create the multitude of phenomena we know from everyday life. Winds can impart their motion to the water of the oceans, and in a far-away place, this motion can be picked up again through the action of the waves. These processes are maintained by the radiation pouring out from the surface of the sun; light flows from there through space, and some of it is intercepted and absorbed by our planet. Both in nature and in machines, heat is produced and transported from place to place. In electrical machines, we make electricity flow in an imitation of its flow in the atmosphere, and in reactors, chemical substances flow while at the same time undergoing change. Today, we even see life as governed by flow processes.

We shall take this observation as the starting point of our investigation of natural and manmade phenomena. It leads to one of the most general description of nature we know today. There are a few physical quantities which can flow into and out of systems, which can be absorbed and emitted, and which can be produced and destroyed. Electrical charge is transported in electrical processes, and mass and substance flow in gravitational and chemical phenomena, respectively. In continuum mechanics, motion is seen as the exchange of linear and angular momentum. Thermal physics is the science of the transport and the production of heat. One of the great advantages of this description of nature is that it relates the different phenomena, which leads to an economical and unified view of physical processes. It turns out that classical continuum physics is a precise method of expressing this point of view for macroscopic systems (see Section P.5).

In this chapter, we shall present some examples of introductory physics, most of which you should be familiar with. We shall use as the main tool the images and the language found in continuum physics. In this way, we hope to prepare the ground for the approach to thermodynamics which you will find in this book. Note, however, that this chapter is a condensed overview, not a text. After reading Section P.1 you may want to venture directly into the main body of the book, in which case you might wish to return to this chapter later on. Either way, we believe you will find it advantageous to draw comparisons between different fields of physics as often as possible during your journey through thermodynamics.

What is this unified approach to physics? In short, it is based on an analogy with continuum physics. First, we have to agree on which physical quantities we are going to use as the fundamental or *primitive* ones; on their basis other quantities are defined, and laws are expressed with their help. Second, there are the fundamental *laws of balance* of the quantities which are exchanged in processes, such as momentum, charge, or amount of substance; we call these quantities *substancelike*. Third, we need particular laws governing the behavior of, or distinguishing between, different bodies; these laws are called *constitutive relations*. Last but not least, we need a means of relating different types of physical phenomena. The tool which permits us to do this is energy. We use the *energy principle*, i.e., the law which expresses our belief that there is a conserved quantity which appears in all phenomena, and which has a particular relationship with each of the types of processes.

To introduce the elements of theories listed above, we shall begin with a comparison of the flow of water and of electrical charge.

P.1 The Flow of Water and Charge

We all are familiar with the flow of water in simple settings, such as the filling or the discharging of containers through pipes (Figure 1). By looking at a special example we will be able to identify the elements of a physical theory which allow us to calculate such things as the current of water through a pipe, the pressure at various points in the fluid, and the time required to discharge a container. The analysis also will tell us that the system and the processes it may undergo are very similar to what we know from electricity. By comparing hydraulic and electrical systems we shall learn about the power of analogies between different fields of physics.

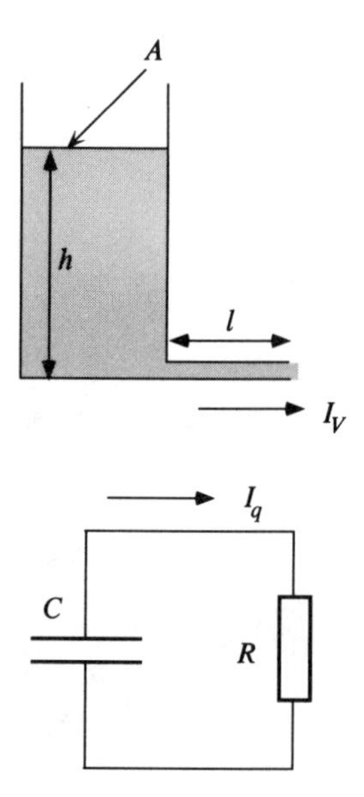

FIGURE 1. A simple container is filled with water up to a certain level. It can be discharged through a long thin pipe fitted at the bottom. This system and the processes it undergoes are comparable to a simple electrical circuit in which a capacitor is discharged through a resistor.

P.1.1 Physical Quantities

Our first question must be which physical quantities we can use as basis for a quantitative description of the flow of water into and out of containers. We shall have to do the same for the electrical system. The choice of fundamental or primitive quantities is not unique. We simply have to begin somewhere, in some way.

We certainly need a measure of the amount of water in a container. There are several possible choices. The simplest of these is the *volume* of the water. Another that comes to mind quickly is the *mass* of the water. Finally, chemists might be inclined to measure the amount of water on the basis of its *amount of substance* (Section P.2.7).

The last of the three measures is the most natural in the case envisioned. If we ask how much "stuff" is in a container, the amount of substance is preferable over the others. Mass (gravitational mass) is the measure of the property of bodies which gives rise to their weight. The volume, on the other hand, is a geometric measure. Therefore we ought to choose the amount of substance as our first primitive quantity. However, as long as there are no chemical reactions there is a simple and direct relationship between a body's amount of substance and its mass that allows us to choose the latter quantity instead of the former (Section P.2.7). Also, if we assume water to be incompressible, we can as well employ its volume as the measure we have been seeking. For this reason we shall express most of what follows in terms of the volume of a body of water. Still, it is important to be aware of the difference between the three quantities introduced so far. Volume is used as a convenient substitute for a body's amount of substance.

The case of electricity is very similar. The physical quantity which measures an amount of electricity is well known: it is the *electrical charge*. A capacitor stores a certain amount of charge, just as a container stores a certain volume of water.

Amount of substance and volume, as well as gravitational mass and electrical charge, are quantities which have a particular property in common. They scale with the system they describe. If a homogeneous body is divided into two equal pieces, each part "contains" half of the original quantity. For this reason these quantities are said to be *extensive*.

If we want to set up a theory of the flow of water we need a primitive quantity which describes its transport. For this purpose we conceive of the rate of flow of water, measured in terms of a new quantity which we call the *flux* of water. This quantity is measured, for example, at the outlet of the pipe shown in Figure 1. The rate at which water flows can be expressed in terms of the *volume flux* or *current of volume*, i.e., the volume of water flowing past a measuring device per time. Alternatively, we may employ the flux of mass or the flux of amount of substance. Again, for practical purposes, we shall choose the first of these measures for most of the following development. In electricity, the quantity analogous to volume flux is the *current of charge*.

Using volume and volume flux we are able to say something about an amount of water, namely the amount of water stored in a system, and the rate at which water is flowing. These quantities, however, do not suffice for a complete theory of the phenomena associated with containers of water and currents flowing in and out. They do not tell us anything about why water should be flowing at all. In electrical circuits as well, we need a quantity which is responsible for setting up currents of charge in the first place.

In addition to quantities measuring amounts of water, we need the *pressure* of the water to explain its flow through pipes. If the same rate of flow is to be sustained through two pipes of different diameters, the pressure difference between the inlets and the outlets of the pipes must be different. Different voltages, i.e. different differences of *electrical potential*, are required if the same electrical current is to pass through two different resistors. These examples demonstrate the nature of pressure and of electrical potential: they are quantities measuring an intensity rather than an amount of something. For this reason they are called *intensive* quantities. In contrast to the extensive quantities, intensive ones do not scale with the size of the system. If a body is divided into two parts, the intensive quantities are the same in both. Summing up, we may say:

To quantify physical phenomena we have to introduce measures of amounts and of intensities.

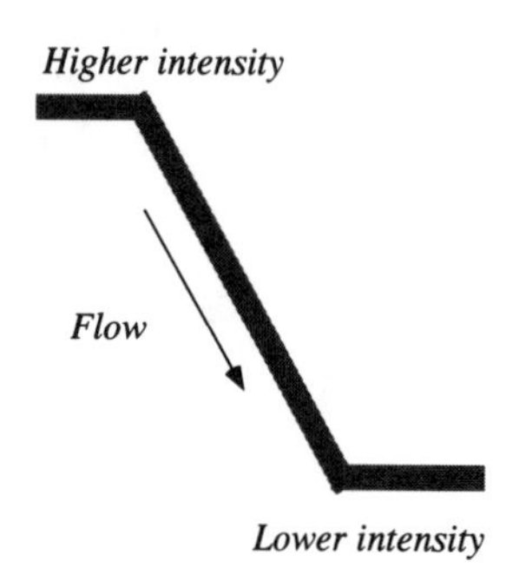

FIGURE 2. Both water in pipes and electrical charge in electrical circuits may be imagined to flow (by themselves) from regions of higher intensities of the potential associated with the phenomenon to regions of lower intensities. Graphically speaking, they flow from regions at high levels to regions at lower levels. This type of diagram is called a waterfall representation of physical processes.

There is a very useful image which may be associated with intensive physical quantities. Since a difference of pressure or of electrical potential is necessary for water or charge to flow we may call such a difference a *driving force* for the process. In the flow of water on the surface of the Earth, a difference of levels commonly is the origin of the transport phenomena. Water falls from higher to lower levels by itself (Figure 2). The comparison of the flow of water or charge with waterfalls suggests that an intensive quantity may be imagined to be a *level*. Pressure therefore takes the role of the hydraulic level, while the electrical potential is visualized as the electrical level.

Note that hydraulics and electricity demonstrate a high degree of similarity — at least in the basic quantities employed. Naturally, only hindsight can tell us if we have chosen the right quantities as the fundamental ones for a given range of phenomena. This means that we have to accept a certain choice, define new quantities on its basis, build a theory, and work out its consequences. If we are satisfied with the results compared to what nature is demonstrating to us, we call the theory a successful one.

P.1.2 Accounting

We want to know how much water is in a container at a given time. Alternatively, we want to say something about how the amount of water can change. For this purpose we shall conceive of a number of quantities which we will define on the basis of the fundamental ones introduced above. The first of these derived variables is the rate of change of the amount of water. Mathematics tells us how to define such a quantity. If we measure the amount of water in a container by its volume V, the rate of change of the volume is the time derivative dV/dt.

There is a simple but fundamental law which allows us to account for amounts of water. If the volume of water is a conserved quantity (as should be the case for incompressible fluids) the volume stored in a given system can change only due to the transport of water across the surface of the system (Figure 3). There must be currents or fluxes of volume with respect to the system, and they alone are responsible for the change of the contents of the system. This case is so simple that we can state right away the following law of balance of volume:

The rate of change of the volume of water in the system must be equal to the sum of all fluxes associated with the currents of water crossing the surface.

If we count the fluxes associated with currents leaving a system as positive quantities the law can be stated formally as follows:

$$\frac{dV}{dt} = -I_{V,net} \tag{1}$$

We shall use the symbol I for fluxes. Since there will be many fluxes for different physical phenomena, indices will be used to distinguish between them. Here, the index V stands for volume. The net flux simply is the sum of all fluxes occurring with respect to a system. Note that the quantity we call *flux* has the dimensions of the quantity which is flowing divided by time. (Often you will find in the literature the term *flux* associated with what we will call *flux density*, namely the rate of flow divided by the surface area through which the current is flowing. Here, we follow the tradition of electromagnetism, where we speak of electric or magnetic fluxes as the surface integrals of the flux densities, which are the quantities **E** or **B**.)

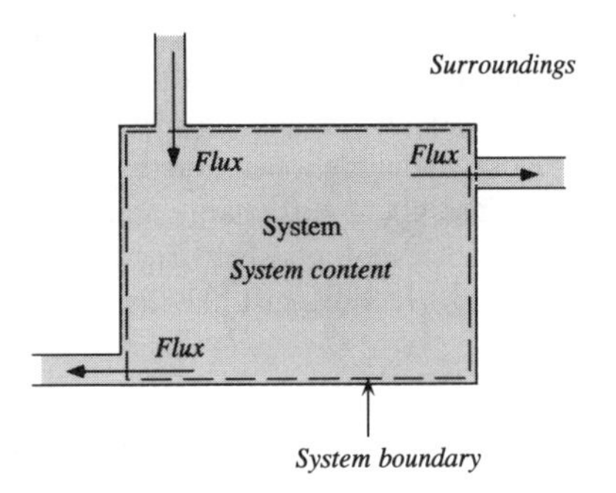

FIGURE 3. A system is a region of space occupied by a physical object. It is separated from the surroundings by its surface. Any physical quantity which we imagine to be stored in the system can change as a consequence of transport across its boundary. The transport is described in terms of currents, and fluxes measure the strength of the currents. Currents leaving a system are given positive fluxes. In some cases transport across the surface may be the only means of changing the contents.

Equation (1) is a fundamental law, namely the formal expression of our assumption that the volume of water is a conserved quantity. It is not a definition of the currents or fluxes of volume. The quantities occurring in Equation (1) are fundamentally different, related only by an interesting property of water. This property serves as one of the basic laws upon which we are going to build the following theory.

If we could no longer assume the volume of water to be a conserved quantity we would have to change the law of balance. We would be forced to account for other means of changing the volume, by introducing other terms in Equation (1). For now, let us assume that this is not necessary.

Often we are interested in the overall change of the volume of water in a container as the result of a process, not just in the rate of change. For this purpose we define two more physical quantities. The first is the *change of volume*, which is simply given by the integral over time of the rate of change of volume:

$$\Delta V = \int_{t_1}^{t_2} \dot{V} dt \tag{2}$$

The second quantity is the measure of how much water has flowed across the surface of the system in a given time. We shall call this quantity the *volume exchanged* in a process. It is defined as the (negative) integral over time of the fluxes of volume:

$$V_{e,net} = -\int_{t_1}^{t_2} I_{V,net}\,dt \quad (3)$$

These two quantities allow us to express the law of balance of volume given in Equation (1) in the following form:

$$\Delta V = V_{e,net} \quad (4)$$

Equation (4) again is a law of nature, while Equations (2) and (3) represent definitions. The law of balance expresses the simple fact that if volume is conserved, a change in the volume of fluid must be equal to the total volume which has crossed the surface of the system.

The laws and definitions are identical in the case of electricity. Since charge is a strictly conserved quantity, the rate of change of the charge of a body must be equal to the sum of all fluxes of charge with respect to this body. In other words, the law of balance of electrical charge looks exactly like Equation (1), with volume replaced by charge. The change of charge and the amount of charge exchanged in a process are defined analogously to Equations (2) and (3). Again, the law of balance in integrated form looks just like the expression in Equation (4).

P.1.3 Constitutive Laws: Resistance, Capacitance, and Inductance

The fundamental law of balance of the quantity we have chosen to represent the amount of water is not of much use by itself. Inspect it, and you will see that we can compute the rate of change of the volume of water in a container only if we have independent information regarding the fluxes. For this reason, a second class of fundamental laws—needed in a physical theory—are laws determining currents in given situations. Clearly such laws depend upon the particular circumstances. Therefore they are called *material laws* or *constitutive laws.*

In our example this means that we will have to state a law governing the flux of water through the particular pipe attached to the container (Figure 1). While Equation (1) is valid as long as the volume of water is a conserved quantity, the law for the flow of water through a pipe depends on many special material properties, and on special circumstances. The same is true for electrical currents as well.

Let us just state an example of a law for currents of water through pipes. Experience tells us that usually we have to force water through a pipe. In other words, there exists a *resistance* to the flow, and we need a driving force to maintain a current. If the fluid is considered to be viscous, and if the flow is laminar, the volume flux through a pipe is governed by the law of Hagen and Poiseuille:

$$I_V = -\frac{\pi r^4}{8\mu l}\Delta P \tag{5}$$

The flux depends linearly upon the difference between the pressures at the inlet and the outlet of the pipe. The factor multiplying the pressure difference depends upon the length l and the radius r of the pipe, and upon the viscosity μ of water. By the way, Equation (5) very much resembles Ohm's law in electricity. This type of relation is found for a number of dissipative constitutive laws, including diffusion and the conduction of heat.[1]

The inverse of the factor multiplying the potential difference is called the *(hydraulic) resistance*:

$$R_V = \frac{8\mu l}{\pi r^4} \tag{6}$$

In terms of this definition Hagen and Poiseuille's law may be expressed in the following simple and intuitive form:

$$I_V = -\frac{\Delta P}{R_V} \tag{7}$$

Constitutive laws specifying currents have to do with transport phenomena. Obviously, we also need a means of saying something about the process of storing water (or electrical charge). It is customary to introduce a quantity which expresses the relationship between a change of the amount of water contained in a system and the change of the associated potential, i.e., the change of pressure. This quantity is constitutive as well, since it depends on the type of system containing the water. It allows us to relate the change of system content to the possibly more easily measured potential. In electricity we are interested in the relationship between the charge contained in a system and the voltage.

Let us now turn to the constitutive quantity used to describe the storage of water in containers. The amount of water depends on the surface level h and the size of the container. For this reason we define the *capacitance* of the storage system: the capacitance is a quantity which indirectly determines the volume of water in the container. To be precise it is the quantity which tells us by how much the content increases if we increase the intensive quantity by one unit. In other words, the capacitance of a container shall be defined as

$$K = \frac{dV}{dP} \tag{8}$$

---

1. For a derivation of the law of Hagen and Poiseuille, and for a comparison of conductive transports of momentum, heat, and mass, see Bird, Stewart, and Lightfoot (1960) or any other book on transport phenomena.

In the case of a container having a uniform cross section it turns out to be

$$K = \frac{\Delta V}{\Delta P} = \frac{Ah}{\rho g h} = \frac{A}{\rho g} \tag{9}$$

It should not come as a surprise that during the process of discharging the water level in the container obeys the same type of equation that describes the discharging of capacitors. By combining the law of balance, Equation (1), and the constitutive laws, Equations (5) and (9), and integrating the resulting differential equation, we obtain:

$$h(t) = h_o \exp\left(-\frac{\pi \rho g r^4}{8 \mu l A} t \right) \tag{10}$$

Here A is the surface area of the container depicted in Figure 1. Together with the resistance introduced above the factor multiplying the variable τ in Equation (10) is the inverse of the product of capacitance and the resistance. The factor $\tau = R \cdot K$ is called the *time constant* of the system consisting of the capacitor (container) and the resistor (pipe). The function describing the discharging of the container takes the form:

$$h(t) = h_o e^{-t/\tau} \tag{11}$$

This is perfectly analogous to what we know from electricity. Actually, the comparison does not stop here. In electrical circuits, currents are determined not only by the resistance of conductors but also by the inductance of electrical elements. In other words, the constitutive law for a current of charge is not just Ohm's law. The potential difference across an element depends not only on the current via the resistance, but also on the rate of change of the current. This phenomenon is known as *induction.*

Consider an open electrical circuit or a container and pipe with a lid over the end of the pipe (Figure 1). If we suddenly close the circuit or open the end of the pipe the currents of charge or water should instantaneously attain a value of U/R or $\Delta P/R_V$, if we calculate processes according to Ohm's law or its equivalent in hydraulics. Naturally this does not happen. Rather, the currents increase from a value of zero. In electricity, the reason can be found in the fact that the magnetic field associated with the current has to be built up, while in hydraulics the water has to be accelerated. Consider the latter case right at the moment when the pipe is opened. Since there is no current yet, there is no momentum transport as a result of friction (Section P.3.4). Rather, the momentum is stored in the water in the pipe, leading to its acceleration. Newton's law (i.e., the balance of momentum; see Section P.2) leads to

$$-A_p(P_2 - P_1) = A_p l \rho \frac{dv}{dt} \tag{12}$$

where A_p is the cross section of the pipe. Note that we have assumed the flow speed to be uniform over the entire cross section of the pipe. Therefore, the pressure difference may be expressed in terms of the rate of change of the volume flux $I_V = A_p v$:

$$\Delta P = -\frac{l\rho}{A_p}\frac{dI_V}{dt} \tag{13}$$

In electricity we find an analogous expression which relates the rate of change of the electric current to the induced potential difference:

$$\Delta\varphi_{el} = -L\frac{dI_q}{dt} \tag{14}$$

Comparison of Equations (13) and (14) suggests that the factor multiplying the rate of change of volume flux is a *hydraulic inductance*:

$$L_V = \frac{l\rho}{A_p} \tag{15}$$

The physical phenomena clearly are comparable. As in the case of electric currents, decreasing a current of water induces a positive potential difference which tends to oppose the change of volume flux. While energy is stored in the magnetic field associated with a current of charge, energy is stored in the flowing water (kinetic energy). The form of the relationship between energy, current, and inductance is the same in hydraulics and electricity.

Normally, dissipative behavior and the phenomenon of induction are combined in a single system. If this is the case, the potential difference across a length of pipe or wire is the sum of the potential differences due to resistance and inductance. We may solve this relation for the current to obtain the following constitutive law:

$$I_V = -\frac{1}{R_V}\Delta P - \frac{L_V}{R_V}\frac{dI_V}{dt} \tag{16}$$

We have written the relation for the hydraulic case, but you may translate it to fit electric phenomena by simply substituting charge for volume. The ratio of inductance and resistance has the dimension of time. Therefore, L/R is the characteristic time scale on which currents change in an inductive circuit (the *inductive time constant*).

If we combine electric or hydraulic capacitors and inductors, the same interesting phenomena occur, namely oscillations. In an electromagnetic LC-circuit, charge oscillates between storage and flow with a frequency which depends only on L and C. Water may be made to oscillate in a U-tube in just the same manner. By calculating the capacitance and the inductance of the container we can find the frequency of oscillation. There is another hydraulic setup which has an electronic equivalent. If the flow of water from an artificial lake to the turbines of an electric power plant has to be stopped abruptly for any reason the pressure may rise to such a level that the pipes rupture. For this reason a hydraulic capacitor is built in parallel to the system, namely a tower which is filled rapidly with the water rushing down the pipes.

P.1.4 Power, Work, and Energy

A number of observations made in different theories of physics have led to the recognition of a powerful new principle. Take, for example, a viscous fluid flowing through a pipe due to the pressure difference between the inlet and the outlet. We know that the pressure difference forces the fluid to flow. Alternatively, we may say that a pump has to do work to make the process happen. The pump, however, does not do this voluntarily. It must be forced to do its job, just as the water has to be forced through the pipe. Work is done *by* some device *on* another physical object.

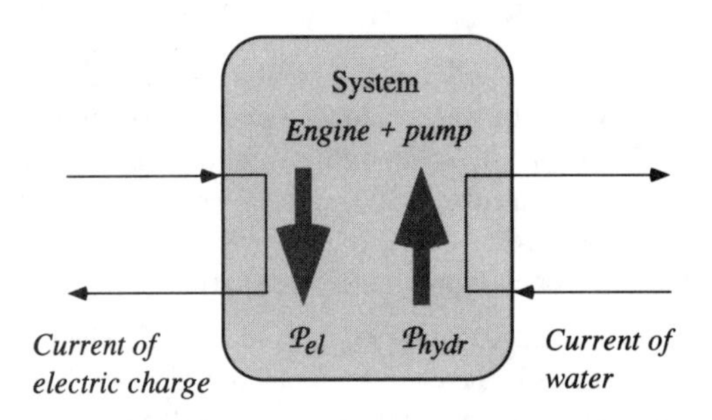

FIGURE 4. Flow diagram of a physical system representing the electrical engine plus the pump. The device receives electrical charge at a high potential. Since the charge "drops" to a lower value of the potential, we say that in this process energy is released. This energy is used to drive the hydraulic process, which consists of pumping water from a point of low pressure to one of higher pressure; the energy released in the first process is used in the second. The rates at which energy is released (depicted by the first vertical heavy arrow) or used (second heavy arrow) are called *power.*

Consider the operation of a device consisting of an electrical engine and a pump (Figure 4). This device takes in water at low pressure and forces it out at the other end at higher pressure. This is what we have called doing work on the fluid. We are accustomed to defining a new quantity, namely *power,* as the formal measure of the rate of working. We may speak of the *hydraulic power* associated with the forcing of water through the pump, and define it as

$$\mathcal{P}_{hydro} = -\Delta P|I_V| \tag{17}$$

Surely the power of a process must depend upon both the difference of levels (the potential difference) and the rate of flow of the quantity under consideration. The integral of the power over time is called *work:*

$$W_{hydro} = \int_{t_i}^{t_f} \mathcal{P}_{hydro}\, dt \tag{18}$$

Now, to function, work has to be done on the device as well. The electrical current, by virtue of flowing from a point of high electrical potential to one of lower potential, does this work. Power and work are defined analogously in hydraulics, mechanics, and electricity. For this reason we say that

$$\mathcal{P}_{el} = -\Delta\varphi_{el}\left|I_q\right| \tag{19}$$

is the *electrical power* of the device made up of electrical engine and pump. (Note that the negative potential difference is the voltage U.) Now comes an interesting observation: it is found that, in ideal engines, the power of the electrical process is equal to that of the hydraulic process it drives:

$$\left|\Delta\varphi_{el} I_q\right| = \left|\Delta P I_V\right| \tag{20}$$

This observation is at the heart of the much wider concept of *energy*. Nowadays we say that the pump in Figure 4 releases energy at a certain rate in the electrical process and then uses it in the following hydraulic process. We also believe that energy neither appears out of the blue nor disappears without a trace; rather, the amount of energy released is transferred to the pump from another system, and the energy used in the hydraulic process is transferred from the pump to another system (Figure 5). Energy enters the device with electrical charge and leaves the device as water flows out of it at high pressure. Talking about "work done by" and "work done on" a system only hides the actual process, namely, the transfer of energy.

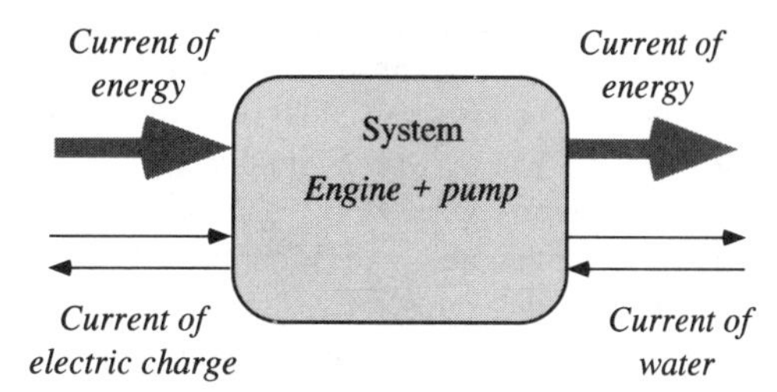

FIGURE 5. If a system works at a steady state, energy released must be supplied to it and energy used must be transferred out of it.

In this image, *power* is a code word for the rate at which energy is released and used; if energy is transferred we shall speak of *currents of energy*. Therefore, work is the amount of energy having been released or used in a process.

A lot more is encompassed by the notion of energy. For example, it is assumed that energy can be stored in physical systems. Also, all known processes suggest that energy is a conserved quantity. Because of its importance, we shall give a more detailed and organized account of the energy principle in Section P.4.

P.2 Transport Processes and Laws of Balance

Processes consist of the transport of one or more substancelike quantities. There exist several distinct modes of transport. The form of the laws of balance is determined by which transports can occur. For this reason, we have to discuss modes of transport and equations of balance together. Basically there are three possibilities, which we will call *conductive, convective, and radiative.* They lead to different terms in the equations of continuity for each of the quantities involved. In addition to transport we will find it possible for some of the quantities to be produced or generated. We shall base our discussion mainly on the example of the transport and the balance of momentum[2] and of amount of substance.

P.2.1 Stress and the Conductive Transport of Momentum

By pushing and pulling on an object we transfer momentum to it. The momentum transferred comes from another system (us), or flows into some other body. In the system under consideration, momentum may be stored. If so, the velocity of the bodies must change. It is possible, however, to leave the motion of the bodies unchanged, if we pull and push equally hard. In this case, the momentum supplied will again flow out of the system.

To avoid having to deal with three components of the vector of momentum, let us consider cases of pure tension or compression only. Consider the process of pushing a wooden block horizontally across a frictionless surface (Figure 6). The block accelerates, which means that you supply momentum to it through your hand. You can feel the flow of momentum through your hand as compressional stress. Naturally, the momentum supplied must be distributed throughout the wooden block. You may convince yourself that this flux has nothing to do with the motion of the body itself: momentum, except for the part stored in the body, does not flow with the body. You may push equally hard on the other side of the block, in which case the block's velocity stays constant, while the momentum supplied at one end leaves at the other end (Figure 6b). In other words, momentum may be transferred even through stationary bodies. Such a type of transport is called *conductive*. This term is known from thermodynamics, where we will encounter it again.

Summing up, we say that through direct contact momentum may be transported across system boundaries and through matter. Conductive transport makes itself felt as stress. The rate of flow of momentum across a part of a system's boundary is measured in terms of the flux of momentum I_p. If conductive transfer is the

2. The attempt to present the equation of motion completely in terms of the balance of momentum has led to the classification of modes of transport used in this section. The terms for the different modes have been taken from the theory of the transport of heat. See Fuchs (1987b).

only mode of transport, the momentum stored in a body can only be changed due to a net conductive flux:

$$\dot{p} = -I_{p,cond,net} \tag{21}$$

This is the *equation of balance of momentum* for this case. Remember that momentum is conserved. The (negative) conductive momentum flux is called a surface force F_s. Using this interpretation, we arrive at

$$\dot{p} = \sum_{i=1}^{N} F_{s,i} \tag{22}$$

which is a form of Newton's law of motion. Thus, Newton's law is a particular case of the general law of balance of momentum. So far it holds only for the case of conductive transport, i.e., only for surface forces acting upon a body.

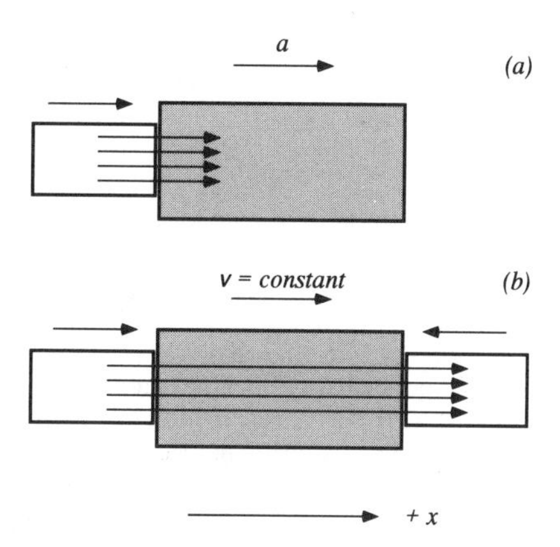

FIGURE 6. A block is pushed in the positive x-direction. It accelerates, which means that momentum is supplied to it (a). If we also push equally hard from the other side, the block's velocity remains constant (b). The momentum supplied from the left side has to leave at the right. In either case, momentum flows through the body itself. Note that in the case of compression, momentum flows through bodies in the positive direction. When momentum enters or leaves a body, we speak of a *flux of momentum* associated with the current.

P.2.2 The Continuous Case: Momentum Flow and Stress in Bodies

If we manage to specify the fluxes of momentum with respect to a body with the help of constitutive relations (Section P.3), we can employ Equations (21) or (22) to calculate the motion of the center of mass of the body. Remember that the momentum of a body is related to the velocity of its center of mass. Equation (21) represents the overall balance of momentum for an entire body subject to surface forces. However, this is certainly not all that could possibly interest us in mechanics. For example, in the case of extended systems, we would like to know things such as the state of stress at every point inside a body. In other words, we view systems as being continuous, having properties that vary throughout space. Here, we shall give only a qualitative description of what is a major subject in continuum physics. The mathematical treatment of simple cases of continuous processes is left to Chapter 3, and mostly, to the Epilogue. The example treated here is rather complex in detail, however, it can be understood on a qualitative level.

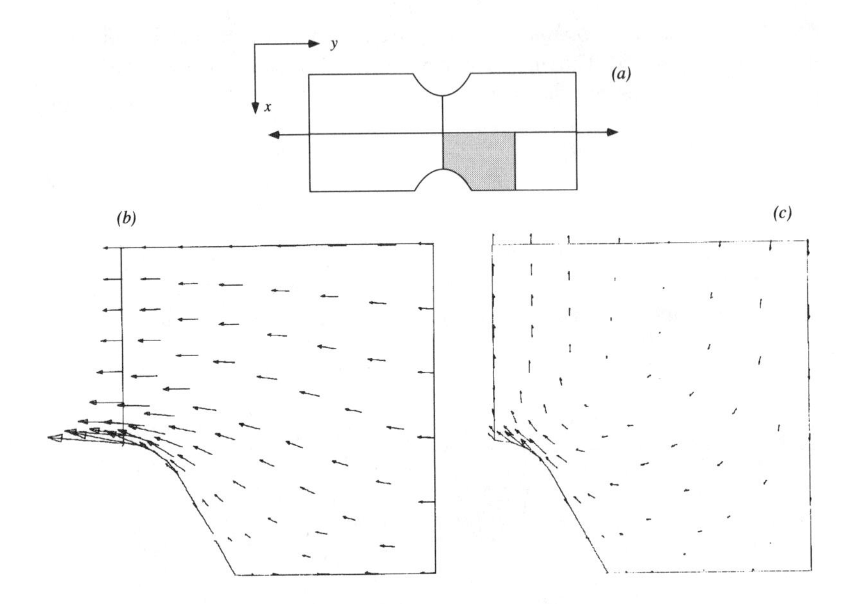

FIGURE 7. A flat strip of metal (a) with a notch on either side is set under tension. The two-dimensional stresses are represented by the flow patterns of two components of momentum (x and y) through the body. The shaded region was been analyzed by the finite element procedure. (b) y-momentum is flowing in the negative direction, demonstrating tension. The notches lead to a channeling of the flow, with higher current density (higher stress) near the notch. (c) Transport of x-momentum has been induced by the sideways flow of y-momentum. The FE computation was performed by K. Bruggisser and interpreted by W. Maurer (1989, 1990a).

The flux of momentum across a system boundary represents only a small part of the information contained in the actual transport process (Figure 6). The current of momentum is distributed over the surface, and there is a spatial flow pattern inside the body. We must somehow be able to specify the spatial distribution of the flow, to fully describe the continuous situation. To this end, a quantity related to momentum currents appropriate for continuum physics is introduced, namely the (surface) *momentum current density* j_p. In the simplest case of a uniform flow of momentum perpendicular to a surface, the product of the current density and the surface area delivers the flux of momentum through the surface:

$$I_p = A\, j_p \tag{23}$$

The (negative) momentum current density is commonly called *stress* in mechanics. Pressure, for example, is the momentum current density in a fluid.

Actually, the current density of every component of momentum is a vector which specifies the spatial distribution of the flow of momentum, just as velocity vectors may be used to visualize the flow pattern of water. In the general case we have to integrate the current density over a surface to obtain the flux through the surface. For every component, a flow pattern has to be drawn. We shall not treat the general theory here. Figure 7 shows an example of a finite element computation which represents the stress inside a body as a flow field of momentum. Momentum current density vectors create the image of this field.

P.2.3 Bodies and Fields: "Radiative" Transport of Momentum

Mechanical interactions of a body with its environment do not just take the form of direct contact with other bodies. Bodies can interact indirectly via fields. In such cases the direct interaction is between bodies and fields.

Consider the simple example of free fall in a homogeneous field, which can tell us much about the form of the transfer of momentum. It is clear that the body receives momentum (if we count the direction of free fall as the positive one). Experience also tells us that every part of a body will experience the same acceleration independently of density and composition. This means that a particular part of a body does not push or pull on other parts while in free fall. In other words, the interior of a freely falling body does not experience mechanical stress. Since stress is associated with conductive flow of momentum *through* matter, the absence of stress tells us that momentum cannot be transported *through* the body in free fall. Rather, every part of a body must receive momentum at the rate necessary for its acceleration.

Since the parts of a body receiving momentum are of arbitrary size and shape, and since there is no exchange of momentum between them, it is clear that momentum simply appears inside the body at every point without being transported there through other parts. This means that there are *sources of momentum* in the body itself. Momentum appears in a body at a particular rate quantified by the *source rate* Σ_p. The case of gravitation tells us that the source rate for a given part depends upon both the strength of the gravitational field and the gravitational mass of the part in question. We shall consider the relevant constitutive relation in Section P.3.3.

How is momentum transported from a field to a body, or vice versa? First of all, momentum must come from, or go to, another body. As a consequence, it must be transported *through the field*. It enters bodies from the field, or bodies emit momentum to the field. Consider a charged body in an electrical field. It is known from electromagnetic theory that momentum flows through the field. Specifically, if there is no charge in a region of the field, momentum entering the region must leave again (Figure 8a). If we place a charged body in this region of space, however, the body changes the field so as to make the net flux of momentum through the region nonzero (Figure 8b). In other words, momentum disappears in a region of a field occupied by a body, but it appears in the body. There are sinks of momentum in the field and sources in the body, and vice versa. The interaction of bodies and fields takes the form of sources or sinks rather than surface currents. This is so since bodies and fields "touch" in three dimensions rather than in two as do ordinary bodies in contact. Both bodies and fields occupy the same regions of space at the same time. By the way, this form of interaction is well known in the case of the transport of heat. Due to radiation, heat may enter or leave bodies directly. For this reason we call this type of transfer of momentum *radiative.*[3]

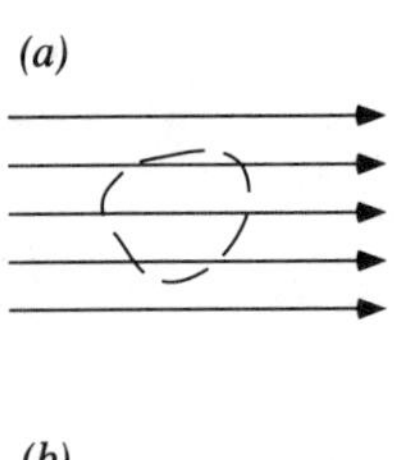

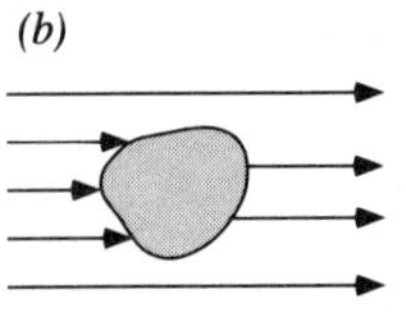

FIGURE 8. The total flux of momentum through a field is zero with respect to a region of space which is occupied by the field only (a). If we place a charged body in the electrical field, however, the situation changes (b). The body alters the field which leads to a net flux with respect to the region now occupied by charge. The change of the external field leads to the interaction of body and field.

The equation of balance of momentum for a body subject to surface and body forces must now include a source term with the fluxes of momentum:

$$\dot{p} + I_{p,cond} = \Sigma_p \tag{24}$$

This is Newton's law of motion including both surface and body forces; it is equivalent to

$$m\dot{v} = F_s + F_b \tag{25}$$

since Equation (24) only holds for a body which cannot change its mass. It is simply impossible to change the mass of a system if the only types of momentum transfer allowed are conductive and radiative in nature. Equation (24) would be wrong in classical mechanics with the derivative of the momentum replaced by the ordinary derivative of the product of mass and velocity. (The resulting equation is not Galilean invariant.[4]) Obviously, cases with systems of variable mass must be treated differently. (See Section P.2.5.)

Equation (25) contains an important additional piece of information: (inertial) mass plays the role of *momentum capacitance*. Just compare the expression $p = mv$ with its electrical analogue $q = CU$. It is clear that inertial mass measures a body's capacity for containing momentum.

P.2.4 The Tides

The example of bodies moving in a gravitational field nicely demonstrates the power of a qualitative solution of a problem in terms of flow patterns of momentum.[5] Consider a thin rod falling freely in a field whose strength increases downwards (Figure 9). We found before that a body in free fall should not experience mechanical stress. This is true only for motion in a homogeneous field.

---

3. Fuchs (1987b,e). See also Herrmann and Schmid (1985) and Heiduck, Herrmann, and Schmid (1987).
4. Consider the equation from the point of view of another observer moving at speed v. The terms representing mass, acceleration, rate of change of mass, and sum of all forces are the same for all observers. The velocity multiplying the time derivative of the mass, however, is not. The problem was posed in this form by R. Resnick (RPI).
5. Fuchs (1987d).

We have concluded that momentum is supplied to every part of a body in a gravitational field. The rate of supply certainly is higher where the field is stronger. If we now consider the rod to be oriented along the field lines, we have to conclude that the source rate of momentum is larger in the lower portions of the rod. Consider the rod to be rigid, in which case every part of it experiences the same acceleration. Therefore, the lower half receives more momentum than is necessary for free fall, while the upper part receives too little. The body solves the problem by rearranging the momentum supplied from the field: it allows the surplus of momentum to flow upwards through the rod to achieve equal distribution. This transport obviously is conductive in nature. Momentum flows through matter in the negative direction (Figure 9), leading to tension in the rod, which may pull it apart. This phenomenon is known as *tides*. In the case of the Earth falling in the gravitational field of the moon, we can see the origin of the two tidal bulges on opposite sides of the fluid sphere. We can even say that the effect of the tides will be largest in the middle of the rod, since all the momentum gathering in the lower parts for upward transport must cross through the center of the body. Beyond this point, the current density decreases.

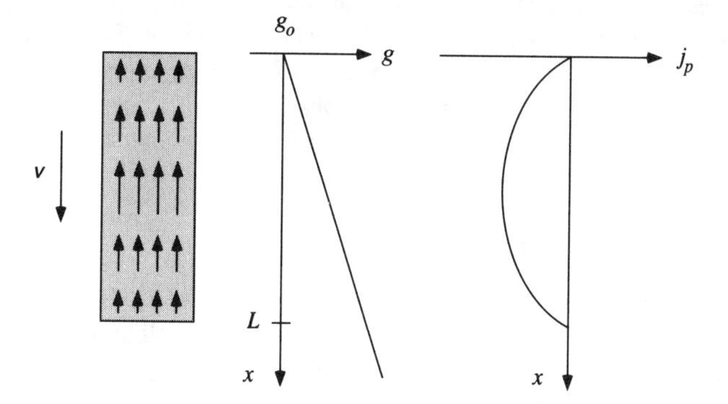

FIGURE 9. A long rigid rod falls in an inhomogeneous gravitational field. Tides are the result of such conductive rearrangement of momentum. Momentum flow is indicated by arrows in the rod. If g varies linearly, the stress, which is measured by the momentum current density j_p, is a quadratic function shown in the second graph on the right.

P.2.5 The Transport of Momentum with Moving Bodies: Convection

The problem mentioned at the end of Section P.2.3 has a simple solution. Momentum also can be transported with moving bodies. Since momentum is stored in moving matter, the flow of water across a system boundary also transfers with it some momentum. This mode of transport is called *convection*.

Consider a homogeneous flow field (Figure 10). The speed of the flow of water is constant over a surface perpendicular to the direction of flow. The flux of volume across the surface is equal to the product of speed and surface area. If we multiply this quantity by the density of the fluid, we obtain the flux of mass (**n** is the vector normal to the surface):

$$I_m = \rho A \mathbf{n} \cdot \mathbf{v} \tag{26}$$

The convective flux of momentum is obtained simply if we multiply this expression by the speed of flow:

$$I_{p,conv} = v I_m \tag{27}$$

If we wish, we can express the convective flux of momentum in terms of the transport of amount of substance, instead of mass. Now we can write the most general expression for the balance of momentum for a region of space such as the one in Figure 10. The rate of change of momentum in this region must be equal to the (negative) sum of all fluxes of momentum, both conductive and convective, and the source rate:

$$\dot{p} + I_{p,cond} + I_{p,conv} = \Sigma_p \tag{28}$$

Here, $\dot{p}$ is the time derivative of the momentum contained in a control volume at a certain instant. It is not the time derivative for a body of constant mass. The problem mentioned after Equation (25) is solved, since we now have a means for changing the mass of a system as a consequence of a particular type of momentum transfer. We simply have to extend Newton's law of motion to include convective momentum fluxes which are not present in the original formulation of the law. This is why it can be applied only to the case of bodies of constant mass. In fact, we shall use the term *body* in the sense of a portion of matter which is always identifiable and of constant mass. Consider a *body* to be a piece of matter surrounded by a surface through which mass may not be transported.

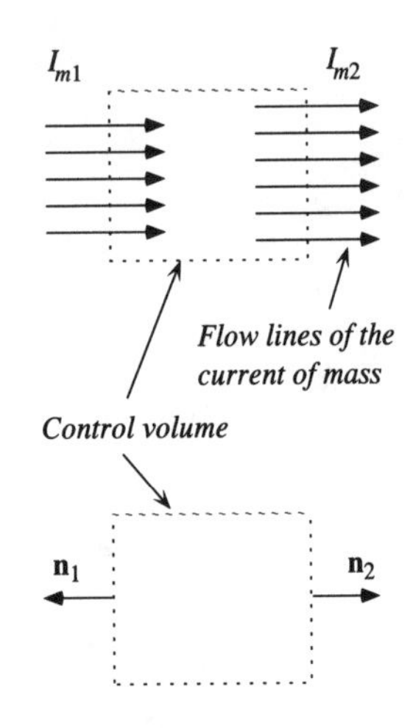

FIGURE 10. If matter is transported across the surface of a control volume, momentum is transferred along with it. This type of transfer is called *convective*.

P.2.6 The Balance of Mass

All of the examples of laws of balance in this section have dealt with momentum. Let us just briefly present the case of another substancelike quantity, namely mass. If we wish to treat convective currents, we obviously have to be prepared to deal with the balance of mass as well.

We may look upon this problem as a special case of the relatively general relation presented in Equation (28) for momentum. We have to ask ourselves two questions if we wish to write a proper equation of balance for mass: How can mass be transported? Is mass conserved?

Now, there can be only one way of transporting mass, namely, convective currents. Mass cannot be conducted or radiated. For this reason, of the three terms in Equation (28) relating to transports, only one will appear in the law to be constructed. Finally, we know that mass is a conserved quantity, leading us to conclude that there is no other way of changing the mass of a system except by convective transport. The proper equation of balance of mass therefore takes the form

$$\dot{m} + I_{m,net} = 0 \tag{29}$$

where it is understood that I_m is the flux of a convective current.

P.2.7 The Balance of Amount of Substance

Chemists measure an amount of stuff not in terms of mass but in terms of *amount of substance*, which has the unit of moles. In the microscopic model of matter this quantity measures the number of particles of a substance. One mole contains $6 \cdot 10^{23}$ particles.

Amount of substance plays a role in processes which have not appeared so far in our description. This quantity may be changed almost at will in chemical reactions. Amount of substance may be created or destroyed in the true sense of the word: in a particular reaction water can be made to disappear while hydrogen and oxygen appear at the same time. If you allow 1 mole of hydrogen gas to react with one mole of oxygen, the amount of hydrogen will change by 1 mole, to 0, while half of the oxygen will be used up, which means that the amount of oxygen reduces by 0.5 mole; at the same time 1 mole of water will appear. Since such reactions may take place inside closed systems, transport across system boundaries is not responsible for changes of the quantity. As a consequence, we will have to extend the equation of balance of amount of substance, with a term allowing for production and destruction of this variable. If we write the symbol Π_n for the rate of production of amount of substance, the equation of balance will take the following form:

$$\dot{n} + \sum_{i=1}^{N} I_{n,i} = \Pi_n \tag{30}$$

The rate of production, measured in mole/s, may be negative, in which case it represents the destruction of the quantity. More details on amount of substance and on the equation of balance it satisfies can be found in Chapter 4.

P.2.8 Conservation and Laws of Balance

The previous examples have given us some insight into how the volume (of incompressible fluids), electrical charge, momentum, mass, and amount of substance are accounted for. All four variables stand for quantities which we may imagine as being contained in regions of space or in bodies, and to be capable of flowing into and out of these regions. We have learned how to describe the rate of transfer for three different types of transport, namely, conductive, convective, and radiative. The sum of all these fluxes and rates must tell us something about the rate of change of the stored quantities. In fact, if the quantities are conserved, the sum of transfer rates and the rate of change of system content must be equal.

While there are different modes of transport, not all need to be present in a given situation. Actually, some of them may not even be allowed in the case of a particular substancelike quantity. Charge, for example, cannot be transferred radiatively, i.e., it cannot hop through fields. Only conductive and convective transport are possible. The volume of fluids, mass, and amount of substance, on the other hand, admit only convective currents.

These behaviors occur in the realm of conserved physical quantities. The equation of balance of momentum, Equation (28), is the most general form of a law of balance in this case. Still, it does not allow for all the possibilities nature has in store for us. It is possible for some quantities to change in a system without transport of any kind having taken place. You have encountered this possibility in the case of amount of substance, which is used to measure amounts of stuff in chemistry. In other words, the term *substancelike* has nothing to do with whether a quantity is conserved. Production and destruction do not rule out the existence of laws of balance. Seen from this perspective, conservation is not a general feature of substancelike or additive quantities. If you review the properties of the variables discussed so far you will notice that they only have two things in common: they may be contained in systems, and they may be exchanged. Every other feature is a particular and not a general one.

P.3 The Properties of Bodies

While the laws of balance introduced above should hold for all bodies, constitutive relations describe the diversity of bodies. Today, intensive research is going on in the field of continuum physics relating to constitutive laws. It would be impossible to give a complete description of material laws in this brief section.

In Section P.1 we encountered a few constitutive laws in the fields of electricity and hydraulics. Here, we will discuss the fundamental role constitutive relations play, and discuss some more examples.

P.3.1 Constitutive Relations and Dynamics

Inspect Newton's law of motion in the form of Equations (24) or (25). It expresses the law of balance of momentum for a body subject to surface and body forces. Since it relates two different concepts, namely properties of the body (its capacity for containing momentum) and interaction of the body and its environment (momentum fluxes or forces), it cannot be used by itself. To compute the motion of a body we need additional information concerning the form of the momentum transfers taking place. Put differently, we need *constitutive laws* for momentum fluxes (forces).

A theory of dynamics relies on constitutive relations. The properties of bodies determine their motion. All bodies obey the same fundamental law of balance of momentum, however, the particular type of motion resulting from interactions depends on material properties, not on the generic properties of momentum. Combining laws of balance with proper constitutive relations lies at the heart of any theory of dynamics, be it mechanical, electrical, or thermal in nature. This also distinguishes statics from dynamics. First of all, statics does not rely on laws of motion. Rather, some variational principle, usually the principle of virtual displacements, is used to find the rest configuration of a system. For example, consider the principle of minimal energy, which tells us that a pendulum at rest should simply hang vertically. While it is possible in some cases to find the static solution for a swinging pendulum from dynamics using some material properties, not all constitutive laws even deliver a solution where the pendulum comes to rest. Without friction, the pendulum will go on oscillating forever.

Consider the following well-known example. The motion of the planets is governed by the same fundamental principle of conservation of momentum as are the motions of all other bodies. What distinguishes the solar system from other cases are the particular constitutive laws appropriate for the sun and the planets. Chief among them is Newton's law of gravitation, which allows us to determine the momentum flux vector for each combination of two bodies. Furthermore, its is asserted that for each pair the flux vectors are of the same size and point along the line joining the centers of the planets. Armed with these laws we can compute the motion of the bodies making up the solar system. In other words, the law of gravitation is a constitutive relation appropriate for the material called *solar system*.

P.3.2 Finding the Properties of Bodies

Actually, much of physics is concerned with the inverse of the problem just presented.[6] Instead of knowing the constitutive relations from which we may calculate processes, we are trying to find the material properties of bodies from our knowledge of processes. The physics of the solar system may again serve as an example. Newton somehow had to find a constitutive relation appropriate to the motion of the planets and the sun. The motion was known (Kepler's laws); what was wanting was a law for the forces governing the motions in the system. In later centuries, the properties of solid bodies and fluids, the electromagnetic field, and thermal systems were investigated. Considering that the phenomena are complex, and that different types of hypotheses might fit the same data, how do we find the proper constitutive laws?

Here, the generic laws, i.e., the laws of balance, come to our rescue. The material properties of systems may not contradict these fundamental laws. In other words, the motion of the planets must obey the law of conservation of momentum in the form of Newton's law of motion. This puts severe restrictions upon the form of possible material laws. The same is true for any other field of physics, including thermodynamics. Summing up we may say that the relation between generic and constitutive laws is twofold. First, constitutive relations are needed, together with the laws of motion, to compute processes. Second, given a set of observations of processes, the generic laws help us in finding out about material properties.

---

6. For an exposition of the problem see Malvern (1969, Chapter 6), and I. Müller (1985).

P.3.3 An Example of Motion

As a concrete example, consider a small metal sphere falling in oil. If you drop such a body into a vertical pipe filled with a fluid of high viscosity you will observe that the body reaches a constant speed rather quickly. In free fall, of course, it would continue to accelerate. Thus there are at least two ways in which the body interacts with its environment. First, it may interact with the gravitational field of the Earth. Second, the metal sphere is in contact with the fluid into which it is dropped. Therefore, we have to consider gravity and fluid flow (around the sphere).

We have described the dual nature of the relationship between generic laws (the laws of balance) and constitutive relations. While the generic laws are used in establishing material laws (the latter may not contradict the former), special examples which rely on particular constitutive relations may suggest the correct form of the fundamental laws. In this example, we shall assume that the constitutive problem has been solved. This will allow us to concentrate upon the task of computing the motion from the known laws.

Since the metal sphere interacts with both the gravitational field and the fluid surrounding it, the appropriate equation of balance of momentum is the following:

$$m\dot{v} + I_{p,fluid} = \Sigma_{p,grav} \tag{31}$$

This is a special form of the expression in Equation (24).

Now we have to discuss the constitutive relations. The law of gravitation tells us that the source rate for a given part of a body is proportional to both the strength of the gravitational field and the gravitational mass of the part. If the field is uniform, the source rate turns out to be equal to

$$\Sigma_{p,grav} = gm \tag{32}$$

if we consider the body to be falling in the positive direction.

The interaction of body and fluid is much more complex. We actually need a theory of hydrodynamics to derive the form of the surface force which is a result of the action of the fluid upon the sphere. We will just quote the result:

$$I_{p,fluid} = g\rho_{fluid}V + 6\pi\mu rv \tag{33}$$

Here, V is the volume of the metal sphere, while μ is the viscosity of the oil. Obviously, the first term represents the buoyancy of the body immersed in the fluid. The second is named after *Stokes*; it stands for the resistive force upon the body moving through the viscous medium.[7] Note that both interactions result in momentum being removed from the body, which makes the momentum flux a positive quantity in the coordinate system chosen. Together, the two parts are the

7. Bird, Stewart, and Lightfoot (1960), p. 59.

result of the momentum transfer across the surface of the body as a consequence of pressure and shear forces. If we knew how to calculate the surface distribution of these transports (i.e., the momentum current density), we could find the momentum flux by integrating over the surface of the sphere.

If all factors have been specified, it should be possible to compute the motion of the sphere moving through the oil. Assembling all the parts leads to

$$m\dot{v} + g\rho_{fluid}V + 6\pi\mu r v = g\rho_{body}V \tag{34}$$

which is our law of balance of momentum for the particular case. Equation (34) can be integrated to deliver the velocity as a function of time.

P.3.4 Viscosity and Momentum Transport

In this section we will discuss a particular example of a constitutive law. We shall encounter relations of the same general form again and again in continuum physics.

Consider a fluid such as water between two large parallel plates (Figure 11). We pull the upper plate horizontally, while the lower one remains stationary. If the fluid were ideal, the plate would simply slip on top of it. Since real fluids are viscous, however, they remain at rest with respect to the surfaces of bodies. In other words, the liquid will be pulled along with the moving plate. Naturally, the velocity of the fluid with respect to the lower plate has to be zero as well. Therefore, the magnitude of the fluid velocity changes from the speed of the upper plate to zero. In other words, there is a velocity gradient dv_x/dy in the y-direction.

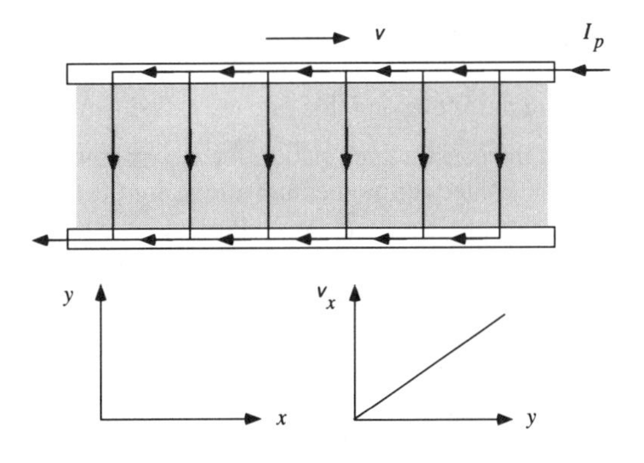

FIGURE 11. A viscous fluid is sandwiched between two large parallel plates. The upper plate is pulled to the right. Because of viscosity, the fluid does not slip at the solid surfaces. In the steady state, a velocity gradient is built up in the fluid. Momentum which is being supplied to the upper plate flows through the fluid from points of higher to points of lower velocity. In the simplest case, a constant gradient is established.

The dynamical situation is as follows: Layers of the fluid move past each other which, because of viscosity, leads to friction in the fluid. We notice that we have to pull the upper plate if we want to maintain a constant velocity. Pulling the body in the positive x-direction implies that x-momentum is being supplied to it. Since in the steady state the plate does not store any more momentum than it needs for its motion, we end up with a flow of momentum through the fluid from the upper to the lower plate. The lower plate conducts the momentum to the Earth. Here, we are confronted with the flow of a component of momentum not in its own direction but perpendicularly to it: x-momentum flows in the y-direction. The fluid experiences *shear stress*. Note that shear stress, like tension or compression, is a case of conductive transfer of momentum. Momentum flows through matter, not with it. In the case considered, the fluid obviously does not move in the direction perpendicular to the horizontal plates.

The flow of momentum is properly described by the current density. Imagine x-momentum to be akin to some kind of "stuff" which may flow in any direction. Momentum is transported down through the liquid from every point of the upper plate. The distribution of the current over a horizontal surface is given by the momentum current density, which represents the shear stress. This quantity is the y-component of the current density vector of x-momentum, which may be abbreviated j_{pxy}. (This is the negative shear stress component τ_{xy}.) Now, a constitutive relation for viscous flow should allow us to relate this flux density of momentum to the kinematics of the flow and the properties of the fluid. Put differently, we are looking for a law relating the velocity gradient, the viscosity of the fluid, and the momentum current density j_{pxy}.

A good number of common liquids and gases exhibit a rather simple constitutive relation for viscous flow. The shear stress, i.e., the momentum flux density, is proportional to the gradient of the velocity in the fluid:

$$j_{pxy} = -\mu \frac{dv_x}{dy} \tag{35}$$

The coefficient μ relating the variables is called the *dynamic viscosity* of the fluid. The minus sign appears since momentum flows from points of higher speed to points of lower speed. The relation is called Newton's law (do not confuse it with the law of balance of momentum, or the law of gravitation, which also carry Newton's name), and fluids obeying this rule are called Newtonian.[8]

The combination of the constitutive relation for viscous momentum fluxes and the generic law of the balance of momentum allows for the computation of flow configurations. The law of Hagen and Poiseuille, for example, follows from the integration of the velocity profile established in steady-state viscous laminar flow through a cylindrical pipe. Qualitatively, the situation is as follows: the fluid is under pressure, which means that momentum is flowing through it in the positive direction; the pressure corresponds to the conductive momentum flux density in the direction of the axis of the pipe. Since the pressure is higher at the inlet, and since the fluid is not accelerated, momentum must be leaving sideways through the fluid. Because of viscosity, the fluid sticks to the walls of the pipe, leading to a velocity profile with a maximum magnitude at the central axis. Due to the radial velocity gradient, momentum is transported toward the wall,

8. Bird, Stewart, and Lightfoot (1960), Chapter 1.

through which it leaves the system. A similar type of momentum flow is established if we pull a crate over the floor. We supply momentum to the crate. However, due to friction, momentum will flow vertically through the crate toward the floor. In a narrow layer between the crate and the surface it slides on, momentum flows from points which possess the speed of the crate to points whose speed is equal to zero (which is the speed of the floor). We know that the fall of a substancelike quantity such as momentum from points of high potential to points of low potential releases energy (Section P.1.4), and we know that this energy is used to drive a process, such as the generation of heat through friction (more about this in Section P.4 and in the chapters on thermal physics).

P.3.5 A Comparison of the Conduction of Momentum and Charge

Viscosity is responsible for establishing velocity gradients in a fluid, and these in turn are responsible for the flow of momentum. We can compare this situation to what happens in an electric conductor such as a piece of wire. Charge is conducted through the wire as a result of the gradient of the electrical potential: charge flows "downhill" from points of higher to points of lower potential. The constitutive law appropriate for this transport is Ohm's law. Commonly, it is written in the form $I_q = U/R$, where R is the electrical resistance. However, this is the definition of the resistance, rather than a constitutive relation. Ohm's law actually establishes the relationship between the flux of charge, the properties of the conductor, and the electrical driving force responsible for the flow. It is found that wires with twice the cross section simply conduct twice the charge. This suggests that the flux per unit surface area, i.e., the charge flux density, is equal in conductors of different cross section as long as all other factors are kept constant. Therefore, the flux density j_q must be expressed by a constitutive law. If the flux density is proportional to the gradient of the electrical potential, the resulting relation is called Ohm's law:

$$j_q = -\sigma \frac{d\varphi_{el}}{dx} \qquad (36)$$

Here, x is measured in the direction of the transport of charge, and σ stands for the electrical *conductivity*.

Compare this to Newton's law of viscosity in Equation (35). Obviously, the viscosity has the meaning of a *momentum conductivity*. The higher the viscosity of a fluid, the better it conducts momentum down the velocity gradient. It is interesting to note that the temperature dependence of the viscosity of water is similar to that of the electrical conductivity of superconductors. Both increase with decreasing temperature, and both make a sudden jump to virtually infinite values at a critical temperature (Figure 12). Electrical conductors may become superconducting below the critical temperature, while water becomes a superconductor for momentum when it freezes! In both cases a phase transition is responsible for the abrupt change of properties. In mechanics this viewpoint might appear unusual, but it agrees with the facts. In electricity and in mechanics, charge or momentum flow, respectively, without a potential difference in the superconducting state. Indeed, in static mechanical situations, momentum is conducted through bodies without the bodies moving at all. And in both cases the transport phenomena are nondissipative: they do not produce any heat.

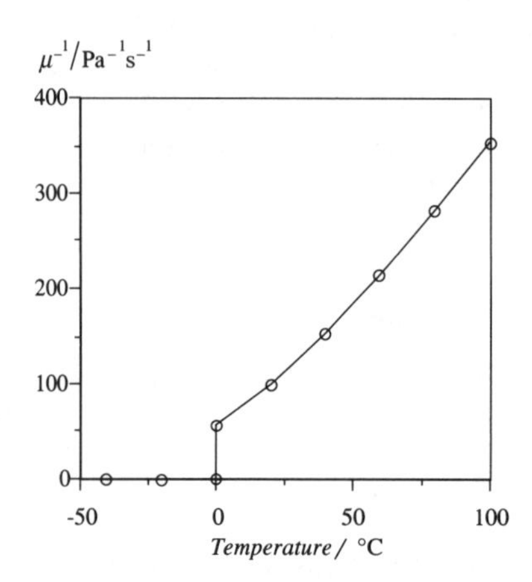

FIGURE 12. Reciprocal of the viscosity of water as a function of (Celsius) temperature. Since the viscosity is the momentum conductivity, its inverse is a resistivity. When water freezes, the resistivity goes to zero, just like the electric resistivity of a superconductor.

P.3.6 Derivation of the Wave Equation

So far we have learned how to set up equations of balance of substancelike quantities, and how to use constitutive relations in conjunction with the laws of balance. Here we will consider an interesting example—the theory of transport of momentum in one direction through a simple fluid. We will see that wavelike transport is the result of the interplay of storage and transport of momentum, where the constitutive relation for the transport is given by an inductance, while the storage is described using the momentum capacitance of the system. In contrast to the case of the flow of water dealt with in Section P.1, we now have a spatially continuous case.

Imagine a fluid which is under pressure. Momentum is transported through it in the positive x-direction (Figure 13). The equation of balance of momentum for this body is rather simple. We consider the boundaries of the body to move with the fluid, which means that there may be only conductive currents of momentum. Therefore, the law of balance takes the form

$$m\dot{v} = -(I_{p1} + I_{p2}) \tag{37}$$

which is a special case of Equation (21). Remember that the mass of the body is the momentum capacitance, and that the momentum current density j_p is the pressure of the fluid. Now, we shall introduce the momentum capacitance per length (mass per length) C^*. If Δx is the length of the body of fluid in Figure 13, the equation of balance of momentum becomes

$$C^*\frac{dv}{dt} = -A\frac{1}{\Delta x}(j_{p2} - j_{p1}) \tag{38}$$

where A is the cross section of the fluid in Figure 13.

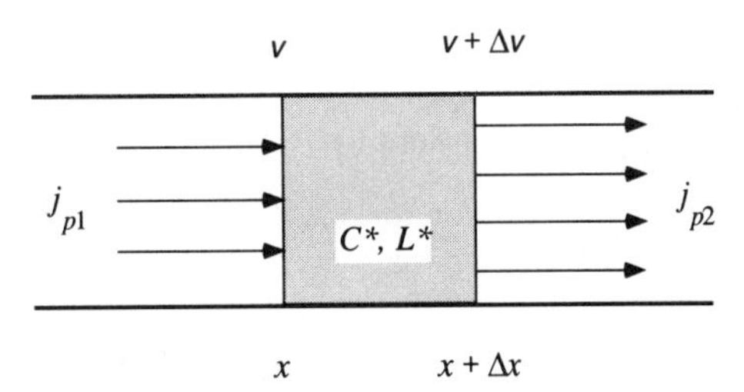

FIGURE 13. A fluid such as a gas is enclosed in a long pipe. If a pressure disturbance is set up, a change of the transport of momentum through the fluid must occur in both space and time. Every part of the fluid possesses a certain momentum capacitance and momentum inductance. As a result, the transport of momentum takes a wavelike form.

The particular form of the transport depends upon the type of constitutive relation satisfied by the momentum current. A body possesses resistance and inductance related to the transport of momentum. Both quantities are introduced in analogy to what we know from electricity or hydraulics. We shall neglect the resistance of the medium to the passage of momentum. (The phenomenon of resistance would lead to an attenuation of the wave traveling through the fluid.) To describe the other effect, the inductance per length L^* will be introduced. The constitutive law of induction relates the difference of speed of the two faces of the body of fluid in Figure 13 to the rate of change of the momentum flux through the body. This phenomenon must appear because of the spring-like nature of the fluid: if there is a difference of speeds at the faces of the body, its density, and therefore its pressure, must change in time:

$$\Delta v = -L^*\Delta x \cdot A\frac{dj_p}{dt} \tag{39}$$

Both Equations (38) and (39) must hold for arbitrarily small parts of the fluid. Therefore, these expressions become partial differential equations:

$$C^*\frac{\partial v}{\partial t} = -A\frac{\partial j_p}{\partial x} \tag{40}$$

$$\frac{\partial v}{\partial x} = -L^*A\frac{\partial j_p}{\partial t} \tag{41}$$

The former represents the balance of momentum for the continuous case, while the latter is the law of (momentum) inductance. We may combine these equations in the following manner. Take the time derivative of Equation (40) and the spatial derivative of Equation (41). In both expressions, derivatives of the momentum current density with respect to both time and space occur. Relating these terms leads to

$$\frac{\partial^2 v}{\partial t^2} = \frac{1}{C^* L^*}\frac{\partial^2 v}{\partial x^2} \tag{42}$$

which is a simple wave equation for transport in one spatial dimension. The term C^*L^* represents the inverse of the square of the speed of sound in the fluid. If we manage to determine the inductance of the fluid, we can also compute the speed of propagation of sound (Chapter 2).

The derivation of the wave equation for one-dimensional compressional waves can be carried over to other fields of physics, such as electricity or heat. In electricity, we may obtain the governing equation for a wave guide (telegrapher's equation). Inclusion of the resistance of the medium leads to the term responsible for dissipation and attenuation of the wave. In thermodynamics, the result equivalent to this particular form of Equation (42) is not generally known. However, the phenomenon of thermal induction exists and leads to interesting results, such as second sound. (See Section E.5.)

P.4 Energy and Physical Processes

The concept of energy provides a link between different types of physical processes. While a particular process may be fully described in terms of the quantities associated with it, energy furnishes the relationships in chains of events. To be specific, while charge and electrical potential, and volume and pressure, are used to model electrical and hydraulic processes, respectively, the amount of energy exchanged in these phenomena leads to a description of how electricity and fluid flow are coupled through an electrically driven pump. In this section we shall detail the ideas behind the energy principle.

P.4.1 Transport Phenomena and (Internal) Processes

We will see that to fully describe the relationship of energy to physical phenomena, we have to distinguish between what we call *transport phenomena* and *internal processes*. This distinction is necessary because very often we want to present discrete models of physical processes rather than a continuum theory. In discrete models, extended objects are described from an overall perspective as physical entities. A water container, for example, is looked upon as a system whose properties may be described by very few variables, such as the volume of water stored, the water level, and its density and temperature. The possibility of some of these variables changing from point to point *inside* the system is neglected. Such models are called *uniform*. An example of a uniform system is a rigid body moving without rotating, in which case every point of the body has the same velocity. Even in the cases of rotating and deforming bodies we may be interested in the overall features of motion as expressed by the motion of the center of mass. Since we neglect the internal life of objects in discrete models, we shall need some way of dealing with internal processes.

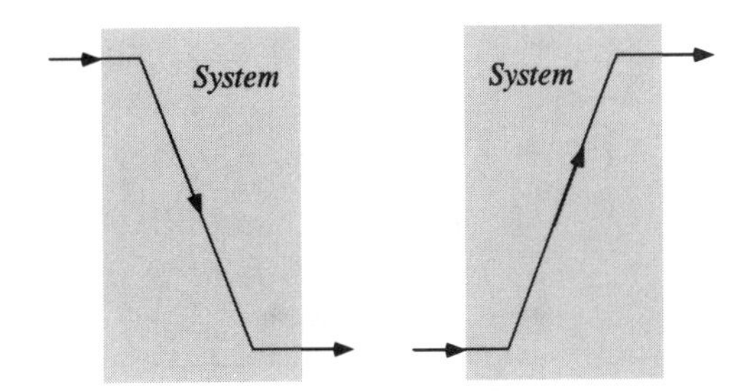

FIGURE 14. Inside a system a substancelike quantity may flow downhill or uphill, i.e., from higher to lower levels, or vice versa. The first case is called a voluntary internal process. The latter is called involuntary.

Examples of internal processes abound. Just think of a turbine taking in and discharging water or steam. An immersion heater belongs in the same category, as does an electrical engine, a wind mill, and a heat engine. You may have noticed that in all of these examples, one or more substancelike quantities are taken in and discharged again by the system. Intake and discharging take place at different levels or potentials. In other words, in systems where internal processes are occurring, one or more extensive quantities are lowered or raised in level. We shall use the term *voluntary* for an internal process in which a substancelike quantity flows from a point of higher potential to a point where it is lower. The term *involuntary internal process* shall be reserved for a phenomenon where substancelike quantities are pumped from a lower to a higher level (Figure 14).

The examples mentioned should clarify what we mean. In an immersion heater, a light bulb, or an electric drill, a current of electrical charge is introduced in the system at a high value of the electric potential. The same current is discharged at a lower potential. A turbine takes in water at high pressure and discharges it at lower pressure. As you probably know, or as you will learn later in this book, a heat engine takes in heat at high temperature and emits it at lower temperature.

Examples of involuntary processes are the pumping of water from points of lower pressure to points of higher pressure; the pumping of angular momentum in a drill or in a wind mill from points where the angular velocity is zero (the environment or the casing of the device) to points of high angular velocity (corresponding to the magnitude of angular velocity of the drill bit or the shaft of the wind mill).

Obviously, internal processes are not all there is in physics. Where, after all, does the water taken in by a turbine come from, and where does it go to? It is clear that physical systems may absorb and emit substancelike quantities. In other words, systems exchange substancelike quantities with the environment (Figure 15). We call these phenomena *external processes* or *transport phenomena*.

In the examples given above, internal processes are always associated with transport phenomena. However, not all internal processes need to be accompanied by transport across system boundaries. We have already mentioned the example of chemical reactions inside a closed system, where processes take place without exchange of quantities. You will learn in the following chapters that thermal physics furnishes more examples of this kind.

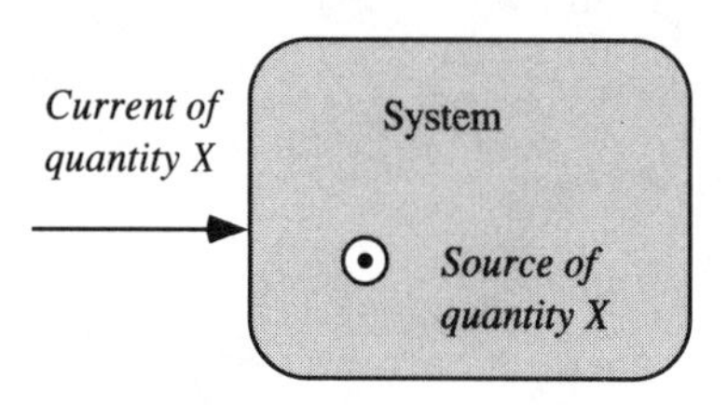

FIGURE 15. In transport phenomena, systems absorb or emit certain substancelike physical quantities, or the quantities are transported across the surface of the system. The arrow is a symbol for two of the three types of transport (conductive and convective) while the small circle inside the system represents radiative transfer (absorption and emission).

Transport does not have to be accompanied by internal processes either. Take the example of pushing a body over a frictionless surface. Momentum is supplied to the body without any of it flowing out again. Momentum simply flows into the system at an instantaneous value of the velocity, and is stored there. In other cases, bodies may emit a substancelike quantity without absorbing it at the same time. Just think of the discharging of water from a container (Figure 1).

To get the full view of natural processes, internal and transport processes have to be combined. In the end we get a single picture of the role of energy in physics. In the following sections we shall describe this role in internal processes and transport phenomena.

P.4.2 Internal Processes and Power

The concept of power was introduced in the examples of discharging of containers and capacitors (Section P.1). We shall use this term in the context of internal processes. By the *power* of a process we mean the rate at which energy is either released or used inside the system. Energy is released if a substancelike quantity falls from a point of higher to a point of lower potential (Figure 16, left side). Conversely, energy is bound in a process in which an extensive quantity is lifted from a lower to a higher level (Figure 16, right side). The examples of electrical, gravitational, and hydraulic processes, in particular, tell us more about the details of the phenomena. We may summarize our experience with these fields of physics by saying that

> *The power of an internal process is proportional to both the rate at which a substancelike quantity X flows through its associated potential difference, and the potential difference $\Delta\varphi_x$:*

$$\mathcal{P} = -\Delta\varphi_x \left| I_x \right| \tag{43}$$

X stands for substancelike quantities while I_x denotes the magnitude of the rate of flow of this quantity. The formulas known for the calculation of electrical, hydraulic, or gravitational power are special cases of this general rule.

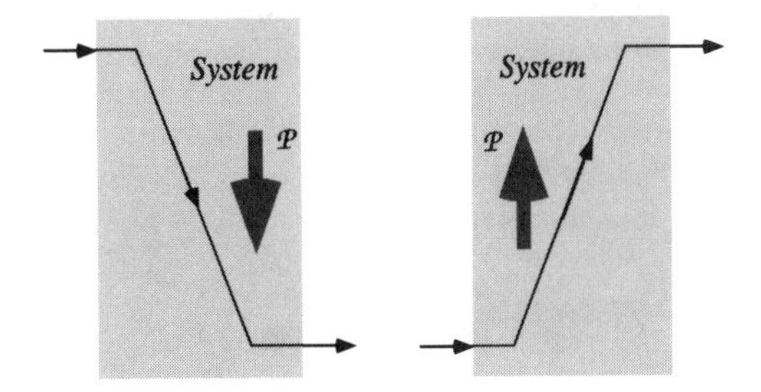

FIGURE 16. Voluntary and involuntary internal processes. The wide arrows pointing down and up symbolize the power of a process, i.e. the release and use of energy as a result of the flow of a substancelike quantity through a potential difference.

P.4.3 Transport Processes and the Flow of Energy

We have interpreted the power of an internal process as the rate at which energy is released (or used) inside a system. If we left it at that, power would be only a defined quantity of little or no importance. However, there is more to the concept of energy than what we have seen so far. Some theories suggest that energy is a quantity which may be transferred. We believe that the energy released in an electric heater or in the fall of water must come from somewhere outside the system. Also, for steady state processes, the same amount of energy delivered to a system must be emitted. This was expressed in Figure 5 for the system of an electric engine and a pump. The energy released in the electric process is delivered to the system, and the energy used to drive the hydraulic process leaves the system. Since the electric and the hydraulic power are equal in the steady state, the fluxes of energy associated with the transfer processes are equal in magnitude.

In the theory of motion, the concept of the transfer of energy is introduced as the power of a force acting upon a body. It is defined as the (scalar) product of two vectors — the velocity of the body and the force. Remember that force stands for the conductive or radiative flux of momentum, and that the velocity is the potential of motion. Therefore, we may interpret the formula as follows:

$$I_{E,mech} = \mathbf{v} \cdot \mathbf{I}_p \tag{44}$$

which means that the product of the potential and the flux of a substancelike quantity is equal to the rate at which energy is transferred into or out of a system as a result of a process. (Note that, in the context of mechanics, the word *power* is used differently than in electricity, where it is employed in the sense of the power of internal processes. We shall use the term *energy current* for the transport of energy, and *power* for the rate of liberation of energy in an internal process.)

This is a different concept than the one used so far. It does not refer to the rate of internal processes; it speaks of the interaction of bodies and their environment. We shall say from now on that in physical transport processes, in which substancelike quantities are exchanged between a body and the environment, energy is exchanged at the same time. Since the transport of energy into or out of a system is always accompanied by the transport of at least one substancelike quantity, we may use a vivid graphical interpretation of transport processes as in Figure 17.

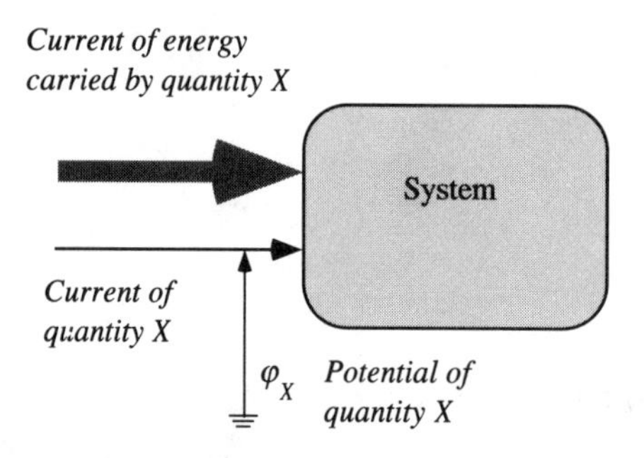

FIGURE 17. Transport processes are accompanied by the transport of energy. The substancelike quantity is called the *carrier* of energy in the process.

In transport processes energy is exchanged at the same time as one or more of the substancelike quantities are transported into or out of a body. The substancelike quantities are called the *carriers of energy*[9] in conductive and radiative transport processes. (Convection is excluded from this image.) The instantaneous value of the potential associated with the extensive quantity is called the *load factor*, and the flux of energy is given by the product of the carrier current and the load factor:

$$I_{E,X} = \varphi_X I_X \quad \textbf{(45)}$$

Here, X refers both to the substancelike quantity and to the type of process it represents: momentum in mechanical processes, charge in electrical processes, etc.

Now we can include the concept of energy in the waterfall picture of internal processes. The energy released in the fall of a substancelike quantity is the difference between the energy fluxes which accompany the transfer of the substancelike quantity into and out of the system (Figure 18). The magnitude of the energy fluxes is determined both by the magnitude of the flux of the carrier and the value of the associated potential. In mechanics, electricity, and gravitation, the value of the potential cannot be fixed absolutely. Only a difference in potentials has any meaning. Therefore, in those cases, the magnitude of a current of energy with respect to a system is arbitrary.

In the discussion of the transfer of energy we have used the example of mechanics. Since we often learn only to deal with the motion of mass points or of rigid

9. Falk, Herrmann, and Schmid (1983).

bodies, the energy flux associated with a transfer of momentum (i.e., with a force) is said to depend upon the velocity of the body. In general, however, there is no such thing as *the* velocity of a body. In a continuous case, every point may move at a different velocity. Therefore, we have to be very careful when stating the relation between energy currents and currents of the substancelike quantities. In fact, the rate at which energy is transferred is equal to the product of the rate of transfer of the extensive quantity and the potential at the point in space where the transfer is taking place. This point lies on the surface of a body in conductive transfers, while in radiative type of transports it is the point inside a body where fields and bodies exchange the quantity of interest.

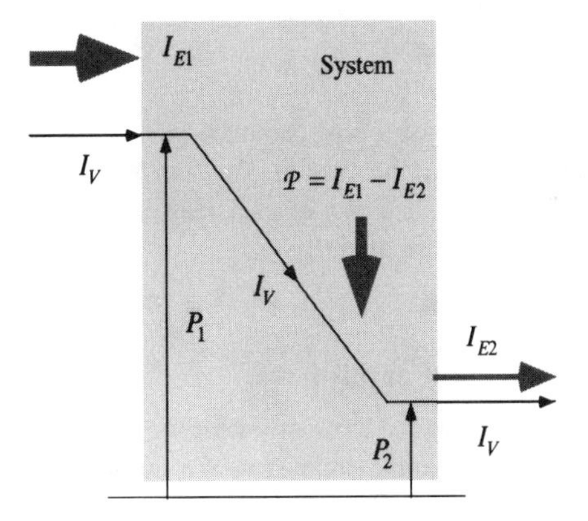

FIGURE 18. In a voluntary internal process, energy is released. The energy released is the difference between the energy supplied to and the energy removed from the system due to the transport of the substancelike quantity. In this figure, a hydraulic process has been chosen to represent other types of phenomena.

P.4.4 The Law of Balance of Energy

One more important feature of energy has to be discussed. Energy not only can be transported, it also may be stored in physical systems. The first example of this fact was found in mechanics. It is possible to derive a new statement from the equation of motion. We simply multiply Equation (25) by the velocity of the body and integrate over time:

$$\int_{t_1}^{t_2} m v \dot{v}\, dt = \int_{t_1}^{t_2} v F_{net}\, dt \tag{46}$$

The quantity on the left-hand side is the change in the kinetic energy of the body, while the term on the right-hand side is the net amount of energy transferred due to the transport of momentum with respect to the body:

$$\Delta\left(\frac{1}{2} m v^2\right) = W_{mech} \tag{47}$$

If we interpret the kinetic energy as the energy of the body due to its motion we may call Equation (47) an equation of balance of a quantity which has been conserved in a process. We may say that the change of the energy content of the body is equal to the net amount of energy transferred.

While this result is interesting, it is not an independent law of physics, since it has been derived from the law of balance of momentum by simple mathematical manipulation. A much more profound result was obtained shortly after 1850 by R. Clausius, who proved the existence of what we now call the internal energy of a fluid, and who derived the law of balance of energy for the thermomechanical processes undergone by simple fluids. After this result was obtained, the concept of energy gained acceptance in all fields of physics. Today we say that energy can be stored and transported, i.e. it satisfies an equation of balance; moreover, energy is conserved:

$$\dot{E} + I_{E,net} = 0 \tag{48}$$

Here, E denotes the energy of a physical system, while $I_{E,net}$ stands for the net flux of energy with respect to the system. The transfer of energy may be a result of a variety of different physical processes. Clearly, Equation (48) is the expression of balance of a conserved quantity.

P.4.5 The Compression of a Fluid

At this point, we will derive the expression for the transfer of energy which results from the compression or the expansion of a simple fluid. (The same derivation leads to the result for the expression of the energy transferred in pulling or pushing on a spring.) Consider a fluid such as air under pressure in a cylinder fitted with a piston (Figure 19). In the rest position, momentum simply flows into and out of the system. Since the gas is not in motion, it does not have any stored momentum. Even though momentum is transferred into and out of the system, energy does not flow, since no part of the surface of the body of fluid is in motion.

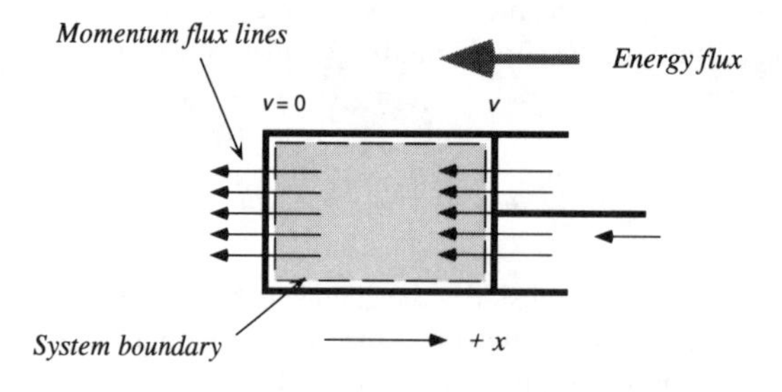

FIGURE 19. When a fluid is compressed, energy is supplied to the system as a result of the mechanical process. During expansion, energy flows out of the system.

If we begin to push the piston, however, the situation must change (Figure 19). First of all, we are upsetting the previous balance of momentum. Now, during acceleration of parts of the fluid, the body will store a certain amount of momentum. If we then hold the speed of the piston steady, the momentum fluxes at the opposing faces of the cylinder will be equal again. Let us consider this situation.

Only one part of the surface of the fluid is in motion, the part touching the piston. Since its speed is not zero, there appears a flux of energy equal to the product of the speed of the surface and the flux of momentum across this part of the surface:

$$I_{E,mech} = v I_p \tag{49}$$

Since the momentum flux is equal to the product of the pressure of the fluid and the surface area of the piston, we may transform Equation (49) as follows:

$$I_{E,mech} = vAP \tag{50}$$

This is equal to the product of the pressure of the fluid and the rate of change of its volume:

$$I_{E,mech} = P\dot{V} \tag{51}$$

This result holds in general for fluids of any shape in three dimensions. We shall have ample opportunity to use it in the following chapters of the book.

P.4.6 The Exchange of Energy in Magnetic Systems

Consider the concrete example of a paramagnetic substance filling the interior of a long straight coil. If we turn on an electric current through the coil, a magnetic field will be set up which leads to the magnetization of the body inside. Naturally, this process involves the transfer of energy to the magnetized body.

From electromagnetic theory it is known that the rate of transfer of energy may be expressed in terms of the product of the *magnetic tension* U_{mag} and the Hertz *magnetic current* I_{mag}:[10]

$$I_{E,mag} = -U_{mag} I_{mag} \tag{52}$$

The magnetic tension and the magnetic current are defined as follows:

$$U_{mag} = \int_{\mathcal{C}} \mathbf{H} \cdot \mathbf{s} dr \tag{53}$$

$$I_{mag} = \int_{\mathcal{A}} \dot{\mathbf{B}} \cdot \mathbf{n} dA \tag{54}$$

These definitions are similar to the quantities known from electricity. $\mathcal{A}$ and $\mathcal{C}$ stand for *surface area* and *curve*, respectively. The former is the path integral of the magnetic field **H**, while the latter is the rate of change of the magnetic flux.

10. Herrmann and Schmid (1986).

Obviously, the magnetic flux plays the role of the extensive magnetic quantity, and its rate of change replaces the rate of flow of electric charge in this analogy.

Let us now derive these quantities for the special example mentioned above. The magnetic tension in the uniform field of the coil is equal to

$$U_{mag} = LH \tag{55}$$

where L is the length of the coil. Since the magnetic flux density **B** is taken to be uniform over the cross section of the coil, the magnetic current turns out to be

$$I_{mag} = A\dot{B} \tag{56}$$

so that the magnetic energy current is equal to

$$I_{E,mag} = -VH\dot{B} \tag{57}$$

With a paramagnetic substance in the field, the magnetic flux density may be expressed as follows:

$$B = \mu_o\left(H + \frac{M}{V}\right) \tag{58}$$

M is the total magnetization of the body. If we consider only the body as the physical system and neglect the field in empty space, the magnetic energy current associated with the magnetization of the paramagnetic substance is

$$I_{E,mag} = -\mu_o H\dot{M} \tag{59}$$

There is an interesting point to be made about the examples just treated. The power involved in the compression of a simple fluid and in the magnetization of a body involves the rate of change of an extensive quantity rather than the transfer of a quantity such as charge or mass. Obviously, there are physical processes in which quantities are not transported. Rather, they change their values at the locations where they are to be found. Such processes may be interpreted in terms of the creation or the destruction of the quantity involved. Production and destruction join transport processes in our description of nature.

P.4.7 Energy and Mass

Finally, let us discuss one of the most fundamental aspects of the nature of energy. Since the beginning of this century it has been known that energy and mass (as a measure of gravity and inertia) are the same. This point will play a crucial role when it comes to the question of the nature of heat (Chapter 1).

The observations made so far about energy and momentum suggest a simple derivation of the relationship between mass and energy. Inertial mass is nothing but the momentum capacitance; i.e., it tells us how much momentum a body has for a given velocity (Section P.2.3). Now we will apply this idea to the phenomenon of light. As you probably know, light also carries momentum. Therefore, we may introduce the concept of momentum capacitance of light, i.e. the quantity measuring the amount of momentum of the system divided by its speed:

$$p = mc \tag{60}$$

The relation between the energy and the momentum added to a body, Equation (44), together with the laws of balance of energy and of momentum, i.e.,

$$\dot{E} = v\dot{p} \tag{61}$$

leads to the following expression for energy and momentum in the case of light:

$$E = cp \tag{62}$$

which is well known from the theory of electromagnetism. (c is the speed of light.) Combining both expressions finally leads to

$$E = mc^2 \tag{63}$$

So far this is not particularly important since it represents only the definition of the momentum capacitance of light. However, if we apply the idea which is embodied by Equation (63) to normal bodies we end up with a very interesting result. If we assume the momentum capacitance (i.e., the mass) of bodies and of light not to be two different concepts we have to conclude that the mass of a body is another expression for its energy: mass, i.e., gravity and inertia, measures the amount of energy and vice versa. We therefore may write the relation between momentum and velocity as follows:

$$p = \frac{E}{c^2}v \tag{64}$$

This result now holds for bodies and light if we set $v = c$ in the case of radiation. We can use it in the calculation of the energy added to a body when its momentum is increased from zero to p. Substituting Equation (64) into Equation (61), we obtain

$$E\dot{E} = c^2 p\dot{p} \tag{65}$$

If we integrate this equation with values of the energy changing from E_o (the rest energy of the body when its momentum is zero) to E (the corresponding amount of energy for momentum p), we end up with the general equation relating energy and the momentum of bodies and light, namely

$$E^2 = E_o^2 + c^2 p^2 \tag{66}$$

Again, this result also holds for light, since the rest energy of radiation is equal to zero. From this equation you may derive the expression for the inertia or gravity (i.e., the mass) of a body as a function of its velocity. Also, the classical expression for the kinetic energy of a body, Equation (47), is the limiting case of the general relation if we allow only for velocities which are small compared to the speed of light.

P.5 Continuum Physics, System Dynamics, and the Teaching of Physics

In this section we will trace some of the roots of what we have presented so far. We have drawn mostly from three sources: the development of continuum physics over the last 40 years, control engineering and system dynamics, and a new model of the teaching of introductory physics. At their cores all three appear to be so similar that it is quite natural to synthesize them into something new. The strength of continuum physics, i.e., its generality, can be put to use even for the beginner if we transfer its ideas to simpler cases by using the approach of system dynamics blended with appropriate didactic tools. The book you are reading is the result of my learning about these various lines of development. I hope to weave these threads into an interesting story of the role of heat in physical processes.

P.5.1 Continuum Physics

You probably have noticed that we have taken the model of *continuum physics* as the basis of our description of physical processes. Even though the continuous case has appeared only at a few points, the form of laws you have seen owes much to the general theory.

Continuum physics might well be the most general way of looking at classical phenomena. It has continued to evolve throughout this century,[11] and continuum thermodynamics is one of its most noteworthy additions to our understanding of nature.[12] Nowadays engineers mostly use classical field theories, and we can see the beginning of a unified presentation of physical processes in their books on momentum, heat, and mass transfer, subjects which are well established.[13] Ba-

11. An extensive reference to continuum mechanics can be found in the review articles by Truesdell and Toupin (1960), and Truesdell and Noll (1965).
12. Truesdell, 1984; Müller, 1985.
13. The text *Transport Phenomena* by Bird, Stewart, and Lightfoot (1960) set the stage for discussing a number of engineering subjects in a unified manner. Later texts for chemical and mechanical engineers include *Momentum, Heat, and Mass Transfer* by Bennet and Myers (1965) and *Transport Phenomena* by Brodkey and Hershey (1988).

sically, all we have done here is to take some of the results of continuum thermodynamics and transform them to an introductory level with the help of some didactic tools (see Section P.5.3). By using the ideas and the language offered by the field theories, we can unify the simpler cases which are presented in introductory courses.

The structure of a theory of continuum physics outlined at the beginning of this chapter constitutes a combination of and a change from the approaches described by Truesdell and by Müller.[14] Commonly, the energy principle is taken with the equations of balance of momentum, mass, electrical charge, etc., while the entropy principle (the balance of heat) is added as a special law. I prefer to add the balance of heat to the list of fundamental laws, where each equation of balance refers exclusively to the phenomena of a particular field of physics. The energy principle is used separately as the principle which unites different types of processes. While formally the change is small, it is important in that it gives a clearer view of the analogies between different fields of physics. Here, thermal physics is placed firmly on the same level as the theories of electromagnetism and motion.

P.5.2 System Dynamics

The "lumped parameter" form of physical laws frequently used in the preceding pages constitutes a simplified version of the general case of continuum physics. Reducing the complexity of the spatially inhomogeneous problem by taking bodies to be uniform leads to an approach which resembles the description of dynamical systems known from *system dynamics.*[15] System dynamics has its roots in control engineering or cybernetics, which was developed around the middle of this century. In systems and control engineering one makes liberal use of analogies of the form discussed here.[16] Unfortunately, in most cases, the comparison of different fields is only carried as far as particular applications demand or suggest. Also, because of this self-imposed limitation, the structure of theories is never developed as fully as possible. Still, control engineering and system dynamics can serve as a basis for a unified description of lumped dynamical systems, and as such these theories are very useful for physics in general and for thermodynamics in particular.[17]

The software tools developed in recent years for system dynamics let us create graphical representations of the models of physical systems and processes in a form directly comparable to the structure of theories outlined above. As an example, consider the flow of a viscous fluid out of a container as in Figure 1. A

---

14. Truesdell, 1984, p. 63–64; Müller, 1985, p. 1–22.
15. J. Forrester, 1961, 1968, 1969; N. Roberts et al., 1983.
16. White and Tauber, 1969.
17. The use of system dynamics goes far beyond physics or engineering, as demonstrated by applications in ecological and social sciences as well as in management.

model prepared using one of the software tools (Figure 20) exhibits the components we have seen so far, namely, laws of balance and associated constitutive laws. Similar models exist for other types of processes, and complex systems can be modeled by combining smaller subsystems.

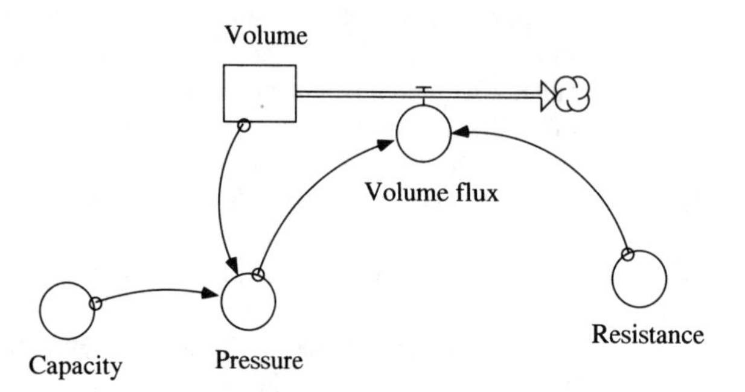

FIGURE 20. System dynamics model of the discharging of a container. This model was built with the program STELLA II™ (High Performance Systems, Hanover, New Hampshire). The rectangle represents a substancelike quantity, while the fat arrow stands for the associated current. Combining a rectangle with one or more currents is equivalent to writing an equation of balance. The circles introduce other variables, and the thin curved arrows allow for constitutive relationships to be built up. The model expresses our assumptions that the volume flux depends on both the pressure and the hydraulic resistance, while the pressure is computed from the volume and the capacitance of the container. Appropriate equations have to be written in the circles. For a complete model numerical solutions can be computed.

P.5.3 The Teaching of Introductory Physics

The last 20 years have seen new interest in generalized approaches to the teaching of introductory physics. One of the models proposed stresses the use of the substancelike quantities,[18] i.e., the additive quantities of continuum physics. These quantities are visualized as "carriers" of energy in transport processes; the image is supported by flow diagrams (similar to those in Figures 17 and 18) depicting the relationships between the fundamental quantities undergoing changes in physical processes. As a consequence, a structure of physics emerges which is centered around the analogies between different fields (Table 1), and which might be called systems physics.[19] At Technikum Winterthur, W. Maurer,[20] and I have combined this model with the fundamental ideas found in continuum physics and with those of system dynamics.

Concrete examples of teaching this type of unified physics demonstrate its usefulness for students all the way from primary school to college. Information theoretical models of learning suggest that the use of analogies should lead to a significant reduction in the amount of information which has to be digested by a student.

18. Falk and Ruppel, 1973 and 1976; Falk and Herrmann, 1977–82; G.B. Schmid 1982, 1984.
19. Burkhardt, 1987.
20. W. Maurer, 1990b.

TABLE 1. A comparison of different physical processes

Subject	Quantity Potential	Power Energy current	Resistance	Capacitance	Inductance
Hydraulics	Volume Pressure	$\Delta P I_V$ $P I_V$	$\Delta P / I_V$	dV/dP	$\Delta P / \dot{I}_V$
Electricity	Charge Electric potential	$\Delta\varphi_{el} I_q$ $\varphi_{el} I_q$	$\Delta\varphi_{el} / I_q$ $j_q = -\sigma d\varphi_{el}/dx$	$q/\lvert\Delta\varphi_{el}\rvert$	$\Delta\varphi_{el} / \dot{I}_q$
Motion (translation)	Momentum Velocity	$\Delta v I_p$ $v I_p$	$j_p = -\mu dv/dx$	p/v	$\Delta v / \dot{I}_p$

Moreover, the perspective outlined in this chapter has led to a critical appraisal of the teaching of physics in the light of students' conceptions of natural processes. In recent years, researchers in the field of didactics have discovered the importance for learning of what may be called students' misconceptions.[21] Since students' ideas about nature often contradict the theories of physics (and of other fields as well), these alternative conceptions are bound to reduce the effectiveness of teaching and learning. The literature in the field of misconceptions therefore abounds with recipes of how we can make students change their minds. It is generally forgotten that the forms of physical theories are not absolute; there would be room for integrating our everyday concepts about nature in a more constructive manner than is normally the case. It is possible that a particular physical theory, rather than our image of processes, is "misconceived".[22] For me, our everyday views of heat furnish one of the main reasons for a new approach to the field of thermodynamics. You will come across this theme in the following chapters.

21. J.D. Novak, 1987.
22. Fuchs, 1986, 1987b,c.

Questions and Problems

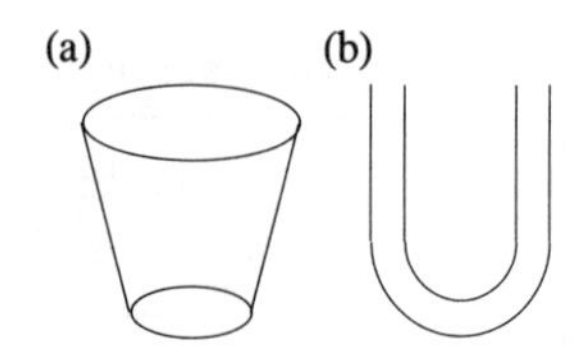

FIGURE 21. Problem 2.

1. Calculate the hydraulic capacitance of a glass tube used in a mercury pressure gauge. The inner diameter of the tube is 8.0 mm.
2. Derive the expression for the hydraulic capacitance of the conical container shown in Figure 21. Do the same for a U-tube.
3. Except for the sign, the hydraulic capacitance of a compressible fluid such as air is defined as in Equation (8). How does the capacitance relate to the compressibility of the fluid?
4. Two currents of water are flowing into a fountain. The first changes linearly from 2.0 liters/s to 1.0 liters/s within the first 10 s. The second has a constant magnitude of 0.50 liters/s. In the time span from the beginning of the 4th second to the end of the 6th second, the volume of the water in the fountain decreases by 0.030 m^3. a) Calculate the volume flux of the current leaving the fountain. b) How much water will be in the fountain after 10 s, if the initial volume is equal to 200 liters?

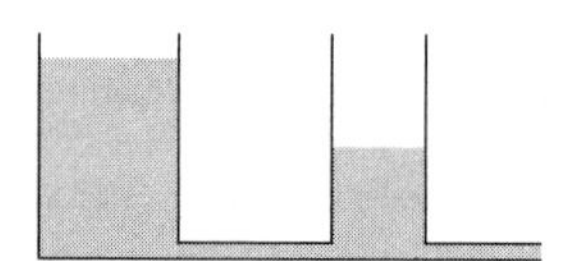

FIGURE 22. Problem 5.

5. Two containers are joined by a pipe as in Figure 22. The second container has both an inlet and an outlet. Assume the flow through the pipes to obey the law of Hagen and Poiseuille. a) Write the equations of balance of volume for the fluid in the containers. b) Derive the relation between volume of fluid and pressure of fluid at the bottom of each of the containers. c) Write the laws for the volume fluxes through both pipes. d) Derive the differential equations for the height of the fluid in each of the containers in terms of the hydraulic capacitance and resistance of the elements of the system.
6. Viscous oil is to be pumped from a shallow container into one lying 10 m higher up. The pipe has a diameter of 5.0 cm and a length of 20 m. If the mass flux is required to be 10 kg/s, how large should the power of the pump be? Draw the energy and carrier flow diagram of the system and the processes. Neglect the acceleration of the fluid.

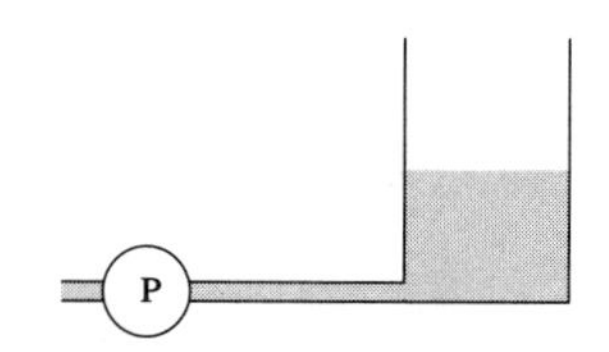

FIGURE 23. Problems 7 and 8.

7. A large oil tank is filled through a pipe at its bottom (see Figure 23). The flow of oil through the pipe is supposed to be laminar. Derive the instantaneous power of the ideal pump in terms of the length and the radius of the pipe, the viscosity and density of the oil, and the height of the oil in the tank.
8. A tank is filled through a pipe at the bottom (as in the previous problem). Assume the flow through the pipe and in the container not to be affected by friction.
a) Express the energy needed to fill the tank up to a certain height in therms of the hydraulic capacitance. b) Where has the energy supplied gone to?
9. A large and shallow lake is going to be filled through a horizontal pipe with a length of 10 km. Initially the lake is empty; in the end it is supposed to contain 10^5 m^3 of water. Assume the hydraulic resistance to be modeled by the law of Hagen and Poiseuille; i.e., take the volume flux to be proportional to the pressure difference across the pipe. The pressure drops by 10^2 Pa per meter of length at a volume flux of 1.0 m^3/s. While the lake is being filled, water evaporates from its surface at a rate of 0.10 m^3/s. a) If the volume flux is constant and equal to 0.50 m^3/s how much energy is required for pumping while filling the lake? b) How large should the (constant) volume flux be for the energy required to fill the lake to be minimal?

10. If you fill the tank of Problems 7 and 8 through a pipe which leads to the top of the tank (Figure 24), how much energy is required? How does this compare to the results of those problems? Has energy been lost?

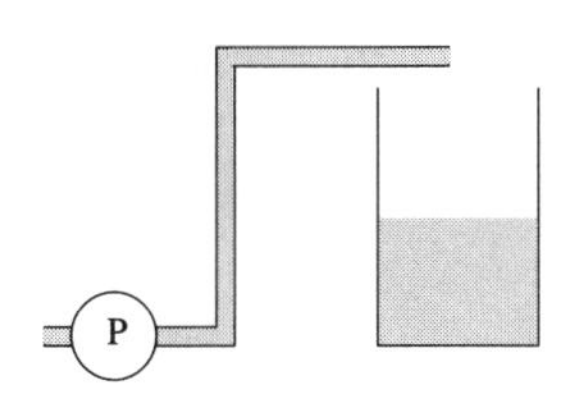

FIGURE 24. Problem 10.

11. Oil with a density of 800 kg/m^3 and a viscosity of 0.60 Pa·s is flowing through a system of pipes as in Figure 25. The pressure at point A is 1.40 bar, while at point C it is 1.20 bar. The thin pipes have a diameter of 1.0 cm while the diameter of the piece leading from A to B has a diameter of 2.0 cm. Neglect the influence of corners in the pipes, and assume the law of Hagen and Poiseuille to apply. a) What is the value of the volume flux through the lower pipes? b) How large is the total volume flux through the pipes? c) What is the pressure at B?

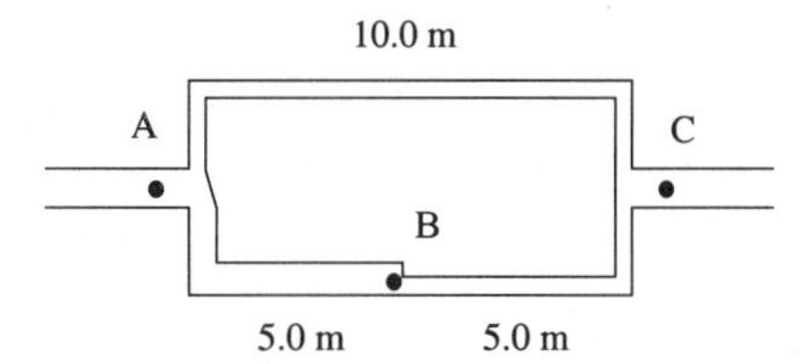

FIGURE 25. Problem 11.

12. Two tanks (see Figure 26) contain oil with a density of 800 kg/m^3 and a viscosity of 0.20 Pa·s. Initially, in the container, which has a cross section of 0.010 m^2, the fluid stands at a level of 10 cm; in the second container (cross section 0.0025 m^2) the level is 60 cm. The hose connecting the tanks has a length of 1.0 m and a diameter of 1.0 cm. a) What is the volume current right after the hose has been opened? b) Calculate the pressure at A, B, C, and D at this point in time. The pressure of the air is equal to 1.0 bar, and C is in the middle of the hose. c) Determine the rate at which energy is released at the beginning as a consequence of fluid friction. d) Sketch the levels in the containers as a function of time. e) Sketch an electric circuit which is equivalent to the system of containers and pipe. f) Sketch a pressure profile (pressure as a function of position) for a path leading from A to D; include a point C* at the other end of the pipe from point B.

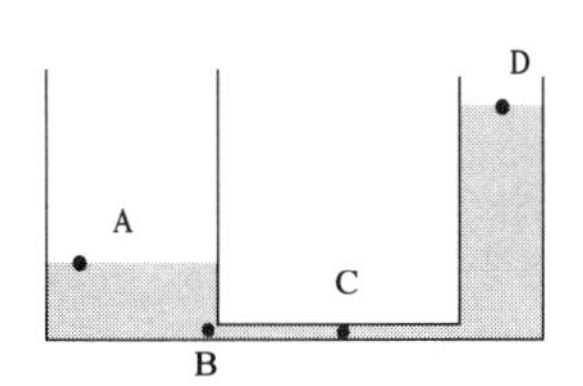

FIGURE 26. Problem 12.

13. What is the hydraulic inductance for a pipe which is channeling water down from an artificial lake to a hydraulic power plant (Figure 27)? The diameter of the pipe is 0.40 m. How large a rise in pressure would you expect if the volume flux (which has a magnitude of 1.0 m^3/s) is reduced to zero within one second?

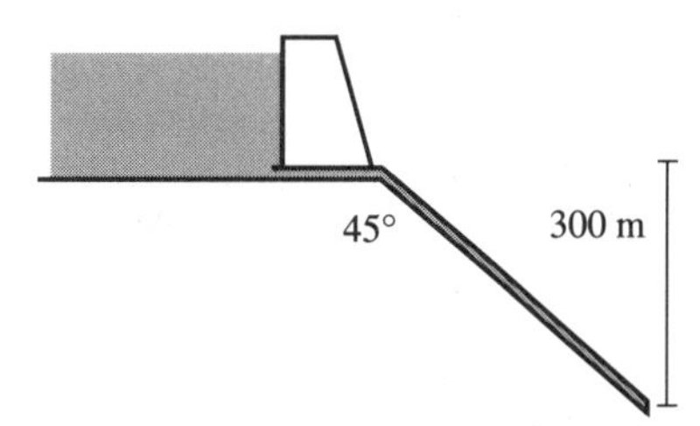

FIGURE 27. Problem 13.

14. A system composed of a straight-walled container with a long thin pipe leading out at the bottom (Figure 1) has a hydraulic capacitance of 10^{-5} m^3/Pa, a resistance of 10^6 Pa·s/m^3, and an inductance of 10^6 Pa·s^2/m^3. Initially, the fluid stands at a level of

1.0 m. Its density is equal to 1000 kg/m^3. a) Calculate both time constants (capacitive and inductive). b) What would be the highest value of the current if the effect of the inductance is neglected? c) Sketch as accurately as possible the volume flux as a function of time for the first 20 s.

15. What is the relationship between Kirchhoff's second law (the loop rule: the sum of all voltages in a closed circuit is equal to zero) and the balance of energy? What is the hydraulic analogue of this law?

16. Derive the expression for the energy stored in a charged capacitor from consideration of the process of charging. Compare the result to the analogous hydraulic expression.

17. Consider two capacitors, one of them charged, connected in a circuit. a) Calculate the final charges and voltages of the capacitors in terms of the initial charge and the capacitances. b) Is the energy of the capacitors conserved? c) Translate the problem into an equivalent hydraulic one.

18. Two identical capacitors are connected in the circuit shown in Figure 28. Initially the voltage of the first is equal to 100 V while the voltage of the second is 50 V. Determine the rates of change of the charge of the capacitors right at the beginning.

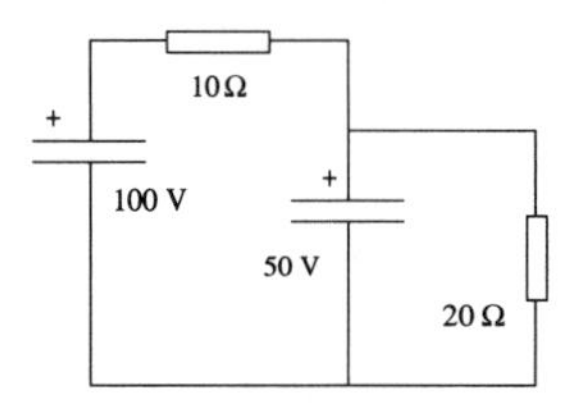

FIGURE 28. Problem 18.

19. A capacitor (capacitance 150 μF) and a resistor (resistance 1500 Ω) are connected in series to a battery (voltage 50 V) at time $t = 0$ s. The initial charge of the capacitor is equal to zero. a) Derive the equation of balance of the charge of the capacitor. From its solution derive the formula for the electric current as a function of time. b) Draw the carrier and energy flow diagrams for the battery, the resistor, and the capacitor. c) What are the values of the electrical power of the three elements after 0.15 s? d) What are the values of the corresponding electrical energy currents at that point in time? e) Calculate the rate of change of the energy of the capacitor. f) How large is the rate of change of the energy of the resistor?

20. An electric resistor with a resistance of 120 Ω is connected in series to a parallel combination of capacitor (4 mF) and a second resistor (60 Ω) (Figure 29). The elements are connected to a battery having a voltage of 24 V. a) Describe qualitatively the voltage across the second resistor as a function of time. b) At a certain point in time an electric current of 150 mA is flowing through the first resistor. What is the value of the charge of the capacitor at this time? c) How large is the rate of change of charge of the capacitor at this point in time? d) Calculate all values of the electrical power with respect to the four elements in the circuit. Explain whether energy is released or used in the electric processes. Explain where energy is coming from and where it is going to.

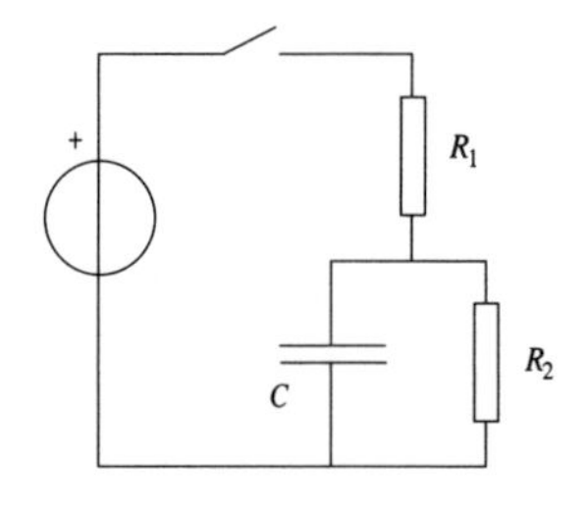

FIGURE 29. Problem 20.

21. In the previous problem a) sketch a hydraulic system made up of containers, pipes, and pumps which would be equivalent to the electrical system shown here; b) write down the equations governing the processes taking place in the circuit. Solve the differential equation for the current through the second resistor and demonstrate that the time constant of the system is equal to

$$\tau = C\frac{R_1 R_2}{R_1 + R_2}$$

22. Derive the expression for the energy contained in an inductive element (consider the process of starting a current flowing through a circuit containing a battery, an inductor, and a resistor). Translate the result for hydraulics. Show that you can obtain the

formula for the inductance of a pipe with fluid, as in Equation (15), by comparing the energy of the inductive element with the kinetic energy of the fluid in the pipe.

23. Derive the equation for the oscillation of water in a U-tube (this can be done by combining the equation of balance of volume and the expression for the current in the presence of inductance and resistance). What is the form of the equation if you neglect the effects of resistance? Demonstrate that the period of oscillation can be calculated from

$$\omega = \frac{1}{\sqrt{L_V K_V}}$$

24. Repeat the previous problem by directly using Newton's law of motion for the water in the U-tube. Do you get the same result?

25. A person is pulling a crate across the floor at constant speed via a rope (Figure 30). Take the positive direction to coincide with the direction in which the body is pulled. a) Identify the closed circuit through which the horizontal component of momentum is flowing. b) Determine the momentum fluxes (and their signs) with respect to the crate, the person, and the earth, i.e., the floor. c) There exist several relationships between the different fluxes identified in (b). Which of these have to do with the action-reaction principle (Newton's third law)? Which condition is expressed by the other relationships?

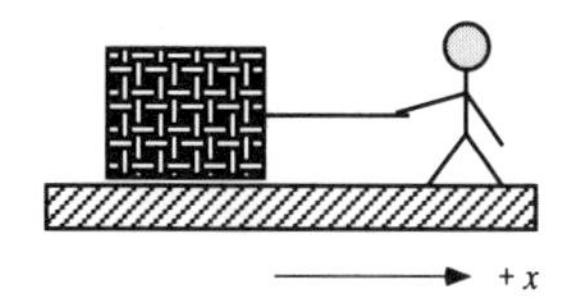

FIGURE 30. Problem 25.

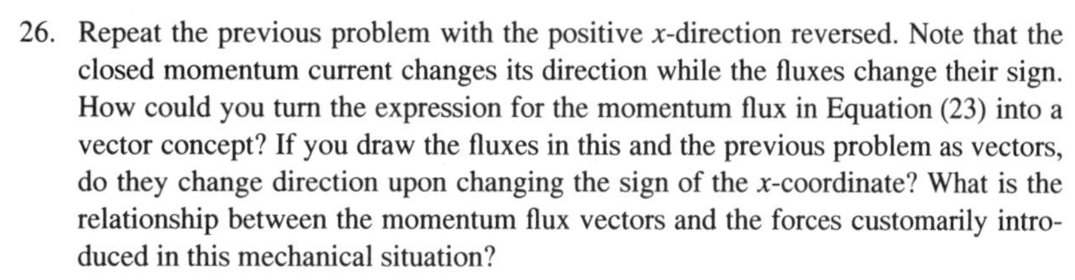
26. Repeat the previous problem with the positive x-direction reversed. Note that the closed momentum current changes its direction while the fluxes change their sign. How could you turn the expression for the momentum flux in Equation (23) into a vector concept? If you draw the fluxes in this and the previous problem as vectors, do they change direction upon changing the sign of the x-coordinate? What is the relationship between the momentum flux vectors and the forces customarily introduced in this mechanical situation?

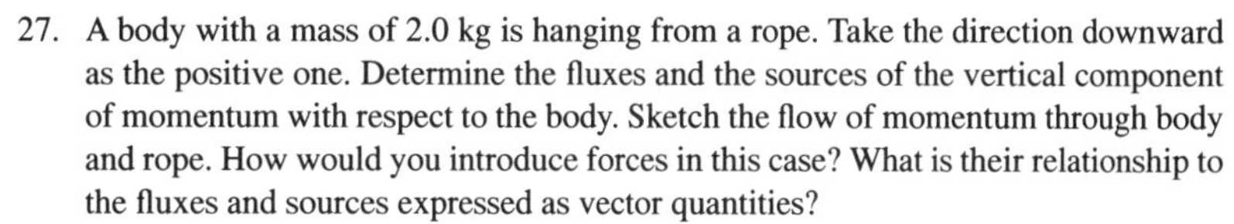
27. A body with a mass of 2.0 kg is hanging from a rope. Take the direction downward as the positive one. Determine the fluxes and the sources of the vertical component of momentum with respect to the body. Sketch the flow of momentum through body and rope. How would you introduce forces in this case? What is their relationship to the fluxes and sources expressed as vector quantities?

28. Consider a body in free fall. a) Determine the momentum fluxes with respect to its surface. How large are the sources of momentum? b) Express the equation of balance of momentum of the body and determine its solution.

29. Solve the equation of balance of momentum for a small sphere falling in a viscous fluid, Equation (34). Set up the corresponding equation if the resistive medium is air.

30. A rope with a given (constant) mass per length is hanging from a hook. Express the equation of balance of momentum for small segments of the rope and derive the appropriate differential equation for the continuous case. Then determine the momentum current density in the rope as a function of position. How does this quantity relate to the mechanical stress in the rope?

31. Consider a long bar falling in an inhomogeneous gravitational field increasing linearly downward (Section P.2.4). The body is assumed to be rigid. a) Calculate the acceleration of the body with the help of the equation of motion of the entire body. b) Compute the momentum current density through the body as a function of posi-

tion in the body. (Hint: set up the equation of balance of momentum for the spatially continuous case).

32. A liquid having a density of 920 kg/m^3 is flowing through a pipe whose diameter decreases from 3.0 cm to 1.5 cm (Figure 31). The speed of flow at the larger entrance is 4.82 m/s. a) Compute the convective momentum fluxes at the inlet and the outlet. b) The pressure of the fluid at the inlet is 1.10 bar. Calculate the conductive momentum flux at the entrance. Compare the magnitude of the convective and the conductive fluxes.

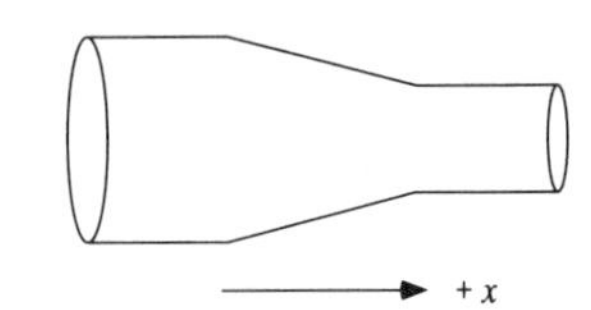

FIGURE 31. Problem 32.

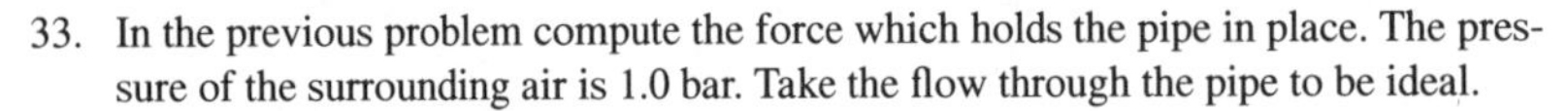

33. In the previous problem compute the force which holds the pipe in place. The pressure of the surrounding air is 1.0 bar. Take the flow through the pipe to be ideal.

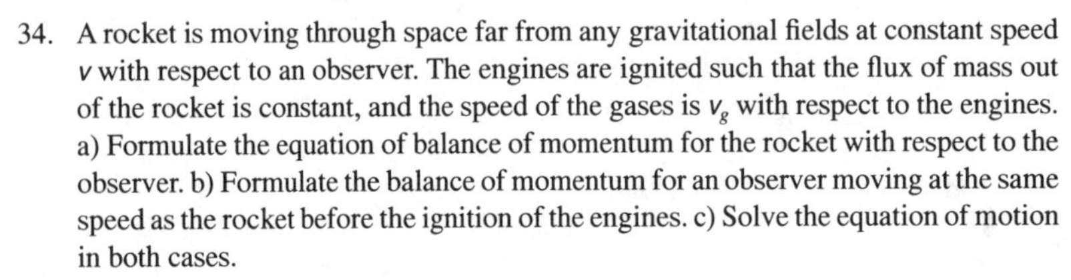

34. A rocket is moving through space far from any gravitational fields at constant speed v with respect to an observer. The engines are ignited such that the flux of mass out of the rocket is constant, and the speed of the gases is v_g with respect to the engines. a) Formulate the equation of balance of momentum for the rocket with respect to the observer. b) Formulate the balance of momentum for an observer moving at the same speed as the rocket before the ignition of the engines. c) Solve the equation of motion in both cases.

35. A balloon is inflated with air and then released. It is observed that it moves horizontally from rest a certain distance in the first fraction of time. How can you estimate the pressure of the air inside the balloon at the moment it is released? Assume the flow of air out of the opening to obey Bernoulli's law.

36. A train engine is pulling an open freight car along a horizontal track. It has to pull with a force of 1000 N. Now coal is dumped into the car from above at a rate of 1000 kg/s (the coal is falling down vertically). What is the force exerted by the engine upon the car if the composition is to continue to move at constant speed? Assume the friction between car and tracks not to change.

37. The following data are given for a rocket-powered car: speed of the gas stream with respect to the car, 2000 m/s; mass flux, 10 kg/s; coefficient of air resistance, 0.32; cross section of the front of the car, 1.40 m^2; frictional force exerted by the road, 700 N. a) How large is the acceleration of the car at a speed of 500 km/h? Its mass at this moment is 1200 kg and the density of the air is 1.20 kg/m^3. b) What is the rate of change of momentum of the car at that moment?

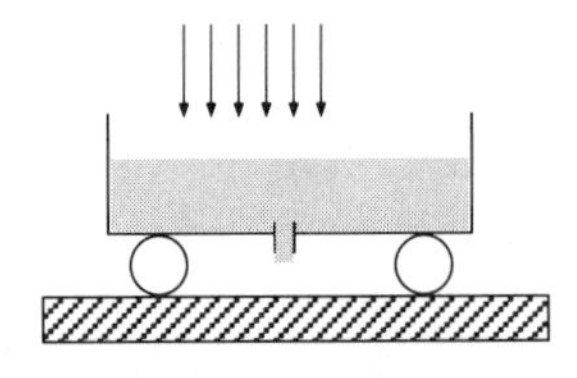

FIGURE 32. Problem 38.

38. An open car moves underneath a vertically falling current of water (Figure 32). At the same time, the car loses water through a hole at the bottom at the same rate at which it picks up water. Set up the equation of motion for the car and determine its speed as a function of time. Assume friction to be negligible.

39. One mole of salt is dissolved in 10 liters of water inside a tank. A current of fresh water of 1.0 kg/s enters the container where it is instantly mixed with the salt-water solution. The solution flows out of the tank at a rate of 0.50 kg/s. a) Determine the concentration of salt inside the tank as a function of time. b) How much salt has left the container in the first 10 s?

40. Two substances A and B are pumped into a reactor at rates of 10 mole/s and 20 mole/s for A and B, respectively. A certain amount of substance B already is present in the reactor. A and B react instantly to form substance C according to A + 2B $\rightarrow$ 3C. C is pumped out of the reactor at a rate of 15 mole/s. Determine the rates of change of the amounts of A, B, and C in the reactor.

41. A flat plate with a surface area of 0.50 m^2 is pulled horizontally over another plate with a film of oil between them. The film has a thickness of 4.0 mm. To move the plate at constant speed of 1.0 m/s, one has to pull with a force of 25 N. How large is the viscosity of the oil?

42. A fluid is confined between the walls of two concentric cylinders (Figure 33). The gap between the cylinders is very narrow. The inner cylinder can be rotated, and the torque upon it and the angular velocity can be measured. Determine the viscosity of the fluid.

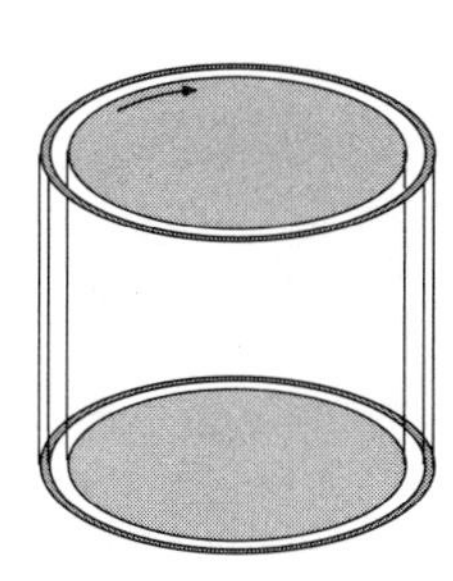

FIGURE 33. Problem 42.

43. Determine the speed of sound in air on the basis of the general wave equation, as in Equation (42), using the hydraulic variables. The hydraulic capacitance of the gas is defined by the relation $K_V = -dV/dP$. Normally, the pressure variations satisfy the adiabatic relation

$$\frac{dP}{dV} + \gamma \frac{P}{V} = 0$$

(See Chapter 2.) Take 1.0 bar for the pressure and 1.2 kg/m^3 for the density of air.

44. Show that if you use momentum as the fundamental quantity, a (linear) spring is an inductor. Determine its inductance. Show that a body hanging from the spring has the property of a capacitor. Now determine the frequency of oscillation from the capacitance and the inductance of the system.

45. Repeat the derivation of the wave equation, Equation (42), for charge flowing through a wire. a) Neglect resistance, b) include Ohm's Law, Equation (36), and demonstrate that you get

$$\frac{\partial^2 \varphi}{\partial t^2} = \frac{1}{L^* C^*} \frac{\partial^2 \varphi}{\partial x^2} - \frac{1}{\sigma A L^*} \frac{\partial \varphi}{\partial t}$$

Does this equation become the simple wave equation if you neglect resistance? What is the form of the equation if you let the inductance go to zero? What does this mean for the speed of propagation of the wave?

46. Set up the equation of motion of a body hanging from a (linear) spring. Solve it and show that you get the same expression for the frequency of oscillation as that found in the previous problem.

47. Explain the difference between power of a process and energy currents with respect to a system.

48. Use the expression for the power of a process for calculating a) the energy required for lifting a body of mass m a certain distance h at the surface of the earth, and b) the energy released from an electric field when an electron moves through a given voltage U in the field.

49. Calculate the source rate of energy with respect to a stone falling freely 2.0 s after it has been released. The mass of the body is equal to 0.20 kg. Where does the energy come from?

50. A car moving horizontally at a constant speed of 120 km/h is using 8.0 liters of gasoline in a distance of 100 km. The mechanical efficiency of the engine is 0.20. Draw a flow diagram for the car as the system, depicting energy carriers, energy currents, and power. How large is the magnitude of the sum of all resistive forces acting upon the body? Repeat the flow diagram with the engine as the system.

51. The car of the previous problem is moving up a slope with an angle of 5°. Draw the flow diagram. At what rate would it be using gasoline (all other quantities remaining the same)?

52. Prove that, in one-dimensional motion, the rate of supply of energy to a rigid nonrotating body is mav as a result of changes of its speed.

53. A linear spring is attached to a wall on one side. As it is stretched, determine all energy fluxes with respect to the spring. Calculate the change of the energy content of the spring as a function of the stretching

54. Show that the energy of a spring looked at as an inductor is equivalent to the common expression found in the previous problem.

55. A mill stone grinds wheat, rotating atop a horizontal surface. Draw a flow diagram for energy carriers, energy fluxes, and power for the stone as the system. The stone rotates once in 2.0 s. If the energy flux supplied to the stone is 1.0 kW, how large is the flux of angular momentum through the shaft of the mill stone? How large is the net torque with respect to the stone?

56. A cylindrical body is suspended from a thin wire (Figure 34). If the wire is twisted it exerts a torque upon the body, making it oscillate. a) Identify fluxes of angular momentum with respect to the wire and the cylinder. b) Twisting the wire by an angle φ leads to a torque $I_L = D^*\varphi$. (I_L is the flux of angular momentum.) Determine the inductance of the wire. c) Which quantity corresponds to the (angular momentum) capacitance of the system? d) Determine the frequency of oscillation of the cylinder.

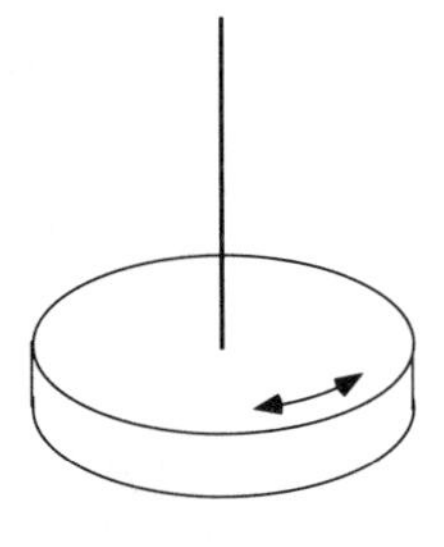

FIGURE 34. Problems 56 and 57.

57. Set up Euler's equation of motion for the rotating cylinder of the previous problem. Solve the equation and show that you get the same expression for the frequency of oscillation as that found on the basis of the systems approach used before.

58. Calculate the dependence of the mass of a body upon its velocity. How large is the percent change of the mass of an electron which has been accelerated through a potential difference of 50 kV?

59. Show that the formula for the kinetic energy of a body moving at speeds much smaller than the speed of light can be obtained as the limiting case of the general relationship between momentum and energy of a body.

60. In Figure 35 you find a system dynamics diagram of a body suspended from a spring and oscillating up and down. Identify the part of the diagram which represents Newton's law. What type of relation is represented by the box and the flow labeled *position* and *velocity*? What is the nature of the relation represented by the flow called net *momentum flux* and the three variables associated with it? Can you identify the feedback loops of the system?

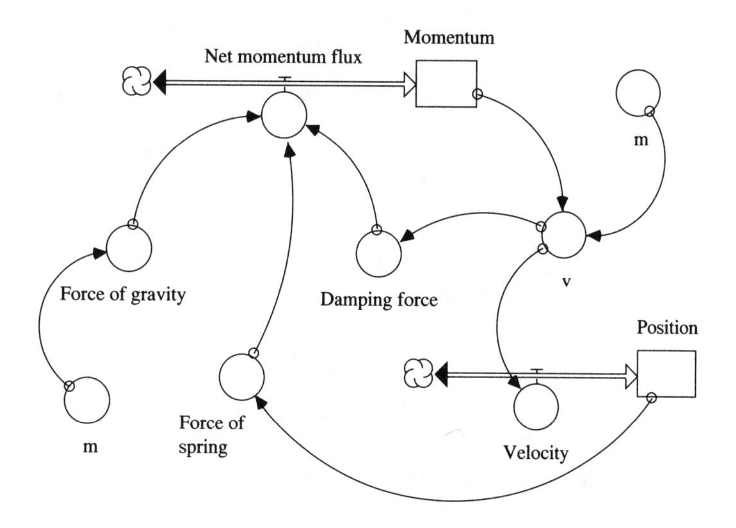

FIGURE 35. Problem 60.

```
momentum(t) = momentum(t - dt) + (net_momentum_flux) * dt
INIT momentum = 0
INFLOWS:
net_momentum_flux = force_of_gravity+force_of_spring+damping_force

position(t) = position(t - dt) + (velocity) * dt
INIT position = 2
INFLOWS:
velocity = v

damping_force = -0.5*v
force_of_gravity = 9.81*m
force_of_spring = -10*position
m = 0.50
v = momentum/m
```

61. Figure 36 shows the system dynamics model of Problem 39. Identify the graphical representation of the differential equations which you have written for the solution of Problem 39. Does the numerical solution presented correspond to what you have calculated?

62. Sketch a system dynamics model for the process of discharging of a capacitor. Repeat the problem for charging with the help of a battery.

63. With the model of Problem 62, include the power of the electrical processes of the battery, the capacitor, and the resistor. Integrate the power over time. What do these quantities represent?

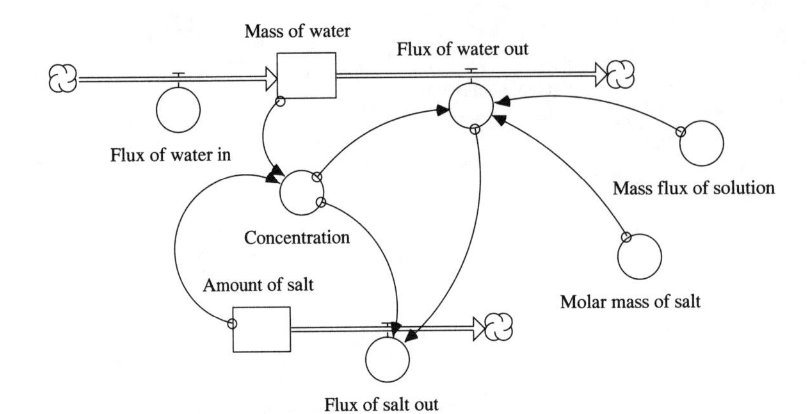

FIGURE 36. Problem 61.

```
amount_of_salt(t) = amount_of_salt(t - dt) + (- flux_of_salt_out) * dt
INIT amount_of_salt = 1
OUTFLOWS:
flux_of_salt_out = concentration*flux_of_water_out
mass_of_water(t) = mass_of_water(t - dt) + (flux_of_water_in - flux_of_water_out) * dt
INIT mass_of_water = 10
INFLOWS:
flux_of_water_in = 1
OUTFLOWS:
flux_of_water_out = mass_flux_of_solution/(1+molar_mass_of_salt*concentration)
concentration = amount_of_salt/mass_of_water
mass_flux_of_solution = 0.50
molar_mass_of_salt = 0.040
```

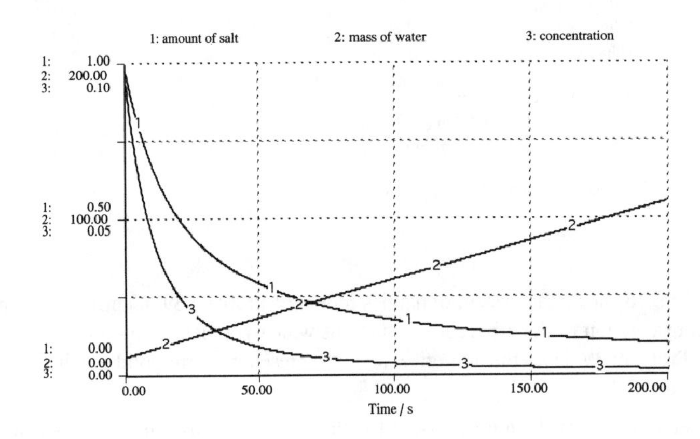

CHAPTER 1

Hotness, Heat, and Energy

Thus the production of motive power is due ... not to any real consumption of caloric, but to its transport from a warm body to a cold body.

S. Carnot, 1824

The goal of this chapter is to introduce the fundamental quantities and concepts of thermodynamics. We shall motivate the generic laws of thermal physics which we are going to use throughout this book. Constitutive laws will be introduced on an informal basis when needed for getting answers to some special problems. They will be treated in great detail in the following chapters. First, we will discuss hotness and the measurement of temperature. Following this, we will study the properties of heat in some detail. Finally, the relationship between heat and energy has to be dealt with. We shall motivate this relationship by considering the principle of operation of heat engines. Paired with the law of balance of energy, we will be able to address some practical problems which are of general interest to engineers and others concerned with energy problems. First however, let us take a look at what kinds of images and concepts can be formed about heat and hotness from our everyday knowledge of thermal processes.

1.1 Thermal Phenomena, Concepts, and Images

Thermal phenomena are part of our everyday experience. They allow us to conceive of the quantities which are necessary for a theory of thermal physics. Nature tells us where to begin and what to look for. On the basis of experience, we construct the concepts which we do not derive from anything else, i.e., those which are truly fundamental. We shall take these to be *hotness*, *heat*, and *energy*. We intuitively know the first two of these quantities, even though our knowledge is not precise in any scientific sense. We have become familiar with energy through our study of other parts of physics. By weaving the elements into a theory of thermal processes, and by comparing the predictions of the theory with new phenomena, we will find out whether our ideas are correct.

Assuming that you probably have had at least a moderate exposure to thermal physics, you might find this section the most demanding of all, even though there will not be a single equation to set up or to solve. It is imperative that you see very clearly what heat is not before we go on with the story.

1.1.1 Moving, Charging, Heating, and Melting Bodies

There are two important points to consider at the beginning of any discussion concerning the nature of heat. First, we would like to know what is responsible for making a stone warm or for melting a block of ice. Second, we have to recognize the difference between a moving body, a charged sphere, and a hot stone, and how this difference can be expressed in a theory of physics.

It is quite clear that *heat* makes a stone warm and melts an ice cube. We put more heat into a body such as a stone to make it warmer; taking heat out of the stone will make it cool down. If we put ice in a warm place, heat will enter the body and make it melt. *Heat* is the perfect quantity for describing what is happening in these situations.

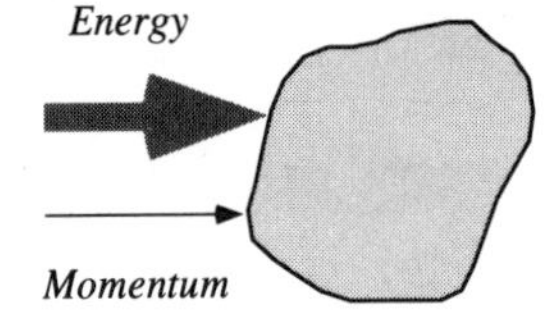

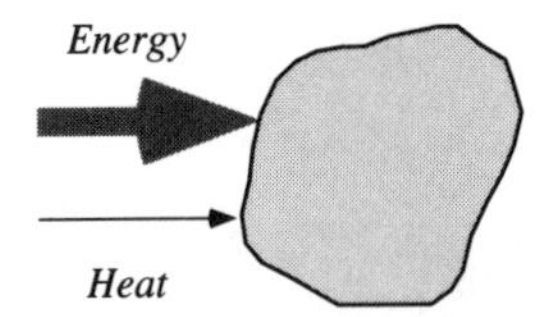

FIGURE 1. Two processes compared. In the first we accelerate a stone. In terms of physics, we say that momentum is added to the body, which makes it move faster. We know that energy is added to the stone at the same time. In the second case, the stone is heated rather than accelerated. The difference between the situations is that now we add heat rather than momentum. Heat is responsible for making the stone warmer.

The second question goes to the core of what heat is not. First try and answer this question: What do the moving, charged, and hot body have in common? Again the answer is quite clear this time for those who have studied physics. In all three situations the body has received energy (Figure 1). In each case, the system contains more energy than if the body were not moving, if it were not charged, or if it were not hot. Seen from the viewpoint of energy, the situations are all the same, and we should ask rather about how a stone which is moving differs from a stone which is hot. We might learn more about how nature works from the answer to this question.

From the perspective of a nonscientist, we could say that the difference lies simply in the fact that the moving body has received an amount of motion while in the case of the hot body we have added some heat (Figure 1). As scientists, we have learned that amounts of motion are measured in terms of momentum. Changing the motion of a body requires that we change its momentum. True, we also change the energy of the body at the same time, but knowing that the energy of a system can be different for many different reasons, we take momentum as the quantity which tells us why the body is moving rather than becoming hot.

Applying this argument to the hot body, we have to conclude that the stone is hot not because it contains energy but because it has more heat than if it were colder. A stone gets hot and an ice cube melts because of heat. Even though energy accompanies all processes, heating included, it cannot distinguish between them. The point is this: *heat is not energy.*

The laws of physics help us to assure ourselves of this point. You know that energy is a different measure of gravity or inertia (Section P.4.7). Increasing the energy of a system means increasing its mass. It does not mean that the body necessarily has become faster, hotter, or both; it might just as well have become electrically charged. Energy, i.e., mass, cannot be used to distinguish between different processes. We absolutely need other quantities such as momentum or charge, if we wish to state what has happened to a body. Applied to thermal processes this means, that if heat is the quantity responsible for warming a stone or melting ice, rather than making the bodies heavier, then heat cannot be energy.

Obviously, then, you will have to look elsewhere for the quantity called *heat*. This is precisely what we will do in the following sections. Please be patient if we use the term *heat* for the quantity which makes your room warm, which its responsible for melting ice or metal, and which drives so many of the other processes going on around us. For the impatient among you, the property emerging from our knowledge of everyday phenomena is *entropy*. We hope to motivate the properties of this fundamental thermal quantity by tying it to the concept of heat before we start using the formal term *entropy*.

1.1.2 Hotness and Temperature

One of our most direct experiences with thermal phenomena is the observation that objects feel hot, warm, or cold. We have a sensation which allows us to place objects in a sequence which we label cold, cool, warm, hot, or very hot. From the *sensation of hotness* we abstract a primitive quantity which we call *hotness*. A primitive quantity is one which we do not derive in terms of other more fundamental quantities.

The concept of hotness goes back to E. Mach.[1] He considered it to be natural and fundamental, like so many other concepts upon which we build the foundations of science. Imagine it to be something like a line on which we organize bodies according to how hot they are, just like beads on a string. Mach wrote:[2]

> Among the sensations by which, through the conditions that excite them, we perceive the bodies around us, the sensations of hotness form a special sequence (cold, cool, tepid, warm, hot) or a special class of mutually related elements … . The essence of this physical behavior connected with the characteristic of sensations of hotness (the totality of these reactions) we call its hotness … . The sensations of hotness, like thermoscopic volumes, form a simple series, a simple continuous manifold … .

The sensation of hotness might serve as a first measure of the physical quantity, albeit not a very reliable one. Naturally, we would like to learn how to determine hotnesses reliably. There are measures of hotness which are more useful than our senses, such as the volumes of bodies which change with the hotness. We can

1. E. Mach (1923) analyzed the development and the logical foundations of thermal physics. An interesting passage in his book compares the potentials associated with different phenomena (velocity for translational motion, electric and gravitational potentials, and temperature).
2. E. Mach (1923), p. 43. With the exception of the choice of a term the translation is from C. Truesdell (1979).

build simple devices to tell how hot an object is. These devices are *thermometers*, with which we measure temperatures. Now, what does *temperature* have to do with the concept of hotness? There is a simple image which explains the relationship. Temperature is like a coordinate on the hotness manifold; it is a numerical indicator of hotness. This, by the way, lets us expect that there may be many possible *temperature scales*, just as there are many possible ways of introducing coordinates along a line. According to Mach:[3]

> The temperature is ... nothing else than the characterization, the mark of the hotness by a number. This temperature number has simply the property of an inventory entry, through which this same hotness can be recognized again and if necessary sought out and reproduced .... This temperature number makes it possible to recognize at the same time the order in which the indicated hotnesses follow one another and to recognize between which other hotnesses a given hotness lies.

There is an interesting and important feature of hotness: it has a lower limit. Experience tells us that this should be the case. Bodies cannot get colder than "really cold": we have never found bodies with a hotness below a certain level, which we call the *absolute zero of hotness*. Put differently, we can say that hotnesses are strictly positive. In fact, the point of absolute zero cannot be reached; it only can be approached. This has been done in experiments to an ever increasing degree. Nowadays, temperatures within a small fraction of one unit (1 Kelvin) from absolute zero can be reproduced.

1.1.3 Heating and Cooling: the Transport of Heat

Clearly, the hotness of objects can change. The sand at the beach is hot at noon, and cool at night. We can get hot water for coffee by heating it. Ice is produced by cooling water.

Therefore there is more to thermal phenomena than just the hotness. We definitely need a means for changing this quantity. Again, on the basis of every-day experience we can introduce the notion of *heating* (or cooling). We can heat bodies slowly or quickly; therefore we often speak of the rate of heating. However, we have used the word *heating* already in the sense of a rate, namely the *rate of transfer of heat* to a body. Cooling stands for the rate of transfer of heat out of a body (Figure 2).

Obviously, heat can be transported. Heat from the sun arrives at the Earth, where it is distributed in the atmosphere before it is radiated back into space. Heat flows out of the depths of the Earth to the surface. A metal bar which is heated at one end also gets hotter at the other end, which demonstrates that heat becomes dis-

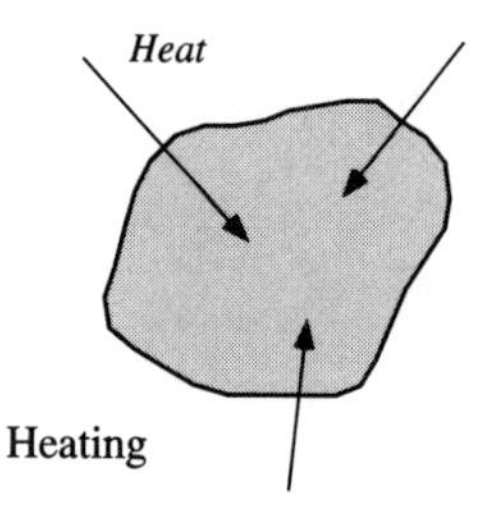

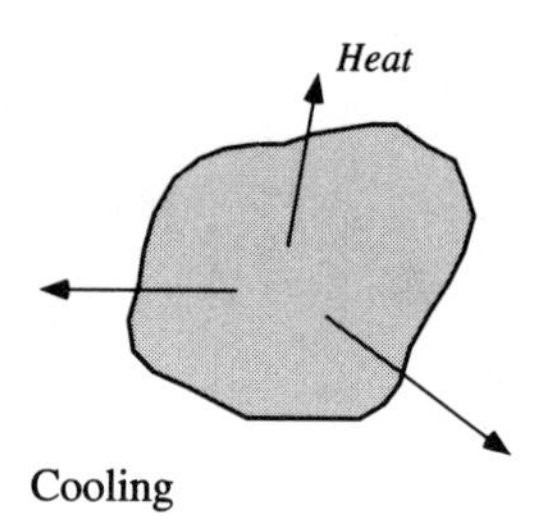

FIGURE 2. Heating and cooling. By *heating* we mean the flow of heat into a body. In the case of *cooling*, heat flows out of the body. We assume that heat is a quantity which can flow. Also, we believe that heat can be stored in bodies.

3. E. Mach (1923), p. 44. In recent years, efforts have been made to base the foundations of thermodynamics upon the hotness as a primitive concept (Truesdell, 1979, 1984; Pitteri, 1982).

tributed throughout the body. Large amounts of heat are transported from the Gulf of Mexico to Europe with the help of the Gulf Stream. Central heating systems transport heat from the burner to the radiators from which the heat flows into rooms. There are many aspects of life which are influenced by the flow, transport, or distribution of heat. Therefore we will study this aspect of thermal physics in detail. Here, we will briefly introduce the three modes of *heat transfer*:

- *Conduction*. There is one class of phenomena in which heat flows through bodies. The bodies do not have to move for heat to get from one place to another. Experience tells us that heat flows only from places which are hotter to those which are cooler. A difference of hotness is needed for maintaining such a flow. Indeed, bodies in thermal contact stop exchanging heat when they are equally hot. This type of transport is called *conduction of heat*.
- *Convection*. Heat can be transported with bodies. There are many examples, such as the Gulf Stream, heated water flowing from a burner to a radiator, or hot air rising up in a chimney. In this case heat does not have to flow through the bodies. Heat which resides *in* bodies flows because the bodies themselves are transported. The driving force of the process must be the one associated with the flow of matter, for example a pressure difference. If heat is transported in this manner we speak of *convective transport of heat*.
- *Radiation*. There must be a third type of transport which, for example, manages to get heat all the way through empty space from the sun to the Earth. You can feel this kind of heat flow when you sit near a fire, or behind a window with the sun shining on you. There is no need for the air to transport heat to you. Moreover, air conducts heat much too little for this to be of any effect. Therefore, heat must flow through another medium, namely a radiation field. This type of transport is called *radiation of heat*.

1.1.4 Heat and Heating: the Storage of Heat

In the previous section, we used still another word for describing thermal phenomena, *heat*. If heating is the rate of transfer of heat, then some *quantity of heat* is transferred in a process. If we heat a body in a particular manner, we transfer more heat the longer we allow the heating to occur. The amount of heat communicated to a body therefore can be calculated simply from the heating. In this sense, heat is a quantity which can be derived from heating; it is not new or independent of what we already have introduced.

Still, there is something new about the notion of heat, something which is not trivial at all. Heating is taken as a term for describing a process or an action, only. The word *heat*, on the other hand, suggests the image of something tangible, something we can measure as it flows past us when heating occurs. There is an amount of something, like an amount of water, or more like an amount of electricity. We might even be tempted to think of heat as something which can be stored in bodies. Where else should the heat be after it has been transferred to an object in the process of heating? Why would a little child ask whether a baked potato is heavier than a cold one, if it were not for the additional heat it contains when it is hot? For this discussion we shall accept the following:

> *Heat is a quantity we can imagine as being stored in bodies, and as being capable of flowing from body to body.*

Consequently, we can say that bodies contain a certain amount of heat, and this amount can change as a consequence of heating or cooling, i.e., as a consequence of the flow of heat. For a physicist, this suggests comparisons with other quantities which have been introduced in other theories, notably those of electricity and motion (Prologue). In this sense *heat is the extensive thermal quantity.*

Remember what we said in the Prologue about substancelike quantities such as electrical charge or momentum. These quantities are abstract; they are not material in any sense of the word. Still, we can profit tremendously from forming simple graphical images such as the ones used in this section. If we think of heat as an abstract "fluid," we will be led easily and directly to the formal mathematical laws that govern it. We just have to be careful not to take the naive picture too literally. Heat does not add weight to a body, it cannot be seen, nor can it be touched. And, most interesting of all, heat can be produced. As we will find out shortly, this quantity is not conserved.

1.1.5 Heat and Hotness

So far we have heat and hotness as the two primitive concepts for building a theory of thermodynamics. Still, many people do not consciously distinguish between the sensation of hotness, and amounts of heat. The term *heat* often is used in the sense of something being hot. Therefore it is important to stress again the difference between these concepts.

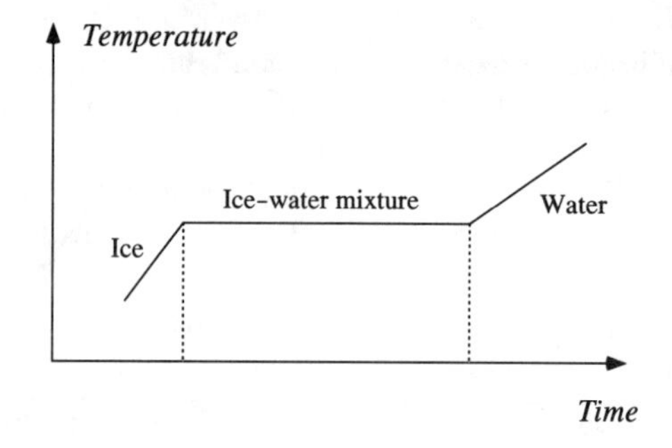

FIGURE 3. We need heat to melt ice; i.e., we have to heat the mixture of ice and water to change all the ice into water. While this process is taking place, it is observed that the hotness of the mixture does not change. (Simply place a thermometer into the ice-water mixture, and observe its reading.) If we record the temperature T read off the thermometer as a function of time, we obtain a horizontal line for as long as there is any ice left in the water. We only have to insure that the heating is slow, and that the water and ice remain well mixed at any moment.

Simple observations tell us that the two quantities cannot be the same. Take an amount of water. We assume that it contains a certain amount of heat. Divide the water into two equal volumes. What can you say about the heat contained in each part, and the sensation of hotness of the parts? Experience tells us that their hotnesses are equal and the same as that of the original body of water. However, the heat content has been divided equally among the parts. Therefore the sensation of hotness and amount of heat clearly are two different things. Hotness is called the *intensive* thermal quantity while heat is the *extensive* one.

Another phenomenon demonstrates beyond any doubt that heating is not just another word for change of hotness (in which case there would not be any need for a new quantity!): when ice is heated, it melts, but its hotness does not change as long as there is a mixture of ice and water. (See Figure 3 and Section 1.3 for more details on this phenomenon.)

Consider the following example, which serves to illuminate the role of hotness even further. Place two bodies at different temperatures in thermal contact and monitor their temperatures. It is quite obvious what will happen. In the course of time the hotter body will cool down while the cooler one must heat up. Experience tells us that this continues as long as the temperatures are different. Finally, after some time, the hotnesses have become equal and the process will stop (Figure 4). Another interpretation of this phenomenon is to say that heat flows by itself from the hotter to the cooler body as long as there is a difference of temperatures.

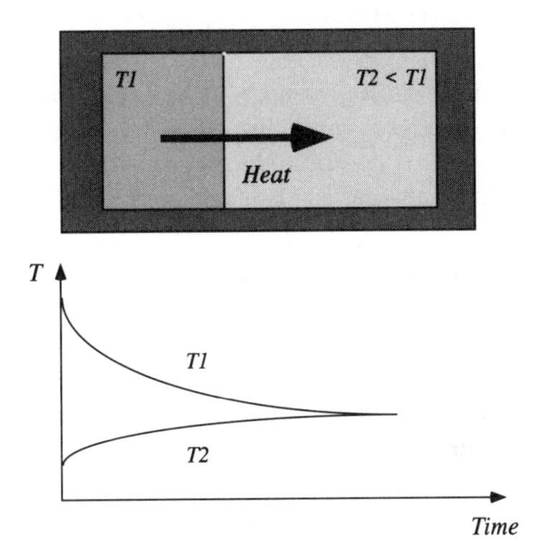

FIGURE 4. Two bodies at different temperatures are in thermal contact but insulated from their surroundings. If we monitor their temperatures, we notice that the hotter body gets colder while the colder one is heated up. This is clearly seen in a plot of temperatures versus time. The process stops when both bodies have reached the same final temperature.

We have seen analogous processes in other fields of physics (Table 1). Two containers filled with water and connected by a hose, or two charged bodies connected by a wire, behave similarly. In those cases we speak of the flow of a certain physical quantity as a consequence of a driving force. We have defined driving forces as the difference of potentials between the physical systems. If we take these phenomena as a guide we can interpret a difference of temperatures as the *thermal driving force*, and hotness as the *thermal potential*. In graphical language, hotness takes the role of the thermal level.

The driving force determines the type of process taking place. In this sense we say that a thermal process is one in which heat flows due to a difference of temperatures. Note that not every mode of heat flow is a thermal phenomenon according to this definition. Witness the following example. You can carry around a bucket of hot water, which means that heat is being transported. But certainly this is not a thermal process. Hotness plays a central role in what we call a thermal phenomenon, and we may not forget the qualifier given in the definition above.

TABLE 1. Comparison of some processes

	Quantity	Potential
Water containers joined by a hose	Water	Water level
Charged spheres in contact	Electrical charge	Electrical potential
Bodies at different temperatures	Heat	Hotness

1.1.6 The Production of Heat

We have not discussed an important question yet: Where does heat come from? Nature tells us that there must be sources of heat. In some cases heat simply flows out of bodies in the process of cooling. However, there are examples which are more interesting. Heat comes from the sun, or out of the earth. Heat is produced in a fire, by rubbing your hands together, or by letting electricity flow through wires.

There appears to be a distinct difference between the first process, and the other ones listed above. In cooling, an object loses the heat it received in a process of heating. The body simply undergoes the reverse of the previous process. In the other cases, the body emitting the heat does not have to change its hotness. As far as we know, the degree of heat of the interior of the sun does not change much as it pours out vast amounts of heat (in fact, it gets even hotter). An electrical heater does not cool as it heats water. And the heat produced by rubbing your hands together has not been put into them previously. So, where does the heat come from in these cases?

We might believe that the amount of heat contained in bodies is always so large that the emission of some of it will not change their hotness. However, this is rather unlikely. Clearly there must be sources of heat in the true sense of the word. Heat which was not there before is pouring out of the body. Why does the electrical heater cool very noticeably when the electricity is turned off? Certainly, after switching off, it emits only a small amount of heat compared to what it emits while working. Why does the hotness change in the former case and not in the latter?

About two hundred years ago two sets of experiments were performed which were believed to demonstrate beyond any doubt that heat cannot be a conserved quantity. The first, the cannon-boring experiments of Count Rumford, have much in common with the electrical heating described above. Heat is produced by friction as long as the boring process is going on; so where does all this heat come from? The second experiment, by H. Davy, was thought to be even more important.[4] We can melt two blocks of ice by rubbing them against each other. The water which results from rubbing the ice certainly contains more heat than the ice it came from, because we need heat to melt ice. If we perform the experiment in an environment whose temperature is lower than that of the melting ice, the heat added to the water cannot have come from outside: heat only flows from hotter to cooler places. We have to conclude that after the process there is more heat in nature than before: the surroundings of the ice and water have not lost any heat, while the system contains more heat than before. Even though Rumford's reasoning is not tight, and Davy's experiment is difficult to perform in any quantitative manner,[5] we shall accept the evidence offered by nature: *heat can be created.*

1.1.7 The Production of Heat and Irreversibility

Heat can be created. Does this mean it also can be destroyed? Since we have had to conclude that it is not a conserved quantity, the question is realistic. If we consider the evidence offered by nature, we come to the conclusion that *heat cannot be destroyed*; it can only be distributed to colder places. Heat which leaves a body goes into the surroundings to heat them. The reason why we often do not notice this effect on the surrounding bodies is simple: they usually are very much larger than the body which is losing heat.

Again, you can use your general knowledge of natural or manmade processes to come to far-reaching conclusions. You know that certain processes run one way but not necessarily in the reverse. Or if you make them run in the reverse, something else will change because of it. Real processes are said to be *irreversible*. If heat did not exist already, you would have to invent a physical quantity with its properties to account for irreversibility.

Left to itself, a rolling car comes to a halt on a horizontal street, and a swinging pendulum will stop swinging after some time. Heat flows only from hotter to colder bodies by itself. The sun radiates heat, it does not suck it up out of space. And living beings get older, never younger. What do these examples have in common, and how do they relate to the properties of heat?

---

4. A critical reappraisal of the experiments shows that a proof of Davy's claim (Davy, 1839) was almost impossible (D. Roller, 1950). The question of where the heat comes from for melting the ice is less than trivial. However, if we could take care of all possible interferences, we would have no doubt as to whether or not we can melt ice by friction; it is certainly possible.

5. D. Roller (1950).

A simple example explains the meaning of irreversibility. A moving wooden block comes to rest on a horizontal surface. The reason is clear: friction hinders the motion and finally lets it "die." Friction also creates heat. Therefore, in our example, the creation of heat lets the block come to rest. This does not explain, however, why the process is irreversible. At first sight, *irreversible* should mean *cannot be reversed.* But the motion of the block can be reversed! We simply have to push the body across the surface, thereby accelerating it. Something, however, cannot be changed here: heat will also be created during the reversed motion; friction makes sure heat will be generated. You have never seen a block absorbing heat from the surroundings and destroying it, thereby starting to move by itself. This is what we mean when we say that a process is irreversible.

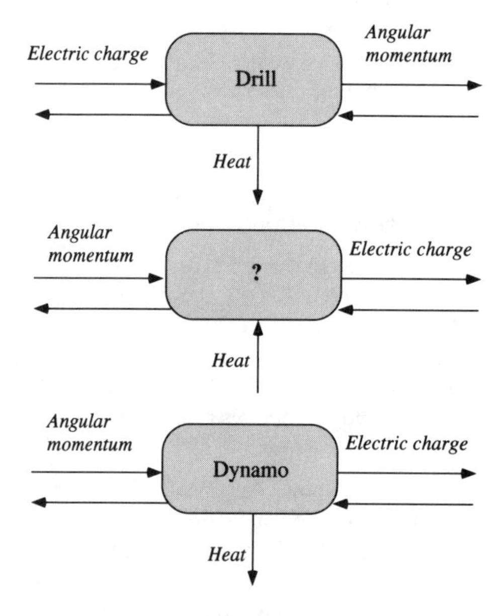

FIGURE 5. A drill is driven by the flow of electrical charge. In a real engine heat is created and emitted. The engine has to be cooled. The complete reverse of the processes depicted in the flow diagram of the drill is given in the middle picture. As far as we know, such an engine cannot exist. The real engine resembling the second one the most is a dynamo whose flow diagram is shown in the third diagram. The dynamo reverses the flows of angular momentum and of electrical charge compared to the drill. However, it cannot reverse the flow of heat. The engine still has to be cooled. This is the hallmark of irreversibility, or of dissipation.

Another good example of irreversibility is that of a drill and a dynamo (Figure 5). A dynamo is the reverse of a drill except for one crucial feature: both produce heat, and they emit it. Engines have to be cooled, not heated, pointing to the equivalence of irreversibility and the production of heat. Again, we can understand the behavior of some aspect of nature on the basis of the properties of heat. We now make the following definition:

> *An irreversible process is one during which heat has been created. A process without the creation of heat is called reversible.*

The term *irreversible* is somewhat unfortunate in the light of the examples given, considering that processes can be reversed with the exception of the heat production. Another descriptive term has been coined for processes which produce heat: they are called *dissipative*.

Now that we have precisely defined the meaning of irreversibility, we can answer the question of what this phenomenon has to do with the properties of heat. You should convince yourself that our assumption that heat can be created but cannot be destroyed is the mirror image of irreversibility. If there are irreversible processes then heat must have the assumed properties; and if heat has the assumed properties then there must be irreversible processes. The "one-way street" represented by irreversible processes must be matched by a one-sided physical quantity. We need heat with its properties.

1.1.8 The Power of Heat

There is another property of heat we have not looked at yet. Heat can be used to do things; it can drive engines; and it is an agent for effecting things. In other words, heat can do work. To quote S. Carnot:[6]

> Every one knows that heat can produce motion. That it possesses vast motive-power no one can doubt, in these days when the steam-engine is everywhere so well known. To heat also are due the vast movements which take place on the earth. It causes the agitations of the atmosphere, the ascension of clouds, the fall of rain and of meteors, the currents of water which channel the surface of the globe, and of which man has thus far employed but a small portion. Even earthquakes and volcanic eruptions are the result of heat.

Since Carnot's time, it has become evident that the action of heat can effect more than just motion. Heat drives a whole range of other processes, such as electric and chemical ones. Today we take these phenomena as a sign of the interrelation between different classes of physical processes.

Does this mean that heat is some sort of work? The answer should be "no." Water can also be used to drive water wheels and turbines. Does this make water some sort of work? Certainly not. Similarly, electricity, i.e., electric charge, can be used to do work, but it is not work. We take heat to be something akin to water or to electric charge, which beautifully explains the phenomena we are familiar with. Work (and with it, energy) is something different. We are used to introducing the physical quantity called *work* in mechanical, gravitational, and electrical processes. We shall do this in the case of heat as well. We will derive a new quantity called *thermal work*, a quantity which is to the physics of heat what mechanical work is to motion.

---

6. S. Carnot (1824), p. 3. It was suggested by Callendar (1911) to resurrect the caloric theory of heat in an extended form. On its basis, he gave a simple description of Carnot's work. A detailed and critical analysis of the development of Carnot's theory has been presented by C. Truesdell (1980).

One of our main tasks will be to decipher the relationships between heat and work, and between heat and energy. As the basis of our exploration we shall use a vivid image proposed by Carnot. He had suggested that we could compare the motive power of heat to that of the fall of water driving a water wheel:[7]

> According to established principles at the present time, we can compare with sufficient accuracy the motive power of heat to that of a fall of water .... The motive power of a fall of water depends on its height and on the quantity of the liquid; the motive power of heat depends also on the quantity of caloric used, and on what may be termed, on what in fact we will call, the *height of its fall*, that is to say, the difference of temperature of the bodies between which the exchange of caloric is made. In the fall of water the motive power is exactly proportional to the difference of level between the higher and lower reservoirs. In the fall of caloric the motive power undoubtedly increases with the difference of temperature between the warm and the cold bodies; but we do not know whether it is proportional to this difference.

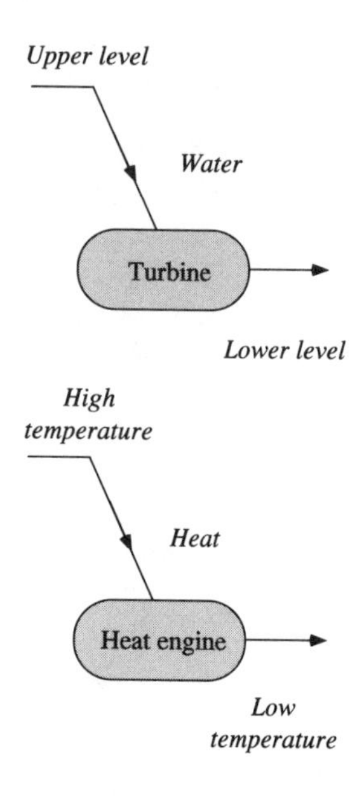

FIGURE 6. Hydroelectric and thermal power plants are structurally comparable. Water drives a turbine by falling from a higher to a lower level. Heat drives a heat engine by "falling" from a higher to a lower thermal level.

Just as water falls from a high level to drive a turbine, after which it flows out of the engine at a lower level, heat is imagined to fall from a high level to a lower one, thereby driving the heat engine (Figure 6). The principle of operation of heat engines is in accordance with this image. Steam takes up caloric (heat) from a burner, and passes through the engine where it effects motion, just to flow out again and to give up its heat to a condenser. As Carnot put it:[8]

> The steam is here only a means of transporting the caloric .... The production of motive power is then due in steam-engines not to an actual consumption of caloric, but to its transportation from a warm body to a cold body.

We know today that we can indeed explain the motive power of heat in terms of these images. There is a deep similarity between different types of physical processes. Hydroelectric power plants and heat engines are two examples which serve to drive home this point (Figure 6).

1.1.9 Heat, Thermal Processes, and the Properties of Bodies

So far we have talked only about the generic properties of heat: it can be stored, it can flow, and it may be produced. These are the things that happen in thermal processes as far as heat is concerned. However, the interesting details of processes depend not on heat so much as on material bodies or systems being heated or cooled. Therefore, we will have to study in detail the special behavior of different types of physical bodies.

Different bodies contain different amounts of heat under given conditions, and different substances require different amounts of heat for melting. The form of

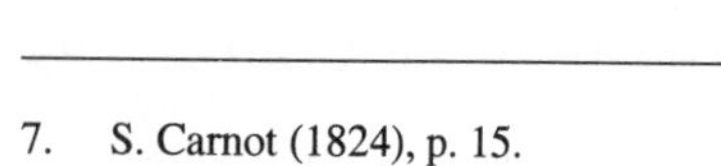

7. S. Carnot (1824), p. 15.
8. S. Carnot (1824), p. 7.

heat flow and heat production, even the question of whether heat is produced in a particular process, will be found to depend upon the properties of the bodies suffering change. If you prefer, the answer to such questions depends upon the *models we construct of the bodies* undergoing processes.

Again and again, we will encounter limitations inherent in a discussion which disregards the properties of bodies. Several examples used in the course of the current chapter in fact depend on special material properties. We will find that the usefulness of the basic law of thermal physics, i.e., the law describing the properties of heat alone, depends on the knowledge of material laws. Therefore results derived from it in this chapter without regard for the properties of bodies cannot be specific or detailed in any way.

1.1.10 Heat as the Substancelike Thermal Quantity

All in all, we form a mental picture of heat as something like a "substance" or a "stuff." We picture bodies as *reservoirs for heat*, we may speak of the *heat content* of an object, which will be described by a function of the independent variables of the body. Also, this "substance" flows into and out of bodies. The process of the *flow of heat* into a body is called *heating*. (The opposite is called *cooling*, or *negative heating*.) By stating such ideas we add heat to the list made up of electrical charge, gravitational mass (gravitational charge), momentum, and angular momentum (Table 2). Heat is another *substancelike* quantity which simply means that *heat can be pictured to be contained in bodies and to flow from one body to another*. No further assumptions are made. Specifically, heat is not conserved!

TABLE 2. Substancelike quantities

Process	Quantity
Electric	Electric charge
Gravitational	Gravitational mass
Translational motion	Momentum
Rotational motion	Angular momentum
Thermal	Heat

We are in good company with these assumptions about heat. The Greeks used a similar picture, and in early modern times up to 1850 physicists used pretty much the same ideas. Back then, heat often was called *caloric*,[9] as you have noticed in

9. In the Interlude, we will give a formal treatment of the thermodynamics of uniform ideal fluids, which starts with the premises of an extended version of the classical caloric theory. We will present the proof that modern thermodynamics can be based upon the caloric theory.

Carnot's description of the operation of heat engines. There are several other terms which aptly describe what we mean by heat: *thermal element* (because of the similarity to Greek thinking), or *thermal charge*[10] (in analogy to electricity). *Heat*, *caloric*, *thermal element*, and *thermal charge*, refer to the same thing, namely, the substancelike thermal quantity which flows into bodies and is stored there if the bodies are heated, and which flows out of the bodies if they are cooled.

It is instructive to list amounts of heat to get a little bit more acquainted with this quantity (Table 3). As always, a physical quantity has a *unit*. Since heat is a new and fundamental quantity which is not associated with anything else we know so far, it should have its own unit. Let us therefore introduce the unit of heat called the *Carnot* (Ct) which we are going to use for now.[11]

TABLE 3. Amounts of heat (rough values in Ct)

Process or state	S / Ct
Heat necessary to melt 1 cm^3 of ice	1
Heat added when heating 1 cm^3 of water from room temperature to boiling	1
Heat needed to vaporize one cm^3 of water	6
Heat content of 1 liter of gas at normal pressure and temperature	10
Heat content of 1 mole of argon gas at room temperature	10^2
Heat content of 1 liter of liquid or solid at room temperature	10^4
Heat generated in burning 1 kg of coal at room temperature	10^5

1.1.11 Questions to Ask About Heat and Hotness

Let us summarize the points discussed in this section by listing a few questions we should answer when we talk about heat and hotness. These questions will underscore the simple and graphical image we may construct for understanding the role of the basic quantities in a theory of the dynamics of heat.

First of all, we need heat inside bodies to make them warm, or to melt them. Second, heat has to be transported so that it gets to where it is needed. In the reverse processes, heat has to be removed from where it is not needed. These features are captured in two questions:

10. McGraw-Hill Encyclopedia of Science and Technology.
11. For historical reasons, the unit of heat is normally given in terms of other units. In Section 1.4 we will find that it should be expressed as 1 Ct = 1 Joule/Kelvin, where Joule and Kelvin are the SI units of *energy* and *temperature*, respectively.

- Where is heat stored?
- How is heat transported, and how does it enter and leave bodies?

In many cases where heat is needed, it is generated rather than moved from places where it already exists. Therefore, we ask:

- How is heat generated in bodies, and how much of it is generated?

When heat resides in bodies, it affects them in different ways; i.e., bodies respond differently to different modes of heating. This means that we should always ask the following question:

- What effects does heat have upon bodies?

Heat, like any other substancelike quantity, may be used to drive other processes; i.e., it may be used to do work:

- How is heat used to do work, and how much work is done by heat?

The last question concerns hotness (not heat), the other quantity without which a description of the nature of thermal processes would be incomplete. Here we ask the simple question:

- How hot are things?

If we keep this list of questions in mind, we will have a guide to the problems of a theory of heat. These questions serve as the bridge between phenomena and theory, between reality and our formal description of it.

1.1.12 Heat, Caloric, and Entropy

The quantity of heat introduced here, which has so much in common with the old concept of caloric, bears the distinct marks of what physicists, chemists, and engineers call *entropy.* It *is* entropy, or rather, it is a generalized version of this fundamental quantity. In fact, our concept of heat is general enough for us to build modern continuum thermodynamics upon its basis. The procedure is very simple. Let us accept that heat (entropy) has the properties of a nonconserved substancelike (i.e., extensive) quantity. This will lead immediately to a law of balance of heat (entropy) as the most important expression of our assumptions about thermal processes.

Normally, we would derive the concept of entropy after introducing thermodynamics. This practice has led to a teaching culture which considers entropy to be a difficult and advanced concept without roots in everyday images of thermal processes. This is not very surprising, considering that it is almost impossible to recognize the true nature of a quantity which is introduced on purely formal grounds and led onto the stage through the back entrance only. If we were to teach basic electricity in a manner analogous to that found in the chapters on thermodynamics of our introductory physics texts, we would never realize that there is a quantity with the properties of electric charge.[12] Therefore it is time to take a second look at our practice and let entropy appear in the first act of the drama as the central figure of thermal physics, namely, as the formal measure of a quantity of heat for which we have very useful everyday images.[13]

To let the image of a quantity of heat take still firmer roots, we will describe simple thermal phenomena using the word *heat* in Section 1.3. The formal term *entropy* will finally be used for this quantity starting with Section 1.4. However, when speaking colloquially, we will always mean a quantity of entropy when referring to heat. In expressions such as "the body has been heated" or "in this process heat has been produced," *heat* will continue to stand for its formal equivalent, namely for entropy. Otherwise we will drop the word *heat* from our list, i.e., we will *not* use it for the only acceptable meaning in traditional presentations of thermodynamics, namely amounts of energy exchanged in heating (Section 1.4.3).

If you still believe that using the concept of entropy from the start is too difficult, consider what J.W. Gibbs once wrote:[14]

> One of the principal objects of practical research ... is to find the point of view from which the subject appears in its greatest simplicity. ... a method involving the notion of entropy ... will doubtless seem to many far-fetched, and may repel beginners as obscure and difficult of comprehension. This inconvenience is perhaps more than counter-balanced by the advantages of a method which makes the second law of thermodynamics so prominent, and gives it so clear and elementary an expression ... . If, then, it is more important for purposes of instruction and the like to familiarize the learner with the second law, than to defer its statement as long as possible, the use of the entropy-temperature diagram may serve a useful purpose in the popularizing of this science.

1.2 Temperature and Thermometry

The properties of heat will be the subject of Sections 1.3 through 1.6. Here we will investigate the concept of *temperature* as the measure of hotness. Temperatures can be measured quite easily. Still, the discussion in this section will show that the concept of temperature is anything but trivial. In particular, we will not be able to find *the* temperature scale, i.e., the one and only scale which reads the "true" temperature. If there is anything like an *absolute temperature* we cannot say yet. For now we have to be satisfied with a couple of empirical scales.

---

12. Assume "electricity" to be a form of energy, formulate the "First law of electricity" and apply it to a cyclic process undergone by a parallel plate capacitor with variable separation. At the end of the development you may introduce a new quantity called *reduced* electricity (which is the ratio of "electricity" and voltage) which turns out to be equivalent to what we know to be a quantity of charge. See Fuchs (1986).
13. Fuchs (1987a,c).
14. Gibbs (1873).

1.2.1 The Celsius Scale of Temperature

We are quite accustomed to measuring temperatures in everyday situations. For example, if you take your temperature you might use a thermometer based on the expansion of mercury in a thin capillary. Mercury expands when it is heated, and the expansion is measured and used to fix the temperature. When bodies are heated, the rise in temperature is accompanied by changes in one or more properties. Change in volume is just one of those possible changes. Others include changes in length, electrical resistance, color, and pressure of gases.

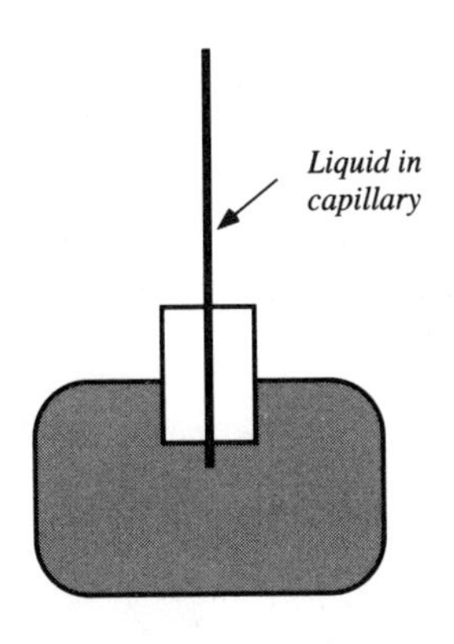

FIGURE 7. Thermometer based on the expansion of a liquid. The small changes of the volume of the liquid are made visible by the capillary.

The change in the volume of a liquid can be observed using a simple device (Figure 7). Fill a small bottle with the liquid and close it with a cork through which a long and thin capillary has been fitted. Allow the liquid to fill the bottom of the capillary. If you hold the bottle in your hands for a while, the liquid in the capillary will climb (in most cases). How does this work? Heat flows from your warm hands to the colder liquid, which expands. The change in volume usually is very slight, but it can be made visible by the capillary: since the capillary is very thin, even small changes in volume translate into large changes in the height of the column of liquid.

TABLE 4. Some temperatures (in °C)

Boiling point of helium	- 269
Boiling point of nitrogen	- 196
Melting point of carbon dioxide	- 79
Melting point of mercury	- 39
Melting point of ice	0
Temperature of human body	37
Boiling point of water (1 bar)	100
Melting point of iron	1535
Surface temperature of sun	5800
Central temperature of sun	$13 \cdot 10^6$

Historically, the first useful thermometers were based on this type of device. Today, we use mercury as the liquid (which we call a *thermometric substance*). On the basis of such thermometers we may introduce a temperature scale, the *Celsius scale*. We measure the length of the column for two different temperatures, e.g., those encountered when water freezes and boils. The first point is assigned the temperature 0°C (zero degrees Celsius); the second corresponds to 100°C. The interval in the length of the column of mercury is divided into 100 equal parts, each part corresponding to a change in temperature of 1°C. The Celsius scale is not the one and only absolute coordinate system on the hotness manifold. Rather, we have arbitrarily set a certain change of temperature to be proportional to the change of length of the mercury column in the capillary.

We cannot be sure that other thermometric substances will deliver scales proportional to the one introduced on the basis of the expansion of mercury. We cannot even be sure that volume or length will always increase when the hotness increases! Water is a beautiful counterexample. Water exists in the range of temperatures spanning 0°C to 100°C. At low temperatures, near its freezing point, the volume of water first decreases, reaching a minimum at 4°C before finally increasing (Figure 8). At temperatures above 4°C, water behaves as we might expect of a "normal" substance. The behavior of water at low temperatures is called an *anomaly*. This anomalous behavior prevents lakes from freezing totally, which saves the fish in them in winter.

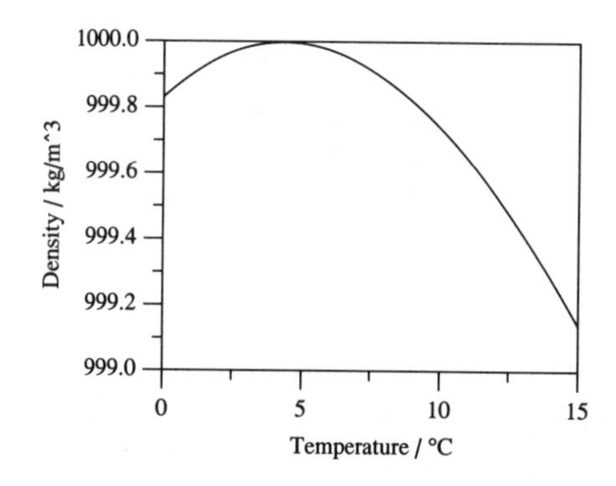

FIGURE 8. Variation with temperature of the density of water. The variation around 4°C is shown in some detail. The pressure is constant and equal to atmospheric pressure.

Water, therefore, could not be used as a thermometric substance. While we would get meaningful results out of a water thermometer in the range of temperatures from 4°C up to 100°C, the same would not be true if we included the density extremum. (The question of why we should not use water as a thermometer in the range between 0°C and 4°C is more subtle.)

1.2.2 Some Thermoscopic Properties

In this section we will discuss some simple material properties. They are needed because we base thermometry on special properties of bodies.

Bodies expand or shrink when their temperature is increased. Figure 8 shows the result of measurements of the thermal expansion of water. In Figure 9 the change of the length of a bar is indicated. To describe the change mathematically, it is customary to introduce a *temperature coefficient of (linear) expansion* α_l. This coefficient is defined as the relative rate of change of length l of a body with temperature θ. This means that the relative change of length is obtained if we integrate the coefficient over temperature:

$$\frac{\Delta l}{l} = \int_{\theta_o}^{\theta} \alpha_l(\theta) d\theta \qquad (1)$$

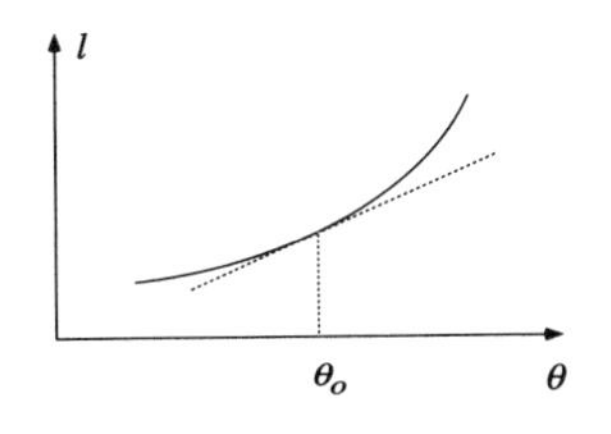

FIGURE 9. Change with temperature of the length of a long bar of a particular material. The relationship between length and temperature usually is a complicated one. Over a small range of temperatures, however, it may be approximated by a linear function.

Over a small range of temperatures we can approximate the length of such a bar by a linear relation. In other words, we may choose for α_l the constant value α_{lo} at θ_o. In this special case, the relative change of length of a body is given by

$$\frac{l-l_o}{l_o} = \alpha_{lo}(\theta-\theta_o) \tag{2}$$

Here, l_o is the length of the body at a reference temperature θ_o (which might be 0°C). In Table A.2, values of the coefficient of expansion for some materials are listed. Note that α_l has the unit 1/K (which is introduced below).

In analogy to the formula for the change of length, the *change of volume* with temperature often is approximated by a linear expression:

$$V(\theta) = V_o(1+\alpha_V(\theta-\theta_o)) \tag{3}$$

Here, α_V is the *temperature coefficient of volume expansion*; values for some materials are listed in Table A.2. This coefficient has the unit 1/K.

Another property of bodies which in many cases changes with the hotness is the *electrical resistance*. Thus it can be used for thermometry as well. A widely used thermometer is based upon the electrical properties of platinum. This thermometer is applied for accurate measurements in the range from 253°C below freezing to roughly 1200°C. Very often, the empirical relationship between temperature and resistance is represented by a quadratic equation of the form

$$R(\theta) = R_o(1+\alpha_R(\theta-\theta_o)+\beta_R(\theta-\theta_o)^2) \tag{4}$$

There is nothing deep or fundamental about the form of this relationship; it is a representation of empirical data just like Equations (2) or (3). However, these equations are important in a different sense: they are examples of constitutive relations, i.e., laws describing aspects of the behavior of special materials. Coefficients for some materials are listed in Table A.3.

EXAMPLE 1. Thermal stress.

A copper bar of constant cross section and length l_o is rigidly attached between walls. Assume it to be free of stress at a given temperature. If the temperature changes, the length of the bar changes, and momentum currents due to compression or tension are set up in the bar. Calculate the thermal stress resulting from a change of temperature of ± 30°C. Young's modulus for copper is equal to $12.3 \cdot 10^{10}10$ N/m^2.

SOLUTION: According to Hook's law, the momentum current density (i.e., the stress) is equal to

$$j_p = E\Delta l / l_o$$

where E is Young's modulus. If the bar gets longer, the stress is compressive and j_p is positive. According to the linear approximation to the law of thermal expansion (1), the change in length of the rod is

$$\Delta l = l_o \alpha \Delta \theta$$

If we combine these expressions we get

$$j_p = E \alpha \Delta \theta$$

With the numbers for Young's modulus and the coefficient of thermal expansion given in Table A.2, the momentum flux density turns out to be $\pm 6.2 \cdot 10^7$ N/m^2.

EXAMPLE 2. Volume expansion of a solid body.

Assume that a solid cube expands by expanding along each of its three axes equally. Show that the coefficient of volume expansion equals $3\alpha_l$ in this case, where α_l is the (linear) coefficient of expansion. What is the relative change of volume per degree Celsius of a cube of 1 kg copper?

SOLUTION: The length of a side of the cube is l. Therefore, $V = l^3$. Now

$$\begin{aligned} \Delta V &= V_2 - V_1 = l_2^3 - l_1^3 = l_1^3 (1 + \alpha_l \Delta \theta)^3 - l_1^3 \\ &= l_1^3 (1 + 3\alpha_l \Delta \theta + \ldots) - l_1^3 \\ &\approx 3\alpha_l V_1 \Delta \theta \end{aligned}$$

Terms involving higher powers have been neglected. The relative change of volume per temperature is

$$\frac{1}{V_1} \frac{\Delta V}{\Delta \theta} = 3\alpha_l$$

For copper this is equal to $50 \cdot 10^{-6}$ K^{-1}. (See Table A.2 for values of the linear coefficient of expansion.)

EXAMPLE 3. Temperature of the filament of a light bulb.

It is found that the current through the filament of a light bulb is 0.010 A at a voltage of 1.0 V. At 150 V the current is equal to 0.50 A. What is the temperature of the filament at the higher reading, if the experiment is performed at room temperature (20°C)? The temperature coefficients are $\alpha = 4.11 \cdot 10^{-3}$ K^{-1} and $\beta = 9.62 \cdot 10^{-7}$ K^{-2} for the filament, which is made out of tungsten.

SOLUTION: At the first reading, the current is so small that the temperature of the filament must be equal to the temperature of the laboratory. In other words, at 20°C the resistance of the filament is equal to $R_o = U/I = 1.0\ \text{V}/0.01\ \text{A} = 100\ \Omega$. Current and voltage also determine the resistance of the wire at the higher reading. It is equal to 300 Ω. The temperature of the filament therefore is determined by the quadratic equation

$$300\Omega = 100\Omega\left(1 + \alpha_R \Delta\theta + \beta_R \Delta\theta^2\right)$$

whose solution is $\Delta\theta = 441°C$. The new temperature therefore is 461°C.

1.2.3 The Ideal Gas Temperature

The Celsius scale of temperature is only one of many we could construct. Now we will introduce another scale which will be of great use in thermodynamics, namely the *Kelvin scale* which is based on the *ideal gas temperature* and which is independent of the particular thermometric fluid. However, we will be able to demonstrate this feature only after we have discussed thermodynamics in Chapter 2.

It is a common experience that the pressure of gases increases if they are heated while the volume is kept constant. For example, consider the build-up of steam in a pressure cooker: the volume does not change, and the pressure increases with increasing temperature. Somehow, the pressure of a gas must be related to its temperature. A simple setup called an *air thermometer* can be used to measure the temperature and pressure of a gas whose volume is kept constant. Figure 10 shows the nature of the results one commonly obtains from measurements made with this apparatus. If we draw the values of pressure as a function of (Celsius) temperature we will find that actual data closely follow a straight line. This behavior is interesting. (It is called the *Law of Gay-Lussac.*) Assume that air will behave in this manner at all temperatures. As a consequence, we can write the relationship between pressure and (Celsius) temperature in the following form:

$$P(\theta) \;=\; P_o(1 + \beta\theta) \tag{5}$$

where P_o is the pressure of the gas at 0°C, and β is called the *temperature coefficient of pressure*. If we know this relationship, we can use the pressure as a measure of temperature in the interval covered by the experiment.

Now there must be a temperature for which the pressure of the gas becomes zero. We find this point by extrapolating the straight line in Figure 10 to lower temperatures. The pressure of a gas cannot vanish, and certainly it cannot take on negative values. We have to conclude that this particular point must constitute a lower limit for the temperature of air.

This is quite surprising in itself. It is more interesting still when we observe that all *dilute* gases show this behavior (in fact this even includes substances dissolved in others, such as sugar in water), and moreover, all extrapolated curves intersect the line of zero pressure at the same point. Measurements put the value of the temperature coefficient of pressure at 1/273.15°C for *all dilute substances.* This means that for gases showing this type of behavior irrespective of their chemical composition, there is a hotness below which the ideal gas cannot exist, and this temperature is the same for all such substances! From the value of β we find that this value must be 273.15°C below the freezing point of water.

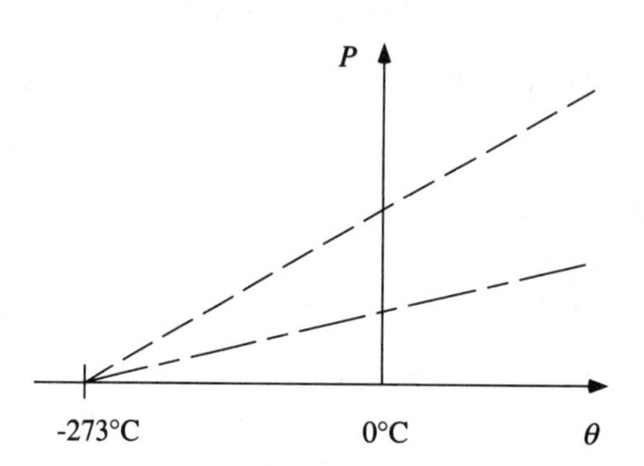

FIGURE 10. For simple dilute gases, the pressure is a linear function of temperature at constant volume. If you were to plot measurements for two different dilute gases, you would obtain two straight lines for which $P = 0$ at the same temperature. This type of behavior is found in all gases which are sufficiently dilute and hot. The straight lines interpolating and extrapolating actual measurements represent the model of the ideal gas.

Our experience with gases which behave in this simple manner (i.e., those whose P-θ curve is a straight line) is a strong indication that the temperature found by extrapolating measurements constitutes a point of "absolute zero" for these kinds of fluids. This does not mean, however, that this temperature has any special meaning for other materials. Only further evidence can show if there are bodies which are capable of attaining hotnesses below the lowest one for dilute substances. Since we have never found such a case we believe that the value of - 273.15°C constitutes the *lowest possible value for hotnesses.*

Therefore, a natural scale of temperature would be one for which the temperature is taken to be zero at this lowest point, i.e., at - 273.15°C, and which is measured by the gas thermometer. This is done in the Kelvin scale which is defined on the following basis: zero Kelvin (0 K) corresponds to the point of absolute zero, and the interval of 1 Kelvin (1 K) corresponds to a change in temperature of 1°C. Therefore, we can convert Celsius temperatures θ into Kelvin temperatures T by:

$$T = \theta\,\mathrm{K/°C} + 273.15\,\mathrm{K} \tag{6}$$

If we use the Kelvin scale, the relationship between the pressure and temperature of gases just demonstrated takes on a particularly simple form:

$$P = P_o \beta T \tag{7}$$

We call the temperature measured by the gas thermometer the *ideal gas temperature* because it is based on the ideal behavior of dilute gases (Section 2.3). The Law of Gay-Lussac is one property of such gases. As we shall see in Chapter 2, the temperature introduced here has an important additional feature: it can be taken as the basis of a scale which is independent of the thermometric substance. It is an *absolute scale*.

EXAMPLE 4. Volume as a function of temperature.

Measurements demonstrate that dilute gases show a linear relationship between volume and temperature, if the pressure is kept constant. a) Express the law in a form equivalent to that of Equation (5). Compare it with the linear approximation to the law of expansion of liquids and solids (Section 1.2.2), and calculate the coefficient of expansion with respect to a temperature of 0°C. Is the coefficient independent of temperature? b) It is found by measurement that $\alpha^*(0°C) = 1/273.15$ K. Show that the ratio of the volumes at different temperatures (for equal pressure) is given by the ratio of the Kelvin temperatures.

SOLUTION: a) The linear relationship can be expressed in the following form:

$$V(\theta) = V_o(1+\alpha^*\theta) \quad \textbf{(E1)}$$

where V_o is the volume of the gas at 0°C. If we compare this to Equation (2) we find that

$$\alpha^* = \alpha_V$$

From Equation (E1) we find that the temperature coefficient of volume is given by

$$\alpha^* = \frac{V - V_o}{\theta \cdot V_o} \quad \textbf{(E2)}$$

$(V - V_o)/\theta$ is a constant. However, if we refer the coefficient of expansion to a different temperature, we divide by a different value of the volume. Therefore, the coefficient depends on the temperature to which it refers.

b) We simply calculate the volumes for two different temperatures according to Equation (E1). Their ratio turns out to be equal to

$$\frac{V(\theta_2)}{V(\theta_1)} = \frac{V_o(1+\theta_2/273.15)}{V_o(1+\theta_1/273.15)} = \frac{\theta_2 + 273.15}{\theta_1 + 273.15}$$

which is equal to the ratio of the Kelvin temperatures; see Equation (5).

1.3 Some Simple Cases of Heating

In this section we will describe a few simple thermal processes. We have introduced the two fundamental thermal quantities, namely heat and hotness, with whose help we will be able to describe roughly what is going on in some situations. The discussion has to remain on a qualitative level since we do not yet have the means of going into the details of thermal phenomena. By looking at some interesting cases we will learn to better understand the nature of heat. We shall choose as examples two classes of thermal processes which have played a major role in the early development of thermodynamics. Using these processes we will be able to discuss a procedure for the measurement of amounts of heat (entropy).

Our first example is *heating or cooling of simple fluids, such as air, whose volume and temperature can be changed.* If we put air in a cylinder having a piston we have a device which lets us compress or expand the gas. At the same time, the air might be heated or cooled. (See Figure 11.) We are interested in the exchange of heat, and the change of the heat content; this means that we want to find out about how much heat flows into or out of a body, and by how much the heat in the body changes. In other words, we will try to *account for amounts of heat.* Naturally we also want to know what happens to the *hotness* of the fluid. We shall describe some processes in terms of these quantities. To make things as simple as possible, we shall conceive of reversible operations, i.e., processes which conserve heat. In Chapter 2 we shall see under which conditions this assumption can be fulfilled.

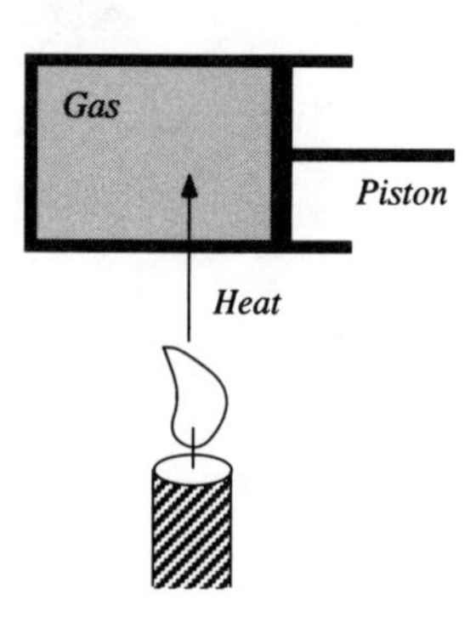

FIGURE 11. A simple device which allows us to put a fluid such as air through thermomechanical processes, i.e., processes which couple thermal and mechanical operations only. The volume may be changed with the help of the piston, and heating and cooling may be present.

There are several special circumstances we should examine. For example, the cylinder might be perfectly insulated, in which case it is impossible for heat to flow across the surface of the system. On the other hand, the volume may be kept constant, or we might try to leave the temperature unchanged. These cases are known as *adiabatic*, *isochoric*, and *isothermal* processes, respectively.

Our second example is *the melting and the vaporization of single substances.* This class of processes is chemical in nature. In melting and in vaporization, substances change. The details of the processes can be rather involved. For the case of melting we shall give a simple mathematical description of the heating taking place. This will serve as an introduction to constitutive relations, a problem which otherwise we will not touch upon much in this chapter.

1.3.1 Adiabatic Compression and Expansion

Let us begin with a phenomenon which might appear rather surprising at first. Consider the special case of air in a cylinder having a piston. The setup is assumed to be perfectly insulated which makes the exchange of heat impossible. We are allowed only to compress or expand the air. If this condition is satisfied, we say that the fluid may undergo only *adiabatic processes.* We know from experience with bicycle pumps that upon sudden compression the hotness of the air rises abruptly. On the other hand, if the gas is allowed to expand under such circumstances it is found that the temperature drops steeply. If you doubt this to be possible, place a small piece of an easily combustible material in a cylinder into whose walls a window has been cut (German tinder in a pneumatic tinderbox, for example). When you suddenly compress the air you will see a flash of light as the material ignites. This situation demonstrates that the temperature of the gas must have increased considerably. In other words:

> *In adiabatic compression or expansion of a fluid such as air, its hotness changes without heating or cooling.*

This is surprising indeed. Would we not automatically believe that some heating must have occurred for the temperature of a body to increase? Heat has not been

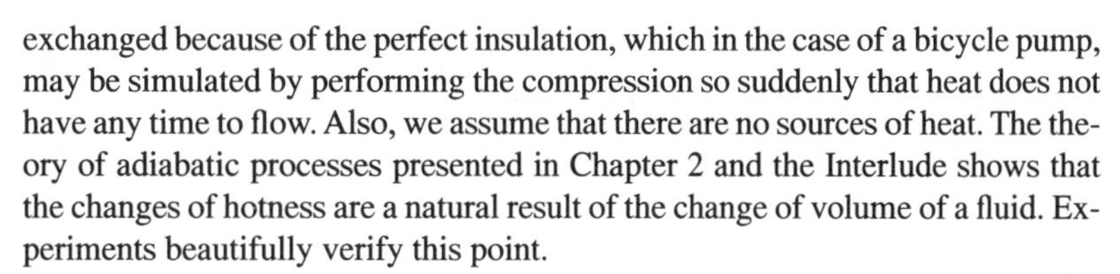

exchanged because of the perfect insulation, which in the case of a bicycle pump, may be simulated by performing the compression so suddenly that heat does not have any time to flow. Also, we assume that there are no sources of heat. The theory of adiabatic processes presented in Chapter 2 and the Interlude shows that the changes of hotness are a natural result of the change of volume of a fluid. Experiments beautifully verify this point.

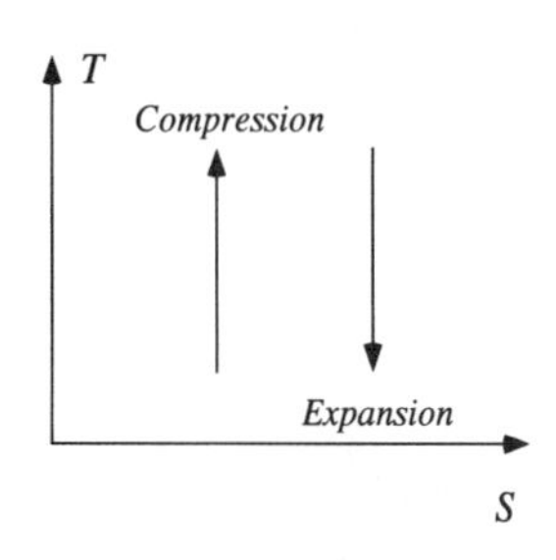

FIGURE 12. Adiabatic compression and expansion in the temperature-heat diagram. This diagram is a valuable tool in thermodynamics as was observed by J.W. Gibbs (see Section 1.1.12).

In summary, the temperature of air changes upon compression or expansion without any heating having taken place, and without any internal sources having supplied heat. In other words, the heat content of the gas must remain constant during adiabatic changes. This is a simple case of accounting for amounts of heat: nothing has gone in, nothing has come out, and nothing has been produced inside. There is a simple and useful tool which allows us to describe this phenomenon graphically, namely the *temperature-heat diagram* (or *T-S* diagram) of the process (Figure 12). We shall use the symbol S for the heat of a body. Since the heat content of the body remains the same, the curve representing the process must be a vertical line in the diagram. The temperature rises upon compression, and it decreases as a result of expansion of air.

If for any reason the process considered is not reversible, i.e., if heat is generated in the fluid, the representation of an adiabatic change in the *T-S* diagram will differ from the one in Figure 12 (see Section 1.6).

1.3.2 Heating at Constant Volume

Next let us discuss the heating of air if its volume is kept constant. We say that under these circumstances the fluid undergoes an *isochoric process*. It is very simple to keep the volume constant in the case of gases heated or cooled. Heat does what we usually believe it does. A body which is heated at constant volume gets hotter:

> *Experience tells us that no matter what kind of body is heated at constant volume its temperature must increase if heat is absorbed, and it must decrease if heat is emitted.*

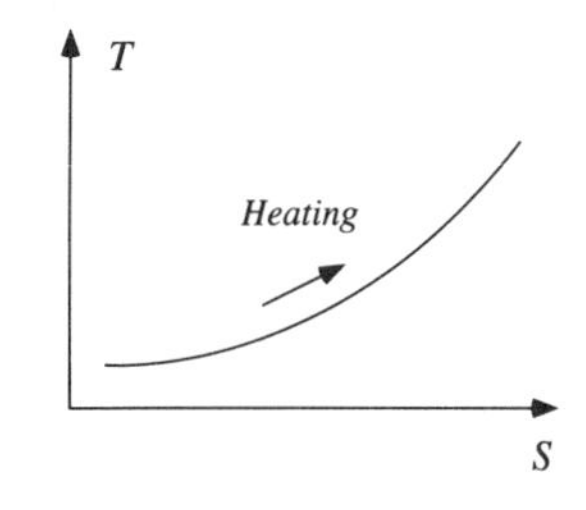

FIGURE 13. Heating at constant volume. If heat is added, the temperature of the body must increase. How it increases depends on the properties of the body.

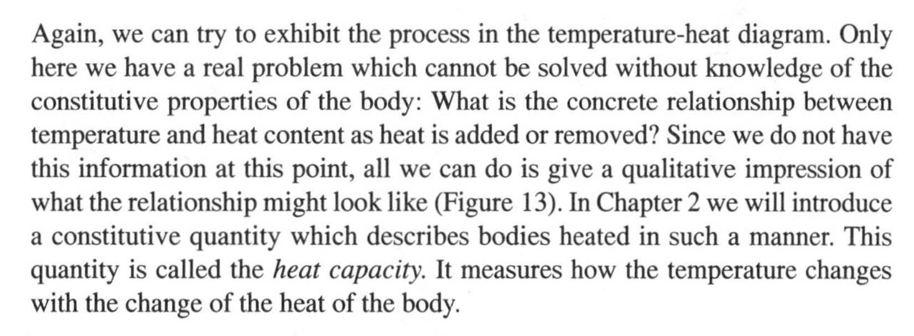

Again, we can try to exhibit the process in the temperature-heat diagram. Only here we have a real problem which cannot be solved without knowledge of the constitutive properties of the body: What is the concrete relationship between temperature and heat content as heat is added or removed? Since we do not have this information at this point, all we can do is give a qualitative impression of what the relationship might look like (Figure 13). In Chapter 2 we will introduce a constitutive quantity which describes bodies heated in such a manner. This quantity is called the *heat capacity*. It measures how the temperature changes with the change of the heat of the body.

Since we assume heat to be conserved we know that the amount of heat communicated to the body is equal to the change of its heat content. Again, this is a particular case of accounting for heat: the heat content of a body can change by transfer only, since nothing is produced inside. The sum of what has been absorbed and emitted (if we take an amount which is emitted as a negative quantity) must be equal to the change of the contents.

1.3.3 Isothermal Processes

Another interesting process is the heating of air at constant temperature. We speak of *isothermal processes* in this case. Since the temperature of air changes if it is heated at constant volume, we obviously have to let the volume change for an isothermal process to take place. On the other hand, we have seen that the temperature changes if the volume is changed without heating (adiabatic processes). Combining these experiences we can come up with the answer of how to perform an isothermal process. We may heat a body which would normally increase its temperature; if we let the fluid expand, its temperature should drop. We only have to combine and fine-tune the rates of heating and of expansion for the temperature to remain constant (Figure 14). On the other hand, the hotness of air rises if it is compressed without heating. Therefore, we must cool it at just the right rate during compression for the temperature not to change. In other words:

> *If a fluid such as air is to undergo an isothermal expansion it has to be heated at the same time. During an isothermal compression it has to be cooled. Therefore, if the volume of air increases isothermally its heat content increases; heat is literally sucked up by the gas. If the volume decreases the heat content decreases; heat is pressed out of the fluid.*

Reservoir
Heat
T = constant

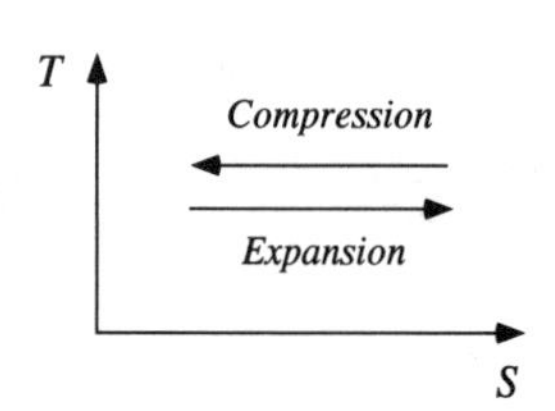

FIGURE 14. If a gas is to expand or contract at constant temperature, it has to be able to exchange heat with the surroundings. The processes of isothermal expansion or compression are represented by horizontal lines in the *T-S* diagram. On expansion, fluids usually absorb heat. However, this does not always have to be the case: water in the range of temperatures between 0°C and 4°C behaves differently!

There is an important exception to the details of adiabatic and isothermal processes just presented. Water in the range of temperatures between 0°C and 4°C behaves differently. Take the isothermal changes first. Water will be found to *emit* heat when *expanding* at a temperature in this range. Also, for part of an adiabatic compression the temperature decreases as the volume is decreased, only to go through a minimum whereupon it increases as is normally expected.[15] Even though there are very few cases of anomalous behavior, the example of water demonstrates that we cannot simply leave it out of consideration. Water is too important a fluid.

We would like to know how much heat is required for a given change in volume of the fluid. Again this question is answered by a particular constitutive quantity, which is called *latent heat*. It will be introduced, with the heat capacity in Chapter 2.

15. Kelvin was one of the first to discuss the significance of this behavior for thermodynamics (see Truesdell, 1980). A theory of classical thermodynamics allowing for the anomaly was first presented by Truesdell and Bharatha (1977).

EXAMPLE 5. Replace an adiabatic process by isochoric and isothermal steps.

Air in a bicycle pump is compressed adiabatically. How can the air be brought back to its original state if we first wait for the air to cool in the pump? Draw the steps performed in a *T-S* diagram.

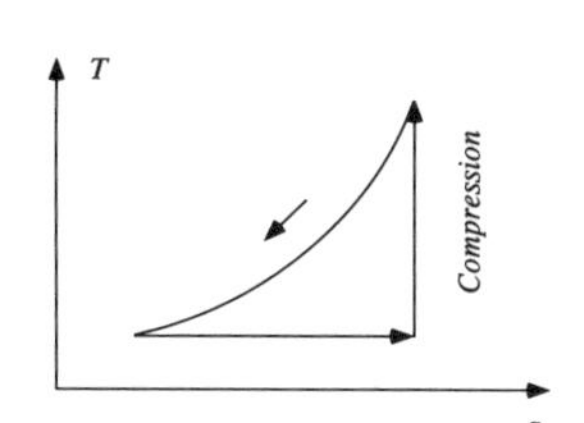

SOLUTION: If we compress air adiabatically, we raise its temperature without changing its heat content (see figure). If we now wait for the air to cool without the piston being moved, the following is taking place: heat is flowing out of the air at constant volume, thereby reducing its temperature. This step restores the initial temperature of the surroundings, but it reduces the heat of the air. Therefore, as a last step, we have to increase the heat content to its original value without changing the hotness of the system. This is achieved by isothermal heating which, also brings the volume of the air back to its initial value.

1.3.4 Melting and Vaporization

Finally, let us take a brief look at *phase changes*. When a block of ice having a temperature below the freezing point is heated, we first observe a rise of temperature of the body. The process is analogous to the heating of a fluid at constant volume (Figure 13). The water which is produced from the ice by melting behaves similarly. We also know that water eventually turns into steam, which can be heated further at constant volume if we like. But what about the processes which turn ice into water, and water into steam?

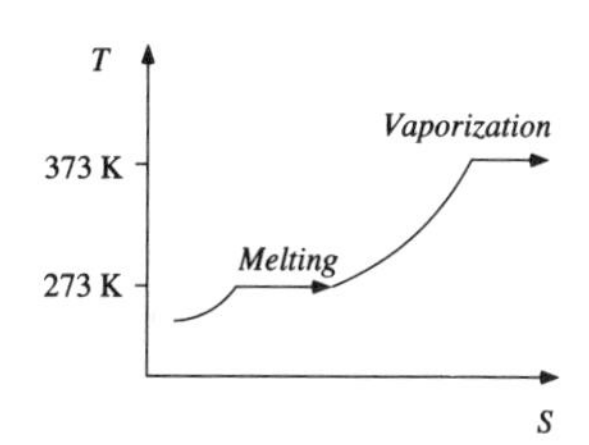

FIGURE 15. Relationship between temperature and heat absorbed for melting or vaporization of ice and water. During the change of phase, the temperature of the body stays constant. The diagram gives only a qualitative representation of the relationship.

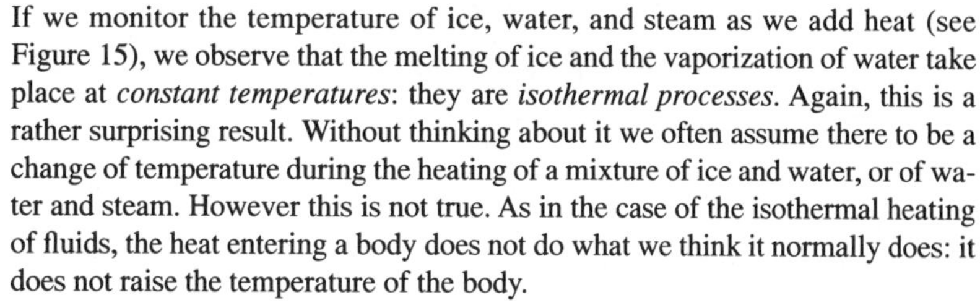

If we monitor the temperature of ice, water, and steam as we add heat (see Figure 15), we observe that the melting of ice and the vaporization of water take place at *constant temperatures*: they are *isothermal processes*. Again, this is a rather surprising result. Without thinking about it we often assume there to be a change of temperature during the heating of a mixture of ice and water, or of water and steam. However this is not true. As in the case of the isothermal heating of fluids, the heat entering a body does not do what we think it normally does: it does not raise the temperature of the body.

The melting of ice has played an important role in the history of the theory of heat. The process served as a means for constructing ice calorimeters (see the following section), and the melting of ice by rubbing two blocks of it against each other provided strong evidence that heat can be created (Section 1.1.6).

The examples of adiabatic and isothermal changes demonstrate clearly that heat and hotness cannot be the same quantities. In one case, the hotness changes even though the heat content does not; in the other, the amount of heat in a body changes while the temperature remains constant.

1.3.5 Measuring Amounts of Heat

Now we can describe a procedure which, at least in principle, allows us to measure amounts of heat exchanged. While not exactly practical, it clearly demonstrates thermodynamic reasoning. The method depends on the availability of bodies which can undergo isothermal and adiabatic processes without producing heat. We need an ice calorimeter and a means of transferring heat between a body and the calorimeter. The calorimeter contains a mixture of water and ice, which as we have seen, remains at a constant temperature. As it emits or absorbs heat, the relative amounts of ice and water change. Since ice and water do not have the same density, the volume of the mixture must change with the heat content. Changes of volume are made visible with the aid of a capillary in the calorimeter (Figure 16).[16]

A body is to be heated with heat taken from the calorimeter (when heat is emitted by the calorimeter the capillary reading goes up since the amount of ice increases). Since the ice calorimeter tells us how much heat has been removed, we will know the amount transferred to the body if the transfer takes place ideally, i.e., without the production of heat. We can accomplish this using an ideal fluid enclosed in a cylinder with piston (Figure 16). As in the case of the calorimeter, we assume the container walls do not contribute to the thermal processes; neither should they store heat. They separate the fluid from the surroundings and let heat pass easily. Let us now bring the fluid in contact with the calorimeter and withdraw some heat from it by letting the gas expand at constant temperature (Figure 16). We then remove the cylinder, insulate it, and compress the fluid adiabatically until its temperature becomes equal to that of the body to be heated.

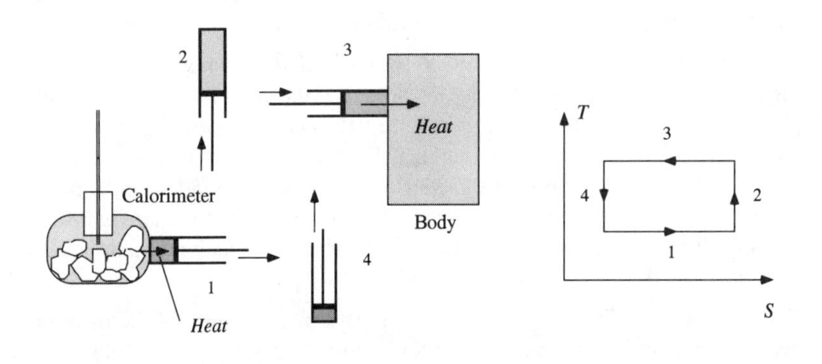

FIGURE 16. The amount of heat transferred to a body can be measured using an ice calorimeter and an ideal fluid. While transferring heat from the cold calorimeter to a warmer body the fluid undergoes a four-step process which is the reverse of a Carnot cycle. The four operations are shown in the accompanying *T-S* diagram. During step 1 the fluid in the cylinder absorbs heat from the calorimeter, and during step 3 it emits heat to the body.

In the third step the fluid is brought in contact with the body and compressed at constant temperature; if the amount of heat communicated to the body is small, its temperature will remain nearly constant. Alternatively, the body may undergo a phase change at constant temperature. The fourth step consists of expanding the gas adiabatically to let its temperature return to the temperature of the calorimeter. Assume, for argument's sake, that not all of the heat taken from the calorimeter is transferred to the body. The gas will be in a different state compared to the beginning of the cycle. We can return to the initial state by simply com-

16. See G. Job (1972), p. 12.

pressing the gas a little more while it is in contact with the ice-water mixture. In the end, the calorimeter reading will tell us how much heat has been transferred to the body.

Consider the fluid which has been used for transferring heat. The process it undergoes is the reverse of a *Carnot process* (Section 1.4), which means that heat has been transferred at constant temperatures only. The cycle the fluid goes through is shown in the *T-S* diagram in Figure 16.

1.3.6 The Law of Heating and of Changes of Heat for Melting

At this point, we shall make an exception and discuss in some detail a constitutive relation, namely one that applies to the melting (or the vaporization) of bodies. Assume a body is melted at constant pressure. The temperature must stay constant. So while heat is absorbed, the only quantity of the body which changes, except for a slight change of the volume, is the amount of substance. It is common, therefore, to state the law of melting in terms of how much heat is needed to melt a unit amount of a given substance. It is found that equal units of heat are needed for melting equal amounts of a substance. The factor $\bar{l}_f$ relating the two numbers is called the *molar latent heat of fusion*:

$$S_e = \bar{l}_f \Delta n \tag{8}$$

Here Δn is the change of the amount of substance resulting from the melting (e.g. water), and S_e is the amount of heat communicated to the body. The index e stands for *heat exchanged.* We count amount of heat exchanged as a positive quantity if heat is added to a body. Values of the latent heat of fusion for different substances can be found in Table A.4. These constitutive quantities have to be determined by experiment. Now that we have learned how to measure amounts of heat, we can apply this knowledge to the process of melting. We begin by defining the unit amount of heat, for example, by stating that it is the amount necessary to melt one cubic centimeter of ice. We translate this into the reading of the ice calorimeter introduced above (Example 7). We know, from now on, how much heat has been withdrawn from or added to the calorimeter. If we transfer heat from the calorimeter to another substance being melted, we can relate the amounts of heat transferred to the body to the change of the amount of the substance involved.

Actually, we are just as interested in the change of the heat content of the body undergoing a phase change as we are in the amount of heat exchanged. This problem is not solved so easily, for the simple reason that we have to take into consideration the possible production of heat as well. Certainly the heat communicated to the substance will be stored there, and it will contribute to the change of the heat content. However, if heat is produced in the body during the process, the change will be larger than the amount absorbed. It is found experimentally that a substance solidifying will emit exactly the quantity of heat it has received in the reverse process, i.e., in melting. In fact, we have already built this observation into the law of phase change, Equation (8). Note that the heat exchanged changes its sign if we reverse the process. We would have been forced to choose another form of the constitutive law if we had found melting and solidifying to be different. Conversely, we can say that by adopting a law of the form of Equation (8) we do not admit irreversible processes. This is what we meant when we said that our models of bodies determine whether heat is produced in a process (Section 1.1.9).

We must conclude from the observations that phase changes are processes which do not produce heat. If they did, a body would have to emit more heat in solidifying than it absorbs as a consequence of melting. For this reason, the *change of the heat content*, denoted by ΔS, is simply related to the amount of heat communicated to the body:

$$\Delta S = S_e \tag{9}$$

This equation is an instance of the fundamental law of thermal physics, namely the *law of balance of heat* (Section 1.6). Here, it simply states that the change of heat of a body is equal to the heat exchanged if there is no production of heat inside the system under consideration. A change can be effected only by heat flowing into or out of the body. This is an application of our general law of balance of any kind of substancelike physical quantity. Finally, by combining the fundamental law, Equation (9), with the material property, Equation (8), we can calculate the change of the heat content in phase changes in terms of the latent heat.

While the most conspicuous variables involved in the melting of a body are heat and amount of substance, we should not forget the change of volume commonly associated with the phase transition. The fact that ice has a smaller density than water plays a vital role in our environment. It determines important details of the formation of sea ice at the Earth's poles, which in turn influences the climate. It also serves as a clear warning signal to anyone who sees an ice cube sink in a drink. Another nice example comes from biology. Sperm whales, for example, control their buoyancy in a most ingenious manner. Much of the head of the whale contains a kind of wax. If the animal takes in cold water through pipes in the head, the wax solidifies; its volume decreases, increasing the average density of the whale. At depths of a couple of thousand meters the whale can float for a long time, waiting for prey. Eventually, warm blood will be brought into the head which melts the wax. The animal will gain positive buoyancy which allows it to surface without much strain.

The law governing the *vaporization* or *condensation* of bodies is analogous to what we have seen in the case of melting. We shall not write down the equations separately. We simply introduce the *molar latent heat of vaporization* $\bar{l}_v$ to cover this phenomenon. (See Table A.5 for values of this quantity.)

EXAMPLE 6. The molar and the specific heats of fusion of ice.

Take the unit of heat as that quantity which melts 0.89 cubic centimeters of ice. Ice has a density of 920 kg/m^3. a) Determine the molar and the specific latent heats of fusion of ice. The specific value l_f is taken with respect to mass rather than to amount of substance.
b) Compute the change of heat of a mixture of ice and water if 200 g of ice are formed.

SOLUTION: a) The specific heat of fusion is defined by the relation

$$S_e = l_f \Delta m = l_f \rho \Delta V$$

which has been transformed with the help of the density of ice. According to the definition of the unit of heat, the factor $l_f \rho$ must be equal to 1 Ct / 0.89 cm^3. This leads to

$$l_f = \frac{10^6}{0.89 \cdot 920} \frac{\text{Ct}}{\text{kg}} = 1220 \frac{\text{Ct}}{\text{kg}}$$

Now we obtain the molar heat of fusion by employing the relationship between mass and amount of substance, i.e., $m = M_o n$:

$$\begin{aligned} \bar{l}_f &= M_o l_f \\ &= 0.018 \cdot 1220 \frac{\text{Ct}}{\text{mole}} = 22 \frac{\text{Ct}}{\text{mole}} \end{aligned}$$

Note that since the unit of heat is traditionally derived in terms of the units of energy and temperature, we are not free to introduce another definition. The one given here has been adjusted so as to agree with the definition introduced in Section 1.4.

b) According to the special form of the balance of heat in Equation (8), the change of the heat content is given by the heat exchanged. Therefore:

$$\begin{aligned} \Delta S &= l_f \Delta m \\ &= 1220 \frac{\text{Ct}}{\text{kg}} \cdot (-0.20 \text{ kg}) = -244 \text{ Ct} \end{aligned}$$

The minus sign tells us that the amount of heat of the mixture has decreased. Heat is emitted when ice forms. Both the change of the heat content and the heat exchanged are negative.

EXAMPLE 7. The ice calorimeter.

How thin does the inner diameter of the capillary of the ice calorimeter in Figure 16 have to be made, if one unit of heat is to translate into a change of level of 2.0 cm? The volume of the calorimeter is equal to 1.0 liters. The densities of water and of ice are taken to be 1000 kg/m^3 and 920 kg/m^3, respectively.

SOLUTION: We calculate the change of volume of the mixture of ice and water from the densities:

$$\Delta V = \frac{\Delta m}{\rho_{ice}} - \frac{\Delta m}{\rho_{water}}$$

which is equal to

$$\Delta V = \left(\frac{1}{\rho_{ice}} - \frac{1}{\rho_{water}} \right) \frac{S_e}{l_f}$$

$$= \left(\frac{1}{920} - \frac{1}{1000} \right) \frac{1}{1220} \text{m}^3 = 7.1 \cdot 10^{-8} \text{m}^3$$

The change of volume of the mixture translates directly into the change of level of water in the capillary. The diameter must therefore be equal to 2.0 mm. The volume of the calorimeter does not matter.

EXAMPLE 8. Vaporization of freon in a refrigerator.

Freon-12 is used in a refrigerator. It is evaporated at low temperature and thereby absorbs heat from the space to be cooled. Assume that freon is evaporating at 0°C. It is to take up a current of heat of 4.0 W/K. a) At what rate must liquid freon be converted to its gaseous form if the latent heat of vaporization at 0°C is equal to 570 Ct / kg. b) At what rate could ice be formed in the refrigerator?

SOLUTION: a) We transform Equation (8) to differential form, and change from the molar latent heat to the specific value. This leads to:

$$I_s = l_v \frac{dm}{dt}$$

Here, I_S stands for the rate at which heat is absorbed; i.e., it represents the heat current. The rate of change of the mass of liquid freon is found to be:

$$\frac{dm}{dt} = \frac{4.0}{570} \frac{\text{kg}}{\text{s}} = 7.0 \cdot 10^{-3} \frac{\text{kg}}{\text{s}}$$

b) According to the results of Example 6, about 3 g of ice could be formed every second if heat is removed at the given rate.

1.4 Engines, Thermal Power, and the Exchange of Heat

So far, we have discussed the properties of heat informally. The discussion can be summarized as follows: we believe there is a substancelike quantity called entropy which is responsible for thermal processes. It is exchanged in heating, produced in irreversible phenomena, and can be stored in bodies. The properties just listed are an informal statement of the law of balance of entropy, i.e., of the second law of thermodynamics (see Section 1.6).

Now we might expect to be able to learn more about entropy if we manage to clarify its relation to energy in thermal processes. In his investigation of the motive power of heat Carnot was the first to see this important point. We shall attack the problem head on by employing the full power of his comparison of heat and water, which was quoted in Section 1.1.8. We will build upon the properties of entropy as they have appeared to us so far; i.e., we shall take for granted the validity of the Second Law. Second, we will assume that the role of energy in thermal processes is the same as that known from other fields of physics as discussed in the Prologue (Sections P.1.4 and P.4); in other words, we shall assume the law of balance of energy (which is called the First Law) to remain valid in the presence of thermal processes. In summary, Carnot's investigation of the principle of operation of heat engines will deliver the following results:

1. A heat engine absorbs entropy from the furnace at higher temperature; if it could be operated reversibly, the same amount of entropy would be transferred to the cooler, i.e., to the environment at lower temperature (Second Law).
2. By lowering entropy from a point of higher temperature to a point of lower temperature, energy is released at a certain rate (thermal power); this energy drives the mechanical process since, in "energy transformations," energy is neither produced nor destroyed (First Law).
3. The energy released in the fall of entropy must be supplied to the engine together with entropy from the furnace. More generally, energy is transferred together with entropy in heating and in cooling.

In short, we simply add thermal phenomena to the list of processes known from other areas of physics. Accepting this, we will be able to directly state the formula for the motive power of heat which will provide for the simplest possible entry into thermodynamics. (A derivation on the basis of some other assumptions can be found in the Interlude.) The consequences of this idea will be developed for ideal engines in the present section. Energy and the production of heat will be discussed in Section 1.7.

Remember that we will use the word *entropy* for the fundamental quantity of thermal processes when it comes to formal relations. Informally, we shall continue to use the word *heat* in the sense of the property which makes stones hot and which lets ice melt.

1.4.1 Heating and Cooling of a Carnot Heat Engine

Heat can do work. This is a simple fact which has been known for a long time. In modern times engineers started to use the *motive power of heat* in steam engines. A number of technical devices based on the effects of heat have become important to our lives. Among them are car engines, refrigerators, thermal power plants, and heat pumps for heating homes and buildings. A few simple but important facts are known about heat engines: First, they need a furnace and a cooling device; i.e., they operate between two environments at different temperatures; they absorb heat from the furnace, and they emit heat to the cooler. Second, the motive power of such engines depends on the temperatures of the furnace and the cooler; indeed, their power increases with increasing difference between the upper and the lower temperatures.

Let us now turn to the first of these observations. Heating the engine first of all means that entropy is taken in from the furnace at the higher temperature. The flow of entropy from the furnace to the engine is denoted by an *entropy flux* I_S which is measured in Ct/s. Since entropy cannot disappear, it has to be emitted if the engine is to be operated steadily. This is accomplished with the help of the cooling device which receives the entropy rejected by the engine. We know from the properties of entropy that the current leaving the engine must be equal to that entering if all the operations are reversible, i.e., if entropy is not produced in the system. Since cooling takes place at a lower temperature, entropy flows from the hot furnace to the cold "refrigerator," and we say that heat engines are driven by the *fall of heat* (entropy). This observation allows us to interpret the principle of operation of a heat engine in accordance with water wheels which are driven by water falling through a certain distance. Temperature therefore plays the role of the *thermal potential.*

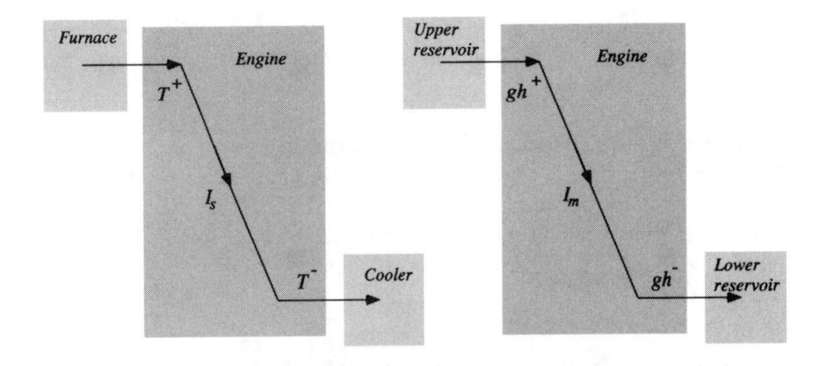

FIGURE 17. "Waterfall diagram" of a thermal process. A current of entropy flows from a higher to a lower thermal level. This image of the action of heat in a heat engine is structurally analogous to what occurs in a hydraulic power plant.

In summary, the engine's task is to transport heat, i.e. entropy, from the furnace to the cooler. How can this be achieved? Consider, as Carnot did, an engine which absorbs entropy at *constant temperature* from a furnace. Then the entropy absorbed is lowered to the temperature of the cooler whereupon *all the entropy (heat)* is emitted, again at *constant temperature*. A heat engine operating in the manner described is called a *Carnot engine*. Actually, for us the *engine* is the working fluid employed, such as air or steam. The fluid operating in such an engine undergoes a cyclic process called a *Carnot cycle*, whose steps we can describe qualitatively in terms of the simple changes discussed in Section 1.3 (Figure 18). To be specific, let us assume the agent of the engine to be air. The first step in the cycle described above must be an isothermal expansion of the fluid. If air expands isothermally, it absorbs entropy at constant temperature. This

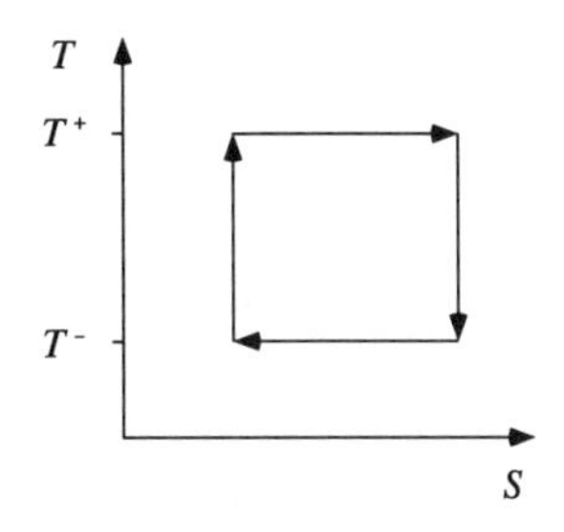

FIGURE 18. Carnot cycle in the *T-S* diagram. It consists of four steps, two of them isothermal, the other two adiabatic. If the steps are performed ideally, i.e., if no entropy is produced in the agent, the cycle has the simple form of a rectangle.

is exactly what we want. Now the temperature of the air has to be lowered to the hotness of the cooler; this process has to take place without any heat being exchanged. Therefore this step must involve an adiabatic expansion. In other words, the air expands during the first two steps of the Carnot cycle.

Now the fluid has to emit the entropy it has absorbed in the first step. Since the emission is to take place at constant temperature this step must be an isothermal compression. In the end, we only have to return the agent to the starting point for it to be able to begin another cycle of operation. The final process is an adiabatic compression, which raises the temperature to the desired level, namely to the temperature of the furnace, without adding or removing entropy. Note that the steps in this cyclic operation are assumed to be ideal, just like those discussed in Section 1.3. (See Figure 16 for comparison.) For this reason the cycle described is that of a *reversible engine*.

1.4.2 Thermal Power and the Balance of Power

On the basis of these observations regarding the operation of heat engines, Carnot was able to suggest a theory relating heat and energy. We already have presented the basic idea of how such engines work, by the usual *waterfall diagram* in Figure 17. Therefore, in analogy to other fields of physics, we should expect the motive power of heat engines to be strictly proportional to the drop of the thermal potential. Compare the process depicted in Figure 17 to the operation of a hydromechanical engine, i.e., a water turbine which is driven by water falling from an artificial lake at a high level.

The gravitational power of the fall of water is measured by the rate at which energy is released in the process. This quantity is known to be equal to

$$\mathcal{P}_{grav} = -(gh^- - gh^+)|I_m| \tag{10}$$

If we want to compare hydromechanical and thermomechanical engines we only have to replace the mass of the water by entropy (or the current of mass by the current of entropy), and the gravitational potential gh by temperature. If we follow the suggestion of this image we have to conclude that the *motive power of heat*, i.e., the rate of energy released by a fall of entropy, is given by

$$\mathcal{P}_{th} = -(T^- - T^+)|I_s| \tag{11}$$

We call this the *thermal power* of the heat engine; fundamentally, it is the same as electrical power or mechanical power.

Heat is used to drive the mechanical process of a heat engine, just as electricity (i.e. electrical charge) is used to turn a motor. To understand the heat engine fully, we only have to compare it to other engines. We know that energy is needed at a certain rate for driving the mechanical process of the heat engine. We have said before that energy is released in the fall of entropy, and this energy drives the engine. Since no other process takes place, and since we assume the device to run in steady state, the two rates—the mechanical and the thermal power—must be equal in magnitude (Figure 19):

$$\mathcal{P}_{th} + \mathcal{P}_{mech} = 0 \qquad (12)$$

This is a direct result of our assumptions concerning thermal processes and energy. We simply include thermal processes with all others. We may then formulate a more general law of balance of power:

In steady state processes, including thermal ones, the sum of all rates at which energy is released and used must add up to zero.

This statement will carry over to all thermal processes, reversible and irreversible. This law allows us to relate thermal processes to other processes, and thus furnishes much needed information about the properties of heat (entropy).

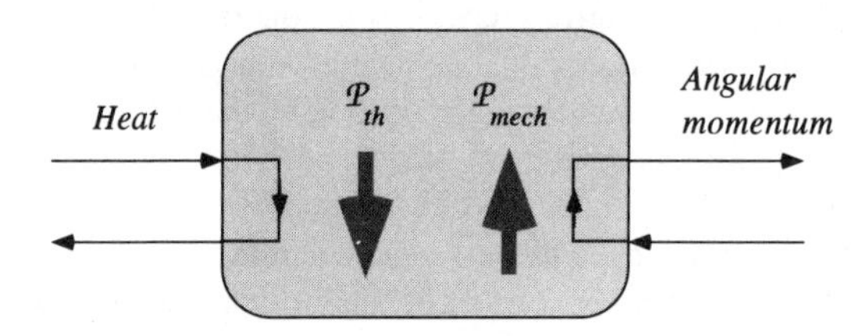

FIGURE 19. The operation of an ideal Carnot engine is analogous to that of many other devices known from physics and engineering. In a thermal process, namely in the fall of entropy, energy is released at a certain rate. This rate is called thermal power. Since the ideal engine drives only a mechanical process, the mechanical power must be equal in magnitude to the thermal power. The energy released in the fall of heat is used to "pump" angular momentum.

EXAMPLE 9. The current of entropy through an ideal Carnot engine.

Consider a Carnot engine working between a furnace at 300°C and a cooler at 40°C. This ideal engine is known to have a mechanical power equivalent to 5.0 MW. a) How large must the current of entropy through this engine be? b) If the current of entropy is unchanged, how large must the temperature difference be for a power of 4.0 MW?

SOLUTION: a) This is a direct application of the relation for the motive power of a heat engine. First we have to conclude that the thermal power of the engine is equal to its mechanical counterpart, namely 5.0 MW. The thermal levels, i.e. the temperatures, are given. Therefore:

$$|I_s| = \left|\frac{\mathcal{P}_{th}}{\Delta T}\right|$$
$$= \frac{5.0}{573-313}\frac{\text{MW}}{\text{K}} = 19.2 \cdot 10^3 \frac{\text{W}}{\text{K}}$$

W/K (= Ct/s) is the SI unit of a current of entropy.

b) The motive power depends linearly upon the difference of temperatures between the furnace and the cooler. For a power which is only 4/5 of the original one, it suffices to have a temperature difference equal to 80% of the original one, i.e. 208 K.

1.4.3 Entropy and Energy in Heating and Cooling

Equation (11) is an example of a general feature of physical processes which we have encountered before in gravitation and in electricity. If quantities such as mass, electrical charge, or entropy fall from higher to lower levels, energy is released at a rate which is proportional both to the flux of the substancelike quantity and the difference in the associated potential. However, there is more to this result than just a formula for the motive power of heat. Since energy cannot just appear out of the blue, we have to look for its origins. The only possible reservoir which can deliver the energy released in the fall of entropy is the furnace (Figure 20). Energy enters the engine with the flux of entropy. Therefore we conclude that a body which receives (loses) entropy also receives (loses) energy. Naturally, the same statement must hold for the cooler as well.

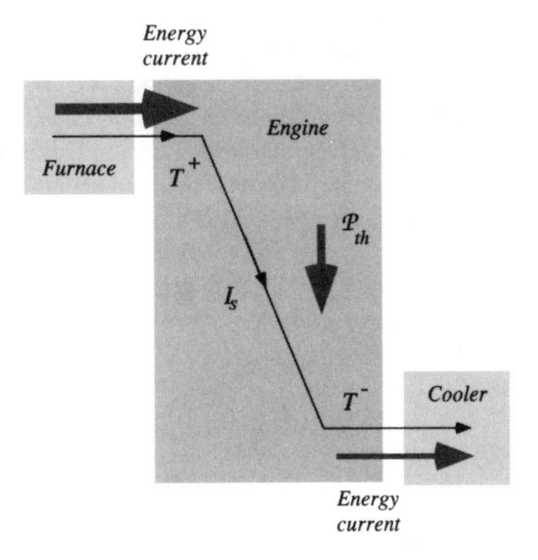

FIGURE 20. The diagram picturing the principle of operation of a heat engine (Figure 17) only has to be extended to include thermal power and fluxes of energy which accompany the fluxes of entropy from the furnace and to the cooler. The thermal power must be equal to the difference of the thermal energy fluxes with respect to the system.

We should expect a straightforward relation to exist between currents of entropy and thermal power in processes involving heating and cooling of bodies. Equation (11) suggests that this relationship is of the same form as that found in other fields of physics. If the motive power is the difference of the thermal powers of heating and of cooling of the engine, then each of the thermal energy fluxes must be determined by the flux of entropy and by the temperature of the body undergoing heating or cooling:

> *If a body at temperature T receives entropy at a given rate, it absorbs energy at a rate which is the product of the temperature and the rate at which entropy is received (Figure 21).*

This is a general law of thermal physics which we are going to use throughout the book. It will serve us well when it comes to determining the relationship between entropy and energy in thermal processes.

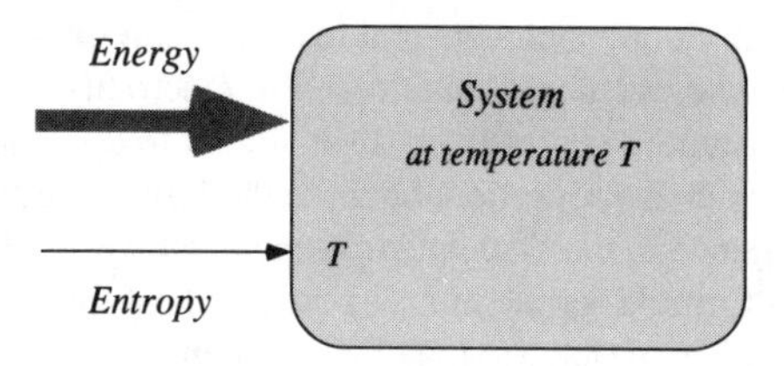

FIGURE 21. The process of heating of a body always consists of the flow of two quantities—entropy, and energy. Energy enters the body with its carrier, i.e., with entropy.

How does a body receive entropy, and therefore energy, in a thermal process? Put differently, how do entropy and energy get from the furnace to the heat engine (Figure 20), from an immersion heater to the water, or from the sun to the Earth's atmosphere? In Section 1.1 we briefly identified three modes of heat transport. Since we are talking about a body receiving entropy, we must exclude the transport of entropy with flowing substances (convection). We are therefore left with radiation and conduction. For now we shall apply the fundamental law relating rates of entropy and energy received (or emitted) to conduction only. In conduction, the transport of entropy takes place with the help of currents across the surface of the body. As a consequence we are led to say

In conduction, the current of energy entering a system at temperature T is given by the product of the current of entropy entering the system and the temperature of the system:

$$I_{E,th} = T I_s \tag{13}$$

Even though this result is less general than the one stated first, it is all we need for the moment. It is the fundamental relationship between currents of entropy and energy in the basic thermal process, namely the transport of entropy by conduction.

Equation (13) can be interpreted in a number of ways. First it introduces the *thermal potential* T in a manner equivalent to the electrical potential. As we will see in Section 1.4.4, this potential can be determined in principle from the operation of ideal Carnot engines, i.e. by measuring their efficiency. This is so since the formula for the motive power of an ideal Carnot engine is independent of the working fluid inside. Steam is just as good as air, or alcohol. Carnot was the first to see this important point. He considered a heat engine and a refrigerator running with different substances to compete against each other. If the heat engine released more energy than the refrigerator needed to pump the entropy back up into the furnace, there would be a net gain of power without any entropy having fallen from the furnace to the cooler. Carnot considered this to be against the laws of physics. We conclude that the potential T appearing in Equation (11) or Equation (13) is independent of thermometric substances, and it can be used as an *absolute temperature* scale. We have to ask whether the thermal potential introduced by Equation (13) is the same as the ideal gas temperature *T*. This is indeed the case, as will be proved in Chapter 2. Therefore, we do not have to distinguish between the temperature we have used so far and the one appearing in this fundamental relation of thermal physics.

We are free to read Equation (13) in still another way: it allows us to *infer currents of entropy from the measurement of energy currents and temperatures.* You will notice the importance of this new principle: it gives additional information needed for the determination of amounts of entropy in particular, and of thermal quantities in general. In particular, we now have the relation between the unit of heat (entropy) introduced in Section 1.1.10, and the units of temperature and energy: 1 Ct = 1 J/K. For currents, or rates, the relation is 1 Ct/s = 1 W/K.

For practical reasons, we often need to know amounts of energy communicated to a body. The energy exchanged in a process is the integral over time of the energy current:

$$Q = -\int_{t_i}^{t_f} I_{E,th}\,dt \tag{14}$$

Using the expression for the thermal energy flux given by Equation (13) we obtain

$$Q = -\int_{t_i}^{t_f} T I_s\,dt \tag{15}$$

Q is the symbol commonly employed for *amounts of energy exchanged in thermal processes*, i.e. in heating and cooling. The sign convention is the usual one: currents entering a system are counted as negative quantities, while a quantity of energy absorbed is counted as a positive one.

Note that the quantity *Q* has received the name *heat* in traditional presentations of thermal physics. Unfortunately, this usage hides the true measure of amounts of heat, i.e., entropy, from our view. Therefore, we shall continue to call *Q* the *amount of energy exchanged in heating or cooling* of bodies.

EXAMPLE 10. The current of entropy absorbed by a river cooling a thermal power plant.

It is known from the operation of a thermal power plant that the water of a river used for cooling carries an energy current of 600 MW. a) How large is the current of entropy entering the environment if the temperature of the water is taken to be equal to 27°C? b) Express the entropy absorbed by a reservoir at constant temperature in terms of the thermal work and the temperature, and c) for the power plant given here calculate the entropy absorbed by the river in one day.

SOLUTION: a) Assume the temperature of the water to remain constant. The amount of water used for cooling is so large that its temperature does not change very noticeably. In this case the temperature remains constant and the current of entropy is given by

$$I_{s,in} = \frac{I_{E,th,in}}{T}$$

$$= \frac{-600 \cdot 10^6 \text{ W}}{(273+27) \text{ K}} = -2.0 \cdot 10^6 \frac{\text{W}}{\text{K}}$$

The law relating currents of entropy and of energy describes an instantaneous situation. It is valid for any temperature, and for nonconstant values as well. In general, therefore, the currents will change in time.

b) Since the temperature of the water is constant during absorption of entropy, the amount exchanged can be computed easily according to Equation (15):

$$Q = -T\int_{t_i}^{t_f} I_s dt = TS_e \qquad \textbf{(E3)}$$

$$S_e = \frac{Q}{T} \qquad \textbf{(E4)}$$

Since heat flows into the body of water at constant temperature, i.e., at a constant level, the energy exchanged is simply equal to the product of entropy exchanged and the temperature at which the exchange is taking place.

c) The amount of energy exchanged in one day is equal to 600 MW · 86400 s = $5.2 \cdot 10^{13}$ J, and the numerical value for the entropy exchanged is $5.2 \cdot 10^{13}$ J / 300 K = $1.7 \cdot 10^{11}$ J/K.

EXAMPLE 11. The thermal power of an immersion heater.

An immersion heater which is hooked up to a battery is heating a body of water. Draw the flow diagram involving the three systems battery, heater, and water. a) Determine the thermal power of the heater in terms of the electrical power of the setup in steady state. b) Compute the current of entropy absorbed by the water for a power of 300 W, and a water temperature of 50°C.

c) Calculate the energy exchanged in the course of 5 minutes. Can you calculate the amount of entropy absorbed by the water in this time span?

SOLUTION: a) From the flow diagram it is clear that the energy released by the operation of the battery is transferred through the heater to the water. In the steady state, the powers must be equal in magnitude, leading to:

$$|I_{E,th}| = |UI_q|$$

b) The current of entropy is determined according to Equation (13) from the momentary values of the thermal energy current and the temperature of the system receiving the entropy:

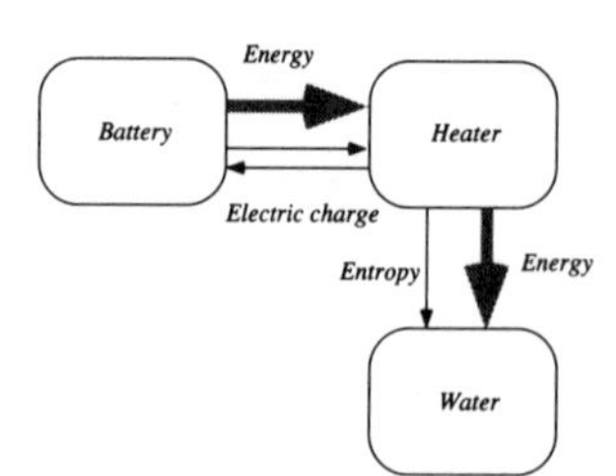

$$I_{s,in} = \frac{I_{E,th,in}}{T}$$

$$= \frac{-300 \text{ W}}{327 \text{ K}} = -0.92 \frac{\text{W}}{\text{K}}$$

c) The amount of energy exchanged in the thermal process within 5 minutes is calculated according to Equation (14):

$$Q = -\int_{t_i}^{t_f} I_{E,th}\, dt$$
$$= -(-300\text{W}) \cdot 5 \cdot 60\text{s} = 90\text{kJ}$$

To compute the amount of entropy absorbed by the water, we would need information about the temperature of the body of water as it is being heated. We do not have this at our disposal at this point. (See Chapter 2 for a solution.)

EXAMPLE 12. Measurement of latent heats of fusion.

a) Demonstrate how the molar or specific entropy of fusion (or of vaporization) can be determined from a measurement of the energy flux of heating, the melting temperature (or the temperature of vaporization), and the rate of change of the amount of substance being melted (or vaporized). b) A mixture of ice and water is heated by an immersion heater and stirred at the same time to insure homogeneous conditions. With an electric power of 50 W, 10 g of ice melt in 67.0 s. Calculate the specific entropy of fusion of ice.

SOLUTION: a) First we write the constitutive law of melting, Equation (8), in terms of fluxes and rates. The amount of entropy transferred to a body is replaced by the current of entropy I_s, while the change of the amount of substance becomes the rate of change:

$$I_s = \bar{l}_f \dot{n} \quad \text{(E5)}$$

Instead of measuring the entropy flux, we determine the thermal energy current entering the system in the process of heating. This quantity is obtained if we multiply (E5) by the temperature T of the process:

$$T \cdot I_s = T \cdot \bar{l}_f \dot{n} \quad \text{(E6)}$$

On the left-hand side we now have the energy current. Since both the temperature and the rate of change of amount of substance of the phases can be measured, we determine the molar latent entropy of fusion from these values according to:

$$\bar{l}_f = \frac{1}{T} \frac{I_{E,th}}{dn/dt} \quad \text{(E7)}$$

The formula for the latent entropy of vaporization is exactly the same. Note that all the currents represent magnitudes. (Signs have been neglected.)

b) With a molar mass of ice of 0.018 kg/mole, the magnitude of the rate of change of the amount of ice is equal to dn/dt = - 0.010 kg / (0.018 kg/mole · 67.0 s) = - 8.3 · 10^{-3} mole/s. The power of the electric process is equal to the thermal power. Therefore, the molar entropy of fusion turns out to be

$$\bar{l}_f = 50\text{W}/(273\text{K} \cdot 8.3 \cdot 10^{-3}\text{mole/s}) = 22\text{J/(K} \cdot \text{mole)}$$

With the help of the molar mass of ice we obtain the specific entropy of fusion: $l_f = \bar{l}_f / M_o$ = 1230 J/(K · kg).

EXAMPLE 13. Thermal work in the ideal Carnot cycle.

Give a graphical interpretation of the energy exchanged in thermal processes during the ideal Carnot cycle and represent the thermal work in the *T-S* diagram.

SOLUTION: If the thermal processes involved are ideal, i.e. if they conserve entropy, the change of the entropy content visible in the *T-S* diagram is the direct measure of the entropy exchanged. Furthermore, entropy is absorbed and emitted at constant temperature only. Therefore, according to Example 10, the energy associated with an amount of entropy exchanged is equal to this amount multiplied by the associated temperature. In our case this is represented in the *T-S* diagram by the areas under the two isothermal steps. The net energy exchanged thermally is equal to the thermal work of the engine, and it is equal to the area enclosed by the Carnot cycle.

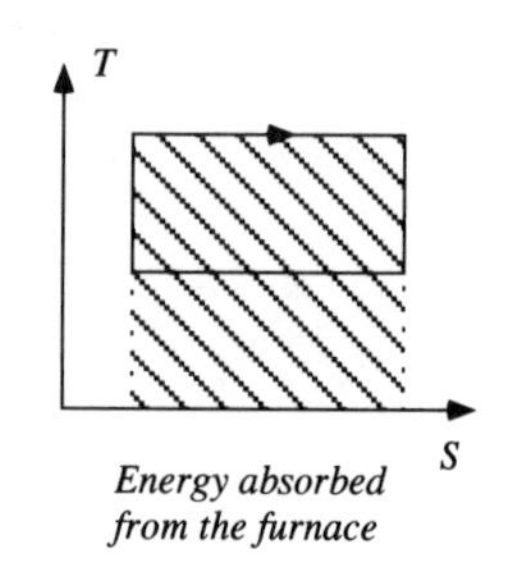

Energy absorbed from the furnace

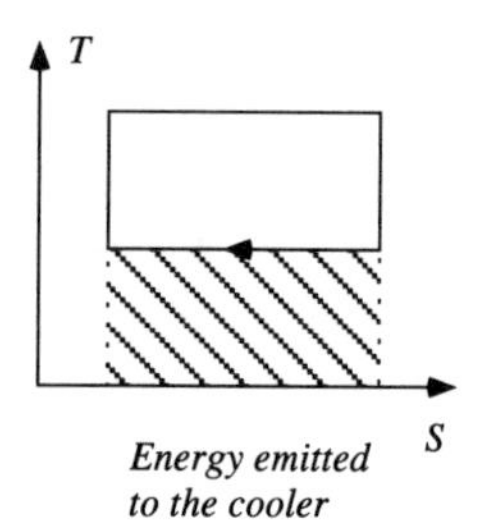

Energy emitted to the cooler

1.4.4 The Balance of Entropy and Energy in Ideal Carnot Heat Engines

For practical reasons, we often are interested in a measure of the efficiency of engines and operations. The efficiency of a heat engine compares the motive power, i.e. the mechanical energy flux, to some other quantity. Since there are different possibilities for comparisons, it is not surprising that two different measures are in use today.

Figure 22 explains these possibilities. First of all, it appears reasonable to compare the mechanical to the thermal power. However, we may also measure the quantity of energy delivered by the engine in relation to the quantity of energy supplied from the furnace. Obviously, under general conditions, the two efficiencies are not the same.

The first of these efficiencies tells us about the fraction of the thermal power of the current of entropy which is delivered by the engine for practical purposes. In other words, it is defined as

$$\eta_2 = \left| \frac{\mathcal{P}_{mech}}{\mathcal{P}_{th}} \right| = \left| \frac{\mathcal{P}_{mech}}{(T^+ - T^-) I_s} \right| \tag{16}$$

(The meaning of the index number 2 will be made clear below.) Put differently, it compares the useful power to the maximum value of the rate at which energy is liberated in the fall of entropy through a given difference of temperatures. Since here we are dealing with ideal engines, the efficiency η_2 is equal to 1. This is not particularly surprising; it just tells us that an ideal Carnot engine does the best possible job of transferring energy for a desired purpose.

On the other hand, it is known that a thermal engine emits entropy, and therefore also energy, to the cooler (Figure 22). Only if the lower operating temperature could be made equal to zero Kelvin would this energy flux be equal to zero. In general, the net thermal energy flux, i.e. the thermal power, with respect to the engine in Figure 22 is

$$I_{E,th} = T^+ I_{s,in} + T^- I_{s,out} = (T^+ - T^-) I_{s,in} \tag{17}$$

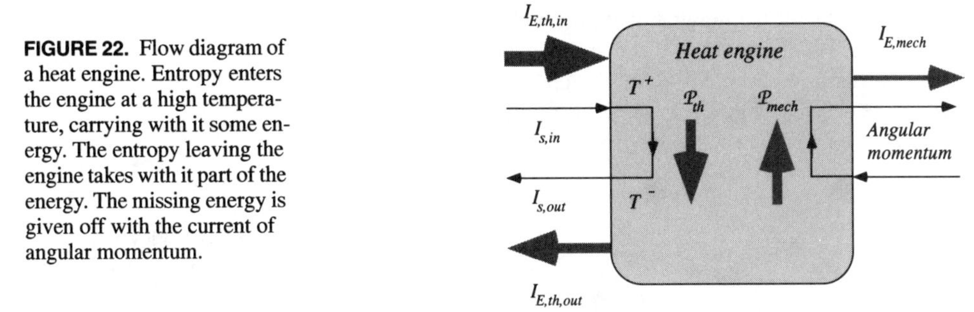

FIGURE 22. Flow diagram of a heat engine. Entropy enters the engine at a high temperature, carrying with it some energy. The entropy leaving the engine takes with it part of the energy. The missing energy is given off with the current of angular momentum.

This is clearly smaller than the energy flux entering the engine from the furnace. The second measure of efficiency compares the mechanical power to the thermal energy current supplied from the burner. Since the mechanical power is equal to the net thermal energy current for an ideal Carnot engine, we obtain

$$\eta_1 = \left| \frac{\mathcal{P}_{mech}}{I_{E,th,in}} \right| = \left| \frac{(T^+ - T^-) I_s}{T^+ I_s} \right| \tag{18}$$

which is equal to

$$\eta_1 = \frac{T^+ - T^-}{T^+} \tag{19}$$

Again, the result only holds for ideal Carnot engines. The value of η_1 is smaller than 1 in general. This second measure of efficiency of a heat engine can approach 100% only if the cooler is maintained at a temperature approaching 0 K. This is so simply because the current of entropy leaving the engine in the process of cooling carries with it a flux of energy according to the associated temperature.

In contrast to the first measure of efficiency given in Equation (16), the latter result suggests that thermal power plants are very wasteful. But then, the efficiency given in Equation (18) is that of an *ideal heat engine*, which according to Equation (16) has an efficiency of 100%. So where is the contradiction? As we will see, there is none. The problem is rather like the one encountered in the case of a hydroelectric plant. Water falls a few hundred meters to drive the turbines. If the plant could be built ideally, it would use 100% of the gravitational power, i.e. all of the energy released in the fall of water. However, the water has been "pumped" to an artificial lake high up in the mountains all the way from the oceans with the help of the sun. From there it is allowed to fall only a fraction of the difference of height measured compared to sea level. Looked at this way, the efficiency of the hydroelectric plant is less than 1 even under ideal conditions (Figure 23). The only difference between the cases of gravitation and heat can be found in the fact that the value of the gravitational potential is arbitrary, while that of the thermal potential is not. However, this does not come to bear in practical situations. Here on Earth, the lower value of the gravitational potential is limited by the level of the oceans, while the lower value of the thermal potential is effectively given by the temperature of the environment.

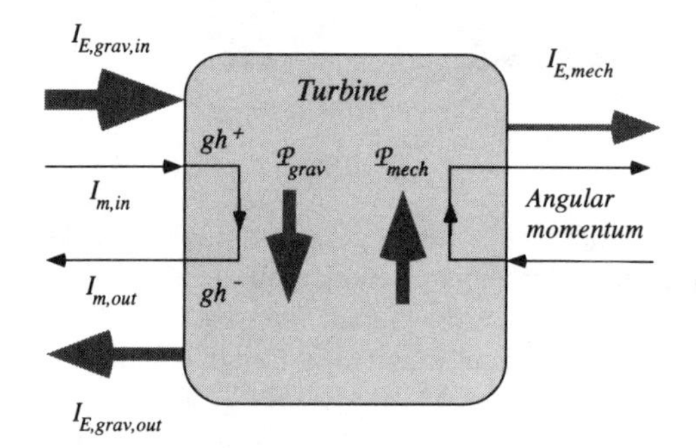

FIGURE 23. Flow diagram of a turbine driven by water falling from a high level. Mass enters the engine at a high gravitational potential, carrying with it some energy. The mass leaving the engine takes with it part of the energy. The missing energy is given off together with the current of angular momentum.

Since we are saddled with two different efficiencies, we might wish to be able to decide which one to use. While both are employed in thermal physics, the first one introduced, i.e. η_2 in Equation (16), appears to be more natural. It compares the result of some process to what nature could possibly deliver. Therefore, it is the measure of the efficiency of a technical device which has been in use in modern energy technology.

You might wonder about the different subscripts used with the efficiencies introduced in Equations (16) and (18). This particular usage can only be explained in historical terms. The efficiency as in Equation (18) was the first to be introduced at the beginning of the development of thermodynamics. For ideal engines, the value given in Equation (19) is often called the *Carnot efficiency* or the *Carnot factor*. The other measure was introduced much later, and only recently has it become established to some extent. In engineering literature, the efficiency η_2 is often called the *second law efficiency*, while the earlier measure η_1 is called the *first law efficiency*. *First law* and *second law* refer to the laws of balance of energy and of entropy, respectively. Their numbering is a result of the historical development; it does not have any deeper meaning.

Note that ideal heat engines are those which work reversibly, i.e. those which do not produce any heat. What happens if heat is generated will be discussed in the following sections.

EXAMPLE 14. A thermal power plant.

The thermal energy current due to burning of coal in a thermal power plant is 1.5 GW, while the mechanical energy current leaving the turbines is 0.6 GW. The steam driving the engine is emitted at a temperature of 50°C. If the turbines operate as an ideal Carnot engine, what are a) the current of entropy flowing through the engine, b) the temperature of the furnace, and c) the Carnot efficiency?

SOLUTION: a) The thermal energy current emitted with the steam is equal to the difference between the energy current entering the engine and the one driving the generator. In other words, it is 0.9 GW. At a temperature of (273 + 50) K, the current of entropy associated with this thermal energy flux is

$$I_s = \frac{I_{E,th,out}}{T^-}$$
$$= 2.79 \frac{\text{MW}}{\text{K}}$$

b) The temperature of the furnace can be calculated in terms of the currents of entropy and energy absorbed by the engine:

$$T^+ = \frac{I_{E,th,in}}{I_s}$$
$$= 538\,\text{K}$$

c) According to Equation (16b), the Carnot efficiency is equal to

$$\eta_1 = \frac{538\text{K} - 323\text{K}}{538\text{K}} = 0.40$$

The same result could also have been obtained with the help of the energy currents:

$$\eta_1 = \frac{I_{E,mech}}{I_{E,th,in}}$$
$$= \frac{0.60}{1.50} = 0.40$$

Note that in these equations the magnitudes of the fluxes have been used. The second law efficiency naturally is equal to 1.0.

EXAMPLE 15. The available power of a current of entropy: *La puissance du feu.*

The efficiency introduced in Equation (16) refers to the power of a fall of heat through a given temperature difference. This latter quantity is often calculated with the temperature of the environment T_o replacing the temperature of the cooler. The power thus calculated is the maximum which can possibly be derived from a current of entropy emitted from a reservoir at temperature T in a given environment. For this reason it is called the *available power* (or the *exergetic power*) of heat. It is simply what Carnot called *La puissance du feu*, the power of heat.

a) Determine the formula for the available power of a current of entropy. b) Express the available power in terms of the thermal energy current at T. c) The fluid of an ideal Carnot engine is assumed to operate between the temperatures of 400°C and 100°C. Furnace and environment are at temperatures of 500°C and 20°C, respectively. Compare the power of the engine to the available power with respect to T and T_o; i.e., compute its second law efficiency.

SOLUTION: a) The current of entropy stems from a reservoir at temperature T. It can only drop to a level T_o which releases energy at a rate:

$$\mathcal{P}_{av} = (T - T_o)|I_s| \quad \textbf{(E8)}$$

Note that this is analogous to the optimal power of water received at a potential gh being allowed to fall to gh_o.

b) The energy current at T is related to the current of entropy by Equation (13). Therefore we obtain:

$$\mathcal{P}_{av} = \left(1 - \frac{T_o}{T}\right)|I_{E,th}(T)| \quad \textbf{(E9)}$$

c) The ratio of the actual power of the engine and the available power can be expressed as follows:

$$\begin{aligned}\eta_2 &= \frac{\mathcal{P}_{mech}}{\mathcal{P}_{av}} \\ &= \frac{(1 - T^- / T^+) I_{E,th}(T)}{(1 - T_o / T) I_{E,th}(T)} \\ &= \frac{0.45}{1 - 293/773} = 0.72\end{aligned}$$

Note that we could not have used the current of entropy leaving the furnace for this calculation. The reason is the following: since entropy drops from the furnace to the fluid operating in the engine without producing mechanical effects, heat is produced (Section 1.5). Therefore, the current of entropy entering the engine is not the same as the one emitted by the furnace. Alternatively, we could have calculated this current first.

1.4.5 Heat Pumps and Refrigerators

So far we have discussed spontaneous or voluntary thermal processes: heat drops from a higher to a lower level, and energy is released. Just as in the case of other phenomena, it is possible to transport the substancelike quantity "uphill." While a spontaneous process gives off energy, an involuntary phenomenon like heat flow against the drop of temperature requires energy (Figure 24). Heat pumps and refrigerators are engines which pump heat out of a cold environment and dump it in a warmer one. Therefore, we say that they make heat "flow uphill" from a lower thermal level to a higher one. They literally pump heat, i.e. entropy, just as water pumps pump water. With their help heat may be transferred into environments where it is needed for heating. In this way we avoid having to produce heat. We only have to supply the energy for accomplishing the task of lifting a certain amount of heat to the desired level.

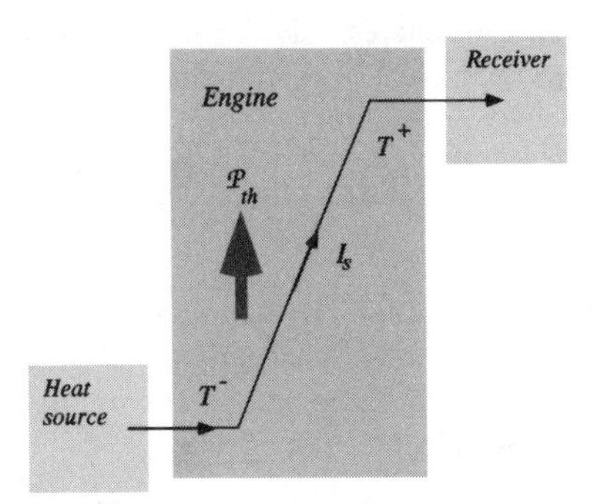

FIGURE 24. A heat pump transfers heat, i.e. entropy, from a lower to a higher potential. This process requires energy; i.e., it has to be driven by an operation which releases energy. The thermal power required is equal in magnitude to the power that would be released in an equal fall of entropy.

The principle of operation of a heat pump is the following (Figure 25). Entropy and its associated energy current enter the engine at the lower thermal potential. If the pump operates ideally (without the production of heat), the same flux of entropy will leave the system at the upper temperature, taking with it some energy. We know that at a higher temperature more energy is flowing with the same current of entropy; see Equation (13). Therefore, more energy is leaving the engine than is entering at the lower temperature. The amount of energy missing has to be provided by the (mechanical) process driving the heat pump. This is the energy needed to pump entropy from a temperature T^- to one of T^+. The required power is equal to the product of the rate of transfer of heat and the difference of the thermal potential. In other words, the formula is perfectly analogous to Equation (11). Since energy is required, the power turns out to be negative.

So far we have not distinguished between heat pumps and refrigerators. In fact, there is no real difference between these devices. However, it is customary to take a slightly different view as to the meaning of the lower and upper operating temperatures. In the case of a heat pump, one commonly pumps entropy from the environment to a reservoir at elevated temperature. A refrigerator, on the other hand, pumps entropy from a reservoir at low temperature to the environment. In other words, the temperature of the environment is the lower thermal potential for heat pumps, and the upper one for refrigerators (Figure 26).

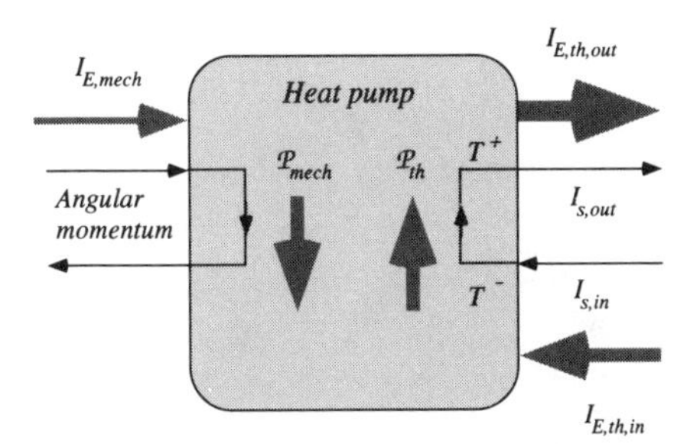

FIGURE 25. Flow diagram of an ideal heat pump or refrigerator. The energy extracted from the cool environment, with the entropy, and the energy added for driving the process leave the device at the higher temperature. The energy added is used to lift the entropy to a higher level.

If we keep in mind the meaning of the operating temperature we can write down the same formula for the *coefficient of performance* (*COP*) of heat pumps and refrigerators. This coefficient is defined as the amount of energy added to or removed from the reservoir of interest, divided by the energy needed to drive the engines. Therefore, the coefficient of performance is given by

$$COP = \left|\frac{I_{E,th}(T^+)}{\mathcal{P}_{mech}}\right| = \left|\frac{T^+}{T^+ - T^-}\right| \tag{20}$$

for heat pumps and refrigerators. The temperatures have been identified in Figure 26. Remember that we have discussed only ideal devices running as (reverse) Carnot engines. Therefore, the particular formula for the coefficient of performance holds only for ideal Carnot engines.

Just as the first law efficiency η_1 of a heat engine, Equation (18), might give the wrong impression of the performance of a thermal engine, the coefficient of performance introduced in Equation (20) tells only half the story. A better way of measuring the efficiency of heat pumps and refrigerators is to compare the thermal power for lifting entropy to the actual mechanical power of the engine:

$$\eta_2 = \left|\frac{\mathcal{P}_{th}}{\mathcal{P}_{mech}}\right| \tag{21}$$

This is a figure analogous to the second law efficiency of heat engines. Note that it is the inverse of what was defined in Equation (16), just as the coefficient of performance is the inverse of the Carnot efficiency, Equation (19). Naturally, for ideal heat pumps and refrigerators, the efficiency η_2 is equal to 1.

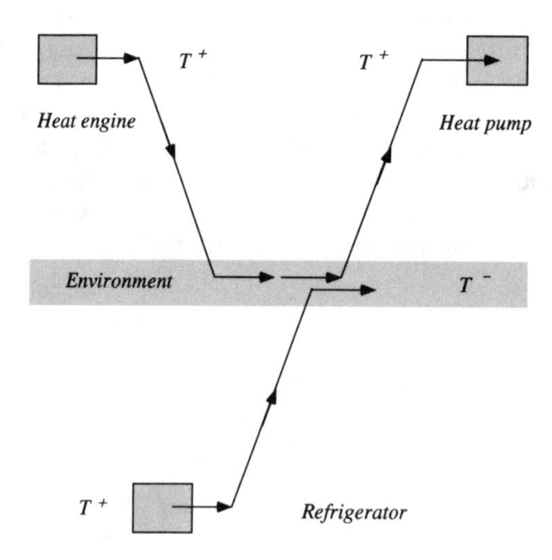

FIGURE 26. For heat pumps and refrigerators, different views are taken as to the meaning of the operating temperature. In the case of a heat pump, we are interested in the entropy transferred into the reservoir at elevated temperature. With refrigerators, one would like to know about the amount of entropy removed from the cold space.

EXAMPLE 16. Heating water with a heat pump.

Assume that a heat pump is installed which requires an energy current of 165 W for operation. The pump takes entropy out of the ground in winter. (The temperature of the ground is 2°C.) a) How large is the entropy current at the beginning, with 20°C water? b) How large is the current when the temperature of the water has reached 100°C? c) How large are the energy currents entering the water in these two cases?

SOLUTION: a) The current of entropy is calculated to be

$$I_{s,in} = \frac{I_{E,mech}}{T^+ - T^-} = \frac{-165\,\text{W}}{18\,\text{K}} = -9.17\frac{\text{W}}{\text{K}}$$

if the water has a temperature of 20°C.

b) If the water has reached a temperature of 100°C the current of entropy diminishes to 165 W/ 98 K = 1.68 W/K. With the same amount of energy less entropy can be pumped through a larger temperature difference. (This result should be compared to the entropy current out of an immersion heater having a power of 165 W, which according to Example 11 is equal to 0.56 W/K at 20°C.)

c) At 20°C the energy current leaving the heat pump (and entering the water) will be

$$I_{E,th,out} = -\left(I_{E,mech} + I_{E,th,in}\right) = -\left(I_{E,mech} + T^- I_{s,in}\right) = -\left(-165\,\text{W} + 275\,\text{K}(-9.17\,\text{W/K})\right) = 2690\,\text{W}$$

In the second case, it decreases to 627 W. Obviously, heating with an ideal heat pump must be more efficient than heating with an immersion heater. (See Example 11.)

EXAMPLE 17. Freezing water in a freezer.

One liter of water (which already has a temperature of 0°C) is frozen in a freezer which we assume to function as an ideal heat pump. The temperature in the freezer has to be maintained at a temperature of 0°C. How much entropy is extracted from the water, and how much energy is needed to emit this entropy to the kitchen at a temperature of 22°C? Use the values for water found in Table A.4.

SOLUTION: The amount of entropy emitted when water is turned into ice is determined by the constitutive law of fusion which was introduced in Section 1.3:

$$S_e = n \cdot \bar{l}_f = m \cdot \bar{l}_f / M_o = \frac{1.0\,\text{kg} \cdot 22.0\,\text{J/(mole} \cdot \text{K)}}{0.018\,\text{kg/mole}} = 1220\,\text{J/K}$$

This amount of entropy flows into the freezer and has to be removed. The energy needed to operate the freezer while the water is freezing is equal to the energy needed to lift this amount of entropy from a temperature of 0°C to one of 22°C:

$$W = (T_2 - T_1)S_e$$
$$= (295\,\text{K} - 273\,\text{K}) \cdot 1220\,\text{J/K} = 26.8\,\text{kJ}$$

Note that in reversible operations we can think of entropy as if it were a substance such as water. The amount of energy just calculated corresponds to an equivalent amount used in pumping water from a lower to a higher place. The formulas involved in the calculations are structurally equivalent in the cases of water and of heat (entropy).

EXAMPLE 18. Different ways of heating using solar power.

Not all types of solar heating are created equal. Consider the following means of keeping a supply of domestic water at 60°C. In a first setup (A), solar radiation is used directly to heat the water. In a second (B), solar radiation is used to heat a furnace to 700°C. The heat from the furnace drives an ideal Carnot engine which rejects the heat to the water at 60°C. The energy released by the engine is used to drive an ideal Carnot heat pump which pumps heat from the environment at 0°C into the water at 60°C. Calculate the ratio of the rates of heating of the two processes. Take I_{E1} to be equal in both cases.

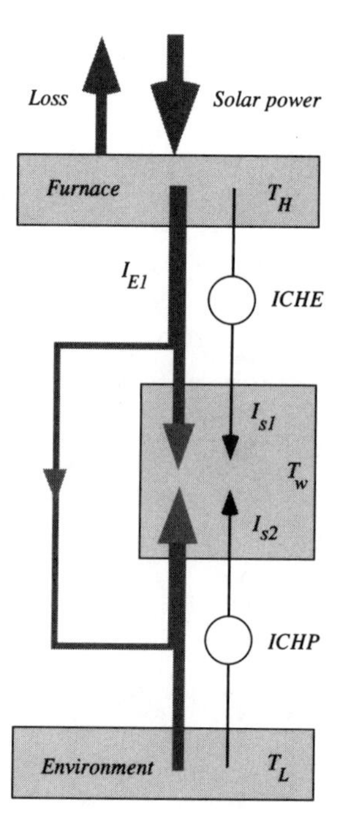

SOLUTION: The water is actually heated by two currents of entropy, one from the high-temperature furnace, the other from the environment. The energy current associated with this heating is given by the sum of the currents of entropy and the temperature of the water:

$$I_E = T_w(I_{s1} + I_{s2}) \quad \textbf{(E10)}$$

All energy fluxes appearing in the equations are thermal, and all are taken to represent magnitudes. The currents of entropy are related according to the following rule. The first determines the rate at which energy is liberated by the ideal Carnot heat engine (*ICHE*). This energy in turn determines the magnitude of the current of entropy pumped by the ideal Carnot heat pump (*ICHP*). Therefore:

$$I_{s1}(T_H - T_w) = I_{s2}(T_w - T_L) \quad \textbf{(E11)}$$

We solve for the second entropy current in (E11) and plug the result into (E10). After some algebra we obtain the following expression:

$$I_E = T_w I_{s1}\left(1 + \frac{T_H - T_w}{T_w - T_L}\right) = T_H I_{s1} \frac{T_H - T_L}{T_w - T_L}\frac{T_w}{T_H}$$

The product of the temperature of the furnace and the current of entropy emitted at this temperature is equal to the thermal energy current entering the heat engine. It is equal to the solar gain I_{E1} in steady state. Thus, the ratio we are looking for is equal to:

$$\frac{I_E}{I_{E1}} = \frac{T_H T_w - T_L T_w}{T_H T_w - T_H T_L} > 1$$

For the numbers given here, case B is more efficient by a factor of 4.0.

1.5 The Production of Heat

So far, we have discussed cases of heat, i.e. entropy, being transported and stored. There is more, however, that can happen to entropy. In Section 1.1 we discussed the possibility of heat being created, which means that entropy must be a nonconserved quantity, unlike electrical charge or momentum. In everyday language, we talk about creating or generating heat. In itself, this does not mean much. We also talk of "generating" electricity. It turns out, however, that electricity cannot be generated; it is known that charge is a conserved quantity. When we "produce" electricity all we do is set up electric currents; the total amount of charge remains constant. In the case of thermal phenomena we now can prove that entropy can be generated in the true sense of the word.

1.5.1 Irreversible Carnot Engines

Real heat engines do not perform like the ideal Carnot engines discussed in the previous section, for two reasons. First of all, they usually are not Carnot engines; i.e., they do not absorb and emit entropy at constant temperatures. This lowers their efficiency compared to Carnot engines. Second, even if we could build engines which absorb and emit entropy at constant temperatures only, we would not get the efficiency calculated according to Equation (19). This has to do with irreversibility, i.e. with the production of heat in the fluid operating in the engine. In the following sections we will discuss only Carnot engines, reversible and irreversible. This will simplify our discussion considerably.

In real life, engines operate below their potentials. They do not use all the energy released in the fall of entropy for mechanical purposes. We say that engines use part of the energy for a different, but special, purpose: *they create entropy!* This statement can be proved if we accept that in real operations the efficiency is lower than the one calculated in Equation (19). This observation is based on experience which shows that engines always operate below the ideal limit, which rests on the assumption that entropy is not generated in the engines.

If an engine absorbs and emits entropy at constant temperatures, we have to conclude that the current of entropy leaving it is larger than the one entering: only then do we get a performance which is less than the one predicted in the previous section. Let us introduce the real efficiency η_1 of an engine, as opposed to the Carnot efficiency η_c. Based on the observation that

$$\eta_1 \leq \eta_c \tag{22}$$

we get

$$\begin{aligned} I_{s,out} &= \frac{I_{E,th,out}}{T^-} \\ &= -(1-\eta_1)\frac{I_{E,th,in}}{T^-} = -(1-\eta_1)\frac{T^+}{T^-} I_{s,in} \end{aligned} \tag{23}$$

or

$$\left|\frac{I_{s,out}}{I_{s,in}}\right| = (1-\eta_1)\frac{T^+}{T^-} \geq \left(1 - \frac{T^+ - T^-}{T^+}\right)\frac{T^+}{T^-} = 1 \tag{24}$$

The real efficiency was replaced by the Carnot efficiency according to Equation (19), which led to the inequality in Equation (22). In other words, the current of entropy leaving a real Carnot engine must be larger in magnitude than the one entering:

$$\left|I_{s,out}\right| \geq \left|I_{s,in}\right| \tag{25}$$

The operation of heat engines in real life tells us that it must be possible to create entropy, and that the reverse, i.e. the destruction of entropy, cannot take place.

The irreversible Carnot engine does not deliver as much energy for mechanical purposes as the hypothetical ideal engine. For this reason, engineers often speak of the *loss of availability*. By availability they mean the energy which can hypothetically be gained from a fall of entropy (Example 15). The loss of availability and the production of entropy are intimately linked. Indeed, the rate at which entropy is produced directly measures the rate at which availability is lost (see Example 19 and Section 1.7). Another point is that the production of entropy occurs inside the engine, i.e. in the working fluid (Example 20). We stress this here since we will later consider a model of a heat engine where the dissipation occurs outside the fluid.

EXAMPLE 19. The production of entropy in a nonideal Carnot engine.

a) For an engine with $\eta_1 < \eta_C$, calculate the rate at which entropy is produced in the engine. Compute the numerical value for an engine operating between reservoirs at 600 K and 300 K, respectively, which has an observed efficiency of 0.30. The thermal energy current entering the engine is 1.0 GW. b) Show that there is a loss of available energy (Example 15), and express the rate of loss in terms of the rate of production of entropy. How large is the loss for the engine described in (a) for one day?

SOLUTION: a) In an engine which is operating in a steady state, the rate at which entropy is produced must be the difference between the rate at which entropy is emitted and the rate at which it is absorbed. If we introduce the symbol Π_s for the rate at which entropy is created, we have:

$$\Pi_s = I_{s,in} + I_{s,out} \tag{E12}$$

This is a particular case of the general law of balance of entropy which we are going to formulate later in Section 1.6. This equation can be transformed using the relations between currents of entropy and energy in heating. If we also use the expression for the efficiency of an ideal Carnot engine, Equation (19), Equation (E12) changes to:

$$\Pi_s = \frac{I_{E,th,in}}{T^+} + \frac{I_{E,th,out}}{T^-} = \left(-(1-\eta_1)\frac{1}{T^-} + \frac{1}{T^+}\right) I_{E,th,in}$$
$$= \left(-(1-\eta_1) + \frac{T^-}{T^+}\right)\frac{I_{E,th,in}}{T^-}$$

or

$$\Pi_s = -(\eta_c - \eta_1)\frac{I_{E,th,in}}{T^-} \quad \text{(E13)}$$

If we introduce numerical values we obtain

$$\Pi_s = -(0.50 - 0.30)\frac{-1.0\,\text{GW}}{300\,\text{K}} = 0.67\frac{\text{MW}}{\text{K}}$$

The value of 0.5 is the Carnot efficiency of the engine operating between 600 K and 300 K. The rate at which entropy is absorbed is

$$I_{s,in} = \frac{-1.0\,\text{GW}}{600\,\text{K}} = -1.7\frac{\text{MW}}{\text{K}}$$

which is 2.5 times larger than the rate of production.

b) The loss of available power is the difference between the hypothetically available power and the actual mechanical power of the engine. According to the definition of the available power in Example 15, and the Carnot efficiency Equation (19), we have:

$$|\mathcal{P}_{av}| - |I_{E,mech}| = \left(1 - \frac{T^-}{T^+}\right)|I_{E,th,in}| - \eta_1 |I_{E,th,in}| = (\eta_c - \eta_1)|I_{E,th,in}|$$

This can be expressed using the result obtained above, Equation (E13):

$$|\mathcal{P}_{av}| - |I_{E,mech}| = T^- \Pi_s \quad \text{(E14)}$$

This result is important. It demonstrates that the loss of power due to dissipation is directly proportional to the production of entropy. At second glance this is not so surprising. After all, the nonconservation of heat leads to the loss in the first place. In one day, the loss is equal to 300 K · 0.67 MW/K · 86400 s = $1.74 \cdot 10^{13}$ J, which is equivalent to $4.8 \cdot 10^6$ kWh of energy. See Section 1.7 for a detailed analysis of the loss of power and production of entropy.

EXAMPLE 20. A nonideal Carnot cycle.

a) Consider a Carnot cycle, i.e. a cycle where entropy is absorbed and emitted at constant temperatures, for which the adiabatic expansion is dissipative. Draw the *T-S* diagram and indicate the energy exchanged in the thermal processes. b) Why is the cycle less efficient than its reversible counterpart? c) Identify the lost available energy in the *T-S* diagram.

SOLUTION: a) Three of the four steps in the Carnot cycle of Example 18 are assumed to be reversible operations. Only the adiabatic expansion produces entropy in the fluid.

Since, in an adiabatic step, entropy is neither added nor removed, the production leads to an increase of the content of entropy. For this reason, the curve representing the adiabatic expansion is no longer a vertical line in the *T*-*S* diagram. Energy is exchanged thermally in two processes, namely during the isothermal expansion and compression. The amount of energy exchanged during these steps is given by the area under the isotherms in the *T*-*S* diagram. The net exchange is equal to the difference of the shaded areas.

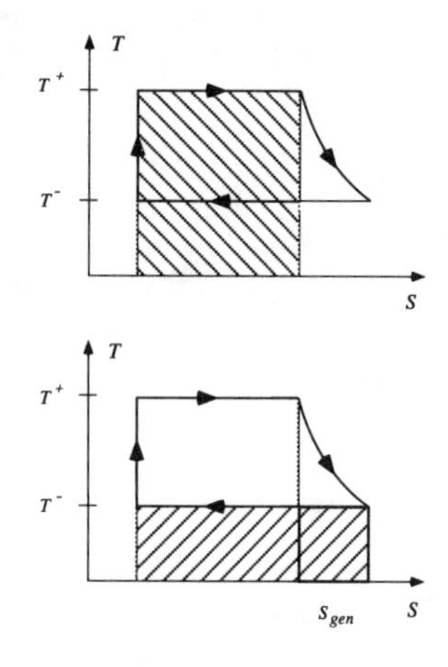

b) Since entropy has been produced during the cycle, more has to be emitted in the isothermal compression compared to what we found in Example 13. This means that more energy is discharged to the cooler as well.

c) The loss compared to a completely reversible cycle is represented by the small rectangle under the lower isotherm with width S_{gen} (entropy generated). The amount lost is equal to S_{gen} multiplied by the temperature of the cooler. (Compare this to Equation (E14)).

1.5.2 The Conduction of Heat

Entropy is produced when it flows through bodies from hotter to cooler places. In other words, *entropy is produced in the conductive transport of heat.* This might come as a surprise to you. We are quite familiar with other processes which create entropy, such as friction and the flow of electricity through wires, but we do not have direct sensory knowledge about what happens in conduction. Therefore we are forced to look at this phenomenon more carefully.

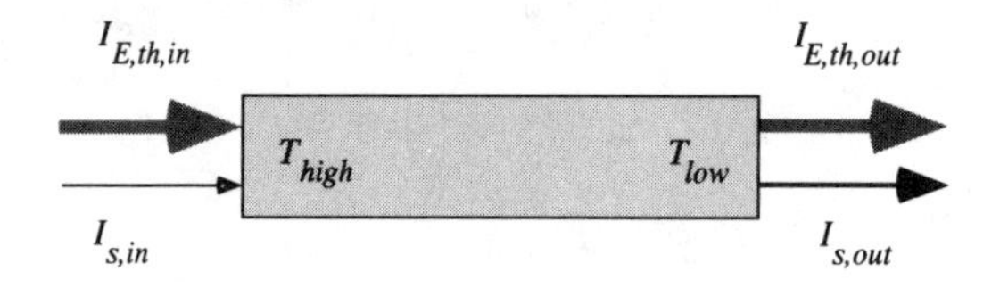

FIGURE 27. If entropy flows through a body from its hotter end to its cooler end; i.e., if entropy is transported conductively, entropy is also produced. This is because the flux of entropy must be larger for a given flux of energy if the temperature is lower. Therefore, more entropy leaves the body than enters at the hotter end.

Consider a long bar of heat-conducting material (Figure 27). Let us keep its ends at constant temperatures, for example by sticking one in boiling water and the other in a mixture of ice and water. Let us wait long enough for steady-state conditions to be established. In other words, we must wait long enough until the currents of entropy and of energy through the bar become steady. Since energy is conserved, we conclude that the flux of energy entering at the hotter end is equal in magnitude to the flux leaving at lower temperature. This immediately tells us that entropy must be produced in the bar: there is a larger current of entropy accompanying the flux of energy at lower temperature. We infer this from the fundamental relationship expressed by Equation (13): the balance of energy tells us that

$$I_{E,th,in} + I_{E,th,out} = 0 \tag{26}$$

Remember that currents entering a body are given negative fluxes. Each of the fluxes of energy will be written as the product of the flux of entropy and the associated thermal potential, according to Equation (13):

$$T_{high} I_{s,in} + T_{low} I_{s,out} = 0 \tag{27}$$

If we solve this equation for the ratio of the fluxes of entropy we obtain

$$\left| \frac{I_{s,out}}{I_{s,in}} \right| = \frac{T_{high}}{T_{low}} > 1 \tag{28}$$

This proves our claim: in steady-state conduction, more entropy leaves the bar than enters at the hot end. Entropy is produced in the body.

If you feel uncomfortable about this result, consider the following arguments. Isn't the flow of entropy the prototype of an irreversible phenomenon? Just look at the process that takes place when two bodies of unequal temperatures are brought in thermal contact (Section 1.1.5). This is irreversibility in its purest form; entropy must be produced as a consequence of the process. Also, you know that we can operate a heat engine between the reservoirs at the hot and cool ends of the bar. We have said that energy can be released for a nonthermal purpose in the fall of entropy. If entropy flows conductively from one point to another, it also drops from higher to lower temperatures. The energy released in the fall of entropy has been used for the only possible process which is a thermal one: entropy is produced continuously. Put in different terms we might say that we have lost available power, according to Example 19, which means that entropy must have been created. No matter how we look at the conductive flow of entropy, it is hard to escape the following conclusion:

Conduction of heat produces entropy; i.e., it is a dissipative process. When entropy is allowed to fall from points of higher temperature to points of lower temperature without causing motion or some other process, entropy is necessarily produced.

In the discussion and the proof of dissipation we have not included any properties of the body conducting heat. We have only talked about general properties of entropy and energy, and we have assumed the fundamental relation between a conductive current of entropy and the associated flux of energy to hold according to Equation (13). As a result, we cannot say anything about how much entropy will be conducted through the bar in Figure 27. Surely, the answer to this question must depend on such quantities as the temperatures at the hotter and cooler ends and material properties of the conductor. Note that the conduction of entropy has much in common with the conductive transport of charge through a metal wire. Again, nature offers us an example of structural analogy between phenomena from different fields of physics. We shall investigate conduction much more thoroughly in Chapter 3, where we will have the opportunity to discuss the role of constitutive relations.

EXAMPLE 21. The rate of production of entropy in a thermal conductor.

A metal bar conducts entropy from a container of boiling water to a mixture of ice and water. It is found that in the latter container, ice melts at a rate of 10.0 g per minute. Compute the rate at which entropy is produced in the conductor.

SOLUTION: Since conditions do not change in the conductor in the steady state, the sum of the fluxes of entropy must be equal to the rate at which entropy is produced. This condition is the same as the one expressed in Example 19:

$$\Pi_s = I_{s,in} + I_{s,out} \tag{E15}$$

On the other hand, the magnitude of the fluxes of energy at both ends must be equal, which leads to the condition given in Equation (26):

$$T_{high} I_{s,in} + T_{low} I_{s,out} = 0 \tag{E16}$$

Since the temperature decreases along the bar, the flux of entropy must increase. We find the rate of production of entropy by first eliminating the flux at the hot end in Equation (E15) using Equation (E16). This yields

$$\Pi_s = I_{s,out} - \frac{T_{low}}{T_{high}} I_{s,out} = \left(1 - \frac{T_{low}}{T_{high}}\right) I_{s,out} \tag{E17}$$

We know the current of entropy which is entering the mixture of ice and water from the rate of change of ice (Example 12):

$$\begin{aligned} I_{s,out} &= l_f \, dm/dt \\ &= 1230\,\mathrm{J/(K \cdot kg)} \cdot 0.010\,\mathrm{kg/60s} = 0.20\,\mathrm{W/K} \end{aligned}$$

If we introduce this value in Equation (E17), we obtain a value of 0.055 W/K for the rate of production of entropy.

1.5.3 A Comparison of Some Dissipative Processes

Nature offers us a whole range of dissipative processes, some obvious, some not so obvious. They have one basic feature in common. Heat is produced when energy is released in the fall of some quantity from a higher to a lower potential without causing motion, electrical, chemical, or other processes.

Consider the obvious case of friction. Imagine a body being pulled at constant speed over a rough surface. Momentum flows through the rope, the body, and into the ground. Since there is a difference of speeds between the body and the ground, we may say that momentum falls from points of higher potential to points of lower potential. In such a process energy is released. Now the only follow-up process is a thermal one: heat is produced. More generally, the flow of momentum through a viscous medium leads to dissipation.

This sounds very much like what happens when electric charge flows through a resistor. Indeed, the analogy is quite perfect. Replace viscosity by resistance, and momentum by charge, and you end up with the same picture of what happens. Charge flows through a resistive medium from points of higher potential to points of lower potential. Again, the energy released in the flow of charge, which might drive another process, serves only to create entropy.

There are some less obvious dissipative processes which nevertheless operate according to the same principles. These have to do with the flow of substances such as in diffusion or the free expansion of air. In these cases, substances flow from points of high chemical potential to points where this potential is smaller. We commonly can measure the chemical potential in terms of the pressure. You certainly are familiar with osmotic pressure arising when a substance is dissolved unevenly in volumes of liquids which are in contact through semipermeable membranes. It does not take much imagination to think of a way to harness the osmotic pressure difference to drive some sort of engine. If instead of doing this we let the dissolved substances diffuse freely, we do not produce motion or any other effect except, again, for the production of entropy.

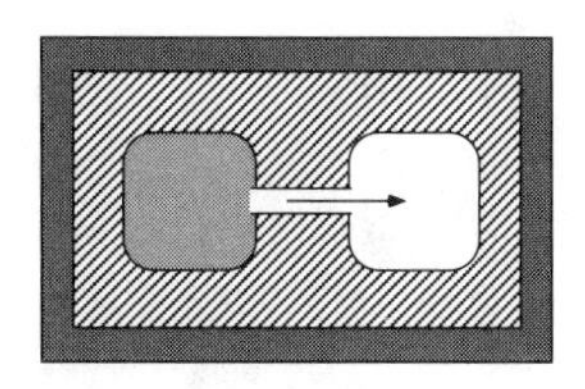

FIGURE 28. Two containers are surrounded by a liquid which is perfectly insulated from the surroundings. One of the bottles is filled with air, the other is empty. It is found that the temperature of the setup does not change if the gas is allowed to become evenly distributed between the two containers.

The free expansion of air from a container into a void falls in the same class of phenomena (Figure 28). Imagine two containers connected by a pipe which may be closed by a valve. The containers are completely isolated from the environment. One of them contains air; the other has been pumped empty. If we open the valve and let the air pass freely into the empty chamber, we find something rather surprising at first sight. If the pressure of the air in the experiment is not too high, and the temperature not too low, the temperature of the expanded gas is the same as that of the compressed air. We might be reluctant to agree that in this process entropy has been produced. However, we cannot escape this conclusion. First, it is clear that we could have used the expansion of the air to drive an engine. Since we have not done so, what could have happened other than the creation of entropy? A second argument should convince you finally that the free expansion of air is dissipative. Since the temperature of the air is the same at the end of the process as at the beginning, the outcome of the complicated flow process is equivalent to the simple isothermal expansion discussed in Section 1.3. There we have concluded that entropy must be supplied to the expanding gas if its temperature is not to decrease. Since in the free expansion we do not supply any entropy from the surroundings, entropy must have been produced in the system.

All in all, we can say now that the conduction of entropy fits neatly into the class of dissipative processes described here. Entropy flows through a resistive medium, which necessarily produces more entropy.

1.5.4 The Problem of Calorimetry

The fact that the conductive flow of heat is dissipative is fundamentally important for our understanding of the measurement of amounts of entropy. Just about all techniques involve the direct contact of bodies at different temperatures, leading to the transport and the production of entropy.

One of the oldest means of measuring amounts of entropy exchanged is a simple water calorimeter such as in Figure 29. Imagine that we have defined (we certainly are allowed to do so) that a certain amount of water has absorbed one unit of entropy if its temperature increases by 1 K. If we now place another object, such as a hot stone, in the perfectly insulated calorimeter, the temperature of the water will rise while the temperature of the stone must drop. The process will come to a halt when the two temperatures have become equal. Assume the hotness of the water to have increased by just 1 K. We therefore know how much entropy has been absorbed by the water. However, we are interested in thermal properties of the stone, such as its entropy capacity (Chapter 2). We therefore want to know how much entropy has been emitted by the stone. How do we solve this problem?

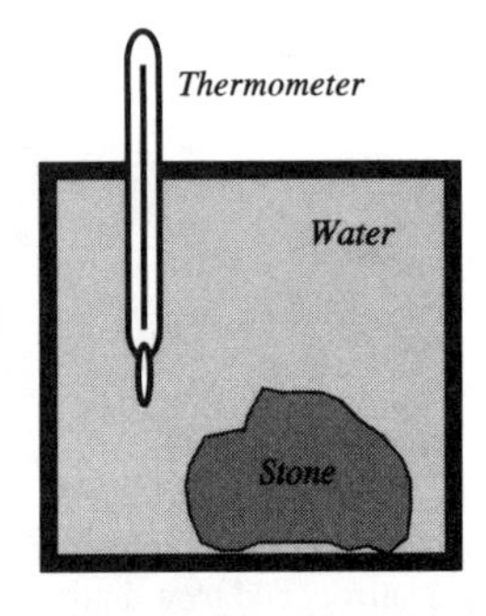

FIGURE 29. The simplest calorimeter consists of an insulated container of water. We place another object, whose properties we wish to investigate, into the container. The stone and the water exchange entropy conductively. Since this process produces entropy, it is impossible to infer quantities of entropy exchanged directly. The device therefore does not measure entropy directly.

In the days of the caloric theory people simply pictured heat to be a subtle fluid which is exchanged between the stone and the water. Most importantly, they assumed heat to be a conserved quantity. This leads to a simple determination of the amount of heat emitted by the stone: it is equal in magnitude to the amount absorbed by the water.

There is only one catch: heat is not conserved in the calorimetric process described. It is produced, and nobody knows by how much. Therefore, we cannot use this simple technique for measuring amounts of heat exchanged. Still, the calorimeter described here, and others of its kind, are alive and in use today. And the theory used for inferring amounts of heat exchanged is the same as the one used by the calorists. Therefore, knowing that heat is created in conduction, we have to conclude that calorimeters do not measure the quantity we have called heat.

There is an element of irony in this story. The quantity for which the calorimeters were built, and for which the theory of conservation was fashioned, namely the old caloric, does not satisfy the assumed law. What do we do in the face of this adversity? We keep the calorimeters and the theory, and by pure luck find another quantity which indeed is conserved in the thermal exchange between the stone and the water. This quantity is energy. Here lies the beginning of the end of the fundamental quantity which we have introduced as heat. Since the new quantity, energy, fits our preconceptions, we loose the single most important thermal concept, namely the old caloric. We loose sight of it so thoroughly that we do not recognize it even when it comes up again in the guise of entropy. Nature gets its revenge, however, by saddling us with a theory of heat, which to put it mildly, is a source of endless confusion. It is as if we had been condemned to create a theory of electricity without the benefit of the quantity called electric charge, or to solve mechanical problems without $F = ma$.

1.5.5 Why Does an Ideal Thermal Engine Need a Cooling Device?

Thermal power plants continue to baffle students by their apparent need for cooling towers or equivalent devices. We all know how much energy is wasted in such plants: roughly two-thirds of the energy released in burning oil or coal, or in splitting uranium nuclei, is emitted to the environment. Only the meager rest makes it to the generator. Ingenious methods have been thought of for circumventing the cooling device. It seems hard to believe that all the energy given up to the environment cannot be channeled back into the burner of the power plant.

Indeed, from the point of view of energy alone we cannot understand why thermal power plants are so wasteful. Compare an electrical device hooked up to a battery, and a turbine which is part of a thermal power plant. We can write down the energy balance of a battery by introducing the energy currents associated with the currents of charge leaving and entering the device. The currents are proportional to the upper and the lower electric potentials, which can be chosen arbitrarily. Only their difference is determined and of interest. Therefore, an ideal battery gives off a net energy current which is equal to the rate at which energy is released by the chemical reactions going on inside. The device hooked up to the battery obtains, and passes on, 100% of the energy released.

Superficially, a thermal power plant operates in a comparable way. We have a closed circuit for water (or steam) flowing from the boiler to the turbine, and back to the boiler via the cooler. However, the quantity which matters, entropy, does not flow in a closed circuit. The reason for this is simple and has to do with the fact that entropy is created in the burning of coal or oil, or in the splitting of atomic nuclei. This first step takes place outside the water circuit. Entropy is injected into the water (or the steam) from the burner, and it continues to be created and added to the water as long as the burner is operating. As a consequence, entropy has to be ejected from the water flowing in the plant. This is done in the cooling device. Without this step the amount of entropy would steadily increase in the circuit, which certainly could not be tolerated. The entropy leaving the system takes with it an amount of energy proportional to the lower operating temperature. Therefore the need for a cooling device is a direct consequence of the irreversibility of the release of energy in the burning process.

1.6 The Balance of Entropy

The most common thermal processes are those in which bodies are heated or cooled. We are all familiar with heating food over a fire or on an electrical heater, heating a stone in the sun, and heating water using an immersion heater. We also know that bodies cool off if left in a cooler environment. We have described these processes in terms of the transfer of heat into or out of a body. Furthermore, we know that heat can be produced inside bodies. These properties of heat tell us that we should devise a theory of accounting for amounts of entropy. We have done some accounting in previous sections. Now we will develop the tools in a slightly more formal manner. The same will be done for energy in Chapter 2.

1.6.1 Currents and Fluxes of Entropy

Thermal phenomena tell us that heat can flow. For this reason, we should introduce a physical quantity associated with the transport of heat. Just as in the case of electricity or motion this quantity is a current—here a *current of entropy.* In the simplest case, heat flows in only one direction through a body. This case can be visualized by flow lines which describe the direction of the transport of entropy as in Figure 30.

Naturally, we are interested in how much entropy flows through a body in a given situation. To answer this question, we imagine a surface perpendicular to the flow lines, and we simply count how much entropy passes through this surface per unit time. We call this quantity the *entropy flux* I_s with respect to the surface. The units of the flux are Ct/s. Actually, we imagine the surface to be part of the surface of a body, such as in Figure 30. In this case, we can give the flux a sign depending on whether entropy flows into or out of the body. The flux is considered to be *positive* if entropy flows *out of* a system.

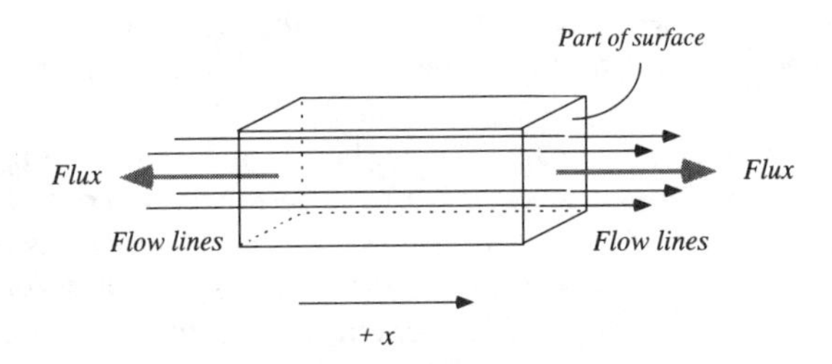

FIGURE 30. Entropy flowing through a body in one direction. The phenomenon of the transport of entropy is visualized by flow lines and is called a *current* of entropy. The rate of transfer of entropy is measured with respect to the surface of a body and is called a *flux* of entropy.

At first, it might seem unnecessary to distinguish between a current of entropy and a flux of entropy. However, we need a means of differentiating between the direction of the transport of entropy, and the direction of the flow with respect to a body. These situations are not the same. In Figure 30 entropy flows to the right, which we might call the positive direction. Therefore we associate a positive quantity with the current. On the other hand, the same current is flowing into and out of the body at the opposing ends, which calls for both negative and positive quantities. The latter quantities are the fluxes I_s with respect to the faces of the body, and they are distinct from the current itself.

Even though we should strictly distinguish between currents and fluxes, we will often use both words in the sense of the flux of entropy if there is no danger of confusion.

1.6.2 The Entropy Content and the Balance of Entropy for Reversible Processes

In addition to the flux of entropy a second quantity is associated with the phenomenon of heat. Since entropy is supposed to be stored in bodies, we need a quantity which describes how much entropy is contained in a system at a given

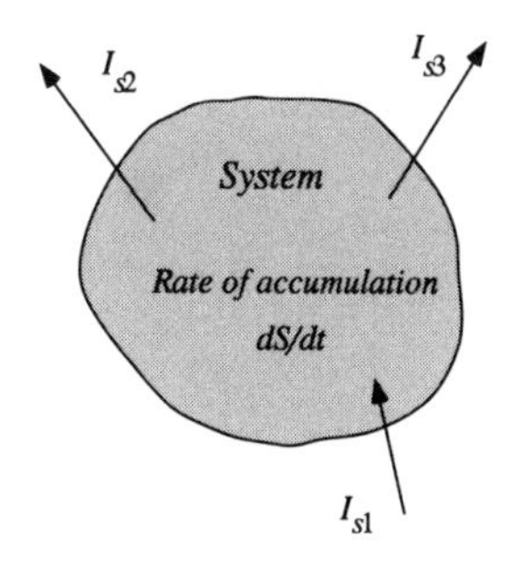

FIGURE 31. Balancing a conserved quantity is quite simple. Only the currents entering or leaving add to the rate at which the system content changes. If the currents or fluxes do not add up to zero, the content either increases or decreases. In fact, the sum of all fluxes determines the rate at which the content changes.

time. This quantity is called the *entropy content of the body*, and it will be denoted by S.

Consider a body having currents of entropy through its surface as in Figures 30 or 31. How do the fluxes relate to the amount of entropy stored in the body? It is clear that entropy flowing into and out of the body will change the entropy content. As in the case of electricity or motion, we sum all fluxes to obtain the *net flux* or net current. The general law of balance of a *conserved* substancelike quantity tells us that the net flux is equal to the rate of change of the quantity: the rate of exchange uniquely determines the rate of change of the stored quantity.

If entropy is neither created nor destroyed, the rate of change of the entropy in the body is equal to the (negative) net flux:

$$\dot{S} + I_{s,net} = 0 \tag{29}$$

We call this the *equation of balance of entropy* in the case of conservation. It is a special form of what is commonly called the *second law of thermodynamics*. For us, the second law is a statement about the balance of entropy.

EXAMPLE 22. Heating and currents of entropy.

A body is heated and cooled at the same time. At one end, entropy flows into the body at a rate of 300 W/K. At the other end, entropy is removed at a rate of 200 W/K. What is the net heating, or the net entropy flux for the body? At what rate does the entropy content of the body change? Assume that entropy is conserved in this process.

SOLUTION: A flux of entropy of I_{s1} = - 300 W/K is associated with the intake of entropy. Remember that currents into a body are considered to be negative. The cooling, therefore, corresponds to a flux of entropy I_{s2} = +200 W/K. The net current is:

$$I_{s,net} = I_{s1} + I_{s2} = -100\,\mathrm{W/K}$$

If entropy is conserved, the rate of change of the entropy content of the body is equal to the negative net flux of entropy across the surface:

$$dS/dt = -I_{s,net} = 100\,\mathrm{W/K}$$

Remember that the choice of sign in the equations is only a matter of convenience or tradition. Mathematicians usually count fluxes out of a body as positive quantities.

1.6.3 Entropy Exchanged and the Change of the Entropy Content

The equation of continuity is an equation relating rates of change of entropy to currents of entropy. Often, however, we are interested in total changes rather than rates. In this context it is convenient to introduce related quantities. We will call the entropy flowing into or out of a body in a given time the *exchanged en-*

tropy, and denote it S_e. The entropy exchanged as a consequence of a current is defined by the integral (Figure 32)

$$S_e = -\int_{t_i}^{t_f} I_s dt \tag{30}$$

Here, t_i and t_f are the initial and the final times of the process of heating, respectively. If more entropy is absorbed than emitted, the minus sign in Equation (30) makes the entropy exchanged a positive quantity. We often speak of the *entropy absorbed* or *emitted* in a process. Both will be taken as positive quantities; i.e., they refer to the absolute value of the integral in Equation (30) for times for which the current enters or leaves the system.

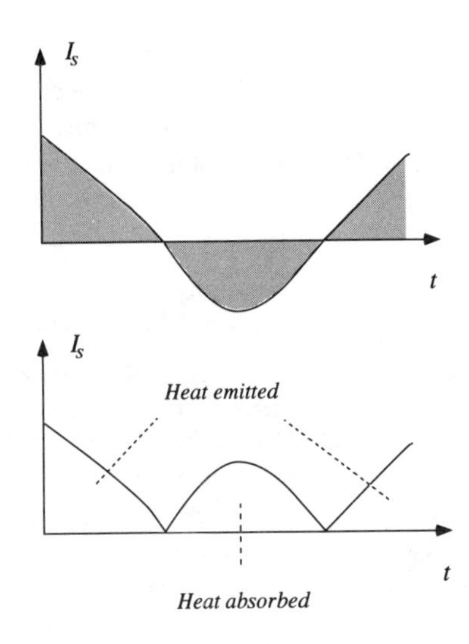

FIGURE 32. Usually the exchanged, absorbed, and emitted entropy are defined in terms of the total or net flux of entropy with respect to a body. The flux is a function of time, and the amount of entropy exchanged in a process can be visualized as the area between the curve $I_s(t)$ and the time axis. Entropy is emitted as long as the net flux is positive, and it is absorbed for negative fluxes. We take both the emitted entropy and the absorbed entropy as positive quantities. Therefore, the exchanged entropy is equal to the difference between them. The same result is obtained if the absorbed and emitted entropy are defined for currents entering or leaving the system, respectively.

In Equation (30) the quantity I_s may stand for individual currents, or for the net flux. It should be clear that the change of entropy in a body is directly related to the net exchanged entropy, or to the difference between the emitted entropy and the absorbed entropy.

If entropy is conserved, the change of the entropy content of a body is equal to the total amount of entropy exchanged

$$\Delta S = S_{e,net} \tag{31}$$

This equation is the time integral of the equation of balance of entropy as expressed by Equation (29). Again, it is a special case of the general law of balance of a conserved quantity in integral form.

EXAMPLE 23. Absorbed, emitted, and exchanged entropy.

A constant flux of entropy of 200 W/K leaves a system, while a current entering the system changes according to - 20 W/(K · s) · *t*. How much entropy is absorbed and emitted in the first 15 s? How much entropy is exchanged? What is the change of the entropy content of the body? Refer the absorbed entropy to the current flowing into the system, and the emitted entropy to the current flowing out of the body.

SOLUTION: Since the current entering the system is constant in time, the amount of entropy crossing the surface of the body can be calculated easily. The emitted entropy is given by

$$S^- = 200\,\mathrm{W/K} \cdot 15\,\mathrm{s} = 3000\,\mathrm{J/K}$$

The absorbed entropy is obtained by simple integration:

$$S^+ = 300\,\mathrm{W/K} \cdot 15\,\mathrm{s}/2 = 2250\,\mathrm{J/K}$$

Therefore, the exchanged entropy is equal to

$$S_e = S^+ - S^- = 2250\,\mathrm{J/K} - 3000\,\mathrm{J/K} = -750\,\mathrm{J/K}$$

The same result would have been obtained if we had calculated the amounts of absorbed and emitted entropy according to the net current (as indicated in Figure 32).

The change of the entropy content is directly given by the amount of entropy exchanged (remember that for now we assume entropy to be conserved):

$$\Delta S = -750\,\mathrm{J/K}$$

EXAMPLE 24. Water being heated by an immersion heater.

A body of water is being heated by an immersion heater. The temperature is observed to rise at a constant rate from 20°C to 90°C within 140 s. The energy current emitted by the heater is constant, and is equal to 300 W. a) Express the current of entropy entering the water as a function of time. b) Calculate the entropy and the energy absorbed by the water due to heating.

SOLUTION: a) The temperature is given as the following linear function of time:

$$T = T_o + 0.5\,\mathrm{K/s} \cdot t$$

In steady state processes the energy current emitted by the heater must be equal to the thermal energy current absorbed by the water. According to the law relating currents of entropy and thermal energy currents, Equation (13), we get

$$I_s = -\frac{300}{293 + 0.5t}\frac{\mathrm{W}}{\mathrm{K}}$$

The negative sign has the usual meaning: the current is entering the body.

b) With the current known, we can compute the amount of entropy added to or removed from a body in a given period. This is done, as always, by integrating the current over time:

$$S_e = -\int_0^t I_s dt$$
$$= 2I_{E,th} \cdot \ln\left(\frac{T_o + 0.5t}{T_o}\right) = 128.5\,\text{J/K}$$

Since the current enters the body, the exchanged entropy is positive. Since the current of energy is found to be constant, the exchanged energy is simply

$$Q = -\int_0^t I_{E,th} dt$$
$$= -(-300\,\text{W}) \cdot 140\,\text{s} = 42.0\,\text{kJ}$$

Note that all of these results hold only for the particular body used in this example—water. This is an example of a constitutive property. We shall study such relations much more extensively in Chapter 2.

EXAMPLE 25. Exchanged energy and the change of entropy in an ideal process.

Assume a fluid is heated without entropy being generated. What is the energy exchanged in a thermal process associated with a change of the entropy content of the body from S_1 to S_2?

SOLUTION: The equation of continuity for entropy is:

$$dS/dt + I_s = 0$$

If applied to the expression for the energy exchanged, Equation (15), we get

$$Q = -\int_{t_i}^{t_f} T \cdot I_s dt = \int_{t_i}^{t_f} T \cdot \dot{S} dt$$

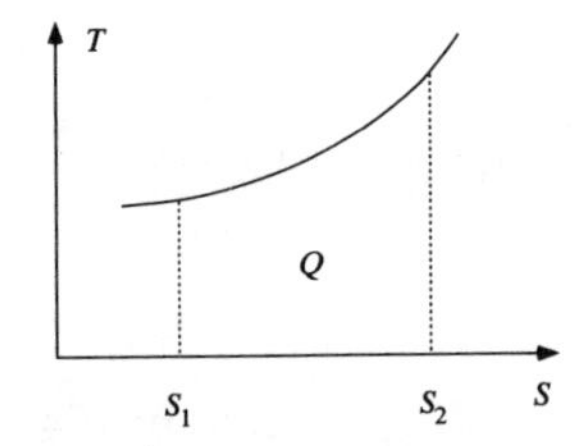

This result holds only in the case of processes which conserve entropy. It can be visualized in a *T-S* diagram as the area under the *T(S)* curve.

1.6.4 The Production of Entropy and the General Law of Balance of Entropy

After having discussed thermal phenomena in some detail in the previous sections, we can list what we mean by entropy. We have found that entropy can be transported, that stored, and created. The last feature mirrors irreversibility, or dissipation, a fundamental property of most processes in nature and in machines. Also, in thermal processes entropy is related to temperature and energy in a form which is analogous to the relationships we know from other fields of physics: entropy is the energy carrier in thermal processes, and temperature is the thermal potential. Therefore, entropy has much in common with electric charge, momentum, angular momentum, or amount of substance. Naturally, there also are important differences which we will have to explore. In summary, we can say that

> *Entropy is a substancelike quantity (like electric charge) which is contained in bodies and which can flow from body to body. It can be created in irreversible processes (burning, electrical currents, friction, heat flow, etc.) but cannot be destroyed.*

Entropy must therefore obey a general equation of balance similar to Equation (29). However, the previous equation of continuity did not contain the possibility of the creation of entropy.

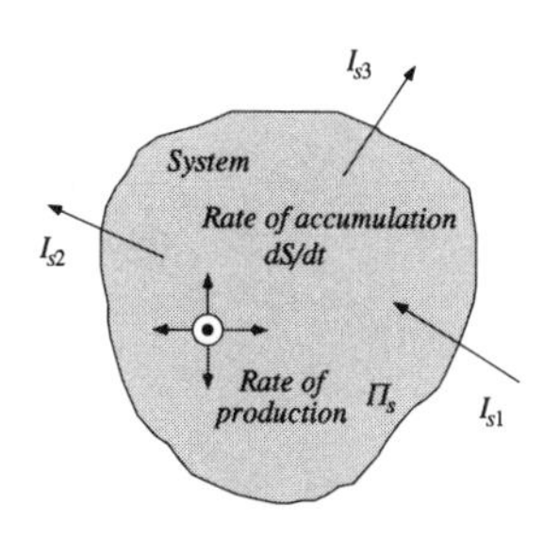

FIGURE 33. Balancing a quantity which can be produced in systems is only slightly more complicated than balancing a conserved quantity. In addition to the currents entering or leaving, the rate of production also adds to the rate at which the system content changes. If the currents or fluxes add up to zero, the content still may increase. The rate at which the content changes is determined by the sum of all fluxes and the rate of production.

The production of entropy inside bodies undergoing irreversible processes is taken into account easily (Figure 33). The equation of continuity must contain a term which describes the rate at which entropy is generated. As before, we shall use the symbol Π_s for this rate. It stands for the amount of entropy generated inside a body per unit time. (It has the units W/K, just like currents of entropy or the time rate of change of the entropy content.) The extended law of balance of entropy now takes the form:

$$\dot{S} + I_s = \Pi_s$$
$$\Pi_s \geq 0 \qquad (32)$$

This equation expresses the fact that the entropy content of a body can change as a consequence of two types of processes: the flow of entropy into and out of the body, i.e., by currents *through* its surface; and the creation of entropy *inside* the body. Equation (32) is the fundamental law of thermal physics comparable to the equation of continuity of momentum in mechanics. As you will see in Chapter 3, it has not yet been written in the most general form. Different modes of the transport of entropy have to be dealt with separately. However, this does not change the fundamental content of the law. The form given in Equation (32) will suffice for us for the moment.

Often, the equation of balance of entropy is used in an integrated form; i.e., it is written in such a way as to express the change of the entropy content of a body. If we use the symbol S_{gen} for the amount of entropy created or generated during a process, the equation takes the form:

$$\Delta S = S_e + S_{gen} \qquad (33)$$

The quantity S_e is the net amount of entropy transferred across the system boundary; Equation (31).

So far, Equation (32) is the most general expression of what is commonly called the *second law of thermodynamics.*

EXAMPLE 26. The rate of change of entropy content in an irreversible process.

A body which is being heated undergoes an irreversible process. The net current of entropy decreases from 100 W/K to zero in 20 seconds. During this time the rate of creation of entropy is equal to 30 W/K. What is the rate of change of the entropy in the body? How large are the exchanged entropy and the change of entropy content in the first 20 s?

SOLUTION: The rate of change of the entropy content is given by

$$dS/dt = -(100\,\mathrm{W/K} - 5\,\mathrm{W/(K \cdot s)} \cdot t) + 300\,\mathrm{W/K} = -70\,\mathrm{W/K} + 5\,\mathrm{W/(K \cdot s)} \cdot t$$

The entropy exchanged in the first 20 seconds is found by integrating the (negative) net current. The emitted entropy is

$$S_e = -\int_0^t (100\,\mathrm{W/K} - 5.0\,\mathrm{W/(K \cdot s)} \cdot t)dt = -1000\,\mathrm{J/K}$$

The change of entropy content during this time is equal to

$$\Delta S = \int_0^t \dot{S}dt = -400\,\mathrm{J/K}$$

This demonstrates clearly that the change of the entropy content does not have to equal the entropy exchanged.

EXAMPLE 27. Irreversible adiabatic processes.

A certain amount of gas is put in a cylinder having piston which is perfectly insulated against the flow of heat. The gas is first compressed and then allowed to expand again to the point where the temperature regains its initial value. The processes undergone by the gas are supposed to be irreversible, possibly because of viscous friction in the fluid. Is the volume at the end smaller than, equal to, or larger than that at the beginning? Write the equation of balance of entropy for adiabatic irreversible processes, and display the operations in the *T-S* diagram.

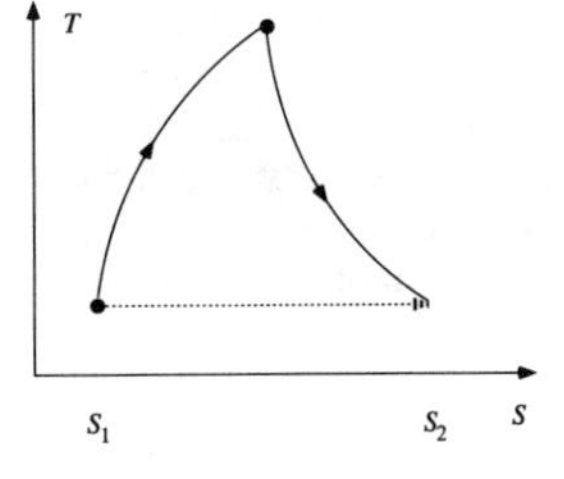

SOLUTION: Because of the ideal insulation the processes are adiabatic; i.e., there is no exchange of entropy (Section 1.3.1). For this reason, the flux term in the equation of balance of entropy is equal to zero. The production term is not equal to zero, though, which leads to the following equation of continuity of entropy:

$$\dot{S} = \Pi_s$$

Thus the entropy of the body can only increase.

In a gas, the temperature increases as the result of adiabatic compression. Because of irreversibility, entropy is produced at the same time. Therefore the amount of entropy in the system increases as the temperature rises, and the process is represented by a curve such as the one shown in the *T-S* diagram.

The second step is an adiabatic expansion, which leads to a reduction of temperature. Again the process is irreversible, which causes entropy to be produced. When the initial temperature is reached, it stops. Compared to the beginning there is more entropy in the gas at the same temperature. We have seen in Section 1.3 that for most substances (with the notable exception of water in the range between 0°C and 4°C) the volume increases if entropy is absorbed at constant temperature. The combined adiabatic irreversible steps actually are equivalent to isothermal heating. Therefore the volume is expected to be larger at the end than at the beginning.

Note that we do not have any information regarding the properties of the gas. For this reason we cannot actually calculate the details of the processes. The curves displayed in the *T-S* diagram are to be taken only as a qualitative description of what is happening.

1.6.5 Systems and Walls

In any physical analysis there is always an important element to be considered. The analyst has to be clear about which system or body is being studied. Only then can the application of a law such as the law of balance of entropy work out successfully. Very often it is quite clear which system has been chosen for investigation, and we do not have to be particularly explicit about our choice. However, nowhere is it more important to be precise and explicit about which part of the world we are going to study than in thermodynamics. The very nature of heat — its tendency to increase through production — calls for careful analysis. We are not allowed to be vague about where the production of entropy occurs in a given situation. Careless treatment of this problem has caused many confusing statements about thermal processes.

The first point to be kept in mind is that the analysis of a situation applies only to the system chosen, and to nothing else. Specifically, this means that the production of entropy takes place *inside the system*. No matter what might happen in the surroundings of the system, dissipation is related to the particular system only. If a body undergoes reversible changes while entropy is being created outside, we have to conclude that entropy has not been produced as far as our equations are concerned. Our analysis cannot make a statement about anything but the body being studied. Conversely, if a process is irreversible, dissipation must have occurred in the body or we would not know about it.

This raises a second point. Consider a body at a uniform temperature in an environment of uniform, but different, hotness. Body and environment touch at the surface of the body. The surface is shared by both the system and its surroundings. So, you might ask, what is the temperature of the surface? Is it the temperature of the body, or that of the environment? The problem becomes more acute if we consider the balance of entropy for a current across this surface. The geometrical surface certainly does nothing to disturb the flux of energy accompanying the current of entropy. In other words, the current of energy must be continuous across the surface (Figure 34). Therefore the current of entropy must be discontinuous; it increases in the direction of flow from the hotter to the cooler body. But then, where has entropy been produced? Where is the seat of dissipation? Unless we are prepared to treat the boundary between body and environment as a physical system in its own right, there is no system which can account for the production. Dissipation must take place in a proper physical system.

Therefore, we are led to introduce surfaces or *ideal walls* across which temperature, entropy flux, and flux of energy are continuous. Such walls do not add to the processes occurring inside a system, particularly to the production of entropy.

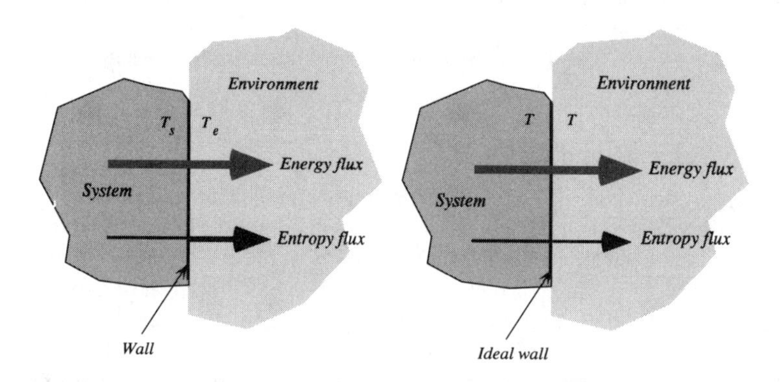

FIGURE 34. If we allow for the system and its environment to be separated by a wall across which the temperature is discontinuous, a current of entropy crossing this boundary will have to be discontinuous as well. Since this brings up the problem of where entropy has been produced, we conceive of ideal walls which do not add to dissipation. Temperature, entropy flux, and flux of energy are all assumed to be continuous across such a wall.

The puzzle presented by two systems at different temperatures can be solved in a number of ways. The ideal system wall may be placed in such a way that the dissipation takes place inside the system. Alternatively, we may exclude the drop of temperature from the system, thereby putting the burden on the analysis of the environment as a physical system. Finally, we may introduce a third system, a finite three-dimensional wall separating the first system and its surroundings. This third body is made responsible for the production of entropy due to the flow of heat from the hotter to the cooler body. The finite wall cannot have a uniform temperature. It exists for communicating between the system and its environment in a physically acceptable way. It is a body with all the physical attributes of the systems we are going to study in this book. Like every proper system it, too, is assumed to be surrounded by an ideal wall.

In essence, then, we consider physical systems having ideal walls. Transfer of entropy and energy across such a wall into or out of the body is governed by the fundamental relationship (13):

> *At the surface of a system the current of entropy and the flux of energy associated with it are related by the temperature at the surface according to Equation (13). Possible production of entropy may take place inside the system, but not at its ideal wall.*

The assumptions stated here are fundamental for thermodynamic analysis. They facilitate the understanding of where and how dissipation may occur. In particular, they do away with statements about thermal energy being added to bodies reversibly or irreversibly. There is no such thing as an irreversible exchange of entropy and energy. The exchange is always governed by Equation (13), and by nothing else.

By the way, if ideal walls did not exist, it would be hard to imagine how temperatures could be measured. A thermometer is a physical system. If its surface were not ideal it might show a different temperature from that of the immediate neighborhood.

1.6.6 Determining the Currents of Entropy: Thermal Constitutive Theories

The equation of balance of entropy is the fundamental law governing the quantity of entropy in all imaginable processes. With this law, and with the qualitative description of some phenomena given above, you might have the impression that thermal processes are independent of such "trivial" details as properties of bodies. However, this is not true. Equation (32) alone would not help us much when it comes to determining thermal processes. We have to be able to specify the fluxes and the rate of production of entropy independently. Only then can we apply the equation of continuity of entropy. The currents and the production rates depend on the properties of bodies. There is no way we can avoid specifying these properties by *constitutive relations.*

Specifying the heating (the current of entropy) by a constitutive relation is analogous to stating a *force law* in mechanics (remember that a force is equivalent to a flux of momentum), or to laying down an expression for an *electrical current.* Take Newton's Law and its application to the solar system. Newton's law is a particularly simple form of the law of balance of momentum; i.e., it is a form of the equation of continuity of linear momentum. With this law alone we could not achieve much. We have to specify properties of the material we want to describe, i.e. the solar system. We have to lay down Newton's law of gravitation, which takes the role of a constitutive relation: it determines the forces for this special system. Only then can we proceed to solve the problem of motion in the solar system.

1.7 Dissipation and the Production of Entropy

In this section, we will discuss how to combine the laws of balance of entropy and of energy for dissipative processes. The combination of these laws will shed light upon the problem of dissipation. The fundamental assumption made here is that the balance of energy is valid in the form given in Section 1.4.2 in the presence of the production of entropy as well. As M. Planck expressed it:[17] "The first law of thermodynamics is nothing more than the principle of conservation of en-

---

17. M. Planck (1926), p. 40.

ergy applied to phenomena involving the production or absorption of heat." We owe much of the early experimental evidence regarding this fact to J. Joule, who performed a series of influential and suggestive experiments more than 100 years ago.

First, we will find a simple relationship between the rate of production of entropy and electrical or mechanical power in such obvious irreversible phenomena as electrical heating or the stirring of a viscous fluid. While entropy is produced out of nothing, it cannot be created by itself. Nature tells us that we have to work to produce entropy; i.e., we have to do things such as rub our hands, burn substances, or let electricity flow through wires. In other words, energy has to be available for driving a process to produce entropy. We will be able to give a graphical answer to the question of how much energy is necessary to create a certain amount of entropy at a prescribed temperature. The amount of energy which is used in the creation of entropy is said to be *dissipated.*

Dissipation reduces the motive power of heat engines, and it leads to less efficient heat pumps. Today we are very much concerned with optimizing engines used in power engineering, refrigeration, or air conditioning. We will find from our analysis that there exists a simple principle governing the reduction of the losses encountered in practice: we have to reduce the production of entropy. Armed with this principle, we will be able to derive the efficiency of a model irreversible heat engine at maximum mechanical power, which comes close to predicting the output of large modern thermal power plants. This example proves the strength of a thermodynamic theory which is built upon the properties of entropy. The following chapters will provide more cases of irreversible processes which are of interest in modern engineering.

1.7.1 The Rate of Production of Entropy in Simple Dissipative Processes

Entropy can be created by letting an electrical current flow through a piece of wire which offers a resistance to its transport. Obviously then, the energy which could have been used for another process (if an ideal electrical engine had been connected to the battery) has been used to create entropy. Consider the heating of water by an electrical immersion heater (Figure 35). An electrical current flows through the device which releases energy for the production of entropy. Both entropy and energy flow into the water. Let us combine the principles of balance of entropy and of energy for the immersion heater as the system undergoing a steady-state process. We need both the second law, Equation (32), and the equation expressing the balance of electrical and thermal power:

$$I_s = \Pi_s \tag{34}$$

$$I_{E,el} + I_{E,th} = 0 \tag{35}$$

If we employ the relationship between fluxes of entropy and of energy in thermal processes according to Equation (13) we arrive at

$$\Pi_s = -\frac{1}{T} I_{E,el} \tag{36}$$

It is common to speak of the dissipation of energy, which is just another way of saying that entropy has been produced. Instead of measuring dissipation in terms of the rate of generation of entropy, we could use the *rate of dissipation of energy* $\mathscr{D}$, which we define as follows:

> *The rate at which energy is dissipated is equal to the product of the rate of production of entropy and the temperature at which the dissipation is taking place:*

$$\mathscr{D} = T\Pi_s \tag{37}$$

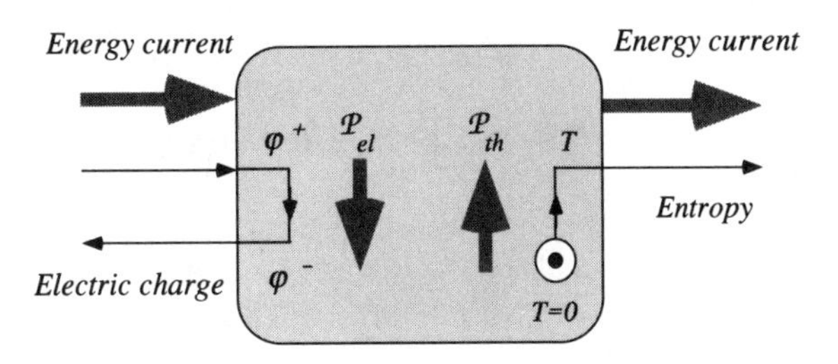

FIGURE 35. Flow diagram of an immersion heater. Charge enters the system at a high potential, and leaves again at a lower one. The energy released in the fall of charge drives the process which produces entropy. Note that there is only one arrow denoting the flow of entropy, namely out of the system. This is in contrast to the situation where either charge, mass, or angular momentum flow. The production of entropy is symbolized by the circle.

If the temperature at which dissipation takes place remains constant, the amount of entropy generated is calculated easily by

$$S_{gen} = \frac{1}{T}\int_{t_i}^{t_f} \mathscr{D}\, dt \tag{38}$$

In other words, it takes TS_{gen} units of energy to produce S_{gen} units of entropy at the temperature T—the higher the temperature, the larger the amount of energy required for creating a certain amount of entropy.

This first example demonstrates how the energy principle, our law of balance of power, can be applied in the presence of dissipation. Note that irreversibility cannot be introduced on the basis of the law of balance of energy. We have taken the law in a form which has not been changed from what was expressed in Section 1.4.2. There is no room in the equation for a dissipative term. Only exchange and storage of energy can be expressed by this law (see Section 2.5). We need the balance of entropy to deal with dissipation.

EXAMPLE 28. Energy dissipated in an immersion heater.

An immersion heater is placed in water and hooked up to a voltage of 110 V. The current is measured to be 1.5 A. How large is the rate of production of entropy in the heater and water combined if the temperature is 20°C? How much entropy is created in 10 s if during this time the temperature changes linearly to 22°C? The electrical quantities remain constant.

SOLUTION: The energy current given off by the heater is equal to

$$|I_{E,el}| = U \cdot |I_q| = 165\,\mathrm{W}$$

which is equal to the rate of dissipation of energy. Therefore:

$$\begin{aligned} \Pi_s &= \frac{\mathcal{D}}{T} \\ &= \frac{165}{273+20}\frac{\mathrm{W}}{\mathrm{K}} = 0.563\frac{\mathrm{W}}{\mathrm{K}} \end{aligned}$$

The amount of entropy produced in this process is

$$\begin{aligned} S_{gen} &= \int_{0s}^{10s} \Pi_s dt \\ &= \int_{0s}^{10s} \frac{\mathcal{D}}{T} dt = |I_{E,el}| \int_{0s}^{10s} \left[T_1 + (T_2 - T_1)\frac{t}{10\mathrm{s}} \right]^{-1} dt \end{aligned}$$

which is equal to

$$S_{gen} = 165\mathrm{W} \cdot 10\mathrm{s} \cdot \ln\left(\frac{T_2}{T_1}\right)(T_2 - T)^{-1} = 5.61\,\mathrm{J/K}$$

Since the temperature is almost constant, the result is very nearly equal to $\Pi_s \Delta t$, with a constant rate of generation of entropy.

EXAMPLE 29. Entropy created in the fall of a body through a resistive medium.

A small steel sphere falls in oil. The frictional force is calculated according to Stokes' law. What is the maximum rate of creation of entropy if the temperature of the medium is 20°C? Take the radius of the sphere to be 1 mm. The densities of steel and oil are 7700 kg/m^3 and 960 kg/m^3, respectively. The dynamic viscosity of oil is 0.99 Pa · s.

SOLUTION: Momentum flows into the sphere from the gravitational field at a rate equal to the weight of the metal sphere. Part of the momentum leaves the system due to friction, with the rate given by Stokes' law:

$$I_p = 6\pi\mu r v$$

Additional momentum leaves as a consequence of buoyancy. Of the energy currents associated with the flow of momentum, only the one due to friction is dissipative. The maximum rate of dissipation is obtained in mechanical equilibrium when the speed of the sphere is largest:

$$mg = V\rho_{oil}g + 6\pi\mu r v_{max}$$

The rate of dissipation of energy is equal to the mechanical power of the force of friction:

$$\mathscr{D} = I_p v_{max} = 6\pi\mu r v_{max}$$

$$= \frac{8p}{27}\frac{r^5(\rho - \rho_{oil})^2 g^2}{\mu} = 4.11 \cdot 10^{-6}\,\mathrm{W}$$

Therefore the rate of production of entropy is equal to

$$\Pi_s = \frac{\mathscr{D}}{T} = 1.40 \cdot 10^{-8}\,\frac{\mathrm{W}}{\mathrm{K}}$$

1.7.2 Pumping Entropy from Absolute Zero

These examples demonstrate that the production of entropy is governed by a simple law. Namely, the amount of energy needed for the creation of entropy in an irreversible process is equal to the product of the amount of entropy produced and the temperature at which the process takes place. If we write the law in a slightly different form we are led to a vivid interpretation of the production of entropy. Instead of using the temperature T in Equation (37) we shall write $(T - 0)$ which is the temperature of the process with respect to absolute zero:

$$\mathscr{D} = (T - 0)\Pi_s \tag{39}$$

Rather than producing entropy in a dissipative process, we might choose to transfer entropy to a body at the same rate. If the body is at temperature T, the current of entropy entering is accompanied by a flux of energy equal to the product of the flux of entropy and the temperature. In this case the thermal power is equal to the rate at which energy is dissipated. The rate of production of entropy is replaced by a current of entropy, and the rate of dissipation of energy corresponds to a current of energy:

$$I_{E,th} = (T - 0)I_s \tag{40}$$

The theory of heat pumps tells us that it takes an energy current of the magnitude just calculated in Equation (40) to lift a current of entropy of magnitude I_s from absolute zero to the temperature T. As a consequence, we may interpret the creation of entropy as follows:

> *Creating entropy at a temperature T corresponds to raising it from absolute zero to this temperature; the energy needed for the production is equal to the energy required to pump entropy to the desired thermal level.*

It is as if heat appears out of nothing at zero hotness, after which it has to be pumped to the temperature at which it makes itself felt.

One more thing can be inferred from the equations just derived, namely that the temperature T must be positive:

$$T > 0 \tag{41}$$

If this were not the case the relationship between the entropy created and the amount of energy needed to produce it would not be definite: the rate of dissipation of energy could be zero or negative for a nonzero, positive, rate of production of entropy. The nonconservation of entropy and the fact that temperature must have an absolute zero are related.

1.7.3 Dissipation and Loss of Available Power

Some of the most common uses of heat are associated with power engineering, heating, and refrigeration. Those who use heat must make the best use of it. In Section 1.5 we saw that the production of entropy leads to a reduction or loss of the power of heat engines. We must quantify the loss occurring in thermal engines as a consequence of dissipation.

As a basis for comparison, we need to know the maximum power which can be derived from heat. This quantity is called the *available power of heat*. The concept of available power holds equally for heat engines, heat pumps, and refrigerators (see Figure 26), but it is best explained in terms of a heat engine operating in an environment at constant temperature. This is pretty much the case for an engine at the surface of the Earth (Figure 36) where the surroundings act as a vast sink for entropy rejected by the engine.

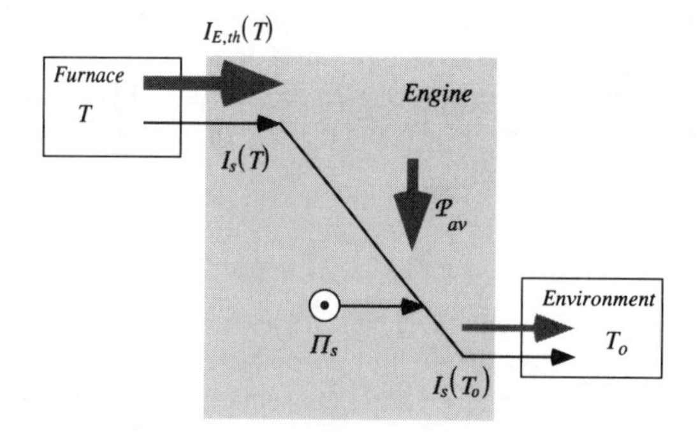

FIGURE 36. A heat engine in contact with a furnace and the environment, both of which act as heat reservoirs whose temperatures do not change. The engine therefore undergoes a *Carnot process*. The model also is applicable to heat pumps and to refrigerators, where the temperatures have to be interpreted in accordance with Figure 26.

The engine cannot receive entropy at a temperature surpassing that of the furnace producing it; let us denote this maximum temperature by T. Furthermore, it is impossible for entropy to be emitted at a temperature below that of the environment, which we set equal to T_o. Let us assume therefore that the engine can come in contact with the furnace and the environment only, and that the temperatures of the engine are T and T_o, respectively, for those times when the reservoirs and the engine are in contact. In other words, we take there to be an ideal wall between the machine and the heat reservoirs (Section 1.6.5). Any dissipation which might occur is thus made a part of the engine. From our discussion of heat engines in Section 1.4 we know that an ideal Carnot engine can make some energy available for other uses. The quantity made available is a definite fraction of the energy transmitted to the engine from the furnace together with entropy, the fraction being determined by the temperature of the furnace, and the hotness of the environment. This is where the term *available power*[18] comes from. It expresses quantitatively the power which may be derived from fluxes of entropy depending on the temperatures involved. What engineers call the *available power* is Carnot's power of heat, i.e. the rate at which energy is released by the fall of entropy. Therefore we define:

$$\mathcal{P}_{av} = -(T - T_o)I_s(T) \tag{42}$$

where $I_s(T)$ is the current of entropy flowing into the engine at temperature T.

There are a few points to be noted about this definition. First of all, it is valid for all three types of thermal engines (Figure 26). T_o is the temperature of the environment, while T is the temperature of the entropy reservoir (the furnace for the heat engine, the receiver for the heat pump, or the cold space for the refrigerator). T is larger than T_o for heat engines and heat pumps, and it is smaller than T_o for refrigerators. $I_s(T)$ is the current of entropy derived from or pumped into the reservoir at temperature T. Its sign is derived from the sense of flow of entropy with respect to the thermal engine in use (not the reservoir). Thus, for heat engines and refrigerators, $I_s(T)$ is negative (the current of entropy enters the engine, Figure 26). For heat pumps, $I_s(T)$ is positive. With the proper sign of the term $T - T_o$, the available power turns out to be positive for heat engines (energy is released in the fall of entropy), and negative for heat pumps and refrigerators (energy is required for pumping entropy).

The next problem is that of finding the *loss of available power* as a result of dissipation in the engine. We shall get the answer indirectly by calculating the actual power of the engine using the laws of balance of entropy and energy. For an engine operating in complete cycles, the rate of change of energy can be set equal to zero, which means that the mechanical power is equal to the sum of the thermal energy fluxes:

$$\mathcal{P}_{mech} = TI_s(T) + T_oI_s(T_o) \tag{43}$$

Note that the mechanical power is equal to the negative of the net mechanical energy current. In contrast to earlier treatments, we now include the rate of pro-

18. In addition to available power, the term *exergetic power* is in use today for the rate of energy liberated in the fall of entropy. The amounts of energy which may be derived from the fall of entropy are called *availability* or *exergy*.

duction of entropy. Since our considerations apply to steady-state operations, the rate of entropy production in the system is equal to the sum of the fluxes of entropy entering and leaving the engine:

$$I_s(T) + I_s(T_o) = \Pi_s \tag{44}$$

If we introduce the latter result into the former we obtain the following expression for the mechanical power of the engine:

$$\mathcal{P}_{mech} = (T - T_o)I_s(T) + T_o\Pi_s \tag{45}$$

The first term on the right-hand side is equal to the negative available power. The second term therefore must represent the *loss of power*:

$$\mathcal{L} = T_o\Pi_s \tag{46}$$

(This term is strictly positive, since entropy can only be produced.) Taken together, we may now write the actual mechanical power in terms of the available power of heat and the loss of power:

$$\mathcal{P}_{mech} = -(\mathcal{P}_{av} - \mathcal{L}) \tag{47}$$

Since the mechanical power of a heat engine is negative, Equation (47) shows that its magnitude is smaller than that of the available power. We cannot derive the full thermal power from the fall of entropy if entropy is produced. On the other hand, for heat pumps and refrigerators, the mechanical power required for pumping entropy is larger than the ideal value derived in Section 1.4.

The result for the loss expressed in Equation (46) can be motivated easily. Since the entropy generated has to be emitted to the environment with the entropy absorbed from the furnace, it carries an additional thermal energy current which will be lost from our application. According to the relationship between fluxes of entropy and of energy, the latter must be equal to the product of the temperature of the body receiving the entropy and the flux of entropy. This directly suggests the form of the result which, in engineering, has been known as the *Gouy-Stodola* rule.[19] Even though the rule has been derived for the simplest possible case, this qualitative discussion suggests it to be valid more generally (Examples 30 - 33).

Finally, we can define the second law efficiency of all three types of thermal engines. For heat engines we have to write

$$\eta_2 = -\frac{\mathcal{P}_{mech}}{\mathcal{P}_{av}} = 1 - \frac{T_o\Pi_s}{(T - T_o)|I_s(T)|} \tag{48}$$

19. See, for example, A. Bejan (1988).

while for heat pumps and refrigerators the inverse of this quantity should be taken:

$$\eta_2 = -\frac{\mathcal{P}_{av}}{\mathcal{P}_{mech}} \tag{49}$$

Since the available power of heat always has the opposite sign of the mechanical power, the minus sign in Equations (48) and (49) make the efficiency a positive number smaller than or equal to 1. Its value directly tells us how well we have used the power of heat, or expressed differently, how badly we have done with producing entropy.

You might wonder why the loss of power is not exactly equal to the rate of dissipation as we have defined it above in Equation (36). Entropy will certainly be produced in the engine at different temperatures which are to be found in the range spanned by those of the furnace and the environment. With the total rate of production fixed by the operation of the engine, the rate of dissipation would be smaller than the loss calculated in Equation (46). This problem has a simple solution. Entropy created at temperatures above that of the environment can contribute to the release of energy for mechanical purposes if it is allowed to fall to T_o without further dissipation (Example 32).

There is an interesting point to be made about losses due to dissipation. Producing entropy for power engineering already constitutes a loss. We have compared real heat engines to their reversible limit, i.e. to the available power of a current of entropy at temperature T. The available power is smaller than the power used for heating the furnace, the loss being determined by the product of the rate of generation of entropy and the temperature of the environment. Just consult Equation (42). The first term, TI_s, is equal to the rate at which we heat the furnace and, through the furnace, the engine. The current of entropy I_s is equal to the rate at which entropy has been produced by the burning of fuel. Therefore the second term in Equation (42) represents the loss due to the production of entropy computed according to our rule expressed in Equation (46). Naturally, there is a loss in a real sense only if there was an alternative to the burning of fuel for driving heat engines. In some cases such an alternative exists in the form of fuel cells (see Example 33). In others, like the Sun, we can do nothing to avoid the loss associated with the production of entropy. We simply accept solar radiation for what it is. As to the production of entropy as a result of the absorption of solar radiation at the surface of the earth, solar engineering is an attempt to temporarily reduce the magnitude of this process.

EXAMPLE 30. Loss of available power in a general thermomechanical process.

Determine the mechanical power of a thermomechanical engine undergoing a steady-state cyclic process. The engine works irreversibly. It is in contact with n heat reservoirs of constant temperatures, and with the environment at temperature T_o. Identify the loss as a result of dissipation, and compute the second law efficiency of the heat engine.

SOLUTION: We start, as always, from the fundamental laws of balance of entropy and energy. The flux of entropy in the equation of balance is the sum of $n + 1$ terms:

$$I_s = \Pi_s \tag{E18}$$

$$I_{E,mech} + I_{E,th} = 0 \tag{E19}$$

with

$$I_s = \sum_{i=0}^{n} I_{si} \tag{E20}$$

The thermal energy current is the result of the transport of heat at $n + 1$ distinct temperatures. Therefore, according to Equation (13), we can write:

$$I_{E,th} = \sum_{i=0}^{n} T_i I_{si} \tag{E21}$$

Let us solve the equation of balance of energy for the mechanical power. Furthermore, we will eliminate the entropy flux to or from the environment:

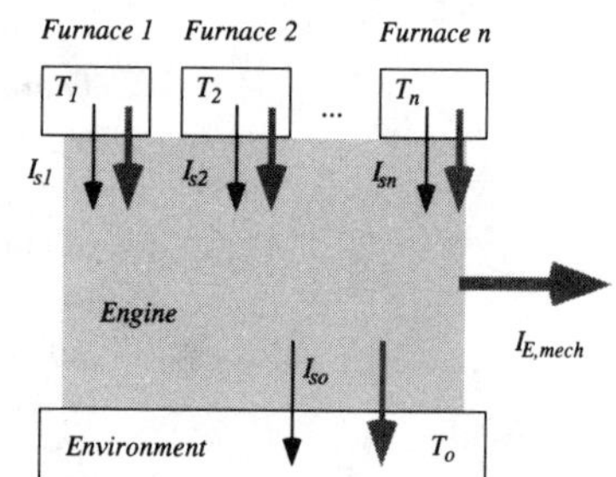

$$\begin{aligned} I_{E,mech} &= -\sum_{i=0}^{n} T_i I_{si} = -T_o I_{so} - \sum_{i=1}^{n} T_i I_{si} \\ &= -T_o\left(I_s - \sum_{i=1}^{n} I_{si}\right) - \sum_{i=1}^{n} T_i I_{si} \\ &= -T_o \Pi_s - \sum_{i=1}^{n} (T_i - T_o) I_{si} \end{aligned}$$

Finally, we identify the last term with the available power according to Equation (42):

$$I_{E,mech} = \sum_{i=1}^{n} \mathcal{P}_{av,i} - T_o \Pi_s \tag{E22}$$

Note that here the mechanical energy current was used instead of the mechanical power. This leads to a different sign compared to Equation (45). The last term in (E22) must be the loss due to dissipation. It follows the Gouy-Stodola rule given in the text:

$$\mathcal{L} = T_o \Pi_s \tag{E23}$$

We can compute the second law efficiency for cyclic processes of a heat engine, which turns out to be:

$$\begin{aligned} \eta_2 &= \left(\sum_{i=1}^{n} \mathcal{P}_{av,i} - \mathcal{L}\right) \Bigg/ \sum_{i=1}^{n} \mathcal{P}_{av,i} \\ &= \left(1 - T_o \Pi_s\right) \Bigg/ \sum_{i=1}^{n} \mathcal{P}_{av,i} \end{aligned} \tag{E24}$$

The definition will turn out to be different for heat pumps or for refrigerators; in fact it will be the inverse of (E24); see Example 31.

EXAMPLE 31. Efficiency of a dissipative heat pump.

a) Determine the coefficient of performance (*COP*) of a dissipative heat pump operating between the environment at temperature T_o and a body to be heated at T. Show that it is smaller than that of an ideal pump. b) Derive the relationship between the *COP* and the second law efficiency.

SOLUTION: a) We start from the definition of the *COP* for heat pumps. It is the ratio of the heating power at temperature T and the mechanical power necessary for driving the pump. Taking into consideration the signs of the fluxes with respect to the pump we have:

$$COP = \frac{I_{E,th}(T)}{-I_{E,mech}} = \frac{I_{E,th}(T)}{-(\mathcal{P}_{av} - \mathcal{L})}$$
$$= \frac{T \cdot I_s(T)}{(T - T_o)I_s(T) + T_o \Pi_s}$$
$$= \frac{T}{(T - T_o) + T_o \Pi_s / I_s(T)} < \frac{T}{T - T_o}$$

The last expression is the *COP* of an ideal heat pump; it is clearly larger than the actual value.

b) According to the definition of the second law efficiency, the actual *COP* must be equal to the product of the ideal *COP* and the second law efficiency:

$$COP = \frac{I_{E,th}(T)}{-I_{E,mech}} = \frac{\mathcal{P}_{av}}{I_{E,mech}} \frac{I_{E,th}(T)}{-\mathcal{P}_{av}}$$
$$= \eta_2 COP(ideal)$$

We could calculate the second law efficiency from the result for the *COP* given in (a). This must agree with the direct definition, which leads to:

$$\eta_2 = \frac{\mathcal{P}_{av}}{I_{E,mech}} = \frac{\mathcal{P}_{av}}{\mathcal{P}_{av} - T_o \Pi_s}$$
$$= \frac{-(T - T_o)I_s(T)}{-(T - T_o)I_s(T) - T_o \Pi_s} < 1$$

Note that the second law efficiency of a heat pump is the inverse of what we defined for a heat engine in Equation (E24).

EXAMPLE 32. Dissipation and loss of available power in conduction.

a) Estimate the rate of dissipation due to conductive heat transfer through a long bar. The temperature can be assumed to vary linearly from a high value at one end to a low one at the other end. b) Calculate the loss of available power for the same bar, and compare the result to the rate of dissipation. Show that the loss agrees with the Gouy-Stodola rule.

SOLUTION: a) The rate of dissipation is given by the rate of production of entropy and the temperature at which it is taking place. Since conditions change through the body, we either need a theory describing the spatial distribution of the generation rate, in which case we integrate to obtain the answer for the body as a whole, or we must assume average values for the variables. Lacking a continuum theory (see Chapter 3 for the solution of this problem), we shall use the latter approach.

The rate of production of entropy in a bar due to conduction was computed in Example 21. In slightly different form, it is given by

$$\Pi_s = \left(\frac{1}{T} - \frac{1}{T_o}\right) I_{E,th}(T)$$

T and T_o are the upper and the lower temperatures, respectively. We choose the average temperature in the bar for estimating the rate of dissipation. This leads to:

$$\mathscr{D} = \frac{1}{2}(T + T_o)\left(\frac{1}{T} - \frac{1}{T_o}\right) I_{E,th}(T)$$

b) The loss of available power is complete in conduction, since there is no mechanical effect at all. Therefore the loss is equal to the available power:

$$\begin{aligned}\mathscr{L} &= -(T - T_o) I_s(T) \\ &= T_o\left(\frac{1}{T} - \frac{1}{T_o}\right) I_{E,th}(T)\end{aligned}$$

This proves that the loss conforms to the rule seen before. It is determined by the rate of production of entropy and the temperature of the environment:

$$\mathscr{L} = T_o \Pi_s$$

The loss turns out to be smaller than the rate of dissipation, as expected. The continuum model given in Chapter 3 will support the estimate given here.

EXAMPLE 33. Comparing efficiencies of different modes of heating water.

We are given a certain amount of hydrogen which we are to use for heating water. The water is to be kept at a constant temperature T which is higher than that of the environment (T_o), and lower than a possible flame temperature if we decide to burn the hydrogen. a) Calculate the heating power for the water in terms of the energy released by the chemical reaction of hydrogen with oxygen for the following three modes of heating: (A) direct heating by burning the hydrogen; (B) heating with the help of an ideal heat engine driven by entropy from a furnace at temperature T_f which receives its entropy from burning of hydrogen, and an ideal heat pump driven by the heat engine (scheme B of Example 18); (C) heating with the help of an ideal heat pump driven by an ideal fuel cell which uses the hydrogen. b) Calculate the loss of power for scheme A, and show that it is equal to the product of the temperature of the water and the rate of production of entropy.

SOLUTION: a) In scheme A, the water is brought in direct contact with the flame. The heating power is equal to the rate at which energy is liberated by the burning of hydrogen:

$$I_{E,th}^{(1)} = I_{E,chem}$$

(There is a small problem with heating via chemical reactions. The energy released for heating usually is not exactly equal to the total energy released by the reaction. However, we may neglect the difference for now.)

The heating power due to scheme B has been calculated in Example 18. In current notation we have the following expression for the energy flux entering the water:

$$I_{E,th}^{(2)} = \frac{T_f T - T T_o}{T_f T - T_f T_o} I_{E,chem} > I_{E,chem}$$

which is larger than the energy flux due to the chemical reactions.

In scheme C, finally, we assume that the entire energy released by the chemical reactions can be made available for driving an electrical current. The electrical power of the fuel cell is set equal to the chemical energy flux. The electrical current is used to drive an ideal heat pump operating between the environment and the water. Therefore,

$$\begin{aligned} I_{E,th}^{(3)} &= COP \cdot I_{E,chem} \\ &= \frac{T}{T - T_o} I_{E,chem} > I_{E,th}^{(2)} \end{aligned}$$

which is larger than result of either of schemes A or B.

b) We should compare the loss of power to the most ideal case, which is represented by the fuel cell. In scheme A, the entropy generated by burning hydrogen is conducted all the way from the temperature of the flame to the temperature of the water with total loss of the chemical (or electrical) power. Since the total rate of production of entropy (due to the burning and the conduction) must be equal to the entropy flux to the water, it is equal to the ratio of the heating power and the temperature of the body receiving the entropy. This makes the loss equal to $\mathcal{L} = T\Pi_S$.

1.7.4 The Influence of Heat Exchangers

Sources of irreversibility are numerous in nature and in machines, but often the most influential factor reducing the power of engines can be found in heat exchangers. (See Section 1.7.6.) They determine the rates of transport of entropy and often dominate the rates of production of entropy. For this reason, we will give a brief and very much simplified description of their action. More about the flow of heat can be found in Chapter 3; details of heat exchangers will be discussed in Chapter 4.

First consider where heat exchangers are needed in conjunction with the operation of heat engines (Figure 37). Heat must somehow pass from the furnace to the engine, i.e. to its working fluid, and from this fluid to the cooler. For this to happen there must be a temperature difference across the heat exchanger: heat enters at the hot end and leaves at the cooler side.

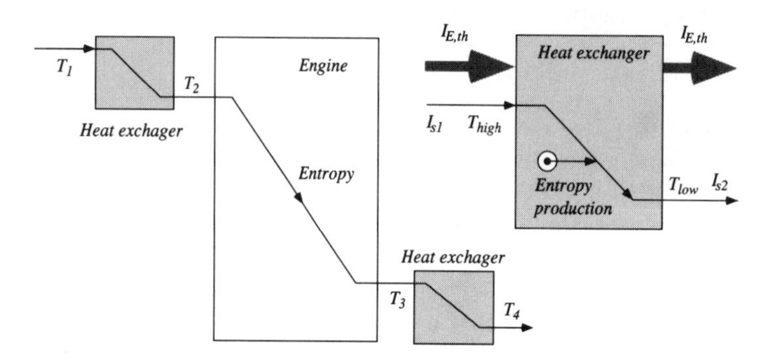

FIGURE 37. Heat exchangers are used for transferring heat to and from an engine. Their basic effect lies in lowering the temperature from one side to the other; combined with the fact that the amount of energy transmitted remains constant across the device, this means that entropy is produced. A simple model of a heat exchanger has entropy entering and leaving at single values of the temperature.

The simplest model of a heat exchanger takes entropy entering at a single given temperature and leaving again at a single, lower value (Figure 37). In general, this is not very realistic since heat is commonly transferred from a hot to a cold fluid. While the hot fluid emits heat its temperature decreases, which means that heat is emitted in a range of temperatures rather than at a single thermal level. The equivalent statement is true about the absorption of heat by the colder fluid. We can approximate the model, however, if we think of the changes of temperature of the two fluids to be small compared to the difference from one fluid to the other. Alternatively, we may think of the temperature difference across the heat exchanger as some mean difference of the actual values. (This is what one does in a theory of heat exchangers; see Chapter 4.)

We now need a simple theory of the behavior of the elements that allow for the transport of heat from one point to another; without it we will not be able to compute the flow of entropy or the rate of generation of entropy. We shall use the simplest model of the combined effect of conduction and convection in a heat exchanger. It is common to approximate the rate of transfer of energy as a linear function of the mean difference of temperatures across the heat exchanger:

$$I_{E,th} = hA(T_{high} - T_{low}) \tag{50}$$

The *transfer coefficient* h is taken to be constant. It includes the effects of the physical properties of the system exchanging entropy. The rate of transfer certainly scales with the surface area A, which has been included separately in Equation (50). The rate of production of entropy is calculated simply for the case of stationary operation. The amount of energy entering the heat exchanger is the same as that leaving. Since the temperatures associated with the flow of entropy are different at the inlet and the outlet, entropy must have been produced at a rate equal to

$$\Pi_s = \left(\frac{1}{T_{low}} - \frac{1}{T_{high}}\right) I_{E,th} \tag{51}$$

Together with Equation (50), this result describes the effect of heat exchangers in our simplified model.

EXAMPLE 34. Ideal heat engine coupled to heat exchangers.

Model a power plant as an ideal Carnot engine running between two heat exchangers communicating heat from the furnace and to the cooler, as in Figure 37. The energy current from the furnace to the first heat exchanger is equal to 1.0 GW at a temperature of 600 K. Waste heat goes from the second heat exchanger to the environment at a temperature of 300 K. The mean heat transfer coefficient in both heat exchangers is equal to 2000 W/(K · m^2). In total, an exchanger surface equal to $2.0 \cdot 10^4$ m^2 is available, which is shared equally among the heat exchangers.
a) How large is the entropy current which passes through the ideal engine? b) What is the magnitude of the useful power of the plant? c) How large is the ratio of this power to the power of a Carnot engine running directly between 600 K and 300 K?

SOLUTION: a) The current of energy through heat exchanger A and the current of entropy leaving this exchanger are given by

$$I_{E1} = (kA)_1(T_1 - T_2)$$
$$I_{E1} = T_2 I_s$$

which leads to the following result for the flux of entropy through the ideal engine:

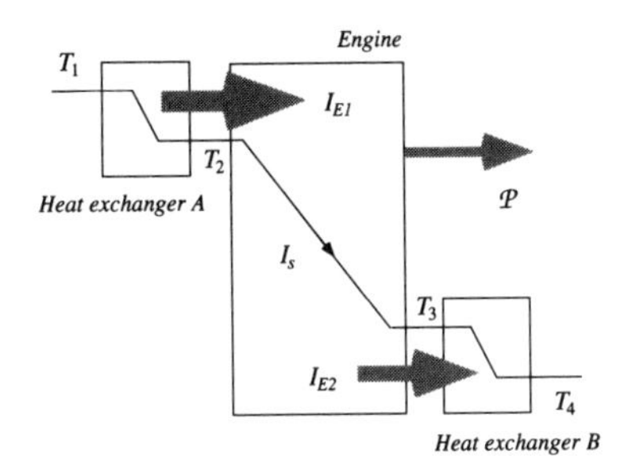

$$T_2 = T_1 - \frac{I_{E1}}{(kA)_1}$$
$$= 600\,\text{K} - \frac{10^9}{2 \cdot 10^7}\text{K} = 550\,\text{K}$$
$$I_s = \frac{I_{E1}}{T_2}$$
$$= \frac{10^9}{550}\text{W/K} = 1.82 \cdot 10^6\,\text{W/K}$$

b) We may now calculate the energy current passing through the second heat exchanger, which yields an expression for T_3. Together with T_2, the power of the engine can be computed:

$$I_{E2} = T_3 I_s$$
$$I_{E2} = (kA)_2(T_3 - T_4)$$
$$\mathcal{P} = (T_2 - T_3)I_s$$
$$\Rightarrow \qquad T_3 = \frac{(kA)_2 T_4}{(kA)_2 - I_s} = 330\,\text{K}$$
$$\mathcal{P} = (550 - 330) \cdot 1.82 \cdot 10^6\,\text{J} = 0.40\,\text{GW}$$

c) The ratio of the actual thermal efficiency to that of the ideal engine running directly between the high and the low temperatures (i.e. the second law efficiency) is equal to

$$\frac{\eta}{\eta_C} = \frac{0.40/1.0}{1 - \frac{T_4}{T_1}} = \frac{0.40}{0.50} = 0.80$$

1.7.5 Minimization of the Rate of Production of Entropy

When entropy is produced, a loss of available power is inevitable. We can conclude this from the results just derived. We were able to quantify the loss in terms of the rate of generation of entropy and the temperature of the body which acts as the final reservoir for entropy; usually this is the environment. For engineers this indicates that a strategy trying to optimize power output or power requirement must try to minimize the production of entropy. While this rule appears to be quite intuitive after all that has been said about the role of entropy in thermal engines, it might be less than obvious in the case of direct heating (without the use of heat pumps). After all, in heating we require the most amount of entropy; so why should we not produce as much as possible of the quantity responsible for making bodies warm? In this section we will demonstrate that the minimization of entropy generation is a general goal we should try to embrace.

Consider the role of entropy production in thermal processes. First of all, if fuels are burned to generate entropy we should ensure the highest possible flame temperature, since this reduces the rate of production of entropy, which is given by the equivalent of Equation (36). In some cases it might be possible to consider fuel cells for harnessing the chemical power of fuels. At least in theory, this should be even better than burning fuels at high temperatures.

The major rule to be followed, however, has to do with the use of entropy once it has been generated. If the goal is, for example, heating of water for domestic consumption, direct heating appears to be utter waste. Instead of harnessing the available power of heat we let it conduct to the desired low temperature, which means pure dissipation. If the entropy is used in engines we likewise should avoid any drop in temperature which is not used for the envisioned purpose. Carnot, without knowing much about rates of production of heat in a modern sense, expressed this point succinctly:[20]

> Since every re-establishment of equilibrium in the caloric may be the cause of the production of motive power, every re-establishment of equilibrium which shall be accomplished without production of this power should be considered as an actual loss.

20. S. Carnot (1824), p. 22.

By *re-establishment of equilibrium*, he meant the fall of caloric (entropy) back to a previous thermal level (temperature). We now know how to measure the actual loss.

It appears that the rule to be given to those who build thermal engines is to avoid any production of entropy (Equation (45) and Figure 36). Indeed, if we set the rate of generation equal to zero in our equations determining the mechanical power, we should achieve the desired theoretical maximum. However, in real life, as a consequence of this requirement nothing moves. *Real* processes which run at finite speed are dissipative.[21] Therefore the rule we are looking for must be expressed in the following form:

To optimize processes involving heat engines, we must minimize the rate of production of entropy under realistic constraints.

We shall demonstrate the usefulness of this idea by calculating the efficiency of a model heat engine at maximum power (Section 1.7.6).

While the rule of minimal entropy production appears to be fairly straightforward for power processes, it might come as more of a surprise that it also holds for heating processes. To understand this, we first must define the goal of heating. An optimal heating process delivers the largest amount of entropy for given heating power to a body at a desired temperature T, where the process takes place in an environment at temperature T_a (see Figure 38). The rule then takes the form:

For fixed heating power, the body to be heated receives the largest amount of entropy if the amount of entropy produced is minimal for the processes involving heating and losses from the furnace.

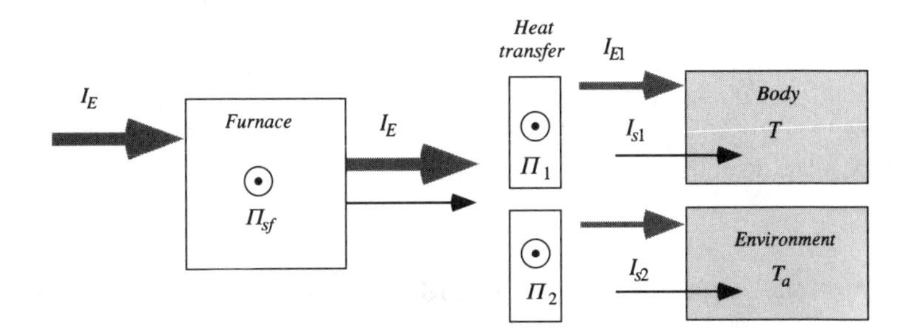

FIGURE 38. A furnace heats a body at a desired temperature T and the environment at T_a (losses). Entropy is produced in the furnace and as a result of heat transfer. The furnace may be of any type, including solar.

The proof goes as follows. The body to be heated at constant temperature T receives an entropy current I_{s1}, while a current I_{s2} is associated with the losses

21. We shall consider *ideal* processes which run at finite speed in Chapter 2; they do not produce entropy.

from the furnace. The production of entropy takes place in the furnace (which receives energy at a fixed rate I_E), and as a consequence of heat transfer from the furnace to the body to be heated and to the environment. Altogether, the rate of production of entropy must be equal to the entropy currents introduced:

$$\Pi_s = |I_{s1}| + |I_{s2}| \tag{52}$$

We can express the second entropy current in terms of the fixed energy current received by the heating system and the temperatures involved:

$$\begin{aligned} \Pi_s &= |I_{s1}| + \frac{|I_E| - |I_{E1}|}{T_a} = |I_{s1}| + \frac{|I_E| - T \cdot |I_{s1}|}{T_a} \\ &= \frac{|I_E|}{T_a} - \left(\frac{T}{T_a} - 1\right)|I_{s1}| \end{aligned} \tag{53}$$

This result demonstrates that the rate of production of entropy decreases as the amount of entropy delivered for heating is increased.

There is a direct relationship between the cases of power engineering and heating. The result demonstrated in Equation (53) can be cast in terms of the available power of heat. If we heat a body at a temperature surpassing that of the environment, we may use its entropy to subsequently drive a heat engine. In other words, we still have some available power which is proportional to both the temperature difference between body and environment and the amount of entropy which can be drawn from the body. Therefore, maximizing the heating of the body in question is equivalent to maximizing the available power of the heating process. Put in these terms, we can appreciate the generality of the rules stated above. Accounting for entropy in all its aspect is not just a convenient theoretical tool, it is the central task of those involved in the use of heat for any purpose.

Consider once again the importance of being able to compute the rate of production of entropy from a constitutive theory. The rule discussed here is nice but completely useless if we do not manage to calculate for concrete cases such quantities as the rate of generation of entropy and currents of entropy. The generic laws which we have been using, i.e. the laws of balance of entropy and energy, alone do not provide this information.

EXAMPLE 35. Optimizing a solar thermal engine with hot water production: the role of availability

We can produce heat in a collector by absorbing solar radiation. Assume that the radiation is completely absorbed; also assume that solar radiation does not bring any entropy with it. (This is very nearly true; see Chapter 3.) The hot collector (simply a hot body) emits entropy to an ideal Carnot heat engine and loses heat to the environment at temperature T_a. The loss is modeled in terms of a law such as in Equation (50) with an overall heat loss coefficient h. The engine rejects the entropy received from the collector to a large body of water at a temperature T_2, which is larger than T_a thus providing both power and hot water at T_2.

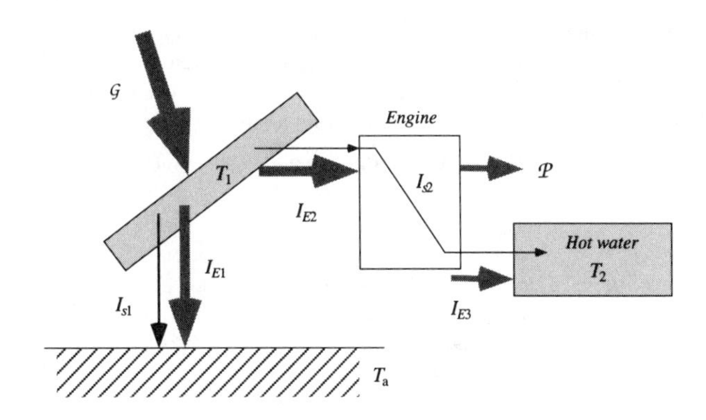

Take values of 290 K and 330 K for T_a and T_2, respectively. The solar radiation is measured in terms of the energy flux per unit area and is taken to be equal to $\mathcal{G} = 800\ W/m^2$. The overall heat loss coefficient is $h = 4.0\ W/(K \cdot m^2)$. a) For which value of the temperature of the collector (T_1) will the rate of production of entropy be minimal? Why would you expect a minimum of entropy production at all? b) How large should the temperature of the collector be for the power of the engine to reach its maximum? c) Why don't the minimum of entropy production and the maximum of power coincide? Is there a quantity related to power that has a maximum coinciding with the minimum of entropy production? Would you run the system at maximum power or at minimum entropy production?

SOLUTION: a) The radiation is absorbed by the collector, leading to production of entropy at the temperature of the collector T_1. The rate of absorption of energy from solar radiation ($A\mathcal{G}$) is equal to the rate of dissipation. (The collector does not do anything else but produce entropy.) Therefore the rate of production of entropy from this source must be equal to $A\mathcal{G}/T_1$. On the other hand, entropy flows from the collector to the environment at T_a, leading to more entropy being produced (see Example 34). The rate of production of entropy in the entire system therefore is made up of these two parts (all other processes proceed reversibly):

$$\Pi_s = \frac{A\mathcal{G}}{T_1} + Ah(T_1 - T_a)\left[\frac{1}{T_a} - \frac{1}{T_1}\right]$$

The minimum of this function is found by differentiating it with respect to the temperature of the collector:

$$\frac{d\Pi_s}{dT_1} = -\frac{A\mathcal{G}}{T_1^2} + Ah\left[\frac{1}{T_a} - \frac{1}{T_1}\right] + Ah(T_1 - T_a)\frac{1}{T_1^2}$$

$$\frac{d\Pi_s}{dT_1} = 0$$

$$\Rightarrow \quad T_{1,min\ S} = \sqrt{\frac{T_a}{h}\mathcal{G} + T_a^2}$$

For the given numerical values the minimum of production of entropy is obtained for a temperature of 377 K. We expect the rate of production of entropy to have a minimum because it should be large for both small and large values of the temperature of the collector but for opposing reasons. At high T_1 the rate of production inside the collector will be small while the losses, and therefore the entropy production due to the flow of heat, must be large. At small values of T_1 we have exactly the opposite conditions. Since the effects are nonlinear we can expect a function with a minimum.

b) The balance of energy for the collector operating at steady-state is given by

$$I_{E2} = AG - Ah(T_1 - T_a)$$

With the help of the relationship between the current of entropy and the current of energy transferred to the heat engine,

$$I_{E2} = T_1 I_{s2}$$

we can now express the power of the engine:

$$\begin{aligned} \mathcal{P} &= (T_1 - T_2) I_{s2} \\ &= \frac{T_1 - T_2}{T_1} I_{E2} \\ &= \frac{T_1 - T_2}{T_1}\left[AG - Ah(T_1 - T_a)\right] \end{aligned}$$

Its derivative with respect to the temperature of the collector will deliver the condition for maximum power:

$$\begin{aligned} \frac{d\mathcal{P}}{dT_1} &= \frac{T_2}{T_1^2}\left[AG - Ah(T_1 - T_a)\right] - \frac{T_1 - T_2}{T_1} Ah \\ \frac{d\mathcal{P}}{dT_1} &= 0 \\ \Rightarrow \quad T_{1,max\ P} &= \sqrt{\frac{T_2}{h} G + T_2 T_a} \end{aligned}$$

The numerical value is 402 K. The expressions for the minimum of entropy production and the maximum power of the engine will only be the same if the "hot water" produced has the same temperature as the environment.

c) The simple rule that minimal production of entropy should deliver maximum output from engines holds only with a single reservoir to which entropy is rejected. In this example, the entropy produced by the absorption of sunlight ends up in two environments which have different temperatures. The loss cannot be calculated by choosing one of the temperatures and multiplying it with the rate of production of entropy. The entropy current rejected to the hot water still represents some available power. We can interpret the power of the engine as being 100% "pure" availability. Therefore, if we add it to the available power of the current of entropy at T_2 we have a quantity whose maximum can be shown to coincide with the minimum of entropy production:

$$\mathcal{P}_{av,tot} = \mathcal{P} + \mathcal{P}_{av}(I_{s2})$$
$$= \mathcal{P} + (T_2 - T_a)I_{s2}$$

If we consider the entire sequence of processes, we should attempt to run at minimum entropy production rather than maximum power.

1.7.6 A Model of an Endoreversible Engine

Only a few years ago, the following model of heat engines was proposed as a more realistic alternative to the ubiquitous ideal Carnot engine.[22] The major drawback of the Carnot engine has to do not so much with the processes undergone by the working fluid but with the rate of transfer of entropy from the furnace to the engine, and from the engine to the cooler. For these rates to be finite there must be finite temperature differences if we think of transferring entropy conductively. No realistic heat exchanger will work without such a drop of temperature. We may then run an ideal Carnot engine between temperatures which are somewhat lower than that of the furnace and somewhat higher than that of the cooler, respectively (Figure 39). Entropy conducted from the furnace and to the cooler causes dissipation in the combined system of heat exchangers plus Carnot engine. We must face the reality that we either have a completely ideal engine at zero power, or a nonideal device at finite power. What we can strive for realistically is an engine which operates with a *minimum rate of production of entropy*. There should be a minimum of entropy generation in the model engine proposed in Figure 39. If we do not allow for any temperature gaps between the heat exchangers and the Carnot engine, the energy supplied from heating simply leaks directly to the environment (cooler) and we have total dissipation. The same situation arises if the upper and the lower operating temperatures of the ideal Carnot engine are made equal. Somewhere between these two extremes must be the optimum condition for the smallest possible rate of production of entropy and the largest mechanical power. Let us look for it.

The model engine works as follows (Figure 39). It is heated from the furnace at a constant rate. In general, the engine can take in part of the current, the fraction being determined by the rate at which entropy can flow through the heat exchanger (represented by the thermal resistor *R1*) to the ideal Carnot engine. The rest of the entropy flux will leak directly to the environment via the thermal resistor *R3*. The core of the engine, i.e. the ideal Carnot heat engine, will reject the entropy at the lower operating temperature, thereby releasing energy in the me-

22. The type of heat engine called *endoreversible* was first proposed and analyzed by Curzon and Ahlborn (1975). They calculated the power of the engine and determined the condition for its maximum. The efficiency at maximum power, Equation (63), has since been called the Curzon-Ahlborn efficiency. A simpler derivation was given by DeVos (1985). The problem was later investigated from the point of view of the minimization of the production of entropy (Salamon et al., 1980; Andresen et al., 1984).

chanical process. The heat exchanger serving the cooler (represented by the thermal resistor *R2*) must be designed to handle this current of entropy. Naturally, once we have found the condition for maximum mechanical power and the flux of entropy associated with it, we will adjust the current of entropy from the furnace to this value to avoid unnecessary leakage through *R3*. Because of the reversible Carnot engine at the core of the system the entire model has been called an *endoreversible engine*.

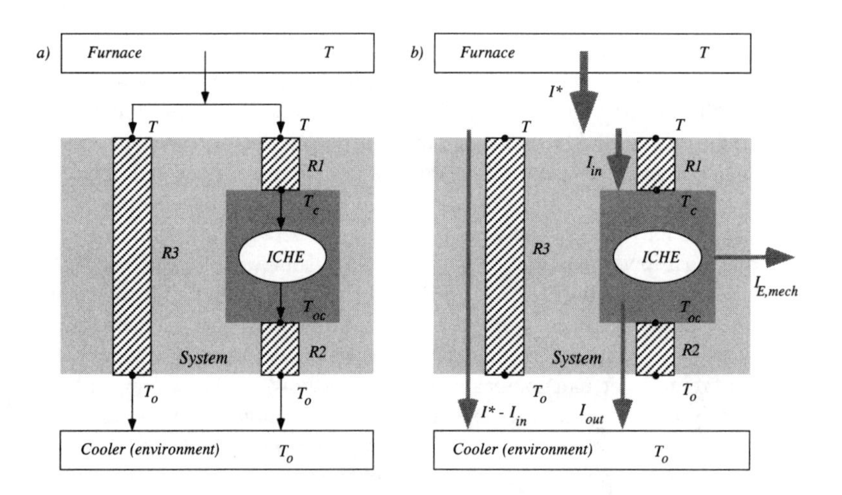

FIGURE 39. (a) The flow of entropy (solid arrows) in a system which contains an ideal Carnot heat engine (*ICHE*) and conductive resistances. The system is heated at a constant rate from the furnace. Due to the resistances of the heat exchangers to the *ICHE*, the flow of entropy in this branch is limited. Entropy is produced in the resistors. (b) Energy flow diagram. The energy currents associated with the currents of entropy are shown as heavy arrows. It is found that minimizing the rate of production of entropy in the composite system will maximize the mechanical power.

We need to be able to calculate the effect of the heat exchangers included with the model engine. It is important for us to know that the form of the simplest constitutive law for heat transfer through an exchanger, Equation (50), is quite applicable to our case. Since the maximum efficiency of our model engine will be found not to depend on the coefficient multiplying the difference of temperatures, we do not have to worry about the actual transport processes.

Now we can turn to the job of first calculating and then minimizing the rate of production of entropy in the system of Figure 39. The only dissipative elements in the engine are the three thermal conductors, i.e. the heat exchangers and the machine itself, which leaks entropy. The rate of production of entropy is equal to the sum of the rates due to the three resistors:

$$\Pi_S = \Pi_S(R1) + \Pi_S(R2) + \Pi_S(R3) \tag{54}$$

As before, we calculate the production rates for steady-state conditions in terms of the fluxes of entropy entering and leaving the bodies. In the following equations, all fluxes represent absolute values, and the index *E* has been dropped from the thermal energy currents for convenience. The overall rate of generation of entropy can be expressed as follows:

$$\Pi_s = I_{in}\left(\frac{1}{T_c} - \frac{1}{T}\right) + I_{out}\left(\frac{1}{T_o} - \frac{1}{T_{oc}}\right) + (I^* - I_{in})\left(\frac{1}{T_o} - \frac{1}{T}\right) \tag{55}$$

Four variables appear in this equation, namely the energy currents entering and leaving the Carnot engine, and the upper and the lower operating temperatures of the ideal engine. The total thermal energy current I^* and the temperatures of the furnace and the cooler are assumed to be fixed. We would like to eliminate three of the variables and leave only the thermal energy current entering the Carnot engine in the equation. The three elements of interest in our model system, namely the Carnot engine and the two heat exchangers, furnish the necessary conditions. Since the engine operates without dissipation, the current of entropy entering is equal to the current which is emitted. This leads to

$$\frac{1}{T_c} I_{in} = \frac{1}{T_{oc}} I_{out} \tag{56}$$

The heat exchangers operate according to the simple constitutive law, Equation (45), which means that

$$I_{in} = a(T - T_c) \tag{57}$$

and

$$I_{out} = b(T_{oc} - T_o) \tag{58}$$

The coefficients a and b shall be taken to be constants. After eliminating these three variables, the production rate of entropy is expressed in the single variable I_{in}:

$$\Pi_s = I^*\left(\frac{1}{T_o} - \frac{1}{T}\right) - \frac{1}{T_o} I_{in}\left[1 - \frac{bT_o}{bT - (1 + b/a)I_{in}}\right] \tag{59}$$

You will notice that the fixed heating power I^* does not contribute to the condition for the minimal generation rate we are going to calculate. Now we are ready to compute the derivative of this expression with respect to the energy flux entering the Carnot engine:

$$\begin{aligned}\frac{d\Pi_s}{dI_{in}} = &-\frac{1}{T_o}\left[1 - \frac{bT_o}{bT - (1 + b/a)I_{in}}\right] \\ &+ \frac{1}{T_o} I_{in} \frac{bT_o}{\left[bT - (1 + b/a)I_{in}\right]^2}\left(1 + \frac{b}{a}\right)\end{aligned} \tag{60}$$

Setting the derivative equal to zero will deliver the condition for the energy flux at minimum rate of production of entropy. After some lengthy algebra we get a quadratic equation for I_{in} whose solution is

$$I_{in}(max) = \frac{ab}{a+b}\left[T - \sqrt{T\,T_o}\right] \tag{61}$$

This should represent the thermal power at the upper operating temperature of the Carnot engine for *maximum mechanical power.* It depends on the temperatures of the furnace and the cooler, and on the physical properties of the heat exchangers which are expressed by the factors a and b. The mechanical power itself is calculated according to the rules for ideal Carnot engines (Section 1.4) which yields

$$I_{mech}(max) = I_{in,max}\left[1 - \frac{abT_o}{abT - (a+b)I_{in}(max)}\right] \tag{62}$$

More significantly, the efficiency of the model engine at maximum power is independent of the dimensions of the heat exchangers. Just as for the ideal Carnot engine it depends only on the temperatures of furnace and cooler:

$$\eta_{1,max} = 1 - \sqrt{\frac{T_o}{T}} \tag{63}$$

This is the result reported by Curzon and Ahlborn.[23] The original derivation was done in terms of the power of the engine whose maximum was to be found. Today we know that engines which are designed for maximum power obviously follow a basic principle, namely that of the minimization of the production of entropy.

It is instructive to compare the actual efficiencies of a number of thermal power plants (Figure 40). It appears that they rather closely approach the efficiency of the model engine in Figure 39. The determining factor in such devices seems to be the transfer of entropy to and from the actual engine, while the behavior of the fluid driving the engine can be regarded as almost ideal.

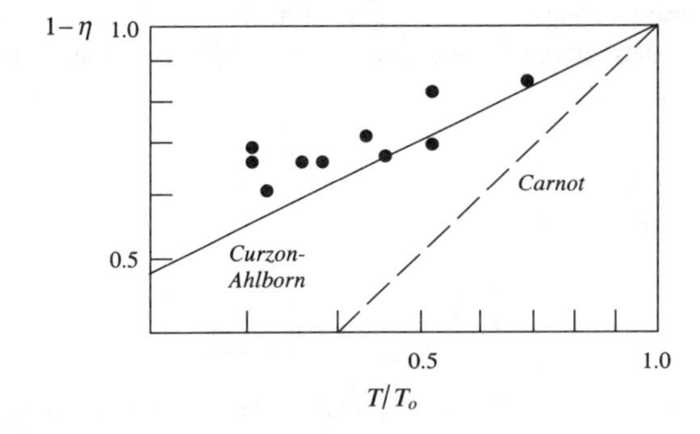

FIGURE 40. Observed efficiencies of thermal power plants are plotted as a function of the ratio of the temperatures of the furnace and the cooler. The broken line on the right represents the efficiency of ideal Carnot engines. The theoretical efficiency according to Equation (63) is drawn as the straight line to the left. The figure has been drawn after Bejan (A. Bejan, 1988, p. 409).

23. Curzon and Ahlborn (1975).

This is an example of what has become the focus of interest in so-called *finite-time thermodynamics* in recent years. Finite-time thermodynamics[24] develops simple aggregate models of dissipative processes. As such, it may be called a part of the modern approaches which deal with processes rather than just states.

EXAMPLE 36. Designing a thermal power plant at maximum power.

The Dungeness nuclear reactor built in Great Britain in 1965 performs very closely to the rule given for thermal plants at maximum power, Equation (63). The temperatures of its furnace and cooler are 663 K and 298 K, respectively. Assume a plant of this type to be designed for 500 MW mechanical power. a) How large must the heating power from the reactor be for the plant to operate at maximum power? b) Assume the heat transfer coefficients of both heat exchangers to be equal (i.e. their physical characteristics are taken to be the same). Furthermore, take the exchangers to be equally large. How large is the effective surface area of each of the heat exchangers if the transfer coefficient has a magnitude of 1000 W/(K · m^2)? c) Determine the upper and the lower operating temperatures of the Carnot engine for the heat exchangers designed according to (b). d) With fixed and equal heat transfer coefficient for both heat exchangers, is there a better way of distributing the available surface area than to make both exchangers equally large?

SOLUTION: a) The efficiency at maximum power of the endoreversible engine is given by Equation (62). Here it turns out to be

$$\eta_{1,max} = 1 - \sqrt{\frac{T_o}{T}}$$
$$= 1 - \sqrt{\frac{298}{663}} = 0.33$$

This figure determines the magnitude of the heating power necessary for optimal operation. If we can avoid direct leakage of entropy, the requirement for the energy flux from the reactor simply is equal to

$$I_{in}(max) = \frac{I_{mech}(max)}{\eta_{1,max}}$$
$$= \frac{500\,\text{MW}}{0.33} = 1.52 \cdot 10^9\,\text{W}$$

b) According to Equation (60), the dimensions of the heat exchangers and the temperatures of furnace and cooler determine the heating power. Alternatively, $ab/(a+b)$ in Equation (60) is determined by the data given so far. With $a = b$ we obtain

---

24. See for example J.M. Gordon (1990), and references therein.

$$a = 2\frac{I_{in}(max)}{T - \sqrt{T \cdot T_o}}$$

$$= 2\frac{1.52 \cdot 10^9}{663 - \sqrt{663 \cdot 298}}\frac{\text{W}}{\text{K}} = 1.39 \cdot 10^7 \frac{\text{W}}{\text{K}}$$

If the heat transfer coefficient has the magnitude stated above, the surface area of each of the heat exchangers should be $1.39 \cdot 10^4 \text{ m}^2$.

c) The operating temperatures of the Carnot engine can be determined from the relations for the rate of transfer of entropy (or energy) according to Equations (57) and (58). The formal solutions are:

$$T_c = T - \frac{1}{a}I_{in}(max)$$

and

$$T_{oc} = \frac{bT - \frac{b}{a}I_{in}}{bT - \left(1 + \frac{b}{a}\right)I_{in}}T_o$$

which lead to numerical values of 554 K and 371 K, respectively.

d) It appears to be reasonable to measure the cost of the heat exchangers in terms of their required surface area. Therefore we are looking for the minimum total surface area which delivers the desired (fixed) output of the engine. In other words, we are looking for the minimum of $x = a + b$ subject to fixed I_{in}. The surface areas are determined by Equation (62):

$$\frac{x}{a(x-a)} = \frac{T - \sqrt{T \cdot T_o}}{I_{in}(max)} \equiv c$$

The minimum of x is found for $a = 2/c$ which yields $x = 4/c$ and $b = a$. In other words, if we have to limit the total surface area of the heat exchangers, we should build them of equal surface area.

Questions and Problems

1. In what sense is hotness the intensity of heat? Why do we have to distinguish it from quantities of heat? To what other quantities in physics may the intensity of heat be compared?
2. Consider a moving body that splits into two halves which continue moving along together. Which mechanical quantity is divided among the bodies? Which other mechanical variable is not divided up, leaving each of the parts with its initial value? Compare electrical and thermal phenomena to this mechanical example. Which electrical or thermal quantities correspond to the mechanical variables?

3. List everyday phenomena which are responsible for our intuitive notion of heat content of bodies. Can you turn the qualitative idea into a physical quantity having a precise meaning?

4. Why shouldn't we think of energy as a mechanical, electrical or thermal quantity? Why would it be particularly wrong to identify stored energy as mechanical, electrical, thermal or other? What consequence does this have for identifying "heat" as stored energy?

5. What happens to all bodies under all circumstances if their energy is increased? Which physical quantity changes if this happens? What kind of conclusions *cannot* be drawn from the statement that the energy of a body has changed?

6. With the help of physical quantities, explain the *difference* between making a body rotate and making it warmer.

7. Which interpretation of "heat" comes close to Carnot's image that heat can do work? How did Carnot compare heat to other quantities?

8. Compare different substancelike physical quantities such as momentum, charge, amount of substance, and entropy. Which two properties do they all have in common? What are possible differences between the quantities listed?

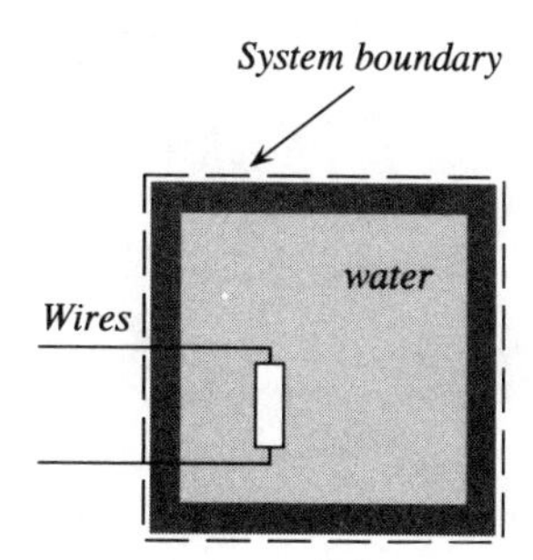

FIGURE 41. Problem 9.

9. Consider a box containing water and an immersion heater (Figure 41). Except for the electrical wires leading to the heater, the system is totally insulated from the environment. Let the heater make the water hotter. If you take the entire setup as the system, why would it be wrong to say that the system has been heated? (Remember the definition of heating, or cooling, in Section 1.1.3.) What does this mean with respect to the question of where the heat has come from which has made the water warmer? What is heat in this case?

10. Rephrase the following expressions in terms of entropy. In which cases would reference to energy be clearly wrong? Do any of the terms and expressions have nothing to do with entropy?
a) heat engine, heat pump; b) heat exchanger; c) heating and cooling; d) heat flow, transfer of heat; e) convective heat flow; f) heat source; g) storage of heat, heat reservoir; h) phenomena in which heat causes motion; i) solar heater; j) production of heat; k) heat transfer coefficient; l) pumping heat from the cold enclosure; k) heating power.

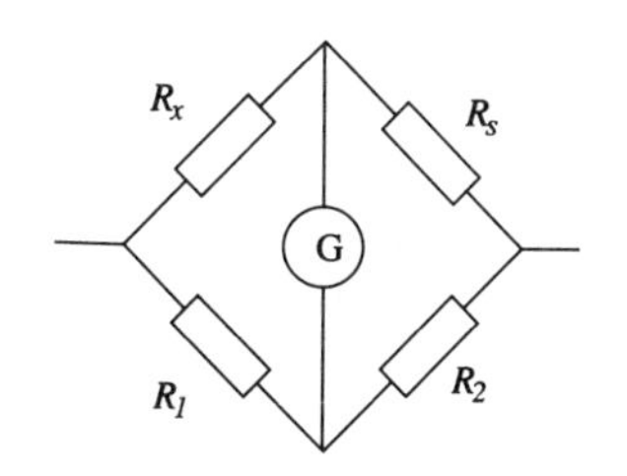

FIGURE 42. Problem 11.

11. An electrical resistor element R_x is used as a thermometer. It is placed as one of the four resistors in a circuit (Figure 42). There is no current through the galvanometer (G) if the thermometer is put in water with a temperature of 20°C and the resistor R_s is given a value of 10.0 kΩ. Now we place the thermometer into a different fluid. Again, electricity does not flow through the galvanometer if we increase R_s by 1000 Ω. What is the temperature of the second fluid? The coefficient of temperature of the platinum resistor is $2.0 \cdot 10^{-3}$ K^{-1}.

12. Water in a cylinder fitted with a piston is evaporated completely by heating at constant pressure. (The piston is moved to compensate for the increase of volume.) In a second step the steam is heated further at constant volume. Finally, an adiabatic expansion decreases the temperature below the initial value at which vaporization took place (without condensation).

a) Sketch the steps in a *T-S* diagram. b) Prepare a rough sketch of the operations in a *T-V* diagram. c) What kind of process can bring the entropy content of the fluid back to its initial value? d) Can you think of a single operation from the list of processes discussed in Section 1.3 which would take the fluid back to its original state?

13. Do any of the simple processes discussed in Section 1.3 lead from the lower right to the upper left corner of the *T-S* diagram? Could any two achieve the goal of changing a fluid's state in such a manner? Will your answer be different for air and for water in a range of temperatures including the latter's anomaly?

14. Phenomena in which heat causes motion, i.e. thermomechanical processes, are probably the most conspicuous ones involving the action of heat. Do you know of electrical, chemical, or other phenomena which are caused by thermal processes?

15. How large must the entropy current through an ideal Carnot engine be if its power is 10 kW, and the upper and the lower operating temperatures are 500°C and 100°C, respectively? How does your answer change if you double the power? If you double the temperature difference?

16. A large power plant of 1.0 GW electrical power emits a thermal energy current to the environment twice as large as the useful one. At what rate does the environment, which has a temperature of 25°C, receive entropy?

17. An immersion heater has a temperature of 120°C as it emits an energy current equal to 0.80 kW. a) How large is the current of entropy flowing across the surface of the heater? b) If the temperature of the water receiving the heat is equal to 80°C, how much entropy flows into the water?

18. Sunlight falling on a window is partly reflected (10%) and mostly transmitted (87%). The rest is absorbed evenly throughout the glass. The energy current associated with the radiation is equal to 900 W. If the temperature of the glass is equal to 30°C, how large is the rate at which entropy is received by this body?

19. In cooling, a body of uniform temperature emits currents of entropy and energy equal to 100 W/K and $6.0 \cdot 10^4$ W, respectively. What is the temperature of the body?

20. As a body is heated, its current of entropy increases linearly from 20 W/K to 40 W/K in 100 s, and its temperature goes from 100°C to 70°C (also linearly). a) Calculate the thermal energy current received by the body as a function of time. b) How much entropy and energy have been absorbed by the body during the process? c) Is this scenario physically possible? Can a body get colder as it receives heat without losing any?

21. A mixture of water and ice is heated in such a way that ice melts at a rate of 0.020 kg/min. a) How large is the current of entropy absorbed by the mixture? b) How much energy does the mixture receive after 10 minutes of heating?

22. It is common to tabulate and use amounts of energy transferred in heating for the melting or the vaporization of bodies. These quantities, referred to unit mass, are called the *specific enthalpy of fusion* (q) and the *specific enthalpy of vaporization* (r), respectively. What is the relationship between q and r and the molar latent entropies of fusion and of vaporization? (See Tables A.4 and A.5.)

23. A Carnot heat engine employing air in a closed cylinder fitted with a piston does not exchange energy at a constant rate during its four-stage cycle. In what sense, then, should we understand Equation (11)? Can you write the equivalent statement integrated over one full cycle?

24. Two or more ideal Carnot heat engines operate in sequence, which means that the entropy rejected by one engine is used by the following one. Each of the engines runs in a distinct interval of temperatures between T_{max} and T_{min}, with the intervals seamlessly covering the entire range of temperatures. a) Show that the power of the sequence of engines is equal to the power of a single engine running between T_{max} and T_{min}. b) Allow for entropy to be added or withdrawn at each inlet to an engine. Show that in this case the power of the sequence of devices should be equal to

$$\mathcal{P} = \sum_{i=1}^{N} I_s(T_i)\Delta T_i$$

$$\Delta T_i = T_i - T_{i+1} \quad , \quad T_1 = T_{\max} \quad , \quad T_{N+1} = T_{\min}$$

c) If the entropy current is a continuous function of temperature (between the maximum and the minimum values), show that the power should be calculated according to

$$\mathcal{P} = \int_{T_{\max}}^{T_{\min}} I_s(T)dT$$

25. How can we infer an absolute temperature scale by measuring the efficiency of an ideal heat engine?

26. Would you treat solar radiation as a high or a low temperature heat source? Discuss the implications of your decision.

27. The *COP* of a refrigerator is defined as the ratio of the thermal energy current extracted from the cold body and the power needed to drive the engine.
a) Derive the formula for the *COP*, and calculate the value for an ideal refrigerator operating between temperatures of - 20°C and 25°C. b) Explain the difference in the viewpoints taken for heat pumps and refrigerators.

28. If the entropy flowing conductively through the metal bar in Example 21 had been allowed to drive an ideal Carnot engine, how large would its mechanical power be? Since no energy has been released for mechanical purposes in conduction, this power is lost. Prove that the loss of power is equal to the product of the rate of production of entropy in the bar and the temperature of the colder end.

29. A mixture of ice and water is placed in a freezer having a constant interior temperature of - 18°C. If the refrigerator works as an ideal Carnot engine, what is the power needed for its operation? Ice is to be formed at a rate of 10 g per minute. The temperature of the environment is taken to be 22°C. What is the difference between this problem and Example 17?

30. Because of imperfect insulation, a thermal energy current of 10 W passes through the walls of the freezer in the previous problem. Answer the questions of Problem 29 for the new situation.

31. In a single stroke of a bicycle pump, air is compressed very rapidly. Compare reversible and dissipative compression. In a dissipative process, would the volume be larger, equal, or smaller than the one found in reversible operation for equal final temperatures? At equal final volume, would the temperature be larger, equal, or smaller in the dissipative case than in the reversible one?

32. Consider the law of balance of entropy and the quantities introduced in this context. a) Can you give a formal definition of entropy absorbed and entropy emitted in terms of integrals of the flux of entropy? Use the definitions to show that the entropy exchanged equals the entropy absorbed minus the entropy emitted. b) Write down the formal definition of an amount of entropy generated; see Equation (33).

33. Over a time span of 100 s, the entropy of a body increases linearly from 300 J / K to 500 J / K. At the same time the rate of generation of entropy decreases from 5 W / K to zero.
a) Compute the net flux of entropy as a function of time. b) How much entropy is exchanged, and absorbed, and emitted?

34. A mixture of ice and water is heated as in Problem 21. a) Calculate the rate of change of entropy of the mixture. c) How large is the rate of change of energy of the volume? Why is it not precisely equal to the energy current in heating?

35. Consider two solid bodies at equal temperatures in thermal contact and completely isolated from the surroundings. What would have to happen if one of the bodies were to get hotter at the expense of the other? Is it possible at all to make one body hotter while the other is cooled?

36. Consider two bodies in thermal contact. Their temperatures are different, which causes entropy to flow from the hotter to the cooler system. Is it possible to set up a model of the process for which the temperatures of each of the two bodies are uniform? Where is entropy produced?

37. Consider water being heated by an immersion heater. a) If you consider the body of water as a system, what is its equation of balance of entropy? (Assume the distribution of entropy through the system to take place reversibly; what does this mean for the conduction of entropy through the system?) b) Answer the question for the case in which you take the system to be made up of water plus heating coil.

38. Usually, the energy added in heating is called “heat.” How much “heat” is added in stirring water with a paddle wheel? What happens to the temperature of the water?

39. If the isothermal expansion or compression of air is dissipative, is the energy exchanged in heating still equal to the area under the isotherm in the *T-S* diagram? Is there a difference in your answers for compression and expansion?

40. Allow for dissipation in the engines of the heating scheme B of Example 18. Assume the temperature of the water to be distinct from the temperatures of both the furnace and the environment. The heat engine and the heat pump are actually running. Is it possible for dissipation to be so large that the ratio of heating power to solar power becomes equal to 1, or even less than 1?

41. Derive the rate of production of entropy for heat transfer which obeys a constitutive law of the form $I_E = a\,(T_1 - T_2)$. a) Write the result in terms of the current of energy. b) Write the formula in terms of the difference of temperatures. c) Two practical problems having to do with heat transfer are thermal insulation and augmentation of transfer. In the former case one wishes to reduce the transfer rate for a given difference of temperatures. In augmentation the rate of transfer is usually prescribed by the problem, and we want to reduce the temperature difference across the heat exchanger. Show that the seemingly contradictory applications are both an exercise in the minimization of the rate of production of entropy.

42. A heat pump is used to heat water at 60°C. Heat is taken from the ground at 2°C. The observed coefficient of performance is 2.2 while the heating power has a magnitude

of 1.0 kW. a) How large is the rate of production of entropy? b) How large is the loss of available power? Show that it is equal to the product of the rate of generation of entropy and the temperature of the environment. c) How large is the second law efficiency of the heat pump?

43. Water is heated by a heat pump driven by the energy from a thermal power plant. How much entropy may be produced by the dissipative engines if this mode of heating is to be competitive with direct heating of water over a burner?

44. Derive the formulas for the coefficient of performance and the second law efficiency of a dissipative refrigerator operating between a cold space at temperature T and the environment at T_a in terms of the rate of production of entropy.

45. One kilogram of ice is formed from water at 0°C in a freezer. The fluid of the engine operates between - 20°C and 30°C, with the temperature of the environment being held at 20°C. Assume the engine to work reversibly. a) Draw a diagram showing the hotness levels involved and the flow of entropy. Identify sources of entropy production. b) How much entropy is generated? c) How much energy is used for driving the engine in excess of what would be necessary if the engine could operate directly between 0°C and 20°C? d) Verify numerically that the excess work (lost available work) is given by the product of the entropy created and the temperature of the environment. e) Verify formally the Gouy-Stodola rule for lost work.

46. A refrigerator is designed to pump heat at a specified rate from the cold enclosure. The heat exchanger at the cold end is given; its properties are fixed. If the fluid operating in the Carnot refrigerator works reversibly, which other factors will determine the performance of the engine?

47. Compare two methods of heating 50°C water. (1) In the first, water at a temperature of 50°C is heated further by a solar collector. (2) The radiation of the sun is used to drive an ideal Carnot heat engine between temperatures of 300°C and 15°C. The energy released by this ideal power plant is used to drive an ideal heat pump which gets its entropy at 15°C to heat the 50°C water. In both cases, 50% of solar radiation is utilized. a) How large is the ratio of the efficiencies of the methods? b) Determine the ratio of the rates of production of entropy for the methods. Assume solar radiation not to supply any entropy.

48. Consider the following model of a real heat pump. Entropy flows from a lake which has a temperature of 4°C through a heat exchanger into the cold fluid of the heat pump at - 3°C. It is pumped up to a temperature of 35°C by an ideal Carnot heat pump from which it flows through a second heat exchanger into a room; there the temperature measures 22°C. The thermal energy current associated with the heating of the room is equal to 3.5 kW.
a) How much power is necessary to drive the heat pump? b) How much entropy is produced every second in the system (heat pump plus heat exchangers)?

49. A liter of water is to be frozen in a freezer. Consider only the process of freezing.
a) How much entropy is emitted by the water during this process? b) What is the minimal amount of energy which you have to supply to the freezer if the entropy is to be rejected to the environment at 22°C?

50. The heat pump of a house is driven by the electrical energy from a thermal power plant. Take the thermal efficiency of the power plant to be equal to 32%. On the way from the power station to the house, 20% of the energy will be lost. a) Draw flow diagrams for carriers, energy, and power with the elements consisting of power plant, transmission, and heat pump. b) What should the minimal value of the coeffi-

cient of performance of the heat pump be for the balance of energy to be positive compared to an oil burner with an efficiency of 80%?

51. An endoreversible engine as in Section 1.7 is to be designed. It consists of the reversible Carnot engine and two heat exchangers serving the furnace and the cooler.
a) Prove that its power at maximum output can be written as

$$I_{E,mech} = \frac{(hA)_f}{1+(hA)_f/(hA)_c} T\left[1-\sqrt{\frac{T_o}{T}}\right]^2$$

where f and c refer to the furnace and the cooler, respectively.
b) If the power is maximized once more by dimensioning relative sizes of the heat exchangers optimally, we get

$$I_{E,mech} = \frac{1}{4} hA \cdot T\left[1-\sqrt{\frac{T_o}{T}}\right]^2$$

where $hA = (hA)_f + (hA)_c$ is the total transfer coefficient multiplying the difference of temperatures.
c) Show that the optimized power of such an engine increases proportionally to $(T - T_o)2/T$ for differences of temperatures which are not too large. This condition is quite applicable to today's range of temperatures. What does this mean for the designer of a power plant?

52. Show that if the heat transfer coefficients of the heat exchangers of an endoreversible power plant are not equal, one should distribute the quantity hA (transfer coefficient times surface area) rather than A equally among the exchangers for the furnace and the cooler.

53. The furnace of a large thermal power plant was designed to deliver energy at a rate of up to 2.0 GW at a temperature of 920 K. Cooling is done at an environmental temperature of 300 K. Model the engine as endoreversible. a) How large is the current of entropy entering the system? b) What is the optimal mechanical power if heat leakage is responsible for a loss of 5% of the heating power? c) What are the magnitudes of the rate of production of entropy and of the loss of available power? Are they related by the Gouy-Stodola rule?

54. Prove that the condition of minimum rate of production of entropy of an endoreversible engine coincides with the condition of maximum power. a) Conduct the proof by directly computing the power of the engine and determining its maximum. b) Can you give a simple argument involving entropy generated and the loss of power to prove the point?

55. Derive a general expression for the second law efficiency of an endoreversible engine. Show that it is given by

$$\eta_2 = \frac{1}{1+\sqrt{T_o/T}}$$

56. According to the model of the endoreversible engine with heat exchangers, an ideal Carnot engine running directly between the temperatures of the furnace and the cooler would have zero power. Is there a contradiction between the conclusions drawn from this model and our earlier treatment of the ideal Carnot engine which

assumed finite power, as in Equation (10)? [*Hint:* Remember the discussion concerning ideal walls and where to place the location of the production of entropy.]

57. Model a refrigerator as an endoreversible engine. Its purpose is to pump heat at a prescribed rate out of the cold enclosure. The heat exchangers at the colder and at the warmer end have been dimensioned so as to make the temperature differences across them roughly equal. You now can add a piece of heat exchanger to only one of the existing exchangers. Which one do you choose? Assume the temperature differences across the heat exchangers to be small compared to the temperatures themselves and to the difference of the temperatures of the cold enclosure and the environment. The added piece of heat exchanger is small compared to the existing ones.

58. Consider a solar furnace with heat engine and heat pump as in Example 18. Water is to be heated at constant temperature T_w. The engines are supposed to be ideal Carnot engines. Heat (entropy) is supplied to and rejected from engines and reservoirs at constant temperatures. Assume the furnace to be an ideal absorber of solar radiation. The losses from the furnace to the environment are taken to be proportional to the difference of temperatures, Equation (50), with constant heat transfer coefficient h. a) Derive the expression for the rate of production of entropy for the entire system. b) How large should the temperature of the furnace be to minimize the production rate of entropy? c) For which value of the temperature of the furnace will the power of the heat engine be a maximum? Why is this value different from the one computed in (b)? Should you try to minimize entropy production or to maximize power output of the heat engine? d) Show that the maximum of the total heating power with which the body of water is being heated occurs at the same temperature of the furnace as that calculated for minimal entropy production.

59. Consider the problem of maximizing the available power of the solar thermal system of Example 35. Show that the expression for the total available power attains its maximum if the least amount of entropy is generated.

60. A low temperature heat engine employing a traditional coolant such as R 134 is designed for use with normal solar collectors. Estimate the efficiency you might expect from such an engine if heat is collected at 90°C and rejected at 30°C. (A detailed calculation carried out for a particular design gives a value of 9% to 10% for the thermal efficiency. T. Koch, Diploma thesis 1993, Technikum Winterthur.)

61. Compare the following traditional statements of the second law with the one given in Section 1.6. Remember that "heat" in these statements is the energy exchanged in heating (or cooling). a) By itself, heat can flow only from hotter to cooler bodies; b) To transfer heat from a place of low temperature to a place of higher temperature, external work must be performed; c) There are no cyclically working engines which continuously withdraw heat from a single heat reservoir, and then transform it into work.

62. Compare the statements "By itself heat only flows from hotter to colder places" and "By itself water only flows downhill." Why would one statement be considered something truly noteworthy (the second law of thermodynamics) while the other is taken to be almost trivial? In your opinion do the statements suggest a comparison between heat and water, or between heat and energy?

63. One part of an isolated isothermal body does not spontaneously get hotter at the expense of another part. In two connected water tanks filled to the same level, water does not fill one container spontaneously at the expense of the other. The first phenomenon often is attributed to irreversibility (production of entropy). The second

case is explained by noting that you would need energy to pump water from one container into the other (lowering water in the first tank does not provide for enough energy for pumping). Do you agree with the interpretations? Shouldn't the explanations be analogous? If entropy could not be produced would this make possible the heating of a part of an isolated isothermal body at the expense of another?

64. Take a fluid in a container which is perfectly insulated to the flow of heat. We still can perform some operations on the fluid such as compression and expansion, letting an electrical current flow through a wire inside the container, or stirring with the help of a paddle wheel. What kind of experience do we have with the behavior of such a system? Is it true that in general it is impossible to reverse a change once it has occurred? Are there examples of reversible behavior? Does our experience call for a physical quantity which can only increase or stay the same, but never decrease?

65. Look at the system described in the previous problem (Problem 64). Assume that entropy has the properties we have ascribed to it. Do the types of behavior of the system as we know them from experience follow from these properties?

CHAPTER 2

The Response of Uniform Bodies to Heating

...the old 'impressive, clear, and wrong' statements regarding latent heat, evolution and absorption of heat by compression, specific heats of bodies and quantities of heat possessed by them, are summarily discarded. But they have not yet been generally enough followed by equally clear and concise statements of what we now know to be the truth.

W. Thomson (Lord Kelvin), 1878

This chapter will present theories of the thermodynamics of some simple materials. While we already possess the generic laws necessary for dynamics (Chapter 1), we cannot really claim to know how certain types of bodies will behave when subjected to heating. Only specific constitutive laws allow for the actual creation of theories of dynamics. The following sections are largely devoted to the thermodynamics of the ideal gas, a simple model fluid which can undergo changes of temperature and of volume. We will be able to compute adiabatic, polytropic, and more general processes undergone by this gas. The simple model chosen allows only for reversible processes, which means that we can actually set up a thermodynamics of nondissipative processes. The ideas will be applied to a couple of other ideal and simple systems, namely magnetized bodies, and blackbody radiation. Finally we shall contrast the results with a brief look at the thermostatics of simple systems.

A more advanced treatment of the subject which rests upon Carnot's Axiom concerning heat engines will be given in the Interlude.

2.1 The Model of Uniform Processes

Without mentioning it explicitly, we made liberal use in Chapter 1 of a model of physical systems which, upon closer inspection, turns out to be anything but trivial. There We spoke of the single temperature of a body as if it were normal for a physical system to have the same temperature throughout. Thermal processes cast a glaring light upon the problem of uniform situations. Normally, when heat flows, temperatures change from point to point, which makes it necessary to set up a continuum theory of nature. So, where does this leave our desire to learn about thermal processes in the simplest settings?

2.1.1 A Continuously Variable World or Eternal Rest?

Objects and systems do not change solely with time. Their properties also vary from point to point in space. The point masses of mechanics certainly are not an example of how things are in nature. The electrical capacitor which we describe in terms of a single value for its capacitance or its electrical field does not exist. While bodies move they also may deform, which can make them nonuniform. When air rushes into vacuum such as in free expansion (Figure 28 of Chapter 1), we are confronted with a situation which makes it impossible to speak of *the* air pressure.

Thermal phenomena present us with more examples. Experience with the world around us demonstrates most clearly that uniform situations do not exist. The temperature never is the same at every point in a body. The Earth's atmosphere is far from a uniform state, and so are our homes and our bodies. When we heat a stone in the Sun or air in a cylinder, heat will gradually spread through the system leaving parts closer to the heat source hotter than those further from it. Therefore the description used in Section 1.3, where a single value was associated with the hotness of air undergoing, variously, heating and compression, seems to be utterly unrealistic.

We might think that it should be possible to select parts of bodies small enough for spatial uniformity to prevail to a significant degree. We could attempt to base our description of nature on such systems, from which we would build the world at large. However, this turns out to be impossible: changes of temperature from location to location are required if thermal processes are to take place at all. Heat does not flow without a temperature gradient, not even in the tiniest part of a body. Inevitably, this leads to the production of heat. Thermal processes are dissipative as a matter of fact, leaving us between a rock and a hard place.

You may object to this stark analysis and insist that situations exist in nature in which physical systems can be described as spatially homogeneous. Again, heat tells the story. If we leave two bodies at different temperatures in contact for a long enough time, their hotnesses will eventually be the same. If we insulate the bodies from the environment we can even maintain this condition over a long period of time without significant change. The air which undergoes free expansion will settle down eventually, making the pressure and temperature uniform throughout. Here, you will say, are cases which we should be able to investigate successfully if we are looking for simple situations.

There only is one problem. The examples provided have nothing to do with *dynamics*. They are cases of eternal rest or, put more prosaically, of *equilibrium*. It seems we must choose between a dynamical world which is too difficult for us to describe, and a simpler, but less interesting, static one.

2.1.2 Uniform Heating in Thermal Superconductors

There must be a way out of this dilemma. After all, we construct theories of mechanical and electrical systems which we describe in simple ways using the notion of spatial uniformity. We calculate the behavior of electrical circuits by assuming them to be composed of discrete elements each of which can be modeled using a few physical variables assumed to have the same values at every point. We model the motion of bodies in the simplest terms, forgetting about spatial inhomogeneity. Ideal pendulums, for example, are points which swing at the end of massless strings through frictionless space. We are quite happy with such simple theories, and we do not let ourselves become unduly worried about the complexities of the real world. After all, the ideal models have an important story to tell despite their simplicity.

Well, then, let us look for and construct a model of spatially uniform bodies which can undergo thermal processes. How could we conceive of *bodies which remain uniform while they are being heated or cooled*? Obviously we require the spatial variation of temperature in a body to vanish while heat is allowed to flow through it. Carnot imagined bodies which let heat pass easily. The situation he described in such simple words is no stranger to us in other fields of physics. In electricity, we build circuits using wires which "let electricity pass easily," and we do not blink an eye when we set their resistances equal to zero. In fact, we know of a perfectly modern phenomenon which lets us support the assumption of ideal wires, namely *superconductivity.* We simply take the wires as being *superconducting*: they let charge pass easily; the potential difference across their length is zero; and they do not produce any heat.

What is the thermal equivalent of electrical superconductivity? Look at a thermal conductor and the simple constitutive relation we have used in Section 1.7; see Equation (50):

$$I_{E,th} = \frac{\Delta T}{R} \tag{1}$$

R is some sort of thermal resistance; it stands for the inverse of the factor hA which was introduced in Section 1.7. Remember that it is common to express the constitutive relation for conduction in terms of the flux of energy associated with the transport of entropy since the flux of entropy increases along the conductor. Note the similarity between Ohm's law and Equation (1). We can obviously make the resistance smaller and smaller while we hold the current through the body constant. This requires making the difference of temperatures across the conductor smaller and smaller in an appropriate way. Therefore, the temperature may very well be uniform in a model system which lets entropy pass easily. Let us calculate the rate of production of entropy in such a conductor. According to the results of Section 1.7, we have

$$\Pi_s = \frac{I_{E,th}\Delta T}{T\,T_o} \tag{2}$$

where ΔT is the difference of temperatures $(T - T_o)$ between the two ends of the conductor. With the law of conduction, this leads to:

$$\Pi_s = \frac{1}{T T_o} R I_{E,th}^2 \tag{3}$$

If we let the thermal resistance vanish while the flux of energy or entropy is kept constant, the rate of production of entropy will go to zero as well. A body working according to this prescription may very well be said to be a *thermal superconductor.*

Other ways of heating bodies may lead us to the same conclusion, namely that it is not forbidden to construct models of uniform heating. Imagine many tiny electrical heaters distributed uniformly through a body of water emitting the entropy they create at an equal rate into every part of the body. Another form of evenly distributed sources of entropy is encountered in the absorption of radiation in an almost transparent body. Radiation from the Sun is absorbed by a few cubic meters of air in our atmosphere at just about a uniform rate. In either case, we may model the actual process as one in which the temperature of the body remains uniform all the time.

Are uniform processes realistic? It is important to realize that it does not matter whether such bodies and circumstances exist in nature precisely as we have described them. They certainly may exist as models in our theories. They are comparable to ideal wires and ideal pendulums. Just like these simple objects, which cannot be found in nature either, thermal superconductors and evenly spread sources of heat are the building blocks of a theory of dynamics, this time of the dynamics of heat in uniform bodies. In fact, despite all the factors which we are ignoring, this model leads to practically important results: bodies undergoing uniform thermal processes approximate many real cases rather well. The idea of the change of a body through homogeneous states is an important ingredient of classical thermodynamics. All we have to do now is to investigate the consequences of such a far-reaching assumption.

2.1.3 Uniform Processes

How, then, are we going to model uniform processes mathematically? The basic idea is simple. We assume that we can describe bodies and the processes they undergo in terms of a few physical variables which have a given value for the entire body. Naturally, prominent among these variables is the temperature. Others have to be added according to the types of systems or materials for which we are going to create theories. In the case of a simple uniform fluid, such as a body of air, an additional variable might be its volume or pressure. In other cases, we might choose the density of an object which, again, is taken to be the same throughout the body. In the case of the heating of magnetized bodies we will have to add a variable pertaining to magnetism. A rubber band can undergo processes which may be described in terms of length and temperature. Electrochemical phenomena may call for amount of substance, electrical potential, and charge as variables.

Each of these quantities is taken to be a *function of time*, with the same value at every point in a body. Thus processes will be modeled using functions such as temperature $T(t)$, volume $V(t)$, magnetization $M(t)$, charge $q(t)$, pressure $P(t)$, and so on. Other quantities will be defined in terms of the basic ones we have chosen for the model, and the laws we believe to govern the processes will be expressed with the help of all of the quantities introduced.

2.2 The Heating of Solids and Liquids: Entropy Capacity

In this section we will discuss a common problem—the heating of solids and liquids—in a greatly simplified manner. We shall construct a model of a body of uniform temperature which is undergoing processes of heating. Indeed, heating and cooling are assumed to be the only phenomena which can be associated with such systems. Their volume will be taken to remain constant. Therefore, in the simplest case, there will be only one independent variable—one function of time—which determines the properties of such a body. We can take this function to be the temperature $T(t)$. Other choices are possible, namely quantities which are related to and change with the temperature. Obviously, these must be the entropy content and the energy of the system. A first question must be asked, namely, how the temperature and entropy content of a body are related. We will answer it using the first constitutive quantity we are going to use, the *entropy capacity*. Then we must formulate constitutive relations which we believe to hold for the processes of heating or cooling of bodies at constant volume. Using the generic laws of balance of entropy and energy, we hope to find the solution of the problem posed, namely the functions $T(t)$, $S(t)$, and $E(t)$, and expressed in terms of these the flux and the rate of production of entropy.

2.2.1 Definition of the Entropy Capacity

When heated at constant volume, a body changes only its entropy content and its temperature. Therefore we are led to introduce a quantity which relates changes of temperature to changes of entropy content. We have encountered this situation before, in electricity and in mechanics. In both cases we introduced a capacity, i.e., a quantity which tells us by how much the electrical charge or momentum of a body will change if the electrical potential or the velocity change by a given amount. In thermal physics we define the following:

> *The entropy capacity K of a rigid uniform body is the ratio of the change of the entropy content and the change of temperature. In differential form:*

$$K(T) = \frac{dS}{dT} \tag{4}$$

It is instructive and useful to write this relationship in terms of time rates of change; the entropy capacity of a body is the factor relating the rate of change of temperature to the rate of change of the entropy content:

$$\dot{S} = K\dot{T} \tag{5}$$

There is a simple graphical interpretation of the meaning of the entropy capacity. Since the entropy of a body is a function of its temperature we can represent this relationship by a curve in the *S-T diagram* (Figure 1). The curve displays the process of heating an ideal body. The entropy capacity at a prescribed temperature is given by the slope of the tangent to the curve at that point. Figure 2 represents the process of heating (or cooling) at constant volume in the *T-V diagram*, which is a natural tool for the description of thermomechanical processes of simple systems. We shall choose the volume of a body as the second independent variable for the theory of thermodynamics of simple fluids such as the ideal gas (Section 2.3).

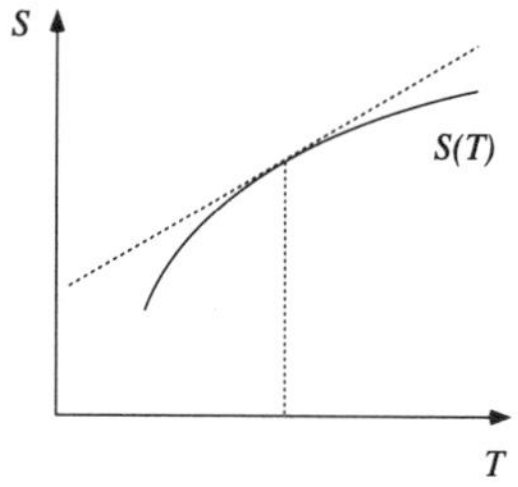

FIGURE 1. *S-T* diagrams are a means for representing the processes of heating in simple cases. If a body is heated at constant volume, the only thermal variables which change are the entropy *S* and the temperature *T*. The entropy capacity of such a body is given by the slope of the tangent to the curve *S(T)*.

The entropy capacity is called a *constitutive quantity* since it describes a particular property of the chosen material. Naturally, we are interested in determining the entropy capacities of different bodies. This can be done if the entropy content is known as a function of the temperature. However, there is a problem: How do we measure the entropy content of a system?

2.2.2 The Heating of Bodies at Constant Volume

We can compute changes of the entropy content of a system if we can measure or determine both the net current of entropy and the rate of production of entropy during a process (Section 1.6). In many important practical cases, nature comes to our rescue: there are bodies which emit as much entropy in a process as they absorb when they undergo the reverse of the process. This has important consequences: the entropy flux must depend linearly on the rate of change of the temperature:

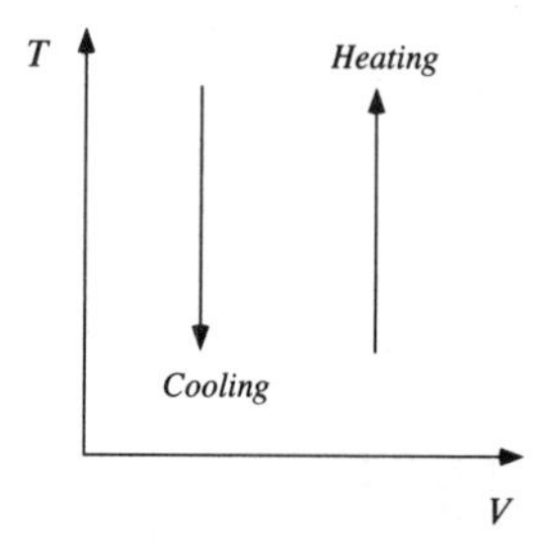

FIGURE 2. *T-V* diagrams are the natural tool for representing thermomechanical processes, i.e., processes for uniform bodies which can change only their volumes and their temperatures. Heating at constant volume is represented by a vertical line in the *T-V* diagram.

$$I_s = -K^*\dot{T} \tag{6}$$

This is the simplest choice which allows for the amount of entropy exchanged to change its sign upon reversal of the process. This relation is not a definition, it is a constitutive law. You can convince yourself that a body obeying this law can undergo only *reversible processes*: the rate of production of entropy must vanish. This is so since upon reversal of a process the entropy exchanged and the change of the entropy of the body reverse their signs. Both conditions can hold only if entropy is not created during the process. For this reason, the rate of change of the entropy content is equal in magnitude to the net current of entropy. Therefore K^* must be identical to the entropy capacity of the body. The formal proof goes as follows. We substitute the definition of the entropy capacity Equation (5) and the constitutive law Equation (6) into the equation of balance of entropy, Equation (32) of Chapter 1, after which we obtain:

$$K\dot{T} - K^*\dot{T} = \Pi_s \tag{7}$$

or

$$(K - K^*)\dot{T} \geq 0 \tag{8}$$

The latter inequality can be satisfied only if $K^* = K$. This is so since the rate of change of temperature in a process can assume any value; in other words, the balance of entropy must hold for any imaginable process. We can find conditions which violate the inequality if K^* and K are not the same. For this reason, the constitutive laws specifying the heating and the rate of production of entropy in our model are:

$$I_s = -K\dot{T} \tag{9}$$

$$\Pi_s = 0 \tag{10}$$

This is a theory of *uniform reversible processes*, which is equivalent to saying that

$$\dot{S} + I_s = 0 \tag{11}$$

for all processes considered in this theory.

The importance of the entropy capacity is quite clear. If we know its value for a given body, we can calculate the rate of flow of entropy or the rate of change of the entropy content during a process of heating or cooling. Knowledge of the entropy capacity allows us to compute the amount of entropy exchanged and the change of the entropy content in such processes. According to Equation (30) of Chapter 1, and Equation (9), the amount of entropy exchanged must be equal to the integral of the entropy capacity over the range of temperatures:

$$S_e = -\int_{t_i}^{t_f} I_s dt = \int_{t_i}^{t_f} K\dot{T} dt \tag{12}$$

or

$$S_e = \int_{T_i}^{T_f} K dT \tag{13}$$

Since the processes admitted are reversible, this quantity also must be equal to the change of the entropy content of the body.

EXAMPLE 1. The current of entropy necessary for heating cold ice.

Measurements put the value of the entropy capacity of 1 kg of ice at 8.1 J/K^2 at a temperature of 13°C below freezing, and at 7.7 J/K^2 at a temperature of 0°C. Calculate the current of entropy which is needed if a lump of ice with a mass of 1kg at a temperature of

- 13 °C is to be heated in such a way that its temperature rises by 1 K per minute. Estimate the total entropy transmitted to the body if it is to be heated up to a temperature of 0°C.

SOLUTION: According to Equation (9) the current of entropy must be

$$\begin{aligned} I_s &= -K \cdot \dot{T} \\ &= -8.1 \frac{1}{60} \frac{\text{W}}{\text{K}} = -0.135 \frac{\text{W}}{\text{K}} \end{aligned}$$

at a temperature of - 13°C. We can estimate the amount of entropy transmitted to the lump of ice by using the average entropy capacity, which leads to

$$S_e = 7.9 \cdot (273 - 260) \frac{\text{J}}{\text{K}} = 103 \frac{\text{J}}{\text{K}}$$

If we knew the entropy capacity as a function of temperature we could find precisely the entropy exchanged by performing the integration as in Equation (13).

2.2.3 Measurement of the Entropy Capacity

The relationship between currents of entropy and currents of energy is of central importance not just in the case of heat engines (Section 1.4). If we multiply Equation (9) by the temperature T, we get a measurable quantity on the left-hand side, the energy current associated with the heating (see Figure 21, Chapter 1):

$$T I_s = -T K \dot{T} \tag{14}$$

or

$$I_{E,th} = -T K \dot{T} \tag{15}$$

Since $I_{E,th}$ and the rate of change of the temperature can be determined quite easily in experiments, we can measure the entropy capacity. The equation just derived shows that

$$K(T) = -\frac{1}{T} \frac{I_{E,th}}{\dot{T}} \tag{16}$$

With the entropy capacity determined we also can calculate the energy exchanged in a thermal process in terms of the entropy capacity. According to Equation (15) and Equation (14) of Chapter 1,

$$Q = \int_{T_i}^{T_f} T K dT \tag{17}$$

The product of entropy capacity and temperature is of particular importance. According to Equation (15) the quantity

$$C = T K \tag{18}$$

which has the dimension of energy divided by temperature, can be determined directly from the measurement of the current of energy and the rate of change of the temperature of the body. We shall call this new quantity the *temperature coefficient of energy* of the body. Its integral over temperature leads directly to the amount of energy exchanged in the heating:

$$Q = \int_{T_i}^{T_f} C dT \tag{19}$$

Due to its direct relationship with the *energy exchanged in heating*, the temperature coefficient of energy, C, is normally reported in tables (see Tables A.6 and A.7). Also, for many substances its value is constant for an appreciable range of temperatures not too close to absolute zero. As a consequence, *in many practical cases the entropy capacity must be inversely proportional to the temperature of the body.* This makes calculations involving entropy capacities rather simple, as we shall see below (Examples 3 and 4).

Note that the quantity C usually is called the *heat capacity.* Since the quantity of energy which is exchanged in a thermal process usually is called *heat*, $C = TK$ has been given a name referring to heat as well. However, it definitely cannot be thought of as a capacity in the ordinary sense of the word, since *heat* in this context may not be thought of as residing in bodies.

EXAMPLE 2. Measuring the entropy capacity of water.

Assume that two liters of water inside an insulated bottle are heated using an immersion heater. The voltage and electrical current are kept constant at 220 V and 1.36 A, respectively. The temperature of the water is monitored. It is found that the temperature is quite nearly a linear function of time, with dT/dt = 0.0356 K/s. Determine $K(T)$ and $C(T)$, and calculate the entropy capacity and the energy capacity per mass for a temperature of 20°C.

SOLUTION: dT/dt is constant, with a value of 0.0356 K/s. The energy current is constant as well:

$$\begin{aligned} |I_{E,th}| &= |U I_q| \\ &= 299\text{W} \end{aligned}$$

Consequently, the temperature coefficient of energy is a constant (or nearly so) for water:

$$(TK)\dot{T} = |I_{E,th}|$$

The experimental results make $C = TK$ = 8400 J/K for this body of water. The entropy capacity of two liters of water therefore is equal to

$$K = \frac{8400\,\text{J/K}}{T}$$

while the energy capacity is constant and equal to $C = TK = 8400$ J/K. The entropy capacity per unit mass at 20°C is $k = 8400/(2 \cdot 293)$ J/(K$^2 \cdot$ kg) = 14.3 J/(K$^2 \cdot$ kg). The temperature coefficient of energy per unit mass is $c = 8400/2$ J/(K · kg) = 4200 J/(K · kg). In Table A.7, the values of the capacity (measured at constant pressure) are listed. Note that the value of TK for water is almost constant over the range of temperatures given (see also Figure 3).

2.2.4 Some Values of Entropy Capacities

Entropy capacities hold for the particular body being heated. It would be very difficult to tabulate such values since we would have to do this for bodies having different amounts of substance or mass. However, the capacities scale with the amount of substance of a body. For this reason we introduce the *molar entropy capacity* $\bar{k}$, or as an alternative, the *entropy capacity per unit mass k* (*specific entropy capacity*). These quantities are independent of the amount of substance (or the mass), and depend only on the type of substance:

$$\bar{k} = \frac{1}{n}K \; , \quad \bar{c} = \frac{1}{n}C = \frac{1}{n}TK \tag{20}$$

$$k = \frac{1}{m}K \, , \quad c = \frac{1}{m}C = \frac{1}{m}TK \tag{21}$$

where n and m are the amount of substance and the mass of the body, respectively. Numerical values for different substances are listed in Table A.6. Commonly, c and $\bar{c}$ are called the *specific heat* and molar *heat capacity*, respectively. They represent the temperature coefficients of the energy of the body, this time divided by its mass or its amount of substance.

For example, the values of k and c of water in the range of temperatures between 0°C and 100°C are given in Table A.7 and in Figure 3 (see Example 2 for how to measure this quantity in a simple experiment). As you can see, the quantity c is almost constant for water.

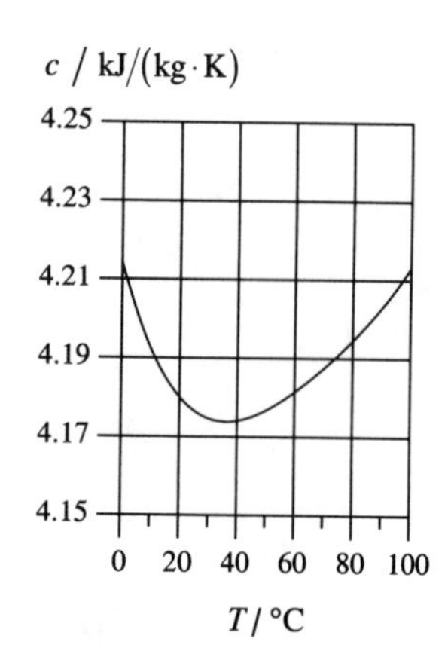

FIGURE 3. $c = Tk$ for water at temperatures between 0°C and 100°C, at a pressure of 1 bar. The data have been computed using thermodynamic property functions implemented in the program EES (Klein et al., 1991).

Similarly, c is almost constant for many other liquids at room temperature. Also, at high temperatures, which often means room temperature or higher, c is a constant for solids. However, if we go to lower temperatures it becomes evident that c changes drastically. Specifically, entropy capacities (and with them the values of c) become zero at 0 K. This finding is equivalent to saying that it is impossible to reach 0 K in any experiment. Either statement is often called the *third law of thermodynamics*. In Figure 3, values of c for some metals are reported. Table A.6 lists values of c for some liquids and solids for a range of temperatures for which they can be considered to be nearly constant.

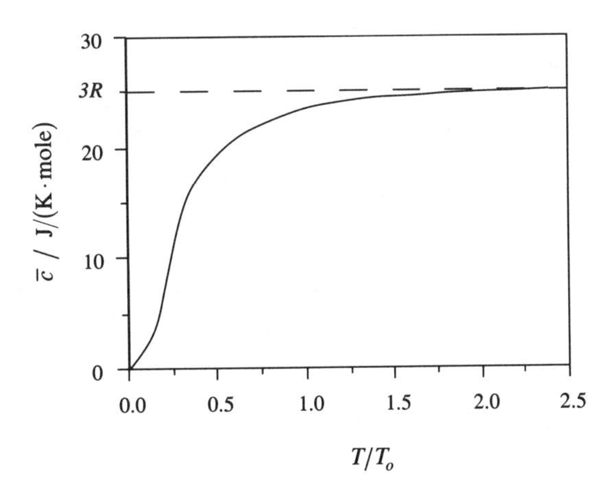

FIGURE 4. $\bar{c} = T\bar{k}$, for several solids. At high temperatures, $\bar{c}$ reaches roughly $3R$ for most solids, where $R = 8.31$ J/(mole · K) is the universal gas constant. T_D is the Debye temperature, which is different for different solids. (For copper and silicon the values are 343 K and 640 K, respectively.)

EXAMPLE 3. The entropy exchanged and the energy exchanged for constant C.

Determine the expressions for the entropy and the energy exchanged in terms of initial and final temperatures for materials having constant C. Calculate the amounts of entropy and of energy exchanged in cooling a piece of iron with a mass of 5 kg from 120°C to 20°C.

SOLUTION: According to Equations (9) and (15), the entropy exchanged in the case of constant energy capacity is

$$S_e = \int_{T_i}^{T_f} K(T)dT = C\int_{T_i}^{T_f} \frac{dT}{T} = C\ln\left(\frac{T_f}{T_i}\right)$$

The expression for the amount of energy which is exchanged in heating is even simpler. According to Equation (19) we have

$$Q = \int_{T_i}^{T_f} C(T)dT = C\left(T_f - T_i\right)$$

In Table A.6 we find that the temperature coefficient of energy per mass of iron is nearly constant, $c = 452$ J/(K · kg). This leads to $S_e = -665$ J/K. The value for the energy exchanged is $Q = -226$ kJ.

EXAMPLE 4. The time required to heat water using a heat pump.

Calculate the times needed to heat the same amount of water a) using an ideal heat pump, and b) using an immersion heater, if both receive the same electrical energy current.

SOLUTION: According to Equation (15), the thermal energy current used for heating is proportional to the time rate of change of the temperature:

$$I_{E,th} = -C\dot{T}$$

with C constant.

a) Consider the energy current to be delivered by an ideal heat engine. According to Equation (20) of Chapter 1, it is given by

$$I_{E,th} = -\frac{T^+}{T^+ - T^-} I_{E,el}$$

If we combine this with the constitutive law stated above we get the differential equation

$$C\dot{T} = \frac{T^+}{T^+ - T^-} I_{E,el}$$

The entropy delivered by the heat pump is taken out of the ground at constant lower temperature, while the upper temperature varies from the initial temperature $T_i \geq T^-$ of the water to the final temperature T_f. Under these conditions, the solution of the differential equation delivers the time for heating using the heat pump:

$$t_p = \frac{C}{I_{E,el}}\left[T_f - T_i - T^- \ln\left(\frac{T_f}{T_i}\right)\right]$$

b) In the case of the immersion heater, the thermal energy current heating the water is equal to the electrical energy current of the immersion heater. Therefore the time required for heating is

$$t_{ih} = \frac{C}{I_{E,el}}\left(T_f - T_i\right)$$

This value is definitely larger than the one obtained for the heat pump.

2.2.5 The Balance of Entropy and the Entropy Content of Bodies

We have considered the exchange of entropy and energy as a consequence of heating. Now we are going to calculate the change of the entropy content of the bodies undergoing those processes. Since the operations considered here are reversible, this will prove to be an easy matter for processes at constant volume. In addition, we will discuss briefly changes of phase.

We have introduced the entropy capacity as the quantity which relates changes of the entropy content to changes of temperature. Since entropy is conserved in the processes admitted in our theory, the entropy capacities measured using the exchange of energy and entropy can be used to calculate the change of the entropy content:

$$\Delta S = \int_{T_i}^{T_f} K(T)dT \qquad (22)$$

For phase changes the result is comparable. According to the laws given in Section 1.3, we calculate amounts of entropy exchanged using the entropy of fusion; S_e must be equal to the change of the entropy content.

Basically, we must know only changes of the entropy content to be able to model the types of thermal processes discussed here. Still, the image of entropy as the substancelike thermal quantity raises an interesting question: How much entropy does a body really contain? In the case of simple bodies such as the ones discussed so far, this question could be answered if we knew the entropy content at the temperature 0 K, and the amount of entropy transmitted to the body as a consequence of phase changes and of raising its temperature from 0 K to the desired level. If the body obeys the theory laid down in this chapter, the latter problem can be solved by measuring the amount of energy absorbed or emitted during heating. Equation (22), and Equation (8) of Chapter 1, allow us to calculate the change of the entropy content with respect to 0 K.

There remains the question of the entropy content at 0 K. The quantity of entropy stored in a body at absolute zero depends on the type of body and the circumstances. Microscopic models of matter (which we are not going to discuss in this book) show that the entropy of a single crystal of a pure element can be taken to be zero at 0 K. Put less formally, in some cases absolutely cold bodies do not contain any entropy. However, in view of the fact that only differences of entropy are of any practical significance, one should not attach too much importance to the problem of *absolute entropy.* It simply is a matter of convenience to set the entropy content of bodies equal to zero at some temperature, which can be taken as 0 K or any other suitable value.

EXAMPLE 5. The entropy of solids.

According to a constitutive law named after P. Debye, the molar entropy capacity of a solid obeys the following relationship at very low temperatures:

$$\bar{k}(T) = \frac{12\pi^4 R}{5T_D^3} T^2$$

Here, T_D is the Debye temperature. The temperature of the solid must be much smaller than this critical value for the formula to hold; i.e., $T \ll T_D$. If we assume the heating to obey the constitutive law of Equation (9), and if the entropy of a solid is equal to zero at 0 K, how large is the entropy content of one mole of such a body at a temperature T? What is the temperature coefficient of energy of this body?

SOLUTION: According to Equation (22) we obtain for the entropy content

$$S(T) = \int_0^T K dT = n\frac{12\pi^4 R}{5T_D^3}\int_0^T T^2 dT = n\frac{12\pi^4 R}{15}\frac{T^3}{T_D^3} \quad \textbf{(E1)}$$

The temperature coefficient of energy is defined as the product of temperature and entropy capacity. Therefore we have

$$C(T) = TK(T) = n\frac{12\pi^4 R}{15}\frac{T^3}{T_D^3} \tag{E2}$$

Note that the quantity calculated in (E1) actually corresponds to the change of entropy content due to heating from absolute zero. In writing (E1) we have set the entropy content equal to zero at 0 K.

2.2.6 The Balance of Energy for Purely Thermal Processes

It is interesting to note that for the processes considered here the law of balance of energy can be derived from the assumptions made so far. This is an example of the more interesting case dealing with the thermomechanics of simple fluids (Sections 2.3 and 2.4). Historically, R. Clausius was able to prove the first law of thermodynamics from other laws he assumed to hold. Looking at the statement of the conservation of energy from this viewpoint may serve as a motivation for the validity of the general law of balance of energy (Section 2.8).

The derivation proceeds as follows: we introduce a quantity *E(T)* whose time derivative is defined to be equal to the product of the temperature and the time rate of change of the entropy content of the body. In other words

$$\dot{E} = T\dot{S} \tag{23}$$

The function E can depend only on temperature since the entropy of a body which cannot change its volume is a function of temperature only. We now introduce the law of balance of entropy into this definition. Since the heating of uniform bodies at constant volume is reversible, the law is given by Equation (11). If we remember that the product of the flux of entropy and the temperature is the thermal energy current, we obtain

$$\dot{E} + I_{E,th} = 0 \tag{24}$$

Integrated over time, this equation takes the form

$$\Delta E = Q \tag{25}$$

Equations (24) or (25) clearly represent the equation of balance of a conserved quantity. We therefore interpret the quantity *E(T)* as the *energy of the body* at temperature *T*. Note that the derivation given here holds only for bodies which are being heated uniformly at constant volume. In no way does this represent the derivation of a general law of nature; that would be impossible.

We can now explain why the quantity $C = TK$ introduced in Equation (18) is called the *temperature coefficient of energy*. If we plug the entropy capacity into the law of balance of energy in Equation (24) we get the following expression:

$$\dot{E} = C\dot{T} \tag{26}$$

This leads to the following relationship between the temperature coefficient of energy and the change of the energy content of a body:

$$\Delta E = \int_{T_i}^{T_f} C(T)dT \tag{27}$$

If C is constant in the range of temperatures envisioned, we obtain the simple case for which the energy increases linearly with the temperature (Example 6).

In the next two sections we shall repeat this type of derivation for the thermomechanical processes undergone by an ideal gas. Afterwards, we shall accept the validity of the law of balance of energy. We could have done this already, in which case Equation (23) would be the derived result for the relationship between entropy, temperature, and energy in pure heating. An equation of the form of Equation (23) is commonly called a *Gibbs fundamental relation* or *Gibbs fundamental form.*

EXAMPLE 6. The change of the energy and the entropy content of water.

Determine the energy of water as a function of temperature. Use the Gibbs fundamental relation to calculate the entropy content of water as a function of temperature.

SOLUTION: For water, the specific temperature coefficient of energy c is constant. The change of the energy content is given by Equation (27). Therefore the energy of a body of water with a mass m is equal to

$$E(T) = mc\left(T - T_o\right) + E_o$$

where E_o is the energy at T_o. With $E(T)$ known we can use the fundamental Equation (23) to calculate the entropy content of the same body of water:

$$S(T) = \int_0^t \frac{\dot{E}}{T}dt + S_o = \int_{T_o}^{T} \frac{mc}{T}dT + S_o = mc\ln\left(\frac{T}{T_o}\right) + S_o$$

This result agrees with the one we could have obtained by integrating Equation (22), or by using the result for the amount of entropy exchanged, which was calculated in Example 3.

EXAMPLE 7. The availability of a body of water.

Take a body of water of mass m at temperature T_i. How large is its availability in an environment at temperature T_a?

SOLUTION: The availability of a body is defined as the amount of energy which may be released by letting its entropy fall to the level of the environment (see Example 15 of Chapter 1, and the figure below). With the expression for the available power,

$$\mathcal{P}_{av} = \left(T(t) - T_a\right) I_s$$

and the balance of entropy and the definition of the entropy capacity,

$$\dot{S} = -I_s \; ; \quad \dot{S} = K\dot{T}$$

we obtain

$$\mathcal{P}_{av} = -\left(T(t) - T_a\right)\frac{C}{T(t)}\dot{T}$$

The availability E_{av} is computed as follows:

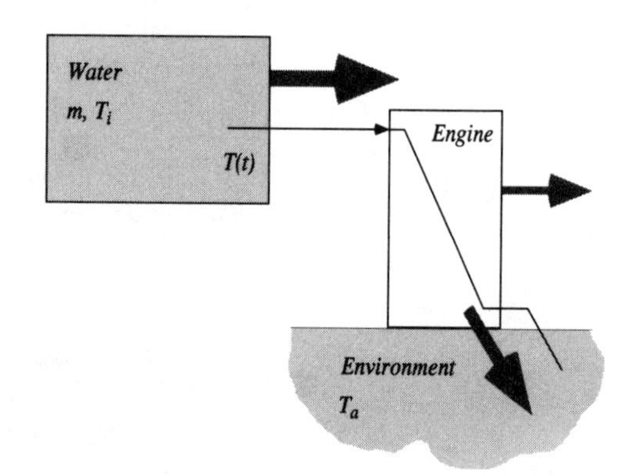

$$E_{av} = \int_{t_i}^{t_f} \mathcal{P}_{av} dt$$

$$= -\int_{t_i}^{t_f} \left(T(t) - T_a\right)\frac{C}{T(t)}\dot{T} dt$$

$$= -\int_{T_i}^{T_a} \left(T(t) - T_a\right)\frac{C}{T(t)} dT$$

$$E_{av} = mc\left[T_i - T_a - T_a \ln\left(\frac{T_i}{T_a}\right)\right]$$

This is how much energy can be released by an ideal heat engine from a body of water at temperature T_i in an environment at temperature T_a. The term proportional to $T_i - T_a$ is equal to the change of the energy of the body while the second term represents the energy rejected to the environment together with the entropy which has been withdrawn from the body. You can calculate the result directly by considering these integrated quantities. (See Example 8 for how to do this.)

EXAMPLE 8. Heat transfer (between two equal amounts of water) using a heat engine.

Consider two bodies having identical entropy capacities, for which C is constant. Their initial temperatures are assumed to be T_1 and $T_2 < T_1$. Entropy is transferred from the hotter to the cooler reservoir using an ideal heat engine. What is the value of the final temperature attained by the bodies, and how much energy is given off by the heat engine in the mechanical process?

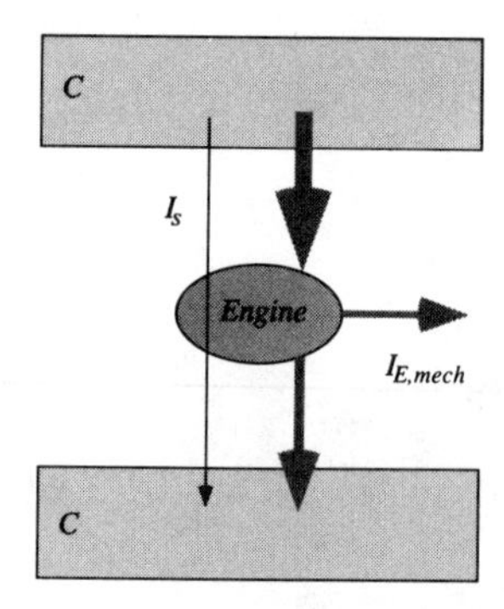

SOLUTION: In this case no entropy is produced, which means that the change of the entropy content of the two bodies is zero. Put differently:

$$\Delta S = \Delta S_1 + \Delta S_2 = 0$$

or

$$\Delta S = \int_{T_1}^{T_f} K\,dT + \int_{T_2}^{T_f} K\,dT$$
$$= \int_{T_1}^{T_f} \frac{C}{T} dT + \int_{T_2}^{T_f} \frac{C}{T} dT$$
$$= C\left[\ln\left(\frac{T_f}{T_1}\right) + \ln\left(\frac{T_f}{T_2}\right)\right] = C\ln\left(\frac{T_f^2}{T_1 T_2}\right)$$
$$\Delta S = 0$$

which leads to

$$T_f = \sqrt{T_1 T_2} \le \frac{1}{2}(T_1 + T_2)$$

The geometric mean is always smaller than or equal to the arithmetic mean. The energy released in the fall of entropy must be equal to the difference of the energy contents of the reservoirs at the beginning and at the end. These are easily calculated according to Equation (27):

$$W = \Delta E_1 + \Delta E_2$$
$$= C(T_f - T_1) + C(T_f - T_2)$$
$$= C\left[2\sqrt{T_1 T_2} - (T_1 + T_2)\right]$$

Since this quantity is negative, the bodies lose energy in the mechanical process.

2.2.7 Exchange of Heat Between Two Uniform Bodies

Let us briefly discuss the problem of how to model a simple process: two bodies which are in thermal contact reach thermal equilibrium if left to themselves. If two bodies at different temperatures are in thermal contact (but otherwise insulated from their surroundings, see Figure 4 of Chapter 1, and Figure 5 below), heat will flow from the hotter to the cooler body, leading to the production of entropy. Once the bodies have reached their final temperatures, which is the same for both, the entropy content of the system will be larger than at the beginning. With what you have learned so far you could easily compute the amount of entropy produced in this process (this is done in Section 2.9): it is equal to the difference of the final and the initial amounts stored in the bodies. Now the problem is this: the model of uniform processes used in this section is one of reversible changes, which does not allow for the production of entropy. If we wish to build a simple model of what is actually happening in nature we will have to add an important ingredient.

This can be done rather simply. We know that the production of entropy is due to heat flow. Let us therefore add a third body to the system (see Figure 5), namely, a conductor placed between the parts exchanging heat. We model the processes of cooling and heating of the first two system parts as reversible changes, as we have done so far, and we let all the entropy be produced in the conductor. Transport of heat through the conductor is modeled as in the case of a heat exchanger, Equation (50) of Chapter 1, with the conductor having a vanishingly small entropy capacity. This simple aggregate model produces the same amount of entropy as that calculated from the changes of state alone. (See Chapter 3.)

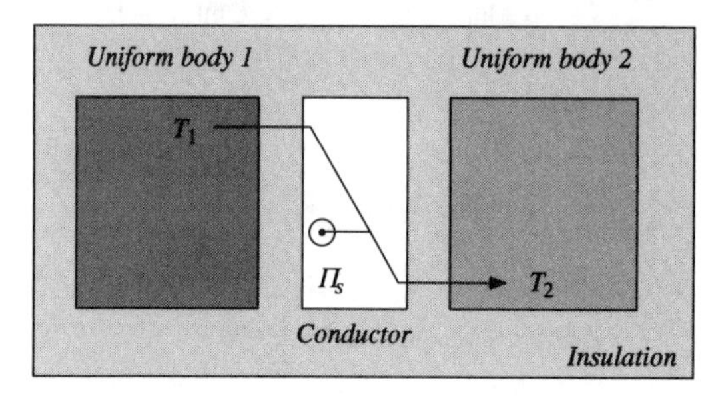

FIGURE 5. Simple "lumped parameter" model of entropy production in the exchange of heat of two uniform bodies. Entropy production is assumed to be limited to a third body, a conductor, while the cooling and the heating of the uniform bodies progresses reversibly.

This problem makes it clear that a full theory of dynamics cannot be built on the simplest homogeneous processes alone. You will discover in this chapter that the constitutive theory of simple fluids either does not cover everyday cases like that of equilibration or allows processes to run at any desired speed; time either seems to be frozen or it does not appear explicitly in the equations of change (Section 2.5.4). The model presented here for the exchange of heat between two uniform bodies represents one little stepping stone in the direction of a more complete theory of the dynamics of heat.

2.3 The Heating of the Ideal Gas

In terms of the concepts introduced so far, the problem of thermodynamics is the determination of the currents and the rates of generation of entropy. We have approached this problem by discussing some very simple constitutive laws, namely those governing the heating of ideal bodies at constant volume and the change of phase of a substance such as water (Sections 2.2 and 1.3). However, by restricting our attention to processes at constant volume we essentially have excluded gases. In this and the following sections we shall extend the analysis to bodies which can change their volumes and their temperatures. The *ideal gas* and *blackbody radiation* (Section 2.6) will serve a examples of such fluids.

We begin by stating a constitutive law, the *thermal equation of state* of the ideal gas, which relates the pressure of the fluid to its temperature and its volume. Heating will be represented by a constitutive relation which introduces the *latent entropy* and the *entropy capacity*. As we will see, the model allows only for *reversible processes* of the ideal gas.

2.3.1 The Thermal Equation of State of the Ideal Gas

The ideal gas is a system which can change its volume and temperature. These two variables are related to a third property of the gas, namely its pressure. In Section 1.2, we encountered the *law of Gay-Lussac*, which is obeyed by dilute gases. It states that the pressure of the fluid is a linear function of temperature if it is heated at constant volume. One finds that such gases have another simple property. Experiments demonstrate that the pressure and volume of a gas like air at room temperature are inversely proportional if the temperature is kept constant:

$$PV = constant \tag{28}$$

This relation is called the *law of Boyle and Mariotte*. The laws of Gay-Lussac and of Boyle and Mariotte together define the *ideal gas* as the fluid having the following *thermal equation of state*:

$$PV = nRT \tag{29}$$

Here, $R = 8.31$ J/(mole · K) is the *universal gas constant*, and n is *the amount of substance* measured in moles.[1] (The proof of this relation is given below).[2]

The *amount of substance* comes into the equation of state of the ideal gas in the following way. It is found that the law holds with a single value of the gas constant if certain mass ratios of different gases are used. For example, 16 g of oxygen gas turn out to be equivalent to 1 g of hydrogen gas. These amounts also turn out to be chemically equivalent in the sense that simple fractions or multiples of these numbers occur in chemical reactions using up all of the reactants. Since the amount of substance is the fundamental measure of how much stuff is

---

1. See Chapter 4 for more details on the subject of the amount of substance.
2. Proof of Equation (29): Since P is proportional to $1/V$ at constant T (Boyle's relation), the product PV is a constant that depends on the temperature:

$$PV = f(T)$$

here, $f(T)$ is an unspecified function. Also, since P is proportional to T at constant volume, P/T is a constant which depends on the volume only. Therefore, P/T is some other function $g(V)$ of the volume:

$$P/T = g(V)$$

If we divide the first equation by T and multiply the second one by V we get

$$PV/T = f(T)/T$$
$$PV/T = Vg(V)$$

This implies that the quantity PV/T is a function of T alone and a function of V alone. Therefore, it cannot be a function of either T or V, which means that it is constant: PV/T = constant.

participating in chemical reactions, the coefficient n appearing in the equation of state also measures the amount of substance.

Gases obeying both the laws of Gay-Lussac and of Boyle and Mariotte do not really exist in nature. Rather, a real gas approximates the behavior of an ideal fluid called the *ideal gas* if its temperature is high and its density low. What can be taken to be sufficiently high or low has to be determined by experience. Basically, gases follow the ideal law for temperatures which are high compared to the point at which they liquefy.

Often, it is useful to write the equation of state of the ideal gas in terms of density instead of volume. We express the volume of the body by its density and its mass: $V = \rho m$. Further, we have the following relationship between the mass and the amount of substance of the fluid: $m = M_o n$. Here, M_o is the *molar mass* (mass per unit amount of substance) of the gas. Using these relationships, the equation of state of the ideal gas becomes

$$P = \frac{R}{M_o}\rho T \tag{30}$$

EXAMPLE 9. Changes of state of air.

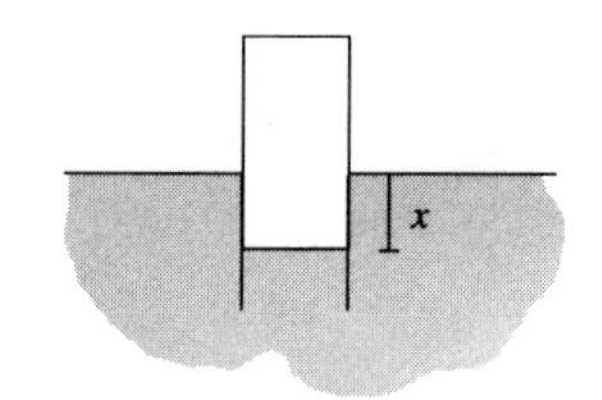

A test tube which is 10 cm long is filled with air at a temperature of 27°C. The air is heated to a temperature of 57°C with the test tube open. Then it is placed upside-down in water so that a 5 cm length of the tube is above the water level. The air in the tube will cool down to 27°C. How far above or below the outer water level will the water rise in the test tube? Atmospheric pressure is set at 10^5 Pa.

SOLUTION: First, there is a relationship between the pressure, temperature, and volume of the gas at T_2 = 273K + 57K:

$$\frac{nRT_2}{V} = \frac{nRT_2}{AH} = P_o$$

Here, n is the amount of gas left in the tube after heating; P_o is the ambient air pressure; and H is the length of the test tube. Later, at T_1 = 273 K + 27 K, the equation of state takes the form

$$\frac{nRT_1}{V} = \frac{nRT_1}{A(x + H/2)} = P(x)$$

where x is the difference between water levels. $P(x)$ is the pressure in the water at depth x:

$$P(x) = P_o + \rho_w g x$$

Here, g = 10 N/kg is the strength of the gravitational field, and ρ_w = 1000 kg/m^3 is the density of water. Combining the three equations, we get

$$\rho_w g x^2 + \left(P_o + \rho_w g \frac{H}{2} \right) x + \left(P_o \frac{H}{2} - H P_o \frac{T_1}{T_2} \right) = 0$$

or

$$x^2 + 10.05x - 0.4091 = 0$$

This equation has solutions x = 0.0405 m and x = - 10.09 m. Only the former makes any physical sense.

EXAMPLE 10. A hot air balloon in an isothermal atmosphere.

A hot air balloon is floating at 2000 m above sea level in an atmosphere whose temperature is taken to be constant at all levels and equal to 0°C. The temperature of the air inside the balloon is 250°C. What is the radius of the spherical balloon, if it has a mass of 450 kg including the passengers? The pressure of the air at sea level is 1 bar.

SOLUTION: Let us call the mass of the balloon and the mass of the air inside the balloon m_b and m_a, respectively. ρ_o and ρ_i are the density of air at 2000 m outside and inside the balloon, respectively. The balloon floats, which means that the buoyancy is equal to the combined weights of balloon and hot air:

$$(m_b + m_a)g = \rho_o V g \tag{E3}$$

where the mass of the hot air is given by

$$m_a = \rho_i V$$

and V is the volume of the balloon, i.e., of the hot air. The density of air can be calculated according to Equation (30), which yields

$$\rho = \frac{M_o P}{RT} \tag{E4}$$

One important condition is the fact that the pressure inside and outside the balloon must be the same; otherwise air would be flowing from the outside to the inside, or vice versa. We can calculate the pressure at 2000 m from the law of hydrostatic equilibrium for our atmosphere:

$$\frac{dP}{dh} = -\rho g$$

With (E4) and T = *constant*, integration of this differential equation leads to

$$P(h) = P_o \exp\left(-\frac{M_o g}{RT} h \right) \tag{E5}$$

We still need the mean molar mass M_o of air. We assume the atmosphere to be composed of 80% nitrogen and 20% oxygen by mass. If x = 0.20 is the mass fraction of oxygen we obtain

$$n_O = \frac{xm}{M_{oO}} \quad , \quad n_N = \frac{(1-x)m}{M_{oN}}$$

which leads to

$$M_o = \frac{m}{n_O + n_N} = \frac{M_{oO} M_{oN}}{x M_{oN} + (1-x) M_{oO}}$$
$$= 0.029 \frac{\text{kg}}{\text{mole}}$$

Equation (E5) yields a value of $P(2000\ \text{m}) = 7.78 \cdot 10^4$ Pa for the air pressure at 2000 m above sea level. This leads to values of $\rho_o = 0.995\ \text{kg/m}^3$ and $\rho_i = 0.519\ \text{kg/m}^3$ for the densities of air outside and inside the balloon, respectively. Solving Equation (E3) with these values gives

$$V = \frac{m_b}{\rho_o - \rho_i} = 946 \text{m}^3$$

which corresponds to a radius of the balloon of 6.1 m.

EXAMPLE 11. Isothermal expansion of the ideal gas.

When a body of ideal gas expands or contracts, it exchanges energy as a consequence of the mechanical process. Determine the energy exchanged for the ideal gas in an isothermal process, and give the values for one mole of an ideal gas whose volume is doubled. The temperature is 300 K.

SOLUTION: In the Prologue (Section P.4.5) we considered the exchange of energy due to the expansion or compression of a simple nonviscous fluid. The mechanical energy flux was found to depend upon the pressure of the fluid and the rate of change of its volume:

$$I_{E,mech} = P\dot{V} \qquad \textbf{(E6)}$$

In isothermal expansion or compression we have to compute the energy exchanged as the result of the mechanical process by integrating (E6) which yields:

$$W = -\int_{V_i}^{V_f} \frac{nRT}{V} dV = -nRT \ln\left(\frac{V_f}{V_i}\right) \qquad \textbf{(E7)}$$

Note that the result does not depend on the absolute values of the volumes involved but rather on the ratio of final to initial volume.

The numerical value turns out to be $W = -1\ \text{mole} \cdot 8.31\ \text{J/(K} \cdot \text{mole)} \cdot 300\ \text{K} \cdot \ln(2) = -1.73\ \text{kJ}$: the gas emits energy as a consequence of the expansion.

2.3.2 Isothermal Expansion of the Ideal Gas

Let us now turn to the constitutive problem of the heating of a fluid such as the ideal gas. In Section 1.3 we discussed the problem of heating or cooling of a fluid at constant temperature. A fluid can be expanded or compressed isothermally if it is heated or cooled at the appropriate rate. The entropy therefore has the effect of changing the volume of the body. It does not do what we expect heat to do: it does not increase the temperature of the body. For this reason the term *latent heat* was coined to denote the heat absorbed in an isothermal process. It is instructive to hear how this name was used more than 150 years ago by J. Ivory:[3]

> The absolute heat which causes a given rise of temperature, or a given dilatation, is resolvable into two distinct parts; of which one is capable of producing the given rise of temperature, when the volume of the air remains constant; and the other enters into the air, and somehow unites with it while it is expanding .... The first may be called the heat of temperature; and the second might very properly be named the heat of expansion; but I shall use the well known term, latent heat, understanding by it the heat that accumulates in a mass of air when the volume increases, and is again extricated from it when the volume decreases.

Here we will make the definition of *latent entropy* formal. It is the factor which relates rates of change of entropy content to rates of change of volume:

$$\dot{S} = \Lambda_V \dot{V} \tag{31}$$

We will call this new quantity the *latent entropy with respect to volume*. For isothermal processes it plays a role comparable to the entropy capacity, which was introduced for processes at constant volume (Section 2.2).

There are ideal bodies which emit as much heat as a consequence of isothermal compression as they absorb during the reverse isothermal expansion. For this reason there must be a simple constitutive law: *currents of entropy are linear functions of the rate of change of the volume of the fluid.* As in the case of heating at constant volume, we can convince ourselves that the processes undergone by a fluid obeying such a law must be *reversible*. The formal proof is analogous to the one that led to Equation (9). Therefore, the coefficient relating currents of entropy and rates of change of volume must be the latent entropy:

$$I_s = -\Lambda_V \dot{V} \tag{32}$$

$$\Pi_s = 0 \tag{33}$$

3. J. Ivory (1827), quoted by Truesdell (1980), p. 17. Much more information regarding the caloric theory may be found in Fox (1971).

The minus sign appears because we count a current into a body as a negative quantity. As in the case of the heating at constant volume, we exhibit the process of isothermal heating in the *T-V* diagram (Figure 6).

We can easily calculate the amount of entropy exchanged in an isothermal process, or the change of the entropy content, by integrating the current of entropy:

$$S_e = -\int_{t_i}^{t_f} I_s dt = \int_{t_i}^{t_f} \Lambda_V \dot{V} dt \quad (34)$$

or

$$S_e = \int_{V_i}^{V_f} \Lambda_V dV \quad (35)$$

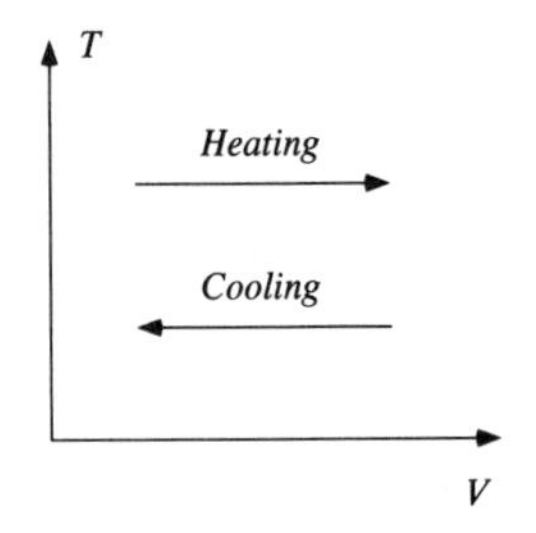

FIGURE 6. *T-V* diagram of isothermal heating or cooling. In most cases the volume of a fluid increases if it is heated at constant temperature. However, there are exceptions, such as water in the temperature range 0 - 4°C.

We have not really solved the constitutive problem yet. The burden of calculating the current has been placed on the determination of a new quantity, namely the latent entropy. We can calculate the process of heating at constant temperature only if we know this quantity. We could try to measure the new constitutive quantity in a manner analogous to what we did in Section 2.2. Multiplying Equation (32) by the temperature at which the process occurs, we get the current of energy associated with the heating. In practice, however, it is very difficult to measure the latent entropy of gases, since these fluids take up very little entropy in comparison to the measuring device. For this reason, it would be nice if we could find new ways of dealing with the constitutive problem. Fortunately, the relationship between entropy and energy, which we have not exploited yet, introduces severe restrictions on the constitutive relations. These restrictions will reduce the burden of measurement: we will be able to determine the latent entropy of the ideal gas using theoretical arguments (see Section 2.3.5).

2.3.3 Heating of the Ideal Gas at Constant Volume

J. Ivory spoke of two effects of heating. If you add heat to a body, either the volume or the temperature (or both) will change. Here, we will investigate the second possibility, namely a change of temperature alone. This can happen only if the volume of the body is forcibly kept constant. With solids or liquids this condition is automatically satisfied to a high degree. Gases have to be put in an airtight container. As before, we shall limit our attention to the ideal gas.

As in the case of heating of solids or liquids, we introduce the *entropy capacity at constant volume* as the coefficient which relates rates of change of the entropy content to rates of change of temperature:

$$\dot{S} = K_V \dot{T} \quad (36)$$

Here we have to refer explicitly to the condition of *heating at constant volume*, since the volume of a gas does not stay nearly constant as in the case of solids or liquids. Again, we assume we are dealing with an ideal fluid for which the heating is a linear function of the rate of change of temperature. According to the reasoning used before, the coefficient in this constitutive relation is the entropy capacity, which leads to:

$$I_s = -K_V \dot{T} \tag{37}$$

$$\Pi_s = 0 \tag{38}$$

In general the entropy capacity is a function of both the volume and temperature of the body. We can determine the amount of entropy exchanged during heating at constant volume:

$$S_e = -\int_{t_i}^{t_f} I_s dt = \int_{t_i}^{t_f} K_V \dot{T} dt \tag{39}$$

or

$$S_e = \int_{T_i}^{T_f} K_V dT \tag{40}$$

Direct measurements of entropy capacities at constant volume of gases are not so simple since the values are rather small compared to those for the measuring apparatus. (We will see how the problem is solved in Examples 18 and 19). We will be able to solve the constitutive problem in a combination of theoretical and experimental steps (Section 2.4.3).

2.3.4 The General Process of Heating of the Ideal Gas

In general, a body which is heated undergoes changes of volume and of temperature simultaneously. For this reason, the entropy content must depend on both the volume and the temperature. Its rate of change must depend upon the rates of change of the independent variables. In other words,

$$\dot{S} = \Lambda_V \dot{V} + K_V \dot{T} \tag{41}$$

The coefficients are the latent entropy and the entropy capacity, respectively. This equation defines the constitutive quantities, so there is nothing new about it. Now we are in a position to state a constitutive law appropriate to this type of process. We assume the current of entropy to be a linear function of both the rate of change of the volume and the rate of change of the temperature. Therefore we write

$$I_s = -\Lambda_V \dot{V} - K_V \dot{T} \tag{42}$$

$$\Pi_s = 0 \tag{43}$$

As stated above, the coefficients must be the latent entropy and the entropy capacity, respectively, and the rate of production of entropy must vanish.[4] We shall see in the following section how this law can be exploited in the case of adiabatic processes undergone by the ideal gas, without even knowing the form of the constitutive quantities.

We would like to introduce the expression for the heating in a form involving pressure and temperature, instead of volume and temperature. Remember that the new independent variable, pressure, can be expressed in terms of the old ones, temperature and volume. In practice, heating often occurs at constant pressure, which makes the new form particularly useful for some applications:

$$I_s = -\Lambda_P \dot{P} - K_P \dot{T} \tag{44}$$

The new constitutive quantities are called the *latent entropy with respect to pressure*, and the *entropy capacity at constant pressure*, respectively. For the ideal gas, it is very simple to relate them to quantities involving the volume of the body. To derive this relationship we need the equation of state of the ideal gas, or its time derivative:

$$P\dot{V} + V\dot{P} = nR\dot{T} \tag{45}$$

---

4. Proof of Equations (42) and (43): We start from J. Ivory's observation (see the quote at the beginning of Section 2.3.2) that the entropy exchanged is resolvable into two parts. Also, the entropy absorbed in a process will be assumed to be emitted in the reverse operation. The simplest form of a constitutive law allowing for these observed features is the following:

$$I_s = -\Lambda_V{}^* \dot{V} - K_V{}^* \dot{T} \tag{i}$$

with the constitutive quantities of heating $\Lambda_V{}^*$ and $K_V{}^*$. Let us now introduce this law for the current of entropy, and the definition of the latent entropy and the entropy capacity, Equation (41), into the equation of balance of entropy, which leads to:

$$\Lambda_V \dot{V} + K_V \dot{T} - \Lambda_V{}^* \dot{V} - K_V{}^* \dot{T} = \Pi_s$$

or

$$(\Lambda_V - \Lambda_V{}^*)\dot{V} + (K_V - K_V{}^*)\dot{T} \geq 0 \tag{ii}$$

Equation (ii) must hold for any value of the rate of change of volume or of temperature. It would be possible to violate the inequality for some values of the rates of change of the independent variables unless we set the constitutive quantities of heating in (i) equal to the latent entropy and the entropy capacity, respectively:

$$\Lambda_V - \Lambda_V{}^* = 0 \quad , \quad K_V - K_V{}^* = 0$$

which leaves us with Equations (42) and (43).

This allows us to replace the time derivative of the volume in Equation (42):

$$\begin{aligned} I_s &= -\Lambda_V \frac{1}{P}\left(nR\dot{T} - V\dot{P}\right) - K_V\dot{T} \\ &= \Lambda_V \frac{V}{P}\dot{P} - \left(\Lambda_V \frac{nR}{P} + K_V\right)\dot{T} \end{aligned} \tag{46}$$

If we compare this to the expression of the heating involving the pressure, Equation (44), we find that

$$\Lambda_P = -\Lambda_V \frac{V}{P} \tag{47}$$

$$K_p = \frac{nR}{P}\Lambda_V + K_V \tag{48}$$

This result demonstrates that knowledge of both entropy capacities is equivalent to knowing the capacity at constant volume and the latent entropy with respect to volume. Since the measurement of the entropy capacities will prove simpler than that of the latent entropy in many cases, the relationships just derived are of particular importance.

2.3.5 The Constitutive Problem of the Ideal Gas

The ideal gas is described by three functions, the pressure, and the latent entropy, and the entropy capacity. We know the first of these three relations, but not the latter two. We can gain additional information on the latent entropy on the basis of theoretical considerations alone. In this subsection we shall give a brief account of the theory along the lines of what Carnot did some 150 years ago. The entropy capacities will be found after considering adiabatic processes in Section 2.4. When we finally have solved the complete constitutive problem of the ideal gas, we will be able to discuss in detail the operations undergone by a fluid serving as the driving agent in a heat engine (Section 2.5).

By applying the laws we know so far to the operation of a Carnot engine (Section 1.4), we will be able to determine the latent entropy of the ideal gas. For this purpose we have to discuss how the ideal gas is used as the driving agent of a Carnot engine. Absorbing entropy from a furnace at constant temperature is achieved by letting the ideal gas *expand isothermally* (step 1, from points 1 to 2 in Figure 7). The second step must be an *adiabatic expansion*, which serves to lower the temperature of the gas to that of the cooler. During step 3 the entropy absorbed must be discharged (Figure 7). This is done by *compressing the fluid isothermally*. Finally, in step 4 (between points 4 and 1), the gas is *compressed adiabatically*. As a consequence, the temperature of the working agent is raised back to its initial value. The four-step cycle just outlined is called a *Carnot cycle*.

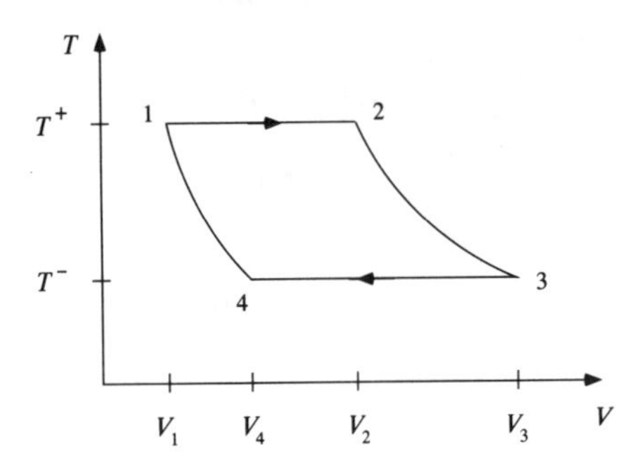

FIGURE 7. Carnot cycle of the ideal gas. The curved lines are the adiabats. The cycle runs between a furnace at high temperature and a cooler at lower temperature.

To obtain a restriction on the constitutive quantities of the ideal gas we proceed as follows. (This pretty much corresponds to what Carnot did in his theory of heat engines; see the Interlude for a more detailed account.) Consider a Carnot cycle operating with a very small difference of the temperatures between the furnace and the cooler:

$$\Delta T = T^{+} - T^{-} << T^{+} \tag{49}$$

In this case the adiabats of the ideal gas in Figure 7 are very short, which means that they do not contribute to the exchange of energy in the *mechanical* process. For this reason we do not have to know more about adiabatic processes at this point. The work done in such a Carnot cycle is determined by the isotherms alone. According to what we have calculated above, Equation (E7), this quantity must be equal to

$$\begin{aligned} W &= -nR(T+\Delta T)\ln\left(\frac{V_2}{V_1}\right) + nRT\ln\left(\frac{V_3}{V_4}\right) \\ &\approx -nR\Delta T\ln\left(\frac{V_2}{V_1}\right) \end{aligned} \tag{50}$$

The last step is a consequence of the fact that for a Carnot cycle having short adiabats the corresponding volumes must be nearly equal (Figure 7), i.e., $V_3 = V_2$, and $V_4 = V_1$. In Section 1.4 we calculated the motive power of a Carnot engine; see Equation (11). The energy released by the engine in one cycle is given by

$$W = -(T^{+} - T^{-})S_e = -\Delta T S_e \tag{51}$$

If we combine Equations (50) and (51), we see that the entropy absorbed from the furnace must be

$$S_e = nR\ln\left(\frac{V_2}{V_1}\right) \tag{52}$$

On the other hand, the entropy absorbed in the isothermal expansion, S_e, is equal to the integral of the latent entropy over the volume, Equation (35). This is possible only if the *latent entropy of the ideal gas* is given by

$$\Lambda_V = \frac{nR}{V} \tag{53}$$

In summary, the theory of thermodynamics—the relationship between entropy and energy—determines one of the constitutive quantities of the ideal gas.

EXAMPLE 12. Isothermal heating of the ideal gas.

Determine the entropy and the energy exchanged for the ideal gas in isothermal heating, and give the values for one mole of an ideal gas whose volume is doubled. The temperature is 300 K.

SOLUTION: The entropy exchanged in this process is calculated according to Equations (35) and (53):

$$S_e = \int_{V_i}^{V_f} \Lambda_V \, dV = \int_{V_i}^{V_f} \frac{nR}{V} dV = nR \ln\left(\frac{V_f}{V_i}\right)$$

The numerical value for doubling the volume of one mole of the gas is + 5.76 J/K (entropy has been absorbed). Since the temperature is constant during the process the amount of energy exchanged in heating is obtained if we multiply the entropy absorbed by the temperature:

$$Q = \int_{V_i}^{V_f} T\Lambda_V \, dV = \int_{V_i}^{V_f} \frac{nRT}{V} dV = nRT \ln\left(\frac{V_f}{V_i}\right)$$

The numerical value is 1.73 kJ for this example. Note that according to this result and Equation (E7), the energies exchanged in isothermal processes of the ideal gas are equal in magnitude but of opposite sign for mechanical and thermal operations.

2.4 Adiabatic Processes and the Entropy Capacities of the Ideal Gas

In this section we will study a phenomenon which appears to be very surprising at first sight. In Section 1.3 we saw that the temperature of a compressible fluid can be raised by compression alone without heating. Processes in which heating is absent, are called *adiabatic*. They play a major role in natural and manmade phenomena, such as the propagation of sound, the transport of heat by convection in the Earth's atmosphere, or cycles in heat engines and refrigerators. In an example, we shall continue the discussion of the propagation of sound which was begun in the Prologue. Since adiabatic processes allow us to measure a constitutive quantity of the ideal gas, namely, the ratio of its entropy capacities, we will be able finally to compute all quantities determining the thermomechanical processes of this fluid. The section will conclude with a short discussion of polytropic processes.

2.4.1 Qualitative Description of Adiabatic Motion of Ideal Fluids

If you suddenly compress air inside a cylinder, its temperature increases; if the air is expanded, its temperature must decrease. These changes occur without heating or cooling. You can insulate the cylinder against the flow of entropy, which does not change the result.

These processes are called *adiabatic*. The term *adiabatic* simply means that *no entropy has been exchanged* between the body and its surroundings. In other words, entropy flow between the body and the environment does not take place. It is convenient to display adiabatic processes in a *T-V* diagram (Figure 8), or in the *T-S* diagram (Chapter 1, Figures 12 and 18). The temperature of the ideal gas drops when the fluid is allowed to expand. Also, the ideal gas can undergo reversible processes only. For this reason, the amount of entropy stored in the gas remains constant while the temperature increases. We have to conclude that the curve representing this process in the *T-S* diagram is a vertical arrow pointing up. The reverse, adiabatic expansion, naturally would result in a drop of temperature at constant entropy. The representative curve in the *T-S* diagram will be a vertical arrow pointing down.

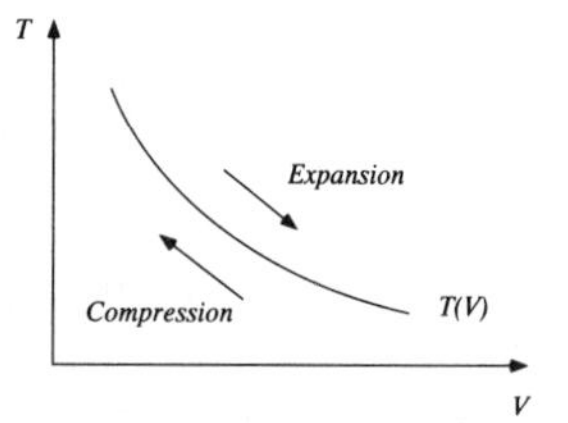

FIGURE 8. Adiabatic compression and expansion. The term *adiabatic* means that entropy cannot be transmitted across the boundaries of the body. The temperature of most fluids, such as the ideal gas, decreases as a consequence of adiabatic expansion. The curve is typical of the ideal gas.

How does this change arise? Our simple descriptions of isochoric and isothermal processes contain the seeds of an understanding of adiabatic operations. In real life, heating at constant volume and heating at constant temperature usually do not occur separately. The entropy content of a body changes when its volume and temperature change. Part of the added entropy increases the temperature, and part of it is changes the volume. In adiabatic changes, however, entropy does not cross the boundary of the body. Therefore the entropy emitted in the isothermal compression of the gas cannot leave the system. As a consequence, it raises the temperature. In somewhat oversimplified terms, we could say that in adiabatic compression some latent heat has been "converted" into sensible heat. (Do not take this literally; there only is one type of entropy inside a body!)

We can give a mathematical statement of the fact that heat does not cross the surface of the system under consideration. The current of entropy vanishes, which means that

$$I_s = 0 \tag{54}$$

which according to Equation (42) is equivalent to

$$0 = -\Lambda_V \dot{V} - K_V \dot{T} \tag{55}$$

or

$$\frac{dT}{dV} = -\frac{\Lambda_V}{K_V} \tag{56}$$

This is the fundamental equation of adiabatic change of simple fluids. In principle we can solve the differential equation if we know the ratio of the latent entropy and the entropy capacity of the fluid.

2.4.2 The Solution of the Problem of Adiabatic Motion of the Ideal Gas

We can derive the theory of adiabatic changes for the ideal gas. As a first step, the expression for the entropy current given in Section 2.3.4 has to be transformed so that it contains the two entropy capacities. The equation of state is used to eliminate the rate of change of the temperature in Equation (42). From Equation (45) we obtain

$$\dot{T} = T\left(\frac{\dot{P}}{P} + \frac{\dot{V}}{V}\right) \tag{57}$$

Now, the entropy current takes the following form:

$$\begin{aligned} I_s &= -\Lambda_V \dot{V} - K_V \dot{T} \\ &= -\Lambda_V \dot{V} - K_V T\left(\frac{\dot{P}}{P} + \frac{\dot{V}}{V}\right) \\ &= -T\left[K_V \frac{\dot{P}}{P} + \left(K_V + \Lambda_V \frac{V}{T}\right)\frac{\dot{V}}{V}\right] \end{aligned} \tag{58}$$

Using the relationship between K_P and K_V, Equations (47) and (48), we conclude that

$$I_s = -T\left[K_V \frac{\dot{P}}{P} + K_P \frac{\dot{V}}{V}\right] \tag{59}$$

If we now apply the condition of adiabatic change, Equation (54), we get a simple differential equation for pressure and volume. With the *ratio of the entropy capacities* defined by

$$\gamma \equiv \frac{K_P}{K_V} \tag{60}$$

this equation takes the form

$$\frac{\dot{P}}{P} + \gamma \frac{\dot{V}}{V} = 0 \tag{61}$$

Naturally, the equation of adiabatic change can be integrated only if the constitutive quantities themselves, or their ratio γ, are known for the ideal gas. Fortunately, we have independent information concerning the ratio of the entropy capacities. Different types of measurements, which do not involve measuring amounts of entropy, all indicate that *for the ideal gas this ratio must be constant*:

$$\gamma = constant \tag{62}$$

Also, this constant is larger than 1, which means that more entropy is needed to raise the temperature of the ideal gas by one degree in an isobaric process than in an isochoric one. An example of how to measure this important ratio is given in Example 13. More important still is the propagation of sound in the ideal gas (Example 14), which gives the same result. Now the solution of the problem of *adiabatic motion of the ideal* gas is very simple. Integration of the differential equation gives the *law of Poisson and Laplace*:

$$PV^{\gamma} = constant \tag{63}$$

Using the thermal equation of state of the ideal gas, $PV = nRT$, this law can be written in different variables:

$$P^{1-\gamma}T = constant \tag{64}$$

or

$$TV^{\gamma-1} = constant \tag{65}$$

Remember that all of these equations hold *for the ideal gas only*. The curves representing adiabatic change in a pressure-volume (P-V) diagram, which are usually found in books on thermodynamics, are derived from the form just calculated. For different types of fluids there may be completely different adiabats which do not even faintly resemble those for the ideal gas.[5] *Existence and form of the adiabats are constitutive properties.* Since $\gamma > 1$, adiabats of the ideal gas in the P-V diagram are steeper than isotherms.

EXAMPLE 13. Adiabatic oscillations for measuring the ratio of the entropy capacities of air.

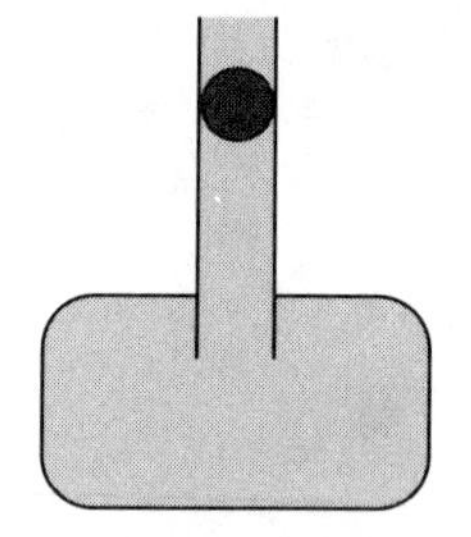

A large bottle is fitted with a thin glass pipe. If we drop a steel ball into the pipe it will oscillate there. (The diameter of the ball has to be slightly smaller than that of the pipe.) The oscillations are slow enough to be observed, yet fast enough to allow for the compressions and expansions of the air to be adiabatic. The bottle has to be large enough so as to leave the pressure and the volume of the air in it virtually unchanged.

5. See Thomsen and Hartka (1962), and the Interlude.

SOLUTION: The oscillations of the ball are driven by the slight pressure changes which we take to be adiabatic. If they are small we can approximate them using Equation (61):

$$\Delta P = -\gamma \frac{P}{V} \Delta V = -\gamma \frac{PAx}{V}$$

where x and A are the displacement in the pipe, and the cross section of the pipe, respectively. The equation of motion of the ball (with mass m) is

$$\ddot{x} - \frac{A \Delta P}{m} = 0$$

or

$$\ddot{x} + \frac{\gamma P A^2}{mV} x = 0$$

If all the factors in the differential equation are constant, its solution is that of a harmonic oscillation of frequency

$$f = 2\pi \sqrt{\frac{\gamma P A^2}{mV}}$$

The frequency and the other quantities can be measured, which allows for the ratio of the specific entropy capacities to be determined. For air we get values of 1.40 for γ, independent of the temperature.

EXAMPLE 14. The speed of sound in the ideal gas.

a) Derive the formula for the speed of sound propagation in a fluid on the basis of the theory given in the Prologue (Section P.3.6). For low frequencies, the wavelength is large, and heat would have to travel very far from areas of compression to those of rarefaction. It turns out that there is no time for this effect, which leads to low frequency sound waves being adiabatic.[6] On the other hand, very high frequency sound waves are isothermal.
b) Derive the result for the speed of sound for the ideal gas for these limiting cases.

SOLUTION: a) We have to compute the capacitance per length and the inductance per length of a fluid to compute the speed of sound according to Equation (42) of the Prologue. The capacitance per length is equal to the mass per length, which is

$$C^* = \frac{\Delta m}{\Delta x} = A\rho$$

6. See Wu (1990).

The inductance per length, on the other hand, may be calculated using Equation (39) of the Prologue. The momentum current density j_p is equal to the pressure P of the fluid. If there is a difference of speeds at the ends of a column of fluid of length Δx and cross section A, its volume is bound to change at a rate

$$\dot{V} = A\Delta v$$

If we plug this into Equation (38) of the Prologue, we obtain

$$\dot{V} = -L^* \Delta x A^2 \dot{P}$$

which may be solved for the inductance per length:

$$L^* = -\frac{1}{\Delta x A^2}\frac{dV}{dP}$$

The square of the speed of sound is equal to the inverse of the product C^*L^*. If we also change the derivative from volume to density we finally obtain

$$c^2 = \frac{dP}{d\rho}$$

for the square of the speed of propagation of sound in a simple fluid.

b) The derivative of the pressure of the fluid with respect to its density has to be evaluated for the cases of adiabatic and isothermal compression. For the former we have the relation

$$P = \text{constant}\,\rho^{\gamma}$$

between pressure and density; see Equation (63). The derivative is computed simply, and we obtain the final result

$$\begin{aligned} c^2_{adiabatic} &= \gamma\frac{P}{\rho} \\ &= \gamma\frac{R}{M_o}T \end{aligned}$$

This expression is a result of the equation of state of the ideal gas, Equation (30), and agrees well with measurements. For very high frequencies, however, the phenomenon is more isothermal, which leads to

$$\begin{aligned} c^2_{isothermal} &= \frac{P}{\rho} \\ &= \frac{R}{M_o}T \end{aligned}$$

This shows that adiabatic waves travel about twenty percent faster than isothermal ones in air. (The ratio of the entropy capacities is 1.40.) The difference between the theory based on isothermal oscillations and actual measurements led Laplace to propose that sound waves are adiabatic.

EXAMPLE 15. The adiabatic temperature gradient in a gravitational field.

Consider a fluid of ideal gas having a pressure gradient as might be set up in the Earth's atmosphere. Calculate the associated temperature gradient if the pressure and density are related by the Poisson-Laplace Law, a condition which can be satisfied approximately if entropy is transferred with the air by convection (Chapter 4). Show that in the Earth's atmosphere the *adiabatic temperature gradient* is constant. Compute its value for air, for which the ratio of entropy capacities is 1.4.

SOLUTION: The Poisson-Laplace Law relates the pressure and temperature of the ideal gas according to

$$P^{1-1/\gamma} = aT \quad , \quad a = \frac{P_o^{1-1/\gamma}}{T_o}$$

The temperature gradient dT/dz can be derived directly from this in terms of the pressure gradient dP/dz:

$$\frac{dT}{dz} = \left(1 - \frac{1}{\gamma}\right)\frac{T}{P}\frac{dP}{dz}$$

This gradient is commonly called the *adiabatic temperature gradient.* We can calculate the pressure gradient in the Earth's atmosphere according to the law of hydrostatic equilibrium

$$\frac{dP}{dz} = -g\rho$$

where $g = 9.81$ N/kg is the gravitational field strength. If we replace the density, using the Poisson-Laplace law

$$P^{1/\gamma} = b \cdot \rho \quad , \quad b = \frac{P_o^{1/\gamma}}{\rho_o}$$

we obtain for the gradient

$$\frac{dT}{dz} = -\left(1 - \frac{1}{\gamma}\right)\frac{g}{a \cdot b} = -\left(1 - \frac{1}{\gamma}\right)\frac{g \cdot M_o}{R}$$

which obviously is constant. The numerical value for air is

$$\frac{dT}{dz} = -\left(1 - \frac{1}{1.4}\right)\frac{9.81 \cdot 0.029}{8.31}\frac{\text{K}}{\text{m}} = -0.010\frac{\text{K}}{\text{m}}$$

This value is not bad, but it is slightly too large. The reason for the discrepancy can be found in the fact that the air contains water vapor. The water vapor condenses during upward motion of the air, and thus injects entropy into the gaseous component, which invalidates the condition of adiabatic motion for this component. A better temperature gradient must therefore lie somewhere between the adiabatic value (which is too large) and the value zero which holds for an isothermal atmosphere. A correction can be made using the theory of polytropic processes (Section 2.4.4). A better value for the actual temperature gradient is - 0.006 K/m.

It is possible, however, to have air temperature differences of - 0.010 K/m between mountains and valleys. This is the case if air ascends on one side of a mountain, gets rid of its moisture through rain, and then descends as dry air on the other side. The air rushing down obtains a larger temperature than the one already present in the valley, leading to a warm, dry wind. (In the Alps this wind is called Föhn.)

2.4.3 Determining the Entropy Capacities of the Ideal Gas

In Section 2.3 we managed to determine the latent entropy of the ideal gas on the basis of theoretical considerations. The missing piece of information, i.e., the entropy capacities of the ideal gas, can be calculated if we know their ratio. We have found that this ratio must be constant and that it can be measured in a number of ways. According to the results of Section 2.3.4, Equations (47) and (48), we obtain

$$K_V = \frac{nR}{\gamma - 1}\frac{1}{T} \tag{66}$$

$$K_P = \frac{\gamma nR}{\gamma - 1}\frac{1}{T} \tag{67}$$

The entropy capacities of the ideal gas are inversely proportional to its temperature. Normally, the products of the entropy capacities and the temperature are used to calculate amounts of energy exchanged in heating. They are defined as follows:

$$C_V = TK_V \tag{68}$$

$$C_P = TK_P \tag{69}$$

We will call the former the *temperature coefficient of energy* (just as in the case of pure heating discussed in Section 2.2), while the latter is the *temperature coefficient of enthalpy*. These names will be explained in Section 2.5. Normally they are called the *heat capacity at constant volume* and the *heat capacity at constant pressure*, respectively. There is a simple relationship between the molar quantities calculated at constant volume and at constant pressure:

$$\bar{c}_P = \bar{c}_V + R \tag{70}$$

See Example 18 for a derivation of this result.

Since the entropy capacities of the ideal gas are inversely proportional to the temperature of the body, its temperature coefficients must be constant. According to quantum theory, we would expect only two or three discrete values for the temperature coefficient of energy:

$$\bar{c}_V = \frac{1}{2} f R \ , \quad f = 3,5,7 \tag{71}$$

The factor f is called the number of degrees of freedom. Monatomic gases have the value f = 3 at all temperatures, while many diatomic molecules exhibit f = 5 at room temperature; f = 7 is found for larger molecules. The gases found in nature agree with this rule to a modest extent (Figure 9). The best agreement is found for the noble gases, which are monatomic molecules. Hydrogen gas displays the transition from f = 3 to f = 5 between 100 K and 400 K, after which a constant value of the temperature coefficient is maintained for several hundred Kelvin. Roughly speaking, this behavior is attributed to the "unfreezing" of internal modes of motion, rotation, and vibration, as the temperature increases. It does not find any explanation in classical physics.

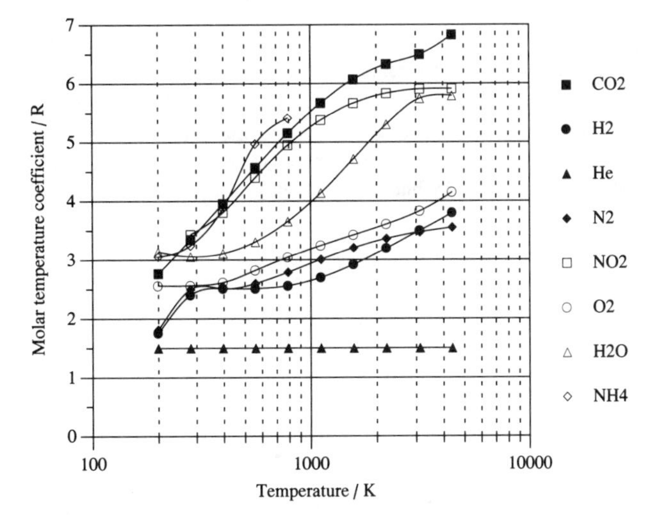

FIGURE 9. Values of the molar temperature coefficient of energy of various gases. Its value, divided by the gas constant, has been displayed as a function of temperature. The data have been computed using thermodynamic property functions implemented in the program EES (Klein et al. 1991).

EXAMPLE 16. Entropy and energy exchanged at constant volume.

Calculate the entropy and the energy exchanged in heating an ideal gas at constant volume in terms of the temperatures of the body. Calculate the numerical values for one mole of a monatomic gas for a change of temperature from 300 K to 600 K.

SOLUTION: Equation (40) tells us that the entropy exchanged is:

$$S_e = \int_{t_i}^{t_f} K_V \dot{T} dt = \int_{T_i}^{T_f} K_V dT = C_V \ln\left(\frac{T_f}{T_i}\right) \quad \textbf{(E8)}$$

for the ideal gas. According to Equation (71) the molar temperature coefficient of energy of a monatomic ideal gas must be $1.5R$ = 12.5 J/(K · mole). Therefore, one mole of the ideal gas absorbs 8.64 J/K of entropy if the temperature is doubled (the actual values of the temperature do not matter).

The energy exchanged in heating at constant volume is found by integrating the thermal energy current over time. The thermal energy current is equal to the product of the current of entropy according to Equation (42) and temperature. If the volume is kept constant during heating, we obtain

$$Q = \int_{t_i}^{t_f} TK_V \dot{T} dt = \int_{T_i}^{T_f} C_V dT = C_V \left(T_f - T_i \right) \qquad \textbf{(E9)}$$

The numerical value turns out to be 3.74 kJ. The results derived here, i.e., relations (E8) and (E9), hold since the temperature coefficient of energy of the ideal gas is independent of temperature.

EXAMPLE 17. Change of the entropy content of the ideal gas.

Calculate the change of the entropy content of the ideal gas as a function of volume and temperature from the knowledge of the latent entropy and the entropy capacity.

SOLUTION: For the reversible processes considered, the change of the entropy content can be computed in terms of the entropy exchanged. In other words, the change of the entropy content can be calculated by integrating either Equation (41) or Equation (42):

$$\Delta S = \int_{t_i}^{t_f} \left[\Lambda_V \dot{V} + K_V \dot{T} \right] dt$$

$$= \int_{V_i}^{V_f} \frac{nR}{V} dV + \int_{T_i}^{T_f} \frac{n\bar{c}_V}{T} dT$$

$$\Delta S = nR \ln\left(\frac{V_f}{V_i} \right) + n\bar{c}_V \ln\left(\frac{T_f}{T_i} \right)$$

The change of the entropy content does not depend on the initial values of volume and temperature, but rather on the ratio of their values at the beginning and the end.

EXAMPLE 18. Temperature coefficient of enthalpy of the ideal gas.

a) Express the molar temperature coefficient of enthalpy of the ideal gas in terms of the temperature coefficient of energy, and calculate its value for a monatomic gas such as helium. b) Compare the result with measurements of the ratio of the entropy capacities, which is 1.66 for helium. c) Show that the integral of the temperature coefficient of enthalpy over temperature is the energy exchanged as a consequence of heating at constant pressure.

SOLUTION: a) According to the definition and Equation (69), the molar temperature coefficient of enthalpy is given by

$$\bar{c}_P = \frac{1}{n} TK_p = \frac{T}{n}\left(\frac{nR}{P}\Lambda_V + K_V \right)$$

If we use the values of the latent entropy with respect to volume and the entropy capacity at constant volume of the ideal gas we obtain

$$\bar{c}_P = \frac{T}{n}\left(\frac{V}{T}\frac{nR}{V} + \frac{n\bar{c}_V}{T}\right) = R + \bar{c}_V$$

The equation of state of the ideal gas was used in the derivation as well. We obtain the numerical value for a gas whose molecules are single atoms from the expression Equation (71) for the temperature coefficient of energy. With $f = 3$, the value is $3R/2 + R = 5R/2 = 20.8$ J/(K · mole), which agrees closely with observation.

b) With the ratio of the entropy capacities measured, we determine the temperature coefficient of enthalpy (or of energy) according to Equation (67). With the definition in Equation (69) we have

$$\bar{c}_P = T\bar{k}_P = \frac{\gamma R}{\gamma - 1}$$
$$= 20.9\frac{\text{J}}{\text{K} \cdot \text{mole}}$$

c) The heating can be related to the flux of energy by

$$I_{E,th} = TI_s = T\left(-\Lambda_P \dot{P} - K_P \dot{T}\right)$$

At constant pressure, the energy exchanged is calculated according to

$$Q = -\int_{t_i}^{t_f} I_{E,th}\,dt = \int_{t_i}^{t_f} TK_P\dot{T}dt = \int_{T_i}^{T_f} C_P\,dT$$

If the temperature coefficient of enthalpy is constant, the energy added is proportional to the change of temperature.

EXAMPLE 19. Measuring the temperature coefficient of enthalpy of air.

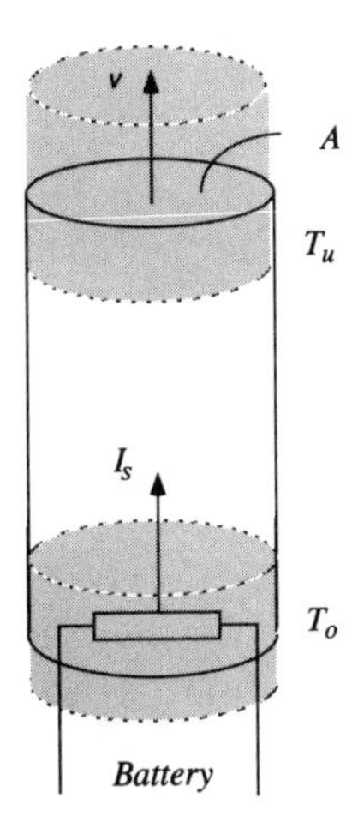

Measuring the entropy capacity at constant volume or the temperature coefficient of energy of gases is difficult, since their values are small. For this reason, one often uses a flow method for measuring the corresponding values at constant pressure. Air is heated electrically at the bottom of a vertical open tube and flows upward. By measuring the speed of air flow and the temperature of the air at the top of the pipe we can infer the value of the temperature coefficient of enthalpy of the gas. (This is a simple experiment which can be built by hand.)

SOLUTION: Consider a certain amount of air which is heated at the lower end and then rises through the tube. We compare the two states of the body of air. If v is the speed of air leaving the tube, the amount of substance of the body of air flowing in time Δt is given by:

$$n = \frac{m}{M_o} = \frac{\rho(T_u)v\Delta t A}{M_o}$$

where A is the cross section of the tube, and T_u is the temperature of the air at the exit. The amount of energy added in heating (Q) is responsible for the change of temperature of the air.

We can calculate the amount of energy transmitted to the air in a time Δt from the value of the power of the electrical heater. Since the process takes place at constant pressure, we have

$$Q = n\bar{c}_P\left(T_u - T_o\right)$$

If P_o is the ambient pressure, the density of air at the outlet can be calculated using the equation of state of the ideal gas, Equation (30):

$$\rho(T_u) = \frac{M_o P_o}{RT_u}$$

Taken together, we arrive at the following expression for the molar temperature coefficient of enthalpy:

$$\bar{c}_P = \frac{RT_u}{AP_o\Delta T v}\frac{Q}{\Delta t}$$

2.4.4 Polytropic Processes

So far we have modeled heating and working of the ideal gas in isothermal and adiabatic processes, where the heating is of a particular kind. We can introduce a more general kind of heating or cooling, of which the processes mentioned are special cases. Remember that in adiabatic processes the heat transfer vanishes while for isothermal changes it takes a special value. In practice, it might take some intermediate value. For this reason, we write the current of entropy in the form

$$I_s = -K\dot{T} \quad \textbf{(72)}$$

An operation conforming to this type of heating is called a *polytropic process*. K is a quantity resembling a generalized entropy capacity. A number of practical processes may be modeled as polytropic changes, including a gas being compressed in a cylinder and being allowed to exchange entropy with the cylinder walls only, or the adiabatic convection of moist air.

This definition shall now be applied to the ideal gas. Remember that this fluid can undergo only reversible processes. Now, with the rate of change of the entropy content expressed in terms of the entropy capacity and the latent entropy, Equation (41), and with the heating assumed for polytropic processes as in Equation (72), we obtain the following equation of balance of entropy:

$$K_V\dot{T} + \Lambda_V\dot{V} = K\dot{T} \quad \textbf{(73)}$$

The latent entropy can be replaced by an expression involving both entropy capacities—see Equation (48)—which leads to the differential equation for polytropic processes:

$$\frac{K_P - K_V}{K_V - K}\frac{\dot{V}}{V} + \frac{\dot{T}}{T} = 0 \tag{74}$$

We can recover adiabatic and isothermal processes if we set

$$\begin{aligned} K &= 0 && \text{for adiabatic processes} \\ K &= \pm\infty && \text{for isothermal processes} \end{aligned} \tag{75}$$

For the general polytropic process, K takes an intermediate value. If we introduce the *polytropic exponent*

$$\gamma^* = \frac{K_P - K}{K_V - K} \tag{76}$$

Equation (74) becomes equivalent to the differential equation of adiabatic motion, which leads to the Poisson-Laplace law with γ replaced by γ^*. The polytropic exponent is a generalization of the adiabatic exponent. We shall always take it to be a constant, just as in the case of the ideal gas. The solution of Equation (74) is a form of the law of polytropic processes of the ideal gas:

$$V^{\gamma^*-1}T = constant \tag{77}$$

which holds only if γ^* is a constant. Its form is equivalent to the laws of adiabatic motion of the ideal gas. Remember that it corresponds to a process in which the fluid is being heated in a particular way.

EXAMPLE 20. Polytropic compression of a body of ideal gas.

Consider a certain amount of an ideal gas (body 1) which is enclosed in a cylinder having a piston (body 2). The cylinder and the gas are insulated from the environment; however, they may exchange entropy with each other. Derive the relation between the pressure and the volume of the gas if it is compressed. The temperatures of the gas and the cylinder walls are assumed to be equal during the process. The cylinder walls have an entropy capacity K_2. (The associated temperature coefficient of energy is taken to be constant.)

SOLUTION: We model both bodies as spatially uniform. Their heating takes the forms

$$\Lambda_{V1}\dot{V} + K_{V1}\dot{T} = -I_{s1}$$

and

$$K_2\dot{T} = -I_{s2}$$

Such bodies can undergo only reversible processes. Since their temperatures are equal at all times, the current of entropy leaving the gas is equal to the flux entering the walls. This leads to the following differential equation:

$$\Lambda_{V1}\dot{V} + K_{V1}\dot{T} = K_2\dot{T}$$

or

$$\frac{nR}{V}\dot{V} + \frac{C_{V1} + C_2}{T}\dot{T} = 0$$

Now we replace the rate of change of temperature by the rate of change of pressure, using the equation of state of the ideal gas. We obtain

$$\frac{nR + C_{V1} + C_2}{C_{V1} + C_2}\frac{\dot{V}}{V} + \frac{\dot{P}}{P} = 0$$

This is the differential equation of a polytropic process with the following polytropic exponent:

$$\gamma^* = \frac{nR + C_{V1} + C_2}{C_{V1} + C_2}$$

Without the temperature coefficient of energy of the walls, i.e., without the effect of exchange of entropy between the gas and the walls of the enclosure, the process would be adiabatic. Here, the polytropic exponent of the gas is smaller than its adiabatic counterpart.

EXAMPLE 21. Adiabatic convection of moist air.

In Example 15, the adiabatic temperature gradient in the Earth's atmosphere was computed for dry air. It was found to be - 0.010 K/m. More often, it is about - 0.006 K/m. The difference is due to the condensation of water vapor due to the decrease of temperature as the air rises. The process may be modeled more closely as polytropic heating of the gaseous component. The entropy comes from the change of phase of the other component. a) Determine the polytropic exponent appropriate to this process. b) Determine the coefficient K appearing in the expression of the heating of the gas. What is the significance of the sign of this quantity?

SOLUTION: a) According to Example 15, the adiabatic temperature gradient in the Earth's atmosphere is given by the expression

$$\frac{dT}{dz} = -\left(1 - \frac{1}{\gamma}\right)\frac{gM_o}{R}$$

where g is the gravitational field strength (9.81 N/kg), and M_o is the molar mass of air (0.029 kg/mole). Here, γ stands for the adiabatic exponent of air, 1.4. If we now model the effect of condensation by applying the theory of polytropic processes, we only have to replace the adiabatic exponent by the polytropic one. The cases are formally equivalent:

$$\frac{dT}{dz} = -\left(1 - \frac{1}{\gamma^*}\right)\frac{gM_o}{R}$$

With the numerical value of the actual temperature gradient we obtain for the polytropic exponent:

$$\gamma^* = \left(1 + \frac{R}{gM_o}\frac{dT}{dz}\right)^{-1}$$

$$= \left(1 + \frac{8.31}{9.81 \cdot 0.029}(-0.0060)\right)^{-1} = 1.21$$

b) From the definition of the polytropic exponent we derive the generalized temperature coefficient K:

$$K = \frac{1}{\gamma^* - 1}\left(\gamma^* K_V - K_p\right) = \frac{1}{\gamma^* - 1}\frac{1}{T}\left(\gamma^* C_V - C_p\right)$$

Its numerical value turns out to be - $0.90C_V/T$. Therefore the heating of the air is given by:

$$I_s = +0.90\frac{C_V}{T}\dot{T}$$

The sign of K tells us that entropy is injected into the gaseous component as the temperature decreases. This must be so since water vapor condenses as the air rises. (See Chapter 4 for more details.)

EXAMPLE 22. Polytropic gas spheres.

Stars are gas spheres which in most cases are pretty nearly in hydrostatic equilibrium. Often, pressure and density are related by a law analogous to the one derived for polytropic changes. This might be due to convection (Chapter 4), or to a particular equation of state (degenerate gases). Derive the relationships between density and temperature, and between pressure and density, and the equation of hydrostatic equilibrium in a gravitating polytrope.

SOLUTION: From Equation (77) we get

$$\rho = bT^n \quad , \quad b = constant \quad , \quad n \equiv \frac{1}{\gamma^* - 1}$$

n is called the polytropic index, and we shall essentially be concerned with values of n which lie between 1 and infinity. The pressure therefore is related to the density in the following way:

$$P = B\rho^{1+1/n} \quad , \quad B = constant$$

For very large n we recover the isothermal relationship between pressure and density. The law of hydrostatic equilibrium in a spherically symmetric star is the same as that derived in the Prologue, with the strength of the gravitational field given by $g = Gm/r^2$, where G is the gravitational constant, and m is the mass contained in a concentric sphere of radius r. We therefore find that

$$\frac{dP}{dr} = -\frac{G \cdot m}{r^2}\rho$$

The mass m is a variable which can be found as the solution to the differential equation

$$\frac{dm}{dr} = 4\pi r^2 \rho$$

In summary, we arrive at a single differential equation of hydrostatic equilibrium in a fluid sphere. Solve the equation giving the pressure gradient for the mass, take its derivative, and set the result equal to $4\pi r^2\rho$:

$$\frac{1}{r^2}\frac{d}{dr}\left(\frac{r^2}{\rho}\frac{dP}{dr}\right) = -4\pi G\rho$$

If we replace the density and the pressure by the polytropic laws given above, and if we introduce the dimensionless variable ξ by defining

$$r = \left[\frac{(n+1)B}{4\pi G}b^{1/n-1}\right]^{-1/2}\xi$$

we get the final form of the law of hydrostatic equilibrium which holds for a spherically symmetric polytrope:

$$\frac{1}{\xi^2}\frac{d}{d\xi}\left(\xi^2\frac{dT}{d\xi}\right) = -T^n$$

This equation is called the *Lane-Emden equation of index n.* (Note that by a particular choice of b, the variable T may be made dimensionless, in which case T is not the ideal gas temperature, but is simply related to the latter.) With appropriate boundary conditions at the center of the sphere it can be solved for $T(\xi)$. It has been used extensively to model stars.[7]

---

7. See S. Chandrasekhar (1967), Chapter 4, for the problem discussed here. Polytropic gas spheres played an important role in the development of the theory of stellar structure. For polytropes, pressure and density are related directly, removing the problem of temperature. If we go to a more general equation of state, we usually have to include the transport of heat in the set of basic differential equations. This considerably increases the size of the (physical and numerical) problem (see Schwarzschild, 1958; D.D. Clayton, 1968).

2.5 Some Applications of the Thermomechanics of the Ideal Gas

In the previous sections we solved the constitutive problem of the ideal gas. In other words, we have the thermal equation of state of this fluid, and we determined its latent entropy and entropy capacity. On the basis of these quantities we have been able to calculate the working and the heating of the ideal gas. In particular, it was found that the fluid can only undergo reversible processes. Now, as you will see in Section 2.5.1, the balance of energy can be proved to hold for this material, and the energy of the ideal gas can be derived. It will be found that the energy of this fluid depends only on its temperature. Our knowledge will then be applied to the computation of some cyclic processes undergone by the ideal gas.

2.5.1 Derivation of the Law of Balance of Energy for the Ideal Gas

If we combine the results obtained so far, we can define a new quantity of the ideal gas which is a function of both independent variables, i.e., temperature and volume. This quantity, which we interpret as the energy of the gas, can then be shown to satisfy an equation of balance for a conserved quantity. We must define the time derivative of the function E as follows:

$$\dot{E} = T\dot{S} - P\dot{V} \tag{78}$$

As mentioned in Section 2.2, a relation of this type is called a *Gibbs fundamental relation* of the system under investigation. We know that the product of pressure and rate of change of volume of the ideal gas is equal to the mechanical power due to compression or expansion. The first term on the right-hand side of the Gibbs relation is the negative of the thermal power, since the ideal gas can undergo only reversible processes. We conclude that the rate of change of the quantity E is equal to the sum of mechanical and thermal energy currents:

$$\dot{E} + I_{E,mech} + I_{E,th} = 0 \tag{79}$$

This clearly is the equation of balance of a conserved quantity. We interpret the function $E(T, V)$ as the energy of the ideal gas which has volume V and temperature T. Equation (79) expresses the fact that the energy of the body can change only as a consequence of the transfer of energy due to thermal or mechanical processes.

Remember that this derivation holds only for a limited case: the thermodynamics of the ideal gas. Naturally, there is a lot more behind this result than just a limited theorem which may be proved by mathematical means.

2.5.2 The Energy and the Enthalpy of the Ideal Gas

Now that we have determined all the constitutive quantities of the ideal gas we should expect to be able to calculate other properties of this simple fluid. We already have found the entropy of the ideal gas as a function of volume and temperature. The energy function is just as interesting. How does the energy of the ideal gas depend upon the independent variables?

Also, other functions derived from the energy function are used to facilitate some computations. They are called *thermodynamic potentials*; among them are quantities such as enthalpy, free enthalpy, and free energy. In this section we will introduce the enthalpy of a fluid, and explain where the names *temperature coefficient of energy* or *temperature coefficient of enthalpy* come from.

The energy function. We can prove an interesting and somewhat surprising result: *the energy of the ideal gas depends only on the temperature*. In other words, the values of the volume and pressure do not matter. As long as the temperature remains constant, the energy remains constant as well, and a change of the energy content of the gas depends only on a change of temperature. To prove this, let us start with the Gibbs fundamental relation for the ideal gas, Equation (78). The rate of change of the entropy of the body is given by Equation (41). Since the processes the ideal gas can undergo are reversible, the heating is equal to the negative rate of change of the entropy. If we plug this into Equation (78) we find that the rate of change of the energy of the fluid is given by

$$\dot{E} = (T\Lambda_V - P)\dot{V} + (TK_V)\dot{T} \quad \textbf{(80)}$$

Because of the form of the latent entropy in Equation (53), the volume dependence of the energy drops out. Therefore,

$$\dot{E} = (TK_V)\dot{T} \quad \textbf{(81)}$$

or in integrated form,

$$E(T) = C_V(T - T_o) + E_o \quad \textbf{(82)}$$

which holds since $C_V = TK_V$ is constant for the ideal gas. This result is important since it allows us to calculate processes undergone by the ideal gas in terms of a change of temperature only. We do not always have to follow the details of a process.

There is an experiment which can throw some light upon this problem, namely the *free expansion of the ideal gas*, which we have discussed in Section 1.5. The free expansion does not allow any exchange of energy between the gas and its surroundings. Thus the energy of the gas does not depend on the volume or pressure. Apparently, it can depend only on the temperature. This is quite surprising, since we might have expected the temperature of the gas to decrease as a consequence of the expansion. Indeed this is the case with real gases. (The volume dependence of the energy of real gases is used in liquefying such fluids.) To be sure, this effect has nothing to do with adiabatic nondissipative expansion, as a consequence of which the temperature of a gas drops dramatically.

Enthalpy. If you heat a fluid at constant volume, the only energy transfer is thermal. Put differently, all the energy transferred as a consequence of heating goes toward the change of the energy of the body. Therefore, we have a simple relation between energy supplied in heating ("heat") and change of energy. By the way, this simplest of all phenomena often leads to the identification of internal energy with the "heat of a body:" the "heat" supplied changes the "heat content" of the fluid. Remember that this usage of terms is not acceptable.

If you heat a fluid in any other way, there also will be an exchange of energy related to the change of volume. Therefore, the energy supplied during heating will not equal the change of the energy of the body. In practice, you will often encounter heating at constant pressure. Take a look at the balance of energy for such a process:

$$\dot{E} + P\dot{V} = -I_{E,th} \tag{83}$$

If you introduce a new quantity defined as follows:

$$H = E + PV \tag{84}$$

you can see that for heating at constant pressure, the rate of change of this new function is equal to the energy current in heating:

$$\dot{H} = -I_{E,th} \quad \text{for} \quad P = constant \tag{85}$$

In other words, the energy added in heating at constant pressure completely goes toward the change of the quantity H, which is called the *enthalpy* of the fluid. Note that in contrast to the energy of a body, there is no simple graphical image as to the meaning of enthalpy. The enthalpy plays a role in the formal development of thermodynamics, and in the formulation of the balance of energy for open systems (flow systems, Chapter 4).

The names *temperature coefficient of energy* and *temperature coefficient of enthalpy* can now be explained. In isochoric or isobaric processes, the energy current associated with heating is written as follows:

$$I_{E,th} = -C_V\dot{T} \quad \text{for} \quad V = constant$$

$$I_{E,th} = -C_p\dot{T} \quad \text{for} \quad P = constant$$

Because of the balance of energy, and Equation (85), the former expression is seen to be the rate of change of the energy, while the latter is the rate of change of the enthalpy of the fluid. Therefore, C_V and C_p play the roles of temperature coefficients of energy and enthalpy. Again, because of the identification of the energy transferred in heating with "heat", the coefficients have acquired the names *heat capacity at constant volume* and *heat capacity at constant pressure*, respectively, leading to considerable confusion as to the meaning of what is supposed to be a quantity of heat. Chemists often call the enthalpy the "heat content of a body."

EXAMPLE 23. Work done in an adiabatic expansion.

Express the energy exchanged in adiabatic expansion of the ideal gas in terms of the temperatures. Remember that the energy of the ideal gas depends only on the temperature.

SOLUTION: Since there is no exchange of energy in a thermal process in adiabatic expansion, the balance of energy takes the form

$$\Delta E = W$$

The change of the energy of the ideal gas is independent of volumes or pressures, and depends only on the initial and final temperatures, according to Equation (82). Therefore,

$$W = C_V\left(T_f - T_i\right)$$

In the following example, we will get the same result by transforming the expression obtained for the work done in adiabatic processes of the ideal gas.

EXAMPLE 24. Energy exchanged in isochoric, isothermal, and adiabatic processes.

Compare the following operations: (1) an adiabatic doubling of the volume of one mole of air from standard conditions, and (2) isochoric cooling followed by isothermal heating which lead from the same initial to final states as (1). a) Draw the processes in the *T-V* diagram. b) Calculate the energy exchanged as a consequence of mechanical and thermal processes for each step. c) Determine the change of the energy content in (1) and (2), respectively.

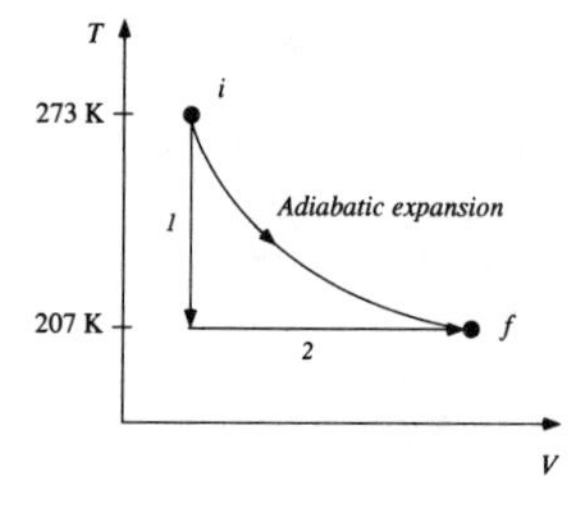

SOLUTION: b1) In the case of an adiabatic process undergone by an ideal gas with constant ratio of the entropy capacities, we use the relationship between volume and pressure according to the Law of Poisson and Laplace:

$$PV^\gamma = P_i V_i^\gamma$$

If we solve this relation for the pressure, we can compute the energy exchanged as a consequence of compression or expansion:

$$W = -\int_{V_i}^{V_f} P dV = -\int_{V_i}^{V_f} \frac{P_i V_i^\gamma}{V^\gamma} dV = \frac{P_i V_i^\gamma}{\gamma - 1}\left[\frac{1}{V_f^{\gamma-1}} - \frac{1}{V_i^{\gamma-1}}\right]$$

For standard conditions, the values of temperature and pressure are 0°C and 1 bar, respectively. The volume of one mole of an ideal gas at standard conditions turns out to be $22.4 \cdot 10^{-3}$ m^3, and the ratio of the entropy capacities is 1.4 for air. Now the energy exchanged must equal

$$W = \frac{1.013 \cdot 10^5 \cdot (22.4 \cdot 10^{-3})^{1.4}}{1.4 - 1}\left[\frac{1}{(44.8 \cdot 10^{-3})^{0.4}} - \frac{1}{(22.4 \cdot 10^{-3})^{0.4}}\right] \text{J} = -1.37\,\text{kJ}$$

The energy exchanged in the thermal process is zero for an adiabatic expansion: $Q = 0$.

b2) Let us number the isochoric and isothermal steps by 1 and 2, respectively. The energy exchanged in the mechanical process is zero for the isochoric change:

$$W_1 = 0$$

For isothermal expansion or compression we have computed the energy exchanged in working in Example 11.The numerical value turns out to be

$$W_2 = -1 \cdot 8.31 \cdot 207 \cdot \ln(2)\text{J} = -1.19\text{kJ}$$

This is so since the temperature drops to 207 K in the adiabatic doubling of the volume, and therefore also in the isochoric cooling. In the latter process the energy exchanged as a consequence of cooling is

$$Q_1 = \int_{T_i}^{T_f} C_V dT = C_V\left(T_f - T_i\right)$$
$$= 1.0 \cdot \frac{5}{2} \cdot 8.31 \cdot (207 - 273)\text{J} = -1.37\text{kJ}$$

The temperature coefficient of energy has been computed according to Equation (71), with $f = 5$.

The energy exchanged in isothermal heating is calculated using Equation (42). Since the latent entropy of the ideal gas is nR/V (see Equation (53) and below), we get

$$Q_2 = \int_{V_i}^{V_f} nRT/V dV = nRT \ln\left(\frac{V_f}{V_i}\right)$$
$$= 1.08.31 \cdot 207 \ln(2)\text{J} = +1.19\text{kJ}$$

c) If we add up all amounts of energy exchanged for the cases (1) and (2), we obtain the same value:

$$\Delta E = -1.37\text{kJ}$$

This is as expected: the change of the energy content does not depend on the details of processes.

EXAMPLE 25. The law of adiabatic change derived on the basis of the balance of energy.

Use the Gibbs fundamental relation to derive the law of Poisson and Laplace, which holds for adiabatic processes of the ideal gas.

SOLUTION: Adiabatic processes of the ideal gas do not lead to changes of the entropy content, since the processes are reversible. Also, for the ideal gas the energy is a function of temperature only. If we introduce these properties into the Gibbs fundamental form we obtain

$$C_V \dot{T} = -P\dot{V}$$

The pressure of the fluid has to be expressed using the thermal equation of state of the ideal gas, which leads to

$$C_V \frac{dT}{T} + nR\frac{dV}{V} = 0$$

According to Equation (70), the product nR is equal to the difference of the temperature coefficients of enthalpy and energy. With the definition of the ratio of the capacities we finally arrive at the following differential equation:

$$\frac{dT}{T} + (\gamma - 1)\frac{dV}{V} = 0$$

which has the solution

$$TV^{\gamma-1} = constant$$

This is the derivation commonly found in texts on thermodynamics. The law, however, was discovered long before the energy principle was known, which makes us wonder how Laplace and Poisson could have derived it. The derivation of the law given in Section 2.4 demonstrates that it is a consequence of the theory of heat alone. It does not depend on the relationship between entropy and energy.

EXAMPLE 26. Measuring the ratio of the entropy capacities.

The following experiment can be used to measure the ratio of the entropy capacities. The air in a large bottle is compressed to slightly above the external pressure, and the pressure inside is measured. Then we let air escape quickly through a valve, such that the pressure inside is reduced to the external value. The temperature inside drops, since this process is adiabatic. Finally, we heat the air slowly until it has the same temperature as at the beginning. The final pressure is measured.[8] a) Draw a T-V diagram of both steps. b) Derive the ratio of the entropy capacities from the values of the pressure. c) Prove that the energy exchanged in the first step must be equal in magnitude to the energy exchanged during the second process.

SOLUTION: b) Even though the amount of air in the container decreases, we shall assume this quantity to remain constant. We simply imagine the air which has escaped takes part in the second step, i.e., in the heating at constant volume.

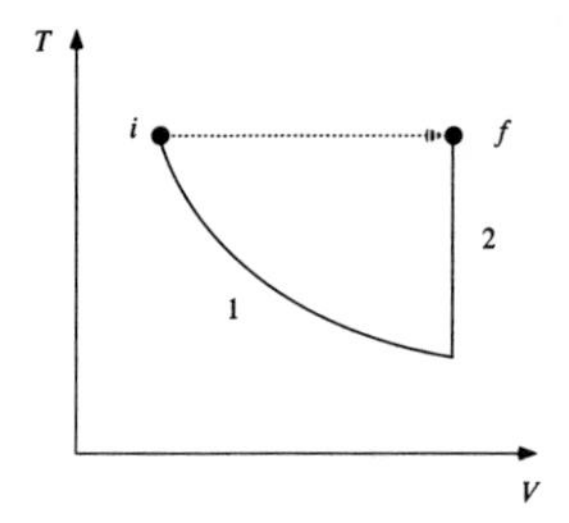

The first step is an adiabatic expansion. Since the changes of pressure and volume are small, we can approximate the differential equation of adiabatic change by

$$\frac{P_i - P_o}{P_o} + \gamma\frac{\Delta V}{V_o} = 0$$

where P_o is the pressure of the outside air, and P_i stands for the initial pressure in the container. Since the temperature first decreases and then increases back to its initial level, these steps are equivalent to an isothermal expansion from the initial to the final state.

8. Method of Clément and Désormes.

(See the figure.) This observation allows us to express the change of volume using the thermal equation of state. Since T is constant we have

$$\frac{P_f - P_o}{P_o} + \frac{\Delta V}{V_o} = 0$$

where P_f stands for the pressure at the end of the second process. If we combine the results, we see that the ratio of the entropy capacities is given by

$$\gamma = \frac{P_i - P_o}{P_f - P_o}$$

c) The energy of the ideal gas must be the same at the beginning and at the end of the sequence of operations since the temperature is the same in both states. For this reason the energy exchanged in the adiabatic expansion is the reverse of the energy absorbed by the air as a consequence of heating at constant volume:

$$\Delta E = 0 \;\Rightarrow\; W = -Q$$

In both steps, only one mode of exchange of energy can take place (mechanical in adiabatic, and thermal in isochoric processes).

2.5.3 The Stirling and Otto Cycles

The thermodynamics of the ideal gas derives some of its significance from the fact that this gas can serve as a model fluid for practical power cycles. Put differently, the actual operations taking place in a variety of thermal engines may be modeled in terms of ideal processes undergone by the gas. Since this fluid may undergo only reversible operations, the analysis of power cycles will deliver upper bounds for their performance. Here we shall discuss two practically important types of engines, the *Stirling* and the *Otto* engines. The former is an example of external combustion processes, while the latter uses internal combustion.

The Reverend R. Stirling invented an engine in which a gas such as air or helium undergoes cyclic processes in a closed environment.[9] Two cylinders, a heater and a cooler, and a regenerator, make up the device. Burning of the fuel for heating takes place externally and continuously, which in practice allows for much better control of the chemical process. This is important for pollution and noise control. Even though it is relatively difficult, and therefore expensive, to build, the Stirling engine has attracted renewed attention in recent years because of its inherent positive properties.

It is convenient to discuss the operation of such an engine directly in terms of the model processes through which the ideal gas runs (Figure 10). Let us begin with the heating of the fluid. The gas is heated, and expands at a constant temperature

9. See for example G. Walker (1973 a,b) and J. Walker (1985).

T. The entropy and the energy absorbed are simply related by this value of the temperature. In the second step, the fluid is cooled at constant volume so that the temperature reaches its lower operating level, T_o. The entropy is discharged to the regenerator, a fact which will become important later in the cycle. Now the ideal fluid is cooled further at constant temperature, which means that its volume must be reduced. In the final step it is heated once again, this time at constant volume, to reach its initial temperature *T*. The entropy needed in this step is supposed to come from the regenerator.

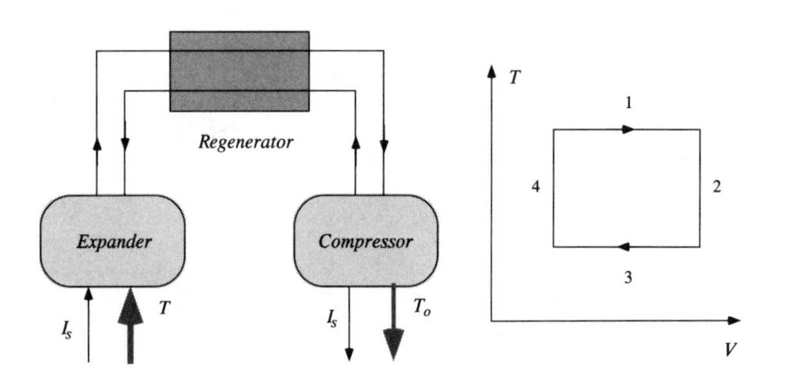

FIGURE 10. The Stirling engine consists of an expander and a compressor, and a regenerator. The fluid in the engine is expanded and heated at constant upper operating temperature *T*. The entropy emitted in the step 2 (constant volume cooling) is given back to the gas in step 4 with the help of a heat exchanger (the regenerator). Before that, the gas is compressed and cooled at constant lower operating temperature T_o. The ideal cycle is depicted in the *T-V* diagram.

The regenerator's significance lies in increasing the thermal efficiency of the cycle. If we look only at the gas, two of the steps are the same as those in the Carnot engine. In the isochoric operations, however, additional entropy is exchanged between the environment and the fluid, thus reducing the efficiency of the engine compared to that of a Carnot cycle. If, however, the isochoric heating and cooling can be made internal to the system, we do not have to supply and waste extra entropy (and energy). In fact, a Stirling cycle having an ideal regenerator will achieve the Carnot efficiency. This is so since the same amount of entropy is emitted during the second step as is absorbed in the fourth (heating or cooling at constant volume between the same initial and final temperatures). If we include the regenerator in the system to be considered, we have to model it as a body which always undergoes reversible operations at the same instantaneous temperature as the gas. In practice, the regenerator will be one of the weak links of the design: it will be difficult to minimize the production of entropy in this device. For modeling purposes we may consider it to be an ideal counter flow heat exchanger (Figure 10).

An example of an internal combustion engine is given by a fluid going through an *Otto cycle*. The gas is assumed to be air, and it is modeled as an ideal gas. In the first step the fluid is compressed adiabatically in a cylinder having a piston (Figure 11). The temperature rises from its lowest value to the highest value achieved in the cycle. During step 2, the air is heated at constant volume as a consequence of the burning of fuel which has been mixed in with it. The final temperature reached is the highest in the complete cycle. Then the gas expands adiabatically, whereupon in a final step it is cooled at constant volume.

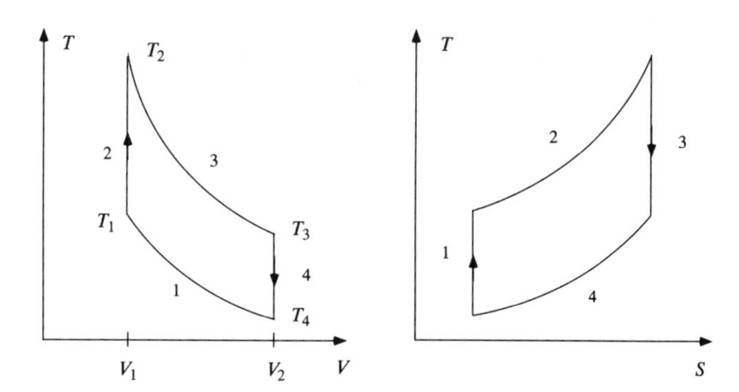

FIGURE 11. A body of ideal gas runs through an Otto cycle. The cycle has been drawn in the *T-V* and in the *T-S* diagrams.

The idealized cycle neglects several features which are a consequence of the internal combustion of the fuel. First of all, the production of entropy is modeled as taking place outside the fluid. We assume that the air absorbs the entropy released by the burning fuel as if it was isolated from it. The heating of the air is then taken to proceed according to the laws laid down in this chapter. Also, in the real engine, air is taken up in each cycle, mixed with fuel, and expelled again from the cylinder with the products of the combustion. The air does not remain enclosed in the engine, and not just air runs through the steps of the Otto cycle. Rather, we have a mixture containing other components as well. In the model we neglect these facts. Still, the ideal Otto cycle is used as a first approximation of the real processes taking place.

EXAMPLE 27. Thermal efficiency of the Otto cycle.

Consider a body of ideal gas going through the Otto cycle described in Figure 11. a) Determine the thermal efficiency of the cycle, and express it in terms of the temperatures involved. b) Express the result in terms of the ratio of the volumes (compression ratio). c) Compare the efficiency to that of a Carnot cycle.

SOLUTION: a) The engine absorbs the same amount of entropy in step 2 as it emits in step 4. Since these steps involve heating and cooling at constant volume, the entropy exchanged is given by

$$S_{e,in} = C_V \ln\left(\frac{T_2}{T_1}\right)$$

$$S_{e,out} = -C_V \ln\left(\frac{T_4}{T_3}\right)$$

Therefore, we have the following relation between the temperatures involved:

$$\frac{T_2}{T_1} = \frac{T_3}{T_4}$$

The thermal efficiency may be expressed as the ratio of the energy delivered in working and the energy absorbed in heating. These quantities are given by

$$Q_{in} = C_V(T_2 - T_1)$$

and

$$W = C_V(T_2 - T_3) - C_V(T_1 - T_4)$$

The amounts of energy exchanged as a consequence of the mechanical operations in steps 2 and 4 just cancel. From this we conclude that

$$\eta_1 = \frac{W}{Q_{in}} = 1 - \frac{T_3 - T_4}{T_2 - T_1} = 1 - \frac{T_4}{T_1}$$

b) The temperatures and the volumes may be related by the law of Poisson and Laplace for the adiabatic steps 1 and 3:

$$T_2 V_1^{\gamma-1} = T_3 V_2^{\gamma-1}$$
$$T_1 V_1^{\gamma-1} = T_4 V_2^{\gamma-1}$$

which means that the thermal efficiency can be expressed as follows:

$$\eta_1 = 1 - \left(\frac{V_1}{V_2}\right)^{\gamma-1}$$

c) The thermal efficiency of the Otto cycle is clearly smaller than the Carnot efficiency. The latter value would be obtained by an engine running between constant temperatures T_2 and T_4 (Figure 11) which are the maximum and the minimum temperatures attained in the Otto cycle. The smaller efficiency is not due to dissipation (the engine is ideal) but simply due to the fact that entropy is absorbed at temperatures smaller than the largest one, and emitted at temperatures larger than the smallest one.

2.5.4 Irreversible Processes, Constitutive Laws, and Time

Processes in nature are irreversible, but the theory laid down so far is one of reversible changes only. This is a direct consequence of the model of the ideal fluid we have chosen as a basis. The model is specified by the constitutive laws describing the body under investigation. Therefore the constitutive laws are responsible for the fact that the theory permits only reversible processes.

If we want to break out of the confinement of ideal processes, we have to change the model of the bodies undergoing thermodynamic processes. In other words, we have to enlarge the class of bodies we are considering. The ideal gas defined by the thermal equation of state derived in Equation (29) and the expression for the heating as in Equation (42), does not allow for anything but reversible changes. However, if we change the laws just a little bit we will find that thermodynamics is different.

An example of a simple body which is capable of irreversible changes is that of a viscous fluid (Epilogue). Viscosity leads to the creation of entropy and to the dissipation of energy. We can introduce viscosity into our models by adding a viscous pressure term to the thermal equation of state of the fluid. This changes everything. Due to this term, the energy exchanged in the mechanical process does not reverse its sign upon reversal of the change of volume. As a consequence, the entropy exchanged will also be different for a process and its reverse. This is in stark contrast to the behavior of the ideal fluid we have investigated up to now. Fluids which behave in such a way produce entropy.

There is another consequence of changing the constitutive laws in the prescribed manner: time appears explicitly in the equations of change. Therefore it matters how fast a process takes place. The equations describing a process will model real initial value problems as we know them from mechanics and other fields of physics. Only the particular properties of ideal fluids might make us believe that time has no place in thermodynamics. Time is right there; we only have to look for it.

2.6 Black Body Radiation as a Simple Fluid

There exists another simple physical system of great interest in thermodynamics: thermal radiation, which can permeate empty space and bodies alike. In its simplest form, the radiation field may be described from a thermodynamics viewpoint using very few variables. Radiation contains entropy. This becomes clear when we consider a body cooling down due to radiation only. Since entropy cannot vanish, the radiation must transport it away from the body. Radiation therefore constitutes a thermal system. In fact, thermal radiation trapped inside a cavity behaves just like a simple fluid, very much comparable to the ideal gas. It is often called a *photon gas*. It possesses a certain amount of entropy and energy, and its pressure and temperature assume well-defined values.

The entropy and pressure of the radiation field turn out to be very small under everyday conditions on Earth, i.e., for small temperatures. For this reason, the properties of the radiation "fluid" are of interest mainly in astrophysics. It contributes considerably to the phenomena inside stars and in the universe as a whole. Transfer of entropy by radiation, however, may very well be appreciable even for low temperatures. Therefore, the radiation of heat plays a major role for engineers and scientists alike. In this section we shall study the photon gas; radiative transfer of heat will be discussed in Chapter 3.

2.6.1 Thermal Radiation and Black Bodies

Bodies can emit and absorb electromagnetic radiation. If these processes occur because a body is warm or hot, we speak of *thermal radiation*. Radiation is a physical system in its own right, apart from the bodies which lead to its creation.

A cavity, for example, may be filled with radiation which is continuously absorbed and emitted by the walls surrounding it (Figure 12). We call the system inside the cavity the *radiation field* or the *photon gas*. Our task will be to decipher the thermal properties of this gas.

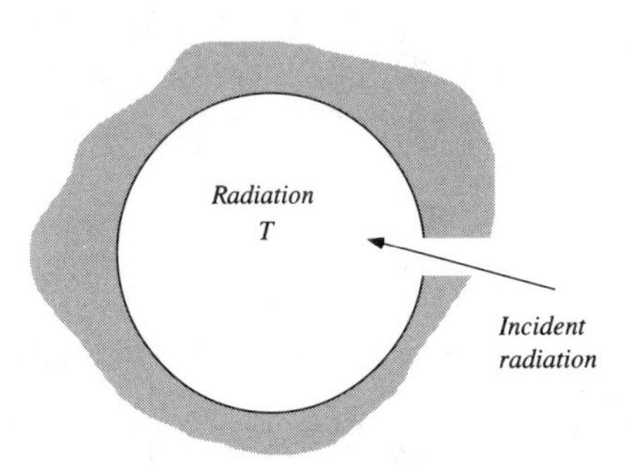

FIGURE 12. A cavity containing thermal radiation. The photon gas may interact either with the walls of the cavity or with a black body at temperature T, which also is the temperature of the photon gas. Radiation falling on a hole in the walls of the cavity will be completely absorbed. Therefore, the hole is said to act as a black body. Conversely, radiation emitted from it is called blackbody radiation.

The radiation inside the cavity may result from emission and absorption processes by the walls, which are kept at a temperature T. If the field is uniform, we can associate the same temperature with the radiation. The radiation filling the cavity is said to be *blackbody radiation*. We denote a body as *black* if it is an ideal absorber of radiation, i.e., if it absorbs all light rays falling upon it regardless of the frequency of the light. Consider a small opening in a wall of the cavity. Any light passing into the cavity from outside will have only a very small chance of ever escaping again. Therefore, we may think of the hole as an ideal absorber. Alternatively, light passing to the outside from the cavity must have the properties of radiation emitted an ideal absorber—black bodies. In fact, the light emitted from a small hole in the walls of the cavity can be used to probe the properties of the radiation inside. (In Section 3.5 we shall investigate the relationship between the properties of the radiation field and the radiation flowing through space.)

We also may make the walls of the cavity completely reflecting in which case we need an ideally absorbing and radiating piece of matter, i.e., a black body, inside the cavity. The black walls or the black body serve as an entropy reservoir for the radiation field. Entropy is added to or removed from the field due to emission or absorption. If we do not wish to heat the photon gas, we can make do without any absorbing and radiating matter. In this case, we still can let the radiation undergo adiabatic processes.

2.6.2 Energy, Pressure, and Heating of Blackbody Radiation

As in the case of the ideal gas, we need to know some constitutive properties of the radiation field. From measurements on the radiation emanating from black bodies we can infer a relation between the temperature of radiation and the energy density of the photon gas. (The energy flux density and the energy density must be directly proportional; see Section 3.5.) It is found that the intensity of light from the opening in the cavity in Figure 12 does not depend on the nature of the walls, or on the size or the shape of the body emitting radiation. The temperature of the body, and therefore of the field, is the only parameter influencing blackbody radiation. Therefore, the energy density of the uniform field, i.e., the quantity

$$\rho_E = \frac{E}{V} \tag{86}$$

depends only on the temperature of the radiation enclosed in the cavity:

$$\rho_E = f(T) \tag{87}$$

Here, V is the volume of the photon gas, while E is its energy. $f(T)$ is an unknown function of the temperature.

Another important piece of information concerns the pressure of the photon gas. We know from electromagnetic theory that radiation transports momentum. While the pressure of radiation is very small under normal circumstances, it nevertheless can be determined experimentally. Both theory and experiment demonstrate that the pressure of the photon gas is equal to one third of its energy density:

$$P = \frac{1}{3}\rho_E \tag{88}$$

Since the energy density is a function of temperature only, the same is also true for the pressure. This is the *thermal equation of state* of radiation; it is comparable to the equation of state of the ideal gas. In the following subsections we shall exploit this information to derive further properties of blackbody radiation. But first we have to specify the heating of the photon gas.

We model blackbody thermal radiation as a spatially uniform system, which means that we associate with it single values of temperature and pressure. We will allow the volume of the cavity in Figure 12 to be variable. Therefore the independent variables used to describe the properties of radiation are volume and temperature, just as in the case of simple fluids. For this reason we can define the *latent entropy with respect to volume* and the *entropy capacity at constant volume* of the photon gas; the procedure is analogous to the one used for the ideal gas in Section 2.3. More importantly, we shall assume the *heating* of the radiation field to be a linear function of the rate of change of volume and of temperature. This reflects the assumption that the uniform photon gas may undergo only reversible operations. Just as in the case of the ideal gas, we can prove that no entropy is produced in a thermomechanical system obeying such a linear constitutive law (Section 2.3.4, Equation (43)), and the factors in the law for the heating are the latent entropy and the entropy capacity, Equation (42).

Since the system can exchange energy only in thermal and mechanical processes, just like any simple fluid, we also obtain the same *Gibbs fundamental relation* as for the ideal gas, namely, Equation (78). This is a direct consequence of the equations of balance of entropy and energy of the radiation field, and of the relationship between the heating and the rate of supply of energy, which is equivalent to Equation (13) of Chapter 1.

2.6.3 The Constitutive Problem of the Photon Gas

We should now determine the latent entropy and the entropy capacity of the photon gas, and with them the density of entropy and of energy. We can derive the missing information on the basis of the constitutive relations (87) and (88) if we consider a Carnot cycle undergone by the radiation inside a cavity. This is in accordance with the derivation carried out for the ideal gas in Section 2.3.5. Let the photon gas go through a Carnot cycle having very short adiabats, i.e., with a very small difference of the temperatures at which entropy is absorbed and emitted (Figure 7). The energy released by an ideal Carnot heat engine is equal to

$$W = -\Delta T \cdot S_e \tag{89}$$

where S_e is the amount of entropy absorbed at the temperature of the furnace, $T + \Delta T$. On the other hand, the work done by the fluid can be computed easily from the pressure, which remains constant for the isothermal steps, and from the change of volume. The adiabatic steps do not contribute significantly to the exchange of energy since they are taken to be very small. For the same reason, the changes of volume are roughly the same for both isothermal operations (Figure 7). If we also take into consideration that the pressure of the photon gas is equal to one-third of the energy density, and if we approximate the difference of pressures, we obtain

$$W = -(P_{T+\Delta T} - P_T)\Delta V = -\frac{1}{3}\frac{d\rho_E}{dT}\Delta T \cdot \Delta V \tag{90}$$

The expressions in Equations (89) and (90) must be equal, which leads to a determination of the entropy added in one step of the Carnot cycle. This quantity is also equal to the integral of the latent entropy over volume. For this reason, the latent entropy with respect to volume of the photon gas must be equal to

$$\Lambda_V = \frac{1}{3}\frac{d\rho_E}{dT} \tag{91}$$

We introduce the rate of change of entropy into the Gibbs fundamental form—see Equation (78)—and observe that the energy E is the product of the energy density and the volume. Since the energy density is a function of temperature only, its rate of change can depend only on the rate of change of T. All these considerations lead to the following result:

$$V\dot{\rho}_E + \dot{V}\rho_E = T(\Lambda_V\dot{V} + K_V\dot{T}) - P\dot{V} \tag{92}$$

or

$$\left(V\frac{d\rho_E}{dT} - T\cdot K_V\right)\dot{T} = \left(\frac{1}{3}T\frac{d\rho_E}{dT} - \frac{4}{3}\rho_E\right)\dot{V} \tag{93}$$

Since volume and temperature can be changed independently, each of the factors appearing in parentheses in Equation (93) must be zero. The right side leads to a differential equation for the energy density as a function of temperature which has the following solution:

$$\rho_E = aT^4 \tag{94}$$

or

$$E = aVT^4 \tag{95}$$

This is called the *law of Stefan and Boltzmann*; a is the *Stefan-Boltzmann constant*, and has a value of $7.56 \cdot 10^{-16}$ J/(m^3 · K^4). It will be of crucial importance not only for deriving quantities related to the photon gas, but also when we calculate the energy associated with the radiation from hot bodies (Chapter 3).

We can now find the latent entropy and the entropy capacity of the photon gas. The former quantity can be evaluated using Equation (92). We obtain the latter if we set the factor multiplying $\dot{V}$ on the left-hand side of Equation (93) equal to zero. The constitutive quantities turn out to be the following:

$$\Lambda_V = \frac{4}{3}aT^3 \tag{96}$$

$$K_V = 4aVT^2 \tag{97}$$

Finally, the rate of change of the entropy content of radiation can be expressed using the constitutive quantities. If we integrate the equation along a simple path in the T-V diagram (Figure 13), we obtain a simple result for the *entropy of black-body radiation*:

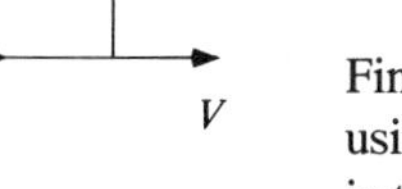

FIGURE 13. We integrate the expression for the rate of change of the entropy of radiation from the state $T = 0$, $V = 0$ to an arbitrary state. First, we integrate over V at $T = 0$, and then over T for a given value of V. The value of the entropy at $T = 0$ is set equal to zero.

$$S(T) = \frac{4}{3}aVT^3 \tag{98}$$

EXAMPLE 28. The ideal gas and radiation inside a star.

Both matter and radiation occupy the same region of space inside a star. If the temperature is high, the gas is completely ionized which makes it an ideal gas even at high densities. Because of the high temperature, radiation can contribute considerably to the pressure.
a) Matter inside a new main sequence star is composed of 70% hydrogen and 30% helium by mass. Calculate the total pressure of matter and radiation at the center of such a star of 15 solar masses, if the temperature and density have values of $34 \cdot 10^6$ K and $6.2 \cdot 10^3$ kg/m^3, respectively. Compute the fraction β of the total pressure which is due to the ide-

al gas. b) Express the heating of the ideal gas plus radiation in terms of the gas pressure P_g and the radiation pressure P_r. c) Calculate the expression for the total entropy capacity in terms of the entropy capacity of the ideal gas and the fraction β of the pressure due to the gas.

SOLUTION: a) We can compute the gas pressure using the equation of state of the ideal gas in the form of Equation (30). For this purpose we need the mean molar mass of the gas inside the star. We proceed in accordance with Example 10. Observe, however, that for each mole of hydrogen nuclei we also have one mole of electrons, and for one mole of helium nuclei we also have two moles of electrons. This is a consequence of ionization. If we write X for the mass fraction of hydrogen, we have

$$
\begin{aligned}
M_o &= \left[\frac{2X}{M_{oH}} + \frac{3(1-X)}{M_{oHe}}\right]^{-1} \\
&= \left[\frac{2\cdot 0.7}{0.0010\,\text{kg/mole}} + \frac{3(1-0.7)}{0.0040\,\text{kg/mole}}\right]^{-1}
\end{aligned}
$$

for the molar mass. The result is $0.615 \cdot 10^{-3}$ kg/mole. The gas pressure turns out to be

$$
\begin{aligned}
P_g &= \frac{R}{M_o}\rho T \\
&= \frac{8.31\cdot 6.2\cdot 10^3 \cdot 34\cdot 10^6}{0.615\cdot 10^{-3}}\,\text{Pa} = 2.8\cdot 10^{15}\,\text{Pa}
\end{aligned}
$$

The pressure of the photon gas is somewhat smaller:

$$
\begin{aligned}
P_r &= \frac{1}{3}aT^4 \\
&= 3.4\cdot 10^{14}\,\text{Pa}
\end{aligned}
$$

The ratio β of the gas pressure to total pressure is 0.89. In other words, 11% of the total pressure is due to radiation. There exist conditions in the universe under which radiation is responsible for an even larger fraction of the total pressure.

b) We have to heat both matter and radiation simultaneously. Therefore, we simply add the expressions for the heating of both components:

$$
I_s = -\left(\Lambda_{Vg} + \Lambda_{Vr}\right)\dot{V} - \left(K_{Vg} + K_{Vr}\right)\dot{T}
$$

This may be changed using the results obtained above

$$
\begin{aligned}
I_s &= -\left(\frac{nR}{V} + \frac{4}{3}aT^3\right)\dot{V} - \left(\frac{nR}{\gamma-1}\frac{1}{T} + 4aVT^2\right)\dot{T} \\
&= -\frac{1}{T}\left(P_g + 4P_r\right)\dot{V} - \frac{1}{T^2}V\left(\frac{1}{\gamma-1}P_g + 12P_r\right)\dot{T}
\end{aligned}
\qquad \text{(E10)}
$$

c) The factor multiplying the rate of change of temperature in (E10) is the total entropy capacity of the mixture of ideal gas and radiation:

$$K_V = \frac{1}{T^2} V \left(\frac{1}{\gamma - 1} P_g + 12 P_r \right)$$

With $P = P_g + P_r$, $\beta P = P_g$, and $(1 - \beta) P = P_r$, we obtain

$$K_V = \frac{1}{T^2} V \left(\frac{1}{\gamma - 1} \beta P + 12 (1 - \beta) P \right)$$
$$= \frac{1}{T} \frac{PV}{T} \frac{1}{\gamma - 1} [\beta + 12(\gamma - 1)(1 - \beta)]$$
$$= \frac{K_{Vg}}{\beta} [\beta + 12(\gamma - 1)(1 - \beta)]$$

For $\beta = 1$ the expression becomes the entropy capacity of the gaseous component only.

EXAMPLE 29. Energy (mass) densities of matter and of radiation in the universe.

The universe is permeated by blackbody radiation of temperature 2.7 K, which is commonly interpreted as a relic of a hot Big Bang. The expansion of space is described by a scaling factor R which is increasing with time. R is set equal to 1 at an arbitrary point in time. A simple model of the universe predicts that the temperature of the background radiation changes inversely with the scaling factor.[10] It is assumed that as the temperature of the universe dropped below a certain value, matter and radiation ceased to interact.
a) Show that as a consequence of this prediction the expansion of the universe must be reversible. b) The present mass density of matter in the universe is estimated to be $4.5 \cdot 10^{-27}$ kg/m^3. Determine the ratio of the mass densities associated with radiation and matter in the universe. c) Express the dependence on R of the mass densities of matter and of radiation. At what value of R compared to today's were both densities comparable?

SOLUTION: a) According to Equation (98) the entropy of radiation contained in the volume V is proportional to this volume and to the cube of the temperature. Now, V is proportional to the cube of the scaling factor R, while T varies inversely with R. For this reason, the entropy of radiation in the universe must be constant. Since the radiation field is not cooled, entropy cannot have peen produced. Therefore, the expansion of the universe is reversible. In fact, the process is nothing but a reversible adiabatic expansion (see the following example).

b) The mass density of radiation is equal to its energy density divided by the square of the speed of light:

$$\rho = \frac{\rho_E}{c^2} = \frac{aT^4}{c^2}$$
$$= \frac{7.56 \cdot 10^{-16} \cdot 2.7^4}{9.0 \cdot 10^{16}} \frac{\text{kg}}{\text{m}^3} = 4.5 \cdot 10^{-31} \frac{\text{kg}}{\text{m}^3}$$

This is just about 10000 times less than the estimated density of matter in the universe.

10. Peebles (1971), p. 121.

c) Naturally, the density of matter varies inversely with the volume which is proportional to the cube of the scaling factor R. The density of radiation, on the other hand, varies as the fourth power of temperature. For these reasons we have:

$$\rho_{matter}(t) = \rho_{matter,o}\left(\frac{R_o}{R(t)}\right)^3$$

and

$$\rho_{rad}(t) = \rho_{rad,o}\left(\frac{T(t)}{T_o}\right)^4 = \rho_{rad,o}\left(\frac{R_o}{R(t)}\right)^4$$

The ratio of the densities follows from these expressions:

$$\frac{\rho_{rad}}{\rho_{matter}} = \frac{\rho_{rad,o}}{\rho_{matter,o}}\frac{R_o}{R(t)}$$

Since the density of radiation decreases faster than that of matter, the two must have been equal sometime in the past. According to the present value of the ratio of densities, they must have been approximately equal when the universe was some 10000 times smaller than it is today. The temperature must than have been roughly 30000 K. It is estimated that this should have occurred some 3500 years after the Big Bang.[11]

EXAMPLE 30. Adiabatic and isothermal processes of the photon gas.

The photon gas may undergo various processes, among them adiabatic and isothermal ones. a) Express the relationship between temperature and volume for blackbody radiation for adiabatic processes. Show that blackbody radiation behaves as an ideal gas with a ratio of entropy capacities equal to 4/3. b) Determine the amounts of energy exchanged as a consequence of isothermal changes of volume, and compute the change of energy of the photon gas.

SOLUTION: a) Since blackbody radiation is capable of reversible processes only, and since the photon gas is not heated in adiabatic processes, the entropy content of the radiation field must remain constant. If we take the expression for the entropy of blackbody radiation contained in a cavity, Equation (98), we see that the following condition must be met for adiabatic processes:

$$VT^3 = V_o T_o^3 \tag{E11}$$

i.e., the product of volume and the cube of the temperature must remain constant. The analogous expression which holds for the ideal gas is

$$VT^{1/(\gamma-1)} = constant \tag{E12}$$

11. J. Silk (1981), p. 340.

Relations (E11) and (E12) are equivalent if we set the ratio of entropy capacities $\gamma = 4/3$ for the photon gas.

b) If the temperature is held constant, the heating is expressed by the latent entropy and the rate of change of volume. For this reason the entropy exchanged in an isothermal expansion or compression undergone by blackbody radiation is given by

$$S_e = \int_{V_1}^{V_2} \Lambda_V \, dV = \int_{V_1}^{V_2} \frac{4}{3} a T^3 \, dV = \frac{4}{3} a T^3 \, \Delta V$$

The temperature stays constant which allows us to compute the energy exchanged together with entropy in a simple way:

$$Q = T S_e = \frac{4}{3} a T^4 \, \Delta V$$

On the other hand, the energy exchanged as a consequence of the change of volume is equal to

$$W = -\int_{V_1}^{V_2} P dV = -\frac{1}{3} a T^4 \, \Delta V$$

since the temperature does not change. These results lead to the expression for the change of the energy of the photon gas in isothermal processes:

$$\Delta E = Q + W = a T^4 \Delta V$$

This is not particularly surprising, since we know that the energy density of blackbody radiation depends only on the temperature (Equation (94)). However, you might note that there is a difference between the ideal gas and the photon gas as far as the dependence of energy on the volume is concerned. In isothermal expansion much more entropy is added than would be expected from an analogy with the ideal gas. This is so since the entropy reservoir supplying entropy to the photon gas also creates more radiation if the volume of the cavity containing the photon gas increases. Radiation does not conserve the amount of substance, while the amount of substance of an ideal gas enclosed by walls does not change.

2.7 The Coupling of Magnetic and Thermal Processes

With the exception of the brief description of melting and vaporization, this entire chapter has dealt with the coupling of thermal and mechanical processes in simple fluids only. The theory which has emerged is called *thermomechanics*. It models only a small portion of what nature has to offer to us: basically all types of phenomena can be coupled.

Some magnetic systems exhibit a coupling of magnetic and thermal properties. In the simplest case, when such a body is heated its *temperature* as can change, as well as its *magnetization* (Prologue). The properties of some paramagnetic substances have enabled physicists to reach very low temperatures in the laboratory by *adiabatic demagnetization* of the magnetic bodies. This is an interesting application, and it is worthwhile to extend thermodynamics to magnetocaloric effects. Since in previous sections we have carefully introduced a number of basic concepts, we can now present the material in a condensed form. You will find the motivation for the ideas developed here in the pages on the thermomechanics of the ideal gas.

2.7.1 Equation of State of a Paramagnetic Substance

Paramagnetic substances exhibit simple coupling of thermal and magnetic properties similar to the coupling of mechanical and thermal quantities which we have found in the case of the ideal gas. Remember that the thermal equation of state of the ideal gas expresses the relationship between the temperature, volume, and pressure of the fluid. In the case of a magnetic substance, the proper extensive and intensive quantities analogous to volume and pressure are the magnetization M and the magnetic field H, respectively (Prologue). It is found that the *thermal equation of state of a paramagnetic substance* can be written in the form

$$MT = C^* H \tag{99}$$

C^* is called the *Curie constant.* (Some values are listed in Table A.8.) This law clearly shows the coupling between magnetic and thermal properties.

In the case of the ideal gas, knowledge of the constitutive law expressed by the thermal equation of state proved to be insufficient for a complete description of the system. We needed additional information, such as measurements of an entropy capacity or of the ratio of the entropy capacities, or knowledge of the form of the energy of the system. The energy of a system expressed in terms of the temperature and some other variable, is called the *caloric equation of state*. For paramagnetic substances it is found that the energy depends only on the temperature:[12]

$$E(T) = \frac{A}{1 + B\exp(D/T)} \tag{100}$$

where A, B, and D are constants. Remember that the energy of the ideal gas also is a function of temperature alone. It will be found that together with the laws of heating and the balance of energy we have enough information to model magnetocaloric processes of paramagnetic substances.

12. Ionic paramagnetism and the production of low temperatures have been discussed in some detail in Zemansky and Dittman (1981), Chapter 18.

2.7.2 The Heating of a Paramagnetic Substance

If a substance exhibits magnetocaloric coupling, its thermal and magnetic properties may change as a result of heating. Again it is possible to have reversible processes, in which case we can write the constitutive law of heating in the following form:

$$I_s = -\Lambda_M \dot{M} - K_M \dot{T} \tag{101}$$

The constitutive quantities have similar meanings as in the case of fluids. The factor multiplying the rate of change of the temperature must be a type of entropy capacity. We call it the *entropy capacity at constant magnetization*. The other factor is the *latent entropy with respect to magnetization*. You will see that the equations of state, together with the energy principle, allow us to determine these quantities. As in the case of fluids we often use the quantity

$$C_M = TK_M \tag{102}$$

which is called the *temperature coefficient of energy* (at constant magnetization). Also, in analogy to thermomechanics where we have used the pressure in place of the volume, it is practical to express the heating in terms of the intensive magnetic quantity, i.e., in terms of the magnetic field H:

$$I_s = -\Lambda_H \dot{H} - K_H \dot{T} \tag{103}$$

Using the thermal equation of state we derive the following relationships between the different constitutive quantities:

$$\Lambda_H = \frac{C^*}{T} \Lambda_M \tag{104}$$

$$K_H = K_M - \frac{M}{T} \Lambda_M \tag{105}$$

2.7.3 Energy and the Gibbs Fundamental Form

The energy principle takes the following form for magnetocaloric processes. The energy of a body can change as a consequence of magnetic and thermal processes. Therefore the equation of balance of energy looks like

$$\dot{E} + I_{E,th} + I_{E,mag} = 0 \tag{106}$$

In the Prologue and Chapter 1 we found that the energy currents are given by

$$I_{E,th} = TI_s \tag{107}$$

$$I_{E,mag} = -\mu_o H \dot{M} \tag{108}$$

If we plug these expressions into the equation of balance of energy, Equation (106), and use the equation of balance of entropy for reversible processes, we arrive at the *Gibbs fundamental form* which is analogous to Equation (78):

$$\dot{E} = T\dot{S} + \mu_o H \dot{M} \tag{109}$$

This equation expresses the following observations: if the entropy of a body is increased at constant magnetization, its energy increases as well. Also, if the magnetization is increased at constant temperature, the energy grows. Equation (109) relates properties of the body only and does not refer to quantities which are exchanged in processes.

The laws listed so far allow us to draw some important conclusions regarding the constitutive properties of paramagnetic substances. First, notice that the energy depends only on the temperature. Therefore the rate of change of the energy of a body is simply related to the rate of change of its temperature:

$$\dot{E} = C_M \dot{T} \tag{110}$$

If we introduce this into the Gibbs fundamental form, and observe that the rate of change of the entropy content is expressed by the entropy capacity and the latent entropy, we get

$$C_M \dot{T} = T\Lambda_M \dot{M} + TK_M \dot{T} + \mu_o H \dot{M} \tag{111}$$

from which we conclude that the latent entropy with respect to magnetization is

$$\Lambda_M = -\mu_o \frac{H}{T} \tag{112}$$

This result is interesting. It tells us that if we lower the magnetization of a substance isothermally it *absorbs* entropy. According to Equations (110) and (100), we can calculate the entropy capacity at constant magnetization. For temperatures which are not too low (practically, this means some tenths of a Kelvin) the expression for the energy of the body given by Equation (100) can be approximated. This leads to the following approximation to the entropy capacity:[13]

$$K_M = \frac{A^*}{T^3} \tag{113}$$

where A^* has a constant value. With these properties derived we can calculate magnetocaloric processes and quantities.

13. Zemansky and Dittman (1981).

2.7.4 Reaching Low Temperatures

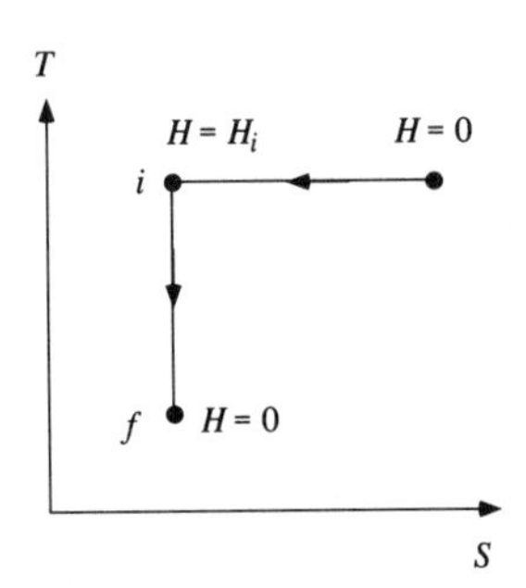

FIGURE 14. Lowering the temperature of a substance using adiabatic demagnetization. First, the field is increased isothermally at relatively high temperature. This leads to a reduction of the entropy in the body. Then the probe is demagnetized adiabatically, which leads to a decrease of the temperature.

Low temperatures can be obtained using liquid helium. However, this does not allow us to go lower than about 1 K. Today one uses paramagnetic substances to lower the temperature even further. This is done in the following manner. A paramagnetic substance whose temperature is near 1 K is slowly magnetized at constant temperature. Because

$$\dot{S} = \Lambda_H \dot{H} = -\frac{C^* \mu_o}{T^2} H \dot{H} \tag{114}$$

for *isothermal processes*, we notice that the body emits entropy during this step (Figure 14). In a second step the magnetization is reduced too quickly for entropy to be exchanged. Therefore this step is an adiabatic change (see the vertical line in Figure 14). According to Equation (103) the mathematical law for this adiabatic process is

$$0 = \Lambda_H \dot{H} + K_H \dot{T} \tag{115}$$

Since the latent entropy is negative, the temperature decreases with decreasing magnetic field. Thus we obtain lower temperatures as a consequence of *adiabatic demagnetization*.

EXAMPLE 31. Adiabatic demagnetization.

Determine the final temperature reached in adiabatic demagnetization in terms of the initial temperature and field. Compute the numerical value for the following example: chromium potassium alum has a Curie constant $C^* = 2.31 \cdot 10^{-5}\ m^3 \cdot K/mole$ and an entropy capacity constant $A^* = 0.15\ J \cdot K/mole$. What temperature is reached if we start with values of 1.0 K and 2.0 A/m for the initial temperature and field, respectively?

SOLUTION: The effect is best determined on the basis of the expression for the heating given in Equation (101), where we have to set $I_s = 0$. In addition, we need the latent entropy and the entropy capacity as given by Equations (112) and (113), respectively. We arrive at the following differential equation of adiabatic demagnetization:

$$-\frac{\mu_o}{C^*} M \dot{M} + \frac{A^*}{T^3} \dot{T} = 0$$

If we change the temperature from T_i to T_f, and simultaneously change the magnetization from M_i to 0, the solution of this equation is

$$\frac{\mu_o}{C^*} M_i^2 = A^* \left(\frac{1}{T_f^2} - \frac{1}{T_i^2} \right)$$

We replace the magnetization by the field H according to the thermal equation of state to obtain

$$T_f = T_i \left[\frac{\mu_o C^*}{A^*} H_i^2 + 1 \right]^{-1/2}$$

The numerical value for this example is

$$T_f = 1.0 \left[\frac{4\pi \cdot 10^{-7} 2.31 \cdot 10^{-5}}{0.15} (2.0 \cdot 10^6)^2 + 1 \right]^{-1/2} \mathrm{K} = 0.036\mathrm{K}$$

The values roughly correspond to those encountered in an experiment by De Haas and Wiersma in 1934.[14]

2.8 The General Law of Balance of Energy

The energy principle, i.e., the statements about the existence and properties of energy, has had a long and interesting history. Bits and pieces of the principle were recognized step by step in the course of the analysis of different applications in physics and in engineering. Ideal mechanical systems offered the first insight into the concept of energy. In some simple mechanical cases we can even prove that there is a quantity, distinct from momentum, which is conserved in the processes considered. Energy can be stored as kinetic or potential energy, and it can be exchanged as work.

A further important step was made in the analysis of heat engines early in the development of thermodynamics. There was supposed to be another form of exchange of energy, in thermal processes. Specifically, it was assumed that the amount of energy released by a steam engine was equal to the difference between the energy absorbed from the furnace and the energy rejected to the cooler. This is in effect the statement we used in the analysis in Section 1.4. While we can recognize in it a special form of the more general energy principle, namely conservation for systems in steady-state processes, it does not constitute anything near the modern concept. R. Clausius was able to show, on the basis of another assumption (Carnot's Principle), that there exists a quantity we may call the *energy content*, or the *internal energy*, of simple fluids, and furthermore, that the change of the internal energy was equal to the sum of amounts of energy exchanged in the various processes undergone by these fluids. This is the beginning of the more general form of the energy principle, which in the context of thermomechanical processes, has been called the *first law of thermodynamics*.

14. Quoted from Zemansky and Dittman (1981), p. 481.

2.8.1 The Law of Balance of Energy

The recognition of the energy principle in Clausius' work was important for a particular reason. In modern language, it demonstrated the principle's validity in the context of phenomena which relate processes of different types, not just mechanical ones. While for dissipationless motion in conservative fields we can prove this principle from the laws of motion, this is no longer possible in the face of the interaction between different fields, such as heat and motion. Here it becomes an independent law, a principle which establishes an *exchange rate*, i.e., a relationship between heat and motion, or between any other phenomena. Therefore we take as our second fundamental law, after the law of balance of entropy, the general *law of balance of energy*:

> *There exists a quantity called energy which we can imagine to be contained in bodies, and to flow from one body to another; it is conserved, which means that the energy content of a system can change only as a consequence of the flow of energy across the surface of the system in mechanical, electrical, thermal, or other processes.*

In mathematical terms, this means that we can write an equation of balance of energy which relates the rate of change of the energy of a system to the net flux crossing the system boundary:

$$\dot{E} + I_{E,net} = 0 \quad \textbf{(116)}$$

Compared to the statement of the law in the Prologue, the only thing new is the appearance in the net energy flux of an *energy current due to thermal processes*:

$$I_{E,net} = I_{E,mech} + I_{E,el} + I_{E,th} + \dots \quad \textbf{(117)}$$

We shall discuss mostly *thermomechanical processes*, which means that we will consider only thermal and mechanical energy fluxes. In this case the equation of balance of energy takes the simple form given in Equation (79). However, from now on we shall assume the balance of energy to be satisfied in all types of processes, be they mechanical, gravitational, thermal, reversible, or irreversible. Remember that in Section 1.4 we used the law of balance of power of engines undergoing steady-state operations. We can now recognize that law as a special case of the balance of energy for which the energy content of the system does not change in time.

EXAMPLE 32. Thermal energy current of an immersion heater.

a) How large is the thermal energy current of an immersion heater in the steady state?
b) How large is it during the heating phase before steady-state conditions are reached?
c) What types of processes are experienced by the immersion heater? In what form is energy exchanged?

SOLUTION: a) In the steady state, the rate of change of the energy of the heater vanishes. Therefore the net flux of energy with respect to the heater is equal to zero. For this reason the thermal energy current must be the negative of the electrical power:

$$I_{E,el} + I_{E,th} = 0 \quad \Rightarrow \quad I_{E,th} = -I_{E,el}$$

b) Before the steady state is reached, energy is stored in the heater, which makes the rate of change of the energy larger than zero. According to Equation (16), with the electrical energy current replacing the mechanical energy flux, we find that

$$I_{E,th} = -I_{E,el} - \dot{E} \quad \Rightarrow \quad \left|I_{E,th}\right| < \left|I_{E,el}\right|$$

Energy enters the system together with a current of electrical charge, which makes the energy flux a negative quantity.

c) In essence, we have already answered this question. Energy is supplied to the system with charge which makes this an electrical process. On the other hand, energy leaves the heater with entropy, which means we are talking about a thermal operation. We say that energy is exchanged in both electrical and thermal processes.

2.8.2 The Integral Form of the First Law

Often we are interested in changes of the energy of a system and in amounts of energy transferred as a consequence of a process. Such an overall balance of energy is very important in thermodynamics. In many cases it is difficult or even impossible to give a detailed account of a process, i.e., a description of how it runs in time. The balance of energy has proved to be an invaluable tool for inferring overall changes. This is very similar to what we know from the collision of bodies. Even if we cannot state the force law for a collision, we still can use the laws of conservation to compute changes of speeds of the bodies involved.

We obtain the overall balance by integrating over the process from the initial to the final time. If we do this for Equation (117), we get

$$\int_{t_i}^{t_f} \dot{E} dt + \int_{t_i}^{t_f} I_{E,mech} \, dt + \int_{t_i}^{t_f} I_{E,th} \, dt + \ldots = 0 \qquad \textbf{(118)}$$

Naturally, the first integral is the change of the energy of the system. The second integral represents the quantity of energy exchanged mechanically which is called *mechanical work*, while the third term is the energy exchanged as a consequence of thermal processes; it might be called *thermal work*.[15] Other terms can be included to represent other types of exchange of energy. Remember that amounts of energy communicated to a body are counted as positive quantities. Equation (118) becomes

15. Calling energy which is exchanged *work* is the practice of mechanics. Remember that we have reserved the term *work* for the amount of energy released in an internal process (Prologue).

$$\Delta E - W_{mech} - W_{th} - \ldots = 0 \tag{119}$$

For purely thermomechanical processes it takes the form

$$\Delta E = W + Q \tag{120}$$

This is how the law of balance of energy is normally written. Q is the preferred symbol for energy exchanged in thermal processes. The equation expresses the following simple idea (Figure 15):

> *The change of the energy content of a body is equal to the sum of the energies exchanged as a consequence of mechanical and thermal processes.*

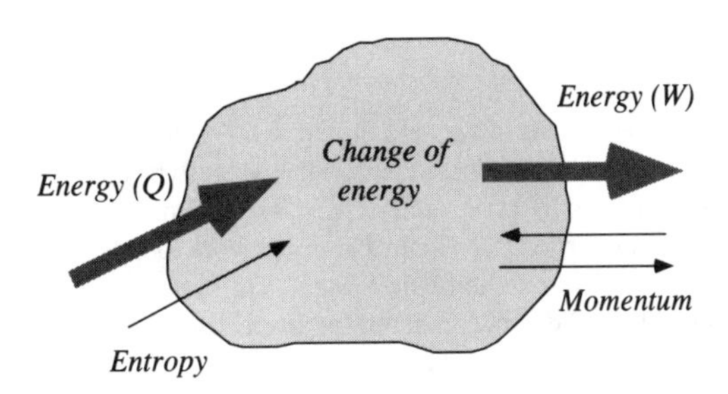

FIGURE 15. Energy enters and leaves a body. Currents are associated with mechanical and thermal processes. Energy is exchanged either with momentum or with entropy. In this case the change of the internal energy of the body is $|Q| - |W|$.

Equation (119) is a form of what often is called the *first law of thermodynamics*. Naturally, it is not a law of thermodynamics but of all of physics. In Equation (120) it has been written in a form applicable to thermomechanical phenomena.

EXAMPLE 33. Balancing energy in a thermomechanical process.

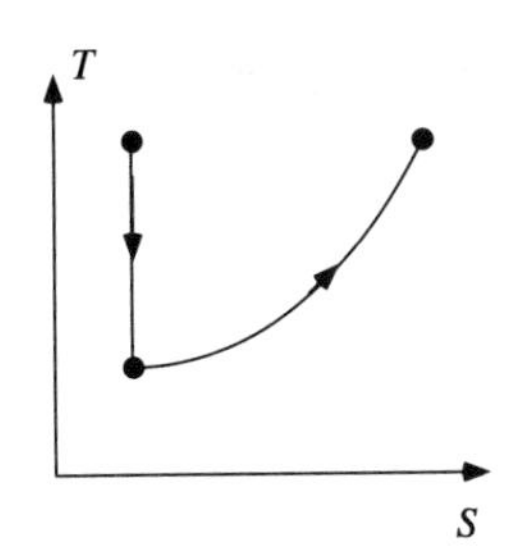

A fluid whose energy content depends only on its temperature is expanded adiabatically; in a second step its temperature is brought back to its initial value at constant volume. It is found that in the adiabatic expansion the fluid exchanges an amount of energy equal to 50 kJ. a) In what type of process is the energy exchanged during adiabatic expansion? What is the sign of the energy exchanged? b) Is energy exchanged in the isochoric step? If so, in what types of processes, and how much? c) Sketch the steps in a T-S diagram (assume entropy to be conserved). Is it possible to visualize ΔE or amounts of energy exchanged in the diagram?

SOLUTION: a) In an adiabatic process of a simple fluid the exchange of energy is due to the change of the volume; heating is excluded. The process is mechanical in nature: a fluid which expands emits energy. Therefore the sign of the energy exchanged is negative: $W = -50$ kJ.

b) Since the energy of the fluid is assumed to depend only on its temperature, and since the temperature returns to its original value, its energy must be the same at the beginning and at the end. According to the first law:

$$0 = W_1 + W_2 + Q_1 + Q_2$$

W_1 and Q_1 are - 50 kJ and 0 kJ, respectively. In an isochoric process energy may flow only in a thermal process, i.e., due to heating. Therefore, W_2 = 0 kJ. This leaves us with

$$0 = -50\text{kJ} + Q_2 \quad \Rightarrow \quad Q_2 = 50\text{kJ}$$

c) Only the amount of energy exchanged during heating in a reversible process can be shown as the area under the *T(S)* curve. Mechanical work does not appear in this diagram; nor does the change of the energy of a system.

EXAMPLE 34. Electrolysis of water.

An apparatus for the electrolysis of water is placed inside a container filled with a liquid. We let a current of electrical charge pass through the water to be electrolyzed. To convert a given amount of water into hydrogen and oxygen, an amount q of charge flows through the battery, which has a voltage U. (Take the battery to work ideally.) Consider water, hydrogen, oxygen, and the electrodes to be the system. The system and the liquid surrounding the apparatus would cool down were it not for an electrical heater built into the liquid. We heat the liquid in such a way that its temperature remains constant, passing an amount of charge q' through a second battery with a voltage U'. a) Write the equation of balance of the system. In what types of processes is energy transferred into or out of the system? b) By how much does the energy content of the system change?

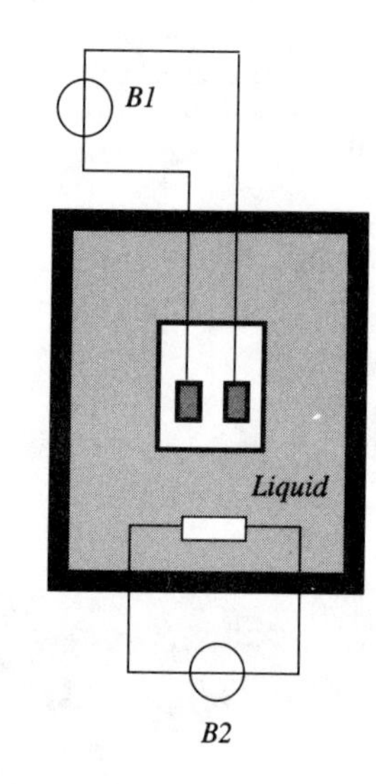

SOLUTION: a) The system *receives* energy both due to the flow of electrical charge used in the electrolysis, and due to heating by the fluid surrounding the system. We have an electrical energy flux which is responsible for maintaining the temperature of the fluid. However, this flux does not reach the system directly. Therefore:

$$\Delta E = W_{el} + Q$$

with $W_{el} > 0$ and $Q > 0$.

b) The energy transferred in the electrical process is calculated simply by the rules we know from the Prologue:

$$|W_{el}| = |Uq|$$

The amount received by the system due to heating from the fluid has to be determined indirectly. Since the properties of the fluid do not change, its energy content can be assumed to remain constant. For this reason the energy transferred from the fluid to the system is equal to the energy received by the fluid from the second battery. Therefore:

$$|Q| = |U'q|$$

The change of the energy of hydrogen and oxygen gas, compared to water, is equal to the sum of the two contributions.

EXAMPLE 35. Change of the energy content of water during evaporation.

Exactly 1.0 g of water is evaporated completely at 100°C and at a constant normal pressure of 1.0 bar to produce 1670 cm^3 of steam. Express the change of the energy of the water in terms of the changes of entropy and volume, and compute its numerical value.

SOLUTION: The change of the energy of the fluid is calculated using the balance of energy, Equation (120). The fluid absorbs energy as a consequence of heating, and it emits some of this energy as a consequence of expansion. The energy exchanged in the thermal process is obtained from the amount of entropy communicated to the body if multiplied by the temperature at which the isothermal process takes place. (See Example 10 of Chapter 1.) The amount of entropy exchanged is calculated according to the constitutive law of vaporization in analogy to Equation (8) in Chapter 1. Since the process of turning water into steam is reversible, the entropy exchanged is equal to the change of the entropy content. Therefore, we arrive at the expression

$$Q = T\Delta S$$

for the amount of energy exchanged in heating. Since the pressure is constant during evaporation, we can compute the energy exchanged as a consequence of the expansion without having to know more about the fluid involved. According to the discussion in the Prologue, the mechanical energy current is the negative product of the pressure and the rate of change of the volume. Therefore, for constant pressure, the energy exchanged in the mechanical process is equal to the negative product of the pressure and the change of the volume:

$$W = -P\left(V_{gas} - V_{liquid}\right)$$

Taken together, we obtain the following expression for the change of the energy of the fluid:

$$\Delta E = T\Delta S - P\left(V_{gas} - V_{liquid}\right)$$

The value of ΔS is calculated using the latent entropy of vaporization from Table A.5 which yields:

$$\begin{aligned} Q &= T l_v \Delta m \\ &= 373\,\text{K} \cdot 6.06 \cdot 10^3 \frac{\text{J}}{\text{K} \cdot \text{kg}} \cdot 0.0010\,\text{kg} = 2.26\,\text{kJ} \end{aligned}$$

The energy exchanged during the expansion at constant pressure is

$$\begin{aligned} W &= -P\left(V_{gas} - V_{liquid}\right) \\ &= -10^5\,\text{Pa}(1670 \cdot 10^{-6} - 10^{-6})\text{m}^3 = -167\,\text{J} \end{aligned}$$

The change of the water's energy content is $\Delta E = Q + W$ = 2.26 kJ - 0.167 kJ = 2.09 kJ.

2.8.3 Combining the Balance of Entropy and of Energy

Let us assume now that the laws of balance of entropy and of energy hold for all processes, including dissipative ones. The equation of balance of energy is not equipped to express dissipation since it refers only to the storage and the exchange of energy. As we have seen in Section 1.7, dissipation of energy is equivalent to production of entropy. Therefore, we need to combine the laws of balance of entropy and of energy if we wish to deal with irreversible processes. To be specific, the latter of these laws will be expressed for thermomechanical operations. If we wish, the results can easily be carried over to more general processes. The laws take the forms:

$$\dot{S} + I_s = \Pi_s \tag{121}$$

$$\dot{E} + I_{E,mech} + I_{E,th} = 0 \tag{122}$$

If we express the thermal energy current in terms of the flux of entropy (Equation 13 of Chapter 1), we can calculate the motive power of an engine from Equations (121) and (122):

$$-I_{E,mech} = \dot{E} + T(-\dot{S} + \Pi_s) = \dot{E} - T\dot{S} + T\Pi_s \tag{123}$$

As in Section 1.7, the product of the rate of production of entropy and the temperature of the body, is called the *rate of dissipation* of energy. The result of Equation (123) may be interpreted graphically in the following manner (Figure 16): the negative mechanical energy current always is larger than the difference between the rate at which energy is stored and the rate at which entropy is stored multiplied by the temperature. Together, these two terms represent the instantaneous motive power of reversible thermal engines. Consider the case for which the mechanical energy current is positive, as it is for a heat engine. The magnitude of the negative energy flux is smaller for dissipative processes than for reversible ones. In other words, the power output of a dissipative engine is smaller than that of an ideal one. Conversely, for a heat pump, the power required increases with the amount of entropy produced. In either case we are faced with a loss due to dissipation. What we derived for steady-state processes in Section 1.7, holds for time-dependent operations as well.

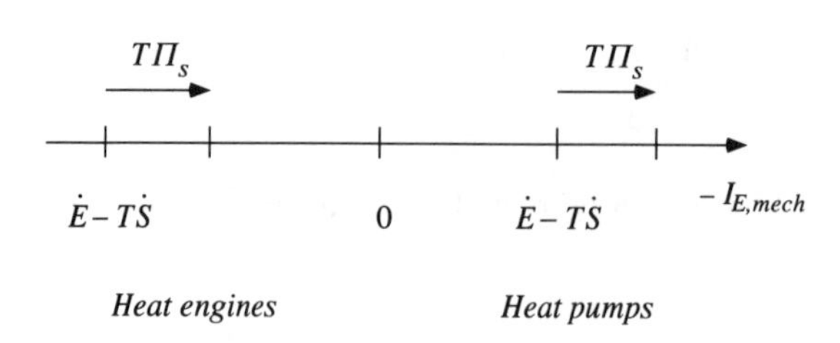

FIGURE 16. Motive power of irreversible thermomechanical processes for power output (heat engine, left side) and power consumption (heat pump, right side). Dissipative heat engines release less energy for mechanical purposes, while real heat pumps require more energy than their reversible counterparts.

Consider the following special case: some water is heated electrically. The heater is included as part of the system, which makes the production of entropy an internal process. An important piece of experimental information concerning dissipation came from Joule's experiments, which demonstrated that the effects of stirring or of electrical heating of a body of water cannot be distinguished from ordinary heating of the same body. We concluded in Section 2.2 that there exists a simple relation between the entropy and the energy of a body undergoing heating at constant volume, namely

$$\dot{E} = T\dot{S} \tag{124}$$

According to experimental evidence, the same relation must hold if the entropy content changes due to production of entropy. If we introduce this in Equation (123) and replace the mechanical energy current by the electrical power, we find that the rate of production of entropy is given by the same relation derived in Section 1.7:

$$\Pi_s = -\frac{1}{T} I_{E,el} \tag{125}$$

(See Equation (36), Chapter 1.) Since no process takes place other than the production of entropy, all the energy supplied to the electrical heater must be dissipated. The rate of production of entropy therefore is equal to the rate of supply divided by the temperature at which dissipation takes place.

EXAMPLE 36. Electrical heating of water.

One liter of water is heated electrically at constant power of 210 W. At a certain moment the temperature is 20°C, and it is found to increase at a rate of 0.045 K/s. a) Determine the rate of dissipation and the rate of production of entropy at that time. b) How large is the current of entropy exchanged between the water and its surroundings, and how large is the current of energy associated with it? c) Calculate the rate of change of entropy and of energy of the system.

SOLUTION: a) Electrical heating in water can lead only to thermal processes, which means that all the energy will be dissipated. Therefore, the rate of dissipation is given by

$$\mathscr{D} = |I_{E,el}| = 210\,\text{W}$$

From this we calculate the rate of production of entropy at temperature T:

$$\begin{aligned}\Pi_s &= \frac{\mathscr{D}}{T} \\ &= \frac{210\,\text{W}}{293\,\text{K}} = 0.72\frac{\text{W}}{\text{K}}\end{aligned}$$

b) The current of entropy with respect to the body of water is computed from the balance of entropy. The rate of production of entropy is known. We can calculate the rate of change of the entropy content from the law of heating at constant volume. Together, these quantities determine the current of entropy:

$$I_s = \Pi_s - \dot{S} = \Pi_s - \frac{C}{T}\dot{T}$$

$$= 0.72\frac{\text{W}}{\text{K}} - \frac{4200 \cdot 0.045}{293}\frac{\text{W}}{\text{K}} = 0.072\frac{\text{W}}{\text{K}}$$

Obviously, entropy is emitted by the water. The rate of production of entropy due to electrical heating is larger than the rate at which entropy is stored in the body. The thermal current of energy flowing out of the water is given by

$$I_{E,th} = T \cdot I_s$$

$$= 293\text{K} \cdot 0.072\frac{\text{W}}{\text{K}} = 21\text{W}$$

c) The rate of change of the entropy of the body has been used above:

$$\dot{S} = \frac{C}{T}\dot{T}$$

$$= \frac{4200 \cdot 0.045}{293}\frac{\text{W}}{\text{K}} = 0.65\frac{\text{W}}{\text{K}}$$

According to the Gibbs fundamental form for water, we find for the rate of change of the energy:

$$\dot{E} = T\dot{S}$$

$$= 293\text{K} \cdot 0.65\frac{\text{W}}{\text{K}} = 189\text{W}$$

This rate of change and the thermal energy current calculated above add up to the electrical power of the heater. Note that we have not used the constitutive relation for heating at constant volume in the form of Equation (6). Rather we have used the definition of the entropy capacity also for this obviously irreversible process. This is justified on the basis of Joule's observations.

EXAMPLE 37. Power and loss of power in Carnot processes.

a) A fluid can be heated and cooled at constant temperatures only (Carnot process). Derive the expression for the motive power associated with the operations undergone by this fluid. The process may be dissipative. Heating occurs at temperature T, while cooling takes place at T_o. The changes of the fluid need not add up to a full number of cyclic operations. b) Show that the loss of power is given by the product of the rate of production of entropy and the temperature of the cooler (the Guoy-Stodola rule of Section 1.7).
c) How is it possible for a fluid to absorb and emit entropy at different temperatures? Can such processes be reversible?

SOLUTION: a) Starting from the laws of balance of entropy and of energy for a fluid undergoing thermomechanical processes,

$$\dot{S} + I_s = \Pi_s$$

$$\dot{E} + I_{E,mech} + TI_{s,in} + T_o I_{s,out} = 0$$

we obtain

$$\dot{E} + I_{E,mech} + (T - T_o)I_{s,in} + T_o(-\dot{S} + \Pi_s) = 0 \qquad \textbf{(E13)}$$

or

$$I_{E,mech} = -(\dot{E} - T_o\dot{S}) + \mathcal{P}_{av} - T_o\Pi_s \qquad \textbf{(E14)}$$

Note that the third term in (E13) is the available power of a current of entropy as defined in Section 1.7 for steady-state processes. Here, for general Carnot processes, we should take both the first and the second terms on the right-hand side of (E14) as the available power. This corresponds to the power available from a fluid undergoing a reversible operation.

b) Look at expression (E14) for the mechanical power in a dissipative process. The difference between the actual power and the ideal one, i.e., the loss of power, is given by the product of the rate of production of entropy and the temperature of the environment (i.e., of the cooler).

c) We know a class of constitutive relations which allow for reversible processes only—heating and working of uniform bodies—which obey the laws discussed in Sections 2.1 - 2.4. To conform to these relations, the fluid must be uniform. Heating and cooling at different temperatures is possible if they do not occur at the same time. Therefore, the expressions occurring in the equations must be suitably averaged over time.

On the other hand, the derivation leading to (E14) does not rely on special constitutive assumptions. For this reason, (E14) is a consequence of macroscopic balances of a body which does not have to be uniform. Heating and cooling may occur at the same time at different parts of the surface of the body where the temperatures are different. In general, such operations are expected to be dissipative. Dissipation is described superficially by the rate of production of entropy in (E14) for which we do not always have a constitutive law.

2.9 Thermostatics: Equilibrium and Changes of State

There are some common irreversible phenomena which we cannot describe using the simple theory laid down so far. Take, for example, the process encountered when two bodies of unequal temperatures are brought into thermal contact. We know that they eventually reach a state in which the temperature is uniform throughout. Simple measurements show that in the case of two identical bodies of water or metal at room temperature the final temperature is just the arithmetic mean of the initial hotnesses. A second example is the free expansion of air. Both phenomena lead to simple results. Still, the processes undergone by the bodies are clearly irreversible. For such operations we lack a theory despite what has been presented so far. We are not able to calculate the processes yet. However, there is another approach to thermal phenomena which will allow us to calculate such quantities as the final temperature reached in thermal contact, or the amount of entropy created in the free expansion of the ideal gas. Instead of calculating the details of actual irreversible processes using "equations of motion" we will compute only the *outcome* of such processes on the basis of a theory of the statics of heat. We shall do this only for some very simple cases. More details can be found in books on classical thermodynamics.[16]

2.9.1 Thermal Equilibrium

While the processes of the flow of heat in thermal contact, and of the free expansion of air, are rather complicated, the initial and the final states are very simple, and conform to the simplifying assumptions made in this chapter. At the beginning and at the end of the processes, the temperature, and possibly the density, are uniform throughout the systems. The state of uniform temperature is of particular interest.

Consider, for example, two blocks of metal which are placed in contact with each other. Assume that they are insulated from the surroundings by a wall which is impermeable to entropy. If the bodies have different temperatures initially, the hotter one will get cooler while the cooler body will get warmer. This continues until the temperatures have become equal. We interpret this process in terms of the transport of entropy from the hotter to the cooler body. The difference of temperatures is called the *thermal driving force*. Using the language introduced in the Prologue, we say that a thermal driving force is needed to maintain the flow of heat from one body to another in thermal contact. Once the driving force has vanished, the process stops, and the temperature is uniform throughout the system. In analogy to mechanical or electrical situations in which the proper driving force is zero, we say that the system is in *thermal equilibrium*. Since equilibrium means the vanishing of the driving force between two systems, or between different parts of a system, thermal equilibrium means that the temperature must be uniform.

2.9.2 The Computation of Changes of State

We have dealt with the situation of uniform temperature before. In fact, we have built the theory of homogeneous processes upon the existence of such simple states. As before, we assume it to be possible to specify such states, which now

16. See H.B. Callen (1985) for an example of an approach to the statics of heat. Callen says: "The single, all-encompassing problem of thermodynamics is the determination of the equilibrium state that eventually results after the removal of internal constraints in a closed, composite system" (p. 26). This obviously is a description of the fundamental idea underlying a theory of the statics of heat, not its dynamics.

are taken to be equilibrium states, by giving just one or two numbers, namely the temperature, and the volume (in the case of systems whose volume may change). The theory of homogeneous processes laid down above has enabled us to calculate processes that carry bodies through states of uniform temperature and density, i.e., through states which are the same as those reached in equilibrium. The hope of calculating the results of irreversible processes now rests on the assumption that we have only to determine the changes from initial to final states irrespective of what happens in between:

In the absence of a proper theory of irreversible processes, we can still compute the outcome of such processes if they lead from one state of equilibrium to another. Under these circumstances we determine changes of state rather than the real processes.

In other words, we assume that a theory of thermostatics of simple systems, which we have not yet developed, and which we are going to discuss only briefly, will lead to essentially the same relations as the theory of uniform reversible processes. There is a distinct difference, however, in what these theories are capable of delivering, not just in their forms. The theory of reversible processes describes how states evolve. A theory of statics cannot do this. All it does is to deliver relations between the variables of systems in equilibrium. On the other hand, the number of materials which exhibit simple equilibrium states might be larger than the number of those which admit models of uniform processes.

Let us discuss the case of thermal contact of two rigid, homogeneous bodies. The total energy of the bodies is conserved since they are insulated from the surroundings:

$$\Delta E_1 + \Delta E_2 = 0 \tag{126}$$

According to the theory describing the heating or cooling of bodies at constant volume, i.e., according to Equation (27), the law of balance of energy leads to the following expression:

$$\int_{T_{i1}}^{T_f} C_1(T)dT + \int_{T_{i2}}^{T_f} C_2(T)dT = 0 \tag{127}$$

This is the proper equation for the *changes of state* of the bodies, rather than an equation describing the real processes undergone. The temperatures T_{i1} and T_{i2} are the initial temperatures of the two bodies, respectively, while T_f is equal to their common temperature at the end of the process. Graphically, this means that the areas under the curves C_1 and C_2 in the appropriate ranges must be equal. For the particular case where $C_1 = C_2$ for all temperatures, this rule is exhibited in Figure 17.

If we can calculate the integrals in Equation (127), we can in principle solve the equation for the common temperature attained by the bodies in thermal contact. The simplest case is the one for C = *constant.* (See Example 38.)

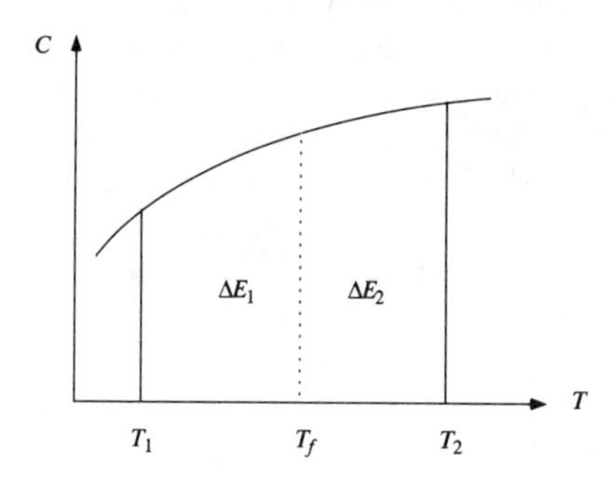

FIGURE 17. In the case of identical bodies the final temperature reached in thermal contact is calculated simply by geometrical means. The energy emitted by the body which is cooling down is the opposite of the energy absorbed by the body being heated.

Assume we have calculated the final temperature reached by the two bodies. We are now in a position to compute the *changes of the entropy content* of either one. If we add the changes, we must get the amount of entropy produced:

$$S_{gen} = \Delta S_1 + \Delta S_2 \geq 0 \tag{128}$$

We expect the sum to be larger than zero. We will not give a general proof. Rather, the amount of entropy created is calculated for a particular case in Example 39.

EXAMPLE 38. The final temperature reached in thermal contact.

Two bodies having *constant* values of $C = TK$ are brought in heat contact. Assume that they are insulated from the environment. What is the final temperature reached?

SOLUTION: We simply use Equation (127) in integrated form:

$$C_1\left(T_f - T_{i1}\right) = -C_2\left(T_f - T_{i2}\right)$$

where T_{i1} and T_{i2} are the initial temperatures of body 1 and 2. We obtain

$$T_f = \frac{C_1 T_{i1} + C_2 T_{i2}}{C_1 + C_2}$$

Note that this is only true if the values of C are constant! If the two bodies are identical or have identical values of C, we conclude that the temperature reached is exactly the arithmetic mean of the initial temperatures.

EXAMPLE 39. The entropy created in the contact of two bodies at different temperatures.

Two bodies having different temperatures T_{i1} and T_{i2} are brought in thermal contact; they are insulated from the surroundings. Their entropy capacities K are equal and are supposed to be inversely proportional to the temperature in the envisioned range; this means that the quantities $C = TK$ are constant. How much entropy is produced in the ensuing process of equalizing the temperatures?

SOLUTION: We can calculate the *changes* of the entropy content of the two bodies occurring as a consequence of the process. Just as in Example 3 we get

$$\begin{aligned}\Delta S &= \int_{T_{i1}}^{T_f} K\,dT + \int_{T_{i2}}^{T_f} K\,dT \\ &= \int_{T_{i1}}^{T_f} \frac{C}{T}dT + \int_{T_{i2}}^{T_f} \frac{C}{T}dT \\ &= C\left[\ln\left(\frac{T_f}{T_{i1}}\right) + \ln\left(\frac{T_f}{T_{i2}}\right)\right] \\ &= C\ln\left[\frac{T_f^2}{T_{i1}T_{i2}}\right]\end{aligned}$$

According to Example 38 the balance of energy leads to

$$\Delta E = C\left(T_f - T_{i1}\right) + C\left(T_f - T_{i2}\right) = 0$$

or

$$T_f = \frac{1}{2}\left(T_{i1} + T_{i2}\right)$$

This we plug into the expression for the change of entropy content to obtain $\Delta S \geq 0$. This is so since the arithmetic mean of the initial temperatures is always greater than or equal to their geometric mean. As a result, the argument of the natural logarithm in the expression for the change of the entropy content is larger than or equal to 1.

EXAMPLE 40. Production of entropy in the free expansion of the ideal gas.

In Section 1.5 the free expansion of the ideal gas was mentioned briefly. Since the temperature of the gas is the same at the beginning and at the end, the result of the process is equivalent to that of an isothermal expansion. Convince yourself that the process is irreversible, and demonstrate that the amount of entropy created can be related to the amount of energy dissipated in the manner discussed in Section 1.7.

SOLUTION: If an ideal gas expands isothermally, it absorbs entropy; i.e., its entropy content increases. For this reason the entropy content of the gas must increase by the same amount. Since entropy has not been exchanged, it must have been produced.

We interpret the process in the following manner. The energy which would have been emitted as a consequence of the mechanical process of isothermal expansion is used internally for the production of entropy. Using a different image, we can say that the energy used to produce entropy stems from the flow of the gas from points of high to low pressure. (See Chapter 4 for more details on flow processes.) We say that the amount of energy which could have been used for other purposes has been dissipated.

From Section 2.5 we know that the energy exchanged in the mechanical process during isothermal expansion of the ideal gas is given by

$$W = -nRT\ln\left(\frac{V_f}{V_i}\right) \tag{E15}$$

On the other hand, we have calculated the amount of entropy communicated to the body in the same process. Since the operation is reversible, the entropy exchanged and the change of the entropy content are equal:

$$\Delta S = nR\ln\left(\frac{V_f}{V_i}\right) \tag{E16}$$

In the free expansion exactly this amount of entropy must be generated internally, since the final states attained by the gas are the same irrespective of the details of the actual process. Therefore, in the free expansion of the ideal gas, the amount of entropy produced S_{gen} is equal to what has been calculated in (E16). Also, the energy delivered for mechanical purposes in the reversible process, (E15), must be equal to the amount of energy dissipated. Therefore we conclude that the relationship between the entropy created and the energy dissipated is

$$\mathscr{D} = T \cdot S_{gen}$$

As a consequence we can state that *the entropy created at a temperature T is equal to the energy needed for the process of production of entropy divided by this temperature*. The same was found to be true in Section 1.7.

2.9.3 Thermostatics and the Maximum Entropy Postulate

In mechanics, equilibrium states are considered in the branch called *statics*. *Thermostatics* is the science of heat which considers the determination of the equilibrium states. In this book we will not deal with statics. However, it is important to take a brief look at the theory.

In nature, equilibrium states can be the outcome of some processes. For example, a pendulum stops swinging after some time because of the effects of friction. Therefore, we could conceivably determine the equilibrium state of a pendulum by considering the mechanical processes which must be described, using proper constitutive laws. Naturally, these laws must include the phenomena which eventually lead the pendulum to stop swinging. Otherwise we will never get the equilibrium state out of our theory. Consider an ideal pendulum. No matter what we do, equilibrium simply cannot be attained if the body is moving at a given moment. This is the approach of dynamics. What we have attempted so far is to transfer the procedure known from mechanics to thermal physics. In this manner, thermodynamics is created.

There is a different approach to statics, however, an approach which has nothing to do with dynamics. It is best explained in the context of thermal phenomena. We shall use the example of bodies in thermal contact, which we have treated above. As you will see, the condition of equilibrium is determined by a new type of principle. The bodies start from equilibrium states and end in another state of uniform temperature. Initially, we assume the bodies to contain given amounts of entropy and energy, and their temperatures to be well defined. During the process, entropy is created. This finally stops when the temperature has become uniform throughout the combined system. In this final state, then, the amount of entropy contained in the system has attained its maximum possible value, which leads us to the *maximum entropy principle*:

> *The values assumed by the quantities specifying the equilibrium states of bodies are those which maximize the entropy of a body.*

When you think about it, this condition must be satisfied in much more general cases than the one discussed. Entropy can be created, but it cannot be destroyed. Therefore, in a composite system which is insulated from its surrounding, the amount of entropy can only increase. In thermal equilibrium the processes which are possible inside the system have come to a standstill. Entropy is no longer produced. Therefore, the amount of entropy contained in the system has reached its maximum value.

The same kind of principle applies to other fields of physics as well. The determination of equilibrium states in mechanics can be built upon a similar variational principle. There is also a proper type of mathematics which deals with this kind of situation. For our purpose it is important just to recognize the difference between thermostatics based on the maximum entropy postulate, and thermodynamics, which follows from equations of balance and constitutive laws. In statics, equilibrium is determined without recourse to the equations of motion.

EXAMPLE 41. The final temperature reached in thermal contact.

Two identical rigid bodies with constant values of $C = TK$ are brought into thermal contact and insulated from the surroundings. Their initial temperatures are T_1 and T_2. Determine the final temperature assuming that the entropy of the bodies is maximized under the condition that the energy remains constant.

SOLUTION: The formulation of the problem tells us that we should try to find the maximum of the entropy of the body for given fixed energy. This is equivalent to the mathematical problem of finding the extremum of the function $S(E)$. Therefore, we need the entropy of the bodies as a function of their energies. According to Equation (27), the energies are given by:

$$\begin{aligned} E_1 &= C(T_1 - T_o) + E_o \\ E_2 &= C(T_2 - T_o) + E_o \end{aligned} \tag{E17}$$

where

$$E = E_1 + E_2 = constant \tag{E18}$$

The entropy contents of the bodies have been calculated before:

$$S_1 = C\ln\left(\frac{T_1}{T_o}\right) + S_o$$
$$S_2 = C\ln\left(\frac{T_2}{T_o}\right) + S_o \qquad \textbf{(E19)}$$

We solve (E17) for the two temperatures, replace E_2 using (E18), and substitute the expressions into the equations for the entropies (E19). If we add the two values of S we get:

$$S = S_1 + S_2 = C\ln\left[\left(\frac{E_1 - E_o}{CT_o} + 1\right)\left(\frac{E - E_1 - E_o}{CT_o} + 1\right)\right] + 2S_o$$

Now we look for the value of the energy E_1 for which the total entropy obtains its maximum. In other words, we have to set the derivative of S with respect to E_1 equal to zero:

$$\frac{\partial S(E, E_1)}{\partial E_1} = C\frac{1}{AB}\frac{1}{CT_o}[B - A]$$

with

$$A = \frac{E_1 - E_o}{CT_o} + 1$$
$$B = \frac{E - E_1 - E_o}{CT_o} + 1$$

This expression is equal to zero, which leads to the simple equation

$$E = 2E_1$$

This means that the entropy of both bodies is maximized if the energy of body 1 is half of the total energy. Naturally, body 2 must have half of the total energy as well. Inspection of (E17) shows that the bodies have half of the total energy if their temperatures are equal to the arithmetic mean of the initial values:

$$T = \frac{1}{2}(T_1 + T_2)$$

This is the expected result. It shows that in the state attained by two bodies in thermal contact, but otherwise insulated from the surroundings, the entropy takes its maximum possible value.

Questions and Problems

1. Explain the meaning of uniform processes. How must heat spread through a body for such a process to be reversible? Is a uniform process necessarily reversible?

2. Consider two bodies in thermal contact. Their temperatures are different, which causes entropy to flow from the hotter to the cooler one. Is it possible to set up a model of the process for which the temperatures of each of the two bodies are uniform? Where is entropy produced?

3. Derive an expression in terms of the electrical current for the production rate of entropy in an electrical resistor operating at a steady state. Do the same for a block sliding over a surface on a viscous film. (Use the momentum current in this case.) Compare the results to Equation (2). How does Equation (2) change if you express it in terms of the current of entropy instead of the energy current associated with it?

4. Why can't a rigid body whose heating is proportional to the rate of change of its temperature undergo irreversible processes?

5. How much entropy and energy are added to 1.0 kg of silicon if the body is heated from 160 K to 640 K? (See Figure 4 for properties of silicon.)

6. A 100 g piece of copper is to be heated at 30 K such that the rate of change of its temperature is 0.10 K/min. How large does the flux of entropy have to be? How large is the current of energy entering the body? (See Figure 4 for properties of copper.)

7. The temperature of a piece of granite lying in the sun is found to change from 20°C to 40°C in 2 hours. Its mass is 0.30 kg. At what average rate does it absorb entropy? Assume its temperature coefficient of energy to be constant over the range of temperatures considered.

8. The entropy of a rigid body changes at a rate of 12.0 W/K at a temperature of 20°C. How large is the rate of change of its energy?

9. Derive an expression for the energy density of a rigid body. (Set the energy equal to zero at room temperature.) Take its temperature coefficient of energy to be constant. What properties should the body have to achieve a high energy density?

10. A rigid body has a constant entropy capacity in a particular range of temperatures. How much energy does it emit if its temperature drops from T_i to T_f?

11. Consider water being heated by an immersion heater. a) If you consider the body of water as a system, what is its equation of balance of entropy? How do you have to formulate the equation of balance of energy? In what form is energy transferred to the system? (Assume the distribution of entropy through the system to take place reversibly; what does this mean for the conduction of entropy through the system?) b) Answer these questions for the case in which you take the system to be made up of the water plus the heating coil.

12. An ideal Carnot engine is driven with the heat from 2000 liters of water at 90°C. Entropy is rejected to the environment at a temperature of 20°C. How much energy does the engine release for mechanical purposes?

13. Two different methods of heat recovery are to be compared. In both cases, 100 kg each of 20°C water and of 40°C water are accumulated. In case A, the samples are mixed in one tank, in case B they are kept separated in two tanks. In both cases an ideal heat pump is supposed to deliver heat at a temperature of 60°C. Entropy is collected from the warm water in the tanks which is cooled to 10°C.

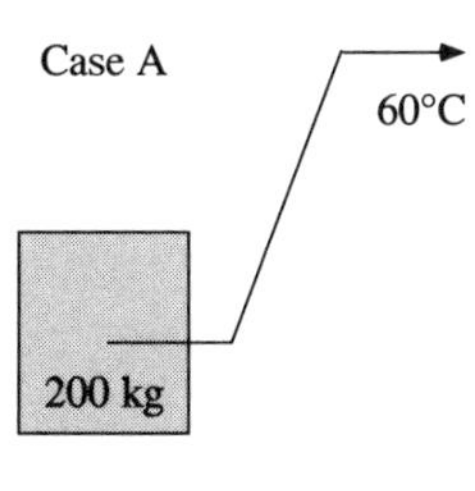

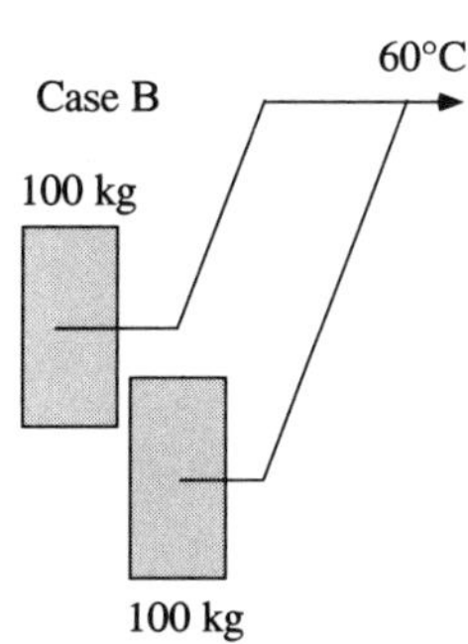

FIGURE 18. Problem 13.

a) In case A, the heat pump takes the entropy from the single tank. How much entropy is emitted by the water as it is cooled to 10°C? b) How much energy has to supplied to the heat pump to pump the entropy to 60°C? c) In case B, the heat pump gets its entropy from both tanks alternatively. How much energy is required for the heat pumps? d) Explain the difference between cases A and B.

14. For the world exhibition in Seville, the architect for the Swiss pavilion suggested building a tower of ice. (The project was abandoned for environmental reasons.) Estimate the minimal amount of energy necessary to produce 1000 tons of ice in a 30°C environment.

15. It is said that the oceanic climate is less extreme with respect to temperature variations because of the thermal buffering effect of the water, which is explained as being a result of the higher "heat capacity" of water compared to that of the land. Compare the entropy capacities per volume of water and soil or rock. If the difference appears to be too small to explain the effect upon climate, how else would you explain the phenomenon?

16. How does the energy content change if water undergoes an isothermal expansion at a temperature of 2°C? Is the answer different for a temperature of 20°C?

17. A body of air of mass 20 kg at 100°C and 1.0 bar is compressed isothermally. a) How much energy has to be supplied in the mechanical process if the volume is to be reduced to 10% of its initial value? b) How large will the pressure of the air be at the end? c) If the compression is to be performed at a constant rate of change of the volume in 10 s, how large does the mechanical power have to be as a function of time?

18. Sketch the *T-S* diagrams for heating of a rigid uniform body whose entropy capacity is a) constant, b) decreasing with increasing temperature, c) increasing with increasing temperature. Which of these cases applies to water?

19. Determine the entropy content of one mole of argon at 300 K and at a pressure of 1 bar. (See the values supplied in Table 5.)

TABLE 5. Values of the molar temperature coefficient of enthalpy of argon[a]

solid		liquid		gaseous	
T K	$\bar{c}_p$ $J \cdot mole^{-1}K^{-1}$	T K	$\bar{c}_p$ $J \cdot mole^{-1}K^{-1}$	T K	$\bar{c}_p$ $J \cdot mole^{-1}K^{-1}$
20	11.76	83.85	42.04	87.29	20.79
40	22.09	87.29	42.05	100	20.79
60	26.59			300	20.79
80	32.13				
83.85	33.26				

a. Values have been taken from Förstling und Kuhn (1983).

20. An ideal fluid undergoes a general process represented by a curve in the *T-V* diagram leading from an initial point (T_i, V_i) to the final state (T_f, V_f). Show that the entropy exchanged along the path is given by

$$S_e = -\int_{V_i}^{V_f} \Lambda_V dV - \int_{T_i}^{T_f} K_V dT$$

21. How do you determine the energy exchanged as a consequence of heating for the general process of Problem 20?

22. a) What type of observation shows that the latent entropy (with respect to volume) of the ideal gas must be a positive quantity? b) What must be the sign of the latent entropy with respect to pressure? What does the latter result mean? c) Prove that the entropy capacity at constant pressure must be larger than the entropy capacity at constant volume. What is the significance of this result?

23. Dry air rushes from the mountains (2500 m above sea level) into a valley (500 m above sea level). The temperature of the air in the mountains is 4°C. Before the arrival of the winds the temperature of the air in the valley is 16°C. By how much will the temperature of the air rise in the valley with the winds blowing?

24. Derive the expression for the pressure as a function of height in the earth's atmosphere if pressure and volume are related by the adiabatic condition for dry air.

25. Measurements of the speed of sound in air at different temperatures give the following values. Determine the adiabatic exponent (the results are presented in the last column).

TABLE 6. The ratio of specific heats for 273.15°C and 101.3 kPa

Gas	c	M_o	γ
Argon	308	0.040	1.67
Helium	971	0.004	1.66
Neon	433	0.020	1.65
Xenon	170	0.131	1.67
Oxygen	315	0.032	1.40
Nitrogen	334	0.028	1.38
Carbon monoxide	337	0.028	1.40
Ammonia	415	0.018	1.29
Chlorine	206	0.071	1.33
Carbon dioxide	258	0.044	1.30
Methane	430	0.016	1.30

26. Why does the entropy content of the ideal gas remain constant as a result of an adiabatic process? Determine the special forms of the equations of balance of entropy and energy for such a process. What happens to the energy of the ideal gas during adiabatic expansion?

27. According to the result obtained for adiabatic processes of the ideal gas, the ratio of the latent entropy and the entropy capacity must be proportional to T/V. As a result, does the entropy capacity depend on T or V? How do C_V and C_P depend on temperature or volume?

28. The results derived for adiabatic motion of the ideal gas require the ratio of the entropy capacities to be constant. Is this requirement realistic for real ideal gases? (Compare with values taken from Figure 9.)

29. Derive the equations of adiabatic change of the ideal gas, using the result of Example 17, i.e., the equation which determines the change of the entropy of the ideal gas.

30. From what you know about the properties of water how would an adiabat look in the *T-V* diagram in the range 0 - 4°C? Try to sketch an adiabat for water for a process in which the temperature changes from 2°C to 6°C.

31. From the values in Figure 9, calculate the specific temperature coefficient of enthalpy for He, Ne, H_2, and N_2 at 300 K.

32. Draw the curve for an isobaric process of air in the *T-S* diagram. Repeat the problem for the same body of air for a process at a higher value of the pressure.

33. A body of air is heated at constant pressure. What fraction of the entropy added remains in the body? What fraction of the energy added as a result of heating remains there?

34. Draw curves for adiabatic, isothermal, isobaric, and isochoric processes of the ideal gas in the *T-S*, *T-V*, and *P-V* diagrams.

35. Replace the process of heating of a body of air at constant pressure by two consecutive processes. The first is heating at constant volume, the latter is an isothermal change of volume. Sketch the steps in the *T-S* diagram.

36. Air having a mass of 5 g, at a pressure of 38 bar and a temperature of 650°C, is heated inside a cylinder by burning some injected fuel. The amount of energy added by the burning fuel is 7.5 kJ. The piston moves in such a way as to leave the pressure of the air constant. (This corresponds to a step in the Diesel process.) Assume that the fuel added does not change the properties of the air in the cylinder. a) How much energy is exchanged as a result of the change of volume of the air? b) Calculate the change of the energy of the gas.

37. A bubble of air with an initial diameter of 5.0 mm starts rising from the bottom of a pond at a depth of 5 m. The temperature of the water is 6°C at the bottom and 15°C at the surface. Assume the bubble to have the same temperature as the surrounding water at all times. Neglect the effects of surface tension. a) Calculate the radius of the bubble shortly before it reaches the surface. b) Approximately estimate the amount of energy exchanged as a consequence of heating while the bubble is rising.

38. Use the law of hydrostatic equilibrium for a column of gas extending from the center of the Sun to its surface to estimate the pressure at the center. The gas at the center of the Sun is ideal. Determine the temperature at the center from a rough estimate of the density. How large is the contribution of radiation to the pressure at the center of the sun?

39. Write the law of balance of energy in differential and in integrated forms for a) adiabatic processes of the ideal gas; b) cooling of rigid bodies; c) isobaric processes of the ideal gas; d) isothermal processes of a fluid; e) isothermal expansion of water at 2°C; f) adiabatic processes of a fluid with internal production of entropy; g) stirring of water by paddle wheels; h) electrical heating of water; i) electrical heating of a heating coil plus water.

40. Is it possible for the energy of systems undergoing the various processes of Problem 39 to remain constant? In which of the processes running at constant energy will the entropy of the system change, and why?

41. Consider a paddle wheel inside a tank containing some viscous fluid. As the fluid is stirred, entropy is produced. If the tank is insulated, the temperature of the systems must rise. Assume the Gibbs fundamental form to hold for the system and derive the relation between the energy dissipated and the entropy produced. Do the same for an immersion heater that is heating up.

42. Suppose we raise the temperature of an amount of water from T_L to T_H by stirring, and then drive an ideal Carnot engine with the entropy of the water released to the environment at temperature T_L. a) Calculate the amount of energy dissipated. b) How large is the amount of energy gained for the mechanical process? c) How large is the loss of availability? d) Why is the loss different from how it is defined in Equation (46) of Chapter 1?

43. Burning hydrogen is supposed to release energy at a rate of 10 kW. How large do you have to make the current of amount of substance of hydrogen to the burner? The burning of hydrogen releases 10.8 MJ of energy per cubic meter of gas supplied at standard temperature and pressure (0°C and 101.3 kPa).

44. In a model of irreversible processes undergone by a viscous fluid, it is found that the Gibbs fundamental form, Equation (78), still holds. Show that the energy current as a result of changes of volume is given by

$$I_{E,mech} = P\dot{V} - T\Pi_s$$

Compare irreversible compression and expansion at constant temperature. Is the mechanical energy current reversible?

45. When ice is melted, is the change of the energy of the system larger or smaller than the energy added in heating? How does the change of the entropy of the system compare to the entropy added during melting?

46. Take a fluid, such as the ideal gas, which may undergo heating and cooling at constant temperatures only (Carnot processes). a) Express the work done by such a fluid in an operation which does not necessarily have to be a closed cycle, in terms of the changes of entropy and energy, and the entropy absorbed. b) Prove that the motive power of a Carnot engine is a special case of this work estimate. c) Calculate the work done in a process in which no entropy is exchanged. d) What is the work done by a fluid in an isothermal process in which entropy is absorbed but not emitted? e) Assume that dissipation cannot increase the energy emitted by the fluid as a consequence of the mechanical process. Show that this assumption leads to the result that the change of the entropy content of the fluid is larger than the entropy exchanged. (Entropy must have been produced.)

47. Consider the following strongly simplified model of a gas of noninteracting point particles. N particles are contained in a cube of side L. Assume one-third of the particles to travel in each of the three directions parallel to the sides of the cube. All particles have the same speed. a) Show that the pressure of the particle gas is given by

$$P = \frac{1}{3}\frac{N}{V} v p$$

where v and p are the speed and the momentum of a single particle, respectively, and V is the volume of the cube. b) Apply this result to an ideal gas of material particles. Derive the relation between the pressure and the energy density of the gas. c) Again for the material ideal gas, derive the expression for the temperature coefficient of energy. Which gases have such a value for the coefficient? d) Apply the idea to the photon gas. Show that the relation between pressure and energy density is given by Equation (88).

48. In geophysics, the thermomechanical behavior of rocks in the Earth's interior is often described in terms of a fluid model. Which essential assumptions about properties of rocks have to be made for this model to apply?

49. Can you explain the difference between thermostatics and thermodynamics? Use Callen's definition of the task of thermostatics (which he calls thermodynamics) found in Footnote 16 for a starting point.

INTERLUDE

Heat Engines and the Caloric Theory of Heat

I suppose implicitly in my proof that if a body has suffered any changes whatever and that if after a certain number of transformations it is brought back to its original state, ... [it will be] found to contain the same quantity of heat as it contained at first, or, in other words, that the quantities of heat absorbed or emitted in its various transformations are exactly compensated.

S. Carnot, 1824

This chapter will present an alternative route to the thermodynamics of simple fluids at a slightly higher mathematical level, allowing for a generalization of the subject of thermodynamics of uniform fluids. The most important difference to the previous development has to do with what we assume to know about the nature of heat. By introducing the law of balance of heat, and the relation between currents of heat and energy in heating (Equation (13) of Chapter 1), we have directly identified heat with entropy, and have made the relation between entropy and energy the cornerstone of our development. This approach has afforded us a great simplification which is important in an introductory course on the foundations of thermodynamics. Still, it somewhat oversimplifies the matter in that it assumes too much about the nature of heat at the start, and it leaves open the question as to the historical development of the subject. In this chapter, we will therefore not assume any knowledge of the relation between heat and energy. Rather, we will start with what is known as Carnot's Axiom (an assumption about the power of heat in ideal engines), and with a statement about the existence of a heat function. On this basis we will be able to derive the relation between heat and energy.

The point about the heat function needs additional explanation; basically, it represents a formal statement of the idea of the existence of a substancelike quantity called *heat*, or *caloric*. Carnot's Axiom, the heat function, and an observation about the ratio of the heat capacities of the ideal gas, lead to a proper theory of thermodynamics of ideal fluids in which heat turns out to be what we, today, call entropy. You may therefore take this chapter as the proof that thermodynamics may be built upon this simple everyday image—of heat as a quantity which can be thought of as residing in bodies, making them warm, letting them expand, or melting them. Therefore, in this chapter, we will use the terms *heat* or *caloric* for the fundamental quantity of thermal physics.

Except for a final step, this is what Carnot had derived by 1824.[1] By demonstrating the difference in assumptions needed for the thermodynamics of Clausius, we will find the tools needed to understand the development of the ideas about the nature of heat. You will see where and why it happened that heat became the energy exchanged in heating, and why the notion of heat as caloric was lost.[2]

The entire derivation will be based upon the model of spatially uniform, ideal fluids, as we have encountered them in Chapter 2. The bodies called *simple* are fluids which obey the following constitutive laws. There is a *thermal equation of state* relating the pressure in the fluid to its volume and its temperature (Section I.1). Second, they are subject to the *doctrine of latent and specific heats*, which postulates the existence of two further constitutive quantities, i.e., the latent heat and the heat capacity (Section I.2). Having defined the materials in this manner, we will postulate how they behave in thermomechanical processes (Section I.3). Put more colorfully, we will state an assumption about the power of heat (Carnot's *puissance du feu*), which is what we need to relate mechanics and thermodynamics. The result will be the identification of heat (caloric) with entropy, and the derivation of the relationship between entropy and energy. Section I.4 derives the existence of the energy function of the fluids, and proves the law of conservation of energy for them. Finally, in Section I.5, we will compare the caloric and the mechanical theories of heat, which should give you a better understanding of the historical development.

I.1 Thermal Equations of State

This section will provide the first step toward a description of the properties of simple fluids. First, an explanation is given of why volume and temperature (and not volume and pressure) should be used as the independent variables of the theory, and paths are defined. In Section I.1.2, the general thermal equation of state is introduced along with a discussion of constitutive inequalities and the problem of invertibility, which is raised by bodies such as water, with its density maximum at 4°C. While other equations of state are not presented in any detail, the theory developed here can be applied to relations other than the one which holds for the ideal gas.

---

1. S. Carnot (1824): *Reflections on the Motive Power of Fire*.
2. In his work, *The Tragicomical History of Thermodynamics*, C. Truesdell gave a detailed and critical account of the historical development of thermodynamics (Truesdell, 1980). He and Bharatha worked out the logical foundations of classical thermodynamics (*The Concepts and Logic of Classical Thermodynamics*, 1977), where the nature of heat is left to be determined almost up to the end. Following their reasoning, you can see what is needed to make heat a quantity of energy. If you take a slightly different approach, however, heat turns out to be entropy. This line of reasoning was demonstrated in a paper by Callendar (1911).

I.1.1 Variables, Processes, and Paths

In Chapter 2 we mainly used the ideal gas as a vehicle for discussing and developing the ideas of thermal physics. We introduced the concept of ideal homogeneous fluids, for which two independent variables are needed. For the ideal gas we were able to derive the relations governing processes such as adiabatic changes. Here, we will discuss why we choose temperature and volume as the independent variables. Also, the meaning of *process* and *path* will be made precise.

The independent variables. In the treatment of the thermodynamics of the ideal gas in Chapter 2, *temperature* and *volume* were used as the independent variables. It is clear that two variables are needed if we are to describe the interaction of thermal and mechanical processes of simple homogeneous bodies, but others might be chosen. The thermal equation of state of the ideal gas relates the pressure of the fluid to its temperature and volume. This equation can be inverted, which gives us the possibility of using, for example, pressure and volume as the independent variables of the theory. However, this is not a general theory but one dealing only with the simplest kind of thermomechanical system, i.e., the ideal gas. Are we free to select any two of the three variables (pressure, temperature, and volume) as the fundamental ones for a more general theory?

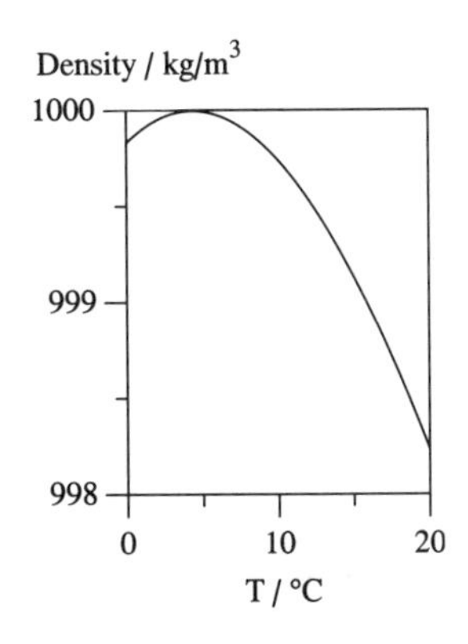

FIGURE 1. The density of water first increases with increasing temperature. Only for temperatures above 4°C does water behave as we normally expect.

The question has a simple answer. If the heating of a fluid is given in terms of temperature and volume, we have to be able to replace the temperature by pressure and volume. In this way we arrive at a description involving only the latter two variables. For this to be possible the temperature has to be replaced using the equation of state. Therefore all depends on whether this equation can be inverted for the temperature. As we shall see in the following section, this is not possible in general. Let us use water as an example, and consider the curve of volume or density versus temperature displayed in Figure 1. The curve has been measured for constant pressure. Therefore, if we allow the temperature to vary between 0°C and 100°C, there may be two values of the temperature associated with one value of the volume at a given pressure. While the function is invertible separately for the ranges from 0°C to 4°C, and from 4°C to 100°C, it is not so globally, and definitely not at the point of maximum density of water. We have to conclude that a general theory of classical thermodynamics which allows for substances like water cannot be built with pressure and volume as the independent variables. Even pressure and temperature do not work under all circumstances. We shall therefore stick with temperature and volume as the independent variables for a general theory.

Variables and state space (the constitutive domain). The physical state of the fluids to be used in this and the following chapter can be described by the instantaneous values of temperature and volume. These two variables assume positive values only. While in the case of the volume this is trivial, it is more subtle as far as temperature is concerned. We have introduced and used the ideal gas temperature as our measure of hotness (Section 1.2.3). This scale is bounded below: temperatures cannot fall below a certain value. All other experience with temperatures points in the same direction: this quantity can only be positive. We shall accept here the existence of such a scale.[3] Therefore an instantaneous state of a fluid can be represented by a point in the upper right-hand quadrant of the *T-V* diagram (Figure 2).

A given fluid, however, cannot take on all possible values of T and V. Natural bodies freeze or melt, or they condense or evaporate. We shall exclude such processes from consideration in this chapter. Therefore, certain areas of the *T-V* quadrant are inaccessible to the fluid. We will call the part of the quadrant covering the possible states of a fluid the *constitutive domain* of the fluid. This domain is a constitutive quantity just like others: it belongs to a body, and it describes how a body differs from other fluids.

Processes and paths. Now we can describe what we mean by processes. A *process* undergone by a body is described by the volume and the temperature as functions of time:

$$V = V(t) > 0$$
$$T = T(t) > 0 \tag{1}$$

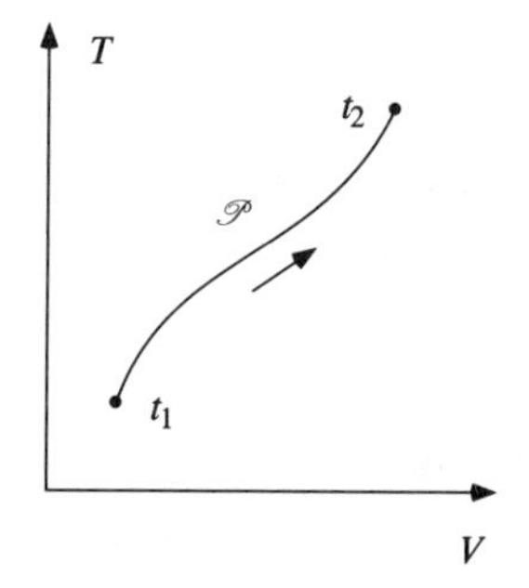

FIGURE 2. A process is exhibited by a curve in the *T-V* quadrant. The process is directed from the initial to the final point. Different processes which can be visualized by the same curve are said to belong to the same path $\mathcal{P}$. The reverse process follows the same path, but in the opposite direction.

It can be visualized by a curve in the *T-V* quadrant (Figure 2). We call t_1 and t_2 the beginning and end of the process, respectively. A process might run up and down the same curve several times, and curves might be complicated and intersect at certain points. We shall always use *simple processes* in which the body does not occupy the same point in the constitutive domain twice. (Later, we will accept an exception to this rule: in a simple cyclic process the body returns to the starting point in the end.) Examples of processes are *isothermal* and *isochoric* changes. They are represented by horizontal and vertical straight lines in the quadrant, respectively.

A process can be visualized by a curve in the *T-V* quadrant. A curve, however, does not represent a unique process. Consider the curve in Figure 2. An infinite number of processes can run along the same line from the initial to the final point. The line does not contain any information about the speed at which it is traversed, nor does it matter which values of time we associate with the end points. Processes which are represented by the same curve in the *T-V* quadrant are called *equivalent*, and the set of all equivalent processes will be called a *path* $\mathcal{P}$. Different processes which belong to the same path correspond to different parameterizations of the same curve. Obviously we have to distinguish between a process and a path generated by it. However, very often this distinction is of no importance, in which case we shall use the terms interchangeably.

The *reverse of a process* runs along the same curve in the opposite direction. The reverse of a path will be denoted by $-\mathcal{P}$.

3. See Truesdell (1979) for a discussion of this matter.

I.1.2 The Thermal Equation of State

The thermal equation of state is a constitutive relation of a fluid body. It expresses a relationship between the pressure, temperature, and volume of the fluid. The best known equation of state is that of the ideal gas, which was introduced in Chapter 2. Derivatives of the quantities occurring in the equation of state satisfy certain inequalities which are of particular importance. For example, they may decide whether the equation is invertible, and whether sound may propagate in the fluid.

If the equation of state of a fluid is know, certain additional useful relations (like the coefficient of thermal expansion, or the compressibility) can be derived.

The equation of state. Simple fluids have a pressure which is a scalar quantity, i.e., the stresses in fluids are such as to make the momentum current density independent of the orientation of the area for any imagined surface area within the fluid. We shall assume here that the fluids obey the following equation of state: the pressure is a function of the volume and the temperature of the body:

$$P = p(V, T) \tag{2}$$

p is called the pressure function. Its domain is the constitutive domain of the fluid. Furthermore, we assume it to possess continuous partial derivatives. Equation (2) is our first constitutive relation of the fluid.

The pressure has a couple of interesting properties. First, if pressure is applied to a fluid, and if the temperature is kept constant at the same time, we always observe its volume to decreases. This leads to the inequality

$$\partial p / \partial V < 0 \tag{3}$$

Second, it is assumed that the pressure takes only positive values:

$$P > 0 \tag{4}$$

This statement and Equation (3) represent *constitutive inequalities*. You should check if they are satisfied by the ideal gas.

In a process, V and T are functions of time: $V(t)$ and $T(t)$. The time rate of change of the pressure of the fluid is given by

$$\dot{P} = \frac{\partial p}{\partial V}\dot{V} + \frac{\partial p}{\partial T}\dot{T} \tag{5}$$

Since the derivative of the pressure with respect to volume does not vanish—according to Equation (3)—the pressure function p may be inverted to obtain V. We can write the time rate of change of V as follows:

$$\dot{V} = \frac{\partial \mathcal{V}}{\partial P}\dot{P} + \frac{\partial \mathcal{V}}{\partial T}\dot{T} \tag{6}$$

Here the volume function $V = \mathcal{V}(P,T)$ has been introduced. We solve Equation (5) for the rate of change of the volume and set the result equal to the previous expression. This leads to two relations which we shall use frequently:

$$\frac{\partial \mathcal{V}}{\partial P} = \left[\frac{\partial p}{\partial V}\right]^{-1}$$
$$\frac{\partial \mathcal{V}}{\partial T} = -\left[\frac{\partial p}{\partial V}\right]^{-1} \frac{\partial p}{\partial T} \tag{7}$$

In a field theory where the density of the fluid is allowed to vary from point to point, the equation of state (2) cannot be applied. In such cases, we should replace the volume by the density. For a homogeneous body this is an easy matter. Since $\rho = m/V$, we can write the pressure as a function of density and temperature:

$$P = p^*(\rho, T) \tag{8}$$

p^* is a pressure function different from p. If the mass m of the body is constant, we get

$$\rho \frac{\partial p^*}{\partial \rho} = -V \frac{\partial p}{\partial V} \tag{9}$$

Furthermore, we can show that

$$\frac{\partial p^*}{\partial \rho} \dot{\rho} = \frac{\partial p}{\partial V} \dot{V} \tag{10}$$

Constitutive inequalities. We have already encountered two constitutive inequalities which express restrictions upon the range of validity of the constitutive relations. First, we have assumed the pressure of a fluid to be positive. This assumption expresses the idea that the fluids discussed here cannot support tension but only compression. Second, we have seen that the partial derivative of the pressure with respect to volume should be negative. This inequality is based upon experience. In Section I.2.3, we will see that it is equivalent to saying that sound may propagate in such a fluid. Expressed in terms of density instead of volume, this inequality reads

$$\partial p^*/\partial \rho > 0 \tag{11}$$

This is a consequence of Equations (7a) and (10). There is another inequality of interest. If the temperature increases at constant volume, we usually observe an increase of pressure. Therefore, we expect that

$$\partial p/\partial T > 0 \tag{12}$$

Equation (7b) demonstrates that this leads to

$$\partial V/\partial T > 0 \tag{13}$$

Equation (13) means that the volume of the fluid will increase if its temperature is increased at constant pressure. This may not be so in nature, however. Water is a beautiful counterexample, showing that we should not simply neglect this case. Its volume *decreases* with increasing temperature between 0°C and 4°C. In this range we have instead of Equation (12) or Equation (13):

$$\begin{aligned} \partial p/\partial T &< 0 \\ \partial V/\partial T &< 0 \end{aligned} \tag{14}$$

The anomalous behavior of water leads to some very unexpected results. However, since the change of sign of $\partial p/\partial T$ causes considerable mathematical difficulties, we shall not generally treat this anomaly.[4] Still we have to be aware of the limitations of the inequalities in Equations (12) and (13).

Since the partial derivative of the pressure with respect to temperature may change its sign, there may exist points at which

$$\partial p/\partial T = 0 \tag{15}$$

Such points will be called *piezotropic points*. The name is borrowed from meteorology, where equations of state often are considered which exclude the dependence of pressure on temperature, i.e., $P = f(V)$. A curve connecting piezotropic points of a fluid is called a *piezotrope*. In the case of water, such a piezotrope should exist at least in a part of the constitutive domain. It is the curve which joins points of minimum volume for given pressures.

Coefficient of expansion, pressure coefficient, and compressibility. Fluid bodies can change their volume when they are heated. Often, such changes of volume occur under conditions of constant pressure. For the purpose of expressing this phenomenon one introduces the *temperature coefficient of expansion*:

$$\alpha_V = \frac{1}{V}\frac{\partial V}{\partial T} \tag{16}$$

This coefficient was introduced in Section 1.2 in connection with measurements of temperature.

Another piece of information is contained in the quantity describing the change of pressure of a fluid during isochoric changes of temperature. We introduce the *temperature coefficient of pressure* of the fluid as follows:

4. For a detailed discussion see Truesdell and Bharatha (1977).

$$\beta = \frac{1}{P}\frac{\partial p}{\partial T} \tag{17}$$

Furthermore, we shall repeatedly use the so-called *isothermal compressibility* of fluids. If a fluid is compressed isothermally,[5] it will offer a resistance and the pressure will increase. This coefficient is defined as follows:

$$\kappa_T = -\frac{1}{V}\frac{\partial \mathcal{V}}{\partial P} \tag{18}$$

A large compressibility means that it is easy to compress the fluid. The compressibility of water is given as an example in Table A.1. There exists a simple relationship between the three coefficients introduced above:[6]

$$\alpha_V = P\beta\kappa_T \tag{19}$$

The coefficient of thermal expansion, the pressure coefficient, and the compressibility of the *ideal gas* are of particular interest. They are calculated easily:

$$\alpha_V = \frac{1}{T}$$
$$\beta = \frac{1}{T} \tag{20}$$
$$\kappa_T = \frac{1}{P}$$

Naturally, the condition of Equation (19) is satisfied by the ideal gas law.

I.2 A Theory of Heat for Ideal Fluids

In Section I.2.1, the concepts of latent and specific heats, which are known from their application in Chapter 2 to the ideal gas, is made precise. A proof is given which shows that the rate of production of heat vanishes in every process: the bodies can undergo only reversible processes. The pressure is introduced as an independent variable, and the problem of bodies having a negative latent heat (water) is discussed. This preparation allows us to derive relationships governing special processes, particularly adiabatic and polytropic ones (Section I.2.2). This

5. In Section I.2, the adiabatic compressibility will be introduced.
6. Proof of Equation (19). Remember the equation of state in the two forms $P = p(V,T)$, and $V = \mathcal{V}(P,T)$, and the definition of α and β. We can express the partial derivatives of the two functions P and $\mathcal{V}$ by Equations (7a) and (7b). This immediately leads to the relationship between the coefficients proposed above.

will lead to an important application, namely the propagation of sound in fluids (Section I.2.3).

Finally, the theory is extended to throw some more light on the constitutive quantities. A few preliminary results can be obtained, but the general solution of the problem has to await the introduction of the concept of energy in Section I.3, where we will discuss the motive power of heat engines.

I.2.1 The Storage and the Exchange of Heat

Thermal processes deal with the storage and the flow of heat.[7] Heat (caloric) is a quantity we account for using an equation of balance (Chapter 1). This equation relates the rate of change of the heat stored in a body, currents of heat crossing its surface, and the rate of production of heat inside the body. If we want to calculate thermal processes we must be in a position to say more about these terms. Here, the problem will be discussed in the spirit of the caloric theory for the fluids which have been introduced in the previous section.

The heat function. The equation of balance of heat contains the time rate of change of the heat of the body. The heat of the body is described by a function S which we will call the *heat function.* To specify it means to determine its functional dependence. For the fluids treated here this is a simple matter: we will assume the caloric S of such a body to be a function of volume and of temperature only:[8]

$$S = \mathcal{S}(V, T) \tag{21}$$

Specifically, it does not depend on such things as the history of the body, or the rates of processes, or other possible effects. At this point we do not ask whether such bodies exist in nature. They simply represent idealized objects, created for the purpose of facilitating the development of a theory. This simplification is in line with what we have discussed so far.

Furthermore, we assume the heat function to have continuous first and second derivatives. This means that the time rate of change of the heat of a body can be written in the following form:

---

7. That heat can be transferred is *the* fundamental assumption underlying both the caloric and the mechanical theories of heat. The caloric theory, however, also assumes the existence of a heat function. We may interpret this graphically as meaning that heat also resides in bodies. Therefore, in the context of the caloric theory, we are justified in speaking of a quantity of heat for which a law of balance must hold. Researchers in Carnot's time never expressed their assumption in this way.

8. Since our assumptions will lead to the identification of caloric with entropy, we shall use the symbol S for heat.

$$\dot{S} = \frac{\partial \mathcal{S}}{\partial V}\dot{V} + \frac{\partial \mathcal{S}}{\partial T}\dot{T} \tag{22}$$

Also, because we assume the heat function to vary smoothly, there exists the following condition of integrability:

$$\frac{\partial}{\partial T}\frac{\partial \mathcal{S}}{\partial V} = \frac{\partial}{\partial V}\frac{\partial \mathcal{S}}{\partial T} \tag{23}$$

This is the well-known statement that mixed partial derivatives of a smoothly varying function should be equal.

At this point, this is all we can say about the heat function. In fact, we only have stated an assumption which a mathematician might call an axiom. Naturally we can hope to learn more about the heat function if we can find out about ways of changing the heat content of bodies. As we know, this is possible by either exchanging caloric or by creating it. Therefore we should investigate currents of caloric and its production rate more carefully. In Chapter 2 we already have considered thermal phenomena which suggest a simple expression of the heating of uniform fluids.

Heating and the latent heats and heat capacities. The simple fluids investigated here can respond to heating (or cooling) in a number of ways: they can change their volume or their temperature. We simply will exclude other responses from consideration.

If the volume of a body is kept constant during heating, its temperature increases. The quantity specifying the amount of heat added for an increase of the temperature, divided by this change, has been called *heat capacity at constant volume.*[9] Since we assume the states of the body to be specified by its volume and its temperature, we have to expect the heat capacity to depend on both variables as well.

If, on the other hand, the temperature stays constant during heating or cooling (isothermal processes), the volume of the body must change. Since the heat added does not do what we normally expect of it, we speak of the *latent heat with respect to volume*, by which we mean the amount of caloric needed to change the volume, divided by this change of volume. In general, the latent heat also depends on both the volume and the temperature.

In summary, we assume the heating I_S to depend both on the rate of change of the volume and the rate of change of the temperature of the body. The coefficients involved are the latent heat and the heat capacity:

$$I_s = -\Lambda_V(V,T)\dot{V} - K_V(V,T)\dot{T} \tag{24}$$

9. In earlier times the term *specific heat* was used for what we now call *heat capacity.* The relationship between the heating and the constitutive quantities expressed below has been called the doctrine of latent and specific heats by Truesdell (1980).

(Section 2.3). Furthermore, the latent heat and the heat capacity are supposed to have continuous partial derivatives. The minus signs have been introduced because we count a current into a system as a negative quantity. Isothermal processes and processes at constant volume are special cases of this relation. The latent heat and the heat capacity are *constitutive relations*; i.e., they are quantities describing the diversities of the bodies satisfying Equation (24). They are assumed to have continuous derivatives. If we can specify them for a given body, we can calculate the processes it can undergo.

As far as we know, we always have to add heat (caloric) to a body to increase its temperature at constant volume. This observation leads to an important inequality:

$$K_V = K_V(V, T) > 0 \tag{25}$$

Bodies which do not obey this inequality are unstable.

The case of the latent heat, however, is more complex. In general we expect the latent heat of a body to be a positive quantity. *If the latent heat is positive, caloric is absorbed in an isothermal expansion.* This is what we expect of simple gases:

$$\Lambda_V = \Lambda_V(V, T) > 0 \tag{26}$$

There are important exceptions to this rule, however, water being the most notable case. Naturally, the opposite behavior is very strange at first sight. It means that a body undergoing an isothermal expansion emits heat. This is the case for water for a certain range of temperatures and volumes.[10] In Section I.3 we will be able to see that the latent heat is indeed negative for water in part of the constitutive domain. All we can say at this moment is that we have to watch out for the possibility of a change of sign of the latent heat. For most of the development in this book, however, we shall neglect this possibility, since it causes considerable mathematical problems.

Equation (24), and the constitutive inequalities following it, are said to describe the *doctrine of latent and specific heats*. This doctrine, together with the thermal equation of state, Equation (2), and its inequalities, are called the *theory of calorimetry* in this text. All investigations on heat up to and including Clausius' (except for those on the conduction of heat) used this theory.[11]

Processes and the absorption and emission of heat. The doctrine of latent and specific heats, Equation (24), lets us determine the *heat exchanged* during a process, which will be denoted by $S_e(\mathscr{P})$. This is the amount of caloric transmitted across the surface of the body under consideration. It is defined as the integral of the heating for the process in the T-V quadrant (Figure 2):

---

10. The proof of this statement is not quite so simple, and it is impossible to give at this point. (It is related to the change of sign of $\partial p/\partial T$ which we discussed in Section I.1.2, but we need another assumption for the proof to be possible.)

11. Truesdell (1980).

$$
\begin{aligned}
S_e(\mathcal{P}) &= -\int_{t_1}^{t_2} I_s dt \\
&= \int_{t_1}^{t_2} \left[\Lambda_V(V,T)\dot{V} + K_V(V,T)\dot{T}\right] dt
\end{aligned}
\tag{27}
$$

In Section I.3 we shall deal with cyclic processes undergone by fluid bodies. In such processes heat will be absorbed and emitted. For this purpose, let us define the *heat absorbed*:

$$
S_e^+ = \frac{1}{2}\int_{t_1}^{t_2} \left[|I_s| - I_s\right] I_s dt \tag{28}
$$

and the *heat emitted*:

$$
S_e^- = \frac{1}{2}\int_{t_1}^{t_2} \left[|I_s| + I_s\right] I_s dt \tag{29}
$$

Both quantities are non-negative. Consequently, the caloric exchanged, Equation (27), can be expressed by these two quantities:

$$
S_e(\mathcal{P}) = S_e^+(\mathcal{P}) - S_e^-(\mathcal{P}) \tag{30}
$$

The heat exchanged, and the heat absorbed and emitted, are quantities which depend on the particular process. This is why the notation $S_e(\mathcal{P})$ was chosen. Equation (24) shows that the heat exchanged on the reverse of the path $\mathcal{P}$ is the negative of the heat added on $\mathcal{P}$:

$$
S_e(-\mathcal{P}) = -S_e(\mathcal{P}) \tag{31}
$$

We call this the *reversal theorem* for the exchange of caloric:

> *If a body absorbs a certain amount of heat (caloric) while traversing a path, it will emit the same amount upon reversing the path.*

The reversal theorem also holds for the heat absorbed and emitted. Common experience with thermal phenomena suggests that there exist bodies which approximate this kind of behavior.

As we shall see shortly, the assumption that the exchange of caloric is governed by the doctrine of latent and specific heats, i.e., by Equation (24), imposes severe restrictions upon the types of processes that simple fluids can undergo.

Reversibility. Does this mean that bodies described by the doctrine of latent and specific heats can undergo only reversible processes? We have to be careful here, since heat can be created in a body. This means that we have to distinguish between the heat exchanged across the surface of the body, and the change of heat content in the body. Caloric satisfies an equation of continuity of the form

$$\dot{S} + I_s = \Pi_s \tag{32}$$

(Chapter 1). Π_S is the rate at which heat is created in the system. The restrictions placed upon the bodies by Equations (21) and (24) make the rate of production of caloric vanish.

Because of Equation (21), we see that the *change of heat content* ΔS is the negative change for the reverse process:

$$\Delta S(-\mathcal{P}) = -\Delta S(\mathcal{P}) \tag{33}$$

Together with Equation (31), this shows that heat may not be created in such bodies. The idea is simple. If the amounts of heat communicated across the surface of the body are the same (except for the sign), and the change of the heat content is the same as well, there may not be any production of heat in the body. Therefore, the *theory of calorimetry admits only reversible processes.*[12]

Thus the equation of continuity for caloric takes a particularly simple form for bodies obeying the theory of calorimetry. Heat cannot be created, which means that there is no creation term in the equation:

$$\dot{S} + I_s = 0 \tag{34}$$

For the simple fluids treated here, the heating is equal to the (negative) rate of change of the heat content.

The heat function and the latent heat and heat capacity. The fact that the bodies described by the theory of calorimetry can undergo only reversible processes has some interesting consequences. Since the rate of change of the heat content is equal to the (negative) net heat current, we can write Equation (24) in terms of the time derivative of the heat function:

---

12. Proof: Integration of the equation of continuity for heat (Equation (32)) over a process $\mathcal{P}$ gives:

$$\Delta S(\mathcal{P}) = S_e(\mathcal{P}) + S_{prod}(\mathcal{P})$$

The reversal theorem, Equation (31), and Equation (33) allow us to transform this into

$$\Delta S(-\mathcal{P}) = S_e(-\mathcal{P}) - S_{prod}(\mathcal{P})$$

For the reverse process $-\mathcal{P}$, integration of the equation of continuity leads to

$$\Delta S(-\mathcal{P}) = S_e(-\mathcal{P}) + S_{prod}(-\mathcal{P})$$

Since $\Pi_S \geq 0$, the last two equations are compatible if and only if $\Pi_S = 0$. The production of heat therefore is zero.

$$\dot{S} = \Lambda_V(V,T)\dot{V} + K_V(V,T)\dot{T} \tag{35}$$

Now the latent heat and the heat capacity can be expressed in different terms. They are no longer related to the exchange of heat, but to the heat content. This means that they are the partial derivatives of the heat function:

$$\begin{aligned} \Lambda_V(V,T) &= \frac{\partial \mathcal{S}}{\partial V} \\ K_V(V,T) &= \frac{\partial \mathcal{S}}{\partial T} \end{aligned} \tag{36}$$

The condition of integrability given in Equation (23) therefore takes the following simple form.[13]

$$\frac{\partial \Lambda_V}{\partial T} = \frac{\partial K_V}{\partial V} \tag{37}$$

The capacity finally obtains the meaning we ascribe to it in other fields of physics: it is the quantity relating changes of the quantity stored in a system to changes of the intensive quantity of that system (Prologue). This seems to be as natural a definition of a capacity (and a quantity analogous to the latent heat) as the one afforded by the doctrine of latent and specific heats. It is important to note that in the other area of physics where capacities are commonly introduced, i.e., in the theory of electromagnetism, the distinction between the two versions does not come into play. Introducing the capacity on the basis of the storage of charge or on the exchange of charge is all the same, since this quantity is conserved in all processes.

In the future, we shall always take the heat capacity to be the partial derivative with respect to temperature of the heat function. When we treat irreversible processes, this will be important. Equation (36) will still hold, and it will express the definition of the latent heat and the heat capacity—no more, no less. However, the expression for the heating as given in Equation (24) will be different (Epilogue).

The introduction of the latent heat and the heat capacity on the basis of the exchange of heat was necessary to ensure the reversibility of the processes we envisage here. However, we could just as well have postulated directly that the production of heat vanishes. Given the existence of a heat function as in Equation (21), the doctrine of latent and specific heats, Equation (24), and the statement that the production rate of heat vanishes, are equivalent.

---

13. This is really what distinguishes the caloric theory from one where "heat" is a quantity related to energy. In the latter, the condition of integrability for the latent "heat" and the "heat capacity" in the form of Equation (37) does not exist.

The pressure as an independent variable. In practice, processes often occur with the pressure of a fluid kept constant. Therefore, it is convenient to express the doctrine of latent and specific heats in terms of pressure and temperature. For this purpose we replace the rate of change of the volume in Equation (24) using the equation of state, Equation (2):

$$\begin{aligned} I_s &= -\Lambda_V \left[\frac{\partial p}{\partial V}\right]^{-1} \left[\dot{P} - \frac{\partial p}{\partial T}\dot{T}\right] - K_V \dot{T} \\ &= -\Lambda_V \left[\frac{\partial p}{\partial V}\right]^{-1} \dot{P} - \left[K_V - \Lambda_V \left[\frac{\partial p}{\partial V}\right]^{-1} \frac{\partial p}{\partial T}\right]\dot{T} \end{aligned} \quad (38)$$

We now introduce the *heat capacity at constant pressure*, and the *latent heat with respect to pressure*:

$$I_s = -\Lambda_P \dot{P} - K_P \dot{T} \quad (39)$$

If we compare this form with the one given above, we see that we can write the new constitutive quantities in terms of the old ones:

$$\begin{aligned} \Lambda_P &= \Lambda_V \left[\frac{\partial p}{\partial V}\right]^{-1} \\ K_P &= K_V - \Lambda_V \left[\frac{\partial p}{\partial V}\right]^{-1} \frac{\partial p}{\partial T} \end{aligned} \quad (40)$$

If we want to determine the heating of our simple fluids, we can specify both heat capacities instead of the capacity at constant volume and the latent heat with respect to volume. These suffice for determining the latent heats. It seems that the specific heats are measured more easily. However, they do not afford a description of thermal processes in as vivid a manner as the latent heat combined with one of the heat capacities.

There are constitutive inequalities equivalent to Equations (25) and (26). We can express them in the following form:

$$\begin{aligned} \Lambda_P &< 0 \\ K_P &> K_V \end{aligned} \quad (41)$$

Equation (40a) demonstrates that the latent heat with respect to pressure always has the opposite sign of the latent heat with respect to volume. The fact that the heat capacity at constant pressure is greater than its counterpart (at least if the latent heat with respect to volume is positive) can be appreciated if you compare heating in a pressure cooker to heating in the open air. We will see later that the capacity at constant pressure should be larger than its counterpart even for substances like water in its anomalous range.

We know that the processes possible in the case of the doctrine of latent and specific heats must be reversible. This allows us to express the latent heat with respect to pressure and the heat capacity at constant pressure in terms of the rate of change of the heat function

$$S = \mathcal{S}^*(P,T) \tag{42}$$

The equivalent to Equation (35) is:

$$\dot{S} = \Lambda_P \dot{P} + K_P \dot{T} \tag{43}$$

The coefficients must be the partial derivatives of the heat function which means that[14]

$$K_P = \frac{\partial \mathcal{S}^*}{\partial T} \tag{44}$$

Frequently, the ratio of the heat capacities with respect to pressure and to volume is used. The second inequality in Equation (41) shows that

$$\gamma \equiv K_P / K_V > 1 \tag{45}$$

---

14. This leads to a useful expression of the calculus of functions of two variables. First we write the time rate of change of both heat functions introduced so far, i.e., $\mathcal{S}^*(P,T)$ and $\mathcal{S}(V,T)$, whereupon we substitute for the time rate of change of the volume, using the equation of state, which leads to

$$\begin{aligned}
\dot{S} &= \frac{\partial \mathcal{S}^*}{\partial P}\dot{P} + \frac{\partial \mathcal{S}^*}{\partial T}\dot{T} \\
&= \frac{\partial \mathcal{S}}{\partial V}\dot{V} + \frac{\partial \mathcal{S}}{\partial T}\dot{T} \\
&= \left(\frac{\partial p}{\partial V}\right)^{-1} \frac{\partial \mathcal{S}}{\partial V}\dot{P} + \left[\frac{\partial \mathcal{S}}{\partial T} - \left(\frac{\partial p}{\partial V}\right)^{-1} \frac{\partial p}{\partial T}\frac{\partial \mathcal{S}}{\partial V}\right]\dot{T} \\
&= \frac{\partial \mathcal{S}(V,T)}{\partial V}\frac{\partial \mathcal{V}(P,T)}{\partial P}\dot{P} + \left[\frac{\partial \mathcal{S}(V,T)}{\partial T} - \frac{\partial \mathcal{S}(V,T)}{\partial V}\frac{\partial \mathcal{V}(P,T)}{\partial T}\right]\dot{T}
\end{aligned}$$

We read off the following identities, which are special cases of well-known rules of calculus:

$$\begin{aligned}
\frac{\partial \mathcal{S}^*(P,T)}{\partial P} &= \frac{\partial \mathcal{S}(V,T)}{\partial V}\frac{\partial \mathcal{V}(P,T)}{\partial P} \\
\frac{\partial \mathcal{S}^*(P,T)}{\partial T} &= \frac{\partial \mathcal{S}(V,T)}{\partial T} - \frac{\partial \mathcal{S}(V,T)}{\partial V}\frac{\partial \mathcal{V}(P,T)}{\partial T}
\end{aligned}$$

These rules hold in general for a function of two variables (the heat function) if one of the variables is replaced by another function of these variables (equation of state).

According to Equations (40b) and (7b) this can be written in the form

$$\gamma = 1 + \frac{\Lambda_V}{K_V} \frac{\partial \mathcal{V}}{\partial T} \tag{46}$$

This ratio can sometimes be measured independently of the specific heats themselves (Section I.2.3). Because of this, we have some information about the constitutive quantities of some fluids which will prove crucial when we have to determine the relationship between heat and energy (Section I.3). In Chapter 2, we saw how difficult a *direct* measurement of the specific heat or the latent heat must be in practice. Therefore, direct information about the ratio in Equation (45) cannot be valued too highly.

Some special expressions for the heating will prove to be useful later. The heating (or the rate of change of the caloric of the body) can be expressed in terms of the two specific heats. We achieve this by multiplying Equation (39) by the difference of the heat capacities, and by using Equations (40) and (2):

$$\left(K_P - K_V\right) I_s = \Lambda_V \left[\left(\frac{\partial p}{\partial V} \right)^{-1} K_V \dot{P} - K_P \dot{V} \right] \tag{47}$$

This equation also can be expressed in terms of pressure and density if we replace the time derivative of the volume by the time rate of change of the density according to Equation (10):

$$\left(K_P - K_V\right) I_s = \Lambda_V \left(\frac{\partial p}{\partial V} \right)^{-1} \left[K_V \dot{P} - K_P \frac{\partial p^*}{\partial \rho} \dot{\rho} \right] \tag{48}$$

These equations will lead directly to the law of adiabatic change, and to an expression for the speed of sound.

I.2.2 Adiabatic and Polytropic Changes

Processes in which the heating has some special form are of practical interest to scientists and engineers. In Chapter 2 we considered *adiabatic processes*, which result if the heating vanishes. Another case leads to isothermal changes. Finally, we can conceive of *polytropic processes*, of which adiabatic and isothermal changes are limiting cases. With their help we can treat a large number of interesting applications, including the calculation of the temperature gradient of the Earth's atmosphere, polytropic stellar models, and the propagation of sound.

The law of adiabatic change. It is well known that the temperature of a gas can be changed without heating or cooling. If you compress a gas rapidly, its temperature is bound to increase. What at first appears strange finds a natural explanation within the theory of calorimetry. In Chapter 2 such processes have been described. They are called *adiabatic changes*. Adiabatic simply means that *no heat is exchanged during the process*, which is equivalent to saying that

$$I_s = 0 \tag{49}$$

This is the general definition of the meaning of adiabatic processes. In our case there also is a special meaning. Since the bodies described by the doctrine of latent and specific heats cannot undergo irreversible processes, caloric cannot be created. Therefore we can also state that

$$\dot{S} = 0 \tag{50}$$

which is equivalent to saying that for the fluids treated here the heat content remains constant during an adiabatic process.

If we apply this to the formulation of the doctrine of latent and specific heats—see Equation (24)—we see at a glance the explanation for the phenomenon of adiabatic compression or expansion:

$$\Lambda_V(V,T)\dot{V} + K_V(V,T)\dot{T} = 0 \tag{51}$$

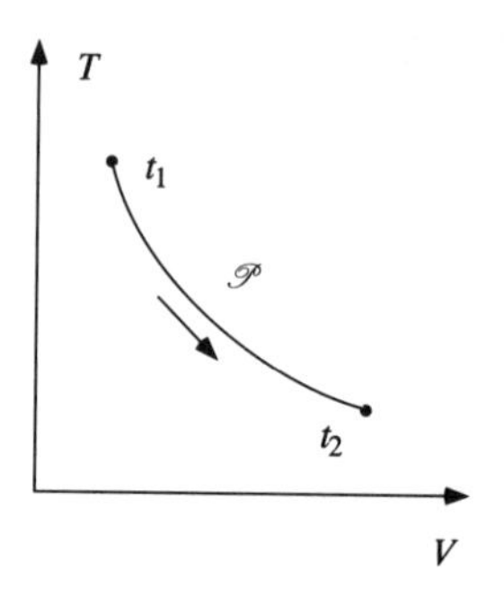

FIGURE 3. An adiabat $\mathcal{P}$. The form shown here roughly corresponds to that of an adiabatic process undergone by the ideal gas. More generally, it is an adiabat for the case where the latent heat with respect to volume is larger than zero. The process can run in the reverse direction as well.

Even though the heat content of the fluid does not change, a change of temperature follows naturally if the volume either decreases or increases. Equation (51) describes the coupling of the rates of change of the temperature and the volume of the gas. Now, because of the inequality Equation (25), and because the time rates of change of volume and of temperature cannot vanish together, Equation (51) can be written in the following form as a differential equation for the volume and the temperature:

$$\frac{dT}{dV} = -\frac{\Lambda_V}{K_V} \tag{52}$$

This is the *differential equation of adiabatic motion.* If the latent and the specific heats of the fluid are known, this equation determines the temperature as a twice continuously differentiable function of the volume. In the *T-V* diagram this is a particular curve (Figure 3 or Figure 4), called an *adiabat* of the body. Since the constitutive functions are supposed to have continuous derivatives, we are assured that only one adiabat runs through any point of the constitutive domain of the fluid.

The nature of the adiabats is a constitutive property of the body. An adiabat depends on the latent heat and the heat capacity. If one body has adiabats of a particular form, another body need not have the same at all. In Figure 3 the curve represents an adiabat for the common case in which the latent heat with respect to volume is larger than zero.

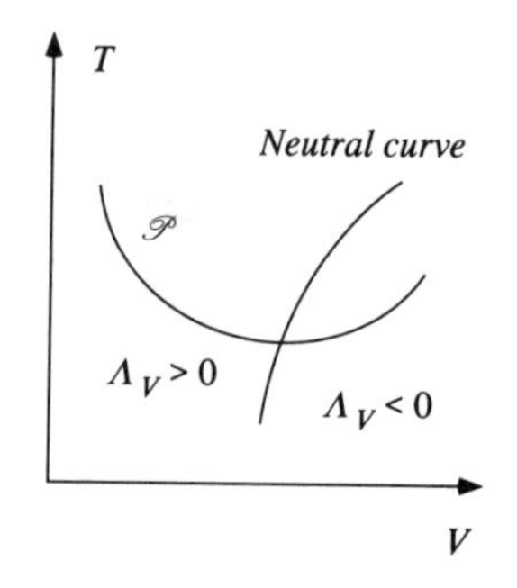

FIGURE 4. An adiabat for a substance for which the latent heat may change its sign. The line crossing the adiabat is the curve for which the latent heat vanishes.

However, we must expect cases in which Λ_V is negative, and there might be points or even parts of the constitutive domain in which the latent heat vanishes. Points at which $\Lambda_V = 0$ will be called *neutral points*, all the other points of the constitutive domain will be called *ordinary points*. Neutral points make the analysis necessary for a general theory of thermodynamics rather difficult. According to Equation (40), at neutral points we have $\Lambda_P = 0$ and $K_P = K_V$

The differential equation of adiabatic motion shows that in a part of the constitutive domain in which the latent heat with respect to volume is negative, the temperature must be an increasing function of the volume, which leads to a totally different form of the adiabats as the one commonly known from books on thermal physics. Finally, if Λ_V changes its sign along a particular adiabat, the curve must have an extremum there. For example, for water near 4°C, the adiabats have a rather unexpected look (see Figure 4).

A different form of the differential equation of adiabatic motion is of particular interest. The expression for the heating developed in Equation (48) shows that

$$\dot{P}/\dot{\rho} = \frac{dp^*}{d\rho} = \frac{K_P}{K_V}\frac{\partial p^*}{\partial \rho} = \gamma\frac{\partial p^*}{\partial \rho} \tag{53}$$

This differential equation will prove particularly useful when we have to discuss the propagation of sound in fluids. For the ideal gas, this reduces to

$$\frac{dP}{d\rho} = \gamma\frac{P}{\rho} \tag{54}$$

The propagation of sound in the ideal gas (Section I.2.3) will show that for the ideal gas the ratio of the specific heats must be constant:

$$\gamma = constant \tag{55}$$

If we accept this for the moment, we can give a specific solution to the problem of *adiabatic motion of the ideal gas having a constant ratio of the specific heats:*

$$PV^{\gamma} = constant \tag{56}$$

This is the *Poisson-Laplace Law of adiabatic change*, which we encountered in Chapter 2 when we derived the consequences of the theory for the ideal gas.

The adiabatic compressibility. Simple experience with gases demonstrates that the compressibility depends on the particular process undergone by the fluid. For example, if we give the gas enough time to cool while it is compressed, its compressibility is larger than in the case of rapid compression. In the latter case, heat cannot flow out of the gas, which leads to a rise in temperature. As a consequence it will be harder to compress the gas: the fluid's compressibility must be lower.

Therefore we should distinguish between the *isothermal compressibility* and the *adiabatic compressibility.* The former has been defined in Equation (18). The adiabatic compressibility, on the other hand, is given by the derivative of the volume with respect to pressure at constant heat:

$$\kappa_s = -\frac{1}{V}\frac{\partial \mathcal{V}^*(P,S)}{\partial P} \tag{57}$$

It is equal to the isothermal compressibility divided by the ratio γ of the heat capacities:[15]

$$\kappa_T = \gamma \kappa_s \tag{58}$$

Polytropic processes. We can introduce a more general kind of heating, of which adiabatic and isothermal processes are limiting cases. Remember that in adiabatic processes the heating vanishes, while for isothermal changes it takes on a very special value. In practice, the heating might take some kind of intermediate value. We start again with a general expression for the heating, namely,

$$-\frac{\partial \mathcal{V}}{\partial T}\Lambda_V \dot{V} - \frac{\partial \mathcal{V}}{\partial T}K_V \dot{T} = \frac{\partial \mathcal{V}}{\partial T} I_s \tag{59}$$

or

$$\left(K_P - K_V\right)\dot{V} + \frac{\partial \mathcal{V}}{\partial T}\left[K_V \dot{T} + I_s\right] = 0 \tag{60}$$

---

15. Proof: We use the two volume functions $V = \mathcal{V}(P,T)$ and $V = \mathcal{V}^*(P,S)$. The rate of change of the volume can be written in the following form:

$$\dot{V} = \frac{\partial \mathcal{V}^*}{\partial P}\dot{P} + \frac{\partial \mathcal{V}^*}{\partial S}\dot{S}$$

Comparing this to Equation (6), and using the doctrine of latent and specific heats, we arrive at the following identifications:

$$\left(1 - \Lambda_V \frac{\partial \mathcal{V}^*}{\partial S}\right)\frac{\partial \mathcal{V}}{\partial T} = K_V \frac{\partial \mathcal{V}^*}{\partial S}$$

$$\left(1 - \Lambda_V \frac{\partial \mathcal{V}^*}{\partial S}\right)\frac{\partial \mathcal{V}}{\partial P} = \frac{\partial \mathcal{V}^*}{\partial P}$$

Using the relationship between heat capacities and latent heats, as in Equation (40), the first of these expressions leads to

$$\frac{\partial \mathcal{V}}{\partial T} = K_P \frac{\partial \mathcal{V}^*}{\partial S}$$

We inert this into the second expression given above to obtain

$$\begin{aligned}
\frac{\partial \mathcal{V}^*}{\partial P} &= \left(1 - \Lambda_V \frac{\partial \mathcal{V}^*}{\partial S}\right)\frac{\partial \mathcal{V}}{\partial P} \\
&= \left(1 - \frac{\Lambda_V}{K_P}\frac{\partial \mathcal{V}}{\partial T}\right)\frac{\partial \mathcal{V}}{\partial P} \\
&= \frac{1}{\gamma}\frac{\partial \mathcal{V}}{\partial P}
\end{aligned}$$

The last step is a consequence of Equations (40) and (7). This proves Equation (58).

which is a consequence of Equation (40) and the transformations in Equation (7). We shall now write the current of heat in the form

$$I_s = -K\dot{T} \tag{61}$$

We take this to be the definition of a *polytropic process*. K is a quantity resembling a generalized heat capacity. With this in mind, the heating turns out to be

$$\left(K_P - K_V\right)\dot{V} + \frac{\partial V}{\partial T}\left[K_V - K\right]\dot{T} = 0 \tag{62}$$

This is the differential equation of polytropic processes. We can recover adiabatic and isothermal processes if we set

$K = 0$ for adiabatic processes

$K = \infty$ for isothermal processes

For the general polytropic process, K takes an intermediate value. These ideas will now be applied to the *ideal gas*. We obtain directly from Equation (62):

$$\frac{dV}{dT} + \frac{V}{T}\frac{K_V - K}{K_P - K_V} = 0 \tag{63}$$

If we introduce the *polytropic exponent*

$$\gamma' = \frac{K_V - K}{K_P - K_V} \tag{64}$$

the last equation becomes equivalent to the differential equation of adiabatic motion which leads to the Poisson-Laplace Law with γ replaced by γ'. The polytropic exponent is a generalization of the adiabatic exponent. For values of K which range from zero to very large ones, the polytropic exponent changes from large negative to large positive values. We shall, however, in general restrict our attention to values of the exponent which lie in the range

$$1 \le \gamma' \le \gamma \tag{65}$$

We shall always take it to be a constant. The solution of Equation (63) is a form of the law of polytropic processes of the ideal gas:

$$V^{\gamma'-1}T = constant \tag{66}$$

This form is equivalent to the laws of adiabatic motion of the ideal gas. Remember that it corresponds to a process using a particular type of heating.

I.2.3 Propagation of Sound in Fluids

The propagation of sound in gases and Laplace's explanation of the speed of sound provide an interesting example of the theory laid down so far. Newton had found that the speed of sound c was given by

$$c^2 = \frac{dP}{d\rho} \tag{67}$$

For a long time it was assumed that the temperature would remain constant during the passage of a sound wave. Accepting this, we would get for the speed of sound in the ideal gas:

$$c^2 = \frac{dP}{d\rho} = \frac{\partial P}{\partial \rho} = \frac{P}{\rho} \tag{68}$$

However, for air this equation delivers a value of the speed of sound which is roughly 20% too low. Where is the source of the error to be found?

The speed of sound. Sound represents a particular type of transport of momentum through a medium which we take to be a fluid here. In the Prologue, it was explained how the interaction of two properties of such a medium could lead to a wavelike propagation of a substancelike quantity. These properties are the capacity to store the substancelike quantity, i.e., momentum, and the ability to react to changing currents. The behavior of a medium is described therefore by a *capacitance* per length and an *inductance* per length. The square of the speed of a wave turns out to be equal to the inverse of the product of the latter two quantities (Prologue).

The momentum capacity is equal to the mass of a body, while the inductance is related to the compressibility of the fluid, i.e., to its springlike behavior. According to the derivation given in the Prologue, the square of the speed of sound in a fluid of this type turns out to be given in Equation (67):

$$c^2 = \frac{1}{L_p^* \cdot C_p^*} = \frac{dP}{d\rho} \tag{69}$$

This confirms the well-known result for the speed of sound in a fluid which obeys an arbitrary equation of state of the form given in Equation (2). Therefore, the general result for the speed of sound given in Equation (67) cannot be responsible for the discrepancy between theory and measurements.

Laplace's theory of sound for the ideal gas. When the discrepancy between the early theory and experiments became known, all possible sources of errors were checked. As we have seen, the basic theory of the propagation of waves in air is not at fault. Finally, Laplace suggested that the propagation of sound in air should be an adiabatic process: each portion of the air which is being compressed by the sound wave changes its volume so rapidly that there is no time for heat to be exchanged. Thus, the amount of caloric in the gas should remain unchanged. In Laplace's theory, therefore, the speed of sound is given by:

$$c^2 = \frac{dP}{d\rho} = \gamma \frac{\partial P}{\partial \rho} = \gamma \frac{P}{\rho} \tag{70}$$

which is a consequence of Equation (53). It also demonstrates that $dP/d\rho$ could be made a negative quantity if the constitutive inequality in Equation (11) was violated. This would make it impossible for sound to propagate in the fluid. The final form in Equation (70) holds for the ideal gas only.

The propagation of sound in the *ideal gas* is of particular interest to us because it delivers specific information about the constitutive properties of the fluid. There are two measurements which throw light upon the ratio of specific heats:

1. The speed of sound is consistently higher than suggested by the theory for isothermal motion. Therefore:

$$\gamma > 1 \quad \Rightarrow \quad K_P > K_V \tag{71}$$

2. The *square of the speed of sound* in the ideal gas is *proportional to the temperature*. Since $P/\rho \sim T$ for the ideal gas, this observation and the theory of sound, Equation (70), show that the ratio of specific heats must be constant:

$$\gamma = constant \ > 1 \tag{72}$$

These determinations of γ for the *ideal gas* will prove to be of great value when we attempt to determine the motive power of heat (Section I.3).

I.2.4 The Heat Function and the Constitutive Quantities

We have not yet made as much progress on the constitutive theory of simple fluids as we might wish. Even in the case of the simplest of all bodies, i.e., the ideal gas, we know only that the ratio of the heat capacities must be constant.

This section exemplifies the difficulty of solving even a simple constitutive problem in thermodynamics. It is the problem faced by the thermodynamicists of the early 19th century. It tells us that only a broadening of the approach will lead to success: without the relationship between heat and work we cannot go much further. Still, a couple of additional points can be gleaned from the theory of heat alone.

The heat capacities and the heat function. Let us go back to the doctrine of latent and specific heats. Together with the assumption that there exists a heat function, this doctrine leads to expressions for the rate of change of the heat content of a body. A particular form is the one which involves the rates of change of the pressure and the density of the fluid, Equation (48):

$$\dot{S} = \left(\frac{\partial p}{\partial T}\right)^{-1} K_V \dot{P} - K_P \frac{\partial p^*}{\partial \rho}\left(\frac{\partial p}{\partial T}\right)^{-1} \dot{\rho} \tag{73}$$

If we introduce the heat as a function of pressure and density, i.e., $S = \mathscr{S}^{\#}(P,\rho)$, we see that its partial derivatives must be the factors in front of the rate of change of the pressure and the rate of change of the density. Therefore the two specific heats take the following forms:

$$K_V = \frac{\partial p}{\partial T}\frac{\partial \mathscr{S}^{\#}}{\partial P} \tag{74}$$

and

$$K_P = -\frac{\partial p}{\partial T}\left[\frac{\partial p^*}{\partial \rho}\right]^{-1}\frac{\partial \mathscr{S}^{\#}}{\partial \rho} \tag{75}$$

Using the expressions for the heat capacities we can derive a differential equation for the heat function. We simply calculate the ratio of the heat capacities according to Equations (74) and (75), and rearrange the terms:

$$\frac{\partial \mathscr{S}^{\#}}{\partial \rho} + \gamma \frac{\partial p^*}{\partial \rho}\frac{\partial \mathscr{S}^{\#}}{\partial P} = 0 \tag{76}$$

Furthermore, the condition of integrability can be given in the following form:

$$\frac{\partial}{\partial \rho}\left[\left(\frac{\partial p}{\partial T}\right)^{-1} K_V\right] + \frac{\partial}{\partial P}\left[\left(\frac{\partial p}{\partial T}\right)^{-1}\frac{\partial p^*}{\partial \rho} K_P\right] = 0 \tag{77}$$

This follows directly from Equations (74) and (75) if we note that the mixed derivatives of the heat function $\mathscr{S}^{\#}$ must be equal.

The heat capacities of the ideal gas. The relations derived above can now be specialized for the ideal gas, and some interesting conclusions regarding the heat capacities of this simple fluid can be drawn. First, the expressions for the heat capacities, i.e., Equations (74) and (75), will look like this for the ideal gas:

$$K_V = \frac{P}{T}\frac{\partial \mathscr{S}^{\#}}{\partial P} \qquad K_P = -\frac{\rho}{T}\frac{\partial \mathscr{S}^{\#}}{\partial \rho} \tag{78}$$

There is one very interesting additional piece of information which we can derive from what we know so far. For the ideal gas, the condition of integrability assumes the form

$$K_P - K_V + \rho\frac{\partial K_V}{\partial \rho} + P\frac{\partial K_P}{\partial \rho} = 0 \tag{79}$$

which follows from Equation (77) if we calculate the partial derivatives using the equation of state. This equation shows that our theory of heat *forbids the ideal gas to have constant heat capacities*, for if they were constant they would have to be equal, which contradicts everything we know about them. This brings up some question which, historically, have played an important role (Section I.5).

What are the heat capacities of the ideal gas if they are not constant? As we have seen in the previous section, their ratio must be constant. This allows us to calculate a general solution of the differential equation (76) for the heat function. As you can verify,

$$\mathcal{S}^{\#}(P,\rho) = \psi\left(\frac{P^{1/\gamma}}{\rho}\right) \tag{80}$$

where ψ is an as yet unspecified function. According to this, the heat capacity at constant volume turns out to be

$$K_V = \frac{P^{1/\gamma}}{\gamma \rho T}\psi'\left(\frac{P^{1/\gamma}}{\rho}\right) \tag{81}$$

If we do not have any other information regarding the specific heats of the ideal gas, we cannot say more about them, or about the heat function in Equation (80).

However, if the specific heat at constant volume is a function of the temperature of the gas only, then we can determine the constitutive quantities and the heat function. At least on an adiabat the capacity must be a function of T only. If we calculate the heat capacity for this case according to Equation (81), we get:

$$K_V = f(T) \quad \Rightarrow \quad K_V = \frac{C_V}{T} \quad \text{with} \quad C_V = constant \tag{82}$$

This is the form of the heat capacity which we have encountered in Chapter 2: the capacities are inversely proportional to the ideal gas temperature T. However, there is nothing in our theory or in experiments which would allow us to derive this result.

I.3 Interaction of Heat and Motion: Carnot's Axiom

In Chapters 1 and 2, we studied the operation of heat engines. Now we will return to the subject and develop it from a more general point of view, all the time following the early historical development.

The use of the "motive power of heat" in steam engines had attracted a young french engineer, S. Carnot. Carnot wrote that "the phenomenon of the production of motion by heat has not been studied from a sufficiently general point of view."

He suggested that a general theory should be developed along the lines of mechanics:[16]

> All cases are foreseen, all imaginable motions obey certain general principles, firmly established under all circumstances. Such is the character of a complete theory. A comparable theory plainly wants for heat engines. We shall have it only after the laws of physics are extended enough, to make known beforehand all the effects of heat acting in a determined manner on a body.

The relationship between motion and heat served as the vehicle for the development of thermodynamics. By studying the heating and mechanical processes of bodies, we can hope to unravel the relationship between heat and energy. This is the importance of heat engines: they are the devices which employ both thermal and mechanical actions in bodies like steam or gases.

Engines work cyclically. The fluid undergoing thermal and mechanical processes must return to its initial state at some point to begin another cycle of operations. For this reason, we will study cyclic processes in this and the following sections, mainly the *Carnot cycle*. Carnot was the first to explore the relationship between thermal and mechanical processes in cyclic operations in some detail. Even though he did not succeed ultimately with his theory, he can be called the founder of thermomechanics. He realized that the laws of thermal physics would provide restrictions upon the constitutive relations for the bodies studied. On the other hand, if one had some information on the thermal properties of only one type of body, one might be able to deduce the relationship between heat and energy. Carnot applied this reasoning to simple homogeneous bodies which can change only their temperatures and volumes, and which are subject to an equation of state relating the pressure of a fluid to its temperature and volume (Sections I.1 and I.2). By limiting our attention to a simple but major example, we will be able to demonstrate clearly the relationship between heat and work, following Carnot's analysis.

I.3.1 Carnot Cycles

Heat engines must operate cyclically. If they do not, they will perform a task and then stop. Consequently we have to apply the ideas developed so far to closed cyclic processes. By a *process* we mean a pair of functions $V(t)$, $T(t)$. The path $\mathcal{P}$ generated by such a process is represented by a curve in the T-V diagram (such as in Figure 2). Here we will discuss the *Carnot cycle* as a particular example before presenting a mathematical tool for the treatment of general cycles.

Description of Carnot cycles. We have seen that for heat engines to work, a *furnace* and a *refrigerator* are needed. The engine operates between these two bodies, drawing heat from the reservoir at the higher temperature, and rejecting heat

16. S. Carnot: *Réflexions sur la Puissance Motrice du Feu et sur les Machines Propres a développer cette Puissance*. Paris, Bachelier, 1824.

to the reservoir at the lower temperature. Carnot composed a cycle of four steps which was to facilitate the desired task. In the first step, the body working as the agent of the heat engine is in contact with the furnace which maintains a constant upper temperature (Figure 5). The agent, usually a gas, expands isothermally by absorbing caloric from the furnace. Then, during the second step, the engine is insulated, and the gas continues to expand adiabatically. As a consequence, its temperature decreases to the lower temperature of the refrigerator. At this point, the fluid is brought in contact with the refrigerator. While caloric is emitted, the gas contracts isothermally. Finally, the gas is insulated again and allowed to contract adiabatically while its temperature rises to the initial value. This four-step cycle is called a *Carnot cycle* and is illustrated in Figure 5.

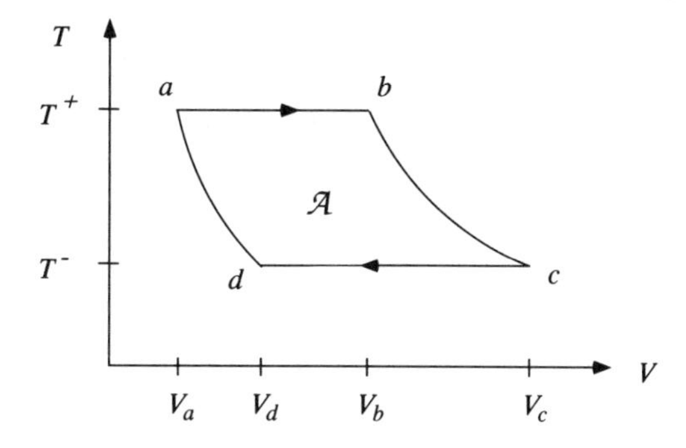

FIGURE 5. Ordinary Carnot cycle. Between *a* and *b*, the body undergoing the cycle absorbs heat. On *cd* it emits as much heat as was absorbed. Parts *bc* and *da* are adiabats. No heat is absorbed or emitted during these steps of the cycle. The cycle is found in a part of the constitutive domain where $\Lambda_V > 0$.

A Carnot cycle is determined uniquely by, for example, the two operating temperatures and the volumes V_a and V_b, i.e., by points *a* and *b* (Figure 5). This is the case because the adiabats follow uniquely from the differential equation (52) for adiabatic motion.

This, however, brings up an interesting point. The adiabats look like the ones drawn in Figure 5 only if the latent heat with respect to volume is larger than zero. In a part of the constitutive domain where the latent heat is negative, the adiabats represent increasing functions of volume, and a Carnot cycle generally looks like the one in Figure 6a. It is possible to have even stranger Carnot cycles,[17] namely, those for which the isotherms lie in parts of the domain having latent heats of different signs where the adiabats cross a neutral curve (Figure 6b). While we should be aware of these problems, we shall not generally treat cases other than those for which the latent heat is of one sign only.

The heat exchanged in a Carnot cycle. The Carnot cycle is chosen in such a way as to make the calculation of the heat exchanged very simple. According to our theory of heat, the caloric absorbed from the furnace and the caloric emitted to the refrigerator are given by

17. See J.S. Thomson and T.J. Hartka (1962); J.E. Trevor (1928); L.A. Turner (1962).

$$S^+(\mathscr{C}) = S(V_b, T^+) - S(V_a, T^+)$$
$$S^-(\mathscr{C}) = S(V_c, T^-) - S(V_d, T^-) \tag{83}$$

respectively (Figure 5). These two quantities must be equal for a cycle:

$$S^+(\mathscr{C}) = S^-(\mathscr{C}) \tag{84}$$

We can prove that this must be the case for general cycles undergone by bodies obeying the theory of calorimetry. As Carnot put it: "... the production of motive power is due *not to any real consumption of caloric, but to its transport from a warm body to a cold body...* ." This image is suggested by the comparison of heat and water.

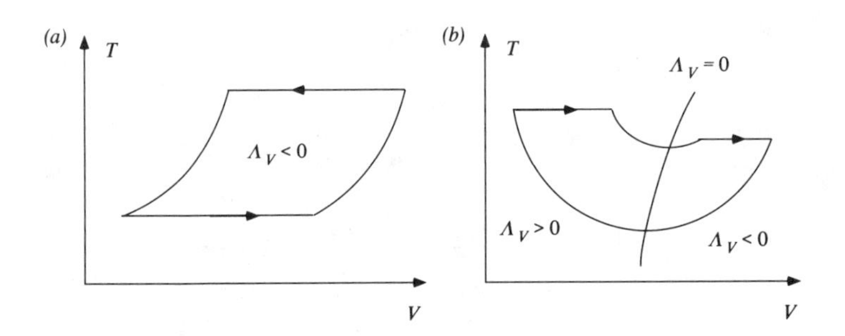

FIGURE 6. Unusual Carnot cycles. If we allow the latent heat with respect to volume to be either negative (a), or both positive and negative (b), we get Carnot cycles which might look like the ones drawn here. The cycle in (a) is like the one in Figure 5. (b) represents a cycle with water as the agent in which the temperatures of the furnace and the refrigerator are 6°C and 2°C, respectively. The curve intersecting the cycle is a neutral curve on which the latent heat vanishes.

I.3.2 Calculation of Cyclic Processes

So far, our approach is lacking in two ways. While we have been able to compute the heat absorbed and emitted in a Carnot cycle, we have not obtained a result for the energy exchanged as a consequence of the working of the fluid (i.e., the motive power), and we do not yet have the means of calculating more general cycles than the one devised by Carnot. Here we will have a brief look at some mathematics necessary for the task.

The exchange of quantities in mechanical and thermal processes. Physical processes are described in terms of the exchange of certain quantities. In the case of motion we talk about the flow of momentum and energy, while in thermal operations heat (caloric) and energy are exchanged. We have encountered simple instances of such problems in previous chapters. Basically all we had to be able to do was to integrate the currents of momentum, heat, and energy, over time.

Note the difference in what we know about mechanical and thermal processes. As far as our limited model goes, the mechanical operation is described completely. The exchange of momentum could be calculated in principle, but this is not necessary since we assume momentum not to be stored in the fluid. The energy exchanged in mechanical processes is known as well from the theory of mechanics. Thermal processes are quite different in this respect. First, the exchange of the quantity which is analogous to momentum, i.e., heat, is nontrivial: caloric is stored and exchanged, which means that the balance of this quantity has to be considered expressly. We have a theory specifying the currents of heat (Section I.2) which allows us to calculate the exchange of heat. The energy exchanged with caloric, however, has to be determined first. This is the goal of this section. In contrast to mechanical processes, the exchange of energy in thermal operations is still undetermined.

Nevertheless, consider what we know. The fluxes of caloric in thermal operations and the currents of energy as a result of mechanical processes are functions of the rates of change of the independent variables of the present theory:

$$I_s = -\Lambda_V(V,T)\dot{V} - K_V(V,T)\dot{T} \tag{85}$$

$$I_{E,mech} = P(V,T)\dot{V} \tag{86}$$

Remember the definition of a process (Section I.1.1). We are dealing with bodies for which temperature and volume are the independent variables. A function like the pressure or the heat capacity of a fluid is defined in general over the constitutive domain in terms of these variables. For a particular process, volume and temperature are variables of time. This makes the functions we are considering, i.e., pressure or heat capacity, functions of time for the process. If we now integrate a current, we have to calculate integrals of functions along the path representing the process. Such integrals are called *path integrals.*

Path integrals and Green's Theorem.[18] Consider a function $A(V,T)$ which, multiplied by the rate of change of one of the independent variables, has to be integrated over time:

$$X_e = \int_{t_1}^{t_2} A(V,T)\dot{V}dt \tag{87}$$

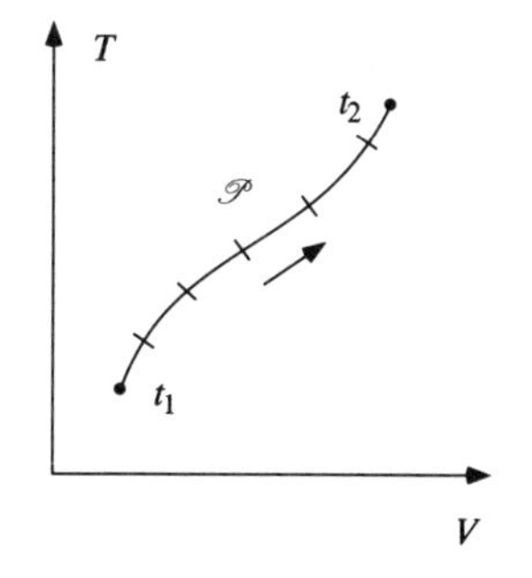

FIGURE 7. The path of a process is divided into N pieces, corresponding to N time intervals.

In other words, we integrate over the range spanned by the initial and final times of the process (Figure 7). Now let us divide the interval of time into N parts. Alternatively, this divides the path $\mathscr{P}$ in the V-T quadrant into N pieces. According to the definition of a simple integral, X_e must be the limit of the following sum:

$$X_e = \lim_{\Delta t \to 0}\left(\sum_{i=1}^{N} A_i(V,T)\dot{V}_i\Delta t\right)$$

This can be interpreted as the sum of products of $A(V,T)$ along the path $\mathscr{P}$ and the differences ΔV between adjacent points subdividing the path in Figure 7. If

18. For more information see, for example, Marsden and Weinstein: *Calculus* III (1985).

we again take the limit, the last form of the sum symbolically is written as the integral

$$X_e = \int_{\mathscr{P}} A(V,T)dV \tag{88}$$

This is called the integral of the function $A(V,T)$ over volume along the path $\mathscr{P}$; it is a path integral. For a function such as the current of caloric, Equation (85), the integral looks like

$$X_e = \int_{t_1}^{t_2} \left[A(V,T)\dot{V} + B(V,T)\dot{T}\right]dt = \int_{\mathscr{C}} \left[A(V,T)dV + B(V,T)dT\right] \tag{89}$$

Now consider a general cycle. A simple cycle is represented by a closed curve in the *T-V* diagram which is traversed only once (Figure 8). We assume that no two points on the path are the same, except for the points representing the beginning and end of a simple operation.

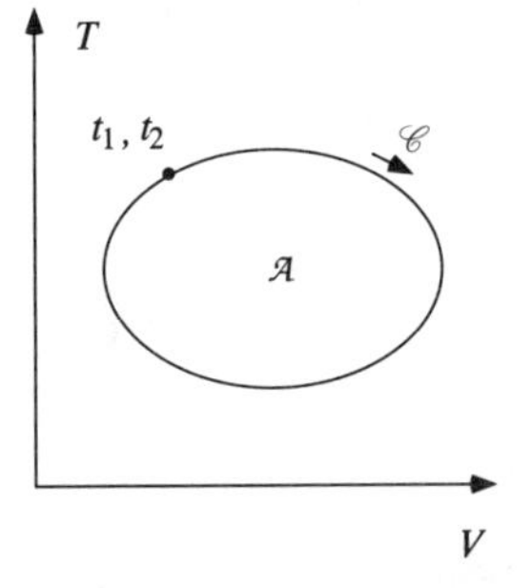

FIGURE 8. A general cycle is traversed clockwise in the *T-V* plane. A simple cycle is traversed exactly once. The area $\mathscr{A}$ is enclosed by the path of the cycle $\mathscr{C}$.

We just saw that quantities like the exchanged heat can be calculated by a path integral. Since we are interested in the interaction of thermal and mechanical processes during a cycle, we will have to calculate such line integrals. Often it will be useful instead to integrate over the area enclosed by the cyclic path. There exists a transformation of integrals taken along a closed path into the corresponding surface integral which is called *Green's Theorem.*

Assume a simple cycle is to be traversed clockwise in the quadrant, as in Figure 8. This means that the area enclosed by the cycle lies on the left side of someone walking along the path. Assume there to be two functions $A(V,T)$ and $B(V,T)$ which are as smooth as necessary. Then there exists the following transformation:

$$\int_{\mathscr{C}} \left[A(V,T)dV + B(V,T)dT\right] = \int_{\mathscr{A}} \left[\frac{\partial A(V,T)}{\partial T} - \frac{\partial B(V,T)}{\partial V}\right] dVdT \tag{90}$$

Here, $\mathscr{C}$ is the path in the *V-T* quadrant, and $\mathscr{A}$ is the area enclosed by it.

Application to mechanical and thermal processes. If we now apply the mathematics of path integrals and of Green's Theorem to the exchange of quantities in mechanical and thermal processes, we get some preliminary results which will be important for the following discussion.

The energy exchanged as a result of the working of the fluid during a closed cycle is calculated according to

$$W_{mech} = -\int_{t_1}^{t_2} P(V,T)\dot{V}dt = -\int_{\mathscr{P}} P(V,T)dV \tag{91}$$

Therefore, W_{mech} is a function of the path of the process. The particular form of the expression for the energy exchanged in the process demonstrates that the energy given off on the reverse path - $\mathscr{P}$ must be the negative of the energy released on $\mathscr{P}$

$$W(-\mathscr{P}) = -W(\mathscr{P}) \tag{92}$$

This is the reversal theorem for the energy exchanged in the mechanical process, which holds for the simple fluids we are using. It is a consequence of the choice of equation of state. It does not hold if the pressure is not related to the volume in the simple way expressed by Equation (2).

For the purpose of treating cyclic operations, we should calculate the mechanical energy exchanged along the path of the cycle. The result is the net energy exchanged in one cycle by a heat engine. It is the motive power of the engine, i.e., the quantity an engineer would like to determine. To this end, we shall use Green's Theorem, which yields

$$\begin{aligned} W(\mathscr{C}) &= -\int_{\mathscr{C}} P(V,T)dV \\ &= -\int_{\mathscr{A}} \frac{\partial P(V,T)}{\partial V} dVdT \end{aligned} \tag{93}$$

Let us now calculate the heat exchanged along the path in Figure 8. We obtain the path integral

$$\begin{aligned} S_e(\mathscr{P}) &= -\int_{t_1}^{t_2} I_s dt \\ &= \int_{\mathscr{P}} \left[\Lambda_V(V,T)dV + K_V(V,T)dT\right] \end{aligned} \tag{94}$$

It was demonstrated in Section I.2 that the fluids chosen here emit as much heat during the reverse of a process as they absorb during the process itself:

$$S_e(-\mathscr{P}) = -S_e(\mathscr{P}) \tag{95}$$

As a consequence, we can prove that the fluids which obey the theory of calorimetry *can undergo only eversible processes.* While the equivalent statement for momentum exchanged is trivial in our case, it is not trivial for caloric exchanged, since caloric is not a conserved quantity.

To calculate the caloric exchanged in a cycle we evaluate Equation (94) along the closed path in Figure 8 and apply Green's Theorem, which leads to the following expression:

$$\begin{aligned} S_e(\mathscr{C}) &= \int_{\mathscr{C}} \left[\Lambda_V(V,T)dV + K_V(V,T)dT\right] \\ &= \int_{\mathscr{A}} \left[\frac{\partial \Lambda_V}{\partial T} - \frac{\partial K_V}{\partial V}\right] dVdT \end{aligned} \tag{96}$$

Now, as a consequence of the condition of integrability for the heat function, Equation (37), we obtain

$$S_e(\mathscr{C}) = 0 \tag{97}$$

Saying that the heat exchanged in a cycle is equal to zero also means that *the amount of heat exchanged in a general process is independent of the path.* This is what we have assumed all along, specifically in Equation (83). Now we have a formal proof of this statement. We shall see that this property is a restriction on the types of constitutive relations we shall be able to accept for the fluids which obey the theory of calorimetry.

I.3.3 Carnot's Axiom

We should now try to answer the question of how much energy will be released for mechanical purposes in one step of a Carnot cycle. Since a Carnot cycle is determined by the caloric it absorbs (among other things), answering this question should furnish the desired relationship between heat (caloric) and energy. By considering experience with heat engines, Carnot came up with a general expression of the work done as a function of the caloric absorbed and the temperatures of the furnace and the refrigerator. We will use his ideas as a stepping stone into thermodynamics.

Motivation of Carnot's Axiom. Carnot made two claims about the cycle he proposed. First, among all cycles operating at the same temperatures and absorbing the same amount of heat, Carnot cycles ought to be the most efficient. Second, he assumed the efficiency of a Carnot cycle to be independent of the agent used: there should be no difference in using steam or air. These claims appear quite sound and logical. If we accept them we can derive as a result *Carnot's Axiom:*

> *The motive power of a Carnot cycle depends only on the two operating temperatures and on the amount of heat absorbed.*

This will be stated below in mathematical terms. However, the claims themselves cannot be proven to be correct. They rest upon assumptions which lie outside the theory being developed here. Therefore it seems to be sensible to accept as our starting point what we derive from them, namely Carnot's Axiom, rather than having to deal with two independent claims.

The first claim sates that Carnot cycles are the most efficient. How could we motivate such an idea? First let us assume that the heat engine operates between two heat reservoirs at fixed temperatures. We could also conceive of cycles running between many, even infinitely many, furnaces and refrigerators which all have different temperatures; however, we will not do this. Under these circumstances it is quite clear that only Carnot cycles achieve the maximum possible motive power. For any other type of cycle, the agent used in the engine would receive at least some heat at temperatures lower than that of the furnace, and it would discharge heat at a time when its temperature must be higher than that of the refrigerator. In any case, we would have to deal with a transfer of heat between bodies of different temperatures without the production of motion. In terms of our discussions in previous chapters, energy would be released in the fall of caloric, but not for mechanical purposes. Rather, in the conduction of heat between the bodies more heat would be generated. Carnot expressed his ideas in the following way:

> Since any re-establishment of equilibrium in the caloric can be the cause of motive power, any re-establishment of equilibrium that occurs without producing this power should be considered a true loss. Now, very little reflection suffices to show that every change of temperature which is not due to a change of volume...can be nothing but a useless re-establishment of equilibrium in the caloric. Thus the necessary condition for the maximum is that in the bodies employed to effect the motive power of heat there be no change of temperature not due to a change of volume. Conversely, whenever this condition is fulfilled, the maximum will be attained.

If we accept Carnot's first claim, we can derive an important result: *any two Carnot cycles which use the same operating temperatures and absorb the same amount of heat release the same amount of energy for mechanical purposes.*[19] In other words, the motive power of a Carnot cycle is independent of the position of the cycle on the two isotherms in Figure 5.

This result has important consequences. It was stated in Section I.3.1 that a Carnot cycle is determined uniquely by the four quantities T^+, T^-, V_a, and V_b (Figure 5). Therefore, the energy delivered by the cycle $\mathscr{C}$ ought to be a function of these four variables:

$$W(\mathscr{C}) = g_{\mathcal{B}}\left(T^+, T^-, V_a, V_b\right) \tag{98}$$

The index $\mathcal{B}$ reminds us that the energy delivered may depend upon the particular body undergoing the Carnot cycle. Now, according to the doctrine of latent and specific heats, the heat absorbed by the heat engine is given by

$$S^+(\mathscr{C}) = \int_{V_a}^{V_b} \Lambda_V\left(V, T^+\right) dV \tag{99}$$

---

19. The proof roughly goes as follows. Consider two competing Carnot cycles which have the same operating temperatures and which absorb the same amount of heat. Any other type of cycle will be less efficient than a Carnot cycle, but it can be made to approach the efficiency of one of the Carnot cycles, say the first. If the second Carnot cycle is less efficient than the first, it will be possible to find a non-Carnot cycle which is more efficient, which contradicts our assumption of maximum efficiency. If, however, the second Carnot cycle is more efficient than the first, we simply exchange their roles and use the same argument just given. Therefore, both Carnot cycles must have the same motive power.

(In a Carnot cycle, the heat is absorbed at constant temperature.) Since we assume that $\Lambda_V > 0$, we can invert this equation to yield V_b as a function of V_a, T^+, and $S^+(\mathscr{C})$:

$$W(\mathscr{C}) = g_{\mathscr{B}}\left(T^+, T^-, V_a, h_{\mathscr{B}}\left(V_a, T^+, S^+(\mathscr{C})\right)\right) \tag{100}$$

According to Carnot's first claim, the argument V_a in the expression for the work done in a cycle must be dropped. The energy released does not depend on the details of the cycle once the temperatures and the heat absorbed are fixed. We can now conclude that the energy given off by a particular body $\mathscr{B}$ undergoing a Carnot cycle is a function of the operating temperatures and the amount of caloric absorbed:

$$W(\mathscr{C}) = G_{\mathscr{B}}\left(T^+, T^-, S^+(\mathscr{C})\right) \tag{101}$$

The mechanical effect of a heat engine may still depend on the particular fluid used in it. Carnot claimed that the work done by heat engines would be independent of the agent used. To support this claim he considered two heat engines with different bodies undergoing Carnot cycles between the same furnace and refrigerator. Body 1 would absorb heat (caloric) from the furnace, and emit just as much to the refrigerator. In so doing it releases energy for mechanical purposes. We could now use this energy to make the caloric given up to the refrigerator return to the furnace again in a cycle undergone by Body 2. If Body 1 delivers more energy than Body 2 needs to pump the caloric back up, the two cycles together would do positive work while neither the furnace nor the refrigerator would have been changed. (The two cycles restore the conditions as they were at the beginning.) According to Carnot this would be contrary to "sound physics" In Carnot's words:

> But, if there were means of employing heat preferable to those we have used, that is, if it were possible by any method whatever to make caloric produce a quantity of motive power larger than we have made by our first series of operations, it would suffice to draw off a portion of this power in order to cause the caloric, by the method just indicated, to go up again … from the refrigerator to the furnace and re-establish things in their original state, thereby making it possible to recommence an operation altogether like the first, and so on. That would not only be perpetual motion but also the creation of boundless motive force with no consumption of caloric or of any other agent whatever. Creation of this kind is entirely contrary to … sound physics; it is inadmissible."

Accepting this claim of universal efficiency, we must conclude that the energy given off by a body undergoing a Carnot cycle must be a *universal function* G of the operating temperatures and the heat absorbed, and it must always be negative for a heat engine:

$$W(\mathscr{C}) = G\left(T^+, T^-, S^+(\mathscr{C})\right) \tag{102}$$

with

$$W(\mathscr{C}) < 0 \tag{103}$$

Carnot had tried to prove this result by using the two claims just discussed. The problem is that both statements cannot be proved. The claim of universal efficiency, for example, rests on arguments involving the bodies undergoing Carnot cycles and the surroundings (the furnace and the refrigerator). For the former we have a theory (the theory of calorimetry); for the latter we do not. The claim of maximum efficiency uses arguments involving irreversible processes occurring between the engine and its surroundings. The agents used in the engine, however, can undergo only reversible cycles, as we have seen here and in Chapter 2. Therefore, proofs are impossible.

It seems to be more natural to accept the result just stated as the most general expression of our experience with heat engines. In mathematical language, we accept as an *axiom* that the energy released in a Carnot cycle for mechanical purposes is a universal function of the operating temperatures and the heat absorbed. Equation (102) is called *Carnot's General Axiom.* On its basis we will be able to derive the motive power of heat together with a proof of the claim that Carnot cycles are the most efficient (Section I.3.4).

Carnot's Axiom and the caloric theory of heat. We shall transform Carnot's Axiom to make it fit our assumptions concerning the nature of heat. We will use it in its final form in Section I.3.4, where we will determine the motive power of heat. First, the energy delivered in a cycle depends linearly upon the heat absorbed:[20]

$$W(\mathscr{C}) = G^*\left(T^+, T^-\right) S^+(\mathscr{C}) \tag{104}$$

---

20. (Truesdell, 1980). A Carnot cycle is divided into two parts as in the figure below. The original cycle $\mathscr{C}$ is the sum of two parts $\mathscr{C}_1$ and $\mathscr{C}_2$. Both parts have the same operating temperatures, which in turn are the same as those of the original cycle. If we apply the definition of the heat absorbed to each cycle, we find that

$$S^+(\mathscr{C}) = S^+(\mathscr{C}_1) + S^+(\mathscr{C}_2)$$

Also, because of the reversal theorem for the energy delivered in the mechanical process, and since the common adiabat is traversed in opposite directions, we have

$$W(\mathscr{C}) = W(\mathscr{C}_1) + W(\mathscr{C}_2)$$

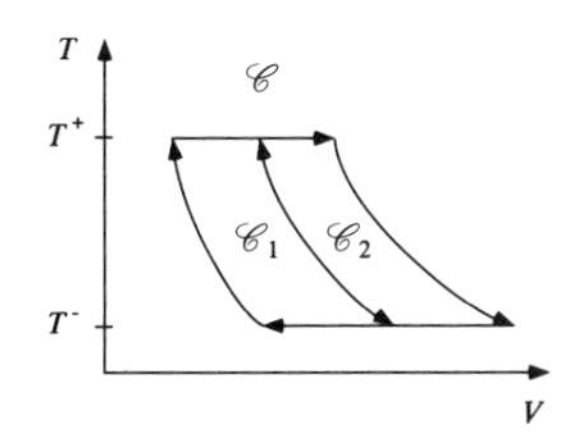

Application of Carnot's General Axiom yields

$$G\left(T^+, T^-, S^+(\mathscr{C})\right) = G\left(T^+, T^-, S^+(\mathscr{C}_1) + S^+(\mathscr{C}_2)\right)$$
$$= G\left(T^+, T^-, S^+(\mathscr{C}_1)\right) + G\left(T^+, T^-, S^+(\mathscr{C}_2)\right)$$

Therefore, the function G is linear in S^+.

Our assumptions regarding the nature of heat suggest that the work done by heat depends on the difference of the operating temperatures in some way. Therefore, we express the axiom in the form

$$W(\mathscr{C}) = -\left[F\left(T^+\right) - F\left(T^-\right)\right]S^+(\mathscr{C}) \tag{105}$$

We can prove that the General Axiom is compatible with our theory of heat if and only if it takes this special form.[21] We introduce the minus sign because we want the energy given *off* by the engine to be counted as a *negative* quantity (as we have done so far for quantities which are exchanged in a process).

We will call this *Carnot's Special Axiom.* The function F in the special axiom is called *Carnot's function.* It must be an increasing function of temperature. If we manage to determine it, we have found the solution to our problem: the relationship between heat and energy.

The sign of the latent heat. We are now in a position to prove that the latent heat of water must be negative in that part of the constitutive domain where the coef-

---

21. (Truesdell, 1980). To prove that the special form is both necessary and sufficient for the General Axiom to be compatible with the theory of heat, i.e., with $S^+(\mathscr{C}) = S^-(\mathscr{C})$, we need the following construction of Carnot cycles (see figure below). A cycle $\mathscr{C}$ is joined with a second cycle $\mathscr{C}_1$ which has the operating temperatures T^- and T_o. The combined cycle is called $\mathscr{C}_2$. From the reversal theorem for heat, Equation (95), we conclude that

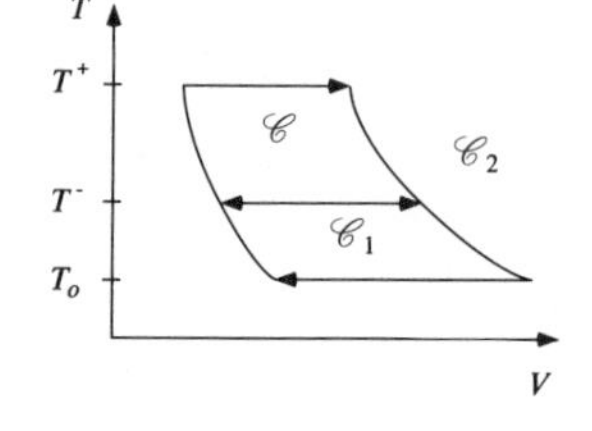

$$S^+(\mathscr{C}_2) = S^+(\mathscr{C})$$
$$S^+(\mathscr{C}_1) = S^-(\mathscr{C})$$

The energy released in the combined cycle $\mathscr{C}_2$ is the sum of the energies emitted in cycles $\mathscr{C}$ and $\mathscr{C}_1$ (reversal theorem for work).

The special form is necessary: if we apply the General Axiom to each cycle, we find

$$G^*\left(T^+,T^-\right)S^+(\mathscr{C}) = G^*\left(T^+,T_o\right)S^+(\mathscr{C}) - G^*\left(T^-,T_o\right)S^-(\mathscr{C})$$
$$= f\left(T^+\right)S^+(\mathscr{C}) - f\left(T^-\right)S^-(\mathscr{C})$$

We have $S^+(\mathscr{C}) = S^-(\mathscr{C})$ for a Carnot cycle, and we conclude that the special form of Carnot's General Axiom holds.

The special form is sufficient: we use the reversal theorems for heat and work as expressed above. If the energy given off in a Carnot cycle is expressed by the special form of Carnot's Axiom, we get

$$\left[F\left(T^+\right) - F\left(T_o\right)\right]S^+(\mathscr{C}) = \left[F\left(T^+\right) - F\left(T^-\right)\right]S^+(\mathscr{C}) + \left[F\left(T^-\right) - F\left(T_o\right)\right]S^-(\mathscr{C})$$

or

$$\left[F\left(T^-\right) - F\left(T_o\right)\right]S^+(\mathscr{C}) = \left[F\left(T^-\right) - F\left(T_o\right)\right]S^-(\mathscr{C})$$

from which we conclude that $S^+(\mathscr{C}) = S^-(\mathscr{C})$.

ficient of thermal expansion is negative. We had suspected this to be the case when we discussed the constitutive inequalities in Section I.2. In fact, we can prove[22] that in the constitutive domain

$$\Lambda_V \frac{\partial p}{\partial T} \geq 0 \tag{106}$$

Also, any piezotropic part of the constitutive domain also is a neutral part. This shows that for water at atmospheric pressures, the latent heat with respect to volume must be negative for temperatures between 0°C and 4°C since there the volume decreases with increasing temperature. According to the discussion in Section I.1 this means that the derivative of the pressure with respect to temperature must be negative as well. Therefore, in this part of the domain, *water absorbs heat when it is compressed isothermally.* Also, if we have a Carnot cycle with one isotherm on either side of the neutral curve, both isothermal operations must be expansive (Figure 6b).

I.3.4 The Motive Power of a Heat Engine

The exploitation of Carnot's Axiom, together with additional information on constitutive quantities (particularly on the ideal gas), will finally allow us to calculate the energy delivered by a heat engine in terms of the temperatures involved and the heat absorbed. We will find that, at least for the fluids treated here, the ideal gas temperature takes the role of the thermal potential, and the relationship between the substancelike quantity of thermal processes (heat, or caloric) and work is the usual one we know from other fields of physics. The result will be applied to some problems which we could not solve before. We can give a proof that Carnot cycles are the most efficient. Therefore, Carnot's claim is not an assumption necessary for thermodynamics; it follows from thermodynamics. You will also see that the ideal gas temperature delivers an absolute scale, i.e., one which is independent of a thermometric substance.

---

22. The proof uses the inequality of Carnot's Axiom, Equation (103). Assume for the moment that the latent heat is positive while $\partial p/\partial T$ is negative at an ordinary point of the domain. Then there must be Carnot cycles in a neighborhood of that point of the type shown in Figure 5. From the expression for the work done in a cycle, Equation (93), we see that in this case $W(\mathscr{C}) > 0$, which contradicts the inequality in Equation (103). Therefore, both the latent heat and $\partial p/\partial T$ must be positive. Now, for a part of the domain in which the latent heat is negative we prove by similar arguments that $\partial p/\partial T$ must be negative as well. In either case the product of latent heat and the derivative of the pressure with respect to temperature must be positive. Also, if we have cycles in a piezotropic part of the domain, the work done must be zero because of $\partial p/\partial T = 0$ and Equation (93). On the other hand, if the latent heat were not zero in this part, i.e., if the part were not neutral, the cycles should deliver energy according to Carnot's Axiom. Therefore, a piezotropic part of the constitutive domain must be neutral.

The Carnot-Clapeyron theorem. We shall now relate the heat absorbed in a Carnot cycle to the energy delivered in the mechanical process. According to the doctrine of latent and specific heats, the heat absorbed in an isothermal process $\mathscr{P}_T$ is

$$S_e(\mathscr{P}_T)=\int_{V_1}^{V_2}\Lambda_V(V,T)dV \quad , \quad V_2>V_1 \tag{107}$$

which holds in this form for positive latent heat. If we set the temperature equal to that of the furnace, Equation (107) yields the heat absorbed in the ordinary Carnot cycle in Figure 5. The energy delivered by the heat engine in one cycle can be expressed as in Equation (93). Now we use Carnot's Special Axiom, Equation (105). Combining the work done in a cycle with the heat absorbed on the upper isotherm yields

$$\int_{\mathcal{A}}\frac{\partial p}{\partial T}dVdT=\left[F(T^+)-F(T^-)\right]\int_{V_1}^{V_2}\Lambda_V(V,T^+)dV \tag{108}$$

for a Carnot cycle. V_1 and V_2 are the volumes at the beginning and end of the isothermal expansion of the Carnot cycle (Figure 5). To get an expression which holds at a point of the constitutive domain, we divide this by $(V_2 - V_1)$ and by the difference between the upper and the lower temperatures $(T^+ - T^-)$, and take the limit as T^- approaches T^+ and as $V_2 \to V_1$. The result is the *Carnot-Clapeyron Theorem*:

$$F'\Lambda_V=\frac{\partial p}{\partial T} \tag{109}$$

where $F' \equiv dF/dT$.[23] This theorem is central to thermodynamics. It shows that the relationship between heat and energy yields a restriction upon the constitutive

---

23. Dividing the right-hand side of Equation (108) by the product $(V_2 - V_1)(T^+ - T^-)$, and taking the limit, yields $F'\Lambda_V$ directly from the definitions of ordinary integrals and derivatives. The case of the double integral on the left-hand side, however, is more difficult. First we rewrite the integral, which in light of the form of an ordinary Carnot cycle (Figure 5), becomes

$$\int_{\mathcal{A}}\frac{\partial p}{\partial T}dVdT=\int_{T^-}^{T^+}\left[\int_{V=\phi_1(T)}^{V=\phi_2(T)}\frac{\partial p}{\partial T}dV\right]dT=\int_{T^-}^{T^+}Y'(T)dT$$

where

$$Y'(T)=\int_{V=\phi_1(T)}^{V=\phi_2(T)}\frac{\partial p}{\partial T}dV$$

Here, $\phi_1(T)$ and $\phi_2(T)$ are the functions describing the two adiabats of the cycle in Figure 5 (on an adiabat, V is a function of temperature alone). We now divide the integral of $Y'(T)$ by the difference of the temperatures $(T^+ - T^-)$ and take the limit which yields relations (upon Λ_V) of the body. If we can find Carnot's function $F(T)$, we can find the specific and latent heats of the simple fluids used here. Or if we have some information about the constitutive relations, we might be able to calculate Carnot's function and thus determine the relationship between heat and energy. This will be demonstrated shortly.

The latent heat of the ideal gas can be determined using the Carnot-Clapeyron Theorem in terms of Carnot's function F. Since

$$P = nRT/V \quad \Rightarrow \quad \frac{\partial p}{\partial T} = nR/V$$

the latent heat turns out to be

$$\Lambda_V = \frac{1}{F'}\frac{\partial p}{\partial T} = \frac{1}{F'}\frac{nR}{V} \tag{110}$$

This will be helpful when we try to determine Carnot's Function below.

The theory of the heat capacities and the determination of Carnot's function. Since we have some specific information regarding the properties of the ideal gas, by using it we can hope to finally determine the relationship between heat and energy which is equivalent to the determination of Carnot's function $F(T)$. If we have F for one body, we have it for all, since the motive power of heat is independent of the agent used (Carnot's Axiom).

Using the relationship between heat and energy as derived above (i.e., the Carnot-Clapeyron Theorem $F'\Lambda_V = \partial p/\partial T$, we get by integration along an isothermal path

$$S_e(\mathscr{P}_T) = \frac{1}{F'}\int_{V_o}^{V}\frac{\partial p}{\partial T}dV \tag{111}$$

For the ideal gas this is equivalent to

$$S(V,T) - S(V_o,T) = \frac{nR}{F'(T)}\ln\left(\frac{V}{V_o}\right) \tag{112}$$

$$\lim_{T^- \to T^+}\frac{1}{T^+ - T^-}\int_{T^-}^{T^+} Y'(T)dT = \lim_{T^- \to T^+}\frac{Y'(T^+) - Y'(T^-)}{T^+ - T^-}$$

which is $Y'(T^+)$. The second step involves dividing $Y'(T^+)$ by $(V_2 - V_1)$ whereupon we take the second limit:

$$\lim_{V_1 \to V_2}\frac{1}{V_2 - V_1}Y'(T^+) = \lim_{V_1 \to V_2}\frac{1}{V_2 - V_1}\int_{V=\phi_1(T)}^{V=\phi_2(T)} Z'dV \quad \text{with} \quad Z' = \frac{\partial p}{\partial T}$$

which yields $\partial p/\partial T$.

From this we determine the heat capacity at constant volume, and its companion, the capacity at constant pressure; see Equation (36):

$$K_V = \frac{\partial \mathcal{S}}{\partial T} = nR(1/F')' \ln\left(\frac{V}{V_o}\right) + S'_o(T) \tag{113}$$

$$K_p - K_V = \frac{V}{T}\Lambda_V = \frac{nR}{TF'} \tag{114}$$

where we have written $S_0(T)$ for $S(V_o,T)$. Remember that these formulas hold only for the ideal gas. The determination of Carnot's function F progresses as follows. We divide the last formula by the heat capacity at constant volume to obtain an expression involving the ratio γ of the two specific heats. The only reliable data on heat capacities we have up to this point concerns this ratio. If we can believe the theory of sound and its experimental determination (Section I.2.3), we have to conclude that this ratio must be constant. Performing the calculation yields

$$\gamma - 1 = \frac{nR}{TF'K_V} \tag{115}$$

Therefore, K_V may depend only on the temperature and not on the volume of the ideal gas (remember that F' is a function of T only). Carnot's determination of the heat capacity at constant volume, Equation (113), shows that F' must be constant; only if this condition holds can K_V be a function of the temperature alone. Therefore, Carnot's function must be a linear function of temperature:

$$F(T) = A \cdot T + B \tag{116}$$

Here, A is a conversion factor for units. From now on we will set it equal to 1 (without units); the arbitrary constant will be set equal to zero.

Interpretation of Carnot's function: the motive power of heat. At this point, let us summarize the results of this chapter in the light of the determination of Carnot's function. We finally know the energy released by a heat engine running through a Carnot cycle:

$$W = -\left(T^+ - T^-\right)S^+ \tag{117}$$

This states that the energy delivered by a heat engine is equal to the product of the caloric absorbed (which is equal to the caloric passing through the engine) and the fall of temperature. This is exactly what we might have guessed on the basis of Carnot's comparison of water wheels and heat engines (Chapter 1). The result is analogous to what we know from other fields of physics. It is interesting to derive the relationship between heat and energy which holds for general cycles. Remember the condition of integrability of the heat function, Equation

(37). Starting with the Carnot-Clapeyron Theorem in the expression for the energy delivered by a general cycle, we obtain

$$\begin{aligned} W(\mathscr{C}) &= -\int_{\mathcal{A}} \Lambda_V dVdT \\ &= -\int_{\mathcal{A}} \left[\Lambda_V + T\left(\frac{\partial \Lambda_V}{\partial T} - \frac{\partial K_V}{\partial V} \right) \right] dVdT \\ &= -\int_{\mathcal{A}} \left[\frac{\partial (T\Lambda_V)}{\partial T} - \frac{\partial (TK_V)}{\partial V} \right] dVdT \\ &= -\int_{\mathscr{C}} T\left[\Lambda_V dV + K_V dT \right] \end{aligned} \tag{118}$$

or

$$W(\mathscr{C}) = \int_{t_1}^{t_2} T I_s dt \tag{119}$$

The last step follows from Green's Theorem. Finally, Equation (119) is a consequence of the doctrine of latent and specific heats. Equation (119) is the general form of Carnot's Special Theorem for arbitrary simple cycles.

This result can be used to derive the motive power of Carnot engines in terms of currents of caloric and energy. Since in a Carnot cycle heat flows only at constant upper and lower temperatures we obtain from Equation (119):

$$I_{E,mech} = \left(T^+ - T^- \right) \left| I_s \right| \tag{120}$$

Here, I_S is the current of heat entering (and leaving) the heat engine. The results show that the product of the temperature of the body and the (net) current of heat with respect to this body is directly related to the rate at which energy is put at the disposal of the mechanical process of the body. Since the energy is delivered by the heating, we shall interpret it as the energy current related to the current of caloric:

$$I_{E,th} = T I_s \tag{121}$$

Remember that this was one of the starting points of the development of thermodynamics in the previous chapters.

Results of the constitutive theory: the constitutive quantities of the ideal gas. The central constitutive restriction of our theory is the Carnot-Clapeyron Theorem, which now takes the form:

$$\Lambda_V = \frac{\partial p}{\partial T} \tag{122}$$

This means that one of the constitutive quantities of the fluids treated here is determined by theory. In fact, with this determination and the measurement of the ratio of heat capacities of the *ideal gas*, all the constitutive quantities of this simple fluid can be calculated:

$$\Lambda_V = \frac{1}{F'}\frac{nR}{V} = \frac{nR}{V} \quad (123)$$

$$\Lambda_p = \frac{\Lambda_V}{\partial p/\partial V} = -\frac{nR}{P} \quad (124)$$

Furthermore, we get

$$K_V = \frac{nR}{(\gamma - 1)TF'} = \frac{nR}{\gamma - 1}\frac{1}{T} \quad (125)$$

$$K_p = K_V + \frac{V}{T}\Lambda_V = \frac{nR\gamma}{\gamma - 1}\frac{1}{T} \quad (126)$$

Finally, the difference of the heat capacities of the ideal gas turns out to be

$$K_p - K_V = nR/T \quad (127)$$

Carnot cycles are the most efficient. Now we can prove that Carnot cycles are the most efficient among all possible cycles undergone by the bodies treated in this chapter. If t^+ and t^- are the parts of the interval of $]t_1,t_2[$ on which heat is absorbed ($dS/dt > 0$) and released ($dS/dt > 0$), respectively, we can write

$$W(\mathscr{C}) = -\left[\int_{t^+} F(T)\dot{S}dt - \int_{t^-} F(T)\left(-\dot{S}\right)dt\right]$$

Since $F(T)$ is an increasing function we can conclude that

$$\begin{aligned} |W(\mathscr{C})| &\le \left|F(T_{max})S^+(\mathscr{C}) - F(T_{min})S^-(\mathscr{C})\right| \\ &= \left|\left[F(T_{max}) - F(T_{min})\right]S^+(\mathscr{C})\right| \end{aligned}$$

The last step is a consequence of the conservation of caloric in reversible cycles. If heat is absorbed only at one temperature (T_{max}), and if it is released at only one temperature (T_{min}), then the equality holds. The converse is true as well. This is exactly the description of a Carnot cycle. Therefore, among all cycles with given extremes of temperature, Carnot cycles are the most efficient.

Absolute temperatures.[24] So far, we have always used the ideal gas temperature as the measure of temperature. This leaves open the question of whether the

24. For more details, see Truesdell (1979).

results obtained depend on this particular body. We would certainly desire this not to be so. Therefore we should discuss whether there is an absolute temperature scale, a scale which is independent of the thermometric fluid used to measure temperature.[25]

The answer to our question is that Carnot's Axiom delivers such scales. Moreover, the ideal gas temperature can be used as an empirical temperature for the absolute scale. Thus, we will not have to change the equations of thermodynamics derived above.

We will briefly sketch the problems involved. Kelvin introduced such a scale in 1848. He used as the absolute temperature τ Carnot's function, which leads to the following relation between heat and energy:

$$F(T) = \tau \quad \Rightarrow \quad W(\mathscr{C}) = \left(\tau^+ - \tau^-\right) S^+(\mathscr{C}) \tag{128}$$

This scale must be independent of the thermometric fluid, since as a consequence of Carnot's Axiom, Carnot's function is universal. As we have seen, Carnot's function can be determined for a particular fluid, namely the ideal gas. It turns out to be equal to the ideal gas temperature T. Therefore, T is an empirical scale for the absolute temperature constructed by Kelvin. We see that none of our equations have to be changed. T can be used as the absolute temperature.[26]

I.4 Internal Energy and Thermodynamic Potentials

Based on the development of the thermodynamics of uniform fluids, R. Clausius was able to prove the existence of the internal energy of the fluids considered. This constituted the first instance of the full blown energy principle in physics. Later, the mathematical development of this branch of classical thermodynamics led to the concept of further quantities related to energy, namely the thermody-

25. Often, *absolute* is taken to mean a scale which has a point of absolute zero. We have taken this for granted by assuming that $T > 0$ all the time. In Chapter 1 we saw that this should be the case because otherwise the relationship between caloric created and the energy used would not be defined.

26. The problem is not quite so simple, however. For example, for a given empirical scale, $F(T)$ is independent of the particular fluid used. We cannot be sure, however, that we get the same function F for other scales as well. Then, the axioms and equations should be independent of the particular scale used; they should be invariant under transformation from one scale to another. (This imposes some restrictions on what kind of fluids can be used for thermometry: water is unsuitable.) Finally, Carnot's function may not vanish; otherwise, the absolute scale chosen above does not make any sense.

namic potentials. In this section we will briefly discuss the concept of internal energy, derive the energy principle for simple fluids, and introduce the thermodynamic potential. The introduction will be very brief; if you wish to find more information on the subject, you should turn to other sources.[27]

I.4.1 The Energy Function and the Balance of Energy

So far, we have considered only the exchange of energy in thermomechanical processes. However, the theory laid down leads to an additional result of great importance. Energy is a quantity which can be stored; therefore we should expect the concept of the energy content of a body to play an important role. The development of thermodynamics along the lines of Carnot's theory enables us to derive the existence of *internal energy*. Together with energy currents, the concept of storage of energy will allow us to formulate the energy principle, the law of balance of energy, for the uniform fluids considered in this chapter.

Internal energy and the Gibbs fundamental relation. We can now prove the existence of a function which corresponds to what we call the energy of the body or its internal energy. If we combine Equations (108) and (118), we obtain

$$\int_{\mathcal{A}} \frac{\partial p}{\partial T} dV dT = \int_{\mathcal{A}} \left[\frac{\partial (T\Lambda_V)}{\partial T} - \frac{\partial (TK_V)}{\partial V} \right] dV dT$$

Therefore:

$$\frac{\partial p}{\partial T} = \frac{\partial (T\Lambda_V)}{\partial T} - \frac{\partial (TK_V)}{\partial V}$$

or

$$\frac{\partial (TK_V)}{\partial V} = \frac{\partial (T\Lambda_V - P)}{\partial T}$$

If the functions involved are sufficiently smooth, there exists a function $E = \varepsilon(V,T)$ such that

$$\frac{\partial \varepsilon(V,T)}{\partial T} = TK_V \tag{129}$$

and

$$\frac{\partial \varepsilon(V,T)}{\partial V} = T\Lambda_V - P \tag{130}$$

This function has all the features of an energy function of the body, as we shall see below. In simple terms, it represents the energy stored in a system as a func-

27. See for example Callen (1985), Chapters 5–7.

tion of volume and temperature. The first step in making clear the importance of this new quantity involves the derivation of an expression for the rate of change of the energy in terms of the rates of change of other variables of the body:

$$\begin{aligned} \dot{E} &= \frac{\partial \varepsilon(V,T)}{\partial V}\dot{V} + \frac{\partial \varepsilon(V,T)}{\partial T}\dot{T} \\ &= \left(T\Lambda_V - P\right)\dot{V} + TK_V\dot{T} \\ &= T\left(\Lambda_V\dot{V} + K_V\dot{T}\right) - P\dot{V} \end{aligned}$$

which is equivalent to

$$\dot{E} = T\dot{S} - P\dot{V} \quad (131)$$

This last form is called the *Gibbs fundamental relation*, which has been derived only for the special fluids treated here. Such relations are constitutive. We have encountered different forms of the Gibbs fundamental relation in the context of the materials discussed in Chapter 2.

Equation (129) demonstrates the root of the name *temperature coefficient of energy* for the quantity

$$C_V = TK_V \quad (132)$$

which was used repeatedly before in Chapter 2. It will become clear in Section I.5, after discussing the mechanical theory of heat, why the product of the heat capacity (i.e., the capacity for caloric) and the temperature of the body has obtained the name *heat capacity* in classical presentations of thermodynamics.

Energy and energy conservation. We can derive the law of conservation of energy for the fluids discussed in this chapter. The term *PdV/dt* is the rate at which energy flows into or out of the body due to mechanical processes, while – *TdS/dt* equals the energy current associated with thermal processes. The sum of the two therefore corresponds to the rate at which the energy flows:

$$\begin{aligned} I_{E,net} &= P\dot{V} - TI_s \\ &= P\dot{V} - T\dot{S} \end{aligned} \quad (133)$$

since caloric cannot be created in the fluids considered. If we substitute this result into the Gibbs fundamental relation Equation (131), we get

$$\dot{E} + I_{E,net} = 0 \quad (134)$$

This is the equation of balance of a quantity which is conserved. We interpret it as the law of balance of energy of uniform fluids undergoing thermomechanical processes. As you already know from the less formal development in Chapter 2, the law of balance of energy can be derived for particular materials if the thermodynamic relations have been obtained on the basis of other information.

The energy function as a "fundamental" relation. In Equation (129) and the subsequent discussion, the energy of an ideal fluid was introduced as a function of volume and temperature. The Gibbs fundamental relation (131) suggests, however, that there exists a more "natural" representation of the energy, namely as a function of the extensive parameters volume and heat (i.e., entropy). If we introduce the energy function $\mathcal{E}(S,V)$, its partial derivatives return the intensive quantities temperature and pressure:

$$\frac{\partial \mathcal{E}(S,V)}{\partial S} = T$$
$$\frac{\partial \mathcal{E}(S,V)}{\partial V} = -P \tag{135}$$

In this functional form, the energy is called a *fundamental relation*. It turns out that all relevant information regarding a particular fluid can be derived from it. To give an example, if you have the energy of the ideal gas as a function of entropy and of volume, you can derive from it the entropy as a function of temperature and volume, or the equation of state, if you wish. For this reason, the function $\mathcal{E}(S,V)$ enjoys a special role in the formal description of the thermodynamics of simple fluids.

I.4.2 Thermodynamic Potentials

It is possible to change the formal representation of the thermodynamic theory of simple fluids in several ways. While we might consider the function $\mathcal{E}(S,V)$ as a very "natural" formulation, considering that entropy and volume are the two extensive fundamental quantities for which equations of balance exist, and energy is the concept which unites different types of processes, there exist other functions which share the property of being "fundamental" in the sense that all other information can be derived from them. Such functions are called thermodynamic potentials. We cannot come up with the same simple graphical images for these potentials that we created to understand entropy and energy. However, they still play an important role in practical computations, since the solutions of concrete problems might be much easier in one representation than in another.[28]

Enthalpy. The following function is the first of the common transformations of the basic representation in the energy form shown above. The enthalpy of a fluid is defined as

$$H = E + PV \tag{136}$$

If we insert the time rate of change of the enthalpy into the Gibbs fundamental form of Equation (131), we obtain

28. For more information on the subject of the formalism of classical thermodynamics, see Callen (1985), Chapters 5-7.

$$\dot{H} = T\dot{S} + V\dot{P} \tag{137}$$

This suggests that there exists a new fundamental form, namely, the enthalpy as a function of entropy and pressure: $\mathcal{H}(S,P)$. Now, the partial derivatives of the enthalpy function deliver the temperature and the volume, respectively:

$$\frac{\partial \mathcal{H}(S,P)}{\partial S} = T$$
$$\frac{\partial \mathcal{H}(S,P)}{\partial P} = V \tag{138}$$

Again, all pertinent information regarding a particular fluid can be derived if the functional relation $H = \mathcal{H}(S,P)$ is known. The enthalpy representation is especially useful in problems involving fluids in contact with a constant pressure environment.

Helmholtz free energy and Gibbs free energy. Two more thermodynamic potentials are in common use. The first is the (Helmholtz) free energy $F = \mathcal{F}(T,V)$, while the second is the Gibbs free energy $G = \mathcal{G}(T,P)$. They are defined by

$$F = E - TS \tag{139}$$

and

$$G = E - TS + PV \tag{140}$$

The Gibbs fundamental form becomes

$$\dot{F} = -S\dot{T} - P\dot{V} \tag{141}$$

and

$$\dot{G} = -S\dot{T} + V\dot{P} \tag{142}$$

respectively. The forms show that the free energy is useful for problems involving fluids in contact with constant temperature environments, while the Gibbs free energy might be best employed for applications where the fluid is in contact with an environment which is both at constant pressure and at constant temperature.

I.5 Caloric and Mechanical Theories of Heat

In this section, we will present the reasoning behind the mechanical theory of heat, i.e., the theory introduced by Clausius, and contrast it with the caloric theory. The comparison demonstrates that both theories lead to a theory of the thermodynamics of ideal fluids. Indeed, their logical development agrees almost up to the final steps, where one or two assumptions have to be changed if one theory of heat is exchanged for the other. This brings up the question of why Clausius chose the mechanical theory over the caloric one, and why we, today, can say without hesitation that the caloric theory must certainly be wrong. We will argue that it was a persistent prejudice as to the nature of heat which decided this turning point in the history of physics. This prejudice has to do with an "atomic" interpretation of thermal phenomena, namely the belief that heat manifests itself in the irregular motion of the particles of which matter is composed, with *motion* being measured as energy.

I.5.1 What is Heat?

Depending on whom you ask, you will get widely differing answers to this question even though the controversy of the early 19th century between followers of the caloric and the mechanical theories of heat has been decided in favor of the latter group. The fact that students, teachers, and writers of textbooks cannot agree on such a simple question today should alert us to a possibly deeper problem.

For children and nonscientists, heat is something akin to a "thermal fluid" residing in bodies, and capable of flowing from one body to another.[29] If you had asked physicists and engineers before 1840, they would have given you very similar answers, using for heat the word *caloric*. Generally, their answers would have been colorfully extended by images of "molecules of heat" and other exotic things.[30] Also, heat was supposed to be a conserved fluid. Since the 1850s it has become clear that this view must be wrong. Heat is a mode of energy transfer (an "interaction")[31], and cannot be thought of as being stored in systems. Students have to give up the images they have formed in everyday life. Today, mechanical engineers are the clearest proponents of this view.

Next, ask chemists. For them, heat still resides in bodies. It is something like internal energy or, quite often, enthalpy.[32] Never mind that in this case we do not have a word for thermal interactions, which actually leaves us impotent when it comes to formulating thermodynamics. Finally, ask modern physicists. They are teaching the mechanical interpretation of thermal phenomena. Little particles and their motions relieve us of the task of being clear about macroscopic thermodynamics. Heat may well reside in bodies, while at the same time being a form of energy transfer.[33]

---

29. Just ask small children before they have been told by teachers that heat is a kind of energy.
30. R. Fox: *The Caloric Theory of Gases.* Clarendon Press, Oxford, 1971.
31. J.B. Fenn: *Engines, Energy, and Entropy.* Freeman and Company, New York, 1982.
32. C.E. Mortimer: *Chemistry. A conceptual approach.* D. Van Nostrand Company, 1979.
33. While a recent text on introductory physics calls heat the energy of the irregular motion of atoms and molecules, its accompanying study guide calls heat a form of energy transfer. (H. Ohanian: *Physics.* W.W. Norton & Company, New York, 1985. Van E. Neie and P. Riley: *Study guide, Ohaninan's Physics.* W.W. Norton, 1985.)

Faced with such an array of interpretations, it may help a student of thermodynamics if the two main contenders of a theory of heat are briefly compared, and their differences explained. Trying to find the source of the differences should take care of many unanswered questions.

I.5.2 A General Theory of Calorimetry

What is heat? This question has had two distinct and so far mutually exclusive answers. Maybe, heat is a subtle "fluid" which can be stored in bodies and which can flow. This is the caloric of the caloric theory of heat.[34] Or perhaps heat is a quantity analogous to work in mechanical processes, and heat and work are interconvertible. This is the heat of the mechanical theory of heat.

There is a common basis on which both theories of heat can be discussed and compared—the theory of calorimetry for ideal uniform fluids, which was presented in Section I.2 with the caloric theory in mind. Now its main points will be rewritten in a neutral form which leaves open the question of what we mean by "heat." Its reduction to either the caloric or the mechanical theory of heat will be performed in the following section.

The bodies used in this theory are those which obey a thermal equation of state relating pressure, volume, and temperature:

$$P = P(V,T) > 0 \tag{143}$$

and for which two further constitutive quantities exist, namely the latent and specific heats (A_V and B_V):

$$\begin{aligned} A_V &= A_V(V,T) > 0 \\ B_V &= B_V(V,T) > 0 \end{aligned} \tag{144}$$

The index V refers to the latent heat with respect to volume, and the specific heat at constant volume. The first inequality in Equation (144) excludes water in the range of temperatures from 0°C to 4°C from our considerations. These assumptions mean that the processes undergone by the fluid bodies can be fully described by the values of two variables, namely volume and temperature.

The latent and specific heats are related to the heating, i.e., to a current of heat I_H:[35]

34. Note that we do not include in the general properties of caloric its conservation or nonconservation. Historically, however, caloric was assumed to be a conserved quantity.

35. Using the symbol *dH/dt* for the heating leaves room for misunderstandings. Is it the rate of flow of heat across the surface of the body, or is it the rate of change of the heat content? In the mechanical theory of heat, there is no "heat content." There, heating may mean only the rate of flow of heat. In the caloric theory of heat, on the other hand, there is a heat content, but heat might not be conserved.

$$I_H = -A_V \dot{V} - B_V \dot{T} \tag{145}$$

The amount of heat exchanged as a result of a thermal process can be calculated as follows:

$$H = \int_{t_1}^{t_2} I_H dt \tag{146}$$

A few consequences of the theory of calorimetry are particularly interesting. First, the bodies described by this theory can undergo only reversible changes (the reasoning is independent of the particular form of the theory of heat; therefore, the results obtained in Section I.2 apply). Second, the latent and specific heats with respect to volume and pressure are related by

$$A_p = A_V \left(\frac{\partial P}{\partial T} \right)^{-1} \tag{147}$$

$$B_p = B_V - A_V \left(\frac{\partial P}{\partial T} \right) A_V \left(\frac{\partial P}{\partial V} \right)^{-1} \tag{148}$$

Finally, we can derive the Poisson-Laplace Law of adiabatic change ($I_H = 0$), which holds for the ideal gas, with constant ratio of the specific heats:

$$P \cdot V^{\gamma} = constant \quad , \quad \gamma = B_p / B_V > 1 \tag{149}$$

The inequality follows from Lapace's explanation of the speed of sound. This speed is always higher than that calculated for isothermal oscillations of the gas. The observation that the ratio of the specific heats must be constant will prove to be crucial when we determine the motive power of heat. Note that nothing has been said about what heat "really" is: the results obtained up to this point are independent of the special form of the theory of heat.

I.5.3 Thermodynamics and the Mechanical Theory of Heat

We already have seen how thermodynamics is derived on the basis of the caloric theory of heat (Section I.3). Two assumptions are necessary to complete our task. First, we take for granted the existence of a heat function; second, we use the observation about the ratio of the heat capacities.

Thermodynamics based on the mechanical theory of heat. We will now talk about a different kind of heat. We shall write I_Q for the flux of heat, and L and C for the latent and specific heats, respectively. Equation (145) becomes

$$I_Q = -L_V \dot{V} - C_V \dot{T} \tag{150}$$

Again, we need some additional assumptions on which to build thermodynamics. The first replaces the caloric theory of heat. In the mechanical theory, heat and work are supposed to be uniformly interconvertible in cyclic processes:

$$\begin{aligned} W &= JQ \\ &= J\left(Q^{+} - Q^{-}\right) \end{aligned} \tag{151}$$

where J is the mechanical equivalent of a unit of heat. This contradicts Carnot's central assumption expressed in Equation (84). In the mechanical theory of heat, a certain amount of heat is "consumed" in the operation of a heat engine. Note that relations equivalent to Equation (37) do not exist here. Heat may not be thought of residing in bodies: there is no heat function. It is clear that in this theory heat is analogous to work in mechanical processes; we might call heat *thermal work.*

The second assumption concerns the heat capacities; one of the specific heats of the ideal gas must be constant. The other one then turns out to be constant as well due to what we know about the propagation of sound in the ideal gas:

$$\begin{aligned} C_V &= constant \\ C_p &= constant \end{aligned} \tag{152}$$

Again we assume Carnot's Axiom to be valid; now, however, we may not use Carnot's Special Axiom because it depends upon the caloric theory of heat. We finally can determine the motive power of heat.[36] Also, we can construct the internal energy, which satisfies:

$$\dot{E} = -J I_Q - P\dot{V} \tag{153}$$

Comparison of the caloric and the mechanical theories of heat. The rival theories of heat do not have to be mutually exclusive, if we accept that *heat* means different things for them. The two versions of thermodynamics are compatible if

$$I_Q = T I_s \tag{154}$$

and

$$\begin{aligned} L_V &= T \Lambda_V \\ C_V &= T K_V \end{aligned} \tag{155}$$

if we set $J = 1$, as is usual today. We conclude that thermodynamics based on the caloric theory of heat is equivalent to thermodynamics based on the mechanical theory if caloric is the entropy of the mechanical theory, or if the heat of the mechanical theory is thermal work. The latent and specific heats of the caloric theory are the latent entropy and the entropy capacity, respectively.

---

36. C. Truesdell (1980), Chapter 8.

I.5.4 The Matter of the Heat Capacities

The caloric theory of heat works. So why did it fail? There is a good reason why heat should not be caloric: our practice of calorimetry can be used to decide between the caloric and the mechanical theories of heat. However, for all its clarity and simplicity, this reason fails to be conclusive. Moreover, as we shall see in the following paragraphs, this most likely was not the point which drove the last nail into the coffin of the caloric theory.

Measuring amounts of heat. Carnot expressed what seems to have been taken for granted all the time, namely that calorimeters measure heat (be it caloric or an energy form): "I regard it as useless to explain here what is quantity of caloric … or to describe how to measure these quantities by the calorimeter."[37]

You will remember that to base thermodynamics on the mechanical theory of heat we had to assume that the specific heats of the ideal gas are constant. Modern calorimetric measurements indeed show this to be the case. At Clausius' time, however, the matter had not been resolved experimentally with sufficient accuracy. Therefore, a prejudice played a major motivating role—the belief in the "motion of the least particles." The simplest kinetic models suggest that the specific heat at constant volume should be constant for the ideal gas. Clausius' thermodynamics could then be constructed.

But why should this have excluded the caloric theory of heat? If the caloric theory is assumed to be valid, we have demonstrated in Equation (79) that

$$K_P - K_V + \rho \frac{\partial K_V}{\partial \rho} + P \frac{\partial K_P}{\partial \rho} = 0$$

for the ideal gas, where ρ is the density of the fluid. This proves that it is impossible for both specific heats of the ideal gas to be constant if we accept the caloric theory of heat. Therefore, if we believe that calorimetric experiments measure caloric, the caloric theory is dead.

However, in the usual calorimetric experiments, caloric is generated. Therefore, the measurements cannot be used for our purpose. All we can say is that

1. If caloric is conserved, the specific heats of water and of the ideal gas are constant.
2. If caloric is not conserved in the exchange of heat which is taking place in the calorimeter, we cannot conclude that the specific heat of water (and of the ideal gas) is constant. Rather, if caloric is generated in the irreversible process, the (average) specific heat is smaller at higher temperatures.

Indeed, the heat capacities of the ideal gas based on the caloric theory turn out to be inversely proportional to the gas temperature; see Equations (125) and (126). The power of the argument against the caloric theory given in this section

37. S. Carnot: quoted from C. Truesdell (1980), p. 81.

rests on an assumption which need not be made, namely that calorimetric measurements directly measure caloric.

What did early experiments on heat prove? The logical chain of reasoning against the caloric theory as presented above is a modern result.[38] It was not really used at the time of Clausius and Kelvin. We usually accept that the series of experiments and investigations on heat conducted by researchers like Rumford, Davy, Mayer, and Joule demonstrated beyond any doubt that heat is energy or an energy form. The following argument shows, however, that these experiments would not have sufficed on the basis of logic alone. They only prepared the ground for already prejudiced scientists for whom heat had to be some kind of "motion." In the end, history chose the mechanical theory of heat because researchers would not give up the idea of the conservation of caloric; and because they were already prejudiced towards the mechanical theory due to their desire to explain thermal phenomena in purely mechanical terms.

Rumford believed to have proved in his famous cannon boring experiments that the heat developed did not come from the iron which had been ground. He determined that the specific heat of the original material was equal to that of the material that had developed so much heat. However, since the specific heat is not the heat content, his experiments did not prove the point he tried to make. He therefore could not prove that heat (caloric) was not conserved, as had been assumed in all earlier investigations. Davy's experiment of melting two blocks of ice by rubbing them was more conclusive. The water contains more heat than the original ice, which means that heat (caloric) has to be created by friction. The logical conclusion from this is that caloric is not conserved, not that the caloric theory is wrong.

Today we accept the notion that Joule's experiments finally proved the caloric theory wrong. However, they did not do anything of the kind.[39] First of all, the values reported for the mechanical equivalent of a unit of heat varied wildly in the series of experiments conducted before the creation of thermodynamics on the basis of the mechanical theory of heat. Second, all but one of the experiments had to do with irreversible processes which had no bearing on the theory of thermodynamics of reversible processes being created then. Finally, since the range of temperatures employed was much too small, the investigations could not contribute to the problem of the nature of heat. It has been known since Carnot's time that heat at one temperature is not the same as heat at another temperature. If experiments were to contribute to the determination of the motive power of heat (and thereby answer the question as to the nature of heat), they would have had to cover a large range of temperatures. Joule's investigations were crucial for supporting the concept which was to open up modern physics: the conservation of energy. They did not answer the question in which we are interested.

---

38. C. Truesdell (1980), Chapter 3.
39. C. Truesdell (1980), Chapter 7.

Also, in the matter of the specific heats of the ideal gas which could have decided between the two rival theories, measurements were too crude up to the time of Clausius and Kelvin. All in all, experiments did not substantially add to the progress of thermodynamics.

Prejudice. As long as experiments did not prove the point, any notion as to the nature of heat would have to be called a prejudice. Certainly, it was hard to think of a "thermal fluid" which could be created but not destroyed. Still, one could not offer more than belief that this should not be so. Here, another belief stepped in: "To this it must be added that other facts have lately become known which support the view that heat is not a substance, but consists in a motion of the least parts of bodies." These are Clausius' words.[40] Since the idea expressed in this sentence did not play the least logical role in the development of thermodynamics at the hands of Clausius,[41] we cannot call it more than a motivating prejudice, a prejudice which probably was necessary to build up the courage to dismiss the caloric theory of heat, Equation (84), and accept Equation (151). The one reason which could have led to a dismissal of the caloric theory without recourse to further experimentation—namely the matter of the specific heats of the ideal gas, combined with accepting that calorimeters measure amounts of heat—was not seen clearly enough.

Questions and Problems

1. Derive the equations for two equivalent processes for which the relationship between temperature and volume is linear. The second process is to traverse the path of the first in half the time. Then calculate the reverse of the first of the processes. The values of temperature and volume at the beginning and at the end of each of the processes are (T_1, V_1) and (T_2, V_2), respectively.
2. Derive the differential equation for the path which represents an isobaric process in the T-V diagram. Calculate the isobars for the ideal gas.
3. A capacitor made up of two (large) parallel plates with variable separation is a simple electromechanical device. Choose independent variables appropriate to the problem and write the "equation of state" for this system analogous to a thermal equation of state. Calculate the curves which correspond to isobars in the thermal case.
4. Check whether the constitutive inequalities (3), (11), and (12), and Equation (13) hold for the ideal gas.
5. Liquids have very small compressibilities. Therefore, their pressures must increase considerably if their temperatures are increased at constant volume. Calculate the

40. R. Clausius: quoted from C. Truesdell (1980), p. 187.
41. Indeed, it is wrong in the mechanical theory of heat, since there heat may not be thought of as residing in bodies. Clausius' results contradict the very prejudice they were built upon.

increase of pressure per 1 K for water at 20°C, and give an approximate formula for the change of pressure with temperature.

6. For the parallel plate capacitor, introduce a quantity which is equivalent to the isothermal compressibility of a fluid. What does *isothermal* mean in this context?

7. Carnot (1824, pp. 29–32) made the following two remarks about thermal processes of a gas. (a) "When a gaseous fluid is rapidly compressed, its temperature rises; on the contrary, its temperature falls when it is rapidly expanded. This is one of the facts best confirmed by experiment." (b) "If, when a gas has been brought to a higher temperature as an effect of compression, we wish to bring it back to its original temperature without causing its volume to change further, we must withdraw some caloric from it." Derive the constitutive inequalities from these observations.

8. For the parallel plate capacitor with variable separation, introduce and derive quantities which are analogous to the latent heat and the heat capacity. Calculate the equation which holds in processes for which no charge is exchanged. These processes are comparable to those of a gas without the exchange of heat (adiabatic processes).

9. Why is a theory of heat sufficient for deriving the relations pertaining to adiabatic processes? Why is knowledge of the relation between heat and energy not necessary? Put differently, why don't we have to know what heat "really" is?

10. We can define a general compressibility as follows:

$$\kappa = -\frac{1}{V}\frac{dV}{dP}$$

Demonstrate that the isothermal and the adiabatic compressibilities follow from this equation by specializing dV/dP to either isothermal or adiabatic processes.

11. According to the derivation in the Prologue, the momentum inductance per length of a column of fluid is given by

$$L_p^* = \left[A\rho\frac{dP}{d\rho}\right]^{-1}$$

Derive the relationship between the inductance and the compressibility.

12. The speed of sound in air has been measured for a constant pressure of 101.3 kPa, and for several temperatures (Table 1). Determine whether the square of the speed of sound is proportional to the temperature of the air, and calculate the value of the ratio of the specific heats. Would the data support the assumption that the speed of sound is proportional to the temperature?

TABLE 1. Speed of sound in air for different temperatures

T / K	c / ms^{-1}	c^2	γ
233	307	94249	1.41
253	319	101760	1.40
273	332	110220	1.41
293	344	118340	1.41
313	355	126030	1.40

13. Derive the relationship between the speed of propagation of sound in a fluid and its compressibility. The speed of sound in water at a temperature of 20°C is 1483 m/s. What is the adiabatic compressibility of water?

14. Show that according to the results of this section the heat content of the ideal gas remains constant during adiabatic compression and expansion.

15. Show that the specific heat at constant volume is a function of temperature only, if it is calculated for adiabatic changes.

16. Prove that the heat function of the ideal gas should take the form

$$\mathcal{S}^{\#}(P,\rho) = \gamma C_V \ln\left(\frac{P^{1/\gamma}}{\rho}\right)$$

if the heat capacity is a function of temperature only.

17. A heat engine using 20 moles of an ideal gas undergoes a Carnot cycle between the temperatures 80°C and 300°C. The volume at the beginning of the isothermal expansion is 0.2 m^3; at the end of this step it has increased to 0.4 m^3. Determine the other corners of the cycle (Figure 5).

18. A Carnot-type cycle is not possible only in thermodynamics. We can let a parallel plate capacitor undergo a cycle which has all the features of the process proposed by Carnot.[42] Consider a capacitor made of two large parallel plates which are separated by a small distance. The surface area of the capacitor plates is assumed to be 1 m^2. Here, voltage, separation, and force between the plates correspond to temperature, volume, and pressure, respectively. There exists a relationship between force, voltage, and separation which is equivalent to an equation of state. a) Describe the four steps which are analogous to the four steps in the Carnot cycle of a heat engine. Specify the direction of the cycle which is to deliver energy for mechanical purposes. b) The engine operates between two batteries which have voltages of 100 V and 20 V, respectively. With the first battery hooked up to the capacitor, the separation of the plates is changed from 0.2 m to 0.1 m. Calculate the remaining two separations corresponding to the other two corners of the cycle. c) Calculate the amount of charge absorbed, and the amount of charge emitted, for the four steps.

19. A gas which satisfies the equation of state of the ideal gas undergoes two different processes which lead from the same initial state to the same final state. Process 1 takes the gas at constant pressure from volume V_1 to volume $V_2 > V_1$. Process 2 consists of two steps: isothermal expansion at T_1 from V_1 to V_2, and isochoric heating at V_2 from temperature T_1 to T_2 (Figure 9). Calculate the energy exchanged in the mechanical processes.

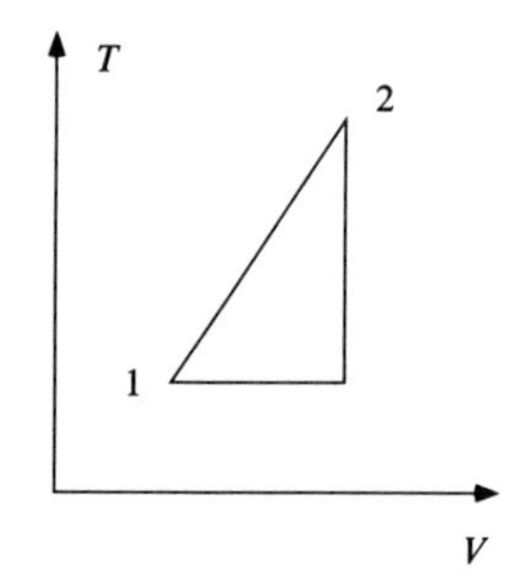

FIGURE 9. Problem 19.

20. If the heat exchanged in a process is to be independent of the path, the constitutive relations of the body undergoing the process *may not be arbitrary.* Calculate the heat exchanged for the processes of Problem 19 using the following two sets of constitutive relations:
(a) $\Lambda_V = a/V$, $K_V = b/T$, where a and b are both constant; (b) $\Lambda_V = aT/V$, $K_V = b$.

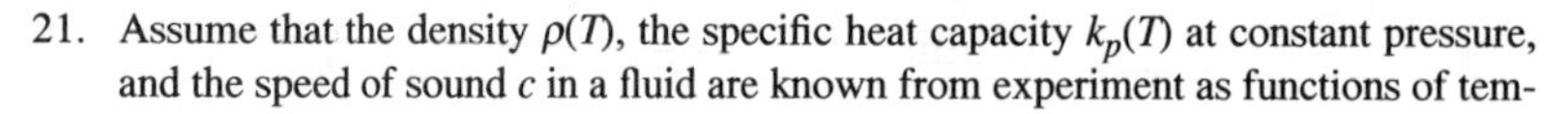

21. Assume that the density $\rho(T)$, the specific heat capacity $k_p(T)$ at constant pressure, and the speed of sound c in a fluid are known from experiment as functions of tem-

42. See Fuchs (1986).

perature for a given pressure. Determine from these values the ratio of the heat capacities γ, the (isothermal) compressibility κ_T, and the latent heat with respect to volume.

22. Show how to obtain the temperature and the pressure of a simple fluid as derivatives of the energy function.

23. Derive the energy of the ideal gas as a function of S and V. Show that you can obtain both the entropy as a function of temperature and volume, and the equation of state of the ideal gas from this information.

24. In Section I.1.2, the compressibility and the thermal coefficient of expansion were defined. Prove the following relationship:

$$K_p - K_V = V\frac{\alpha^2}{\kappa_T}$$

Transform this relation to show that the specific temperature coefficients of enthalpy and of energy are related by

$$c_p = c_V\left(1+\gamma^* \alpha_V T\right)$$

where

$$\gamma^* = \frac{\alpha_V}{\rho c_V \kappa_T} = \frac{\alpha_V}{\rho c_p \kappa_s}$$

is called the Grüneisen ratio. (α_V and κ_T are the temperature coefficient of volume and the isothermal compressibility, respectively.) Show that for the ideal gas the Grüneisen ratio is

$$\gamma^* = \frac{c_p}{c_V} - 1 = \gamma - 1$$

Hint: Use the relation between the different coefficients defined according to Section I.1.2 and remember the Carnot-Clapeyron law derived in Section I.3.4.

25. Show that the law of adiabatic change of an ideal fluid written with temperature and density as the independent variables is given by

$$\left.\frac{dT}{d\rho}\right|_{ad} = \frac{T}{\rho}\gamma^*$$

where γ^* is the Grüneisen ratio defined in Problem 24.

26. For convective motion in the mantle of the Earth it is found that the Grüneisen ratio is not constant, but rather varies inversely as the density of the material.[43] Show that in this case the solution of the differential equation of adiabatic motion can be written as follows:

43. See Stacey(1992), p. 305 and appendixes therein.

$$\frac{T_1}{T_2} = \exp\left[\gamma_1 * \left(\frac{\rho_1}{\rho_2} - 1\right)\right]$$

27. Show that the adiabatic temperature gradient of a fluid in hydrostatic equilibrium can be expressed by

$$\left.\frac{dT}{dz}\right|_{ad} = -\frac{T\alpha_V g}{c_p} \quad \text{or} \quad \left.\frac{dT}{dz}\right|_{ad} = -\gamma * T\rho g \kappa_s$$

28. Why should the prejudice that the "heat" of an ideal gas is the energy of the motion of its particles lead to the prediction of constant "heat capacity" of a monatomic gas? Why should we reject the caloric theory of heat if the "heat capacity" of the ideal gas were indeed constant? Why does the notion of the motion of the least particles being the "heat" of the gas contradict the very theory (the mechanical theory of heat) which was proposed in place of the caloric theory?

CHAPTER 3

The Transport of Heat

It is almost always emphasized that thermodynamics is concerned with reversible processes and equilibrium states, and that it can have nothing to do with irreversible processes … . But this verbalism gets nowhere; physics is not thereby absolved from dealing with irreversible processes.

P.W. Bridgman, 1953

In this chapter simple aspects of the transport of heat will be introduced, extending the treatment of thermal processes into the realm of phenomena which are missing from the theory of the thermodynamics of ideal fluids. Many texts on thermodynamics and on heat transfer sharply distinguish between the two subjects, which only emphasizes that a unified presentation of all thermal phenomena is called for. While we will not achieve the stated goal in this chapter, the ground will be prepared for a theory of continuum thermodynamics of which we will get a first glimpse in the Epilogue.

The first section of this chapter provides a qualitative description of the three types of heat transport: conduction, convection, and radiation. It introduces the formulation of the law of balance of entropy for an entire body. Then, simple applications of all three forms of heat transfer are discussed, giving an overview of some practical problems, and irreversible processes involved with heat transfer are treated in Section 3.3.

The flow of heat leads to spatially inhomogeneous situations which call for a description of natural processes in terms of fields. The following treatment of conduction and of radiation serves as a first introduction to a field theory of thermodynamics. Conduction is an irreversible process which forces us to provide a constitutive law for the rate of production of entropy in addition to an expression for the fluxes of entropy and energy. Sources of entropy and energy are discussed, and the combined action of supply and conduction are treated in a simple setting (Section 3.4). The transport of entropy through the radiation field, and the interaction of bodies and fields, including the irreversibility of emission and absorption of radiation, provide the backbone of the description of radiative transfer (Section 3.5).

While conduction and radiation are discussed at a level beyond that of the simple examples of Section 3.2, additional topics in convection are left to Chapter 4.

3.1 Transport Processes and the Balance of Entropy

In this section we will describe qualitatively the basic phenomena underlying the transport of heat. Simple observations tell us that entropy can flow in three different ways: *conduction*, *convection*, and *radiation*. Consideration of these types of transport will lead to the formulation of the law of balance of entropy in a more general form than previously encountered, and will yield a better understanding of the role of hotness in thermal processes. In the end, the equation of balance of entropy will contain terms describing the different modes of transport.

These types of transfer processes are found not only in thermal physics, but in other fields of the natural sciences as well. Momentum transports have been classified in the same manner in the Prologue. For this reason alone, it is important to have a clear understanding of the nature of heat transfer.

3.1.1 The Conductive Transport of Entropy

Heat one end of a piece of metal over a flame; in a very short time the other end will feel hot as well. If you throw a hot stone in cold water, it will cool down while the water gets warmer. In a heat exchanger a hot fluid might be flowing through pipes, heating a cooler fluid which flows around the pipes. In all of these examples entropy is removed from some bodies and added to others. Why else should some objects become colder while others heat up? The possibility of changing the temperature by compression, i.e., adiabatic processes, does not occur in these examples. Therefore we say that entropy has been transferred. Obviously, entropy flows from hotter to colder bodies.

How is entropy transported in these examples, and what are possible conditions for this process to occur? First, we observe that material transport cannot be involved. A piece of metal heated at one end retains its integrity. A hot stone does not dissolve in water, thereby spreading the entropy it contains. In the case of the heat exchanger, it is true that the fluids move; however, entropy must be transferred through the walls of the pipes. Heat therefore flows *through* bodies *without the help of a body transporting it*, and it flows from one body to another if the two are brought in *direct contact*. These are examples of heat conduction.

An example which we studied in Chapter 1 tells us something about the role of temperature in the conductive transport of entropy. Two bodies which have different temperatures are brought in thermal contact, and their hotnesses are monitored. It is found that the temperatures of the bodies change until they have become equal. As long as they are changing, entropy must be flowing: one of the bodies is cooled, the other is heated. In the end, however, the exchange stops. We conclude that entropy flows conductively as long as there is a difference of temperatures between the bodies which exchange heat, and that by itself, entropy flows only from hotter to colder objects.

Driving forces. This type of behavior is well known from a number of different physical phenomena. Connect two containers having different cross sections which are filled with water up to different levels; let the water flow between them. As a different example, connect two electrically charged spheres with a wire and monitor the electrical potential of each of the spheres. We know what will happen in both cases: the water levels in the containers will reach the same height, and the electric potentials of the two spheres will be the same after the process ends. In each case, something flows as long as there is a difference of potentials, i.e., a *driving force*. In analogy to these well-known phenomena, we shall interpret the conductive transport of entropy as follows:

> *In conductive transport, entropy flows by itself through bodies from points of higher to points of lower temperature. In other words, entropy flows as long as there is a difference of temperatures, i.e., a thermal driving force.*

The balance of entropy. The conductive transport of heat is a prime example of an irreversible process. (See Section 1.5). A body conducting entropy produces more entropy at the same time. This must be so, since in a steady-state process, the same amount of energy which enters the body at high temperature leaves it at a lower thermal level. Therefore, the current of entropy leaving the body must be larger than the one entering. Clearly then, the equation of continuity of entropy must include the production term first introduced in Section 1.6 in addition to the term describing conductive transfer of heat into and out of the body:

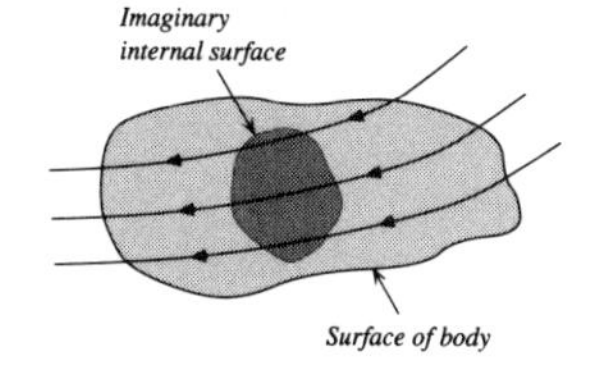

FIGURE 1. Conductive currents of entropy cross surfaces. Such surfaces may be real surfaces of bodies or imaginary surfaces, such as those which we introduce to separate different parts of bodies. The flow lines in this figure do not reflect the fact that entropy is produced in conduction.

$$\dot{S} + I_{s,cond} = \Pi_s \tag{1}$$

Here, $I_{s,cond}$ is the net current of entropy transported conductively with respect to the body in question. We call it the *conductive flux of entropy.* The balance of entropy in conduction will be discussed in much more detail in Section 3.4. There, our task will be to determine the conductive currents and the production rate of entropy.

Flow across surfaces. The conductive current is our way of describing a phenomenon in which we picture entropy to flow *across the surfaces of bodies.* If we are interested in the flow through a body we simply introduce imaginary surfaces inside. Again entropy flows across a surface where one part of a body touches another (Figure 1). In this sense, conduction is a surface phenomenon, and it is rendered formal by a physical quantity, namely a flux I_s, whose distribution over a surface is of prime interest (Section 3.4). We stress this point since a body can pick up or lose entropy in other ways (i.e., by radiation and sources of heat).

3.1.2 The Transport of Entropy with Moving Bodies: Convection

Northern Europe would be a pretty cold place to live were it not for the Gulf Stream, which transports huge amounts of heat from the Gulf of Mexico to the west coast of Europe. And our weather would be pretty dull were it not for the currents of hot or cold air in our atmosphere. These are just two important examples of a different mode of heat transfer. It is quite clear that in these cases entropy is transferred with the help of a material medium, like air or water. You can find examples all around you. Heated air rises from a radiator in a room; hot water which is pumped through the pipes of a central heating system delivers entropy to the radiators; water begins to boil at roughly 100°C, transporting entropy via a material current. If entropy is carried by a material which is flowing we speak of *convective entropy transport.*

The examples demonstrate that convection is a very important phenomenon in our daily lives. We shall deal with some simple aspects of convection in Section 3.2. (More details will be provided in Chapter 4.) We are interested in a particular question at this point, the problem of the *driving force* of this type of heat transport.

The driving force. A difference of temperatures drives the conduction of entropy. You can easily see that this cannot be the driving force in the case of convection. The reason why hot water flows through pipes to your shower definitely cannot be found in a difference of temperatures: a pump drives the flow of water. The fact that the water is hot is immaterial to this transport phenomenon. We have to conclude that the cause of convective heat flow has to be sought in the driving force which lets the material substance (water, air, etc.) move: we know that this is a *pressure difference* set up by a pump or through some other device or process:

> *Entropy can be transported via a flowing substance. In this case the flow of entropy is accidental. The driving force of the process is the difference of pressure which lets the material substance flow.*

There are some important examples of convection which might make us believe that a temperature difference must be the driving force of the process. Think of air rising above a hot radiator in a room. Also, the water circulating in a central heating system does so apparently because it is heated at one end (in the boiler). Indeed, the water does not flow if the heating is stopped.

Still, the immediate driving force for the flow of water (which is responsible for the transport of entropy in the system) is not a difference of temperatures but a *pressure difference caused by the heating*. The hot water in the boiler is slightly less dense than the surrounding liquid; therefore, it begins to rise as a consequence of buoyancy, which is a consequence of a pressure difference. The heating is responsible for the flow only in an indirect way. Another example is presented by our atmosphere: air can easily flow into a region where the temperature is higher. We call this phenomenon *free* or *natural convection* to distinguish it from convection induced by a pump, which is called *forced convection.*

The balance of entropy. Since convection and conduction are obviously different types of entropy transport, we should distinguish between them. For this reason we also introduce *convective currents* in the equation of continuity of entropy:

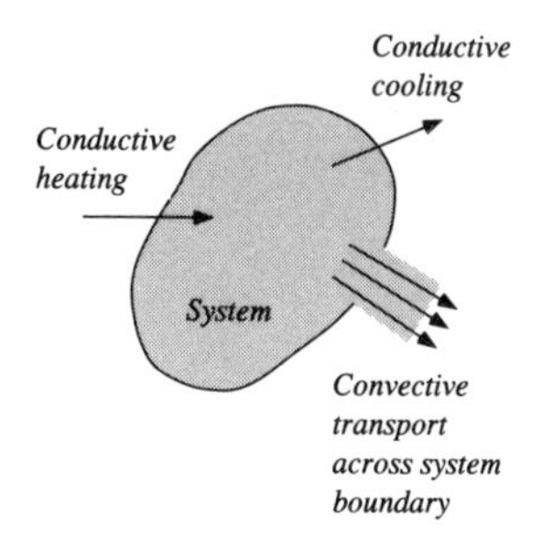

FIGURE 2. Entropy may flow across the boundary of a body either by conduction or by convection. In the former case, matter does not cross the surface and entropy flows through matter. In convection, a substance flows across the surface whereby entropy it contains is transported into or out of the system as well.

$$\dot{S} + I_{s,cond} + I_{s,conv} = \Pi_s \tag{2}$$

This equation tells us that the entropy content of a body can change as a consequence of two types of flow and the production of entropy (Figure 2). Again, we are confronted with a surface phenomenon. Substance flows into and out of regions of space across surfaces, real or imagined. Just as in the case of conduction, we introduce fluxes to describe mathematically what is going on.

There is an important point to note. The transport of substance leads to changes of this quantity in regions of space influenced by the flow. Therefore we have to be extremely careful to state what we are talking about, i.e., to identify the system for which we are performing a balance of entropy. So far, we have always used an identifiable material body as the physical system under consideration (Chapters 1 and 2). Such a body is assumed to retain its material integrity; i.e., it is not allowed to exchange matter with the surroundings. We shall continue to use the term *body* in this sense—an aggregate of matter which can always be identified and separated from the rest of the world. For a body such as a stone this identification is quite simple and clear. It is still simple in the case of air enclosed by rigid walls. In situations where matter flows, however, this becomes more difficult. Still, we may think of an identifiable amount of water moving with the flow in a river. The body of water is thought to be separated from the rest of the water by an imaginary surface which moves and deforms with the body (Figure 3). If we have a body in mind, the balancing of quantities such as entropy always refers to this piece of matter. The time derivative of the entropy function (or of other functions) in Equation (1) is taken with respect to *the entropy of the body*. For this reason it is sometimes called a *material derivative.*

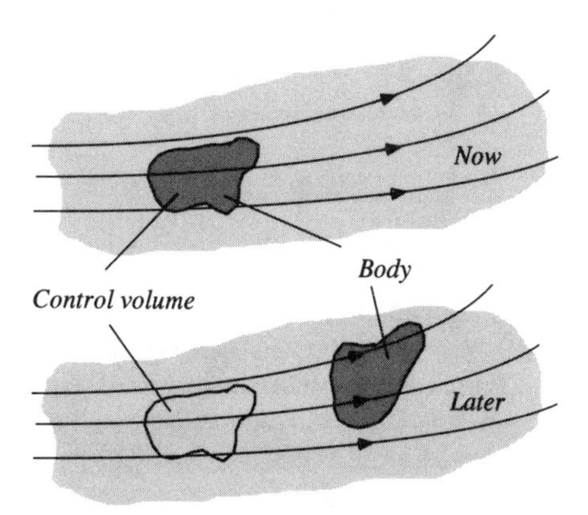

FIGURE 3. A body and a control volume in a general situation involving the flow of substance. A body moves and deforms with the flow. A control volume is any region of space, itself moving or stationary. In general, a control volume is penetrated by convective currents. In this example, the body and control volume occupy the same region of space initially.

Control volumes. Often it is more convenient to do the accounting with respect to a region of space rather than an identifiable body (Figure 3). This is particularly true in cases where matter flows. Imagine a region of space surrounded by an imaginary surface. We often speak of a *control volume* and a *control surface* to distinguish it from bodies. A control surface may easily be penetrated by flows of matter, which leads to changes of the amount of substance in the control volume. This is the case if we consider convective currents, as we have done above. The time derivative of the entropy in Equation (2) is not taken with respect to a body, but with respect to some control volume (which may be stationary or moving). The derivative therefore is of *the entropy of the control volume*. We will learn later how to distinguish mathematically between this derivative and a material one. By the way, systems which may exchange matter are called *open*, while those which do not are called *closed*. Bodies are closed systems by definition.

EXAMPLE 1. Conductive and convective fluxes of entropy.

Consider hot water flowing through a metal pipe as in the figure below. a) Consider the interior of a part of the pipe as the system, and assume this part not to move or deform. Account for all fluxes of entropy penetrating the surface of this control volume. b) Consider the water which is in the control volume at a particular instant to be the system. Follow this body of water in its motion and repeat problem (a).

SOLUTION: a) We are dealing with a stationary control volume through which water and entropy are flowing. (See the upper figure; water is flowing from left to right.) We have to find the currents flowing through the control surface and determine the fluxes associated with them.

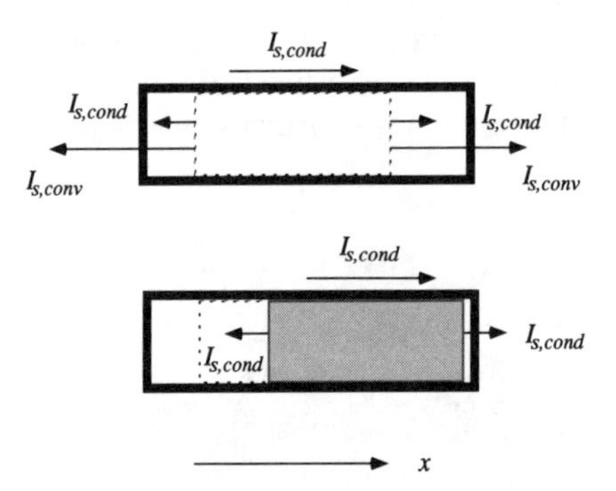

First of all, entropy must be flowing radially outward through the pipe if the surroundings are cooler than the water. This means that we have a conductive current of entropy penetrating the cylindrical surface. Since the flow is outward, its flux is positive (arrow in the positive direction).

Second, because of the loss of heat through the walls of the pipe, the water entering the control volume will be warmer than the water leaving. We have a thermal driving force in the direction parallel to the axis of the pipe. At the control surface, there must be conductive currents of entropy through the water in its direction of flow. Therefore, we have a negative flux associated with the conductive current at the entrance to the control volume, and a positive flux due to the current leaving the system.

Finally, two convective fluxes are associated with the flow of water into and out of the control volume. Entropy stored in the water is carried across the surface of the system.

Again the flux is negative at the inlet, and positive at the opposite side. The flux at the entrance is larger in magnitude than the one at the outlet.

b) If we follow a certain body of water in its motion, the system wall moves with it (lower figure). No water flows across the surface of the body, which is represented by the shaded area. This means that there are no convective currents of entropy to be considered. The conductive currents still exist, and they are the same as the ones identified in (a).

3.1.3 Transport of Entropy with Radiation

In some cases it is obvious that entropy is transported neither by conduction nor by convection. Take the heat of the Sun, which travels to us through empty space, covering a distance of 150 million kilometers. It is clear that the Sun must radiate heat since it produces vast amounts of entropy all the time without changing noticeably. The transport cannot be via conduction. Also, there is no material substance which can act as a carrier of entropy in a convective process.

Heat which is emitted by warm bodies can even be photographed. You can see objects on infrared films. They look unfamiliar, but the process must be similar to photography with normal light. This suggests that there is a medium which transports heat in these cases after all. This medium would be similar to light. Indeed, this is the accepted picture: electromagnetic radiation (X-rays, ultraviolet, visible, infrared, or radiofrequency) carries heat. Hot bodies emit electromagnetic radiation which then transports heat.

One group of phenomena is so pervasive that it makes us think that some bodies must radiate heat (entropy). (On closer inspection, however, you may realize that these phenomena are not the kind of proof we are looking for.) You can sit behind a glass window and feel the heat of the Sun's radiation. You can observe the same phenomenon when you sit at a fire; while all the heated air might go up the chimney, you still get hot sitting there. Meals can be kept hot by lamps, and again conduction or convection are not responsible for the flow of heat. The problem with these cases is that the heat which is felt by the bodies absorbing radiation may be produced inside them. Indeed, in the case of solar radiation, almost all the entropy which appears in a body absorbing the Sun's rays is created (see Section 3.5).

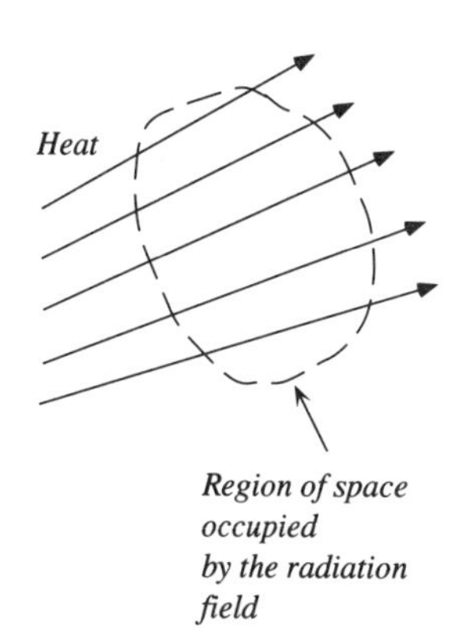

FIGURE 4. Imagine an empty region of space between the Sun and the Earth. Entropy is transported with radiation through this control volume. There is no difference between temperatures across the region.

Transport through the radiation field. We call this type of transport the *radiation of heat*. It is interpreted as the flow of entropy (and other quantities) through a physical system different from normal bodies, namely the *electromagnetic field*. Nevertheless, this field has some features in common with bodies. It can store and transport such quantities as entropy, momentum, and energy, just like ordinary bodies. Therefore we can write down an equation of balance of entropy for the electromagnetic field in an otherwise empty control volume:

$$\dot{S}_{field} + I_{s,rad} = 0 \quad \textbf{(3)}$$

The flow of entropy through the field is a surface phenomenon with currents flowing across imaginary surfaces drawn around regions of space. The amount of entropy in a region of space occupied by a radiation field changes as a consequence of the transport of entropy together with radiation into and out of the region (Figure 4). The flow of heat through empty space is not dissipative. As a simple example, consider two imaginary spheres drawn concentrically around the Sun, the first near its surface, the second much further out. Later in this chapter we will learn how to compute the flux of entropy through surfaces cutting through the radiation field. We will find that the same amount of entropy flows through both spheres in the same time span. Therefore the rate of production of entropy for a region of space which contains only the radiation field is zero.

It is interesting to ask whether we need a difference of temperatures for entropy to flow radiatively through the electromagnetic field. In fact this is not the case. We associate the same temperature with the radiation which has just left the Sun and with the radiation that arrives at the Earth. In this sense, radiative transfer of entropy has much in common with convective transport. The driving force for the transport, if one is needed at all, is not the thermal driving force responsible for conduction. This is of profound importance for the determination of the relationship between fluxes of entropy and of energy (see Sections 3.2.4 and 3.5). Put simply, there is a great difference between entropy flowing by itself in conductive transport, and entropy being carried by something else, be it water or radiation.

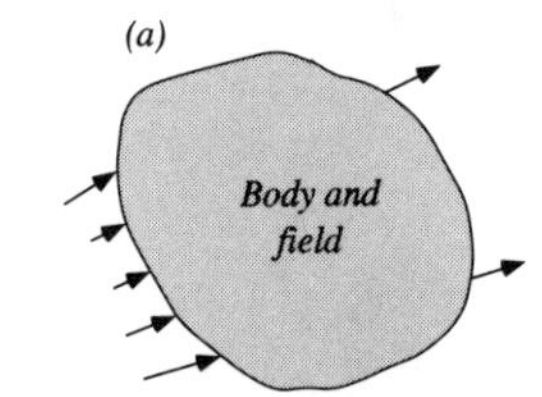

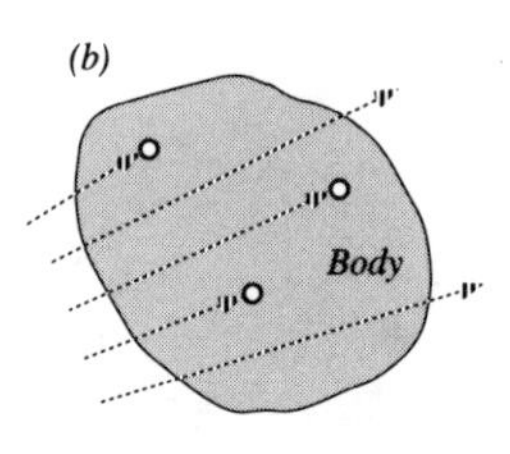

FIGURE 5. The same region of space as in Figure 4 is now filled with matter such as air. Field and body can occupy the same space at the same time. For this reason their interaction takes place at every point inside the system. If we consider the material body only, we have to introduce sources of entropy where the body absorbs radiation from the field, and sinks where it emits entropy to the field.

3.1.4 Interaction of Bodies and Fields

Often we are not interested in the transport of heat through the radiation field itself, but in the interaction of fields and bodies. The examples of the Sun emitting radiation and of the radiation penetrating the Earth's atmosphere can tell us much about the form of this interaction. The radiation which is not reflected back into space enters the atmosphere, where part of it is absorbed along the way to the surface of the Earth. We know from experience that only part of the radiation is absorbed; the rest reaches the surface. At the same time, the air must emit entropy since it cannot continually absorb radiation without getting hotter and hotter. Absorption and emission take place inside every part of the atmosphere. This means that the radiation field pervades the air; it does not stop where the layer of air surrounding our planet begins. In other words, the *radiation field and the atmosphere occupy the same region of space* at the same time (Figure 5).

The balance of entropy for body and field. To motivate the law of balance of entropy in the case of radiative transfer, we shall proceed in two steps. First consider the combined system of matter and field occupying some region of space (Figure 5a). In the case of solar radiation interacting with the Earth's atmosphere, this system absorbs some of the radiation flowing through the field. As far as the region of space is concerned, we have only radiative fluxes of entropy with respect to its surface. (Neglect for the moment that entropy may be conducted through air, and that air may flow through the system.) The entropy of the system may change only due to such radiative currents and the production of entropy in case of dissipation:

$$\dot{S} + I_{s,rad} = \Pi_s \tag{4}$$

Indeed, as we shall learn in Section 3.3.4, the absorption and emission of entropy are irreversible processes. For this reason, we may not neglect the production term in Equation (4).

Often, however, we are interested only in the balance of entropy with respect to the body alone (Figure 5b). In this case, we have to consider the interaction of matter with the part of the field which occupies the same space. The interaction between the two, if it takes place at all, takes the form of absorption of radiation from the field by the body, or emission from the body to the field. Absorption and emission take place at every point in space occupied by the two systems. The properties of the body and of the field determine the amount of entropy which is absorbed or emitted.

Sources due to absorption and emission. Absorption and emission of radiation are not surface phenomena, but rather *volumetric processes*. In the absorption of radiation by the Earth's atmosphere there is no flow of entropy through this material body. Entropy enters the material system via the field, which means that there are *no currents through matter* associated with this type of transport. It simply appears at every point according to the degree of interaction. If we write an equation of balance of entropy for the material body only, we have to represent the interaction using a *source term* instead of currents:

$$\dot{S}_{body} = \Sigma_{s,body} \tag{5}$$

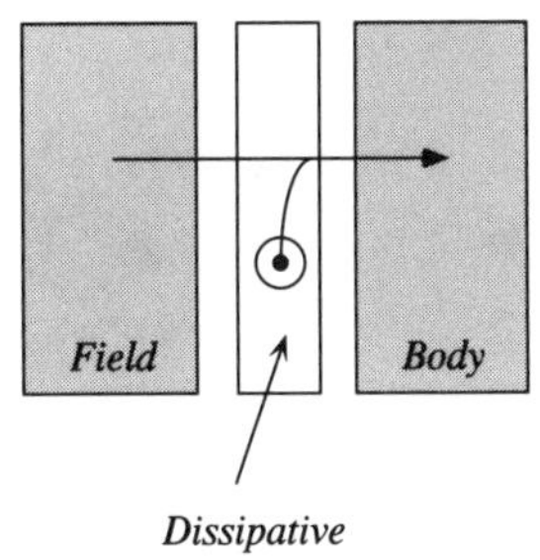

FIGURE 6. If we model the result of the transfer of radiation from the field to the body as uniform heating of the body, we have the problem of deciding where to include the source of irreversibility. The solution presented in the equations corresponds to introducing a dissipative component between the field and the body.

Here, $\Sigma_{s,body}$ is the *entropy supply* or *source strength of entropy*, which is the net time rate at which entropy enters or leaves the body as a result of the interaction. The equation of balance of entropy of the field, on the other hand, must take the form

$$\dot{S}_{field} + I_{s,rad} = \Sigma_{s,field} \tag{6}$$

If we now combine the two equations of balance, we obtain a different relation from Equation (4):

$$\dot{S}_{body} + \dot{S}_{field} + I_{s,rad} = \Sigma_{s,body} + \Sigma_{s,field} \tag{7}$$

Comparison of these expressions tells us that as a result of emission (or absorption) of radiation by the field and absorption (or emission) by the body, entropy must have been produced. In other words, more entropy is absorbed by the body than is emitted by the field. The relation between the two source rates and the rate of production of entropy must be given by

$$\Sigma_{s,body} + \Sigma_{s,field} = \Pi_s \tag{8}$$

with the entropy being produced as a result of the transfer between the field and the body (Figure 6).

The general law of balance of entropy. If we include the source term with the conductive and convective fluxes and the production of entropy in Equation (2), we finally obtain the most general case of the equation of balance of entropy for a body:

$$\dot{S} + I_{s,cond} + I_{s,conv} = \Sigma_s + \Pi_s \tag{9}$$

This equation includes all the processes we are going to discuss. It expresses the fact that the entropy of a body may change as a result of three distinct types of transport: conductive, convective, and radiative, including the effects of irreversibility.

EXAMPLE 2. Absorption of radiation by a body.

Sunlight is passing in one direction through a gas inside a long cylinder. The flux of entropy at the surface where the light is entering has a magnitude of 5.0 W/K. At the opposite end, the flux of the current of entropy leaving the body is 4.0 W/K. a) Determine the net flux of entropy with respect to the region of space occupied by the body. b) At what (minimal) rate is the entropy of the body changing? c) What is the value of the source rate of entropy for the field? How large is the flux of entropy with respect to the material body?

SOLUTION: a) The surface of the region occupied by the body and the field is penetrated by currents of entropy. The net flux of entropy with respect to this system simply is the sum of the single fluxes associated with the different currents:

$$\begin{aligned} I_{s,rad} &= I_{s,rad,in} + I_{s,rad,out} \\ &= -5.0\,\mathrm{W/K} + 4.0\,\mathrm{W/K} = -1.0\,\mathrm{W/K} \end{aligned}$$

The currents are radiative, since entropy is transported through the radiation field only.

b) The entropy which seems to disappear from the field in the region occupied by the gas is actually stored in the body. As for every irreversible process, entropy must be produced, and the rate of change of the entropy content of the body must be larger than the rate at which entropy disappears from the field. Since we do not know the constitutive law of production of entropy as a consequence of absorption of radiation, we can only say that

$$\dot{S}_{body} \geq 1.0\ \mathrm{W/K}$$

c) The only type of entropy transport with respect to the body is radiative in nature and it takes the form of sources. For this reason, the equation of balance of entropy in the form of Equation (6) must hold for the field. Therefore, the source rate of entropy is - 1.0 W/K. There is no flux of entropy with respect to the body.

3.1.5 The Balance of Energy

One of the most important practical problems in the theory of heat transport is the determination of the fluxes and source terms of entropy in the equation of balance. (See Equation (9).) We have to find the constitutive laws which let us calculate these quantities in concrete situations. At this point, the energy principle will come to our rescue. All three types of entropy transport are accompanied by the flow of energy. For this reason we should consider the law of balance of energy as well. Since energy is a conserved quantity, the amount stored in a body can change only by way of transfer to or from another system. The type of transfer of energy depends on the type of entropy flow. In the cases of conduction and convection, energy flows with entropy across system boundaries. This means that in these cases it is accounted for in terms of conductive or convective currents. If entropy is transferred radiatively, however, the interaction of bodies and fields leads to sinks or sources of energy in the body (or in the field). As a result of entropy transfer, energy either flows across system boundaries, or it pours into bodies via a radiation field. Therefore we distinguish between two types of currents and a source term of energy for material systems:

$$\dot{E}_{body} + I_{E,cond} + I_{E,conv} = \Sigma_{E,body} \tag{10}$$

For the radiation field alone, the equation of balance of energy must take the form

$$\dot{E}_{field} + I_{E,rad} = \Sigma_{E,field} \tag{11}$$

The term on the right-hand side of Equation (10) is the *source rate* or the *supply of energy*. Actually, in Equation (10), we have neglected the transport of energy due to other processes such as mechanical ones. In the case of convective currents we cannot always do this. However, for the purpose of this chapter we shall regard such contributions as negligible compared to the other terms.

As we shall see, the relationship between entropy and energy in thermal transport phenomena will help us greatly in resolving the constitutive problem. To be specific, we are interested in a number of relationships, namely those between:

1. Fluxes of entropy and energy in conduction.
2. Fluxes of entropy and energy in convection.
3. Fluxes of entropy and energy through the radiation field.
4. Sources of entropy and energy.
5. The production of entropy.

The following sections will deal with different modes of transport in turn. The one type of relationship between entropy and energy in thermal processes which we have considered so far (remember Equation (13) in Chapter 1) is not of a general nature for transport. Rather, convection and radiation must lead to different expressions relating fluxes of entropy and energy.

3.2 Some Simple Applications of the Flow of Heat

In this section we shall introduce some simple aspects of all three modes of heat transfer to gain some idea of the breadth of applications. We will encounter a simple version of heat conduction, a discussion of the radiation of heat from surfaces and of absorption and emission of radiation inside bodies, and an introduction to heat transfer from solid bodies to fluids (or vice versa). Time-dependent problems will be presented for discrete uniform systems only. These cases require only a modest amount of mathematics. Later in this chapter, conduction and radiation will be presented in more depth.

3.2.1 Currents of Entropy in Conduction: Fourier's Law

Upon what factors does the current of entropy in conductive transport depend? If the temperature of a body changes from place to place, there must be *temperature gradients*. This is one factor upon which we expect the rate of flow of heat, the entropy current, to depend. Second, the material through which the entropy flows must play a role in the determination of the current. The influence of the material will be described by its *conductivity*. If the current of entropy depends on the temperature gradient and the conductivity in the simplest possible way, we say that it obeys *Fourier's law*.

Transport of entropy in conduction. We can motivate the form of Fourier's law in a simple manner. The idea is borrowed from electricity, where we also have encountered phenomena which have to do with conduction, namely the conduction of charge (Prologue). Remember that we found the current to depend linearly upon the difference of the electrical potential which allowed us to introduce the electrical conductivity of a conductor. Here we will proceed analogously.

Consider the conduction of entropy through a slab of material as in Figure 7. Assume that entropy flows only in one direction, and that the distribution of the current of entropy does not vary in a plane perpendicular to the flow. In other words we will consider only the simplest possible case of a flow field. Now we introduce a measure of the distribution of the current over the surface of a body, namely the current density j_S. In our case it is related to the magnitude of the flux I_s as follows:

$$|I_s| = A|j_s| \qquad (12)$$

The meaning of the density of a current of entropy will be explained in more detail in Section 3.4.1. Obviously, the unit of the current density of entropy is W/(K · m^2).

The basic question is the following: how does the current density of entropy depend on the circumstances? Experience tells us that this phenomenon is comparable to the conduction of charge. Two quantities play a central role in these cases—the spatial rate of change of the potential, and material properties. In the case of entropy, the former is the *temperature gradient dT/dx*, while the latter is called the *thermal conductivity* k_S. It is clear that the current of entropy must vanish if the temperature gradient is zero. In the simplest case the current density will depend linearly upon the gradient, just as in electricity. Also, entropy is not conducted if we deal with a perfect insulator whose conductivity is zero. For these reasons we are led to propose the following law governing conductive transport of entropy:

$$j_s = -k_s \frac{dT}{dx} \tag{13}$$

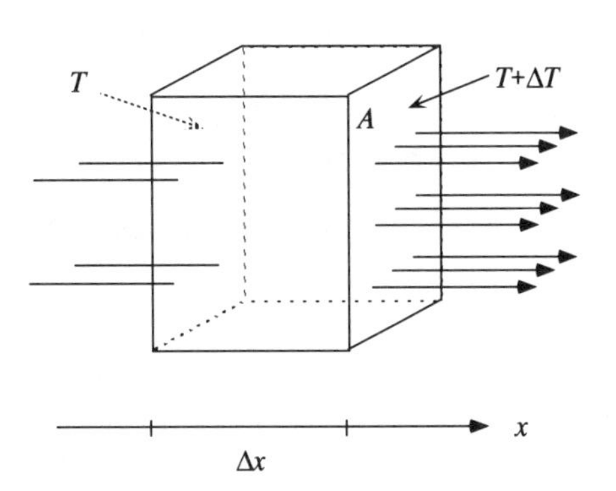

FIGURE 7. Heat flows in one direction only through a slab of matter. We assume that the distribution of the current does not vary in planes perpendicular to the x-direction. There is a difference of temperatures between front and back faces which serves as the driving force of the flow of entropy.

This is *Fourier's law* of conduction. The minus sign is introduced since entropy flows from places of higher temperature to places of lower temperature. Equation (13) clearly is the simplest form of a constitutive law for the conductive current of entropy. The only variables appearing are the temperature gradient dT/dx, which enters the formula linearly, and the thermal conductivity k_S. Naturally, the thermal conductivity is expected to depend on the material of which the body is made, and on temperature (Figure 8). Some values of thermal conductivities are given in Table A.3.1.

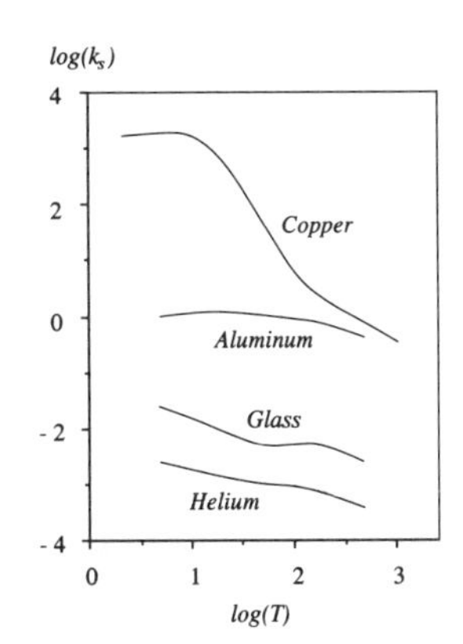

FIGURE 8. Some thermal conductivities k_S as functions of temperature. The values of the conductivities with respect to energy, i.e., those which are commonly listed in tables, are equal to k_S multiplied by the temperature.

Note that we have not yet solved the constitutive problem of the conductive transport of heat. While we now have a relation for the current of entropy appearing in the equation of balance, we still do not have an expression for the rate of generation of entropy. This problem will be taken up in Section 3.4.3.

The flow of entropy and energy in conduction. A current of energy is always associated with a current of entropy in conductive transport. Naturally, we would like to know the relationship between these fluxes. This part of the problem was stated in Section 3.1.5. Obviously, conduction is one means of heating a body. In Chapter 1 we concluded that there must be a simple relationship between the entropy and the energy received in heating. The flux of the latter is obtained by multiplying the flux of the former by the temperature of the body. There and in Chapter 2 we considered bodies having the same temperature throughout. Now we are in a position to write a relationship for the more realistic situation in which the temperature varies from point to point within a body. We imagine a surface—either the surface of a body or one inside—which is crossed by the currents of entropy and of energy. If the temperature at the surface is T, the currents are related by the well-known law we have used so often, Equation (13) of Chapter 1. It directly carries over to the current density:

$$j_{E,th} = T j_s \tag{14}$$

Just as there is a current density of entropy, there also must be a current density of energy. Equation (14) holds for every point in a body through which entropy flows conductively. The validity of the generalization of the law $I_E = TI_s$ can be proved more rigorously. For now, however, let us accept it as intuitively clear.

Fourier's law for the energy current. If we introduce the definition of the *conductivity with respect to energy*

$$k_E = T \cdot k_s \tag{15}$$

Fourier's law can be expressed in terms of the thermal energy current:

$$j_{E,th} = -k_E \frac{dT}{dx} \tag{16}$$

Since we consider pure conduction of heat, only the thermal energy current appears in a process. Therefore, the expression for the balance of energy will take a particularly simple form. This fact will help us in the following applications.

EXAMPLE 3. Fluxes and flux densities of entropy inside a copper bar.

A copper bar of length 0.50 m and cross section 10.0 cm^2 has a temperature of 500 K at one end and 300 K at the other. As heat is flowing through the bar in steady state, measurements indicate that the temperature varies linearly along the bar. a) Determine the temperature gradient. Take the direction of entropy flow to be positive. b) Estimate the current densities of entropy and of energy for the center of the bar using the values read from Figure 8. How large is the conductivity with respect to energy? c) Divide the bar into two equal parts. With this current of entropy flowing, what is the flux of entropy at the surface where the parts touch with respect to the part from where the entropy is flowing?

SOLUTION: a) The temperature gradient is obtained simply. Since the temperature changes linearly, we can write:

$$\frac{dT}{dx} = \frac{\Delta T}{\Delta x}$$
$$= \frac{300 - 500}{0.5} \frac{\text{K}}{\text{m}} = -400 \frac{\text{K}}{\text{m}}$$

b) For the center of the bar the temperature is 400 K. According to Figure 8, the value of the conductivity is about 2 W/(K^{2}m). This gives a value of

$$j_s = -2 \cdot (-400) \text{W}/(\text{K} \cdot \text{m}^2) = 800 \text{W}/(\text{K} \cdot \text{m}^2)$$

The value for the current density of energy turns out to be

$$j_E = T j_s$$
$$= 400\text{K} \cdot 800\,\text{W}/(\text{K} \cdot \text{m}^2) = 3.2 \cdot 10^5\,\text{W/m}^2$$

The value of the conductivity with respect to energy is $k_E = T\,k_s = 800$ W/(K · m).

c) The magnitude of the flux of entropy at the midpoint is

$$I_s = A j_s$$
$$= 10 \cdot 10^{-4}\,\text{m}^2 \cdot 800\,\text{W}/(\text{K} \cdot \text{m}^2) = 0.80\,\text{W/K}$$

It is a positive quantity since it corresponds to a current leaving the system considered.

3.2.2 Application of the Balance of Energy to Conduction

Before we go on to the constitutive problem of conduction in Section 3.4, we will discuss some simple examples. We can do some calculations on the basis of the balance of energy alone, without detailed knowledge of the production of heat. Indeed, because of the conservation of energy, the treatment of pure conduction usually is given in terms of the energy current rather than the entropy current. Some formulas turn out to be simpler this way. Since we consider pure conduction, the only energy fluxes are thermal ones. In one-dimensional steady-state conduction, the flux of energy through a body is independent of the position in the body (as well as of time); i.e., it is constant in the direction of flow. This is demonstrated by the equation of continuity of energy, which we derive in the following manner. The sum of the fluxes of energy entering and leaving the slab of material in Figure 7 must be equal to the rate of change of the energy content, which vanishes in the steady state. The law of balance of energy in steady-state conduction therefore leads to

$$I_{E,in} + I_{E,out} = 0 \tag{17}$$

or

$$A(j_E(x) - j_E(x + \Delta x)) = 0 \tag{18}$$

in the case depicted in Figure 7. Equation (17) also holds for different geometries. (This is used in Example 5.)

Let us now assume that *the conductivity with respect to energy is constant* for a given body. This leads to a particularly simple form of Fourier's law for an extended body. Just consult Equation (16). Since both the current density and the conductivity are taken to be constant, *the temperature gradient must be constant* as well. For this reason we can replace the gradient by its average value, which leads to:

$$\left|I_{E,th}\right| = k_E A \frac{|\Delta T|}{\Delta x} \tag{19}$$

This suggests that we can write Fourier's law in terms of the thermal driving force and a *thermal resistance* R_E:

$$|I_{E,th}| = \frac{|\Delta T|}{R_E} \tag{20}$$

where

$$R_E = \frac{\Delta x}{k_E A} \tag{21}$$

In this form the law is analogous to what we know from electricity (Ohm's Law). Note, however, that here we compare energy, and not entropy, to electrical charge. We can apply it also to bodies of different geometries if we generalize the definition of the thermal resistance (Example 5). Obviously the same relations as those found in electricity must hold for combinations of "thermal resistors." If two bodies conducting entropy are set up *in series*, we simply add the respective resistances. On the other hand, for bodies placed *in parallel*, the inverse of the total resistance is given by the sum of the inverse resistances:

$$R_E = \sum_{i=1}^{N} R_{E,i} \tag{22}$$

and

$$\frac{1}{R_E} = \sum_{i=1}^{N} \frac{1}{R_{E,i}} \tag{23}$$

EXAMPLE 4. Heating of an integrated circuit[1].

A silicon chip is attached to an isothermal surface which is called a *header*. The top of the chip is covered uniformly by a power device which dissipates energy at a rate of 50 W. Thirty thin gold wires connect the top with the header. Assume the entropy to be conducted down through the chip in one direction only. What will the steady-state temperature be at the top of the chip, if the header is kept at a temperature of 25°C?

The chip has a surface area of 0.51 cm by 0.51 cm. It is composed of three layers. The first is made out of silicon with a thickness of 0.051 cm. The chip carries a thin layer of gold at the bottom; its thickness is 0.010 cm. Between these two a thin layer of silicon dioxide forms, which has a thickness of 0.00013 cm. The thermal conductivities with respect to energy are 88 W/(K · m), 312 W/(K · m), and 0.157 W/(K · m), respectively. The gold wires are 0.130 cm long and have a diameter of 0.0254 cm.

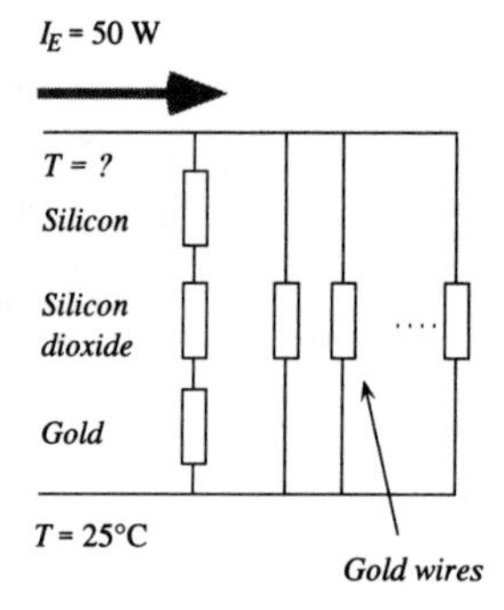

SOLUTION: The device represents a thermal circuit with elements in parallel and in series (see figure). A constant thermal driving force is maintained over the circuit. We must

1. P. Ridgely (1987).

figure out the total thermal resistance offered by the circuit to the flow of entropy and energy. The chip has a resistance of

$$R_{E,chip} = \sum_{i=1}^{3} \frac{1}{k_{E,i}} \frac{\Delta x_i}{A}$$
$$= \frac{1}{0.0051^2}\left(\frac{0.051 \cdot 10^{-2}}{88} + \frac{1.3 \cdot 10^{-6}}{0.157} + \frac{0.010 \cdot 10^{-2}}{312}\right)\frac{\text{K}}{\text{W}} = 0.554 \text{ K/W}$$

This device is connected in parallel with thirty identical wires. Therefore, the total resistance is calculated to be

$$\frac{1}{R_{E,tot}} = \sum_{i=1}^{31} \frac{1}{R_{E,i}} = 30\frac{k_E A}{\Delta x} + \frac{1}{R_{E,chip}}$$
$$= 30\frac{312 \cdot \pi(1.27 \cdot 10^{-4})^2}{0.0013}\frac{\text{W}}{\text{K}} + \frac{1}{0.554}\frac{\text{W}}{\text{K}} = 2.17\frac{\text{W}}{\text{K}}$$

This corresponds to a resistance of 0.461 K/W. With the energy flux given, we can calculate the thermal driving force, i.e., the difference of temperatures between the top and the bottom of the chip:

$$\Delta T = R_E I_{E,th}$$
$$= 0.461 \text{K/W} \cdot 50 \text{W} = 23.0 \text{K}$$

The temperature at the top of the chip therefore is 48°C. Without the gold wires it would be 53°C.

EXAMPLE 5. The flow of heat through the mantle of the Earth.

The total flux of energy from the interior of the Earth through its surface can be estimated from the values of the temperature gradient in the crust and the conductivity with respect to energy. Their values are 0.06 K/m and 1 W/ (K · m), respectively. Assume that the entire flux is conducted from the core at a depth of 3400 km through the solid mantle (the radius of the Earth is 6400 km). Take as an average thermal conductivity the one found for the upper crust. According to these assumptions how large would the temperature of the core of the Earth be?

SOLUTION: First we have to calculate the energy flux at the surface of the Earth. According to Equation (16) it must be

$$I_{E,th} = 4\pi R^2 k_E \frac{dT}{dr}$$
$$= 3.1 \cdot 10^{13} \text{ W}$$

R is the radius of the Earth. The energy flux out of the Earth is about 5000 times smaller than the one we receive from the Sun. We have applied Fourier's law, which was motivated for flat geometry. This is certainly allowed in the case of purely radial flow. We only have to replace the normal temperature gradient by its radial counterpart. The following development, however, changes because of the differences in geometry.

If we knew the thermal resistance R_E of the Earth's mantle we could easily calculate the temperature difference necessary to conduct this current from the bottom of the mantle up to the surface. Since the conducting body is not flat, the surface area through which conduction is taking place varies constantly. Therefore let us write Fourier's law in the form

$$|I_{E,th}| = \frac{dT}{dR_E}$$

with

$$\frac{dR_E}{dr} = \frac{1}{k_E A}$$

This means that we have to calculate the thermal resistance by integration. For a spherical shell with inner and outer radii r_i and r_o, respectively, and with constant k_E we get:

$$R_E = \int_0^{R_E} dR_E = \int_{r_i}^{r_o} \frac{dr}{k_E A} = \frac{1}{4\pi k_E} \int_{r_i}^{r_o} \frac{dr}{r^2} = \frac{1}{4\pi k_E}\left(\frac{1}{r_i} - \frac{1}{r_o}\right)$$

Note that the thermal resistance is of the form given in Equation (21), with $\Delta x = r_o - r_i$ and $A = 4\pi r_i r_o$. According to Equation (20), the difference of temperatures between the core-mantle boundary and the surface must be

$$\begin{aligned}\Delta T &= R_E \cdot I_{E,th} \\ &= 4.36 \cdot 10^5 \mathrm{K}\end{aligned}$$

This value is rather far off from the estimated temperature difference of some 3000 K. From seismic measurements we know that the mantle is solid, which limits the temperature below the melting point of rocks. A good number of reasons can be given to explain this huge discrepancy. The value of the conductivity might be wrong. (However, it will not be all that far off.) The entropy flowing out through the surface of the Earth might not come from the core; it might be produced in the mantel and the crust by radioactive decay; this is indeed the case (see Section 3.4.5 and Example 27). The flow of entropy varies with time; in our case, however, this does not change the result much because of the long time scale. Finally, the entropy might be transported not by conduction but by radiation and convection. This is true as well; it is mostly convection which transports entropy through the mantle, even though the mantle is solid! Over very long time scales the material of the mantle is deformable and it moves; this process apparently is responsible for the continental drift.

3.2.3 The Nature of Heat Transfer at a Solid–Fluid Boundary

Now we will introduce some aspects of the heat transfer across interfaces separating different types of bodies. Consider a hot solid body submersed in some fluid. Entropy is conducted through the body to its surface from where it enters the fluid and is carried away convectively. The transport of entropy from the solid to the fluid, or vice-versa, is of interest in the design of heat exchangers, in the loss of heat from a building, and in many other situations. Another important case is heat transfer between liquids and gases. Just think of the interaction between the Earth's atmosphere and the oceans, which has received much attention recently. Questions concerning the balance of energy and entropy, and of carbon dioxide and other trace gases, are of vital interest in environmental, atmospheric, and oceanographic sciences.

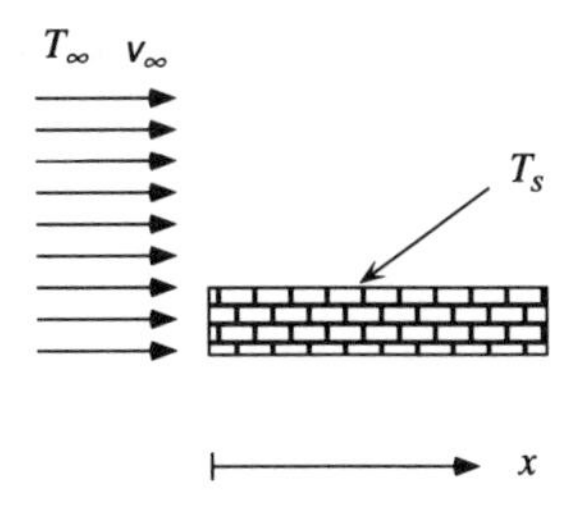

FIGURE 9. A fluid flows past a long flat plate. The fluid has free stream values of velocity and temperature far from the plate. The surface temperature of the solid body is assumed to be constant and different from the free stream value of the fluid. As a result, entropy and energy will be carried across the interface by the combined action of conduction and convection.

Boundary layers. For now, let us limit our attention to the flow of entropy from solids to fluids. The transport mechanism usually is a mixture of conduction, convection, and radiation. (The latter will be treated in Section 3.2.4.) Entropy flows through a hot body to its surface from where it somehow enters the fluid. As an example, consider a viscous fluid flowing along a flat plate as in Figure 9. The hydrodynamic phenomenon is described by the velocity of the fluid in the vicinity of the plate. The conditions in the undisturbed fluid are given by the free stream values of velocity and temperature far from the plate. Because of viscosity the speed of flow is reduced to zero at the surface of the body. It is found that the velocity changes in a direction perpendicular to the surface from the value of zero to the free stream velocity further away in the undisturbed flow. The velocity gradient is confined to a thin *hydrodynamic* or *velocity boundary layer* in which all the interesting action takes place (Figure 10a). The thickness of the boundary layer is zero at the leading edge of the plate, and it increases with increasing distance along the surface. The boundary layer is defined to extend to points where the velocity has reached 99% of the free stream value. Typically, in the situation described, it has a thickness of the order of only a few millimeters.

Now consider the temperature of the fluid. At the surface of the solid the fluid is at rest and its temperature is that of the solid surface, which in general, is different from the free stream value. Therefore temperature gradients must develop perpendicularly to the surface; i.e., the temperature changes from the surface value to the free stream value, this time in a thin *thermal boundary layer* (Figure 10b).

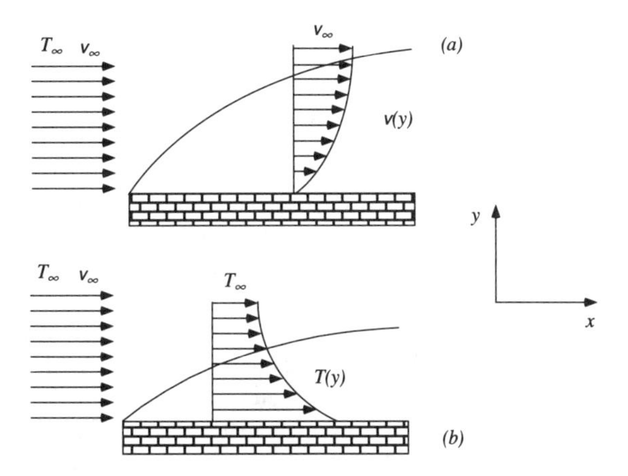

FIGURE 10. Velocity (a) and temperature (b) boundary layers develop at the surface of the solid body. The velocity is zero right at the surface, from where its value increases to the free stream velocity. The distance over which the quantity changes marks the extent of the boundary layer which increases along the plate. The temperature is equal to the surface temperature for $y = 0$. It decreases (or increases) from the surface to obtain the free stream value. The thicknesses of the velocity and temperature boundary layers are not the same.

Again the thickness of this boundary layer increases along the plate from a value of zero at the leading edge. We can understand the importance of the conditions in the boundary layer for the transport of entropy. At the surface of the solid, entropy is transferred into the fluid in the conductive mode only. This allows us to write the energy flux in terms of the conductivity of the fluid and the temperature gradient in the fluid at the surface:

$$j_{E,th} = -k_{Ef} \left.\frac{dT}{dy}\right|_{y=0} \tag{24}$$

The convective heat transfer coefficient. Entropy and energy which enter the fluid conductively will be carried away with the flow of matter. The entropy and energy currents crossing the interface must depend in some way on the physical state of the fluid and the temperatures at the surface of the body and far from it. The process is a rather complex phenomenon. For this reason it is commonly described in a strongly simplified manner. First of all, the rate of energy transfer is used in the formulas instead of the entropy flux because of the production of the latter. Second, the energy flux density is expressed in terms of the difference of temperatures between the surface of the solid body and the fluid far from the surface, and a coefficient which summarizes the complexity of the physical state of the fluid:

$$j_E = h(T_s - T_\infty) \tag{25}$$

h is called the (local) *convective heat transfer coefficient with respect to energy*, while T_s and T_∞ represent the temperature of the surface and of the undisturbed fluid, respectively. The coefficient depends on the details of the fluid flow. It has to be either calculated on the basis of a complete hydrodynamic theory, or measured in experiments. The energy flux is continuous across the interface, which allows us to equate the flux densities in Equations (24) and (25). This leads to an expression for the heat transfer coefficient:

$$h = -\frac{k_{Ef}}{T_s - T_\infty} \left.\frac{dT}{dy}\right|_{y=0} \tag{26}$$

Both the conductivity of the fluid and the temperature difference can be taken to be constant. Therefore the convective transfer coefficient depends on the temperature gradient at the surface of the solid, which is determined by the conditions in the boundary layer. Experience tells us that a hot body submersed in a flowing medium cools much faster than in a still fluid. Therefore, the rate of transfer of entropy from a solid into a liquid or gas crucially depends on the state of motion of the fluid. The type of flow plays an important role as well. We have to distinguish between *laminar* and *turbulent* flows on the one hand, and forced and free convection on the other. The rate of entropy transfer is very different in these cases. It is clear that we have not really solved the problem of convective heat transfer; we have simply shifted it to the task of determining the transfer coefficient from a theory combining motion and heat transfer.

The temperature gradient at the surface of the solid obviously diminishes with increasing thickness of the layer, which leads us to conclude that the local value of the transfer coefficient decreases along the plate. Often the coefficient is replaced by an average value. In this case we can relate the entire energy flux to the change of temperature and the *average transfer coefficient* h_a:

$$I_E = h_a A(T_s - T_\infty) \tag{27}$$

A is the total surface of the body. In this simplified form the constitutive law of convective entropy transfer commonly serves as a boundary condition for the conductive transport of entropy through the solid body. For concrete applications we need to know the average transfer coefficient. Such values are listed in Table A.10 for a few practical situations.

The exchange of entropy at the interface. The energy current which is expressed by Equation (27) is carried across the surface of the solid body by conduction alone. For this reason it is possible to give a simple form of the entropy flux entering or leaving the solid. In conduction the entropy and energy currents are related by the local temperature. Therefore the entropy flux at the surface is equal to

$$I_{s,surface} = \frac{1}{T_s} h_a A(T_s - T_\infty) \tag{28}$$

Naturally, the entropy flux varies across the boundary layer. Entropy will be produced in the fluid due to both conduction and viscous friction. These are two of the possible dissipative processes taking place in the general type of fluid considered here.

The total heat transfer coefficient. How large is the flux of entropy or energy through the wall of a building or through the insulation of a pipe? Obviously we are dealing with multilayer situations in which both conduction through solids and convection at solid–fluid boundaries occur. Consider as an example the transfer of heat through a wall of a house (Figure 11). The wall may be made up of a sequence of solid layers touching one another. We have learned to deal with the conduction of heat through such a sequence by adding up the thermal resistances of each layer (Section 3.2.2). Now, there is air touching the inside and the outside of the wall. Each of these transition zones represents another thermal resistor with a resistance

$$R_{conv} = \frac{1}{h_a A} \tag{29}$$

The difference of temperatures occurring in Equation (28) is taken as the thermal driving force of the transfer of heat. The combined effect of the wall and the two transition zones is obtained by adding their resistances. Often the transfer of energy is described in terms of the total difference of temperatures between the inside and the outside of the building, and an *overall heat transfer coefficient* h_{tot}:

$$I_E = h_{tot} A(T_1 - T_2) \tag{30}$$

From the continuity of the flux of energy this coefficient is calculated to be

$$\frac{1}{h_{tot}} = \sum_{i=1}^{N} \frac{1}{h_{ai}} + \sum_{j=1}^{M} \frac{\Delta x_j}{k_{Ej}} \tag{31}$$

where the Δx_i are the thicknesses of each of the conducting layers which have conductivities k_{Ei}. There are N transition layers and M conductive ones. Equation (31) allows for the calculation of the sum of all the resistances belonging to the different layers.

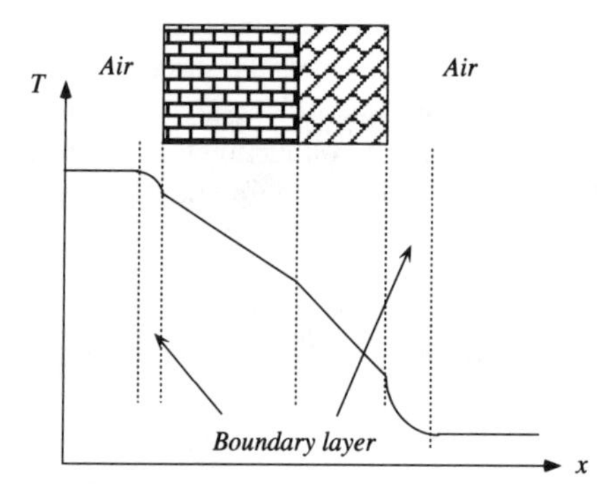

FIGURE 11. The wall of a house forms a sequence of heat conducting and convective transition layers. The overall effect of the wall and the outer and inner boundary layers is described by the sum of all thermal resistances of the single layers. The temperature drops by a certain amount in every part of the sequence.

EXAMPLE 6. Surface temperature of a central heating radiator.

A radiator with an effective surface area of 0.40 m^2 emits a thermal energy current of 500 W. A reasonable value of the heat transfer coefficient in this situation (air rising on a vertical metal surface) is 40 W/K · m^2. a) How large must the surface temperature of the radiator be, if the temperature of the air in the room is 20°C? b) What is the magnitude of the entropy flux emitted by the surface of the radiator?

SOLUTION: a) The energy current is expressed by Equation (28), which we solve for the temperature of the surface:

$$T_s = \frac{I_{E,th}}{A h_a} + T_\infty$$

$$= \frac{500}{0.40 \cdot 40} \text{ K} + 293 \text{ K} = 324 \text{ K}$$

b) The flux of entropy and of energy at the surface of the radiator are related by the temperature at that location:

$$I_s = \frac{I_E}{T}$$

$$= \frac{500}{(273+51)} \frac{\text{W}}{\text{K}} = 1.54 \frac{\text{W}}{\text{K}}$$

EXAMPLE 7. Surface temperatures of a single pane window in winter.

Consider a window having a metal frame. The window measures 1.20 m by 2.00 m. The glass has a thickness of 3.0 mm, and a conductivity with respect to energy of 1.0 W/K · m. Take the convective transfer coefficients inside and outside to be 8.0 W/K · m^2 and 12.0 W/K · m^2, respectively. The metal frame is 3.0 cm wide around the window, and 5.0 mm thick. The conductivity is 220 W/K · m (aluminum), and the transfer coefficients inside and outside are taken to be 30 W/K · m^2 and 50 W/K · m^2, respectively. The temperature on the inside is 20°C; on the outside it is - 10°C. a) What are the temperatures of a single pane window inside and outside in winter? b) Calculate the flux of energy through the window if it has a metal frame. c) How large is the total transfer coefficient of the window?

SOLUTION: a) The energy flux through the glass is given by Equations (31) and (32):

$$I_{E,th} = h_{tot} A \Delta T$$
$$= \left[\frac{1}{8.0} + \frac{0.0030}{1.0} + \frac{1}{12.0}\right]^{-1} 2.4 \cdot 30 \text{ W} = 340 \text{ W}$$

We use this value to calculate the temperature drop from the inside to the surface of the window:

$$\Delta T = \frac{I_{E,th}}{h_{a1} A} = \frac{340}{8.0 \cdot 2.4} \text{K} = 17.7\text{K}$$

which makes the temperature on the inside of the window 2°C. The same consideration for the thermal boundary layer outside delivers a temperature drop of 11.8 K. This means that the change of temperature through the glass is very small, and the outside surface approximately has the same temperature as the surface on the inside.

b) The metal frame adds to the energy current. (It is in parallel with the window pane.) The surface area of the frame is roughly 0.19 m^2. Just as above, we calculate the energy current:

$$I_{E,th} = \left[\frac{1}{30} + \frac{0.0050}{220} + \frac{1}{50}\right]^{-1} 0.19 \cdot 30\text{W} = 110\text{W}$$

The total energy current turns out to be 450 W which is very large. Note that the metal frame has a strong influence despite its small surface area.

c) The total transfer coefficient is given by Equation (31):

$$h_{tot} = \frac{I_{E,th}}{A \Delta T}$$
$$= \frac{450}{2.6 \cdot 30} \frac{\text{W}}{\text{K} \cdot \text{m}^2} = 5.8 \frac{\text{W}}{\text{K} \cdot \text{m}^2}$$

Good windows achieve a much smaller value of this coefficient (by as much as a factor of 10).

EXAMPLE 8. A surprising effect of insulation.

A metal pipe is to be insulated. It is found that, at least in principle, the insulation can have the opposite effect of what we would expect: the current of heat through the walls and the insulation of the pipe increases! How is this possible? Determine the conditions for the maximum heat flow.

SOLUTION: The thermal resistance of the insulation is made up of the resistance of the layer of insulation itself, and of the effect of convection at its surface. While the resistance of the insulating cylindrical shell grows with increasing thickness, the resistance due to the thermal boundary layer decreases because of the increase of surface area. There will be a minimum value of the total resistance at a certain outer radius of the insulation, depending on the material properties.

First we need an expression for the thermal resistance of a cylindrical shell. We proceed as in the case of a spherical shell (Example 5):

$$R_E = \int_{r_o}^{r} \frac{1}{2\pi k_E L} \frac{dr}{r} = \frac{1}{2\pi k_E L} \ln\left(\frac{r}{r_o}\right)$$

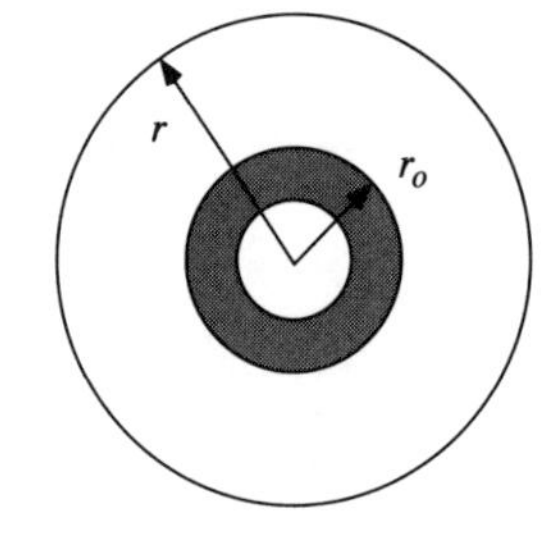

The total resistance of insulation and boundary layer is

$$R_{tot} = R_{insulation} + R_{conv} = \frac{1}{2\pi k_E L} \ln\left(\frac{r}{r_o}\right) + \frac{1}{2\pi h_a L r}$$

Its minimum is found by setting its derivative with respect to the radial variable equal to zero. We obtain

$$r_{min} = k_E / h_a$$

The value of r for which the thermal energy current becomes largest does not depend on the radius of the pipe. However, the latter radius certainly must be smaller than the quantity just calculated. For normal values of the constitutive quantities the pipe (and the insulation) must be rather thin. One might imagine the effect to play a role, for example, when ice starts to build up around thin branches or fibres in plants. An interesting suggestion has been made concerning the improvement of heat transfer through the air–water or air–air heat exchanger of a heat pump. At the cold end of the device, ice tends to build up at the surface, normally reducing the effectiveness of the pump. Possibly, the geometry of the device could be such that frost building up at its surface would lead to an increase of the rate of heat transfer.

3.2.4 Blackbody Radiation from Opaque Surfaces

Next let us consider how a body radiates heat into its surroundings. Even though radiation is a rather complex phenomenon, one case can be treated fairly simply—the emission (and absorption) of radiation by an opaque body. Emission and absorption are volumetric processes as discussed in Section 3.1; still the emission from an opaque body looks like the flow of radiation from a surface. Let us describe briefly how this happens.

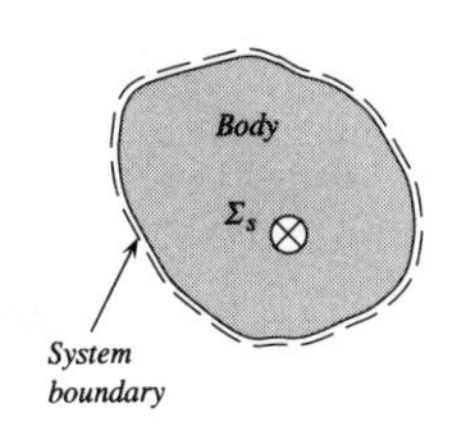

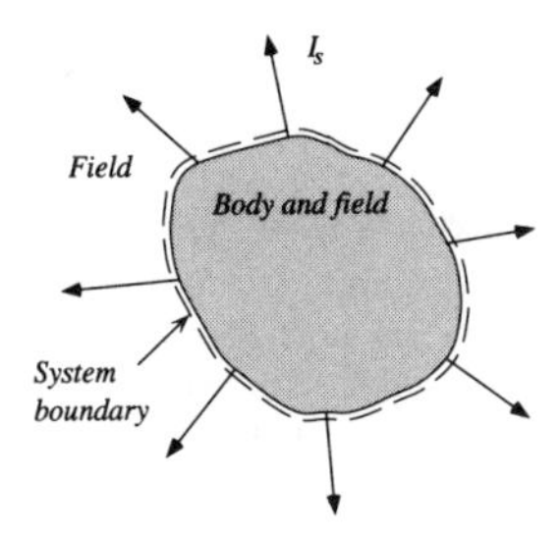

FIGURE 12. Radiation apparently from the surface of an opaque body has been emitted to the field inside the body itself. There is a constant exchange of radiation between matter and field in the space occupied by both (top). Outside the body we simply see the radiation which is traveling through the field. Since we cannot see inside the body, the radiation seen outside effectively comes from a very thin layer at the surface of the material body (bottom).

A warm body emits radiation to the field occupying the same region, leading to a sink of radiation with respect to the body (and a source with respect to the field). Radiative transport through the field inside the system boundary in Figure 12 is rather complicated. Radiation is emitted and reabsorbed constantly at such rates that the net effect is a flow of heat from hotter to cooler points (Section 3.5.4). Since the body is assumed to be (just about) opaque to the radiation, what we see outside the system must originate from a relatively thin layer at the surface of the body. Outside the space occupied by matter, however, we have a simpler situation. There, radiation is traveling away from the region where it was emitted. If we surround the system by a surface, radiation effectively flows through the field through this boundary: we have normal fluxes of entropy and energy with respect to this surface.

Hemispherical emission by blackbody surfaces. We are interested in an expression for the fluxes of entropy and energy through the surface of a body which is emitting heat to its surroundings. In this section we will discuss only the simplest cases, starting with radiation from the surfaces of *black bodies*. A black body is defined as one which absorbs all radiation falling upon it. In Section 2.6 we studied blackbody radiation inside a cavity. On the basis of what is known about such radiation, we can motivate the form of the law for radiative transfer from the surface of a body such as the one in Figure 13. If the surface layers of an opaque body have properties which lead to blackbody radiation, the radiation will be the same as if it had originated from a cavity deep inside the system. Since entropy and energy are carried away by radiation, their flux densities must be related to their (volume) densities inside the radiation field. From Chapter 2, Equations (94) and (98), we know that the density of entropy of blackbody radiation is proportional to the third power of its temperature, while the energy density depends on the fourth power of the temperature. For this reason the fluxes of these quantities from the surface of a body have the same dependence on temperature. It is customary to introduce the *rate of emission of energy of a black body* or the *hemispherical emissive power of a black body* $\mathcal{E}_b$ (also abbreviated by j_{Eb}'), which is defined as the amount of energy emitted by the surface of a black body per unit time and per unit surface area. We expect a law of the form:

$$\mathcal{E}_b = \sigma T^4 \quad (32)$$

Its counterpart relating to entropy j_{sb}' is expressed by

$$j_{sb}' = \frac{4}{3}\sigma T^3 \quad (33)$$

This is again valid for blackbody radiation only. These expressions will be derived in Section 3.5. The constant σ introduced in these relations is called the Stefan-Boltzmann constant, and has the value $5.67 \cdot 10^{-8}$ W/(m$^2 \cdot$ K^4).

You should notice an important point: the relation between fluxes of entropy and of energy which applies to the heating or cooling of a body (Equation (13) of Chapter 1) does not hold in the case of radiative fluxes through the radiation field. Heating and cooling of the material body take the form of sources and sinks of entropy (and energy) for which the simple and direct relation between source rates of entropy and energy holds; see Equation (48).

The fluxes of entropy and energy from the entire surface of a body are obtained simply by multiplying the emissive powers by the surface area A:

$$I_{s,rad} = \frac{4}{3}\sigma A T^3 \tag{34}$$

$$I_{E,rad} = \sigma A T^4 \tag{35}$$

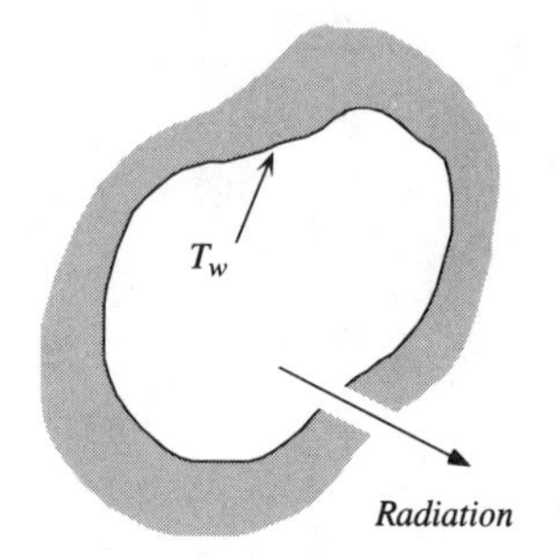

FIGURE 13. Blackbody radiation from the surface layers of an opaque blackbody is the same as that originating from the inside of a cavity in a body (Section 2.6).

The net energy flux between a small body and a black body surrounding it. In general these are not the net fluxes, since the body might absorb heat from another piece of matter radiating towards it. It is instructive to derive the net energy flux for a black body totally surrounded by another black body at a different temperature T_w (such as the small piece of matter in the cavity in Figure 14). Remember that the radiation field set up by the walls of the cavity is isotropic and the same at every point inside. This means that a point at the surface of the small body surrounded by the walls sees blackbody radiation coming at it at the same rate from all directions. (See Equation (39) for a more formal expression of this statement.) Therefore the amount of energy per second and per unit surface area radiated toward the body is σT_w^4. Since a black body absorbs all the radiation falling upon its surface, the net flux of energy with respect to the chunk of matter inside the cavity, which is the difference between the rates of emission and absorption, is given by

$$I_{E,rad,net} = \sigma A\left[T^4 - T_w^4\right] \tag{36}$$

The relations derived here will be extended to bodies other than black bodies in Section 3.2.5.

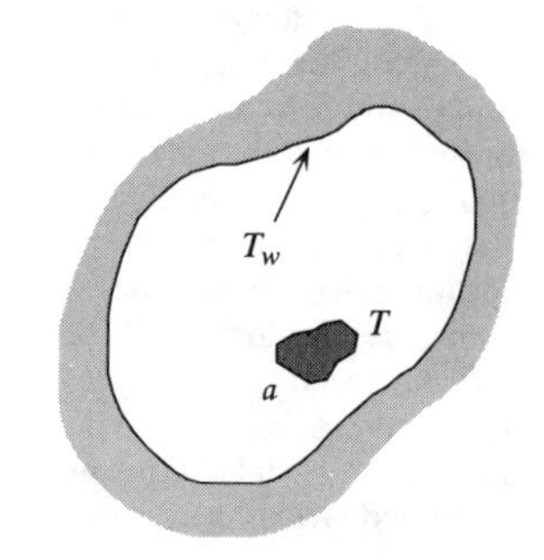

FIGURE 14. A cavity serves as a container of blackbody radiation. The walls are at a temperature T_w, while the small body inside the cavity has a temperature T. a is the absorptivity (Section 3.2.5), which is 1 for blackbody surfaces.

EXAMPLE 9. Surface temperature of the Sun.

The radiation originating in the thin surface layer of the Sun almost has blackbody properties. Using the solar constant, the distance from the Earth to the Sun, and the Sun's radius, derive the temperature of its surface. Also calculate the rate at which entropy is emitted by the total surface.

SOLUTION: The solar constant G_{sc}, i.e., the value of the energy flux per unit normal area at the distance of the Earth, is 1370 W/m². From this value we calculate the emissive power of the Sun (called the luminosity L):

$$L = 4\pi d^2 G_{sc}$$
$$= 4\pi \cdot \left(1.5 \cdot 10^{11}\,\text{m}\right)^2 \cdot 1370\,\text{W/m}^2 = 3.9 \cdot 10^{26}\,\text{W}$$

The distance between Sun and Earth is $150 \cdot 10^6$ km. The Sun approximates a black body which does not receive radiation from the surrounding space. Therefore, Equation (35) applies to the relation between emissive power and temperature, leading to a value of

$$T = \left(\frac{L}{4\pi R_S^2 \sigma} \right)^{1/4}$$

$$= \left(\frac{3.9 \cdot 10^{26}\,\text{W}}{4\pi \cdot \left(7.0 \cdot 10^8\,\text{m}\right)^2 \cdot \sigma} \right)^{1/4} = 5770\text{K}$$

for the surface temperature of the Sun. The Sun's radius is 700000 km. The entropy flux flowing away from the Sun through the field is given by

$$I_s = \frac{4}{3} A \sigma T_s^3$$

$$= 8.95 \cdot 10^{22}\,\text{W/K}$$

EXAMPLE 10. Surface temperature of the Earth as a uniform black body.

Model the Earth as a black body of uniform temperature, absorbing radiation from the Sun and emitting radiation to outer space. How large is the value of the temperature attained by the surface of this body in steady state?

SOLUTION: In steady state, the fluxes and the rates of absorption of energy with respect to the Earth must be equal:

$$I_{E,rad} = \Sigma_E$$

We may model the absorption of energy by a source term, while the emission might be described in terms of the flux of energy through the radiation field set up by the Earth. As a black body, the Earth absorbs all the radiation falling upon its surface, leading to a rate of absorption

$$\Sigma_E = \pi R^2 \mathcal{G}_{sc}$$

where R is the radius of the planet. The factor multiplying the solar constant is the cross section of the Earth. The surface of the Earth attains a particular temperature, which leads to a flux due to emission equal to

$$I_{E,rad} = \sigma 4\pi R^2 T^4$$

The balance of energy finally tells us that the surface temperature of our planet would be

$$T = \left(\frac{\mathcal{G}_{sc}}{4\sigma} \right)^{1/4}$$

$$= \left(\frac{1370\,\text{W/m}^2}{4\sigma} \right)^{1/4} = 279\text{K}$$

were it a black body. The mean surface temperature of our planet is more like 288 K, leaving us with the problem of how to explain the difference. (See Sections 3.2.5 and 3.5.6.)

EXAMPLE 11. The radiative heat transfer coefficient.

Write the equation for the exchange of energy between a black body and its surroundings, Equation (36), in a form which resembles the equation of convective heat transfer at a solid-fluid boundary. How would you write the overall heat transfer coefficient, including convection?

SOLUTION: It is possible to transform the term involving the difference of the fourth powers of the temperatures in such a way that the difference of temperatures occurs in the equation:

$$\begin{aligned} I_{E,rad,net} &= \sigma \cdot A\left[T_1^4 - T_2^4\right] \\ &= \sigma \cdot A\left(T_1^2 + T_2^2\right)\left(T_1^2 - T_2^2\right) \\ &= \sigma \cdot A\left(T_1^2 + T_2^2\right)\left(T_1 + T_2\right)\cdot\left(T_1 - T_2\right) \end{aligned}$$

Comparison with the desired form, Equation (27), shows that

$$h_{rad} = \sigma\left(T_1^2 + T_2^2\right)\left(T_1 + T_2\right)$$

Obviously, the radiative heat transfer coefficient strongly depends upon the temperatures involved.

If convection is present as well, we are dealing with a case of parallel flow of heat. The flux of energy is equal to the sum of the radiative and the convective fluxes. Therefore, the overall heat transfer coefficient must be equal to the sum of the radiative and convective transfer coefficients.

EXAMPLE 12. Surface temperature of a star, estimated from its angular diameter.

Normally, the surface temperatures of stars are derived from their colors or their spectra. However, it is possible to calculate this quantity from the intensity of their light (i.e., from the irradiance at the surface of the Earth), and from their angular diameter as seen from the Earth. Angular diameters of some nearby stars can be determined with the aid of interferometric methods. In the case of the star Sirius in the constellation of Canis Majoris, these values are $8.6 \cdot 10^{-8}$ W/m^2, and $6.12 \cdot 10^{-3}$ arc seconds, respectively.

SOLUTION: First we relate the intensity of the star's light to its radiant power or luminosity L. If the distance of the star from us is d, the luminosity and the irradiance $\mathcal{G}$ are related by

$$L = 4\pi d^2 \mathcal{G}$$

We may assume that stars radiate isotropically. Furthermore, the distance and the radius r of the star are related by the angular diameter α:

$$r = d\tan(\alpha/2)$$

Finally we assume that stars radiate as black bodies, which is quite nearly true. For this reason, the luminosity can be expressed by

$$L = 4\pi r^2 \sigma T^4$$

We can eliminate the unknown distance, radius, and luminosity from these equation to obtain a relationship between the surface temperature, the angular diameter, and the intensity of the star's light:

$$T = \left(\frac{\mathcal{G}}{\sigma \tan(\alpha/2)^2} \right)^{1/4}$$

In the case of Sirius, this relation leads to a surface temperature of 9100 K, which agrees very well with values obtained from other methods.

3.2.5 Radiative Properties of Gray Surfaces

So far we have limited ourselves to the case of blackbody radiation. Now we should take a closer look at the emission and absorption of heat from surfaces which do not have blackbody properties. Since a black surface was defined as one which absorbs all incident radiation, we will now be concerned with *gray* surfaces, which absorb only a fraction thereof.

The absorptivity of gray surfaces. There are three quantities (relating to energy) which we introduce for a description of the radiative processes at surfaces. The first, which has already been used above, is the total rate per unit area at which energy is emitted by a surface; it is called the *total hemispherical emissive power* $\mathcal{E}$, or the *emittance* (units W/m^2). The second is the *irradiance* $\mathcal{G}$, which is the rate of energy incident upon the surface per unit area from other sources; again the units are W/m^2. Finally we need the *absorptivity* a, which is defined as the ratio of absorbed to incident power. Obviously, the rate of absorption of energy per unit area (or *absorptance*), which is abbreviated by $\mathcal{A}$, is given by

$$\mathcal{A} = a\mathcal{G} \quad \textbf{(37)}$$

For a black body $a = 1$. Sometimes, the *reflectivity* ρ of an opaque surface is introduced in place of the absorptivity. The energy not absorbed by such a surface is reflected, which means that the reflectivity and the absorptivity are related by

$$\rho = 1 - a \quad \textbf{(38)}$$

Kirchhoff's Law. The interesting question now concerns the emissivity of non-black surfaces. How does it compare to blackbody radiation? Let us consider once more the radiation inside a cavity. Since the field in the cavity is that of blackbody radiation, the irradiance in this enclosed space is the emissive power of a black body:

$$\mathcal{G} = \mathcal{E}_b \tag{39}$$

In other words, the irradiance is the hemispherical power in a blackbody field, Equation (32). Whether the walls are black surfaces does not matter. If they are not, the combination of emittance and reflectance still leads to radiation with blackbody properties. This is the reason why the radiation in a cavity is called *blackbody radiation*. Since the material of which the walls of the cavity are made does not play a role, the only factor determining the radiation in the cavity is the temperature of the walls.

Now, an arbitrary body with absorptivity a is introduced in the cavity (Figure 14). After some time stationary conditions will have been reached, the temperature of the small body will be the temperature of radiation in the cavity, and the emissive power of the body will be the fraction of the irradiance absorbed by the body. Since the radiation in the cavity is that of a black body, the emissive power of the body inside the cavity must equal the product of absorptivity and blackbody emissive power:

$$\mathcal{E} = a\mathcal{E}_b \tag{40}$$

This is another form of *Kirchhoff's law*. It states that

> *The emissive power of a body is a fraction of the emissive power of a black body at the same temperature, where the fraction is the absorptivity.*

If we introduce the *emissivity* e of the body as the fraction of the emissive power of a black body at the same temperature,

$$\mathcal{E} = e\mathcal{E}_b \tag{41}$$

we can state Kirchhoff's law by saying that *the emissivity is equal to the absorptivity*. Values of the absorptivity (or emissivity) of some materials are listed in Table A.11.

The net radiant flux for a gray body surrounded by blackbody radiation. If the small body in the cavity has a different temperature, then the rate at which energy is absorbed by the body is not equal to the rate at which it is emitted. The net source rate of radiant energy is calculated to be

$$I_{E,rad} = A(\mathcal{E} - a\mathcal{G}) \tag{42}$$

A is the surface area of the body in the cavity. Since the emissive power is given by Kirchhoff's law, and since the irradiance in the cavity is equal to the emittance of the walls, we finally obtain the following expression for the flux of radiant energy from the surface of the body at temperature T irradiated from a distant surface at temperature T_w which completely surrounds it:

$$I_{E,rad} = aA\left[\mathcal{E}_b(T) - \mathcal{E}_b(T_w)\right] \tag{43}$$

The rate is the difference between the emission and absorption rates as calculated for a blackbody surface, multiplied by the absorptivity of the surface. The blackbody emissive power is the same as that given by Equation (40). Taken together, the laws stated here allow us to calculate the flux from a surface area A which has a temperature T, and which is subject to radiation from surroundings at a temperature T_w:

$$I_{E,rad} = aA\sigma(T^4 - T_w^4) \tag{44}$$

Remember that this equation holds only for the particular geometry used in the example: the body is completely surrounded by the walls of the cavity. As a result, all the radiation emitted by the walls will be incident upon the body and vice versa. For different geometries, where only part of the radiation emitted by either of the bodies strikes the other surface, the result is much more complicated. In such cases it is customary to write the result in just about the same form with an additional factor (the *shape factor*) taking care of the difference (see Section 3.5).

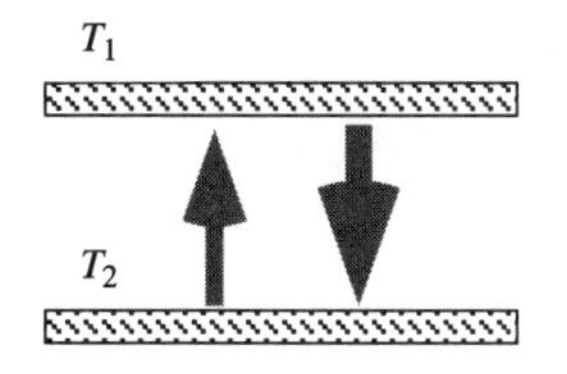

FIGURE 15. Radiant heat exchange between two extended parallel plates. If their temperatures are different, there will be a net flux of energy and entropy from the hotter to the cooler of the plates.

Radiant exchange between extended parallel plates. Here, we will derive the relation for the case of two extended gray surfaces facing each other in such a way that all the radiation originating from one of the bodies is intercepted by the other (Figure 15). This geometry is found, for example, in flat-plate solar collectors. The two plates will be distinguished by indices 1 and 2. Their radiative properties will be expressed using the emissivities (absorptivities), and the reflectivities.

In the course of the derivation, we will need an expression for the total flux of energy per unit area emanating from each of the plates. Since the plates have gray surfaces, they will not absorb all the radiation falling upon them; rather, part of the radiation will be reflected. It is common to call the total flux per unit area, i.e., the sum of what is emitted and what is reflected, the *radiosity* $\mathcal{B}$ of the surface. For the plates, the radiosities are

$$\begin{aligned} \mathcal{B}_1 &= e_1\sigma T_1^4 + \rho_1\mathcal{G}_1 \\ \mathcal{B}_2 &= e_2\sigma T_2^4 + \rho_2\mathcal{G}_2 \end{aligned} \tag{45}$$

The reflectivities are related to the absorptivities (emissivities) of a surface, and the irradiance of one of the plates is the radiosity of the other. Therefore,

$$\begin{aligned} \mathcal{B}_1 &= e_1\sigma T_1^4 + (1-e_1)\mathcal{B}_2 \\ \mathcal{B}_2 &= e_2\sigma T_2^4 + (1-e_2)\mathcal{B}_1 \end{aligned}$$

If we insert the radiosities expressed by Equation (45) into this result, we obtain the following relations for the radiosities of the parallel plates:

$$\mathcal{B}_1 = e_1\sigma T_1^4 + (1-e_1)(e_2\sigma T_2^4 + \rho_2\mathcal{G}_2)$$
$$\mathcal{B}_2 = e_2\sigma T_2^4 + (1-e_2)(e_1\sigma T_1^4 + \rho_1\mathcal{G}_1)$$

Now, the net flux density of energy radiated from plate 1 to plate 2 is the difference of the radiosities:

$$\mathcal{B}_1 - \mathcal{B}_2 = e_1 e_2\sigma T_1^4 - e_1 e_2\sigma T_2^4 + (1-e_1)(1-e_2)(\mathcal{G}_2 - \mathcal{G}_1)$$

which leads to

$$(\mathcal{B}_1 - \mathcal{B}_2)[1-(1-e_1)(1-e_2)] = e_1 e_2\sigma(T_1^4 - T_2^4)$$

A little algebra finally yields the expression for the net energy flux flowing from the hotter to the cooler of the two parallel plates:

$$I_{E,rad,net} = \frac{\sigma A(T_1^4 - T_2^4)}{\frac{1}{e_1} + \frac{1}{e_2} - 1} \quad \textbf{(46)}$$

For blackbody surfaces the emissivities are equal to 1, and we regain the simpler expression already derived in Equation (36).

EXAMPLE 13. Using a wire as a fuse.

A copper wire is placed in an evacuated tube. How large does its radius have to be if it is to melt when an electrical current of 10 A passes through it? The resistivity at 20°C and the temperature coefficient of resistance of copper are $1.8 \cdot 10^{-8}\ \Omega$ and $3.9 \cdot 10^{-3}\ K^{-1}$, respectively.

SOLUTION: The wire will melt if its temperature surpasses the temperature of melting T_m, which is 1356 K for copper. The temperature of the wire also determines the rate of emission of energy from its surface:

$$I_{E,rad} = a\sigma(T_m^4 - T_o^4)2\pi r L$$

Radiation is the only mode of heat transfer occurring in this case. Here, L and r are the length and the radius of the wire, respectively. In the steady state this rate must be equal to the rate at which energy is released in the wire:

$$I_{E,rad} = R I_q^2$$
$$= \rho_{el}\frac{L}{\pi r^2} I_q^2$$

The electrical resistivity is denoted by ρ_{el}. We have to take into account that the resistivity changes with temperature according to

$$\rho_{el} = \rho_o\left(1+\alpha_R\left(T_m - T_o\right)\right)$$

Taken together, we find that the radius of the wire must satisfy the following relationship:

$$r^3 = \frac{\rho_o\left(1+\alpha_R\left(T_m - T_o\right)\right)}{2\pi^2\sigma\left(T_m^4 - T_o^4\right)} I_q^2$$

where we have set the absorptivity of the surface of the wire equal to 1. If we take the temperature T_o of the surroundings to be 293 K, the numerical result for the radius is 0.14 mm. Note that the change of the resistivity of copper has been modeled by a linear law even though this might not be all that accurate.

EXAMPLE 14. Surface temperature of the Earth as a gray body.

Repeat the calculation of Example 10. This time, model the Earth as a gray body of uniform temperature absorbing the radiation from the Sun and emitting radiation to outer space. How large is the value of the temperature attained by the surface of this body in steady state?

SOLUTION: In steady state, the fluxes and the rates of absorption of energy with respect to the Earth must cancel:

$$I_{E,rad} = \Sigma_E$$

We model the absorption of energy by a source term, while the emission is described in terms of the energy flux through the radiation field set up by the Earth. As a gray body, our planet absorbs a fraction a of the radiation falling upon its surface, leading to a rate of absorption

$$\Sigma_E = \pi R^2 a \mathcal{G}_{sc}$$

where $\mathcal{G}_{sc}$ is the solar constant. The factor multiplying the solar constant is the cross section of the Earth. The surface of the planet attains a particular temperature, which leads to a flux due to emission equal to

$$I_{E,rad} = e\sigma 4\pi R^2 T^4$$

Remember that the absorptivity and the emissivity must be equal, which leads to an interesting result. The balance of energy tells us that the surface temperature of our planet should be

$$T = \left(\frac{\mathcal{G}_{sc}}{4\sigma}\right)^{1/4}$$

$$= \left(\frac{1370\,\mathrm{W/m^2}}{4\sigma}\right)^{1/4} = 279\mathrm{K}$$

This is precisely the same result calculated in Example 10. Black and gray bodies of the same geometry attain the same temperature in the light of the Sun, independent of the value of the absorptivity. We still have to explain the difference between the value of the temperature calculated in the current model and the actual mean temperature of the Earth's surface (Section 3.5.6).

3.2.6 Cooling and Heating of Bodies by Emission and Absorption of Radiation

As discussed in Section 3.1.4, bodies can emit and absorb radiation. These processes are the result of the interaction of bodies and fields which occupy the same region of space. Therefore, emission and absorption are volumetric phenomena calling for source rates for their formal description. Here, we will motivate the relationship between the source rates and the rates of production of entropy on the one hand, and the source rates of energy accompanying the processes on the other hand.

Take the model of a uniform body at temperature T. For the sake of argument, let the body emit entropy and energy to the field occupying the same region of space. (The derivation also applies to the case of absorption of radiation.) The rates of emission of these two quantities are equal to the rates of change of the entropy and energy of the body if there are no other modes of transfer present:

$$\begin{aligned} \dot{S} &= \Sigma_s \\ \dot{E} &= \Sigma_E \end{aligned} \tag{47}$$

Remember that processes are reversible in the model of uniform processes. Since the rates of change are related by the Gibbs fundamental relation for a simple body (Equation (23) of Chapter 2), the rates of emission satisfy the equation

$$\Sigma_E = T\Sigma_s \tag{48}$$

which means that the rate of emission of entropy to the field inside the system is equal to the rate of energy emitted divided by the temperature of the body. This result holds for the chunk of matter occupying the region of space in Figure 16. It neglects the fact that entropy is produced as a result of the emission of radiation. However, as far as the body is concerned, this point is immaterial; it does not affect the balance of entropy for the chunk of matter, since in our model, we associate the irreversibility with an additional element between the body and the field (see Figure 6). For the body, emission of heat to a field has the same effect as cooling by conduction as a result of direct contact of a uniform body with its surroundings. This carries over to the relationship between entropy and energy source rates, which formally looks like the relation that applies to currents of entropy and of energy in heating (Equation (13) of Chapter 1).

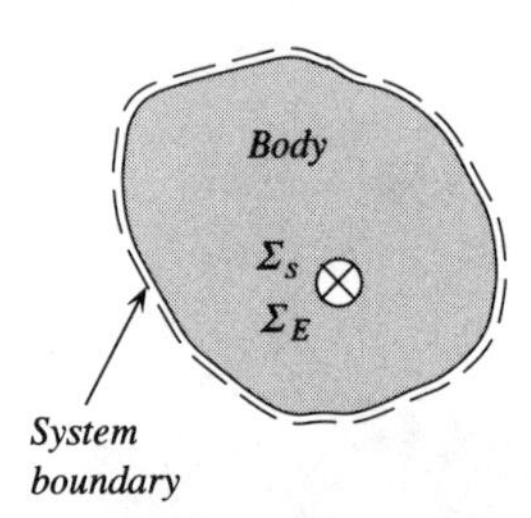

FIGURE 16. A material body and a radiation field occupy the same region of space (only the body is shown). Emission of radiation means that the body loses entropy and energy (and other quantities) at every point inside the field. In this view, the body and field are two separate physical systems.

We should be interested in the rate of production of entropy as a result of the emission or the absorption of radiation. If the processes of emission and absorption are irreversible, the source rates of entropy for the body and the field are not the same. We have expressed this point in Equation (8):

$$\Sigma_s = -\Sigma_{s,field} + \Pi_s \quad (49)$$

Together with Equation (48), we can express this in the form

$$\Sigma_E = T\left(-\Sigma_{s,field} + \Pi_s\right) \quad (50)$$

Consider the case of no entropy being supplied by radiation which is absorbed by a body. Under these circumstances all the entropy leading to the heating of the body must have been produced, the process being completely irreversible. Put differently, all the energy supplied to the body via the field has been dissipated, and Equation (50) is formally equivalent to the expression for the relationship between the rate at which energy is dissipated and entropy is generated (see Equation (37) of Chapter 1).

In summary, we may interpret the results of emission and absorption of entropy in terms of the cooling or heating of bodies. The only difference from the case treated so far, i.e., heating by conductive surface currents, is that we have to deal with source rates of entropy and energy, Equation (48), instead of fluxes (Equation 13 of Chapter 1). In this model, irreversibility is associated with an extra element placed between the field and the body (Figure 6).

EXAMPLE 15. Absorption of solar radiation: the balance of entropy.

A body absorbs a fraction f of the energy current associated with solar radiation I_E intercepted by it. Represent the losses to the environment in terms of a total heat transfer coefficient h. Assume that solar radiation does not carry any entropy. (Because of the high temperature associated with solar radiation, this assumption is quite applicable here.)
a) Calculate the sum of the rates of entropy generation due to absorption of radiation and losses. b) Show that you obtain the same result using the balance of entropy for the body if you take the system boundary to coincide with the environment at temperature T_a.
c) Compare the magnitude of the effects for a body with a surface area of 1.0 m^2 at a temperature of 50°C absorbing 80% of an energy flux of 1000 W/m^2 in an environment of 20°C. The heat transfer coefficient has a value of 10 $W/(K \cdot m^2)$.

SOLUTION: For the solution of the problem we will need the equation of balance of energy for the body:

$$\dot{E} = \Sigma_E - I_{E,loss} = f I_E - hA\left(T - T_a\right)$$

a) Entropy production is due to two distinct irreversible processes, the absorption of radiation and heat transfer to a colder body (the environment). According to Equation (50), the rate of production due to absorption of radiation is

$$\Pi_{s,1} = \frac{\Sigma_E}{T}$$

The rate of production of entropy as a result of heat transfer, on the other hand, is given by

$$\Pi_{s,2} = hA(T - T_a)\left(\frac{1}{T_a} - \frac{1}{T}\right)$$

b) If we consider the body as our system and draw the system boundary at the location of the environment at temperature T_a, we include in the system the part responsible for heat transfer. In this case the equation of balance of entropy takes the form

$$\dot{S} + I_s = \Pi_s$$

Remember that the radiation is assumed not to deliver any entropy, so there is no source term. Now we have

$$\Pi_s = \frac{\dot{E}}{T} + \frac{I_{E,loss}}{T_a} = \frac{fI_E - hA(T - T_a)}{T} + \frac{hA(T - T_a)}{T_a}$$

This result is equivalent to what we obtained by calculating the rates of production independently.

c) Inserting the numbers into the expression obtained in a) gives values of 2.5 W/K and 0.095 W/K, respectively.

3.2.7 The Heat Loss Coefficient for Flat–Plate Solar Collectors

Solar collectors provide a very nice application of heat transfer as we have discussed it so far. Basically, all three modes of transport take place in the process, which leads to loss of heat from a collector to the surroundings, with radiation and convection at interfaces being the most important.

Solar collectors receive radiation from the Sun; they lose heat to the surroundings when heated above the level of ambient temperature; and they remove heat via fluid flow through the device. Here, we will consider the problem of calculating heat loss. Removal of heat will be dealt with in Section 4.6.1, while the absorption of radiation will be the subject of Section 3.6.4.

To define the problem of the exchange of entropy and energy with the surroundings, take a closer look at Figure 17. A typical collector consists of an absorber for solar radiation, a duct for the fluid below the absorber which is insulated at the back, and possibly one or two glass covers to reduce top heat loss. The latter process will be the subject of interest in this section. Naturally, heat also may be lost to the back and to the sides, but these effects will not be considered here.

Assume a collector having a single cover made of a sheet of glass. Heat loss from the absorber plate to the environment is the result of the combined effects of radiation and convection from the plate to the cover, and from the cover to the air surrounding the collector. Radiation and convection act as parallel modes of transfer from one body to the next, while the transports from the absorber to the cover, and from the cover to the surroundings are in series. Therefore, the combined effect of all modes of transfer can be described by the simple equivalent circuit also shown in Figure 17.

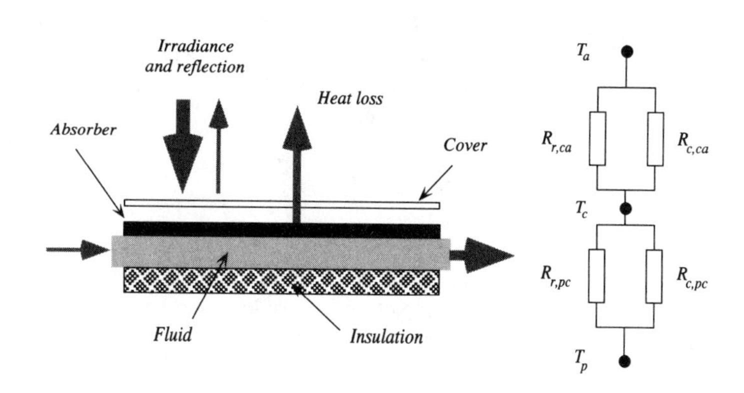

FIGURE 17. Heat loss of a solar collector occurs mostly through the top. The entropy and energy not carried away by the fluid will be transferred to the surroundings. The network on the right symbolizes the combined effect of radiation and convection from the absorber plate to the cover (which here is a single sheet of glass), and from the cover to the surroundings. Subscripts p, c, and a denote the absorber plate, the cover, and the ambient air, respectively, while r and c stand for radiation and convection. R is a heat transfer resistance.

If we use heat transfer coefficients h instead of resistances R to calculate the total energy flux due to loss, we can write

$$I_{E,loss} = A_c U_t \left(T_p - T_a\right) \quad (51)$$

where

$$\frac{1}{U_t} = \frac{1}{h_{pc}} + \frac{1}{h_{ca}} = \frac{1}{h_{r,pc} + h_{c,pc}} + \frac{1}{h_{r,ca} + h_{c,ca}} \quad (52)$$

A_c is the net surface area of the collector (essentially the surface area of the absorber), and U_t symbolizes the total heat transfer coefficient for top loss. Note that conductive transport through the thin glass cover has been neglected in this analysis. If we wish to compute the heat loss coefficient we have to be able to quantify the heat transfer coefficients for convection and for radiation. While the former pose a problem which we cannot solve at this point,[2] the latter can be expressed in terms of what we have learned so far.

The radiative heat transfer coefficient from the absorber to the cover can be written in the form used in Example 11. If we apply the result derived for radiation between two parallel plates, Equation (46), we obtain

$$h_{r,pc} = \frac{\sigma\left(T_p^2 + T_c^2\right)\left(T_p + T_c\right)}{1/e_1 + 1/e_2 - 1} \quad (53)$$

Obviously, to calculate this heat transfer coefficient, we need to know the temperatures of both the absorber and the cover. While the former has to be speci-

2. We simply have to assume reasonable convective heat transfer coefficients for both transfer from the cover to the surrounding air, and from the absorber to the cover. See Duffie and Beckman (1991; p. 160–176) for a detailed discussion of the problem.

fied, the latter has to be obtained as part of the solution of the problem. Now we still need the radiative transfer coefficient for transport from the cover to the surroundings. Radiation occurs between the cover, which has a particular emittance, and the sky for which we must use an equivalent blackbody temperature[3] T_S. Therefore, the coefficient turns out to be

$$h_{r,ca} = \sigma e_c\left(T_c^2 + T_s^2\right)\left(T_c + T_s\right) \tag{54}$$

See Equation (44). The temperatures strongly depend upon operating conditions, while the convective heat transfer coefficient from the cover to the ambient air is a function of wind speed. Typical values for heat loss coefficients of collectors of the type described are around 5 W/(K · m^2).

3.2.8 Time–Dependent Heat Transfer in Discrete Models

So far, we have treated only steady-state processes, i.e., cases where the conditions do not change with time. As soon as we allow the amount of entropy in a body to vary with time we have to deal with dynamics. Dynamical processes abound in nature, so we will approach them here in a simple setting.

Basically, bodies are both conductors and storage elements of entropy. Expressed differently, while entropy is flowing through bodies, it is stored there at the same time. This leads to spatially inhomogeneous situations which change in time. The combination of both effects necessitates the use of partial derivatives and partial differential equations. However, it is also possible to model these processes using spatially separate parts for storage and flow. Naturally, the treatment of such *discrete* systems is much simpler than that of the *continuous* ones. We shall consider discrete systems before turning to the continuous case (Section 3.4).

Spatially uniform systems. We can conceive of models of heat transfer and storage processes in which the elements are separated: while storage occurs in one body (say, the liquid in a bottle), transfer of entropy takes place through a system which is spatially separated from the former (the walls of the bottle). This allows us to treat the system which contains entropy as spatially uniform in the manner of Chapter 2. The transfer of entropy, on the other hand, takes place under steady-state conditions, since we assume the conductor to be unable to act as a storage element. In this manner we have effectively separated the phenomena whose interplay leads to time-dependent processes. (In Chapter 2, the description of the heating of uniform bodies was cast in the form of a theory of dynamics as well; however, since nothing was said about the details of heat transfer, the resulting theory appeared to be one which does not give time a proper role.)

3. See Duffie and Beckman (1991), p. 158.

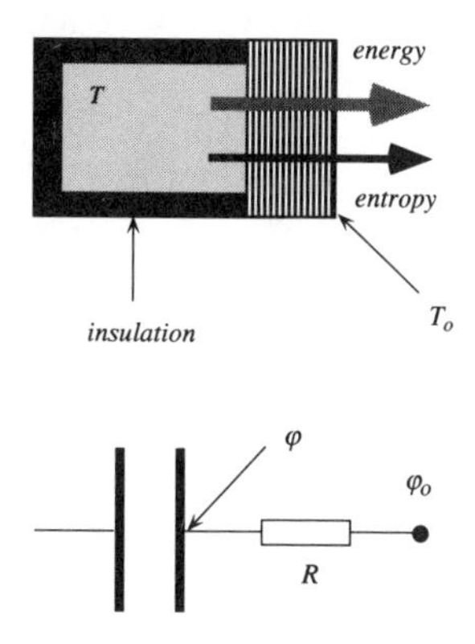

FIGURE 18. Comparison of discharging of a capacitor with the cooling of a homogeneous body. In electricity we commonly treat the simple model of a capacitor without resistance, and a resistor without capacitance: the two elements are separated in our model. In thermal physics this means that we consider a body which cools uniformly; in other words, we assume the body to be a perfect conductor.

Consider a body of some substance having a high thermal conductivity enclosed by the walls of a container (Figure 18). Effectively, we take the storage element to be an ideal conductor. The conductivity of the walls, on the other hand, is taken to be much smaller than that of the system enclosed by them. However we do not want to make it so low that it effectively prevents the flow of entropy out of or into the enclosure. The high conductivity of the material in the container leads to an almost uniform temperature at all times, which allows us to treat the body as homogeneous in space, subject to uniform processes. If we exclude changes of volume of the enclosed system, the only variable which can change is its temperature. Alternatively, we might consider a phase change as the only process the substance can undergo.

Entropy and energy which flow into or out of the enclosed body have to pass through the walls of the container and the boundary layers developing on the walls. We attribute the transport processes to these elements. They conduct, convect, and radiate entropy according to the rules investigated in the previous sections. However, they do *not* store entropy, or at least we assume this not to be the case. For this reason the entropy fluxes obey the constitutive laws of heat transfer discussed above even though they may change in time. Therefore we are permitted to treat the heat transferring system as a simple resistor incapable of storing entropy and energy.

The storage of entropy in a superconducting system combined with a transmitter which cannot store entropy is analogous to our treatment of simple electric circuits (Figure 18). A capacitor commonly is treated as having no resistance, while the resistor is said not to be a capacitor at the same time. (Let us exclude inductors and inductances. It is an interesting question whether such elements exist in thermal processes as well; see Section 3.4.6 and the Epilogue.) In this manner two phenomena which are linked in nature are separated in our models of discrete systems.

The balance of entropy. The mathematical description of the dynamical processes occurring in discrete systems is rather simple. The equations of balance of entropy or energy expressed for homogeneous systems, combined with the appropriate constitutive laws for the currents suffice for our purpose. Let us base the derivation of the equations on entropy. For a uniform body entropy is conserved:

$$\frac{dS}{dt} = -I_s \tag{55}$$

The currents of entropy are counted at the boundary of the system, which in our model, is the place where the storage element and the thermal resistor join. This point has to be made plain, since entropy is created in the resistor.

Constitutive relations. There are two more steps in the analysis which have to be performed. First, we have to express the rate of change of the entropy stored in the system in terms of other variables which are subject to change. Two cases are commonly treated. If the systems are cooled or heated at constant volume without a phase change occurring, the change of entropy is related to the change of temperature of the system. In this case the equation of continuity of entropy takes the form

$$K\frac{dT}{dt} = -I_s \tag{56}$$

K is the entropy capacity of the system in the container. The other possibility often considered is that of melting or solidification as the result of heating or cooling:

$$\bar{l}_f\frac{dn}{dt} = -I_s \tag{57}$$

Here, $\bar{l}_f$ is the molar entropy of fusion of the substance. Both equations represent ordinary initial value problems rather than partial differential equations.

The second step in our analysis has to do with the formulation of an expression for the net flux of entropy. There is nothing new about this. We have analyzed the conditions for heat transfer in previous sections. We simply apply the rules found there for calculating conductive, convective, and radiative fluxes, or their combination. The fluxes of entropy can be expressed in terms of the energy flux and the temperature at the surface of the system. If, for example, the flux of energy is given by the difference of temperatures between the body's surface and the surroundings, and by a thermal resistance with respect to energy, we may write for the flux of entropy:

$$I_s = -\frac{1}{T}\frac{T_o - T}{R_E} \tag{58}$$

where R_E is the thermal resistance with respect to energy. Finally, we introduce the result in the differential equations (56) or (57). If we take the former, we obtain the following differential equation:

$$C\frac{dT}{dt} = -\frac{T_o - T}{R_E} \tag{59}$$

$C = TK$ is the temperature coefficient of energy of the storage element. We could have obtained this equation from the consideration of the balance of energy as well. The following examples give an impression of the cases encountered.

EXAMPLE 16. Cooling a bottle of wine.

A bottle of white wine is placed in a refrigerator whose inner temperature we take to be constant at 0°C. How long will it take for the temperature of the wine to decrease from an initial value of 20°C to the desired 8°C?

Treat the wine as a uniform system of mass 0.75 kg. Use the constitutive quantities of water. The bottle is made out of glass with a thickness of 5.0 mm. The height and the diameter of the main body of the bottle are 25 cm and 8 cm, respectively; neglect its bottom and its neck, and treat the geometry as plane. The convective transfer coefficients inside and outside are 200 W/(K · m^2) and 10 W/(K · m^2), respectively.

SOLUTION: The problem has much in common with the discharging of an electrical capacitor. We will introduce a time constant in analogy to the electrical case. Here the time constant must be the product of the energy capacity and the thermal resistance (with respect to energy). The thermal resistance turns out to be

$$R_E = \frac{1}{h_{tot} \cdot A} = \frac{1}{A}\left[\frac{1}{h_{a1}} + \frac{1}{h_{a2}} + \frac{\Delta x}{k_E}\right]$$
$$= \frac{1}{2\pi \cdot 0.040 \cdot 0.20}\left[\frac{1}{10} + \frac{1}{200} + \frac{0.0050}{1.0}\right]\frac{\mathrm{K}}{\mathrm{W}} = 2.2\frac{\mathrm{K}}{\mathrm{W}}$$

Now the equation of balance looks like:

$$C\frac{dT}{dt} = -\frac{T - T_o}{R_E}$$

As you can verify, its solution is

$$T - T_o = (T_i - T_o)\exp\left(-\frac{t}{\tau}\right)$$

with the time constant

$$\tau = R_E C = 2.2 \cdot 0.75 \cdot 4200\mathrm{s} = 6900\mathrm{s}$$

T_i is the temperature of the wine at the beginning of the cooling process. If we solve the differential equation for the time t, we obtain a value of 6300 seconds, or 105 minutes. (A colleague of mine tells me that he measured the time constant of a bottle of Swiss wine to be some 100 minutes.)

EXAMPLE 17. Formation of ice on the surface of a lake.

While the temperature of the air is - 10°C, ice forms on the surface of a lake. How long does it take from the time ice begins to form for the sheet to reach a thickness of 20 cm? Take the temperature of the air to be constant. The convective transfer coefficient from ice to the air is 10 W/(K · m^2). The conductivity with respect to energy of ice is 2.2 W/(K · m). Neglect the transfer from the water to the ice.

SOLUTION: For ice to form at the surface of the lake, the water there must have reached a temperature of 0°C. Heat flows from the water into the air, first directly, and later through the ice; therefore water will freeze. If we can calculate the rate of formation of ice, we can determine the rate at which the thickness of the sheet grows.
The proper equation of balance describing the formation of ice is Equation (57) where m stands for the mass of the water which is freezing. If we take the mass of ice instead, we have to change the sign appearing in front of the energy current. This current is determined by the rules discussed in Section 3.2.3:

$$I_{E,th} = -\frac{\Delta T}{R_E} = -\left(T_{air} - T_{water}\right)A\left[\frac{1}{h_a} + \frac{x}{k_E}\right]^{-1}$$

If ice has not formed yet, the total transfer coefficient is the convective transfer coefficient alone. Furthermore, we can express the rate of change of the mass of the ice in terms of the rate of change of its thickness:

$$q\frac{dm}{dt} = q\rho_{ice}A\frac{dx}{dt}$$

Taken together, we get the following differential equation for the thickness of the sheet of ice:

$$q\rho_{ice}A\frac{dx}{dt} = -\left(T_{air} - T_{water}\right)A\left[\frac{1}{h_a} + \frac{x}{k_E}\right]^{-1}$$

Separating the variables and integrating, we get

$$\int_0^t dt = -\frac{q\cdot\rho_{ice}}{h_a\cdot k_E\left(T_{air} - T_{water}\right)}\int_0^x \left(k_E + h_a x\right)dx$$

For the sheet of ice to grow to a thickness of 0.20 m we have to wait for a time

$$t = \frac{q\cdot\rho_{ice}}{h_a\cdot k_E\left(T_{water} - T_{air}\right)}\frac{1}{2h_a}\left[(k_E + h_a x)^2 - k_E^2\right]$$
$$= \frac{3.3\cdot 10^5\cdot 920(4.2^2 - 2.2^2)}{2.0\cdot 10^2\cdot 2.2\cdot(10-0)}\text{s} = 8.8\cdot 10^5\text{s}$$

This corresponds to about 10 days, which seems to be a pretty reasonable time span.

EXAMPLE 18. Heating of a current carrying wire.

A tungsten wire of given length and radius is heated by a constant electrical current. The voltage is assumed to be constant as well. The wire is to be placed in an evacuated container whose walls are kept at constant room temperature T_o. Even though the absorptivity of tungsten varies strongly with temperature, assume its value to be constant. Derive the differential equation for the temperature of the wire as a function of time.

SOLUTION: Entropy is created in the wire as a consequence of dissipation of energy, and it is radiated from the surface of the wire. Therefore the equation of balance of entropy takes the following form:

$$\dot{S} = \Sigma_s + \Pi_s$$

The production rate of entropy in the wire is the rate of dissipation of energy divided by the instantaneous temperature:

$$\Pi_s = \frac{1}{T}\left|U I_q\right|$$

while the source rate of entropy is expressed in terms of the source rate of energy due to radiation from the surface and the temperature:

$$\Sigma_s = -\frac{1}{T} a\sigma 2\pi r l \left(T^4 - T_o^4\right)$$

T_o is the temperature of the surroundings. The rate is negative since entropy is emitted. If we now express the rate of change of the entropy of the wire in terms of its entropy capacity and the rate of change of the temperature:

$$\dot{S} = mk\dot{T} = \frac{mc}{T}\dot{T} = \frac{\rho\pi r^2 lc}{T}\dot{T}$$

we arrive at the following differential equation:

$$\frac{\rho\pi r^2 lc}{T}\dot{T} = -\frac{1}{T} a\sigma 2\pi r l \left(T^4 - T_o^4\right) + \frac{1}{T}\left|U I_q\right|$$

We could have found the same equation on the basis of the balance of energy. This differential equation can be separated and integrated, which yields a rather complex expression.

3.3 Heat Transfer and Entropy Production

Heat transfer generally is irreversible, which means that it is accompanied by entropy production. Entropy is generated in conductive heat flow, it is produced in convection as a result of viscous fluid flow, and it is created when radiation is absorbed and emitted. The following sections discuss irreversibility in heat transfer. Examples demonstrate that conditions often can be found which minimize entropy production.

3.3.1 A Model of Two Uniform Bodies in Thermal Contact

The first example deals with heat transfer between two rigid bodies in thermal contact. So far, we have dealt with this problem in the context of uniform bodies (Section 2.9) which did not allow for calculating time rates of change; all we could do was to compute the final temperature reached, and the total amount of entropy produced.

Now we shall extend the model to include the effect of conductive heat transfer in the simplest possible setting. This involves the introduction of a thermal conductor between two bodies treated as uniform. (Figure 6 can be used to represent the model if we replace the field by the second body.) This leads to the roughest possible "finite element" approximation of the continuous case. (How to calculate the effect of spatially continuous conductive heat transfer can be seen in Section 3.4.6.) We will see how this simple model leads to processes with a real time evolution, how it compares to a better finite element approximation, and how much entropy is produced until both bodies have reached the same final temperature.

The model can be described using the entropy or the energy balances for the bodies; remember that the conductor is treated as an element which does not store entropy. Let us use the balance of energy:

$$C\dot{T}_1 = -I_E \quad , \quad T_1(0) = T_{1i}$$
$$C\dot{T}_2 = I_E \quad , \quad T_2(0) = T_{2i} \tag{60}$$

The bodies are taken to be identical, with different initial temperatures. The energy current flowing from the hotter to the cooler body is expressed by

$$I_E = \frac{k_E A}{\Delta x}(T_1 - T_2) \tag{61}$$

See Equations (20) and (21). A is the cross section of the bodies (the surface area where they are touching), k_E is the thermal conductivity, and Δx is a distance which might be taken to be the distance of the centers of the two bodies. We now divide the differential equations by the temperature coefficients of energy, and then subtract the second from the first. This leads to a differential equation for the difference of temperatures $\Delta T = T_1 - T_2$:

$$\frac{d}{dt}\Delta T = -2\frac{k_E A}{\Delta x \cdot C}\Delta T \tag{62}$$

whose solution turns out to be

$$\Delta T = \Delta T_i \exp\left(-\frac{t}{\tau}\right) \quad , \quad \tau = \frac{\Delta x \cdot C/2}{k_E A} \tag{63}$$

where ΔT_i is the initial temperature difference of the bodies (see Figure 19).

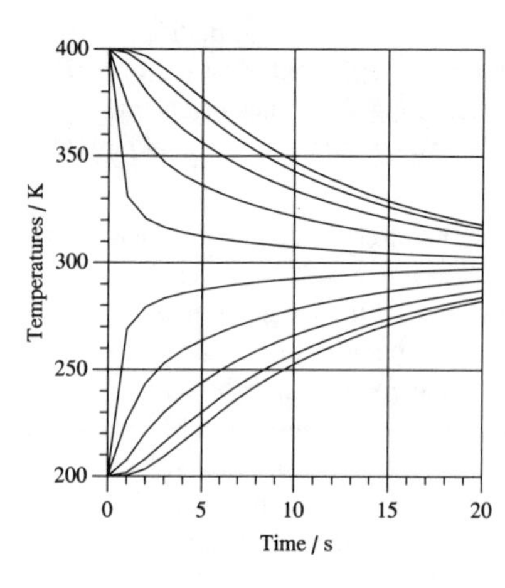

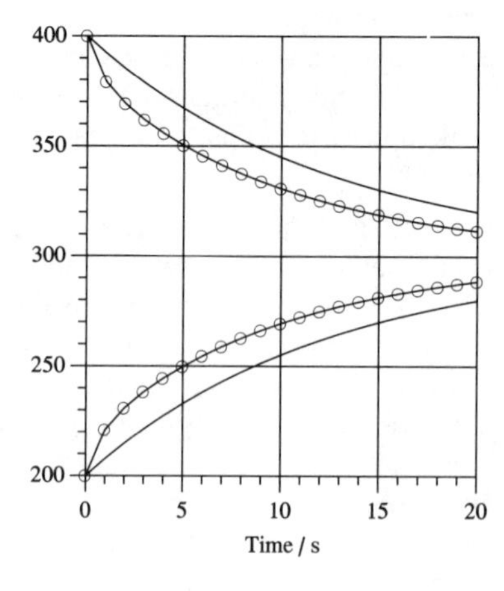

FIGURE 19. A long thin bar has an initial temperature distribution with its left half at 400 K and its right half at 200 K. The first of the diagrams depicts the temperatures at evenly spaced points throughout the bar as a function of time. The second graph shows the average temperature of the two halves of the bar (circles) and the solution computed according to Equation (63) with Δx equal to half of the length of the bar. (Values are: $k_E = 1$, $C = 5$, $\Delta x = 5$, $A = 1$.) A judicious choice of Δx can make the solutions quite similar, and the rough model can serve as an estimate of what is happening in the bar.

The irreversibility is measured in terms of the production of entropy. The rate of generation of entropy is expressed by

$$\begin{aligned} \Pi_s &= I_E\left(\frac{1}{T_2} - \frac{1}{T_1}\right) = \frac{k_E A}{\Delta x}(T_1 - T_2)\frac{T_1 - T_2}{T_1 T_2} \\ &= \frac{k_E A}{\Delta x}(T_1 - T_2)^2 \frac{1}{(T_f + \Delta T/2)(T_f - \Delta T/2)} \end{aligned} \quad (64)$$

If we introduce the solution into this formula, we obtain

$$\Pi_s = \frac{k_E A}{\Delta x}\frac{4\exp(-2t/\tau)}{4(T_f/\Delta T_i)^2 - \exp(-2t/\tau)} \quad (65)$$

Integration of this expression over time (from zero to infinity) leads to[4]

$$S_{prod} = C\ln\left(\frac{4(T_f/\Delta T_i)^2}{4(T_f/\Delta T_i)^2 - 1}\right) = C\ln\left(\frac{4T_f^2}{4T_f^2 - \Delta T_i^2}\right) = C\ln\left(\frac{T_f^2}{T_{1i}\cdot T_{2i}}\right) \quad (66)$$

The last form is equivalent to the one obtained from thermostatic considerations alone (Example 40 in Chapter 2).

3.3.2 Fluid Flow and Minimal Production of Entropy

Many cases of heat transfer lead to situations with two (or more) sources of irreversibility, one of which increases as a parameter is changed, while the other decreases. With such counteracting influences, we may expect a condition to exist for which the total rate of production of entropy becomes minimal.

Consider as an example the flow of hot oil through a pipe. For a given volume flux, if the radius of the pipe is decreased, the effect of fluid friction increases, while the loss of heat decreases. Now calculate the rate of production of entropy from the two sources. If we model the viscous flow of the oil using the law of Hagen-Poiseuille, the first contribution leads to

$$\Pi_{s,friction} = \frac{1}{T}\mathcal{P}_{hydro} = \frac{1}{T}R_V I_V^2 = \frac{1}{T}\frac{8\eta l}{\pi r^4}I_V^2 \quad (67)$$

4. The solution is obtained from

$$\int \frac{\exp(ax)}{b + c\exp(ax)}dx = \frac{1}{ac}\ln\left(b + c\exp(ax)\right)$$

where T is the temperature of the fluid (taken to be uniform over the length of the pipe). Heat loss leads to a rate of generation of entropy equal to

$$\Pi_{s,hl} = I_E\left(\frac{1}{T_a} - \frac{1}{T}\right) = h2\pi rl(T - T_a)\left(\frac{1}{T_a} - \frac{1}{T}\right) \tag{68}$$

A graph of the contributions to entropy production (Figure 20) demonstrates the claim made before: there exists a particular value of the radius of the pipe for which the total rate of generation is minimal. The formal calculation

$$\frac{d\Pi_s}{dr} = \frac{d}{dr}\left[\frac{1}{T}\frac{8\eta l}{\pi r^4}I_V^2 + h2\pi rl(T - T_a)\left(\frac{1}{T_a} - \frac{1}{T}\right)\right]$$

$$= -\frac{4}{T}\frac{8\eta l I_V^2}{\pi}\frac{1}{r^5} + h2\pi l(T - T_a)\left(\frac{1}{T_a} - \frac{1}{T}\right) = 0$$

leads to

$$r_{min} = \left[\frac{4}{T}\frac{8\eta l I_V^2}{\pi}\left(h2\pi l(T - T_a)\left(\frac{1}{T_a} - \frac{1}{T}\right)\right)^{-1}\right]^{0.2} \tag{69}$$

for the radius of the pipe yielding minimal value of the rate of generation of entropy. The particular law of fluid friction used here can be replaced by more general expressions known from the theory of fluid flow.

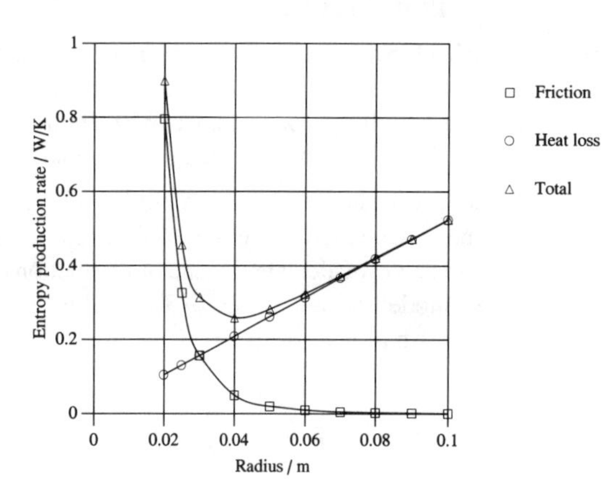

FIGURE 20. The rate of production of entropy shows different behavior as a function of the radius of the pipe for heat loss and for fluid friction. The net result is such that there exists a radius for which the rate of generation is minimal. Values taken are 0.2 Pa · s for viscosity, 1m for the length of the pipe, 10 W/(K · m^2) for the heat transfer coefficient, 0.01 m^3/s for the volume flux, and 400 K and 300 K for the temperatures, respectively.

EXAMPLE 19. Minimizing "total loss" in fluid flow with heat loss.

Define the total loss of power in the case of hot oil flowing through a pipe in terms of energy loss due to heat loss and of friction. Calculate the radius of the pipe for which the loss is minimal. Why do you get a different result than the one calculated above?

SOLUTION: The loss incurred because of pumping is calculated in terms of the hydraulic power

$$\mathcal{L}_{friction} = \mathcal{P}_{hydro} = R_V I_V^2 = \frac{8\eta l}{\pi r^4} I_V^2$$

The loss of energy due to heat loss is given by

$$\mathcal{L}_{hl} = h 2\pi r l (T - T_a)$$

The sum of these contributions has a minimum for a value of the radius of the pipe given by

$$r_{min} = \left[4 \frac{8\eta l I_V^2}{\pi} \frac{1}{h 2\pi l (T - T_a)} \right]^{0.2}$$

A loss expressed directly in terms of energy loss is not equivalent to the expression for the total irreversibility (computed in terms of the rate of generation of entropy). If you wish to work with quantities having to do with energy, you have to compute the minimum of the loss of available power rather than the loss of power; only then will you get the same result as the one obtained on the basis of the consideration of entropy production. The former is what you should calculate if you want to quantify the losses to a power plant supplied with heat from a hot fluid.

3.3.3 Charging and Discharging a Heat Storage System

The problem of minimizing entropy production also occurs in time-dependent problems involving the storage of entropy. Again, rather than delivering a piece of theory, we will treat an interesting example.

Consider a building having a heat storage wall. In fact, just consider the wall itself, which obtains heat from the Sun and discharges heat both to the surroundings and to the room (Figure 21). Take the control volume to include the wall, and let the surfaces touch the environment on the outside and the air of the room inside; i.e., include in the control volume all three sources of entropy production. Now, the law of balance of entropy can be written in the form

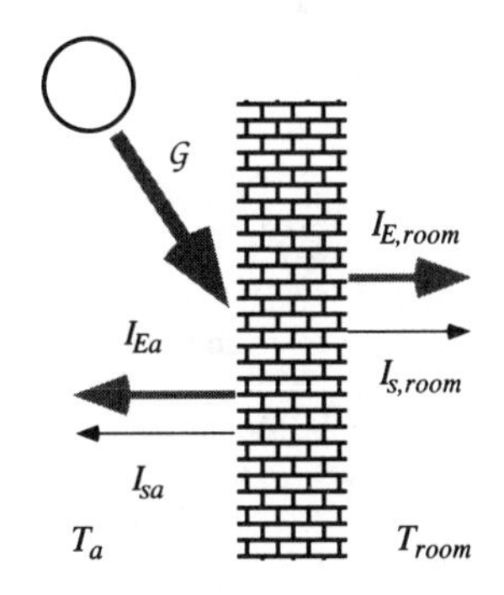

FIGURE 21. Solar radiation absorbed by a wall leads to three sources of irreversibility. One is associated with the absorption of radiation. The other two have to do with heat flow into the environment and into the room.

$$\dot{S} = -I_{sa} - I_{s,room} \tag{70}$$

where we have assumed solar radiation not to deliver any entropy. The currents and the rate of change are expressed by

$$\begin{aligned} I_{sa} &= \frac{1}{T_a} h_a A (T - T_a) \\ I_{s,room} &= \frac{1}{T_{room}} h_{room} A (T - T_{room}) \\ \dot{S} &= \frac{1}{T} \dot{E} = \frac{1}{T} \left(-I_{Ea} - I_{E,room} + a\mathcal{G} \right) \end{aligned} \tag{71}$$

The last expression is a consequence of the balance of energy, which takes the form

$$\dot{E} = -I_{Ea} - I_{E,room} + a\mathcal{G} \tag{72}$$

Here, $a\mathcal{G}$ is the product of absorptivity of the wall and solar irradiance. The problem is to find expressions for the heat transfer coefficients from the wall to the surroundings and the room, and one for the relation between energy (or entropy) and the temperature of the wall. If we treat the wall as spatially uniform, and if we take the model used in Section 3.3.1 for the computation of the transfer coefficients, we get

$$\begin{aligned} &\dot{E} = C\dot{T} \\ &\frac{1}{h_a} = \frac{1}{h_{wall-air}} + \frac{d/2}{k_E} \\ &\frac{1}{h_{room}} = \frac{1}{h_{wall-room}} + \frac{d/2}{k_E} \end{aligned} \tag{73}$$

where d is the thickness of the wall, and heat is assumed to be transferred from the middle of the wall out into the environment or into the room. Now it turns out that the total amount of entropy produced over a cycle of charging and discharging of the storage element has a minimum for a particular value of the thickness of the wall (Figure 22). Not surprisingly, this value also coincides with the maximum of the entropy delivered to the room.

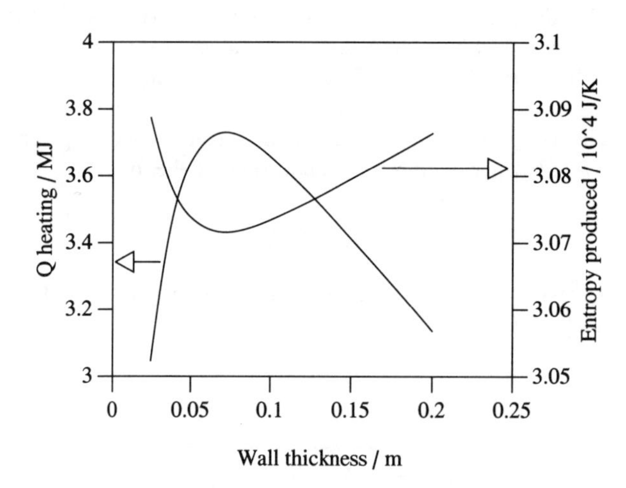

FIGURE 22. Numerical integration of the equations of balance for the storage wall. The wall is considered to be charged for 8 hours, during which time it also loses heat to the surroundings. (It starts at a temperature of 20°C.) After charging, it is allowed to lose heat to the room and to the environment. With the environment and the room at temperatures of 0°C and 20°C, respectively, the calculation is performed up to the time when the wall has again reached a temperature of 20°C. The rate of absorption of energy from radiation is set equal to 300 W, while the temperature coefficient of energy of the wall is assumed to be proportional to the wall thickness.

3.3.4 Radiative Transfer and the Production of Entropy

Next, we will discuss the rate of production of entropy in the emission and absorption of radiation. Consider the case of emission: the rate at which entropy is emitted to the field is smaller than the rate at which it flows away from the surface of the body (Figure 12). This can be shown quite easily. Consider the model of a uniform body at temperature T for which

$$\Sigma_E = T\Sigma_s \tag{74}$$

On the other hand, the flux of entropy through the field at the surface of the body is four-thirds this quantity; see Equations (34) and (35):

$$\begin{aligned} I_{s,rad} &= \frac{4}{3}\frac{I_{E,rad}}{T} \\ &= \frac{4}{3}\frac{\Sigma_E}{T} \\ &= \frac{4}{3}\Sigma_s \end{aligned} \tag{75}$$

We have to conclude that more entropy leaves the space occupied by the body than has been emitted by the body to the field. Therefore, entropy must have been produced in the volume occupied by radiating matter. In the same manner, we can prove that it is impossible for a body just to absorb entropy from a source at the same temperature. Entropy would have to be destroyed, which we know to be impossible. Therefore, it is impossible for a body to absorb entropy at the same temperature without emitting entropy at the same time.

Let us discuss the example of emission and absorption in a quantitative manner. Take two bodies having geometries such as in Figure 14: the walls of a cavity completely surround a smaller body. Assume both surfaces to be black bodies. Let the smaller body be the hotter one, with temperature T_1 and surface area A. Then the net flux of energy from the smaller to the larger surface is

$$I_E = \sigma A\left(T_1^4 - T_2^4\right) \tag{76}$$

See Equation (36). T_2 is the temperature of the enclosure. Now we can compute the rate of production of entropy in the two bodies combined:

$$\begin{aligned} \Pi_S &= \dot{S} + \sum I_{s,i} \\ &= \dot{S} \end{aligned} \tag{77}$$

The sum of all fluxes with respect to both bodies combined is zero. Now the rate of production of entropy is the sum of the terms for each of the bodies. Each of these values is the rate of absorption or emission of energy divided by the respective temperature. Therefore we find

$$\begin{aligned} \Pi_S &= \dot{S}_1 + \dot{S}_2 \\ &= \frac{1}{T_1}\dot{E}_1 + \frac{1}{T_2}\dot{E}_2 \\ &= \left(\frac{1}{T_2} - \frac{1}{T_1}\right) I_E \end{aligned} \tag{78}$$

Finally, we combine Equations (76) and (78), which leads to

$$\begin{aligned} \Pi_s &= \frac{\sigma A}{T_1 T_2}(T_1 - T_2)(T_1^4 - T_2^4) \\ &= \frac{\sigma A}{T_1 T_2}(T_1 - T_2)(T_1^2 - T_2^2)(T_1^2 + T_2^2) \\ &= \frac{\sigma A}{T_1 T_2}(T_1 - T_2)^2 (T_1 + T_2)(T_1^2 + T_2^2) \end{aligned} \tag{79}$$

This expression is larger than zero as long as one of the bodies is hotter than the other, and it vanishes if they have the same hotness. Therefore, it does not matter which of the bodies we take to possess the higher temperature. Emission and absorption of entropy by bodies at different temperatures is necessarily dissipative. We could have performed the computation of the balance of entropy for each of the bodies separately. Using the result for one of them, you can convince yourself that the particular statements made above regarding the irreversibility of emission and the impossibility of absorption without simultaneous emission are correct.

The irreversibility of radiative processes is not limited to the absorption and emission of (blackbody) radiation. Conversion of monochromatic radiation into blackbody radiation and the scattering of radiation have to be added to the list of irreversible processes. The fact that irreversibility necessarily accompanies radiative transfer is of importance for power engineering.

EXAMPLE 20. Absorption of entropy by the Earth's atmosphere.

Consider the Earth to be a uniform body. a) How large is the rate at which entropy appears in the atmosphere, biosphere, and the oceans of the Earth if we take their temperature to be 300 K? The solar constant outside the atmosphere is 1.36 kW/m^2. 30% of the radiation is directly reflected back into space. b) How large is the flux of entropy through the radiation field just before radiation is absorbed? c) How large is the rate of production of entropy on the planet as a result of absorption? d) How large is the rate of entropy generation overall?

SOLUTION: a) The rate at which energy is absorbed at the surface of the Earth is 70% of the solar constant multiplied by the projected surface area of the sphere:

$$\Sigma_E = 0.7 \cdot 1360 \text{ Wm}^{-2} \pi (6.4 \cdot 10^6 \text{ m})^2 = 1.2 \cdot 10^{17} \text{ W}$$

According to Equation (48) the rate of absorption of entropy by the Earth is

$$\Sigma_s = \frac{\Sigma_E}{T} = \frac{1.2 \cdot 10^{17}\ \text{W}}{300\ \text{K}} = 4.1 \cdot 10^{14}\ \frac{\text{W}}{\text{K}}$$

b) The flux of entropy through the field is computed according to Equation (34). With a value of 5770 K for the surface temperature of the Sun, the result is

$$I_s = \frac{4}{3T_s} 1360\pi \left(6.4 \cdot 10^6\right)^2 \text{W/K} = 0.40 \cdot 10^{14}\ \text{W/K}$$

of which 70% is emitted from the field to the planet. Clearly the absorption of solar radiation by the Earth is highly dissipative: the rate of absorption of radiation is much larger than the rate at which entropy flows toward our planet.

c) The rate of entropy production as a consequence of absorption is the difference between the entropy absorbed by the planet and the entropy emitted by the field to the planet:

$$\begin{aligned} \Pi_s &= \Sigma_s - 0.70 \cdot I_s \\ &= 4.1 \cdot 10^{14}\ \text{W/K} - 0.70 \cdot 0.40 \cdot 10^{14}\ \text{W/K} = 3.82 \cdot 10^{14}\ \text{W/K} \end{aligned}$$

d) The total rate of production of entropy due to exchange of radiation must be larger than the value just calculated. This is so because the emission of radiation again leads to the production of entropy. For the radiation flowing from the Earth, we have the following relation between fluxes of energy and of entropy:

$$I_{s,emitted} = \frac{4}{3} \frac{1}{T_E} I_{E,emitted}$$

The balance of entropy for this part of the overall process leads to an additional rate of production of $1.35 \cdot 10^{14}$ W/K. In summary, the outgoing flux of entropy is about twenty times larger than that absorbed from the field of solar radiation.

3.3.5 Maximum Power of a Solar Thermal Engine

A question of practical interest is how much of the energy radiated to us by the Sun could possibly be released for mechanical purposes by a heat engine. On the one hand, we are used to very low efficiencies of commercial photovoltaic cells (typically less than 10%); on the other hand we know that solar radiation transports entropy at a very high temperature, which suggests a large value of the Carnot efficiency. At 5800 K for the temperature of the entropy of radiation arriving from the Sun, we obtain $\eta_c = 1 - 300\text{K}/5780\text{K} = 0.95$ for the thermal efficiency of an ideal Carnot engine (see Chapter 1). Here we have taken the temperature of the environment on Earth to be 300 K. In other words, the available power of solar radiation is very high, which tells us that we should use solar radiation as a high temperature source in thermal engines.

It is clear that it is impossible to achieve this kind of efficiency even in theory. Radiative transfer is dissipative by nature. To achieve finite rates of transfer, we need differences of temperatures between the Sun and the collector on Earth.

Even if we could build an ideal Carnot engine between these elements of heat transfer, it would achieve a thermal efficiency smaller than the 95% calculated above.

The statement of the problem reminds us of the model engine discussed in Section 1.7 which we constructed to run at maximum power. We can do just the same in the case of a solar thermal engine (Figure 23).

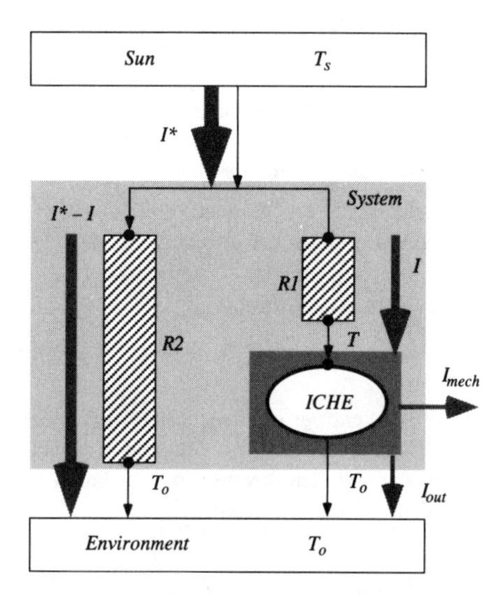

FIGURE 23. An ideal Carnot heat engine (*ICHE*) absorbs entropy from the light concentrated by mirrors. Because of the concentration, the radiation serves as an entropy reservoir at T_S, the surface temperature of the Sun. Since the upper operating temperature is smaller than T_S, part of the entropy will be radiated to the environment. Both elements of heat transfer constitute thermal resistors in which entropy is produced. The engine will achieve maximum power for minimal rate of production of entropy.

Consider the following simplified model (Figure 23). First of all, we need a concentrator for the sunlight to recreate conditions of the radiation field found at the surface of the Sun. In other words we have to use mirrors which concentrate light in such a manner as to make the solid angle from which radiation flows towards the absorber 2π (equivalent to radiation from a hemisphere). Under these conditions, a blackbody absorber could reach the same temperature T_S as that of the surface of the Sun. However, if this were the case, the net rate of transfer of entropy between the absorber and the Sun would drop to zero. For a finite rate, the absorber's temperature must be smaller than the maximum T_S. For the same reason, the engine will not be able to absorb the entire flux of entropy from the Sun which has been concentrated by the mirrors. A part of the current will be radiated into the environment, which has a temperature of about 300 K (T_o). Since the heat engine is assumed to be ideal, it rejects the same amount of entropy to the environment that it absorbs from the concentrators. Let us neglect the irreversibility of the transfer of heat taking place at the cooler; we simply assume the lower operating temperature of the engine to be T_o. Entropy will be produced due to radiative transfer from the concentrators to the absorber, and from the mirrors to the environment (Figure 23).

The power of the engine will depend on two factors—the size of its absorber and the upper operating temperature T. We expect the maximum power per unit absorber area to coincide with the minimal rate of entropy production. Therefore we have to express this rate in terms of the temperatures involved. Just as in examples treated before, the rate of entropy production in a heat transfer element is given by the energy flow through the element, and by the upper and the lower temperatures, respectively. If I^* is the energy current from the collectors, a part I will be absorbed by the engine. As a result, the rate of entropy production is

$$\Pi_s = (I^* - I)\left(\frac{1}{T_o} - \frac{1}{T_s}\right) + I\left(\frac{1}{T} - \frac{1}{T_s}\right) \tag{80}$$

The flux of energy absorbed by the engine depends on its surface area and on the temperatures T_s and T:

$$I = \sigma A\left(T_s^4 - T^4\right) \tag{81}$$

If we insert this expression into Equation (80), we obtain the rate of production of entropy in terms of the unknown temperature T. We only have to determine the derivative of Π_s with respect to T and set the result equal to zero, which yields

$$\frac{4}{T_o}T^5 - 3T^4 - T_s^4 = 0 \tag{82}$$

The solution of this nonlinear equation may be obtained by numerical methods. For values of T_s = 5762 K and T_o = 300 K, the upper operating temperature of the heat engine turns out to be 2465 K which corresponds to a thermal efficiency of η_1 = 0.88. This value is smaller than the one calculated above for a completely reversible engine. However, it still is very large, which again stresses the point that solar radiation should not in every case be wasted for low temperature applications. Some existing solar thermal power plants in the US reach efficiencies above 20%, a number which compares favorably with current photovoltaic elements.[5]

3.3.6 Winds Powered by the Sun

We can view the Earth's atmosphere as a simple heat engine which is powered by the Sun. The motive power of this engine drives the winds.[6] Obviously, the processes involved are highly dissipative, and we might ask what the efficiency of the engine would be at maximum mechanical power. As you will see, the re-

---

5. These SEGS power plants running in California are a commercial success. They use linear parabolic concentrators, which reach a concentration factor of about 40. Steam is produced at a temperature approaching 400°C. Currently, plants with a combined power of roughly 150 MW have been installed, and several more plants of the same basic design are being constructed.
6. This problem was first treated in a different manner by Gordon and Zarmi (1989).

sults of a simple model of a heat engine at maximum power yields a good upper bound on the energy of the winds on the surface of the Earth.

As a first approximation, the winds are the result of differential heating of the surface air which leads to giant cells of rising and descending air. Let us model the entire atmosphere as moving above a flat surface in one big cell (Figure 24). To minimize any problems we shall only consider spatial and temporal averages. In other words, we shall model a spatially uniform engine working in a steady state.

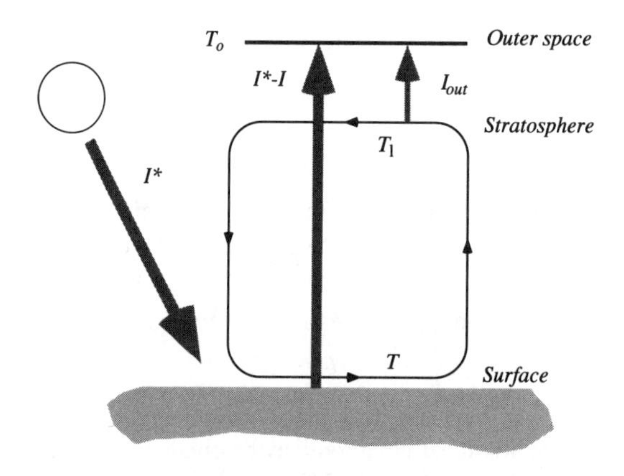

FIGURE 24. The surface of the Earth is heated by solar radiation. Part of the entropy is radiated back into space while the rest heats the air. As a result the air begins to rise and cool down. At high altitudes, the entropy absorbed is radiated into space whereupon the air descends back to the surface. In its simplest form this process constitutes a Carnot cycle which releases energy for a mechanical purpose. In our case, this is the generation of the winds. The gray arrows represent the currents of energy associated with the fluxes of entropy.

We shall consider the air as the heat engine. This engine receives entropy at an unknown temperature T near the surface. Naturally, T must be smaller than the temperature of the Sun if entropy is to be transferred radiatively from the Sun to the Earth. A fraction of the radiation will be accepted by the engine, which we consider to run ideally. The rest of the entropy absorbed by the surface will be radiated directly into space, which we assume to have a temperature T_o near 0 K. The heated air will rise adiabatically, after which the entropy it received will be radiated away into space. The temperature of the air at higher altitudes where radiation into space takes place will be T_l. After this step the air will sink back down to the surface, again without exchanging any entropy. The model presented here obviously is a relatively crude approximation of actual conditions. Nevertheless, we still can hope to obtain some interesting results.

While we have taken the model engine to work as an ideal Carnot heat engine, the entire system, which includes the radiative transfer of heat, is dissipative. Emission of entropy from a body at higher temperature and absorption of it by a different body at lower temperature leads to the production of entropy. It seems reasonable to assume that the entire system will maximize power output by minimizing the rate of production of entropy. Therefore we shall have to consider in detail the dissipation occurring as a result of the absorption and emission of entropy. For this reason we must include in the thermal system to be modeled elements allowing for the transfer of heat, which we may call heat exchangers (Figure 25). The first heat exchanger is between the Sun and the surface of the

Earth, which are at temperatures T_s and T, respectively. This element is responsible for the transfer of a certain average energy current for which we shall take a fixed value I^*. (We shall drop the index E for energy currents for convenience.) The value of this current, or rather the associated average current density over one hemisphere of the Earth, will be calculated below. This first dissipative element leads to a rate of production of entropy equal to

$$\Pi_{s1} = I^* \left(\frac{1}{T} - \frac{1}{T_s} \right) \tag{83}$$

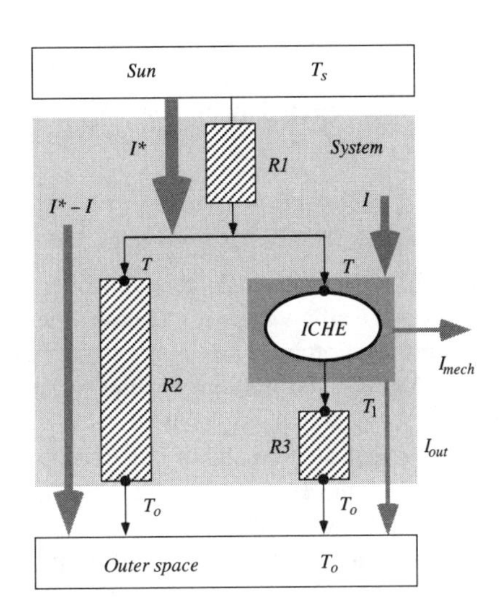

FIGURE 25. The model of the wind machine includes the sources of dissipation, namely the elements which are responsible for the radiative transfer of heat. Entropy is emitted by the Sun and (partly) absorbed by the Earth. This leads to a first source of entropy production. The heat engine composed of the atmospheric cell can take up a certain fraction of this entropy, depending on the temperatures involved. The rest is directly radiated to space, again causing entropy to be created. The entropy taken up by the wind machine is radiated into outer space from high altitudes at a lower temperature. This constitutes the third element of irreversibility.

The Carnot engine will absorb a fraction I of the total current of energy I^*. The fraction will be determined by the temperatures between which the engine will operate at maximum power. The rest will be radiated directly into space, leading to another term in the production of entropy:

$$\Pi_{s2} = (I^* - I) \left(\frac{1}{T_o} - \frac{1}{T} \right) \tag{84}$$

Finally, the entropy rejected by the Carnot engine at high altitudes is responsible for the third contribution to the production rate:

$$\Pi_{s3} = I_{out} \left(\frac{1}{T_o} - \frac{1}{T_l} \right) \tag{85}$$

We must add up the three terms and determine the minimum of this function.

The elements which make up our model in Figure 25 are governed by particular constitutive laws. First of all, the Carnot engine conserves entropy, which means that the fluxes of entropy entering and leaving it are equal in magnitude. Therefore, we have

$$\frac{1}{T}I = \frac{1}{T_l}I_{out} \tag{86}$$

The current of entropy not entering the heat engine, i.e., the one associated with the energy current density $j^* - j$, is determined by the law of Stefan and Boltzmann. If we assume the temperature T_o of outer space which receives the entropy to be much smaller than T, the law will take the form

$$j^* - j = \sigma T^4 \tag{87}$$

Finally, the current of entropy rejected by the heat engine is determined according to its associated energy current:

$$j_{out} = \sigma T_l^4 \tag{88}$$

Several unknown quantities contribute to the total rate of entropy production. We assume that they will take specific values in the case where the engine runs at maximum mechanical power. Foremost among these we have the temperatures at which the Carnot engine receives and rejects entropy. At first sight it appears advantageous to make T as high as possible and T_l as low as possible; in other words, near the temperature of the Sun and of outer space, respectively. In this case the Carnot efficiency of the engine would take its largest possible value. However, this would lead to a very small rate of transfer of entropy to the engine, or from the engine to space. In fact, radiation would simply short circuit the engine. Also, if we make the upper temperature T of the engine equal to its lower temperature T_l, we have a complete short circuit with maximum rate of entropy production. Therefore, we expect the winds to be produced at a rate which somehow minimizes the unavoidable dissipation. The Earth's atmosphere chooses the values of T and T_l so as to make this happen.

The total rate of production of entropy will now be written in a form involving current densities. For this reason, we divide the production rate of entropy by the surface area A. If we use the constitutive laws (86)-(88), we arrive at

$$\frac{\Pi_S}{A} = j^*\left(\frac{1}{T_o} - \frac{1}{T_S}\right) + \frac{1}{T_o}j\left(\frac{T_l}{T} - 1\right) \tag{89}$$

where the lower temperature of the heat engine T_l is given by

$$T_l = \left(\frac{j^* - \sigma T^4}{\sigma T}\right)^{1/3} \tag{90}$$

The derivative of Equation (89) with respect to T has to be set equal to zero, which leads to the following nonlinear expression with T as the only unknown:

$$-4\sigma T^3\left(\frac{T_l}{T}-1\right)+\left(j^*-\sigma T^4\right)\left\{-\frac{T_l}{T^2}+\frac{T_l^{-2}}{3T}\left[-4T^2-\frac{T_l^3}{T}\right]\right\}=0 \tag{91}$$

If we wish to compute the value of T for which the rate of production of entropy becomes minimal, we need to know the average energy current density j^* incident upon the surface of the Earth. The time average of the energy current density of solar radiation at the distance of the Earth is 1373 W/m^2. The planet cuts an area equal to its cross section out of the flow of entropy. The radiation, however, is distributed over the surface area of the hemisphere, which is twice the cross section. If we note that 35% of the incoming radiation will be reflected back into space without being absorbed, we arrive at a value of

$$j^* = 1373\mathrm{W/m}^2 \cdot (1-0.35)/2 = 446\mathrm{W/m}^2 \tag{92}$$

for the rate of absorption of energy by the Earth's surface. Using this value we arrive at the following solution for our problem (which has to be obtained by numerical methods since the equations are nonlinear). The rate of production of entropy, Equation (89), has a minimum in the allowed range of temperatures (roughly between 3 K and 300 K) for T = 277 K, and T_l = 193 K, which leads to energy flux densities j = 111 W/m^2, and j_{mech} = 34.2 W/m^2.

The heat engine operates only over one hemisphere, but the winds are distributed over the entire surface of the Earth; therefore we arrive at an estimate of 17.1 W/m^2 for the power of the winds averaged over the planet. This number should be compared to the measured value, which is about 7 W/m^2 averaged over time and surface. The result is very interesting in that it demonstrates that not just man-made engines may be designed according to the rule of minimal rate of production of entropy. The Earth's atmosphere seems to follow the same rule quite naturally.

3.4 The Balance of Entropy and Energy in Conduction

In this section we shall turn to the discussion of the constitutive theory of conduction. As you know, entropy can flow through bodies. However, it does so only if different parts of a body have different temperatures. Entropy flows from warmer to cooler places. Therefore it is impossible for bodies which conduct heat to be homogeneous as far as temperature is concerned. We have finally arrived at the point where we no longer can associate a single temperature with a body. Bodies through which entropy flows are *inhomogeneous*. This means that we have to describe the conduction of heat in a completely new way. We have to give every point of a body its own temperature which may differ from temperatures at arbitrarily close neighboring points. Such an approach is called a *field description* of the phenomenon.

A body which conducts entropy undergoes changes of shape and volume, and parts of it may flow relative to other parts. The description of these processes is rather complex. Therefore we will treat only the simplest possible case of conductive transport at this point—the flow of entropy through an otherwise unchanging body. We will call this situation *pure conduction* of heat. Thus the only independent variable of our theory, apart from time, will be the temperature, or rather the temperature field.

The first sections will deal with pure one-dimensional steady-state conduction without sources of entropy. One by one, the conditions will be relaxed, and we will treat sources, and nonsteady state processes. (Multidimensional flow will be discussed in the Epilogue.) Finally, we will look at a phenomenon which does not seem to have a place in everyday conductive heat transfer, namely the action of what in other fields of physics would be called inductive behavior. The normal theory of conduction predicts infinite speed of propagation of thermal pulses. Only if we extend the constitutive law of conduction to include time rates of changes of the currents of heat, will we get a theory which predicts finite speed of propagation of thermal disturbances.

3.4.1 Currents, Current Density, and Flux

In Section 1.6.1, it was pointed out that we should distinguish between currents and fluxes. There we used the term *current* to denote the phenomenon of the flow of entropy. A current would be represented by an image of flow lines. A flux, on the other hand, is a signed number which tells us how much entropy flows across a given part of the surface of a body per time. The flux is positive or negative depending on whether it belongs to a current out of or into the body, respectively.

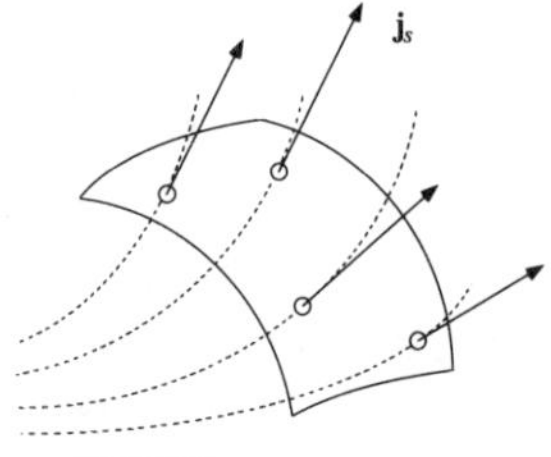

FIGURE 26. A current of heat always is distributed over a surface. The fundamental quantity describing this distribution is called the current density (or flux density) $\mathbf{j}_S$. At every point of the flow field we can imagine a vector $\mathbf{j}_S$ which determines the direction and the magnitude of the current density. The current density is a quantity rather like the vectors **E** and **B** of the electromagnetic field which represent the electric and magnetic flux densities, respectively.

A current of entropy is distributed over a surface (as in Figure 26). A natural way of describing a current is by means of a quantity which specifies this distribution. Since this quantity may vary over the surface, we want to be able to tell how much entropy flows per time at a given point of the surface. Such a quantity is the *current density* or *flux density* which is the amount of entropy crossing a surface per time and per surface area.

With the concept of current density, the distinction between current and flux can be rendered more precise. The current density actually is a vector $\mathbf{j}_S$. We take it to be the quantity which at every point in the flow field determines both the direction and the magnitude of the density of the current (Figure 26). In this chapter we will consider purely one-dimensional cases only (in which case the vector points in one of two opposing directions only).

In one dimension, the component of the current density vector obviously has a distinct sign depending on the direction of flow with respect to the choice of positive coordinate (Figure 27). This sign is independent of the body and the surface through which entropy is flowing. As you know, we also need a quantity related to the current which allows us to do the accounting for entropy with respect to a chosen body. We have to be able to tell how much entropy is flowing across the entire surface or parts thereof. For this reason we introduce the *entropy flux* I_s

which is a signed scalar quantity defined with respect to a body and (a part of) its surface. In the simple one-dimensional case depicted in Figure 27 the magnitude of the entropy flux must be given by the product of the current density and the surface area, just as in Equation (12). To make the definition precise we have

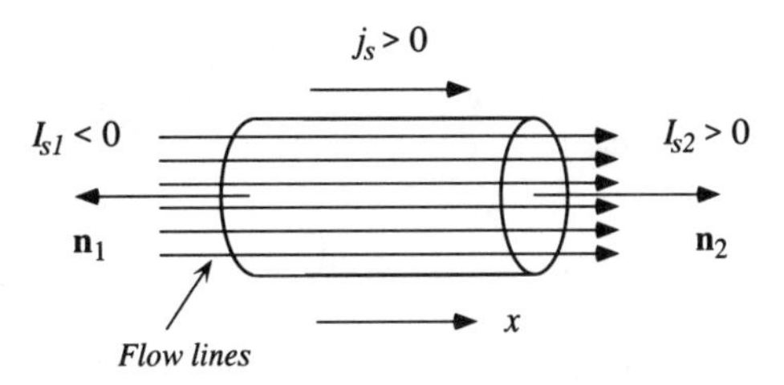

FIGURE 27. A one-dimensional current cuts through a body with plane surfaces perpendicular to the flow of heat. A flux I_s is associated with each surface. Its sign tells us whether the current is flowing into or out of the body.

to introduce the orientation of the surface of a body, using a unit vector **n** perpendicular to the surface at every point and pointing outward, i.e., away from the body under consideration (Figure 27). With this vector, the flux of entropy through either of the two surfaces of the body cut by the current in Figure 27 is given by the scalar product

$$I_s = A\,\mathbf{j}_s \cdot \mathbf{n} \tag{93}$$

In other words, we count the flux as a positive quantity if the current is flowing out of the body. The net flux is computed by the sum over discrete parts of the surface:

$$I_{s,net} = \sum_{i=1}^{N} A_i\,\mathbf{j}_{si} \cdot \mathbf{n}_i \tag{94}$$

In the case depicted in Figure 27, the net flux is

$$I_{s,net} = A(j_s(x + \Delta x) - j_s(x)) \tag{95}$$

Even though we clearly distinguish between currents and fluxes, we shall continue to use the term *current* in the sense of flux if there is no danger of confusion.

3.4.2 The Balance of Entropy and Energy for a Continuous Body

Let us derive the differential form of the law of balance of entropy in conductive transport. This is necessary since we want to obtain conditions for every point inside a body as befits a theory describing a continuum. We start from the equation of balance of entropy for a *body*. In the steady state, the net current must be equal to the production rate of entropy:

$$I_s = \Pi_s \tag{96}$$

This much we read off from the equation of balance of entropy for conduction, as in Equation (1), by setting the rate of change of the entropy content equal to zero. The derivation, which in one dimension is a very simple example of the more general case, proceeds as follows: We express the quantities in Equation (96) in terms of integrals of densities over the body. Let us start with the flux of entropy. It expresses the net rate of flow of entropy across the surface of the body. Above, we introduced the *current density*, which describes the distribution of the current over the boundary. Obviously, we will obtain the net flux if we integrate the current density over the surface of the system:

$$I_s = \int_{\mathcal{A}} j_s dA \tag{97}$$

$\mathcal{A}$ is the closed surface of the body. In the one-dimensional case the current density is a function of the position variable x only. Since entropy flows only in the x-direction, the integral over the surface is very much simplified. There are contributions only from the flow across the faces of the body (Figure 27).

Now we proceed to derive the rate of production of entropy for the entire body in terms of a *production density*. Since the production of entropy takes place inside a body, it is a volumetric process. A density is the proper quantity for expressing such a phenomenon. The production rate for the body is given simply by the integral of the associated density over the volume:

$$\Pi_s = \int_{\mathcal{V}} \pi_s dV \tag{98}$$

Here, π_s is the *density of the rate of generation of entropy* in the body, and $\mathcal{V}$ is the volume. Next we change the variable of integration to the volume in Equation (97). This is achieved by replacing the current density by its derivative with respect to x, and integrating over x as well. Because of the balance of entropy we set the result equal to the expression in Equation (98):

$$\int_{\mathcal{V}} \frac{dj_s}{dx} dx\, dA = \int_{\mathcal{V}} \pi_s dV \tag{99}$$

or

$$\int_{\mathcal{V}} \left[\frac{dj_s}{dx} - \pi_s \right] dV = 0 \tag{100}$$

Note that the expression $dx\, dA$ is equal to dV. This equation lets us draw an important conclusion. The integral can be equal to zero under all circumstances only if the integrand is identical to zero. This means that

$$\frac{dj_s}{dx} = \pi_s \tag{101}$$

This is *the differential form of the equation of balance of entropy in the steady state* for one-dimensional flow. It holds at every point of the body, and not just as an overall expression for the entire body. It expresses a simple idea, namely, that the spatial rate of change of a current of a nonconserved quantity depends on the rate at which this quantity is produced at every point. Obviously then, the spatial rate of change of a conserved quantity in one-dimensional flow should be zero.

This is exactly what happens for energy. If the rate of change of energy vanishes, the equation of balance becomes

$$I_{E,th} = 0 \tag{102}$$

for the body. The net flux will be expressed as the integral over the surface of the body of the flux density of energy. This expression must be equal zero under all circumstances, which is possible only if the integrand is identically equal to zero:

$$\frac{dj_E}{dx} = 0 \tag{103}$$

This is the differential form of the law of balance of energy, which now replaces the overall balance in Equation (102). We will see how this law, together with the relationship between fluxes of entropy and energy (Equation (14)), leads to a determination of the density of the production of entropy in Equation (101).

3.4.3 The Generation of Entropy in Conduction

So far we have solved only one part of the problem of the balance of entropy. Since entropy is produced in conductive transport, we still need an expression for the production term in the equation of balance. The problem will be solved for steady-state processes. Actually, as we shall see in Section 3.4.6, the result holds for general processes as well.

The result is a consequence of the combination of the laws of balance of entropy and energy. We will consider purely one-dimensional flow of heat, as before (Figure 27). What we need is the relationship between the fluxes of entropy and energy. This is given by Equation (14), which leads to

$$\frac{d}{dx}(Tj_s) = T\frac{dj_s}{dx} + j_s\frac{dT}{dx} \tag{104}$$

According to the balance of energy, this expression is equal to zero. Therefore we have

$$\frac{dj_s}{dx} = -\frac{1}{T}j_s\frac{dT}{dx} \tag{105}$$

Comparison to Equation (101) demonstrates that the density of production of entropy in conduction takes the form:

$$\pi_s = -\frac{1}{T} j_s \frac{dT}{dx} \tag{106}$$

Let us develop a graphical interpretation of this result. Where does the energy come from for the production of entropy in the case of conduction? If we take seriously the interpretation of conductive transport of heat given in Section 3.1, we can use the image of entropy flowing from a higher level to a lower one just as in the case of the operation of a heat engine. The waterfall picture used for heat engines also applies to our current case: entropy falling from a higher to a lower temperature releases energy for a follow-up process. Something, however, must be vitally different in the cases of conduction of heat and the operation of a heat engine. No directly visible process is driven by the energy released in the fall of entropy if heat is transported conductively. There are no mechanical, electrical, or other phenomena occurring. So what is happening to the energy which is released when entropy flows from points of higher to points of lower thermal potential? The energy must be used for something, even if we cannot see it right away.

We know the answer to this question. In the conduction of heat, the energy is used to produce entropy; i.e., the energy drives a thermal process, not a mechanical or an electrical one. You might say that the energy flowing through a body has been "enriched" with more entropy; that is all. This is not as unusual as might appear at first. We can compare the conduction of heat to a number of different processes which share the same fundamental property. The flow of electrical charge through a resistor, the flow of gas as in the free expansion of air, or the flow of momentum between two bodies rubbing against each other are examples of the class of processes in which the energy released drives entropy production. Since we often say that heat is produced by friction, we can interpret the conduction of heat as a case of "thermal friction." Just as in the other cases, the rate of entropy production is given by the ratio of the rate at which energy is dissipated, and the temperature at which the process is taking place. The term $j_s\, dT/dx$ in Equation (106) is the density of the dissipation rate of energy. The entropy production rate is obtained if we divide this quantity by the temperature. Analogous equations are derived for the entropy produced by electrical currents, by dissipative currents of momentum (viscous friction), or by diffusion.

EXAMPLE 21. Currents and production densities in a metal bar.

The ends of a copper cylinder length 0.50 m and diameter 0.05 m have constant temperatures of 373 K and 273 K. Entropy is conducted in the direction of the axis of the cylinder only. The thermal conductivity with respect to energy of copper is 384 W/(K · m), which we take to be constant. Calculate the energy current density, the energy flux, and the production densities of entropy at the ends of the cylinder. From the production density as a function of position determine the production rate of entropy in the copper bar. Show that the result is equal to the net current of entropy.

SOLUTION: We have concluded that the temperature gradient must be constant if the conductivity with respect to energy is constant. Therefore the current density of energy is calculated easily according to Equation (16):

$$\begin{aligned} j_E &= -k_E \frac{dT}{dx} \\ &= -384 \frac{273-373}{0.5} \frac{\text{W}}{\text{m}^2} = 7.68 \cdot 10^4 \frac{\text{W}}{\text{m}^2} \end{aligned}$$

The magnitude of the energy flux is

$$|I_E| = A|j_E| = 603 \text{ W}$$

It is identical at both ends of the metal bar. Using Equation (106) we calculate the production densities at the hotter and cooler ends:

$$\pi_s = -\frac{1}{T} j_s \frac{dT}{dx} = -\frac{1}{T^2} j_E \frac{dT}{dx}$$

The numerical values are 110 W/(K · m^3) and 206 W/(K · m^3) at the hotter and cooler ends, respectively. The production of entropy is higher at lower temperatures; first the rate of dissipation of energy increases along the cylinder in the direction of decreasing temperature, and second, this rate is divided by the temperature. The net production rate for the bar is calculated by integrating the production density:

$$\begin{aligned} \Pi_s &= \int_0^L A\pi_s dx = -A\int_0^L \frac{1}{T^2} j_E \frac{dT}{dx} dx \\ &= -Aj_E \int_{T_u}^{T_l} \frac{1}{T^2} dT = I_E \left(\frac{1}{T_l} - \frac{1}{T_u} \right) \end{aligned}$$

The numerical value is 0.59 W/K. The result is indeed equivalent to the net current of entropy. (See also Example 21 of Chapter 1.)

3.4.4 The Field Equation for Temperature

The conduction of heat has forced us to change the description of thermal processes from single values of temperature to fields of temperature. In a body which transports heat conductively, the temperature changes from point to point. (If we do not consider steady-state conditions, it also changes in time.) Here, we will derive an equation governing the field of temperature in steady-state conduction. Such an equation is a field equation, which as we shall see, follows from the fundamental law of balance of entropy combined with the particular constitutive law describing conduction (i.e., Fourier's law).

Fourier's law for the current density of entropy will be cast in a form using the conductivity with respect to energy:

$$j_s = -\frac{k_E}{T} \frac{dT}{dx} \tag{107}$$

First we need the spatial derivative of this expression:

$$\frac{dj_s}{dx} = -\frac{d}{dx}\left(\frac{k_E}{T}\frac{dT}{dx}\right)$$
$$= -\frac{k_E}{T}\frac{d^2T}{dx^2} - \frac{1}{T}\frac{dT}{dx}\frac{dk_E}{dx} + \frac{k_E}{T^2}\left(\frac{dT}{dx}\right)^2$$

which according to the equation of balance of entropy, Equation (101), must be the production density of entropy given in Equation (106). With the current density replaced by Fourier's law, the latter quantity becomes

$$-\frac{1}{T}j_s\frac{dT}{dx} = \frac{1}{T}\left(\frac{k_E}{T}\frac{dT}{dx}\right)\frac{dT}{dx}$$
$$= k_E\frac{1}{T^2}\left(\frac{dT}{dx}\right)^2 \tag{108}$$

According to the equation of balance of entropy we get the following result:

$$\frac{k_E}{T}\frac{d^2T}{dx^2} + \frac{1}{T}\frac{dT}{dx}\frac{dk_E}{dx} = 0 \tag{109}$$

which is equivalent to

$$\frac{d}{dx}\left(k_E\frac{dT}{dx}\right) = 0 \tag{110}$$

This equation is the one-dimensional field equation for temperature in steady-state heat conduction.

EXAMPLE 22. The temperature gradient in a bar.

a) Assume the thermal conductivity with respect to energy, i.e., the factor k_E, to be independent of temperature. Determine the temperature in a long bar as a function of the position in the bar. The values of the temperature are T_u and T_l at the upper and the lower faces of the bar, respectively. b) What is the solution if the conductivity k_s is constant?

SOLUTION: a) If the conductivity with respect to energy is constant, the field equation for temperature, Equation (110), becomes

$$k_E\frac{d^2T}{dx^2} = 0$$

for constant k_E. This demonstrates that the temperature must be a linear function of the position in the bar, which leads to

$$T(x) = T_u + (T_l - T_u)\frac{x}{L}$$

where L is the length of the bar, and x is measured from the hotter end. Note that the temperature decreases linearly along the bar, a result which was derived before.

b) If the conductivity k_S is independent of temperature, the conductivity with respect to energy must be a linear function of temperature, namely

$$k_E = Tk_s$$

The field equation for temperature tells us that in this case the product of the temperature gradient and the temperature must be a constant:

$$Tk_s \frac{dT}{dx} = constant$$

Integration of this differential equation yields the following result:

$$T^2(x) = T_u^{\,2} + \left(T_l^2 - T_u^{\,2}\right)\frac{x}{L}$$

The boundary conditions are the same as in the first case.

EXAMPLE 23. The field equation derived from the balance of energy.

Derive the field equation for temperature in steady-state conduction on the basis of the balance of energy instead of the balance of entropy.

SOLUTION: Since energy is not created, its current is constant in the direction of flow if there is no change of the energy stored in the conducting body. This means that the current entering part of a body is equal to the current leaving this part, from which we conclude that

$$\frac{dj_E}{dx} = 0$$

We simply insert Fourier's Law expressed for the current of energy into this equation to obtain the field equation:

$$\frac{d}{dx}\left(k_E \frac{dT}{dx}\right) = 0$$

This equation is equivalent to the field equation obtained on the basis of the balance of entropy.

EXAMPLE 24. The field equation in spherical symmetry.

Derive the field equation for temperature in steady-state conduction for spherically symmetric heat flow. Determine the temperature as a function of radius in a spherical shell for constant thermal conductivity with respect to energy (k_E).

SOLUTION: As in the previous example, we use the balance of energy. For steady-state conditions, the equation of balance for a spherically symmetric shell reads

$$A(r+\Delta r)j_E(r+\Delta r)-A(r)j_E(r)=0$$

If we insert the expression for the surface area of a sphere, and divide the resulting equation by the volume of the shell, which is equal to

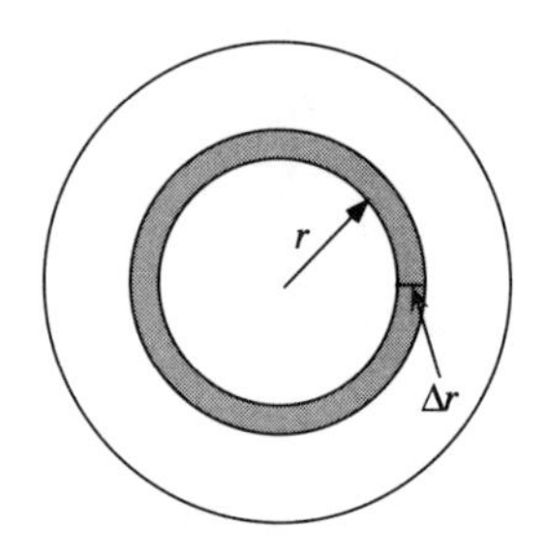

$$V=4\pi r^2\Delta r$$

we get the following difference equation:

$$\frac{1}{4\pi r^2\Delta r}\left[-4\pi r^2 j_E(r)+4\pi\left(r^2+2r\Delta r+(\Delta r)^2\right)\left(j_E(r)+\Delta j_E\right)\right]=0$$

If we neglect all higher order terms in Δr and Δj, the differential equation of balance for energy in spherical coordinates becomes, after taking the limit for $\Delta r \rightarrow 0$:

$$\frac{dj_E}{dr}+\frac{2}{r}j_E=0$$

or

$$\frac{1}{r^2}\frac{d}{dr}\left(r^2 j_E\right)=0$$

To obtain the field equation for temperature, we simply introduce Fourier's Law in the form of Equation (16) into the equation of balance, which leads to

$$\frac{1}{r^2}\frac{d}{dr}\left(r^2 k_E\frac{dT}{dr}\right)=0$$

If k_E is independent of the temperature, this differential equation yields the following result. The first integration tells us that the product of the radial variable squared and the temperature gradient must be a constant, which will be called $-B$. Therefore the temperature is inversely proportional to the radial variable r. The second integration delivers another constant A, and the final result turns out to be

$$T(r)=A+\frac{B}{r}$$

in the shell. The constants A and B are determined by the appropriate boundary conditions.

3.4.5 The Equation of Balance of Entropy for Conduction and Supply

In Section 3.2.6 we first came across the mode of transfer in which entropy appears inside bodies without flowing over system boundaries via currents, and without being created in the system. This is the case in radiative transfer, which we have modeled using sources (or sinks) of entropy inside a body.

Another important class of sources of entropy are chemical and nuclear reactions. Normal fires, and the fires inside stars which are driven by nuclear reactions, create vast amounts of entropy. If we do not count the entropy and energy they release to be part of the system under consideration, entropy and energy are effectively added from outside. Looked at this way, entropy is not created in the system, which means that there must be source rates, rather than production rates of entropy which account for the processes.

In this section we will discuss the role of source terms in the equation of balance of entropy. At the same time we will ask about the role of energy in the processes just listed. The description will be limited to one-dimensional steady state cases. We will assume entropy to appear or disappear at every point of a body, while it is transported through the body by conduction.

Entropy supply. We already have seen instances of how the increase of entropy inside a system without transport across surfaces is described mathematically. One way of adding entropy is by production. Irreversibility appears in the equation of balance of entropy by way of the production rate, as in Equations (1) or (2). The particular case associated with the conduction of heat is described by a production rate per volume π_s, which was derived in Equation (106). The second possibility, i.e., sources and sinks of entropy, is modeled in an analogous manner. This is evident in the case of radiation, where a source rate was first introduced in Equation (5).

Mathematically, there is no difference between production and source terms in the equation of balance of entropy. Physically, this means that either type of process leads to the appearance (or possibly the disappearance) of entropy in a way which cannot be described by currents across surfaces, but which must be modeled as sources or sinks. How then, do we distinguish between the production of entropy and sources or sinks of entropy?

A simple difference lets us state clearly which is which. It all depends on how we treat energy in a given case. If we say that energy has been supplied to a system, we shall treat the appearance or disappearance of entropy associated with this process as due to sources or sinks in the proper sense. In the example above, we do not count the energy stored in some chemicals before a reaction as belonging to the system, then the energy released in the reaction is supplied to the system, and the entropy associated with it is said to have been supplied as well. If, on the other hand, the energy released is counted as part of the energy of the system to begin with, then the energy is not supplied. Rather it might be dissipated, which leads to the production of entropy in the system. Therefore we say that:

Entropy appearing in a system due to the dissipation of energy which already is counted as part of the system is said to have been produced. Entropy appearing (or disappearing) together with energy which is flowing from or to other systems is said to have been supplied.

Balance of entropy. For a formal derivation of the laws governing conduction with supply, we can start with the equation of balance, which holds for the entire body under steady-state conditions:

$$I_s = \Pi_s + \Sigma_s \tag{111}$$

The term Σ_s describes all effects of the supply of entropy. Proceeding as we did in Section 3.4.2, this becomes

$$\int_{\mathcal{A}} j_s dA = \int_{\mathcal{V}} \pi_s dV + \int_{\mathcal{V}} \sigma_s dV \tag{112}$$

Transformation of the integrals leads to

$$\frac{dj_s}{dx} = \pi_s + \sigma_s \tag{113}$$

If we assume conduction to be the only dissipative process in the body, we get

$$\frac{dj_s}{dx} = -\frac{1}{T} j_s \frac{dT}{dx} + \sigma_s \tag{114}$$

The spatial rate of change of the current density (left side of the equation) is the result of the production of entropy by conduction (first term on the right) and the sources of entropy which are described by a *source rate density* or *supply rate density* σ_s (second term on the right).

Supply of energy and entropy. A key point in thermodynamics concerns the relationship between sources of entropy and of energy. When entropy is supplied to a body which has a particular temperature, a definite amount of energy must be supplied as well. Our discussion so far suggests that the constitutive relation between currents of entropy and of energy, and between dissipation rate and production rate should carry over to the supply of entropy and energy as well:

$$\sigma_E = T\sigma_s \tag{115}$$

σ_E is the density of the rate at which energy is supplied to the body from some other system. We can motivate this relation on the basis of the constitutive theory of simple fluids provided in Chapter 2. In a homogeneous body, the Gibbs fundamental relation tells us that the energy increases at T times the rate of increase of entropy. If there are only sources of entropy, then Equation (115) directly follows for such fluids.

The balance of energy takes a rather simple form. In the absence of energy sources, the spatial rate of change of the energy current density is zero (Example 23), since energy is not generated. Now, however, we have sources which lead to the following *equation of balance of energy* in the steady state:

$$\frac{dj_E}{dx} = \sigma_E \tag{116}$$

The field equation for temperature. We can prove Equation (115) to be correct by deriving the field equation for temperature from the equation of balance of entropy. Fourier's law and Equation (114) yield

$$\frac{d}{dx}\left(-\frac{k_E}{T}\frac{dT}{dx}\right)=\frac{1}{T^2}k_E\left(\frac{dT}{dx}\right)^2+\sigma_s \quad (117)$$

This becomes the *field equation for temperature* in the presence of entropy sources:

$$\frac{d}{dx}\left(k_E\frac{dT}{dx}\right) = -T\sigma_s \quad (118)$$

The same field equation derived from the balance of energy yields

$$\frac{d}{dx}\left(k_E\frac{dT}{dx}\right) = -\sigma_E \quad (119)$$

which is equivalent to Equation (118) if and only if the relationship between sources of entropy and energy is given by Equation (115).

EXAMPLE 25. The source density of entropy and energy in a radioactive sample.

Calculate the source density of energy and the production (or source) density of entropy in a 1-kg sample of enriched uranium at a temperature 300 K. Uranium has a density of 18950 kg/m^3. Assume there to be 97% U-238 and 3% U-235. The energy released in the decay of one nucleus is roughly 4.2 MeV and 4.6 MeV, respectively. The half-lives of the two isotopes are $4.5 \cdot 10^9$ years and $7.1 \cdot 10^8$ years, respectively.

SOLUTION: In 1 kg of enriched uranium there are roughly $2.5 \cdot 10^{24}$ nuclei of U-238, and $7.6 \cdot 10^{22}$ nuclei of U-235. Using the half-lives we calculate the activities, which are

$$A=\lambda N=\frac{\ln(2)}{T_{1/2}}N$$

or $1.2 \cdot 10^7\ \mathrm{s}^{-1}$ and $0.23 \cdot 10^7\ \mathrm{s}^{-1}$ for U-238 and U-235, respectively. This yields an energy generation rate of $6.1 \cdot 10^7$ MeV/s (for the entire sample), which is $9.8 \cdot 10^{-6}$ W.

The entropy supplied to the system is produced as a consequence of the decay of nuclei. Calculating the generation rate therefore yields the source rate. The generation rate of entropy is equal to the rate at which energy is dissipated, divided by the temperature. Therefore the density of the generation rate is

$$\begin{aligned}\pi_s &= \frac{\Sigma_E}{T\cdot V}=\frac{\Sigma_E\cdot\rho}{T\cdot m}\\ &=\frac{9.8\cdot 10^{-6}\cdot 18950}{300\cdot 1}\frac{\mathrm{W}}{\mathrm{K\cdot m^3}}=6.2\cdot 10^{-4}\frac{\mathrm{W}}{\mathrm{K\cdot m^3}}\end{aligned}$$

The density of the source rate of energy is 0.19 W/m^3.

EXAMPLE 26. The temperature in a bar which is heated internally.

Consider a bar of length L and constant conductivity with respect to energy which can conduct heat only in the direction of its axis. It is attached to entropy reservoirs at constant temperatures T_o and T_L at $x = 0$ and $x = L$, respectively. Let us take $T_o \geq T_L$. It is heated internally with a constant energy source rate. (This might be due to radioactive heating, as in the previous example.) a) Determine the steady-state temperature of the bar as a function of position. b) Would it be possible for the temperature to have a minimum in $0 \leq x \leq L$? c) Determine the condition for the temperature to decrease monotonically from T_o to T_L. d) What is the alternative?

SOLUTION: a) Under the given conditions the differential equation for the temperature is

$$\frac{d^2T}{dx^2} = -\frac{\sigma_E}{k_E} \quad , \quad \frac{\sigma_E}{k_E} = constant \tag{E1}$$

with boundary conditions

$$T(x=0) = T_o$$
$$T(x=L) = T_L$$

The solution of Equation (E1) is a quadratic function which turns out to be

$$T(x) = -\frac{1}{2}\frac{\sigma_E}{k_E}x^2 + \left(\frac{1}{2}\frac{\sigma_E}{k_E}L - \frac{T_o - T_L}{L}\right)x + T_o \tag{E2}$$

It is easy to see that for a vanishing source rate the solution turns out to be identical to the one given in Example 22, in which case the temperature is a linear function of position.

b) A quadratic function has an extremum. Here, the second derivative of the temperature with respect to position, i.e., the value of $-\sigma_E/k_E$ is negative which tells us that the extremum is a maximum, not a minimum. A minimum would be possible only if there were an energy sink instead of a source in the bar.

c) For the temperature to decrease monotonically from its value at the hotter end, the maximum of the quadratic function obviously must lie outside the range of the bar, i.e., outside $0 \leq x \leq L$. We find the position of the maximum by setting the first derivative of Equation (E2) equal to zero, which yields

$$-\frac{\sigma_E}{k_E}x_{max} + \frac{1}{2}\frac{\sigma_E}{k_E}L - \frac{T_o - T_L}{L} = 0$$

Since $T_o - T_L$ is a positive quantity, the position of the maximum may not be larger than $x = L/2$. For this maximum to lie outside the bar it must therefore be negative. This is the case if

$$\frac{\sigma_E}{k_E} < 2\frac{T_o - T_L}{L^2}$$

We can understand intuitively that the source rate may not be too large for entropy still to be transported from the hotter to the cooler reservoir.

d) The alternative to the situation just described is the following. The maximum of temperature lies in the range $0 \le x \le L/2$. In this case the temperature gradient is zero or *positive* at the end of the bar which is in contact with the hotter entropy reservoir. Entropy will not flow from the reservoir into the bar; rather, it will flow from the point of maximum temperature inside the bar towards both ends! This happens if

$$\frac{\sigma_E}{k_E} \ge 2\frac{T_o - T_L}{L^2}$$

i.e., if the source rate of energy becomes large. For increasing values of the source rate the point of maximum temperature moves towards the middle of the bar. Remember that these considerations hold only for steady state conditions.

EXAMPLE 27. The flow of heat through the Earth's mantle, with sources of entropy.

Assume that just about all the entropy which flows out of the surface of the Earth is produced by radioactive decay in the mantle. Calculate the temperature at the base of the mantle if all the entropy is transported conductively. The mantle-core boundary is at a depth of 3400 km. The Earth's radius is 6400 km. Take the conductivity with respect to energy to be constant and equal to 1 W/K · m. The temperature gradient at the surface is 0.06 K/m. Take the surface temperature to be 0°C.

SOLUTION: We can derive the differential equation for temperature in spherical symmetry by going through the same development as in Example 24. If we allow for sources, the equation of balance of energy turns out to be

$$\frac{dj_E}{dr} + \frac{2}{r} j_E = \sigma_E$$

If we use Fourier's law we obtain the field equation:

$$k_E\left(\frac{d^2T}{dr^2} + \frac{2}{r}\frac{dT}{dr}\right) = -\sigma_E$$

if the conductivity with respect to energy is constant. This equation can be changed into a simpler one using the following transformation of variables:

$$u = Tr$$

We obtain

$$\frac{d^2u}{dr^2} = -\frac{\sigma_E}{k_E} r \qquad \textbf{(E3)}$$

The particular solution of this equation is proportional to the third power of the radius, while the solution of the homogeneous equation is a linear function. Taken together, the solution of Equation (E3) is given by

$$u(r) = Ar^3 + Br + C$$

Together with the boundary conditions

$$T(R) = 0$$

$$R\left.\frac{dT}{dr}\right|_R = D$$

we arrive at the following result for the temperature as a function of radius:

$$T(R) = -\frac{1}{6}\frac{\sigma_E}{k_E}r^2 + D + \frac{1}{2}\frac{\sigma_E}{k_E}R^2 - \left(\frac{1}{3}\frac{\sigma_E}{k_E}R^3 + Dr\right)\frac{1}{r}$$

We now need a value for the density of production of energy due to radioactive decay. This rate must be the flux of energy through the surface of the Earth, which can be calculated from the temperature gradient and the conductivity. If we divide this quantity by the volume of the crust we obtain

$$\sigma_E = \frac{4\pi R^2 k_E \left|dT/dr\right|_R}{\frac{4}{3}\pi\left(R^3 - R_{core}^3\right)} = 3.14 \cdot 10^{-8}\,\frac{\mathrm{W}}{\mathrm{m}^3}$$

Now the numerical values in the temperature function look like

$$T(R) = -5.2 \cdot 10^{-9} r^2 + 2.6 \cdot 10^5 - \frac{2.8 \cdot 10^{11}}{r}$$

If we insert the value for the radius of the core–mantle boundary, we obtain a temperature of 120,000 K. Even though it is smaller than what we got in Example 5, this value is still much too high for the mantle to remain solid. We have made some strongly simplifying assumptions regarding the conductivity and the distribution of the entropy sources. Still it seems rather improbable that entropy can be transported through the mantle of the Earth by conduction alone. In fact, there must be a mechanism much more efficient than conduction. Today we believe this mechanism to be convection.

3.4.6 The Balance of Entropy in Time-Dependent Conduction

The interaction of storage and flow of entropy is the most general problem we are confronted with in thermodynamics. In steady-state transport processes, bodies lose as much entropy by outflow as they gain by production and inflow. If the transport processes are to change in time, it must be due to the change of the amounts of entropy and energy which are stored in systems. In other words, the properties of bodies (which might be described by the entropy capacity and the latent entropy) are responsible for the existence of dynamical processes.

Consider bodies which conduct heat. If the amount of entropy stored is allowed to change, we are dealing with time-dependent conduction. The proper equation of balance of entropy for an extended body conducting heat is

$$\dot{S} = -I_s + \Pi_s \qquad \textbf{(120)}$$

The difference between this and Equation (96) is the appearance of the time rate of change of the entropy of the body.

The differential form of the balance of entropy. Now we will derive the differential form of the equation of balance. (Recall the procedure of Section 3.4.2.) All three quantities appearing in Equation (120) are integrals of associated quantities which vary from point to point in a body. Remember that a surface density is associated with the flux of entropy I_s. This surface density is the flux density. If this quantity is integrated over the surface, the flux is recovered. The rate of production of entropy, on the other hand, is obtained if we integrate the density of the rate of production over the volume of the body. (See Equation (98).)

In the same manner, a *density of entropy* ρ_s is associated with the entropy of a body. Just as the mass of a body is obtained by integrating the mass density over the volume, we recover the entropy of the body by evaluating the integral of the density of entropy over the volume:

$$S = \int_{\mathcal{V}} \rho_s dV \tag{121}$$

If we insert the integral expressions listed above into the law of balance of entropy in Equation (120), we arrive at

$$\frac{d}{dt}\int_{\mathcal{V}} \rho_s dV = -\int_{\mathcal{A}} j_s dA + \int_{\mathcal{V}} \pi_s dV \tag{122}$$

This equation must be transformed somewhat to yield the final differential form we are looking for. Since we are treating conduction through stationary bodies, the volume under consideration does not change with time. Therefore, we may place the time derivative in the left-hand side under the integral. The first term on the right is transformed as in Equation (99). All of these changes lead to

$$\int_{\mathcal{V}} \frac{\partial \rho_s}{\partial t} dV = -\int_{\mathcal{V}} \frac{\partial j_s}{\partial x} dx dA + \int_{\mathcal{V}} \pi_s dV \tag{123}$$

Now all three terms may be combined:

$$\int_{\mathcal{V}} \left[\frac{\partial \rho_s}{\partial t} + \frac{\partial j_s}{\partial x} - \pi_s \right] dV = 0 \tag{124}$$

For the same reason given after Equation (100), the expression in the integral must be zero for all values of the independent variables. The resulting equation is the differential form of the equation of balance of entropy for time-dependent, one-dimensional conduction:

$$\frac{\partial \rho_s}{\partial t} + \frac{\partial j_s}{\partial x} = \pi_s \tag{125}$$

Since we have two independent variables, namely time and position in the x-direction, the equation is a partial differential equation. The first term is due to the storage of entropy, the and second one describes conduction, while the third (on the right-hand side) is responsible for the production of entropy.

Constitutive relations. Naturally, constitutive relations are needed if we attempt a solution of Equation (125). Here, we must obtain three relations, namely for the density, the flux density, and the density of the rate of production of entropy. The latter two are known already. Fourier's law determines the flux density, Equation (13), while the rate of production of entropy is given by Equation (106):

$$j_s = -k_s \frac{dT}{dx} \tag{126}$$

$$\pi_s = -\frac{1}{T} j_s \frac{dT}{dx} \tag{127}$$

The latter result is obtained if the laws of balance of entropy and of energy for conduction are combined as in Section 3.4.3. The form of the constitutive relation for storage of entropy is new. Naturally, this feature is determined using the entropy capacity K of the rigid thermal conductor. We simply have to transform the original definition of the entropy capacity. For a uniform body we may write Equation (5) of Chapter 2 as follows:

$$V\dot{\rho}_s = V\frac{K}{V}\dot{T} \tag{128}$$

The entropy capacity divided by the volume of the body is the product of the entropy capacity per mass and the mass density. Therefore we have

$$\dot{\rho}_s = \rho k \dot{T} \tag{129}$$

If this determination is introduced with Equations (126) and (127) in the equation of balance of entropy, Equation (125), we obtain, after some algebra, the *field equation for temperature* for time-dependent conduction:

$$\rho c \frac{\partial T}{\partial t} = \frac{\partial}{\partial x}\left(k_E \frac{\partial T}{\partial x}\right) \tag{130}$$

Instead of the quantities referring to entropy, the temperature coefficient of energy and the conductivity with respect to energy have been introduced.

The result is a partial differential equation for temperature as a function of time and position. Its solution may be very difficult or even impossible to obtain in analytical form. Most often, in practical cases, numerical methods are used to compute a solution. Naturally, the solution is subject to proper boundary and initial conditions. Theories of analytical and numerical solutions of equations such as Equation (130) are beyond the scope of this book. The derivations performed in this section are important simply as examples of how we can deal with the equation of balance of entropy and constitutive relations in somewhat more complicated situations.

The speed of propagation of thermal disturbances. The result for heat penetrating the surface layer of soil or rock in the course of a year, presented in Example 28, suggests the image of waves propagating downward as time passes. However, this appearance is deceptive: the theory of conduction which builds upon Fourier's law does not allow for real waves spreading through a medium.

We can see this by comparing the time-dependent field equation for temperature for the case of conduction, to the wave equation obtained from electricity. The latter result was derived in Problem 45 of the Prologue. Starting with the equation of balance of charge (which is a conserved quantity)

$$C^* \frac{\partial \varphi}{\partial t} = -A \frac{\partial j_q}{\partial x} \tag{131}$$

and a combination of Ohm's law and the law of induction

$$\frac{\partial \varphi}{\partial x} = -\frac{1}{\sigma} j_q - L^* A \frac{\partial j_q}{\partial t} \tag{132}$$

we obtain the equation for the propagation of electrical disturbances

$$\frac{\partial^2 \varphi}{\partial t^2} = \frac{1}{L^* C^*} \frac{\partial^2 \varphi}{\partial x^2} - \frac{1}{\sigma A L^*} \frac{\partial \varphi}{\partial t} \tag{133}$$

Obviously, the difference between this and the conduction of heat has to be found in the constitutive law of Equation (132), which includes the effect of induction. If this phenomenon did not play a role, we would obtain an equation resembling the one for conduction. Let the inductance in Equation (133) go to zero, which leads to

$$\frac{C^*}{A} \frac{\partial \varphi}{\partial t} = \sigma \frac{\partial^2 \varphi}{\partial x^2} \tag{134}$$

This then should be compared to

$$\rho c \frac{\partial T}{\partial t} = k_E \frac{\partial^2 T}{\partial x^2} \tag{135}$$

The latter equation follows from Equation (130) for constant values of the conductivity with respect to energy.

In summary, the equation representing conduction alone is obtained from more general laws if induction is neglected. Doing so means setting the inductance equal to zero, which in turn means that the speed of propagation of the waves approaches infinity. (Remember that $1/(L^*C^*)$ is equal to the square of the speed of propagation.) In other words, the classical theory of conduction predicts infinite speed for thermal pulses. This is a totally unphysical result; the resolution of this problem will be presented in the Epilogue.

EXAMPLE 28. Penetration of heat into the upper layers of soil.

What is the effect of periodic changes of temperature at the surfaces of the Earth? How far down from the surface can one still notice daily or yearly changes having amplitudes of 7.5 K and 15 K, respectively?

SOLUTION: We need a simplified model of the penetration of heat into the upper layers of the ground. Assume the soil to have a constant temperature (in space and in time) in the absence of the changes at the surface. Model the ground as a body with a plane surface extending infinitely into the vertical direction downward. The material is assumed to have constant properties. These assumptions reduce the problem to one of purely one-dimensional conductive heat transfer, which can be described using Equation (130):

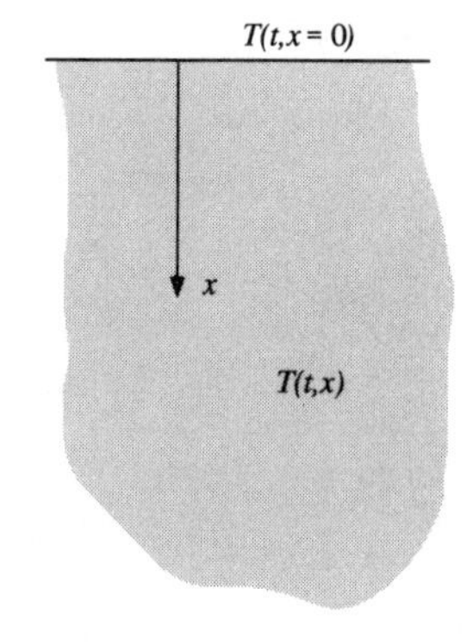

$$\rho c \frac{\partial T}{\partial t} = \frac{\partial}{\partial x}\left(k_E \frac{\partial T}{\partial x}\right) \quad \Rightarrow \quad \frac{\partial T}{\partial t} = \frac{k_E}{\rho c}\frac{\partial^2 T}{\partial x^2}$$

The second form holds for constant conductivity. The factor multiplying the spatial derivative of the temperature is called the *thermal diffusivity* α. The field equation for temperature can therefore be written as follows:

$$\frac{\partial T}{\partial t} = \alpha \frac{\partial^2 T}{\partial x^2}$$

with

$$\alpha = \frac{k_E}{\rho c}$$

The boundary condition at the surface takes the form

$$T(t,0) = T_o \cos(\omega t)$$

This partial differential equation can be solved as follows. Assume the solution to be separable, which means that we can write

$$T(t,x) = f(t)g(x)$$

Now the differential equation becomes

$$\frac{1}{f(t)}\frac{\partial f(t)}{\partial t} = \frac{\alpha}{g(x)}\frac{\partial^2 g(x)}{\partial x^2}$$

Since we have functions of only one variable on the left and functions of the other on the right, this equation can be satisfied only if each side is a constant K, which means that

$$\frac{\partial f(t)}{\partial t} = K\, f(t) \quad \text{and} \quad \frac{\partial^2 g(x)}{\partial x^2} = \frac{K}{\alpha} g(x)$$

These are ordinary differential equations with simple solutions:

$$f(t) = a\exp(Kt)$$

$$g(x) = b\exp\left(\sqrt{\frac{K}{\alpha}}x\right)$$

In summary, the solution of the partial differential equation can be written

$$T(t,x) = T_o \exp\left(Kt + \sqrt{\frac{K}{\alpha}}x\right)$$

The boundary condition at the surface tells us that

$$T_o \exp(Kt) = T_o \cos(\omega t)$$

Now, it makes sense to change to the complex domain. Both the real part and the imaginary part of a complex function are solutions of the differential equation. Instead of $K = \omega$, we write $K = i\omega$, which leads to

$$T(t,x) = \text{Re}\left[T_o \exp\left(i\omega t \pm \sqrt{\frac{i\omega}{\alpha}}x\right)\right]$$

for the real part. This can be transformed into

$$T(t,x) = T_o \exp\left(-\sqrt{\frac{\omega}{2\alpha}}x\right)\cos\left(\omega t - \sqrt{\frac{\omega}{2\alpha}}x\right)$$

The minus sign must be chosen, since otherwise the temperature would increase with increasing depth.

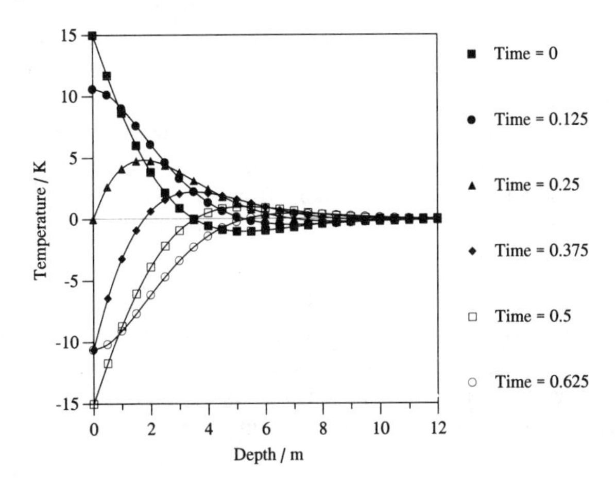

The graph shows the solution for a periodic change of temperature at the surface with a period of one year and an amplitude of 15°C. The curves are for different times, where the time is given as a fraction of a year. The thermal diffusivity of soil was taken to be $5.0 \cdot 10^{-7}$ m^2/s. It is seen that the disturbance decreases to a few degrees some 5 m into the ground. A daily disturbance is felt only to a depth which is about 20 times smaller than for the yearly temperature fluctuation. A periodic climate change with a period of 1 million years, on the other hand, would be felt to a depth of several thousand meters (the curves are similar if the depth variable is scaled as the square root of the period of variation).

3.5 Radiative Transport of Heat

In this section we will turn to the third mode of entropy transfer, radiation. This will complete the list of possible ways of transporting entropy. The most important point for us to understand about radiation is that we have to distinguish between the transport of entropy and energy through the radiation field, and absorption and emission of radiation by bodies. The latter phenomenon is the result of the interaction of bodies and fields. Since radiative transfer of entropy is at least as complicated in detail as the transport of entropy by convection, we shall deal only with its most fundamental aspects. Some topics in solar radiation will be treated in Section 3.6.

3.5.1 The Flow of Entropy and Energy Through the Radiation Field

Entropy and energy are transported through space by electromagnetic radiation. Since visible light is the most obvious part of the spectrum of radiation, we shall often use the term *light* instead of the word *radiation*. As a first step, we will discuss the flow of entropy through the radiation field in the absence of bodies, examples of which are furnished by radiation in a cavity such as in Figure 13, or in the space between the Sun and the Earth. Radiative transport is more complicated than conduction or convection in the sense that it is not sufficient to specify a radiative flux density vector of entropy (or energy) at every point of space. There certainly exists such a vector quantity which describes the net rate per unit area at which entropy is transported at a given point (Figure 26). This net rate, however, is the result of entropy being carried by radiation in every direction of space at every point. Put differently, light may be traveling in any direction of space from a point, and it may arrive at this point from any imaginable direction. Only if we manage to describe the flow of radiation in detail will the condition of the field be specified completely.

Distribution functions and intensity. To describe mathematically the radiation field, one commonly introduces a *distribution function* which is used to specify two fundamental properties of this system. First of all, radiation possesses a *density of entropy* and energy at every point in space and time. Second, at every point in space radiation is traveling in different directions which cover at least a thin cone of finite solid angle, however small (Figure 28). Rays of perfectly parallel light do not exist. Radiation may even be flowing in all directions of space in equal amounts, in which case the field is called *isotropic*. To capture this feature, we determine the *distribution of directions* of flow, i.e., the fraction of the density of entropy (or energy) which belongs to radiation moving in a cone pointing in a given direction (Figure 28).

Both properties of the field, namely density and distribution of directions, are included in the distribution functions for entropy and energy (in the following, x will stand for either quantity). The function $f_x(t,\mathbf{r},\Omega)$ is said to represent the density of entropy or energy of radiation at time t and at point $\mathbf{r}$ traveling in a cone about the direction Ω, divided by the solid angle covered by the cone. The definition of this distribution function tells us that we obtain the density of quantity x by integrating the distribution over the complete sphere (solid angle 4π):

$$\rho_x(t,\mathbf{r}) = \int_{(4\pi)} f_x(t,\mathbf{r},\Omega)d\Omega \tag{136}$$

In the theory of heat transfer, we are interested in the rate of flow of entropy or energy in given directions. The entropy contained in the field is transported at the speed of light. In this respect, radiative transfer resembles convection where entropy stored is transported with the speed of the flowing medium. Therefore, the same relation holds between the density of entropy (or of energy) and the flux density in the case of radiation as well. In other words, the quantity

$$i_x(t,\mathbf{r},\Omega) = c f_x(t,\mathbf{r},\Omega) \tag{137}$$

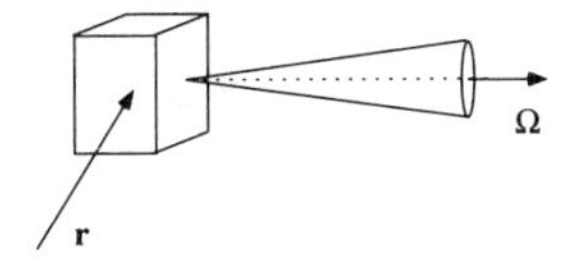

FIGURE 28. Radiation contained in a small volume about the point **r** travels in all possible directions. A certain fraction of this radiation will be flowing in the direction within the solid angle shown in the figure. The flux density per unit solid angle is called the intensity of the radiation.

represents the flux density of quantity x at point $\mathbf{r}$ of radiation traveling within the cone of Figure 28 which is pointing in direction Ω, divided by the solid angle of the cone. This quantity is called the *intensity of radiation*, measured either in terms of entropy or of energy (the units are $W/(K \cdot m^2 \cdot sr)$ and $W/(m^2 \cdot sr)$, respectively, where sr is the unit of the solid angle). To be specific, for the measurement of the intensity at point $\mathbf{r}$ we imagine a small area perpendicular to the direction Ω. We compute the flux of the radiation flowing within a narrow cone pointing in the direction of the vector Ω, and divide this quantity by the surface area to obtain the flux density of the fraction of radiation under consideration. Finally, we divide the result by the solid angle of the cone. Note that the condition of the radiation field is described completely in terms of either the distribution functions or the intensities.

Now we may express the density of quantity x of the radiation field using the intensity. By combining Equations (136) and (137) we obtain:

$$\rho_x(t,\mathbf{r}) = \frac{1}{c}\int_{(4\pi)} i_x(t,\mathbf{r},\Omega)d\Omega \tag{138}$$

The intensity of blackbody radiation. For us, the special case of isotropic radiation is of particular importance. Just consider the radiation contained in the cavity of Figure 13. Black body radiation does not depend on the properties of the cavity walls. The only parameter influencing this type of radiation is the temperature T. Since T is the same everywhere, the entropy arriving from different directions in the cavity must be the same. In other words, the intensity is independent of direction. Therefore, the integral in Equation (138) may be evaluated simply. The solid angle of the complete sphere is 4π. For this reason we obtain the following expression for the *density of isotropic radiation:*

$$\rho_x = \frac{4\pi}{c} i_x \tag{139}$$

We may now express the intensities of *blackbody radiation* in terms of the densities of entropy and of energy which have been derived in Section 2.6:

$$i_{sb} = \frac{c}{3\pi} a T^3 \tag{140}$$

and

$$i_{Eb} = \frac{c}{4\pi} a T^4 \tag{141}$$

In these relations, the index b refers to blackbody. These equations support the claim about the temperature dependence of entropy and energy in radiation made in Section 3.2.4.

The flux density. One of the problems encountered in radiative heat transfer calculations is the question of how much entropy and energy penetrate a given surface in a certain amount of time. Related to this problem is the computation of onesided fluxes, i.e., the rates at which entropy or energy flow from one side of a surface to the other. Consider a part of an imaginary or real surface of size ΔA in a radiation field, such as in Figure 29. The orientation of the surface is determined by the unit vector **n** normal to the surface. The rate at which entropy or energy flow through this area is equal to the corresponding flux through an imaginary sphere of radius r centered on it. Consider the portion of the radiation which travels through the surface within a narrow cone in the direction of the vector Ω. The latter is tilted at an angle θ with respect to **n**. The cone cuts a circular area out of the imaginary sphere, and the radiation flux I_x through the cone is proportional to this area; furthermore, it is inversely proportional to the square of the distance from ΔA. In other words, the flux is proportional to the solid angle of the cone. Finally, it is proportional to the projection of ΔA perpendicular to the direction of the flow of radiation, which is $\Delta A \cos(\theta)$. The flux density of radiation contained within the cone is found by dividing the flux by ΔA. The total flux density, finally, is obtained by integrating over the entire sphere, i.e., over the complete solid angle. Combining all of this information, we can express the flux density of quantity x as follows:

$$j_x = \int_{(4\pi)} i_x \cos(\theta)\, d\Omega \tag{142}$$

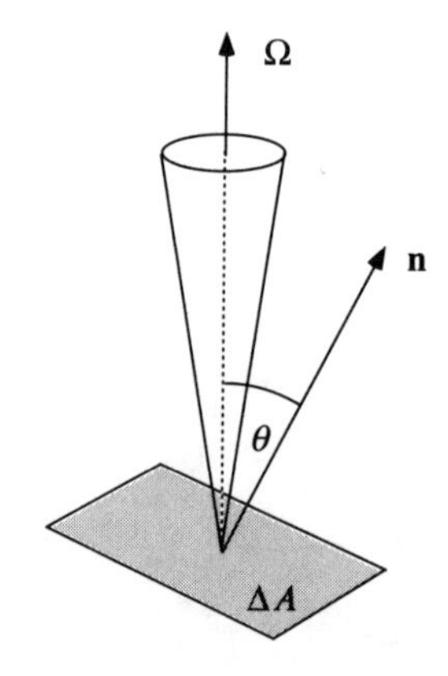

FIGURE 29. Part of an imaginary (or real) surface is penetrated by radiation. The vector **n** determines the orientation of the surface. Consider entropy flowing within a cone centered on the direction of the vector Ω (which represents part of the total entropy flux through the surface ΔA). The flux through the surface due to radiation in the cone depends upon the orientation of the cone with respect to the surface. The problem is to compute either the total flux with respect to ΔA, or the one-sided or hemispherical flux.

This quantity is not a vector, as would be a proper flux density. Rather, it is the component of the flux density vector in the direction of **n**. However, this point does not have to concern us here. Note that the fluxes given by Equation (142) are zero for isotropic radiation: equal amounts of entropy or energy flow from one side to the other, and back again. This condition is satisfied, for example, by blackbody radiation inside a cavity, or by radiation in an extended body where the radiation and body have the same temperature.

The flux density of radiation flowing from one side of a surface to the other is called the *hemispherical flux density* j'_x. It is of practical importance since it represents, for example, the entropy or the energy radiated per time and per unit area by the surface of a hot body (Section 3.2.4). The hemispherical flux density is obtained by integrating the expression in Equation (142) over a hemisphere instead of over the entire sphere. For *isotropic* radiation the result is particularly simple:

$$j_x' = \pi i_x \tag{143}$$

Details of the derivation of this equation are provided below. The result for the hemispherical flux density of entropy and energy of *blackbody* radiation is of special interest:

$$j_{sb}' = \frac{4}{3}\sigma T^3 \tag{144}$$

and

$$j_{Eb}' = \sigma T^4 \tag{145}$$

where

$$\sigma = \frac{1}{4}ca \tag{146}$$

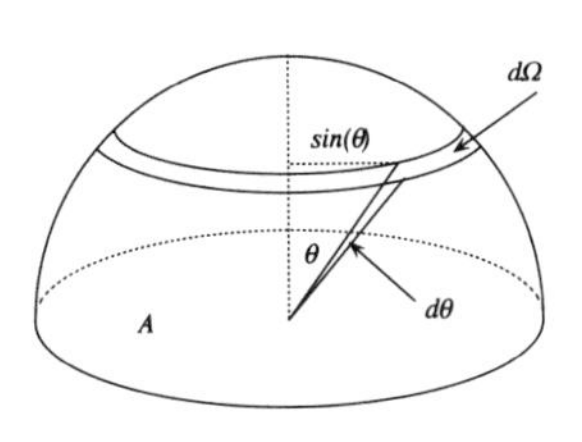

FIGURE 30. Radiation flows isotropically from the lower toward the upper side of surface A. The total entropy or energy carried through the surface from one side to the other per unit time is called the hemispherical flux density.

The constant σ is called the *radiation constant*. It is a combination of the Stefan–Boltzmann radiation constant and the speed of light, and has the value of $\sigma = 5.67 \cdot 10^{-8}\ \mathrm{W \cdot m^{-2} \cdot K^{-4}}$.

Hemispherical flux density for isotropic radiation. To obtain the rate of transfer of entropy or energy flowing through a surface towards one side, we need to evaluate Equation (142) for the flux density for a hemisphere; as before the variable x stands for either entropy or energy of radiation:

$$j_x' = \int_{2\pi} i_x \cos(\theta) d\Omega \tag{147}$$

The intensity has to be integrated over one side of the area A (Figure 30). Remember that we are dealing with isotropic radiation. This means that the quantities appearing in the integral are independent of the longitudinal angle. We can divide the surface of the hemisphere into small circular rings whose areas are

$$d\Omega = 2\pi \sin(\theta) d\theta$$

The hemisphere has a radius of unity. Therefore, the area of a ring translates directly into the solid angle associated with it. Now we can perform the integration indicated in Equation (147), which leads to:

$$j_x' = \int_0^{\pi/2} i_x \cos(\theta) 2\pi \sin(\theta) d\theta = \pi i_x \int_0^{\pi/2} \sin(2\theta) d\theta = \pi i_x \tag{148}$$

This result holds for isotropic radiation in general. j_x' represents the amount of quantity x transferred from one side of an area A to the other, per time and per unit area. It includes all the radiation traveling in all directions seen from one side of the surface A. This is the result presented above in Equation (143).

EXAMPLE 29. The intensity of solar radiation.

a) Derive the intensity with respect to energy of solar radiation near the Earth. The energy flux density of the Sun's radiation at the Earth is 1360 W/m2 (the *solar constant*). The Sun's radius is $7.0 \cdot 10^8$ m, and its distance from the Earth is $1.5 \cdot 10^{11}$ m (one astronomical unit). b) If you consider this to be the intensity of blackbody radiation, what is its equivalent temperature? Compare the result to the surface temperature of the Sun, which from spectroscopic measurements, is known to be 5780 K. What is the importance of this result? c) Derive the relation between the intensity with respect to entropy and with respect to energy for blackbody radiation. Compute the entropy flux density of solar radiation at the Earth's distance.

SOLUTION: a) The radiation arriving at a point on Earth travels in a cone having a solid angle equal to that of the Sun's disk as seen from here (see the figure). This angle is nearly equal to the cross sectional area of the Sun divided by the square of its distance from Earth. Therefore, the intensity of light is given by

$$i_{Eb} = \frac{\mathcal{G}_{sc}}{\Omega} = \frac{\mathcal{G}_{sc}}{\pi R^2/d^2}$$

$$= \frac{1360(1.5 \cdot 10^{11})^2}{\pi (7.0 \cdot 10^8)^2} \frac{\text{W}}{\text{m}^2\text{sr}} = 2.0 \cdot 10^7 \frac{\text{W}}{\text{m}^2\text{sr}}$$

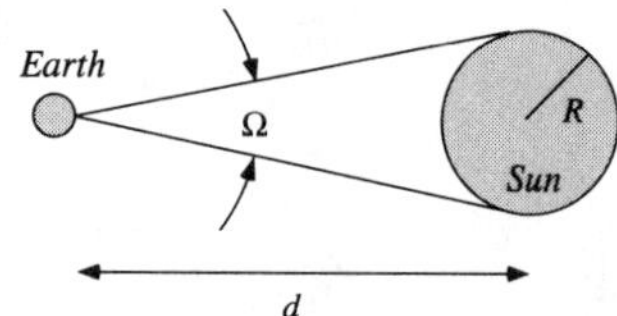

b) According to Equation (141), blackbody radiation with such an intensity has a temperature

$$T = \left(\frac{4\pi}{ca} i_{Eb}\right)^{1/4} = \left(\frac{4\pi \cdot 2.0 \cdot 10^7}{3.0 \cdot 10^8 \cdot 7.56 \cdot 10^{-16}} \text{K}^4\right)^{1/4}$$

The numerical value turns out to be 5770 K. This is equal to the surface temperature of the Sun. Therefore, the intensity of solar light is the same on the surface of the Sun as here on Earth. While the flux density is much lower here than on the Sun, the flux density per solid angle is the same. This suggests that by concentrating the Sun's light we can recreate the conditions found for radiation at its surface. This is important for solar energy engineering, since it shows that we may heat bodies here on Earth to temperatures approaching that of the Sun.

c) The relation between the intensity with respect to entropy and to energy in the case of blackbody radiation is derived simply by using Equations (140) and (141). We obtain:

$$i_{Eb} = \frac{3}{4} T i_{sb}$$

Note that this is *not* equal to the energy flux density divided by the temperature of the radiation. The relation between entropy and energy transported by radiation is not given by Equation (13) of Chapter 1. Rather, the relation depends upon the densities of entropy and energy of the radiation field, as for convective transfer. With a value of $2.0 \cdot 10^7$ W/($m^2 \cdot$ sr) for the intensity with respect to energy for solar radiation, the intensity of entropy is $4.6 \cdot 10^3$ W/(K $\cdot$ $m^2 \cdot$ sr). The flux density of entropy is obtained if we multiply the intensity by the solid angle of the Sun's disk. The value turns out to be 0.32 W/(K $\cdot$ m^2).

EXAMPLE 30. Hemispherical flux density of blackbody radiation.

a) Apply the result for the hemispherical flux density of isotropic radiation to blackbody radiation. b) Compute the numerical values of the hemispherical energy flux density of radiation at the surface of the Sun and inside an oven which is kept at a temperature of 300°C.

SOLUTION: a) If we apply the result to the transfer of entropy in a field of blackbody radiation, we obtain the following relation:

$$j_{sb}' = \pi \cdot i_{sb} = \pi \frac{ca}{3\pi} T^3 = \frac{1}{3} caT^3$$

which can be written as follows:

$$j_{sb}' = \frac{4}{3} \sigma T^3$$

Since this expression holds for the entire surface A (the temperature is the same everywhere), integration over the surface leads to the expression for the total entropy flux:

$$I_s = \frac{4}{3} A \sigma T^3$$

The derivation is the same for energy. Using Equation (141) we obtain the expression in Equation (145).

c) Even though the surface of the Sun is not a cavity, the one-sided flux still is the same as that for a black body. (The reasoning goes as follows: Consider matter to fill all of space radiating to and receiving radiation from a field occupying the same space. Matter and field are assumed to be in thermal equilibrium, and the field has the characteristics of blackbody radiation. If you now take away half of the matter on one side of a dividing plane, the radiation field left on that side still has blackbody properties.) Therefore we have in the case of the Sun:

$$j_{Eb}{}' = \pi i_{Eb} = \frac{ca}{4} T^4 = \sigma T^4$$
$$= 5.67 \cdot 10^{-8} (5780)^4 \frac{\text{W}}{\text{m}^2} = 6.3 \cdot 10^7 \frac{\text{W}}{\text{m}^2}$$

For the interior of an oven, where we have blackbody radiation, the numerical result is $6.1 \cdot 10^3$ W/m^2. Note the strong dependence of these values on the temperature of radiation.

3.5.2 The Radiation Shape Factor

The formulas derived for the fluxes of entropy or energy exchanged between surfaces hold only for the special case where all the radiation emitted by one of the bodies is intercepted by the other, as with the small piece of matter inside the cavity in Figure 14. In many practical cases we do not have such geometrical arrangements. The particular geometry is taken into account via a *radiation shape factor* (which also is called the *view factor* or the *configuration factor*).

As a more general case consider two black surfaces as in Figure 31. The orientation of each of the surfaces is given by a unit normal vector. Let us calculate the flux of energy emanating from surface 1 which is intercepted by a small part of the second area. This quantity is the intensity of radiation emerging from A_1, multiplied by the direction cosine of angle θ_1, and by the solid angle subtended by the small area on surface 2. Finally we have to multiply this expression by the area on surface 1 from which the radiation originates. The flux of energy flowing from surface 1 to surface 2 is then

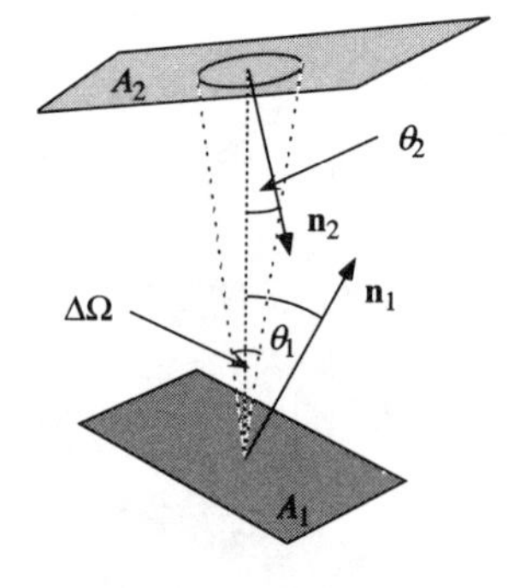

FIGURE 31. Radiative exchange of heat between two black surfaces. Their sizes and orientations are arbitrary. The figure shows a part of the radiation flowing from surface 1 toward surface 2.

$$I_{E12} = \int_{A_1} \int_{\Omega} i_{Eb} \cos(\theta_1) d\Omega dA_1 \tag{149}$$

The solid angle subtended by a small area on surface 2 may be expressed in terms of the distance r between the surfaces and the projection of this area perpendicular to r. Using the direction θ_2, Equation (149) becomes

$$I_{E12} = \int_{A_1} \int_{A_2} i_{Eb} \cos(\theta_1) \frac{\cos(\theta_2) dA_2}{r^2} dA_1 \tag{150}$$

If we now invoke the law of Stefan and Boltzmann, the final result is

$$I_{E12} = \frac{\sigma T_1^4}{\pi} \int_{A_1} \int_{A_2} \frac{\cos(\theta_1)\cos(\theta_2)}{r^2} dA_2 \, dA_1 \tag{151}$$

The same kind of expression emerges for the flux of radiation flowing from surface 2 to A_1, with the temperature T_2 substituted for T_1. The net flux therefore must be

$$I_{E,net} = \frac{\sigma}{\pi}(T_1^4 - T_2^4)\int_{A_1}\int_{A_2}\frac{\cos(\theta_1)\cos(\theta_2)}{r^2}dA_2\,dA_1 \qquad (152)$$

It is customary to write this in an abbreviated form which resembles the original expression for the law of Stefan and Boltzmann:

$$I_{E,net} = \sigma A_1 F_{12}(T_1^4 - T_2^4) \qquad (153)$$

Therefore, the *radiation shape factor* F_{12} is related to the surfaces exchanging radiation by

$$A_1 F_{12} = \frac{1}{\pi}\int_{A_1}\int_{A_2}\frac{\cos(\theta_1)\cos(\theta_2)}{r^2}dA_2\,dA_1 \qquad (154)$$

Note that we could have started by considering the radiation from surface 2. We would have obtained the same result, only with the surface area A_1 replaced by A_2. Therefore, F_{21} is related to F_{12} by

$$A_1 F_{12} = A_2 F_{21} \qquad (155)$$

In practice, the computation of the integral may only be possible using numerical methods. Moreover, the result of Equation (154) is not general. First of all, we have considered only two surfaces exchanging heat; second, the surfaces have been taken to be black. Relaxing either of these conditions introduces additional complexities.[7]

EXAMPLE 31. The radiation shape factor for two disks.

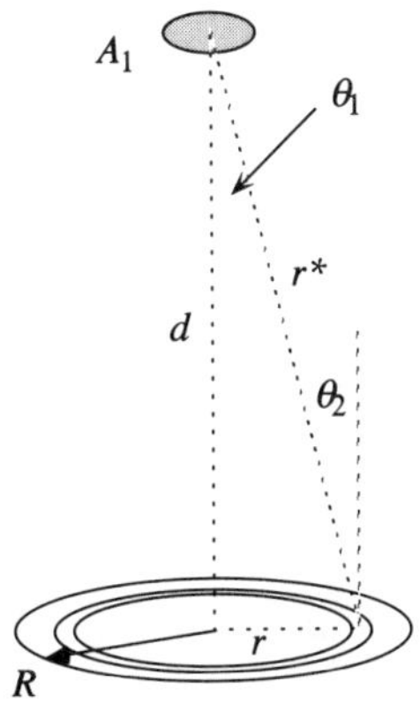

Compute the radiation shape factor F_{12} for two disks perpendicular to the same center line. Assume the first disk to be very small, having a surface area A_1, while the second disk has radius R. How large is the shape factor F_{21}?

SOLUTION: We can calculate the product of the shape factor and the surface area A_1 according to Equation (154). Since the first disk is very small, integration over it effectively reduces to multiplication by the area:

$$A_1 F_{12} = \frac{A_1}{\pi}\int_{A_2}\frac{\cos(\theta_1)\cos(\theta_2)}{r^{*2}}dA_2$$

As indicated in the figure, the integration over A_2 is performed by dividing the surface area into rings of radius r. The distance r^* is computed simply according to $r^{*2} = d^2 + r^2$. Both angles occurring in the equation are the same, and their cosines are d/r^*. Since the

A_1 θ_1 r^* d θ_2 r R

7. For more details see Siegel and Howell (1992), Chapter 6.

surface area of a small ring on A_2 is given by the product of its thickness and its circumference, we obtain

$$A_1F_{12} = \frac{A_1}{\pi}\int_0^R \frac{d^2}{\left(d^2 + r^2\right)^2} 2\pi r dr$$

which yields the final result:

$$F_{12} = \frac{R^2}{R^2 + d^2}$$

The shape factor F_{21} is equal to this expression, multiplied by the ratio of the surface areas of the disks.

EXAMPLE 32. Radiation networks: gray surfaces and shape factors.

Assume the disks of Example 31 to have gray surfaces with absorptivities a_1 and a_2, respectively. Calculate the rate of transfer of energy between the bodies if their temperatures are T_1 and T_2, respectively. Show that the surface characteristics and the geometrical arrangement may be written in the form of resistances. Their combined influence is represented by the sum of the three resistances. Compute a numerical result for the energy flux for two disks 1.0 m apart, the first with a surface area of 0.50 m^2, the second with an area of 0.010 m^2. Take their temperatures to be 1000°C and 20°C, respectively. The larger of the surfaces has an absorptivity of 0.95, while that of the smaller one is 0.30.

SOLUTION: If we compare the flux of heat or energy between two arbitrary gray surfaces with the expression provided by the law of Stefan and Boltzmann for black bodies, Equation (36), we see that the results for gray surfaces are smaller than those for black bodies for given temperatures of the surfaces. We may interpret this situation as being the consequence of three resistances hindering the flow of heat. First, there is a surface resistance involved with surface 1 which is due to the less than perfect emissivity (or absorptivity) of the gray body. Second, the fact that bodies of arbitrary shapes do not intercept all the radiation emitted by one of the surfaces, may again be described in terms of a resistance. Finally, a resistance is associated with the second gray surface. It is customary to interpret emittances and *radiosities* (see below) as potentials. Therefore we can compare the situation to a radiation network as in the diagram.

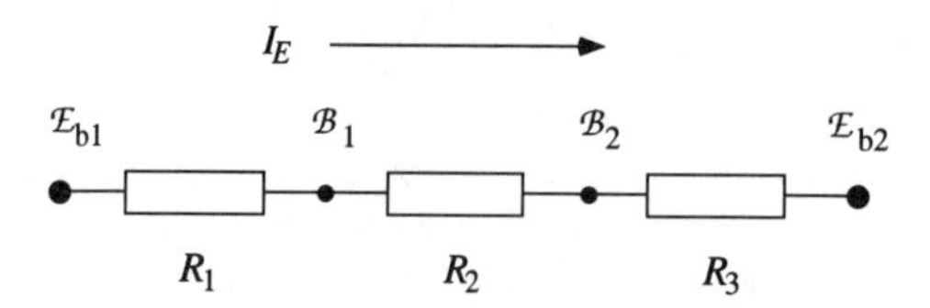

The *radiosity* $\mathcal{B}$ is defined as the total flux of energy per unit area leaving a surface. With a gray surface, it is a combination of the radiation emitted and the radiation reflected. The former is the product of absorptivity and blackbody emittance, while the latter is the product of reflectivity (which is $1 - a$ for nontransmitting surfaces) and the irradiance:

$$\mathcal{B} = a\mathcal{E}_b + (1-a)\mathcal{G}$$

On the other hand, the net flux of energy with respect to surface 1 is determined by the difference between the radiosity and the irradiance:

$$I_E = A_1(\mathcal{B} - \mathcal{G})$$

We may express the flux in terms of the radiosity and the blackbody emittance, which yields

$$I_E = \frac{a_1}{1-a_1} A_1 (\mathcal{E}_{b1} - \mathcal{B}_1) \tag{E4}$$

We interpret the difference between black body emittance and radiosity as a potential difference (which depends on the temperature of the surface), and the factor multiplying it as the inverse of a resistance:

$$R_{rad,1} = \frac{1-a_1}{a_1 A_1} \tag{E5}$$

The flux of the current of energy leaving surface 1 which is intercepted by the second surface is given by the surface area, the radiation shape factor, and the radiosity:

$$I_{E12} = A_1 F_{12} \mathcal{B}_1$$

An analogous expression holds for the flux from surface 2 to surface 1. Therefore, the net flux computed in (E4) may also be calculated in terms of the geometrical factors:

$$I_E = A_1 F_{12} (\mathcal{B}_1 \mathcal{B}_2)$$

The corresponding resistance is

$$R_{rad,2} = (A_1 F_{12})^{-1}$$

if we interpret the radiosities as the potentials. The resistance associated with the second gray surface is given by an expression analogous to (E5):

$$R_{rad,3} = \frac{1-a_2}{a_2 A_2}$$

Finally, the net flux of energy passing between the surfaces may be computed in terms of the sum of the resistances and the difference between the blackbody emittances:

$$I_E = \frac{\sigma}{R_{total}} (T_1^4 - T_2^4)$$

In our example, the numerical value of the total resistance is

$$
\begin{aligned}
R_{total} &= \frac{1-a_1}{a_1 A_1} + \frac{r_1^2 + d^2}{r_1^2 A_2} + \frac{1-a_2}{a_2 A_2} \\
&= \frac{1-0.95}{0.95 \cdot 0.50} + \frac{0.40^2 + 1.0^2}{0.40^2 \cdot 0.010} + \frac{1-0.30}{0.30 \cdot 0.010} \\
&= 958 \mathrm{m}^{-2}
\end{aligned}
$$

r_1 is the radius of the large disk. The net flux of energy passing between the disks is $I_E = 5.67 \cdot 10^{-8}(1273^4 - 293^4)/958$ W = 155 W. The current flows from the larger (hotter) to the smaller disk.

3.5.3 Bodies and Fields: Absorption, Emission, and Kirchhoff's Law

So far we have discussed the flow of entropy through the radiation field only in the absence of bodies. Now we shall consider the interaction of a body with a radiation field in the space which is occupied by both systems.[8] In general, a body is bounded while the field associated with it may extend beyond the space filled by matter, in which case radiation penetrates the surface of a material body (Figure 32).

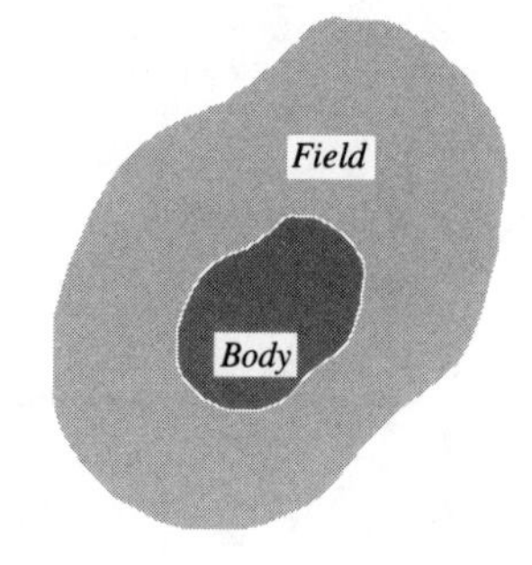

FIGURE 32. For a bounded body interacting with a radiation field, the field may extend beyond the volume occupied by the body.

Interaction of matter and radiation. Matter emits light, and it absorbs light. It also may scatter radiation. *Emission*, *absorption*, and *scattering* are the processes which make up the interaction of bodies with radiation. From the viewpoint of a body, emission and absorption are volumetric processes. In other words, light is emitted and absorbed by every part of a body. Entropy is not transported through matter; rather it appears in, or disappears from, every arbitrarily small part. Entropy appearing in a body disappears in the radiation field, and vice-versa. Obviously this type of transport is comparable to the transport of momentum from fields to bodies due to their interaction. This is not to say that entropy does not flow through the space occupied by a body. However, if this is the case, entropy is transported through the radiation field. While radiation is created or destroyed in the processes of emission and absorption, scattering changes the direction of a ray of light without absorbing it. Even though this process may be important in some cases (just think of the scattering of light in the Earth's atmosphere), we shall neglect it in the following discussion. Still, one point deserves to be mentioned: while absorption and emission depend on the frequency of radiation, this influence is particularly pronounced in scattering. In general, the magnitude of scattering increases strongly with increasing frequency of the light. This fact explains the blue color of the sky. Thus we have to take into account

8. Here, we can only touch upon this subject. If you wish to know more about it, you should read the book by Max Planck: *The Theory of Heat Radiation* (1906), which still is one of the best accounts of the physics of these phenomena. A modern and detailed text on the subject is *Thermal Radiation Heat Transfer* by Siegel and Howell (1992).

the spectral dependence of the quantities introduced so far. However, for the moment we shall leave this point out of consideration and return to it in Section 3.6.2.

Processes at surfaces. Radiation which travels through a medium undergoes the three types of interaction mentioned. When a ray of light reaches the surface of a body, however, we have to deal with additional processes, namely reflection and refraction at the surface, and transmission into the medium adjacent to the body. These phenomena are known from optics, and they satisfy the laws of propagation of electromagnetic radiation at interfaces. Again, we shall not deal with these laws in any depth. Nevertheless, we need to know a few facts about the processes occurring at interfaces (see Figure 33). In general, a surface separating two media will reflect at least part of the light incident upon it. The fraction of light reflected is called the *reflectivity*. The reflection may either be *specular*, in which case we shall call the surface smooth; or *diffuse*, which is the case for rough surfaces. The reflectivity usually depends on the angle of incidence of the ray of light, and on the wavelength. Also, a surface may be smooth for radiation of a particular frequency, and rough for light of another wavelength. If the reflection is specular and complete, we call the surface a *mirror*; if the reflection is complete and totally diffuse, the surface is called *white*. The opposite of a white surface is a *black* one, whose reflectivity is zero. The fraction of light incident on a surface and absorbed by the second body is called the *absorptivity*. By definition, the absorptivity of a black body is equal to 1. Black surfaces do not really exist in nature; for example, a smooth surface separating two optically different media cannot transmit all radiation incident upon it. In particular, if the speed of propagation of light is different in the bodies touching each other, the reflectivity cannot be zero. In general, therefore, a black surface should be rough, the body admitting the light should be thick enough for the radiation not to emerge from the other side, and it should not scatter radiation; otherwise the light transmitted will flow back towards the first body. With a smooth interface, the optical properties of the media must be the same for radiation to be transmitted completely.

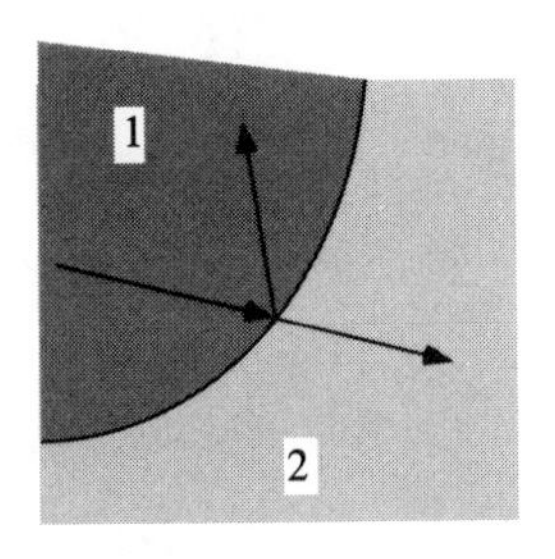

FIGURE 33. Surface phenomena involving radiation occur wherever different bodies touch. A "body" may also be vacuum.

Sources and sinks of radiation. Let us now turn to a more formal discussion of the interaction of matter and radiation. If we disregard scattering, the interaction is described using the source rates of entropy and energy which were introduced in the laws of balance in Equations (5) and (6). These determine the net rate at which an entire body exchanges entropy or energy with the field. If conditions inside a body vary from point to point, we have to introduce the *density of the source rate*, or simply, the *source density* of entropy or energy. These quantities have the units W/(K · m³) and W/m³, respectively. For a body, the source rates are obtained by integrating the densities over its volume $\mathcal{V}$:

$$\Sigma_x = \int_{\mathcal{V}} \sigma_x dV \qquad \textbf{(156)}$$

Again, x stand for entropy or energy. These densities play the same role as quantities introduced previously, such as mass and entropy densities, the density of entropy production, the distribution function introduced above, or flux densities and the intensity of radiation. They all describe the spatial distribution of physical quantities, a feature we cannot do without in a continuum.

Physically, the source rate or its density are the result of emission and absorption of radiation by matter. It is customary to introduce *emission and absorption coefficients of energy* to describe the interaction of bodies and fields in more detail. These coefficients incorporate information concerning the distribution of directions of radiation. At every point inside a body, matter may emit radiation in any direction, and radiation which is absorbed may stem from rays traveling in different directions. The *emission coefficient* ε_E is defined as the rate of emission of energy per volume and per solid angle. In other words, the expression

$$\sigma_E^- = \int_{(4\pi)} \varepsilon_E(\Omega) d\Omega \tag{157}$$

represents the *density of the rate of emission* of energy. Absorption is introduced in terms of a coefficient α which plays a role comparable to ε. It is the rate at which energy is absorbed from the field per volume and per solid angle. The radiation absorbed depends on the intensity of radiation present in the field; in fact, for a short distance in the direction of a ray the absorption density is proportional to the intensity. For this reason one defines the *absorption coefficient* κ_E, which when multiplied by the intensity, yields the density of the rate of absorption per solid angle:

$$\alpha_E(\Omega) = \kappa_E i_E(\Omega) \tag{158}$$

The absorption coefficient κ_E has units 1/m. It describes the fraction of energy absorbed by the body from the field over a small distance, divided by this distance. Obviously, the *density of the rate of absorption* of energy is obtained if we integrate α_E over the solid angle:

$$\sigma_E^+ = \int_{(4\pi)} \kappa_E i_E(\Omega) d\Omega \tag{159}$$

Armed with these quantities we can finally express the source density of radiation in terms of emission and absorption coefficients. With

$$\sigma_E = \sigma_E^+ - \sigma_E^- \tag{160}$$

we obtain for a body contained in volume $\mathcal{V}$:

$$\Sigma_E = \int_{\mathcal{V}} \int_{(4\pi)} (\kappa_E i_E - \varepsilon_E) d\Omega dV \tag{161}$$

Remember that all of these quantities have been defined with respect to the body, and not with respect to the radiation field. The signs appearing in Equations (160) and (161) reflect this assumption. Since both the rate of absorption and the rate of emission are taken to be positive quantities, emission leads to a loss of entropy in a body, while the body will gain entropy through absorption.

Kirchhoff's Law. The foregoing discussion involves the definition of quantities which are used in a continuum theory. Now, we will motivate an important relation of the theory of radiative transfer—a simple relationship between the coefficients of absorption and of emission called *Kirchhoff's law.*

To appreciate this law we first have to figure out the condition of the radiation field generated by a body. Initially, we shall assume this body to be very large and isotropic. It is supposed to be enclosed by an adiabatic wall. If conditions are not uniform at the beginning, they will be soon because of conduction and radiative exchange of heat. In a fluid, differences in pressure and density will quickly be smoothed. After some time we will have a uniform isotropic body in steady state. In the end, the only free parameter describing the state of a body of given composition will be the temperature. Under these circumstances, the radiation field will be isotropic and uniform as well. First of all, the intensity of radiation must be isotropic and the same for points of the body far enough from the surface. These points are not influenced by the wall, and there is no reason why conditions should be different at different locations. For every bundle of rays propagating through a small surface inside the body within a small solid angle, there exists light flowing with equal intensity in the opposite direction converging toward the imaginary surface. In other words, the intensity must be isotropic and uniform and may depend only on the temperature and the composition of the medium.

Actually, the state of the radiation field cannot be any different near the surface of the body under consideration. A ray flowing from the surface toward points inside the medium must have the same intensity as a ray traveling exactly in the opposite direction from interior points towards the wall. If this were not the case we would end up with a net transport of entropy and energy in a particular direction. This would disturb the uniform and steady conditions which have already been reached. As a consequence, the body whose field we are investigating does not have to be large at all. Indeed, it may be arbitrarily small and of any shape, and the radiation field will be uniform and isotropic throughout.

We may now consider small bodies of differing composition in contact with each other. Each body shall be taken to be uniform and isotropic. If they have the same temperature, entropy will flow from one to the other only by means of radiation. Assume the speed of propagation of light to be same in all of them, and let each ray be transmitted directly through the surfaces. (Actually, these conditions can be relaxed, and the problem of the speed of light can be taken care of;[9] however, we shall not discuss this point here). The same kind of reasoning tells us that the intensity of radiation must be the same in two adjacent media. A ray traveling in a particular direction from one of the bodies will be transmitted by the wall separating the different regions of space. It should not matter to the second body

9. M. Planck (1906, 1921).

where the ray comes from. Again, its intensity must be the same as that of a ray propagating in the opposite direction. Therefore, we conclude that the intensity of radiation found in a large body made up of small regions of different materials cannot depend on the composition. All in all, only the temperature may influence the state of the radiation field. This conclusion is very important: it means that *if we know the intensity as a function of temperature for one medium, we know it for all media in the system.*

Finally, consider one of the different parts of the body to be a cavity surrounded by matter. For the reasons just listed, the radiation inside this cavity must be the same as that found inside the volume occupied by matter. Remember that for simplicity the optical properties of all media, and therefore also of empty space, have been taken to be the same. Since we know the field in the cavity to be that of *blackbody radiation*, the same must be true for the field of the entire uniform and isotropic body. In other words, the intensity of radiation is given by Equations (140) and (141).

Now we are ready to state *Kirchhoff's law*. Under stationary conditions in a body of the type discussed, the net source rate of energy must be zero. If this were not so, the body would heat up and cool down at different points, contrary to our assumption. Therefore, the expression in Equation (161) must vanish. Since this is true for bodies of any size and shape, the integrand, which itself is an integral over solid angle, is equal to zero at every point inside the body. Finally, since the radiation field is isotropic, we find that

$$i_{Eb}(T) = \frac{\varepsilon_E}{\kappa_E} \tag{162}$$

In other words, while the coefficients of absorption and of emission depend on the composition of bodies (and on temperature), their ratio is a universal function of temperature only. If one of the coefficients is known, the other may be calculated from Kirchhoff's law. Note that Equation (162) holds at every point in a body.

We have found Kirchhoff's relation by considering uniform conditions with the radiation field having the same temperature as matter. At first sight it may appear as if this is true only if the radiation has blackbody characteristics. However, the coefficients of absorption and emission depend only on the properties of matter, not on those of the field. Their ratio must therefore be independent of the actual intensity of radiation. If the radiation inside a body is not black, the net source rates of entropy and energy simply do not vanish. Indeed, if i is the intensity of radiation in a volume occupied by matter, the source rate of energy turns out to be

$$\Sigma_E = \int_V \int_{(4\pi)} \kappa_E (i_E - i_{Eb})\, d\Omega dV \tag{163}$$

This result, which is a consequence of Kirchhoff's law, has been derived from the expression in Equation (161).

Remember that scattering has not been considered in deriving Equation (163). Under conditions found here on Earth its influence may often be left out of consideration;[10] in astrophysical applications, however, it may be much more important.[11] Also, the phenomenon of *stimulated emission* of radiation has been ignored. The emission coefficient introduced above is that of *spontaneous emission*. Stimulated emission results from radiation present in a medium, and its effect is proportional to the intensity of light passing through the medium. Since this is true for absorption as well, stimulated emission tends to reduce the value of the absorption coefficient. We may still work with Equation (161) with the absorption coefficient replaced by a somewhat different expression.[12]

EXAMPLE 33. Extinction of a ray of light.

a) Consider a cube of length Δx and surface area A filled with a fluid. A narrow cone of light passes through the surface A. Show that the coefficient of absorption is the fractional change of the intensity of radiation per unit length. b) The attenuation (extinction) of a ray of light is due to both absorption and scattering. The scattering coefficient β is defined in the same manner as the coefficient of absorption. Derive the differential equation for the intensity of light as a function of position x along the ray. c) The intensity measured in terms of energy of a ray of light is attenuated to half its value when passing through 20 km of air. Compute the average of the sum of coefficients of absorption and scattering (with respect to energy).

SOLUTION: a) For a narrow ray of light the flux density of energy is the intensity with respect to energy multiplied by the solid angle. The flux through surface A is the flux density multiplied by the area A. If the indices 1 and 2 refer to the faces of the cube where the ray enters and leaves, respectively, the rate at which energy is absorbed is

$$\Sigma_E = -\left(i_{E2} - i_{E1}\right)\Omega A$$

On the other hand, the same quantity must be given in terms of the coefficient of absorption:

$$\Sigma_E = A\Delta x \sigma_E = A\Delta x \kappa_E i_E \Omega$$

σ_E is the source density, and $A\Delta x$ is the volume of the cube. The source density is given by Equation (159) where we only have to integrate over the small solid angle of the narrow cone. If we combine these pieces of information, we find that the coefficient of absorption may be expressed as follows:

$$i_{E2} - i_{E1} = -\kappa_E i_E \Delta x$$

10. However, this certainly cannot be done if the flow of solar radiation through the Earth's atmosphere is to be calculated (Section 3.6).
11. S. Chandrasekhar (1960).
12. Zel'dovich, Raizer (1966), vol. I, p. 118.

or

$$\kappa_E = -\frac{1}{\Delta x}\frac{\Delta i_E}{i_E} \qquad \textbf{(E6)}$$

b) Since extinction is the result of both absorption and scattering, we can introduce a coefficient of extinction which is the sum of the coefficients of scattering and of absorption:

$$\mu = \kappa_E + \beta_E$$

This represents the fractional change of the intensity per unit length, replacing the absorption factor in (E6). In other words, the differential equation for extinction along a narrow ray of light is the following:

$$\frac{di_E(x)}{dx} = -\mu i_E(x) \qquad \textbf{(E7)}$$

This follows from a transformation of Equation (E6).

c) Assume the coefficient of extinction to be constant along the path of the ray of light. If this is the case, the differential Equation (E7) is solved easily. We obtain

$$i_E(x) = i_{Eo}\exp(-\mu x)$$

for the intensity as a function of position x. If 20 km of air absorb and scatter 50% of a ray, the numerical value of the coefficient of extinction turns out to be

$$\mu = \frac{1}{\Delta x}\ln\left(\frac{i_{Eo}}{i_E}\right) = \frac{\ln(1/0.5)}{20\cdot 10^3\ \text{m}} = 3.5\cdot 10^{-5}\,\text{m}^{-1}$$

3.5.4 Radiative Transfer Through a Medium with a Temperature Gradient

How does the interior of the moon cool, or how is entropy transported through a star? It might surprise you that in each of these cases radiative transfer must play an important role, even though the bodies appear to be virtually opaque. Still, if transport by means of conduction or convection is not efficient enough, then entropy must flow through the radiation field inside the bodies while it is continually absorbed and emitted.

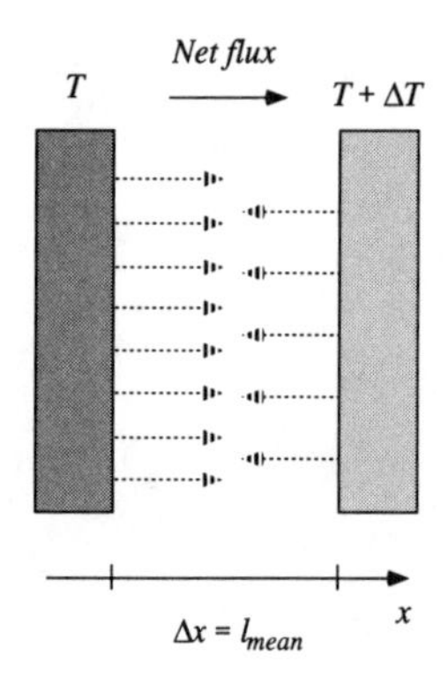

FIGURE 34. Radiative transfer through a planar medium having a temperature gradient. If the two surfaces have different temperatures, there results a net flow of entropy and energy through the field from the hotter to the cooler surface.

Consider one-dimensional heat flow through a planar slab of matter as in Figure 34. Consider two surfaces facing each other at a distance equal to the mean distance between consecutive events of emission and absorption of radiation, which is sometimes called the *mean free path* of the radiation. We model absorption and emission of entropy inside the body by the absorption and emission of these surfaces. Each radiates entropy and energy toward the other, according to its temperature. If they are at the same temperature, there is no net flux of entropy or energy between them. However, if one of them has a lower temperature than the other, a net flux results. If we assume conditions to be such that the radiation field is that of a black body, the net current density in the direction of decreasing temperature can be approximated by the temperature gradient. The expression

$$j_E = \sigma\left[(T+\Delta T)^4 - T^4\right] \approx \sigma 4T^3 \Delta T$$

can be written

$$j_E = -4\sigma l_{mean} T^3 \frac{dT}{dx} \qquad (164)$$

In this form the transfer of entropy looks very similar to that of conduction: the flux of entropy or energy is proportional to the temperature gradient of the material. The factor multiplying the gradient can be interpreted as a kind of thermal conductivity (with respect to energy):

$$k_{E,rad} = 4\sigma l_{mean} T^3 \qquad (165)$$

Because of this behavior the radiative transfer of heat through matter with a temperature gradient is often called *diffusion of radiation*. Its efficiency depends on the properties of the body expressed by the mean free path. In astronomy, in particular, the mean free path is replaced by the *opacity*, which is defined as the fraction of radiation which is absorbed by a slab of matter (Example 34).

Note that here we have introduced the flux density of entropy as in conduction and convection, even though entropy is transported radiatively. However, as you can tell from this discussion, we can easily interpret the net result of this transport in terms of the flow of entropy *through* the body which is permeated by the radiation field. In fact, we do not even consider the field anymore in our description.

EXAMPLE 34. Mass absorption coefficient and the diffusion of radiation.

The mass absorption coefficient (opacity) κ_ρ of matter is defined as the fraction of radiant energy absorbed divided by the mass density. To be precise, it is defined by the law of absorption:

$$\frac{dj}{dx} = -\kappa_\rho \rho j$$

In other words, the opacity and the mean free path must be related by

$$l_{mean} \sim \left(\kappa_\rho \cdot \rho\right)^{-1}$$

The numerical factor in this relation is of the order of unity. Compute a rough value of the mean free path and the opacity of stellar material. Take a star, such as our Sun, with a central temperature of 15 million K, a radius of 700000 km, a mass of $2 \cdot 10^{30}$ kg, and a net energy flux of $4 \cdot 10^{26}$ W. Estimate the quantities for a median point inside the star.

SOLUTION: The quantities given for the Sun lead to an average temperature gradient of 0.02 K/m, a median temperature of about 1 million K, and a density of 1400 kg/m^3. Since all the energy of the star is released near the center, the energy flux through a concentric surface midway between the center and the surface is equal to the total flux. The flux density roughly equals $2.6 \cdot 10^8$ W/m^2. Equation (164) yields a value of

$$l_{mean} = 0.06\text{m}$$

The definition of the opacity leads to a value of

$$\kappa_\rho = 0.01\text{m}^2 / \text{kg}$$

This means that inside the Sun radiation is absorbed on average every few centimeters. The interior of the Sun is virtually opaque.

3.5.5 The Spectral Distribution of Radiation

In the preceding sections we have neglected an important feature of radiation, namely its spectral dependence. Just about all the quantities introduced so far depend on the frequency of the light transporting entropy and energy. For example, Kirchhoff's law cannot hold generally in the forms stated above, with a single value of absorptivity and emissivity independent of frequency. With the model of gray surfaces we were not able to account for the temperatures attained by bodies in the light of the Sun (Examples 10 and 14). The difficulty can be resolved only if we accept that absorption and emission depend on the frequencies of the light absorbed or emitted.

Therefore, we need to know the spectral distribution of radiation. The problem of how to obtain it from first principles was solved for blackbody radiation by Max Planck.[13] He was able to calculate the distribution of energy and entropy in the *normal spectrum*, i.e., in the spectrum of blackbody radiation. Moreover, he argued how, on the basis of Wien's displacement law, the results can be extended and the temperature of rays of monochromatic radiation derived. Since we often have to deal with nonblackbody radiation, this is an important result, which we are going to use in calculating the entropy of solar radiation here on Earth (Section 3.6).

In the following paragraphs we will first explain how the spectral dependence of radiative quantities is introduced. Then, Kirchhoff's law will be stated in its general form, and Planck's result for the distribution of energy and entropy in the normal spectrum will be presented. Finally, the reasoning which leads to a generalization of the results to nonequilibrium radiation will be outlined. As before, a number of simplifications will be made. The influence of the speed of light in different media will not be considered; and we will neglect scattering of radiation until we treat the special case of solar radiation in the Earth's atmosphere.

---

13. M. Planck: *The Theory of Heat Radiation* (1906).

Dependence of radiative quantities upon frequency. The fundamental terms used in the description of the radiation field are the distribution functions f_x and the intensities i_x. Obviously these must depend upon the frequency of a ray of light. To capture this property of radiation, we introduce the *spectral distribution functions* $f_{x\nu}$ and the *spectral intensities* $i_{x\nu}$. (As before, x stands for both entropy and energy.) Again these represent some sort of density, this time with respect to frequency. If we integrate these densities over the entire range of frequencies of the electromagnetic spectrum, we recover the overall quantities introduced before:

$$f_x(t,\mathbf{r},\Omega) = \int_0^\infty f_{x\nu}(t,\mathbf{r},\Omega,\nu)\, d\nu \qquad (166)$$

and

$$i_x(t,\mathbf{r},\Omega) = \int_0^\infty i_{x\nu}(t,\mathbf{r},\Omega,\nu) d\nu \qquad (167)$$

The volume densities of entropy and energy of the radiation field therefore have a spectral dependence as well. The *spectral density of entropy* and the *spectral density of energy* of the field are denoted by $\rho_{s\nu}$ and $\rho_{E\nu}$, respectively. They are obtained in terms of the spectral distribution function or the spectral intensity if we replace the original quantities in Equation (138) by the spectral ones. As a consequence, the densities are computed by integration of the spectral functions over frequency:

$$\rho_x = \int_0^\infty \rho_{x\nu} d\nu \qquad (168)$$

In fact, if appropriate, all the quantities introduced in the previous paragraphs relate to their spectral counterparts in just this way. There is one exception to the rule, namely the absorption coefficient κ_E defined by Equation (158). The spectral coefficient $\kappa_{E\nu}$ exists, but it cannot be integrated by itself to give the original quantity. Rather, Equation (158) has to be replaced by

$$\alpha_{E\nu}(\Omega) = \kappa_{E\nu} i_{E\nu}(\Omega) \qquad (169)$$

This equation has to be integrated over the frequency to yield the expression found in Equation (158).

The relation between the density and the intensity of radiation derived for isotropic flow carries over to the spectral quantities. For entropy and energy, we have

$$\begin{aligned} \rho_{s\nu} &= \frac{4\pi}{c} i_{s\nu} \\ \rho_{E\nu} &= \frac{4\pi}{c} i_{E\nu} \end{aligned} \qquad (170)$$

Often, the spectral densities and intensities are written with respect to wavelength λ instead of frequency. Since the total intensity i_x must be the same if calculated from the spectral intensity with respect to frequency or from the spectral quantity $i_{x\lambda}$, we have

$$i_x = \int_\nu i_{x\nu} d\nu = \int_\lambda i_{x\lambda} d\lambda = \int_\nu i_{x\lambda} \frac{c}{\nu^2} d\nu$$

The last form is a consequence of the relation between the speed of propagation of radiation, its frequency, and its wavelength, i.e., $c = \nu \lambda$. Comparing the first and the third integrals we obtain

$$i_{x\lambda} = \frac{\nu^2}{c} i_{x\nu} = \frac{c}{\lambda^2} i_{x\nu} \quad (171)$$

for the law of transformation from frequency-dependent quantities to those referring to wavelength.

The spectrum of solar radiation. An example of the spectral distribution of radiation is provided by the solar spectrum. The Sun radiates mostly in the visible part of the spectrum, but its light also contains important contributions from ultraviolet and infrared radiation. Figure 35 shows the intensity of solar radiation outside the Earth's atmosphere, which should correspond to the light produced in the photosphere of the Sun. Note that according to Equation (171), the peak of the distribution is not at the same wavelength if the spectral values are measured with respect to the frequency of radiation.

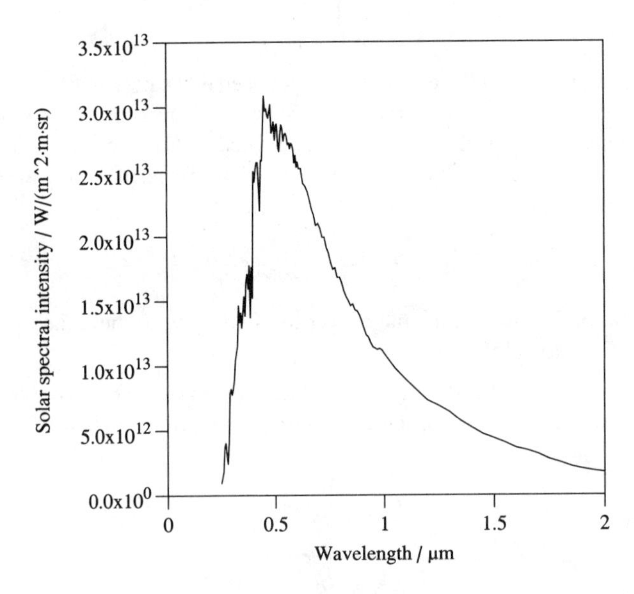

FIGURE 35. Spectral intensity (with respect to wavelength) of solar radiation. The spectrum is similar to that of radiation from a black body at a temperature of about 5800 K (see Figure 37). The integral of this quantity taken over wavelength must yield the total intensity, which in this case, is $2.01 \cdot 10^7$ W/($m^2 \cdot m \cdot sr$). Data represent the World Radiation Center (WRC) spectrum and were taken from Iqbal (1983).

Kirchhoff's law. We can now express the source rate of energy which was first calculated in Equation (161) more generally. The quantities relating to the absorption and emission of radiation have to be integrated over frequency, solid angle, and volume:

$$\Sigma_E = \int_0^\infty \int_V \int_{(4\pi)} \left(\kappa_{E\nu} i_{E\nu} - \varepsilon_{E\nu}\right) d\Omega dV d\nu \tag{172}$$

This leads to an important result, *Kirchhoff's law*, which can now be stated in its general form:

$$i_{E\nu b}(T,\nu) = \frac{\varepsilon_{E\nu}}{\kappa_{E\nu}} \tag{173}$$

Kirchhoff's law results from Equation (172) if we accept that the source rate must vanish separately for each frequency in a uniform body under steady-state conditions for which the radiation field is that of blackbody radiation. (Therefore, the blackbody intensity appears in Equation (173).) If the source rate did not vanish for every color of light separately, the spectral distribution of radiation would change due to selective absorption and emission. This contradicts our assumption that the spectral intensity of the radiation field should not change with time. Remember that the intensity of blackbody radiation must be independent of the medium; the reasons are the same as those given in Section 3.5.3. Therefore, while the spectral coefficients of absorption and of emission in general depend upon the material absorbing and emitting radiation, their ratio is a unique function of temperature.

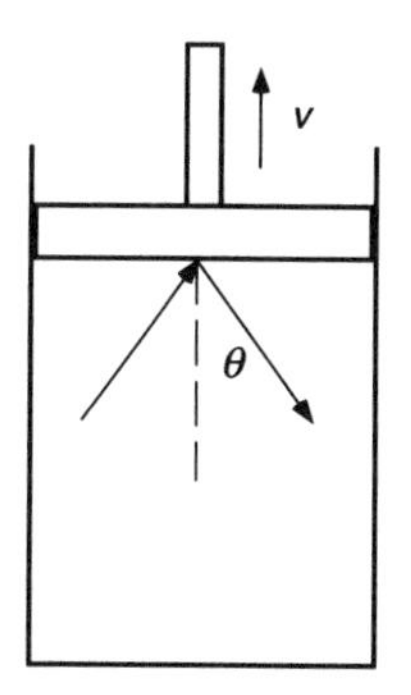

FIGURE 36. Radiation of an arbitrary spectral distribution is expanded adiabatically inside a cavity having reflecting walls and a piston. The radiation striking the moving piston changes its frequency as a result of the Doppler effect.

Wien's displacement law. Recognizing that radiation quantities depend on frequency raises the question of the form of the spectral distribution of blackbody radiation. A first step toward solving this problem can be made by deriving Wien's displacement law. The idea is to calculate the change of the frequency distribution of radiation which results from reversible adiabatic compression or expansion of radiation contained in a cavity with perfectly reflecting walls, and with no trace of an absorbing or emitting medium inside (Figure 36). It is important to recognize that the following derivation holds for radiation of any spectral distribution, not just for blackbody radiation;[14] however, we will assume the radiation in the cavity to be isotropic.

Let us now calculate the rate of change of the spectral energy of the radiation inside the cavity as the piston is pulled or pushed. The spectral energy is the product of the instantaneous volume of the cavity and the spectral energy density. In other words, we wish to calculate the quantity

$$\frac{d}{dt}\left(V\rho_{E\nu}\right)$$

14. If the cavity contains absorbing and emitting matter, the radiation within it would become blackbody radiation. This is why the walls must be perfectly reflecting.

The frequency distribution will change as a result of radiation hitting the moving piston, i.e., as a consequence of the Doppler effect. Let us consider a particular frequency ν. If radiation of that frequency bounces off the piston, it will have a different color. This means that the spectral energy at frequency ν will be reduced by just the amount of energy contained in the radiation flowing toward the piston. With isotropic light, the spectral energy flux striking the piston of surface area A must be

$$j_{E\nu}{}'^{-} = \pi A i_{E\nu}$$

which is the rate at which the spectral energy at frequency ν is being depleted. The second effect will add radiation of frequency ν to the spectrum. If light of a frequency ν_1 which is larger than ν strikes the receding mirror, it will become radiation of frequency ν. The formulas for the Doppler effect let us calculate the relation which must hold between the rays of the two different frequencies ν and ν_1[15]

$$\nu_1 = \nu\left(1 + \frac{2v\cos\theta}{c}\right) \tag{174}$$

θ is the angle between the direction of a ray and the normal to the surface of the piston, and v is the speed of the piston. The spectral intensity of the light which will add to the rays of frequency ν is $i_{E\nu 1}$. This quantity can be approximated by $i_{E\nu}$ as follows:

$$i_{E\nu 1} = i_{E\nu} + (\nu_1 - \nu)\frac{\partial i_{E\nu}}{\partial \nu} + \ldots$$

If we calculate the difference between frequencies according to the Doppler effect and neglect higher order terms in the expansion, we obtain

$$i_{E\nu 1} = i_{E\nu} + \nu\frac{2v\cos\theta}{c}\frac{\partial i_{E\nu}}{\partial \nu}$$

for the intensity of the radiation which will be changed into light of the desired frequency. We again have to integrate this quantity over the hemisphere to get

15. For relative speeds u of observer and light source which are small compared to the speed of light, the relation should be

$$\nu = \nu_1\left(1 - \frac{u}{c}\right)$$

which reduces to Equation (174); ν is the frequency of the reflected radiation while ν_1 is that of the original light. Note that the light reflected from the receding mirror seems to come from a source which recedes at twice the speed of the mirror. Naturally, a ray parallel to the mirror will not experience any Doppler shift.

the energy flux which is responsible for increasing the spectral energy at frequency ν:

$$j_{E\nu}{}'^{+} = \pi A i_{E\nu} + \pi\nu \frac{4}{3c}\frac{\partial i_{E\nu}}{\partial \nu}\dot{V}$$

Note that the product of the speed of the piston and its surface area is equal to the rate of change of the volume of the cavity. Now, the balance of energy for frequency ν is given by

$$\frac{d}{dt}\left(V\rho_{E\nu}\right) = j_{E\nu}{}'^{+} - j_{E\nu}{}'^{-}$$

which, according to Equation (170b), is equivalent to

$$V\dot{\rho}_{E\nu} + \rho_{E\nu}\dot{V} = \frac{1}{3}\nu\frac{\partial \rho_{E\nu}}{\partial \nu}\dot{V}$$

This finally yields the differential equation

$$\dot{\rho}_{E\nu} = \left(\frac{1}{3}\nu\frac{\partial \rho_{E\nu}}{\partial \nu} - \rho_{E\nu}\right)\frac{1}{V}\dot{V} \tag{175}$$

for the spectral energy density during an adiabatic change. Since this quantity must be a function of volume and frequency, Equation (175) can be changed to

$$V\frac{\partial \rho_{E\nu}}{\partial V} = \frac{1}{3}\nu\frac{\partial \rho_{E\nu}}{\partial \nu} - \rho_{E\nu} \tag{176}$$

which has the general solution[16]

$$\rho_{E\nu} = \nu^3 f\left(\nu^3 V\right) \tag{177}$$

This result is called *Wien's displacement law*, and holds for every frequency independent of the particular spectral distribution of the radiation. It reduces the

---

16. This may be shown by inserting the result back into the differential equation. The two derivatives of the spectral energy density with respect to volume and to frequency are

$$\frac{\partial \rho_{E\nu}}{\partial V} = \nu^3 \frac{\partial f}{\partial\left(\nu^3 V\right)}\frac{\partial\left(\nu^3 V\right)}{\partial V}$$

$$\frac{\partial \rho_{E\nu}}{\partial \nu} = 3\nu^2 f + \nu^3 \frac{\partial f}{\partial\left(\nu^3 V\right)}\frac{\partial\left(\nu^3 V\right)}{\partial \nu}$$

This shows that Equation (176) is satisfied by the solution given in Equation (177).

complexity of the problem by reducing the spectral densities and intensities to functions of a single argument. (Before, we had to take them as functions of frequency and volume, or as will be shown, of temperature.) The law shows that if we know the distribution for a specific value of the volume, the function f may be found which then lets us calculate the spectral energy distribution for any other volume.

If we assume the original spectrum of the radiation in the cavity to be that of blackbody radiation, we can apply the relation between its volume and temperature during adiabatic changes as calculated in Section 2.6:

$$T^3 V = constant \tag{178}$$

Using this relation, Equation (177) can be transformed into a relation which shows that the spectrum should depend upon the ratio of temperature and frequency:

$$\rho_{E\nu} = \nu^3 f_2\left(\frac{\nu}{T}\right) \tag{179}$$

The concrete law found to hold for the spectral distribution of blackbody radiation indeed is of the form suggested by Wien's displacement law; see Equation (185).

Entropy and temperature of radiation. Recall what we know about the thermodynamics of blackbody radiation. From the results of Section 2.6 we can derive the Gibbs fundamental relation for the integral quantities. The total entropy and energy of radiation inside volume V satisfy

$$\dot{S} = \frac{1}{T}\dot{E} + \frac{P}{T}\dot{V} \tag{180}$$

According to Equation (88) of Chapter 2, the radiation pressure is one-third the energy density. If we introduce the densities of entropy and of energy, the Gibbs fundamental form becomes

$$\frac{d}{dt}(\rho_s V) = \frac{1}{T}\frac{d}{dt}(\rho_E V) + \frac{1}{T}\frac{1}{3}\rho_E \dot{V}$$

Assembling the terms which depend either upon the volume or upon its rate of change, and noting that the latter two quantities are independent, we end up with the following two results:

$$\rho_s = \frac{4}{3}\frac{1}{T}\rho_E \tag{181}$$

$$\dot{\rho}_s = \frac{1}{T}\dot{\rho}_E \tag{182}$$

The second of these is the Gibbs fundamental relation for the integral radiation densities. Now, while the total entropy density is a function only of the energy density, its spectral counterpart is a function of both spectral energy density and frequency:

$$\rho_{s\nu} = g(\rho_{E\nu}, \nu)$$

We therefore can write the rate of change of the spectral entropy density in terms of the rates of change of the independent variables:

$$\dot{\rho}_{s\nu} = \frac{\partial \rho_{s\nu}}{\partial \rho_{E\nu}} \dot{\rho}_{E\nu} + \frac{\partial \rho_{s\nu}}{\partial \nu} \dot{\nu}$$

Integration over the entire range of frequencies yields

$$\begin{aligned}
\dot{\rho}_s &= \int_\nu \left(\frac{\partial \rho_{s\nu}}{\partial \rho_{E\nu}} \dot{\rho}_{E\nu} + \frac{\partial \rho_{s\nu}}{\partial \nu} \dot{\nu} \right) d\nu = \int_\nu \frac{\partial \rho_{s\nu}}{\partial \rho_{E\nu}} \dot{\rho}_{E\nu} d\nu + \int_\nu \frac{\partial \rho_{s\nu}}{\partial \nu} \dot{\nu} d\nu \\
&= \int_\nu \frac{\partial \rho_{s\nu}}{\partial \rho_{E\nu}} \dot{\rho}_{E\nu} d\nu + \dot{\nu} \int_\nu \frac{\partial \rho_{s\nu}}{\partial \nu} d\nu = \int_\nu \frac{\partial \rho_{s\nu}}{\partial \rho_{E\nu}} \dot{\rho}_{E\nu} d\nu + \dot{\nu} \int_{\rho_{s\nu}} d\rho_{s\nu} \\
&= \int_\nu \frac{\partial \rho_{s\nu}}{\partial \rho_{E\nu}} \dot{\rho}_{E\nu} d\nu + \dot{\nu} \left[\rho_{s\nu}(\nu = \infty) - \rho_{s\nu}(\nu = 0) \right] \\
&= \int_\nu \frac{\partial \rho_{s\nu}}{\partial \rho_{E\nu}} \dot{\rho}_{E\nu} d\nu
\end{aligned}$$

Here, we have used the facts that the rate of change of the frequency is independent of the frequency itself, and that the spectral densities of entropy must vanish both for small and for very large frequencies, because otherwise, its integral would not be finite. If we apply the result to steady-state blackbody radiation, we note that the rate of change of the entropy density should be zero. Since this is true for the rate of change of the energy density as well, the relation

$$0 = \int_\nu \frac{\partial \rho_{s\nu}}{\partial \rho_{E\nu}} \dot{\rho}_{E\nu} d\nu$$

yields the result that the derivative of the spectral entropy density with respect to the spectral energy density must be a constant. As a consequence of the Gibbs fundamental relation for the densities, Equation (182), this constant must be the inverse of the temperature of the radiation under consideration:

$$\frac{\partial \rho_{s\nu}}{\partial \rho_{E\nu}} = \frac{1}{T} \tag{183}$$

This equation states the law of the spectral distribution of blackbody radiation; if we manage to find the entropy of radiation, we will be able to compute the energy distribution. (This is what Planck set out to do; see below.) While this result was derived for blackbody radiation, it must be of more general importance since, as Planck noted:[17]

> For since $\rho_{s\nu}$ depends only on $\rho_{E\nu}$ and ν, monochromatic radiation, which is uniform in all directions and has a definite energy density $\rho_{E\nu}$, has also a definite temperature given by $\partial\rho_{s\nu}/\partial\rho_{E\nu} = 1/T$, and, among all conceivable distributions of energy, the normal one is characterized by the fact that the radiations of all frequencies have the same temperature.

Planck therefore extended the notion of radiation temperature to monochromatic radiation. We may think of this as a result of both the existence of a clear relation between energy and entropy for every color of light, and the fact that rays having different frequencies and traveling in different directions in a cavity without absorbing substances do not interact. Therefore, radiation of any spectral distribution will stay undisturbed inside such a cavity. If you start with blackbody radiation of a given temperature, you may change the spectral values of energy and entropy of all frequencies but one by adding or withdrawing some radiation. The spectral entropy and energy densities of the particular frequency considered will not change, their relation will still be the same, and we should still associate with it the temperature belonging to the original blackbody radiation.

We can now derive Wien's displacement law for the spectral entropy density of radiation. Equation (179) may be inverted to yield

$$\frac{1}{T} = \frac{1}{\nu} f_3\left(\frac{\rho_{E\nu}}{\nu^3}\right)$$

Since the inverse of the radiation temperature is given by Equation (183), we obtain by integration

$$\rho_{s\nu} = \nu^2 f_4\left(\frac{\rho_{E\nu}}{\nu^3}\right) \tag{184}$$

This law has a significance for every frequency, i.e., for every ray of a given color, and therefore also for radiation of arbitrary spectral distribution.[18]

Planck's law of the spectral distribution of blackbody radiation. At the beginning of this century, Planck managed to derive an expression for the spectral intensity of blackbody radiation on the basis of the newly introduced quantum hypothesis. We shall begin by presenting the result for the spectral energy intensity:

---

17. M. Planck (1906), paragraph 93.
18. The general forms of the laws should contain reference to the speed of light which may be different in different media. Since we have assumed c to be the same in all media, its appearance is not required. See Planck (1906), paragraph 94, for the general forms of the laws.

$$i_{Evb}(T,\nu)=\frac{2h\nu^3}{c^2}\frac{1}{e^{h\nu/kT}-1} \tag{185}$$

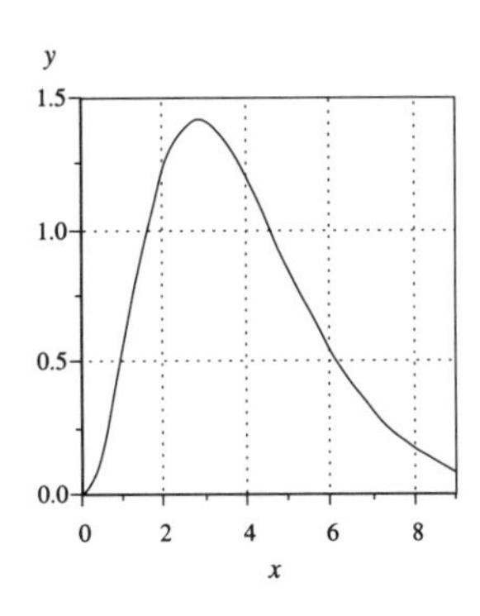

FIGURE 37. The Planck function $x^3/(e^x-1)$, where $x = h\nu/kT$, is a dimensionless representation of the result shown in Equation (185). The spectral intensity of blackbody radiation as a function of frequency has the same form. Higher temperatures mean higher maxima of the curves and higher values for the frequency of the maxima.

h and k stand for Planck's constant and Boltzmann's constant, respectively. Their numerical values are $h = 6.63 \cdot 10^{-34}$ J · s and $k = 1.38 \cdot 10^{-23}$ J/K. Planck's formula beautifully reflects measurements made of the spectral intensity of blackbody radiation (see Figure 37). Equation (141) for the overall intensity of blackbody radiation may be obtained from Equation (185) by integrating over the entire spectrum. Note that the spectral energy density, which is obtained by virtue of Equation (170), i.e.,

$$\rho_{Evb}(T,\nu)=\frac{8\pi h\nu^3}{c^3}\frac{1}{e^{h\nu/kT}-1} \tag{186}$$

indeed has the functional dependence suggested by Wien's displacement law given in Equation (179).

Spectral entropy distribution and temperature of radiation. The fundamental relationship between the entropy and the energy of radiation which is expressed by Equation (183) lets us calculate the spectral entropy density of radiation as a function of the energy density and of frequency. As you can verify by taking the derivative, $\rho_{s\nu}$ must be given by[19]

$$\rho_{s\nu}=\frac{8\pi k\nu^2}{c^3}\left\{\left(1+\frac{c^3\rho_{E\nu}}{8\pi h\nu^3}\right)\ln\left(1+\frac{c^3\rho_{E\nu}}{8\pi h\nu^3}\right)-\frac{c^3\rho_{E\nu}}{8\pi h\nu^3}\ln\frac{c^3\rho_{E\nu}}{8\pi h\nu^3}\right\} \tag{187}$$

The index b for blackbody has been removed from the entropy and energy densities. If we assume Equation (186) to hold individually for every frequency independently of whether the distribution of radiation is that of a black body, then

---

19. Write Equation (187) in the form

$$\rho_{s\nu}=\frac{1}{b}\frac{k}{h\nu}\left\{(1+b\rho_{E\nu})\ln(1+b\rho_{E\nu})-b\rho_{E\nu}\ln(b\rho_{E\nu})\right\} \ , \quad b=\frac{c^3}{8\pi h\nu^3}$$

The derivative $\partial\rho_{s\nu}/\partial\rho_{E\nu}$ is

$$\frac{\partial\rho_{s\nu}}{\partial\rho_{E\nu}}=\frac{k}{h\nu}\left\{\ln(1+b\rho_{E\nu})-\ln(b\rho_{E\nu})\right\}=\frac{k}{h\nu}\ln\left(1+\frac{1}{b\rho_{E\nu}}\right)$$

which by Equation (186) is the reciprocal of the absolute temperature of the radiation at frequency ν. If we wish to extend the concept of temperature to monochromatic radiation, as we have done implicitly, we can do so by defining a temperature of light at frequency ν which the ray would have if it were part of an entire blackbody spectrum, i.e., by using Equation (186).

$$\rho_{E\nu} = \frac{8\pi h \nu^3}{c^3} \frac{1}{e^{h\nu/kT} - 1} \tag{188}$$

also defines a temperature of radiation at frequency ν according to

$$T_\nu = \frac{h\nu}{k} \frac{1}{\ln\left(\frac{8\pi h \nu^3}{c^3 \rho_{E\nu}} + 1\right)} \tag{189}$$

We interpret this as the temperature of monochromatic rays. The result will be used in a practical application when we attempt to compute the entropy and the temperature of solar radiation passing through the Earth's atmosphere (Section 3.6).

Polarization of radiation. Electromagnetic radiation consists of transverse waves. We may construct an image of a wave in which the vector of the electric field vibrates in a definite plane that also contains the direction of propagation of that wave, as in Figure 38. Such a wave is said to be *plane polarized.* A ray of unpolarized light, such as undisturbed blackbody radiation, consists of waves with a random distribution of planes of vibration of the electric field. Viewed head on, completely polarized and unpolarized light would appear as at the bottom of Figure 38.

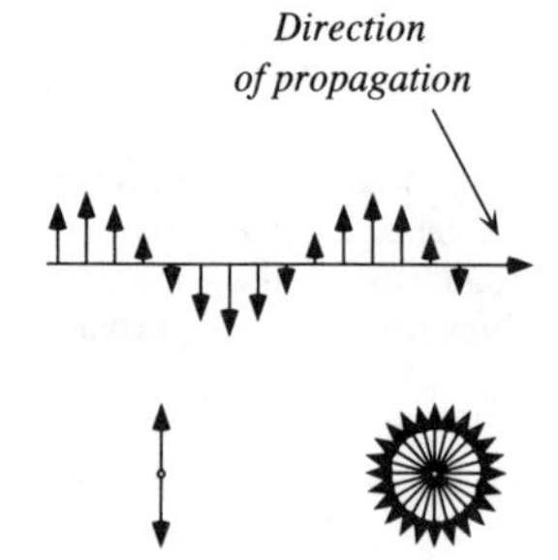

FIGURE 38. Electromagnetic radiation consists of waves with the electric field vibrating in particular planes. Plane polarized light consists of waves which have only a single plane of vibration, while unpolarized radiation consists of waves with randomly oriented planes.

Originally unpolarized radiation may become (partly) polarized as a result of several types of processes. Reflection, double refraction in calcite crystals, passage through polarizing sheets, and scattering all are sources of polarized light. It is found that the intensity of a beam of radiation in any state of polarization can be calculated as the sum of two plane polarized components where the planes have to be perpendicular to each other but, otherwise, can have an arbitrary orientation. If we use the notation i_ν for the intensity of a plane polarized component, we can say that

$$\begin{aligned} \mathrm{i}_\nu' &= \mathrm{i}_\nu^{max} \cos^2\varphi + \mathrm{i}_\nu^{min} \sin^2\varphi \\ \mathrm{i}_\nu'' &= \mathrm{i}_\nu^{max} \sin^2\varphi + \mathrm{i}_\nu^{min} \cos^2\varphi \end{aligned} \tag{190}$$

and

$$i_\nu = \mathrm{i}_\nu' + \mathrm{i}_\nu'' = \mathrm{i}_\nu^{max} + \mathrm{i}_\nu^{min} \tag{191}$$

The intensity of the beam is independent of the orientation φ of the planes of polarization. i^{max} and i^{min}, which are the maximum and minimum values attained by two perpendicular plane-polarized components, are called the principal values of the intensity, and their respective planes are called the principal planes of vibration. These relations hold for the energy intensity, and for the entropy intensity for independent (noncoherent)[20] components. For example, the integral intensities of entropy and of energy are calculated in terms of

$$i_S = \int_\nu \left(\mathrm{i}_{S\nu}^{max} + \mathrm{i}_{S\nu}^{min}\right) d\nu$$
$$i_E = \int_\nu \left(\mathrm{i}_{E\nu}^{max} + \mathrm{i}_{E\nu}^{min}\right) d\nu \tag{192}$$

Now, the spectral values of the single (principal) components are computed using the following relations:

$$\mathrm{i}_{E\nu} = \frac{h\nu^3}{c^2} \frac{1}{e^{h\nu/kT} - 1} \tag{193}$$

$$\mathrm{i}_{S\nu} = \frac{k\nu^2}{c^2} \left\{ \left(1 + \frac{c^2 \mathrm{i}_{E\nu}}{2h\nu^3}\right) \ln\left(1 + \frac{c^2 \mathrm{i}_{E\nu}}{2h\nu^3}\right) - \frac{c^2 \mathrm{i}_{E\nu}}{2h\nu^3} \ln \frac{c^2 \mathrm{i}_{E\nu}}{2h\nu^3} \right\} \tag{194}$$

where i stands for either of the two principal components. For an arbitrary state of polarization, we introduce the degree of polarization P and calculate

$$\mathrm{i}_{E\nu}^{min} = i_{E\nu} \frac{1}{2}(1 - P)$$
$$\mathrm{i}_{E\nu}^{max} = i_{E\nu} \frac{1}{2}(1 + P) \tag{195}$$

These values have to be inserted in Equation (194), from which we can calculate the entropy and the temperature of polarized radiation.

EXAMPLE 35. The blackbody spectrum in terms of the wavelength.

a) Derive the formula for the spectral energy intensity of blackbody radiation in terms of wavelength and present a nondimensional form of the result, as was done for the frequency distribution in Figure 37. b) Determine the location of the maximum of the spectral distribution and give the numerical result for the radiation of the Sun if it is interpreted as blackbody radiation at 5780 K. c) Derive the expression for the spectral entropy intensity with wavelength as the independent variable.

---

20. If the light is (partly) coherent, the entropy intensity calculated from arbitrary components is larger than the sum of the entropy intensities associated with the principal planes of polarization. See Planck (1906), p. 100–102.

SOLUTION: a) We start with the formula for the spectral distribution of the energy intensity in terms of frequency as given in Equation (185):

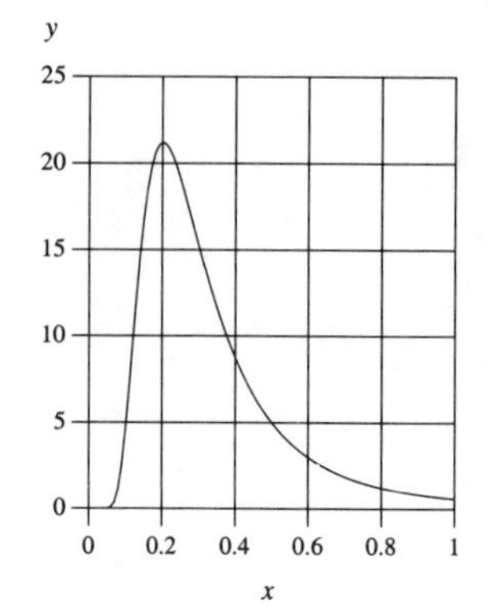

$$i_{E\nu} = \frac{2h\nu^3}{c^2} \frac{1}{e^{h\nu/kT} - 1}$$

Application of the transformation from frequency to wavelength according to Equation (171) leads to

$$i_{E\lambda} = \frac{2hc^2}{\lambda^5} \frac{1}{e^{hc/k\lambda T} - 1} \tag{E8}$$

If we introduce the variable $x = kT\lambda/(ch)$, the form of this equation can be written

$$y = \frac{1}{x^5} \frac{1}{e^{1/x} - 1}$$

This function is displayed in the figure. Note that its maximum does not coincide with the maximum of the distribution in terms of frequency (Figure 37).

b) The maximum of the spectral distribution can be calculated from the dimensionless form just derived. The derivative

$$\frac{dy}{dx} = -\frac{5}{x^6} \frac{1}{e^{1/x} - 1} + \frac{1}{x^5} \frac{-e^{1/x}}{\left(e^{1/x} - 1\right)^2} \left(-\frac{1}{x^2}\right)$$

is set equal to zero, which yields the transcendental equation

$$-5x_m\left(e^{1/x_m} - 1\right) + e^{1/x_m} = 0$$

Its solution is $x_m = 0.2014$, which yields

$$(T \cdot \lambda)_m = 0.2014 \frac{ch}{k} = 2.897 \cdot 10^{-3}\,\mathrm{K \cdot m}$$

This equation often is called Wien's displacement law, instead of the general results derived in Equations (179) and (184). If we use the temperature of the surface of the Sun, we obtain a value of 501 nm for the maximum of the spectral distribution.

c) The results derived so far allow us to calculate the desired quantity by simple derivation. First, we calculate the spectral entropy intensity from the density given in Equation (187). Using Equation (170) we obtain

$$i_{s\nu} = \frac{c}{4\pi} \rho_{s\nu} = \frac{2k\nu^2}{c^2} \left\{ \left(1 + \frac{c^2 i_{E\nu}}{2h\nu^3}\right) \ln\left(1 + \frac{c^2 i_{E\nu}}{2h\nu^3}\right) - \frac{c^2 i_{E\nu}}{2h\nu^3} \ln \frac{c^2 i_{E\nu}}{2h\nu^3} \right\}$$

Multiplying this quantity by c/λ^2 and replacing the frequency by the wavelength yields the desired relation:

$$i_{s\lambda} = \frac{2kc}{\lambda^4} \left\{ \left(1 + \frac{\lambda^5 i_{E\lambda}}{2hc^2}\right) \ln\left(1 + \frac{\lambda^5 i_{E\lambda}}{2hc^2}\right) - \frac{\lambda^5 i_{E\lambda}}{2hc^2} \ln \frac{\lambda^5 i_{E\lambda}}{2hc^2} \right\}$$

EXAMPLE 36. Converting monochromatic light to blackbody radiation.

Consider sunlight which is passed through a filter that blocks out radiation except for a narrow band of frequencies $\Delta\nu$. The light of the Sun has a temperature T_s = 5800 K. The filtered ray is allowed to pass into a cavity where it is converted to blackbody radiation. a) Calculate the temperature of the resulting blackbody radiation assuming the frequency interval of the filtered light lies between 490 nm and 500 nm. b) How much entropy has been produced in the process?

SOLUTION: a) The original radiation stems from the surface of a black body at temperature T_S. In other words, by filtering the Sun's radiation we simply cut a narrow part out of its spectrum. (See the figure.) The radiation which passes through the filter possesses a certain energy. This is the energy of the original blackbody radiation in the frequency band considered here.

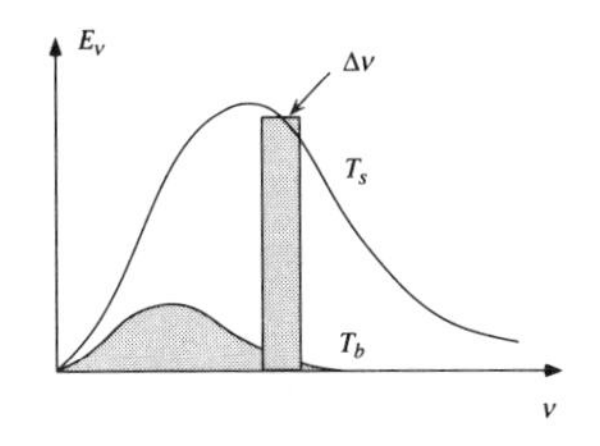

The spectral energy density is given by Equation (186). If the frequency interval of the radiation is narrow, we may calculate the energy contained in a volume V as follows:

$$E = \frac{8\pi h\nu^3}{c^3}\frac{1}{e^{h\nu/kT_s}-1}\Delta\nu V \tag{E9}$$

In the end, the light will have been converted into blackbody radiation with a temperature of T_b inside the cavity:

$$E = aVT_b^4 \tag{E10}$$

Combining (E9) and (E10) we can solve for the unknown temperature T_b:

$$T_b = \left(\frac{8\pi h\nu^3}{ac^3}\frac{1}{e^{h\nu/kT_s}-1}\Delta\nu\right)^{1/4}$$

To give a numerical example, the temperature attained by light filtered in the frequency range 490 - 510 nm turns out to be 2330 K. Obviously, the temperature of radiation resulting from the redistribution of frequencies is lower than the original value. In terms of utilization of solar radiation this means that less can be achieved with the new light.

b) The spectral entropy density of radiation according to Equation (187) will allow us to compute the entropy of the solar radiation contained in the narrow band of frequencies. For the light considered in this example, the quantity

$$x = \frac{c^3}{8\pi h\nu^3}\rho_{E\nu}$$

is much smaller than 1. Therefore, the expression in Equation (187) can be approximated by

$$\rho_{S\nu} = \frac{8\pi k\nu^2}{c^3}\{(1+x)\ln(1+x) - x\ln(x)\} \approx \frac{8\pi k\nu^2}{c^3}x(1-\ln(x))$$

For the same reason, x is approximately equal to

$$x = e^{-h\nu/kT} \quad \Rightarrow \quad \ln(x) = -h\nu/kT$$

The spectral entropy density can therefore be written as follows:

$$\rho_{s\nu} \approx \frac{1}{T}\rho_{E\nu}\left(1+\frac{kT}{h\nu}\right)$$

Now, the entropy associated with the solar radiation filtered into the cavity can be obtained using average values for the narrow range of frequencies:

$$S = \frac{1}{T_s}\rho_{E\nu}\left(1+\frac{kT_s}{h\nu}\right)\Delta\nu V$$

Since the light inside the cavity has been transformed into blackbody radiation, we can use the result known for the total entropy density of blackbody radiation to calculate the entropy S_b of the converted radiation:

$$S_b = \frac{4}{3}\frac{1}{T_b}\rho_{E\nu}\Delta\nu V$$

These results allow us to compute the ratio of the new amount of entropy to the amount delivered by the incoming filtered light:

$$\frac{S_b}{S} = \frac{\frac{4}{3}\frac{1}{T_b}}{\frac{1}{T_s}\left(1+\frac{kT_s}{h\nu}\right)} = \frac{4}{3}\frac{T_s}{T_b}\frac{1}{\left(1+\frac{kT_s}{h\nu}\right)} \approx 1.1\frac{T_s}{T_b}$$

Note that the result is larger than 1, which we expect from the fact that entropy has been produced. The numerical value of the ratio of the entropies turns out to be about 2.8.

3.5.6 Selective Absorbers

Experience shows that bodies having different surface characteristics attain different temperatures if they are exposed to the light of the Sun. Gray bodies which have the same geometrical properties should, on the other hand, all have the same temperature if placed in the same radiation field (see Examples 10 and 14). For this reason we should in general expect the rates of absorption and emission to depend upon the frequency of the radiation involved in the processes. The most striking case of selective absorption and emission of radiation is the greenhouse effect, which heats up the surfaces of the Earth and of Venus and the interior of glass-covered spaces well beyond levels expected from an oversimplified analysis.

Calculating the temperature of absorbers exposed to the light of the Sun. Consider a body in sunlight. If the light is not concentrated, it will most likely attain a temperature of a few hundred Kelvin, and radiate with a spectrum which bears the mark of this hotness. On the other hand, solar radiation which is absorbed has a much higher temperature. Therefore solar radiation and radiation emitted by bodies here on Earth essentially occupy two distinct regions of the spectrum. Consider now a body with a surface which has a high absorptivity for radiation at short wavelengths (solar radiation) but is a poor emitter for long wavelength radiation. The simplest model for dealing with this case is to treat the absorptivity of surfaces as having two distinct but constant values for the two separate frequency intervals of interest. Each value represents some average for radiation in each of the two sections of the spectrum (see Figure 39).

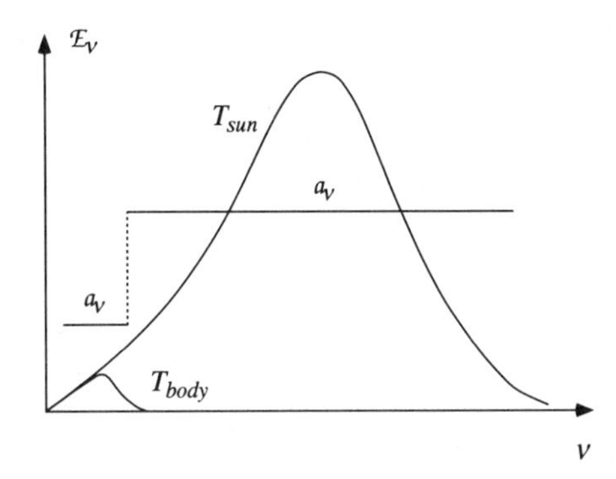

FIGURE 39. Spectrum of solar radiation and of infrared radiation emitted by a body exposed to the light of the Sun. The absorptivities (and therefore the emissivities) are taken to be different in the two distinct ranges of frequencies occupied by solar light and by infrared radiation.

To find surface temperatures we may perform a balance of energy analysis for a selectively absorbing and emitting body exposed to solar radiation. In this analysis we will neglect the angular dependence of radiation properties. In other words, let us assume all surfaces to radiate isotropically in one hemisphere. If A_a is the effective surface absorbing solar radiation, and if $\mathcal{G}_\nu$ is the spectral irradiance of solar radiation at the surface of the absorber, the rate of absorption of energy must be

$$I_E^+ = A_a \int_0^\infty a_\nu \mathcal{G}_\nu \, d\nu \tag{196}$$

The absorptivity depends upon frequency and is denoted by a_ν. The emission from the surface obeys the relation derived in Equation (40), again with the absorptivity (emissivity) taken as a function of frequency:

$$I_E^- = A \int_0^\infty a_\nu \mathcal{E}_{\nu b}(T) d\nu \tag{197}$$

A fraction a_ν of the spectral blackbody intensity is emitted at each frequency. Here, A is the emitting surface area, and T is the surface temperature of the body exposed to the Sun's radiation. This factor determines the spectral distribution of the radiation being emitted. Since $\mathcal{G}_\nu$ depends on the spectrum of solar radiation, the absorptivity does not drop out of the equations as in the case of gray bodies. As a result, the temperature attained by a selective surface strongly depends upon the average values of the absorptivity (or the emissivity) for the respective range of frequency of radiation (visible for the Sun's light, infrared for bodies at around room temperature).

A daisy world. There is a nice example of a model of a planet where the presence of life leads to a self-regulating mechanism which keeps the temperature within a narrow range even though the sun is getting brighter all the time. The model is called a *daisy world*,[21] and it addresses the question of how the surface of the Earth could have had a relatively stable temperature over the course of billions of years, even though the radiation of the Sun must have increased considerably during the same period, which should have led to a steady increase of the surface temperature of our planet.

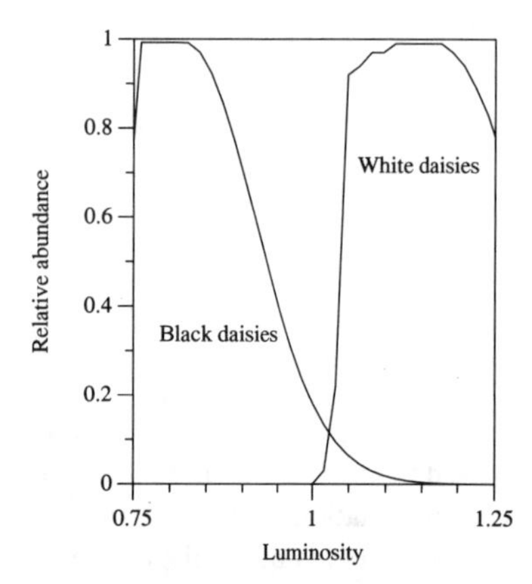

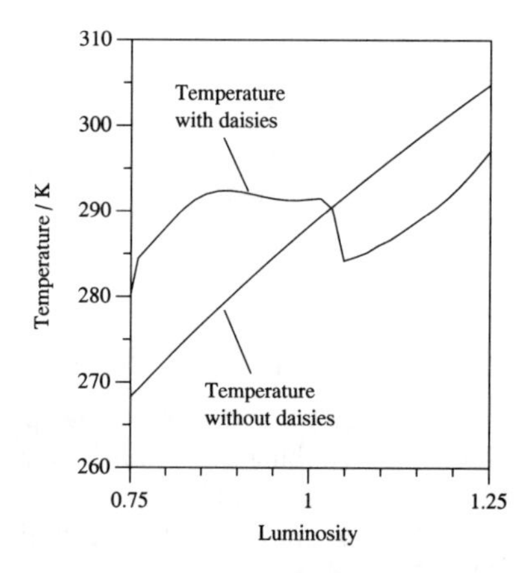

FIGURE 40. Simulation results of a simple model of a self-regulating daisy world. The radiative properties of the planet without life were taken as constant. The absorptivity in the visible part of the spectrum is taken to be larger for black daisies than for white ones, with the absorptivity of the dead surface somewhere in between. The emissivity at infrared wavelengths has been assumed to be the same for all three materials (black and white daisies, and rocks). As the central Sun gets brighter, first black daisies start to grow and cover the planet, making the surface warmer than it would otherwise be. As the luminosity of the Sun grows further, the black daisies get too hot, which lets them slowly die. White daisies, which do not get that hot in the same light, start growing later. As their number increases sharply, due to proper temperatures and increasing space on the planet, the average surface temperature first drops. Finally, it also gets too hot for the white daisies, leading to their demise. The temperature of the planet finally approaches that of the dead surface again. (This is an example of a simple system dynamics model created with Stella™. It consists of two first-order differential equations representing the laws of balance of the number of black and white daisies, respectively. The constitutive laws used are those for the temperature of the different types of surfaces, and laws governing the reproductive rates of the flowers.)

Imagine a planet without life circling the central star of the planetary system. The luminosity of the star is supposed to increase as a function of time. If the planet's

21. A.J. Watson and J.E. Lovelock (1983).

surface has constant radiative properties, its temperature will increase in accordance with the change of luminosity of the Sun. Now assume that black daisies start growing, leading to an increase of the absorptivity at visible wavelengths: the planet will grow hotter than it otherwise would. Now, as the central star continues brightening, white daisies, which have a lower absorptivity, start replacing the black ones. This will tend to lower the temperature of the surface of the planet. All in all, it is possible to envisage a relatively simple model of a planet whose interaction with life leads to a surface temperature which stays within narrow bounds for quite a while.

EXAMPLE 37. Temperature of absorbers on Earth.

Consider a flat disk and a piece of white paper both facing the Sun. The disk is made out of cast iron, for which the average absorptivity for solar radiation is 0.94. For the radiation emitted by the surface (at a temperature not far from 300 K) the value is 0.21. The sheet of paper has values of 0.15 and 0.90 for the absorptivities for solar radiation and for its own radiation, respectively (Table A.3.3). Assume the bodies to emit radiation evenly from both surfaces. Consider radiation from the environment to be blackbody radiation at a temperature of 20°C. The irradiance of solar radiation is taken to be 800 W/m^2. What are the temperatures reached in the steady state by the surfaces if you neglect convection?

SOLUTION: The balance of energy takes a simple form. Since the absorptivities are constant (but different for absorption and emission), we get

$$A a_s G_{sun} = a_b \sigma 2A\left(T_b^4 - T_o^4\right)$$

Since the temperature of the body will not be very different from that of the environment, we can assume the absorptivity to be the same for this component as for the radiation from the body itself.

For the surface made out of cast iron we obtain the following numerical values:

$$T_b = \left(\frac{a_s \cdot G_{sun}}{2a_b \cdot \sigma} + T_o^4\right)^{1/4} = \left(\frac{0.95 \cdot 800}{2 \cdot 0.21 \cdot 5.67 \cdot 10^{-8}} + 293^4\right)^{1/4} \mathrm{K} = 445\mathrm{K}$$

For the piece of paper the result is quite different, due to the different values of the absorptivities:

$$T_b = \left(\frac{0.28 \cdot 800}{2 \cdot 0.95 \cdot 5.67 \cdot 10^{-8}} + 293^4\right)^{1/4} \mathrm{K} = 312\mathrm{K}$$

As we might have expected, the paper does not get as hot as the cast iron disk. Note that the environment has a relatively large influence upon the result, at least in the second case. Naturally, convection due to current of air flowing over the surfaces would prevent the bodies from getting as hot as calculated.

EXAMPLE 38. The Earth as a selective absorber.

a) Model the surface of the Earth as a uniform selective absorber. With an average temperature of 15°C and an absorptivity for the Sun's light of 0.70, how large is the emissivity (absorptivity) at infrared wavelengths? b) What is the expected effect of the increase of the amount of greenhouse gases in the atmosphere? c) Different latitudes receive different amounts of radiation from the Sun in the course of the year. The following formula gives a rough representation of the actual values observed:

$$s(x) = 1 - 0.26\left(3x^2 - 1\right)$$
$$x = \sin(latitude)$$

$s(x)$ represents relative values such that the integral of $s(x)$ over the range of x is equal to 1. Model the temperature of the Earth as a function of latitude if you assume a constant absorptivity of 0.70 for the Sun's light, and a constant emissivity of 0.61 (calculated from the average values according to the first problem).[22] Assume that only radiation is responsible for the temperatures attained.

SOLUTION: a) We can repeat the analysis of Example 10, this time with differing values for the absorptivities for emission and absorption. The balance of energy yields

$$a_V \pi R^2 G_{sc} = a_{IR} 4\pi R^2 \sigma T^4$$

where V and IR stand for visible and infrared, respectively; G_{sc} is the solar constant. The absorptivity (emissivity) in the infrared therefore must be

$$a_{IR} = \frac{a_V G_{sc}}{4\sigma T^4} = \frac{0.70 \cdot 1360}{4 \cdot 5.67 \cdot 10^{-8} \cdot (273+15)^4} = 0.61$$

b) The effect of the greenhouse gases can be modeled by lowering of the emissivity at infrared wavelengths.

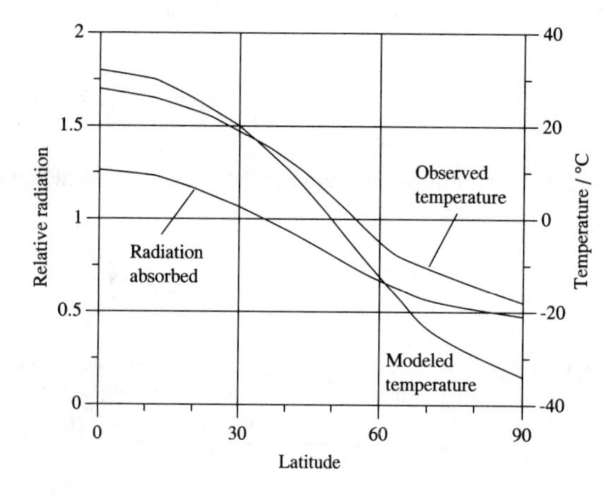

22. See R.S. Lindzen (1990) for a slightly different model and more details.

c) If there is no other type of heat transfer in the latitudinal direction, there must be a balance of energy for each strip of constant latitude which takes the form

$$a_V s(x) \mathcal{G}_{sc} = a_{IR} 4 \sigma T^4$$

We can solve this equation for different values of x to obtain the hypothetical latitudinal distribution of temperature (see the diagram).

If the modeled temperature distribution is compared to the measured average temperature as a function of latitude, it is observed that the actual values do not change as much over the globe as the numbers obtained from the model. Convective currents from the equator to the poles redistribute the entropy, making the temperature gradient from the equator to the poles much smaller.

3.6 Solar Radiation

Our Sun is a rather average main sequence star[23] of spectral type G2[24] which means that it has a surface temperature of a little less than 6000 K and pours out radiation with an energy flux of $4 \cdot 10^{26}$ W. At the distance of the Earth, this flux has thinned out to a flux density (normal irradiance) of 1360 W/m^2. This is the radiation which penetrates the Earth's atmosphere, sets in motion many of the processes in the air and in the oceans, and makes possible the existence of life.

In the last section of this chapter, we will consider the origin and the form of solar radiation found above our planet, before discussing radiative processes in the atmosphere. This should lead to an assessment of the entropy and the temperature of the radiation found at the surface of the Earth. This includes a process which was left out of consideration so far, namely, scattering of radiation. Finally, two subjects of interest in solar energy engineering will be addressed: the concentration of sunlight to achieve high temperatures, and the absorption of radiation by simple flat solar collectors.

3.6.1 The Origin of Solar Radiation

Stars such as our Sun are spheres made of hydrogen and helium (plus a small fraction of the heavier elements), which under the influence of their own gravity attain high temperatures at their centers that allow nuclear reactions to occur. It is believed that all the energy released as a consequence of these reactions is car-

23. Main sequence stars are those in the first and longest phase of their life, during which they burn hydrogen at their centers. On astrophysics in general, see F.H. Shu (1982); on stars in particular, see C. Payne-Gaposhkin (1979), or I.S. Shklovskii (1978).

24. The spectra of normal stars are put into a sequence representing decreasing surface temperature which corresponds to changing the apparent color of the star from blue to red. The sequence is labeled O B A F G K M, and it represents surface temperatures from some 20000 K to about 3000 K.

ried outward and radiated away from the surface. Since solar material is fairly opaque, the radiation we receive essentially originates in a thin layer at the surface of the star. This radiation then spreads through space, a process which preserves the essential characteristics of the light, namely, its entropy and energy intensity and, therefore, also its temperature.

The interior of the Sun. Our central star is a sphere of gas with a radius of about 696,000 km and a mass of $1.989 \cdot 10^{30}$ kg, made out of hydrogen (70% by mass), helium (27%), and a sprinkling of heavier elements (3%).[25] The composition of stars is derived from spectral observations of their surfaces. Naturally, to take this as the composition of the interior as well, we must have reason to believe that a star is well mixed. Indeed, it is assumed that during formation the material of a star is subject to heavy mixing, leading to a uniform composition. Now, since few stars, including our Sun, are found to be precisely at the beginning of their life, we moreover have to assume that the changes of composition due to nuclear reactions in the interior do not reach the surface. Models of the structure of main sequence stars show that convection occurs in these objects either deep inside the core, or near the outer layers only, which prevents mixing of the interior and the surface.

Therefore, new stars can be considered to be uniform spheres of gas. For a given composition, there essentially is only one parameter which distinguishes between different stars and influences their evolution from newly born to old and highly evolved—the mass of the object. Masses of stars span a range of about one hundredth of the mass of our Sun to maybe 100 times its mass. Since, during their first stage of evolution as main sequence stars changes occur rather gradually, one is justified in approximating their structure as spherically symmetric fluid bodies in hydrostatic equilibrium. A simple calculation[26] shows that the pressure at the center of a sphere with the properties of our Sun should be around $6 \cdot 10^{14}$ Pa. Assuming ideal gas properties we then estimate the central temperature of the Sun to be some 10 million K. This result justifies the initial assumption that the matter inside a star must behave very much like an ideal gas[27] even at densities of 100 times that of water, a value which comes close to that at the center of the Sun. The temperatures found at and near the center of a star are high enough to allow for hydrogen burning to occur (hydrogen fuses into helium). These nuclear reactions release the vast amounts of energy which, in the end, are carried away with the radiation from the surface of the star. Before this can happen, however, entropy and energy have to be transported from the interior to the surface.

---

25. Numbers can be found in K.R. Lang (1980).

26. Consider a column of fluid extending from the center to the surface and calculate its weight from an average value of the gravitational field inside the star. See Schwarzschild (1958) for a simple discussion of the interior state of stars.

27. The atoms inside a star must be completely ionized, leading to a gas made out of nuclei and electrons, both behaving as ideal gases for the states found inside main sequence stars.

It is found that, again for properties encountered in main sequence stars, conduction of heat is negligible. This leaves convection and radiation as modes of transfer. If the temperature gradient required for the diffusion of radiation (see Section 3.5.4) is not too large, radiation will prevail and the gas inside the star will be stable against convective disturbances (Section 4.6.3). If, on the other hand, the temperature gradient required for transporting all the heat becomes too large, convection will set in and totally dominate the transport mechanisms. Now, numerical models of the interior of well mixed spheres of gas show that stars of solar mass are stable against convection in the inner parts. Further out toward the surface, however, radiation can no longer carry the fluxes of entropy and energy; convection starts and takes over the transport all the way to the surface. With stars whose mass is larger than about twice that of the sun the interior state is just the opposite. Because of the much higher rate of release of energy at the center compared to solar-type stars, convection is required for the inner parts while radiation occurs in the outer layers.

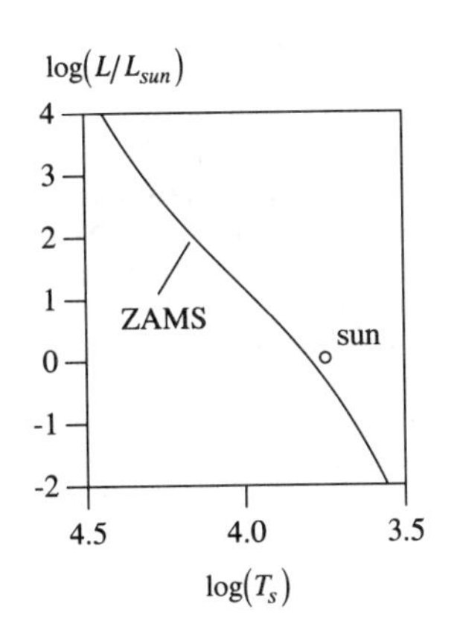

FIGURE 41. Hertzsprung-Russell diagram of main sequence stars. The diagram shows the logarithm of the ratio of the luminosity of a star to that of the sun as a function of the logarithm of the surface temperature. The result holds for models of zero age main sequence stars. Results of model calculations have been taken from Iben (1967).

Changes of the initial structure of stars essentially are a consequence of nuclear reactions near the center which slowly change the composition of the gas.[28] As a result, the interior structure changes quite considerably. It is interesting to see that during the hydrogen burning phase, the exterior appearance of the main sequence stars does not change nearly as much.

Radiation from the surface of stars. Models calculated on the basis of the processes just described yield the radius and the luminosity of the star in addition to details on the interior state.[29] These values allow for the temperature of the surface of the star to be calculated if we assume the radiation to be that of a black body (Figure 42). As we know from the laws of hemispherical emission from the surface of a black body, the relation between hemispherical flux, surface area, and temperature yields

$$L = 4\pi R_S^2 \sigma T_S^4 \tag{198}$$

In astronomy, the energy flux carried away from the surface of a star is called the luminosity L. Currently, the luminosity of the Sun, as measured from the radiation above the Earth's atmosphere,[30] is $3.844 \cdot 10^{26}$ W. With a radius of $6.96 \cdot 10^8$ m, the equivalent blackbody temperature for the surface of the Sun is 5777 K. If we do this calculation for the one-parameter sequence of models of gas spheres for varying mass, we may plot the result in a luminosity-temperature diagram which is called the Hertzsprung-Russell diagram (Figure 41). The most interesting result of these computations shows that the surface properties put the model

28. Stars of solar mass have a luminosity which allows them to burn hydrogen at the center for some 10 billion years. A star of 5 solar masses, however, burns so fast that hydrogen will be depleted in the inner regions in about 70 million years.
29. See D.D. Clayton (1968) for an account of stellar structure and evolution.
30. The value of the luminosity of the Sun is the result of the integration of the WRC spectrum over wavelength (Figures 36 and 42).

stars along a line from the upper left to the lower right in the diagram precisely where observations place the socalled main sequence stars. Therefore, main sequence stars are interpreted as stars in the first phase of their life, during which they transform hydrogen into helium. They stay in a narrow band along the main sequence in the Hertzsprung-Russell diagram as long as they have enough hydrogen at the center for this type of nuclear reaction to occur.

The radiation field at the surface of the sun and at the distance of the Earth. As you can see by comparing the measured spectrum of solar radiation with that of a black body of equivalent temperature (Figure 42), for the purpose of solar energy engineering solar radiation quite nicely approximates the ideal case.

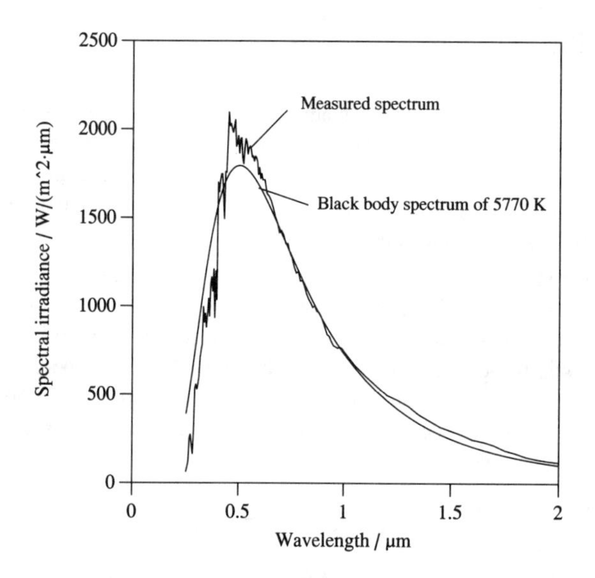

FIGURE 42. Black body spectrum for a temperature of 5770 K superimposed on the spectrum of solar radiation observed above the surface of the Earth's atmosphere. Since the properties of the radiation do not change on their way from the Sun to the Earth, the spectrum is equal to that found at the surface of the Sun, multiplied by the (constant) solid angle of the Sun as seen from the Earth.

The radiation emitted by our Sun originates in the uppermost layers at the surface. This part of the Sun is called the photosphere; obviously, conditions there are more complicated than envisioned by the model of a black body. What you cannot see in the spectrum in Figure 42 are the many narrow absorption lines which result from atomic absorption in the outermost layers of the photosphere. These lines provide most of the detailed information about the conditions at the surface of our star. In particular, through spectroscopy we determine the composition of the Sun's surface; in this way, in the last century, a new element was found in the solar spectrum which received the name of the Sun, helium.

Following the discussion of the laws for monochromatic radiation in Section 3.5.5, we may derive the values of the entropy intensity and the temperature for different wavelengths for the radiation at the surface of the Sun. As calculated in Example 35, the entropy intensity is given by

$$i_{s\lambda} = \frac{2kc}{\lambda^4}\left\{\left(1+\frac{\lambda^5 i_{E\lambda}}{2hc^2}\right)\ln\left(1+\frac{\lambda^5 i_{E\lambda}}{2hc^2}\right) - \frac{\lambda^5 i_{E\lambda}}{2hc^2}\ln\frac{\lambda^5 i_{E\lambda}}{2hc^2}\right\} \quad (199)$$

while the monochromatic temperature is computed from Equation (E8) which yields

$$T_\lambda = \frac{hc}{k\lambda}\frac{1}{\ln\left(\frac{2hc^2}{\lambda^5 i_{E\lambda}}+1\right)} \quad (200)$$

Calculation of the monochromatic temperature and the entropy intensity for both the measured values of the solar spectrum and its equivalent blackbody spectrum shows (Figure 43) that the difference between the actual radiation and the ideal one is small. Integration of the spectral entropy intensity yields a value which is only slightly larger for the blackbody spectrum than for the actual one. (Since turning solar radiation into an equivalent blackbody spectrum should produce entropy, this result is to be expected.) Especially in the range where the Earth's atmosphere lets most of the radiation pass, the deviation is very small. We should expect changes in the radiation from blackbody conditions to result mostly from the interaction with our atmosphere.

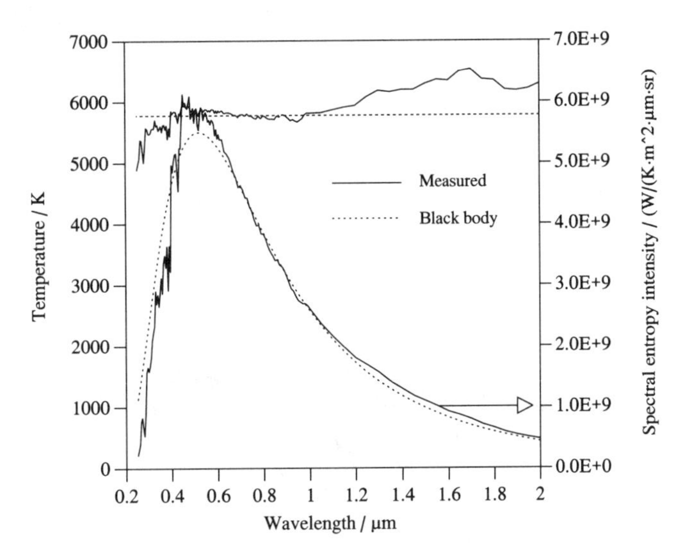

FIGURE 43. Monochromatic temperature of the radiation from the surface of the sun both for measured values and for a blackbody spectrum at 5777 K. Also shown are the entropy intensities for measured values and for the equivalent blackbody spectrum.

3.6.2 Absorption, Scattering, and Polarization in the Atmosphere

The Earth's atmosphere considerably changes the physical state of solar radiation which arrives at the surface of our planet. Different molecules absorb some of the light, and molecules and aerosols scatter part of it. As a result, for cloudless skies, we have both direct and diffuse radiation, where the diffuse part is composed of the light which has been scattered. There still is another effect which should be taken into account: scattering polarizes the rays, which before they hit the atmosphere, are essentially unpolarized.

Naturally, the magnitude of the effects depends both on the properties of the atmosphere and its thickness, i.e., the amount of air solar rays have to penetrate. The latter quantity is called relative *air mass*, where an air mass equal to 1 means that the sun is precisely overhead. If we give the position of the sun in the sky in terms of its zenith angle θ, we can calculate the distance solar rays have to travel through the atmosphere to the observer. For angles which are not too close to 90°, the relative air mass m_a is

$$m_a(\theta) = \frac{1}{\cos(\theta)} \qquad (201)$$

Note that the solar irradiance for the horizontal surface at the top of the atmosphere depends upon the zenith angle of the Sun as well:

$$\mathcal{G}_{ho} = \cos(\theta)\mathcal{G}_{sc} \qquad (202)$$

where $\mathcal{G}_{sc}$ is the solar constant of 1367 W/m^2 which is obtained from integration of the WRC spectrum (Figure 42). The following points will be discussed for a cloudless atmosphere only.

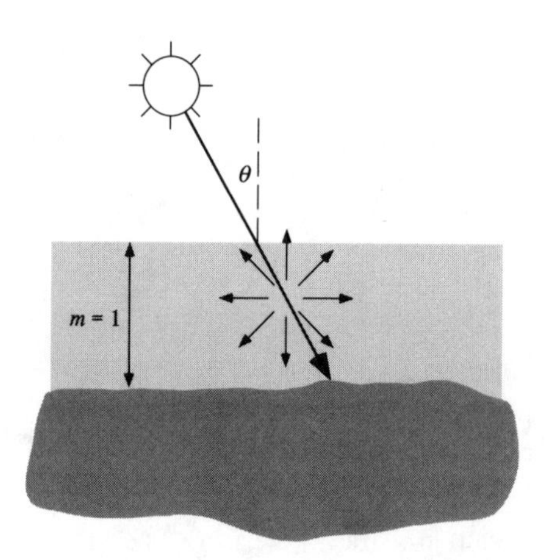

FIGURE 44. When solar radiation penetrates the Earth's atmosphere, it is both absorbed and scattered. In addition, while extraterrestrial radiation is more or less unpolarized, scattering polarizes the light arriving at the surface of the planet. All these effects change the relative amounts of entropy and energy of radiation, therefore leading to changes of the monochromatic temperatures.

Attenuation of solar radiation. The influence of the Earth's atmosphere is commonly described in terms of *attenuation or extinction coefficients* which should be given for every wavelength. An extinction coefficient is the sum of absorption coefficient and scattering coefficient, i.e.,

$$\mu_\lambda = \kappa_{E\lambda} + \beta_{E\lambda} \qquad (203)$$

(see Example 33), and it is defined such that

$$\frac{di_{E\lambda}(s)}{ds} = -\mu i_{E\lambda}(s) \tag{204}$$

which, for constant attenuation coefficients along a path turns out to be equivalent to

$$i_{E\lambda}(s) = i_{E\lambda o} \exp(-\mu s) \tag{205}$$

In these equations, s is the length of the path travelled by a ray in the atmosphere, and the index o refers to the radiation above the atmosphere. In solar radiation computations it is common to refer the extinction coefficients to relative air mass, which means that you have to replace s by m_a, according to Equation (202), in the equations. Since the term containing the exponential function gives the ratio of the transmitted light to the incident light, it is called the *transmittance* τ of the atmosphere:

$$\tau_\lambda = \exp(-k_\lambda m_a) \tag{206}$$

where k_λ replaces the normal extinction coefficient. If the coefficients and the original undisturbed spectrum are known, the effect of the atmosphere can be computed for every wavelength upon which the total influence is obtained by integration over the spectrum.

Absorption and scattering. Absorption of a ray of light by a clear atmosphere is due to molecular effects, and strongly depends upon wavelength, leading to absorption bands in the solar spectrum at the surface of the Earth. In the part of the radiation extending from short wavelengths up to 0.35 μm, the most important contribution to molecular absorption comes from ozone; absorption bands due to water vapor influence the radiation mostly between 1 μm and 4 μm; and the uniformly mixed gases mainly absorb at wavelengths above 2 μm. Aerosols add a little bit to absorption but their main influence is upon scattering and will be considered later.[31]

The spectral transmittance of ozone is calculated in terms of the attenuation coefficient $k_{O\lambda}$, the amount of ozone, which is given as an equivalent length l in cm, and the air mass m_a:[32]

$$\tau_{O\lambda} = \exp(-k_{O\lambda} l m_a) \tag{207}$$

The attenuation coefficient has been measured and is given in tables. For uniformly mixed molecular absorbers such as CO_2 and O_2, the combined effect is

31. The discussion in this section is based upon Iqbal (1983), Chapter 6.

32. We should use the relative air mass for ozone which differs from the normal value for large zenith angles of the Sun (Iqbal, 1983, Chapter 5).

$$\tau_{g\lambda} = \exp\left(-\frac{1.41 k_{g\lambda} m_a}{\left(1 + 118.93 k_{g\lambda} m_a\right)^{0.45}} \right) \tag{208}$$

while water vapor absorption is calculated according to

$$\tau_{wa\lambda} = \exp\left(-\frac{0.2385 k_{wa\lambda} w m_a}{\left(1 + 20.07 k_{wa\lambda} w m_a\right)^{0.45}} \right) \tag{209}$$

where w is the amount of precipitable water given in cm.[33] Note that the form of these laws for the transmittance differ somewhat from the simple form presented in Equation (206); they are the result of parameterization of more detailed absorption calculations.

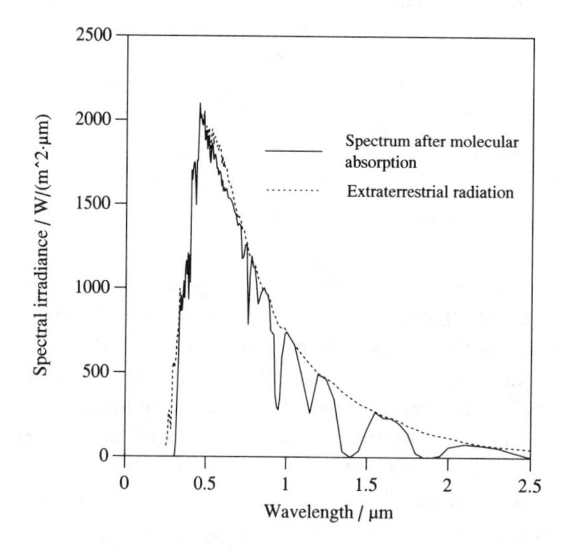

FIGURE 45. Molecular absorption changes the extraterrestrial radiation in some areas of the spectrum. If scattering is neglected, the radiation changes according to the data shown in this graph. The amount of ozone used in the calculation corresponds to 0.35 cm, and that for water vapor to 2 cm, while the air mass was taken to be 2. Spectral absorption coefficients have been taken from Iqbal (1983).

Let us now turn to a brief description of scattering. Light traveling along a beam through the atmosphere is partly absorbed and partly scattered; only the contribution which is not influenced by either mechanism is transmitted to the ground. Scattering removes radiation from a beam by changing the direction of incoming radiation; the frequency and energy of the scattered component are not changed, but the change of directional distribution certainly leads to changes in the intensity of the light. The latter process has important consequences for the entropy and the temperature of scattered radiation (see below); scattering is an irreversible process.

33. See Iqbal (1983), Chapter 5.4.

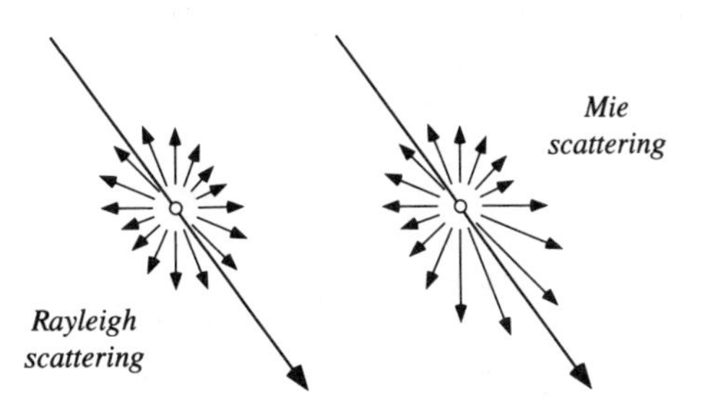

FIGURE 46. Scattering of a ray of light leads to a redistribution of directions of radiation without changing the energy and the frequency. In Rayleigh scattering, forward and backward scattering are favored equally over sideways scattering. In Mie scattering, due to larger particles, forward scattering is dominant.

There are two distinct effects to be taken into account, namely scattering from molecules (*Rayleigh scattering*) and from larger particles (*Mie scattering*). The distribution of the directions of scattered radiation is not generally isotropic. While in Rayleigh scattering equal amounts are scattered in forward and backward directions, forward scattering is preferred in Mie scattering. There is a clear difference between scattering and absorption in that the former is a continuum effect while the latter is selective with respect to wavelength.

Both effects are described in terms of the part of the incident ray which is transmitted without being scattered. In the case of Rayleigh scattering, the transmittance is calculated as follows:

$$\tau_{r\lambda} = \exp\left(-\frac{8.735 \cdot 10^{-3} m_a}{\lambda^{4.08}}\right) \tag{210}$$

Note that in this and the following formula the wavelength must be given in µm. Mie scattering may be expressed in terms of two factors α and β which describe the turbidity of the atmosphere:[34]

$$\tau_{a\lambda} = \exp\left(-\frac{\beta m_a}{\lambda^{\alpha}}\right) \tag{211}$$

The factor β varies from 0 to 0.4 for clean to very turbid atmosphere (visibility ranging from 340 km to less than 5 km), while α is around 1.3.

Note that scattering strongly depends upon the frequency of light, short wavelengths being preferred. Rayleigh scattering in particular leads to the filtering out of blue light from a beam. The transmittance due to this effect is zero at 0.3 µm, while at 0.6 µm it is already above 90%; this explains the blue color of the sky.

Calculation of direct and diffuse radiation on the ground. The formulae presented for absorption and for scattering now allow us to calculate the solar radi-

34. Angström's turbidity formula for aerosols; see Iqbal (1983), p. 117–119.

ation expected at the ground for the case of cloudless atmospheres. The direct ray is attenuated as a result of both absorption and scattering, which means that the part transmitted can be calculated from the radiation incident upon the atmosphere and the product of all transmittances:

$$\mathcal{G}_{\lambda hb} = \tau_{r\lambda}\tau_{a\lambda}\tau_{g\lambda}\tau_{O\lambda}\tau_{wa\lambda}\cos(\theta)\mathcal{G}_{\lambda o} \tag{212}$$

$\mathcal{G}_{\lambda o}$ is the spectral irradiance at the top of the atmosphere, measured for a surface normal to the rays, while $\mathcal{G}_{\lambda hb}$ is the transmitted spectral direct (beam) irradiance for a horizontal surface. The amount of radiation scattered is then calculated according to

$$\mathcal{G}_{\lambda hs} = \tau_{g\lambda}\tau_{O\lambda}\tau_{wa\lambda}\left(1-\tau_{r\lambda}\tau_{a\lambda}\right)\cos(\theta)\mathcal{G}_{\lambda o} \tag{213}$$

where s stands for scattered. There is a problem with calculating the diffuse radiation reaching the ground from the total amount scattered unevenly in all directions. The simplest possibility, which still neglects the effect of light reflected from the ground and scattered back by the atmosphere, is to assume that exactly half of the originally scattered radiation flows downward from the sky to the ground. (This is the assumption made in the calculations which led to the results presented in Figure 47.) The sum of the direct and the diffuse radiation is called global radiation.

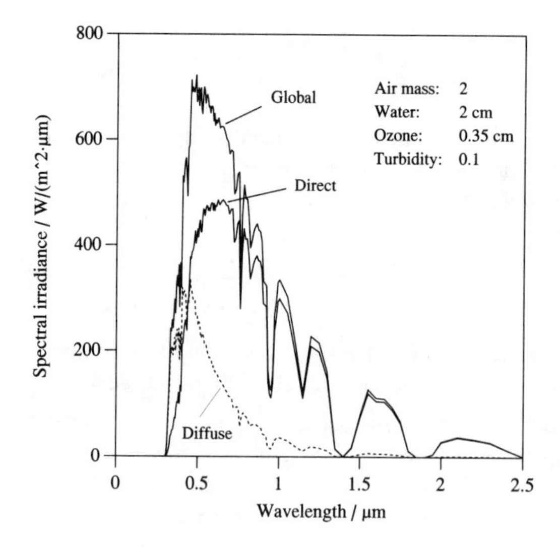

FIGURE 47. Computed spectral distribution of direct and diffuse solar radiation for the horizontal surface for a cloudless sky. It is assumed that precisely half of all the radiation scattered flows in the direction of the ground. The sum of direct and diffuse irradiance is called *global* irradiance. The computation starts with the WRC spectrum and assumes absorption and scattering according to the relations presented in this section. Values of absorption coefficients have been taken from Iqbal (1983); see Table A.12.

The entropy and the temperature of solar radiation. According to Planck's theory presented in Section 3.5.5, we can associate a spectral entropy intensity and a monochromatic temperature with radiation of a given spectral energy intensity. To find the entropy and the temperature according to Equations (199) and

(200) we therefore have to calculate the energy intensity for direct and diffuse radiation from Equations (212) and (213), respectively. For the former we obtain

$$i_{E\lambda b} = \frac{1}{\cos(\theta)} \frac{G_{\lambda hd}}{\Omega_s} \tag{214}$$

where Ω_s is the solid angle of the Sun as seen from the Earth, which is equal to

$$\Omega_s = \frac{\pi R_S^2}{d_{ES}^2} = \frac{\pi\left(6.960 \cdot 10^8\right)^2}{\left(1.496 \cdot 10^{11}\right)^2} = 6.80 \cdot 10^{-5}\,\text{sr}$$

The cosine of the zenith angle of the Sun appears in Equation (214) since we have to take a surface normal to the direction of the direct beam. For the diffuse component of solar radiation, on the other hand, we take the light scattered to the horizontal surface. If we assume half of the radiation scattered in all direction to reach the ground, and if we take this component to have an isotropic distribution over the hemisphere, the energy intensity is given by

$$i_{E\lambda d} = \frac{0.5 G_{\lambda hs}}{\pi} \tag{215}$$

This follows from Equation (148) for isotropic hemispherical radiation. The values obtained from Equations (214) and (215) are plugged into the relations for the entropy intensity and the monochromatic temperature. The temperature of solar radiation is presented in Figure 48 for the spectrum shown in Figure 47.

Obviously, absorption and scattering will change black radiation into nonblack light. Both effects reduce and redistribute the spectral intensities leading to a spectrum of temperatures for radiation which originally had only a single temperature. We may introduce the "effective" temperature of a component of radiation by comparing the integral values of entropy and energy intensity. According to Equation (140) and its counterpart for energy we may write

$$i_E = \frac{3}{4} T_{eff} i_s \tag{216}$$

This quantity is shown in Figure 48 for both direct and diffuse solar radiation. Calculations show that the effective temperature of direct light is only slightly smaller than the temperature of the surface of the Sun, even for relatively low elevation of the Sun in the sky. (For the spectrum calculated in Figure 47 the value turns out to be roughly 5100 K.) The temperature of diffuse light from the sky still has a surprisingly high temperature of around 1700 K. Detailed studies of the entropy of solar radiation[35] demonstrate that the main assumption made here with regard to diffuse light, namely, that it is supposed to be isotropic, is acceptable. Assuming realistic angular distributions leads to changes in the results of only a few percent.

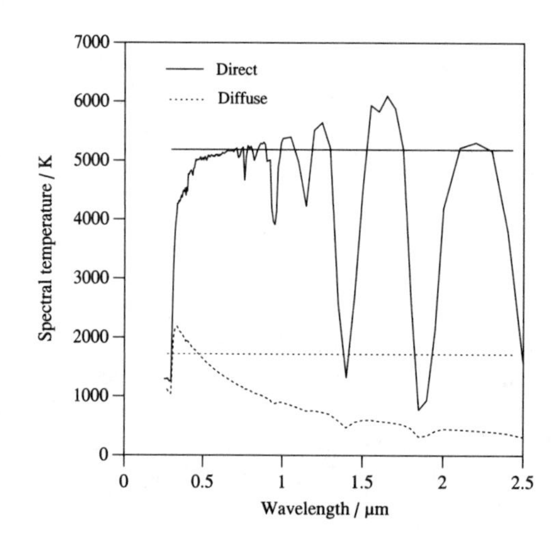

FIGURE 48. Spectral temperatures of direct and diffuse radiation according to the spectra given in Figure 47. The horizontal lines represent effective temperatures of overall radiation calculated according to Equation (216). For visible light, the temperature of direct radiation is close to 5100 K while that for diffuse light is between 1500 K and 2000 K.

Polarization of solar radiation. There is another effect which could be of interest because of its influence upon the entropy and the temperature of radiation, namely polarization. For the case of solar radiation in the Earth's atmosphere we need models of the degree of polarization as a function of the angles of incidence of diffuse light (scattered and reflected). Again, for realistic models of solar radiation, the effect of polarization upon the temperature of radiation is relatively minor. Allowing for complete polarization, however, reduces the entropy of radiation by about 20% compared to the value for unpolarized light.[36]

EXAMPLE 39. Calculating direct and diffuse spectral irradiances.

Calculate the spectral irradiances of direct and diffuse sunlight for a wavelength of 600 nm for the following atmospheric conditions. The solar elevation is taken to be 30°. The amount of ozone and of water vapor are set to 0.35 cm and 2.0 cm, respectively. The Angstöm coefficients for the turbid atmosphere are set equal to $\beta = 0.10$ and $\alpha = 1.3$, respectively.

SOLUTION: According to Table A.12, the values of the extraterrestrial spectral irradiation and the molecular absorption coefficients are given by

35. Kabelac and Drake (1992), p. 239–246.
36. Kabelac and Drake (1992), p. 239–246.

λ / µm	G_λ / W/m²µm	$k_{O\lambda}$	$k_{g\lambda}$	$k_{wa\lambda}$
0.600	1720.00	0.125	0.00E+00	0.00E+00

We have to calculate the transmission coefficients due to absorption and to scattering according to Equations (207) through (211). Since the absorption coefficients for the uniformly mixed gases and for water vapor are equal to zero at the chosen wavelength, we only have to take into account Equation (207) for absorption by ozone:

$$\tau_{O\lambda} = \exp\left(-k_{O\lambda} l m_a\right) = \exp(-0.125 \cdot 0.35 \cdot 2) = 0.916$$

Since the elevation of the Sun is 30°, the zenith angle is 60° which yields a value for the relative air mass of 2.0. Scattering leads to following two transmittances:

$$\tau_{r\lambda} = \exp\left(-8.735 \cdot 10^{-3} m_a / \lambda^{4.08}\right) = \exp\left(-8.735 \cdot 10^{-3} \cdot 2/0.600^{4.08}\right) = 0.869$$

$$\tau_{a\lambda} = \exp\left(-\beta \cdot m_a / \lambda^{\alpha}\right) = \exp\left(-0.1 \cdot 2/0.600^{1.3}\right) = 0.678$$

All in all, the direct spectral irradiance on the horizontal must be

$$\begin{aligned} \mathcal{G}_{\lambda hb} &= \tau_{r\lambda}\tau_{a\lambda}\tau_{g\lambda}\tau_{O\lambda}\tau_{wa\lambda}\cos(\theta)\mathcal{G}_{\lambda o} \\ &= 0.869 \cdot 0.678 \cdot 1 \cdot 0.916 \cdot 1 \cdot 0.5 \cdot 1720\,\mathrm{W/(m^2 \mu m)} = 464\,\mathrm{W/(m^2 \mu m)} \end{aligned}$$

The diffuse radiation, on the other hand, is

$$\begin{aligned} \mathcal{G}_{\lambda hs} &= 0.5 \cdot \tau_{g\lambda}\tau_{O\lambda}\tau_{wa\lambda}\left(1-\tau_{r\lambda}\tau_{a\lambda}\right)\cos(\theta)\mathcal{G}_{\lambda o} \\ &= 0.5 \cdot 1 \cdot 0.916 \cdot 1 \cdot (1-0.869 \cdot 0.678) \cdot 0.5 \cdot 1720\,\mathrm{W/(m^2 \mu m)} = 162\,\mathrm{W/(m^2 \mu m)} \end{aligned}$$

where we have assumed that half of the scattered light strikes the ground. These values can also be read from the curves in Figure 47.

EXAMPLE 40. Entropy produced in scattering.

Consider solar radiation penetrating the Earth's atmosphere, and assume it to be undisturbed blackbody radiation of temperature 5777 K. Calculate the entropy generated if all the radiation in the wavelength band 595 - 605 nm is scattered isotropically in all directions.

SOLUTION: We first have to calculate the entropy intensity of the incident radiation for the spectral range chosen:

$$j_s = \frac{2kc}{\lambda^4}\left\{\left(1+\frac{\lambda^5 i_{E\lambda}}{2hc^2}\right)\ln\left(1+\frac{\lambda^5 i_{E\lambda}}{2hc^2}\right) - \frac{\lambda^5 i_{E\lambda}}{2hc^2}\ln\frac{\lambda^5 i_{E\lambda}}{2hc^2}\right\}\Delta\nu\,\Omega_s$$

where

$$i_{E\lambda} = \frac{2hc^2}{\lambda^5}\frac{1}{e^{hc/k\lambda T} - 1}$$

Plugging in values yields

$$i_{E\lambda} = \frac{2 \cdot 6.62 \cdot 10^{-34} \cdot \left(3 \cdot 10^8\right)^2}{\left(6 \cdot 10^{-7}\right)^5 \left(\exp\left(\frac{6.62 \cdot 10^{-34} \cdot 3 \cdot 10^8}{1.38 \cdot 10^{-23} \cdot 6 \cdot 10^{-7} \cdot 5777} \right) - 1 \right)}$$
$$= 2.45 \cdot 10^{13}\,\mathrm{W/(m^2 m \cdot sr)}$$

and

$$j_s = \frac{2kc}{\lambda^4}\{(1+0.016)\ln(1+0.016) - 0.016\ln(0.016)\} 10 \cdot 10^{-9} \cdot 6.80 \cdot 10^{-5}\,\mathrm{W/(K \cdot m^2)}$$
$$= 5.25 \cdot 10^9 \cdot 10 \cdot 10^{-9} \cdot 6.80 \cdot 10^{-5}\,\mathrm{W/(K \cdot m^2)}$$
$$= 3.57 \cdot 10^{-3}\,\mathrm{W/(K \cdot m^2)}$$

Now, half of the irradiance is scattered into the hemisphere. Therefore, the energy intensity of the scattered light is

$$i_{E\lambda,scatt} = \frac{\Omega_s}{\pi} 0.5 \cdot i_{E\lambda} = 2.65 \cdot 10^8\,\mathrm{W/(m^2 m \cdot sr)}$$

The scattered entropy flux density turns out to be

$$j_{s,scatt} = \frac{2kc}{\lambda^4}\left\{\left(1+1.73 \cdot 10^{-7}\right)\ln\left(1+1.73 \cdot 10^{-7}\right) - 1.73 \cdot 10^{-7}\ln\left(1.73 \cdot 10^{-7}\right)\right\}$$
$$\times 10 \cdot 10^{-9} \cdot \pi \;\; \mathrm{W/(K \cdot m^2)}$$
$$= 1.83 \cdot 10^5 \cdot 10 \cdot 10^{-9} \cdot \pi \;\; \mathrm{W/(K \cdot m^2)}$$
$$= 5.76 \cdot 10^{-3}\,\mathrm{W/(K \cdot m^2)}$$

Comparison with the original result shows that the rate of entropy production per square meter is $2.19 \cdot 10^{-3}$ W/(K · m^2).

EXAMPLE 41. Monochromatic temperature of solar radiation.

Calculate the monochromatic temperature of direct and diffuse solar radiation at 600 nm for the conditions used in Example 39.

SOLUTION: Monochromatic temperatures are calculated according to Equation (200):

$$T_\lambda = \frac{hc}{k\lambda}\frac{1}{\ln\left(\dfrac{2hc^2}{\lambda^5 i_{E\lambda}} + 1\right)}$$

Now, the direct and the diffuse irradiances have been computed in Example 39. The normal direct spectral irradiance is 464 W/(m^{2}μm)/cos(60°) = $9.28 \cdot 10^8$ W/(m^{2}m), while the value for the diffuse light is 162 W/(m^{2}μm) = 1.62 · 10^8 W/(m^{2}m). The spectral energy intensities therefore are equal to

$$i_{E\lambda,dir} = \frac{1}{\Omega_s} g_{\lambda n,dir} = \frac{9.28 \cdot 10^8}{6.80 \cdot 10^{-5}} \mathrm{W/(m^2 m \cdot sr)} = 1.37 \cdot 10^{13} \mathrm{W/(m^2 m \cdot sr)}$$

$$i_{E\lambda,diff} = \frac{1}{\pi} g_{\lambda,diff} = \frac{1.62 \cdot 10^8}{\pi} \mathrm{W/(m^2 m \cdot sr)} = 5.16 \cdot 10^7 \mathrm{W/(m^2 m \cdot sr)}$$

The appropriate temperatures then turn out to be

$$T_{\lambda,dir} = \frac{hc}{k \cdot 6 \cdot 10^{-7}} \frac{1}{\ln\left(\dfrac{2hc^2}{\left(6 \cdot 10^{-7}\right)^5 1.37 \cdot 10^{13}} + 1\right)} = 5072\mathrm{K}$$

$$T_{\lambda,diff} = \frac{hc}{k \cdot 6 \cdot 10^{-7}} \frac{1}{\ln\left(\dfrac{2hc^2}{\left(6 \cdot 10^{-7}\right)^5 5.16 \cdot 10^{7}} + 1\right)} = 1394\mathrm{K}$$

which also can be read off the curves displayed in Figure 48. For a less turbid atmosphere and for higher elevation of the Sun, the temperature of the direct beam approaches that of the surface of the sun. Diffuse radiation has a surprisingly high temperature which might be useful in solar energy engineering.

3.6.3 Concentrating Solar Radiation for Power Engineering

Considering that direct solar radiation has a temperature nearly as high as the surface of the Sun suggests that we should be able to exploit sunlight as a high-temperature "fuel" with which bodies can be heated to hotnesses approaching that of the Sun. Solar energy engineering[37] and materials research could benefit from such a source.

To heat a body to the temperature of the surface of the Sun we must place it in a radiation field similar to the one found there. The difference between direct solar radiation on Earth and the radiation at the surface of our star is not to be found in the intensity but rather in the angular distribution: on Earth, solar radiation comes from a small solid angle, while at the surface of the Sun it strikes the body from all directions of a hemisphere. The radiation field found there can be recreated here at the focus of an ideally concentrating mirror, and if you place the body at that focus, it will receive the kind of radiation needed for attaining the same high temperature. A simple argument shows how we can calculate the maximum concentration necessary for the desired application. Without concen-

37. Detailed information on concentration of sunlight and on related optical problems in solar energy engineering can be found in the books by A. Rabl (1985, Chapter 5) and by J. Duffie and W. Beckman (1991, Chapter 7).

tration, a body having a surface perpendicular to the direct rays of the Sun receives an energy flux equal to $\Omega_s i_E$. Ideally, the energy flux should be $\pi \, i_E$. Therefore, we need a concentration C calculated according to

$$C_{max}\Omega_s i_E = \pi i_E \quad \Rightarrow \quad C_{max} = \frac{\pi}{\Omega_s} = 46200 \tag{217}$$

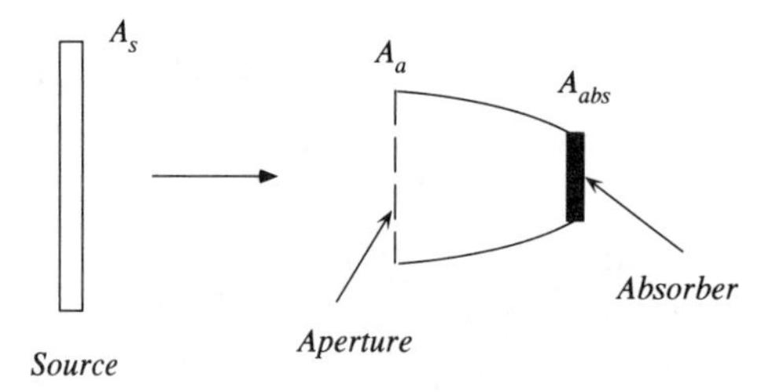

FIGURE 49. Radiation from a source can be concentrated by a device having an aperture and an absorber. The ratio of the surface areas of the aperture and the absorbing body is called the *concentration factor.*

Let us give a more careful thermodynamic argument of why the number just calculated corresponds to the maximum value of concentration of sunlight permitted by the law of balance of entropy.[38] Consider a setup as in Figure 49, which concentrates the radiation from a source onto an absorber. First, the concentration factor of the concentrating device is defined by

$$C = \frac{A_a}{A_{abs}} \tag{218}$$

where A_a and A_{abs} are the surface areas of the aperture and the absorber, respectively. The largest possible concentration factor must be related to the highest possible temperature attained by a body receiving radiation from the sun. Now from a balance of entropy performed for the bodies exchanging radiation we know that if radiation is transferred from body 1 to body 2, the rate of production of entropy is larger than zero—see Equation (79)—and the temperature of the second body is smaller than that of the first. The highest possible value of the hotness of body 2 is reached when the rate of production of entropy vanishes, which is equivalent to saying that the temperatures of the bodies must have become equal and the net rate of transfer of energy must have dropped to zero.

We can express the energy currents flowing from one body to another in terms of the fraction $\mathcal{F}_{12}$ of the radiation emitted by the first of these bodies which is

38. Note that the result depends upon the index of refraction of the medium through which the light travels before hitting the absorber. Therefore, higher values of concentration can be attained which do not violate the laws of thermodynamics.

intercepted by the second.[39] If the surfaces radiate like black bodies, the energy current flowing from body 1 to body 2 will then be

$$I_{E,12} = \mathcal{F}_{12} A_1 \sigma T_1^4$$

Conversely, the energy flux leaving body 2 which is intercepted by body 1 will be expressed by

$$I_{E,21} = \mathcal{F}_{21} A_2 \sigma T_2^4$$

The condition of maximum concentration leads to equality of these expressions, which yields the important result that

$$A_1 \mathcal{F}_{12} = A_2 \mathcal{F}_{21} \quad (219)$$

We can apply this relation to the surface of the arrangement of source, aperture, and absorber shown in Figure 49, to obtain

$$A_s \mathcal{F}_{s-a} = A_a \mathcal{F}_{a-s}$$
$$A_s \mathcal{F}_{s-abs} = A_{abs} \mathcal{F}_{abs-s}$$

from which we conclude that the concentration factor can be calculated in terms of

$$C = \frac{\mathcal{F}_{s-a} \mathcal{F}_{abs-s}}{\mathcal{F}_{a-s} \mathcal{F}_{s-abs}}$$

In the case of maximal concentration, all the radiation entering the aperture should flow to the absorber; this requires the factors $\mathcal{F}$ for radiation from the source to the aperture and from the aperture to the absorber to be equal. This results in

$$C = \frac{\mathcal{F}_{abs-s}}{\mathcal{F}_{a-s}}$$

For this ratio to be maximal, the factor $\mathcal{F}_{abs\text{-}s}$ which describes the flow of radiation from the absorber back to the source, must have its largest possible value which, obviously, is 1. This yields the result for the largest possible concentration factor:

$$C_{max} = \frac{1}{F_{a-s}} \quad (220)$$

39. If the radiation emitted by body 1 and received by body 2 gets there directly without intermediate reflection(s), this fraction is equal to the radiation shape factor introduced before in Section 3.5.2.

Note that the factor $\mathcal{F}_{a\text{-}s}$ must be the radiation shape factor $F_{a\text{-}s}$ since there are no intermediate reflectors between the source and the aperture. This shape factor has been calculated in Example 31; the arrangement used there is equivalent to what we have in the case of the Sun aligned with a (small) circular surface directly facing our star. Obviously, the result is equal to the square of the sine of the half-angle θ_s subtended by the Sun for an observer on the Earth. Since we have to have a concentrator whose aperture sees only the Sun if we wish to obtain maximal concentration, this angle is equal to the acceptance half-angle θ_a of the optical setup. Therefore:

$$C_{max} = \frac{1}{\sin^2(\theta_a)} \tag{221}$$

This is indeed what we had derived before on the basis of the simple argument presented in Equation (217).

EXAMPLE 42. Maximum concentration of line focus concentrators

Use arguments like those which led to Equation (217) to show that the maximum concentration reached in a line focus concentrator is given by

$$C_{max} = \frac{1}{\sin(\theta_a)}$$

where θ_a is the acceptance half-angle of the concentrator. Calculate the corresponding value for direct sunlight.

SOLUTION: We should consider a thin long cylinder receiving direct sunlight from the narrow angle (0.5°) subtended by the sun. How much more light would it receive if we had isotropic radiation with the same intensity as that of direct solar light coming from the part or all of the upper half of a cylindrical dome as in the figure?

In linear concentrators we have to deal with angles instead of solid angles. Reverting the flow of radiation, we can argue as follows: the light emitted by the cylindrical pipe in the center spreads into the upper half-space and over the dome. The flux of energy or entropy associated with this flow must be

$$j_x' = \int_{\cap} i_x \cos(\theta) d\omega$$

The factor $\cos(\theta)$ is a consequence of the projection of the radiation onto the plane receiving it. According to the geometry in the accompanying figure, this can be transformed into

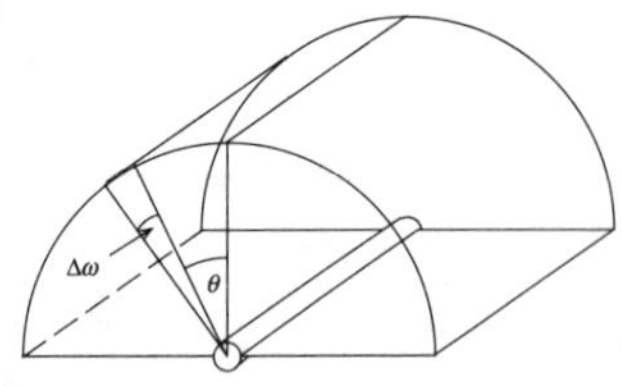

$$j_x' = 2i_x \int_0^{\pi/2} \cos(\theta) d\theta = 2i_x$$

for isotropic radiation. If we have sunlight which comes only from a narrow angle ω_S flowing from all over the dome towards the pipe, the energy flux just calculated would correspond to C times the actual flux:

$$2i_E = C\omega_s i_E$$

Now, since the angle (not the solid angle) subtended by the Sun is $2R_S/d = 2\sin(\theta_a)$, where R_S and d are the radius and the distance of the Sun, respectively, we conclude that the concentration factor is indeed given by the formula presented above. For direct sunlight, the maximum possible concentration factor turns out to be 214.

EXAMPLE 43. Steady-state temperatures reached in a parabolic trough concentrator.

Consider an uncovered metal pipe having a radius of 4.0 cm at the center of a line focus parabolic concentrator. Take the concentrated light (with a concentration ratio 40) to be perfectly intercepted by the cross section presented by the pipe. a) If only radiation is considered to cause heat loss, calculate the temperature reached by the pipe in the steady state. The pipe absorbs a fraction $(\tau\alpha) = 0.95$ of the incoming light, and its emissivity is 0.90 at infrared wavelengths. The irradiance onto the aperture of the concentrator is 900 W/m^2, and the ambient temperature is 20°C. b) Calculate the width of the parabolic trough concentrator.

SOLUTION: a) A simple balance of energy will deliver the temperature reached by the pipe absorbing concentrated solar light. In the steady state, we have

$$2r\,L(\tau\alpha)C\mathcal{G} = 2\pi r\,Le\sigma\left(T^4 - T_a^4\right)$$

which delivers

$$T = \left(\frac{(\tau\alpha)C\mathcal{G}}{\pi e\sigma} + T_a^4\right)^{1/4}$$

$$= \left(\frac{0.95\cdot 40\cdot 900}{\pi\cdot 0.90\cdot 5.67\cdot 10^{-8}} + 293^4\right)^{1/4} = 685\text{K}$$

b) Since the concentration is 40, the aperture, i.e., the width of the parabolic trough, must be 40 times as wide as the pipe. Therefore, the width is 3.2 m.

3.6.4 Transmission and Absorption in Flat-plate Solar Collectors

The concepts of absorption, reflection, and transmission of radiation can be applied to a nice example in the field of solar energy engineering, namely the computation of the amount of light absorbed by a flat-plate collector. Consider a flat piece of metal (the absorber) covered by one or more sheets of glass, as in Figure 50. The Sun's light is partly reflected, absorbed, and transmitted by the cover. Part of the transmitted radiation is reflected back through the cover, and the rest is absorbed, which is what we would like the collector to do. In general, the radiative properties needed here, i.e., the reflectance, absorptance, and transmittance,[40] all depend upon the frequency of the light; moreover, reflection and absorption properties also depend upon the angles of incidence. In the following development, we shall take these properties to be independent of wavelength, which for reflection and transmission of the cover, is quite correct for normal glass.[41] Also, the absorptance of the absorber will be taken to be independent of the angles of incidence of radiation. Finally, for diffuse light we shall be working with effective angles of incidence, to deal with the problem in a simple manner.

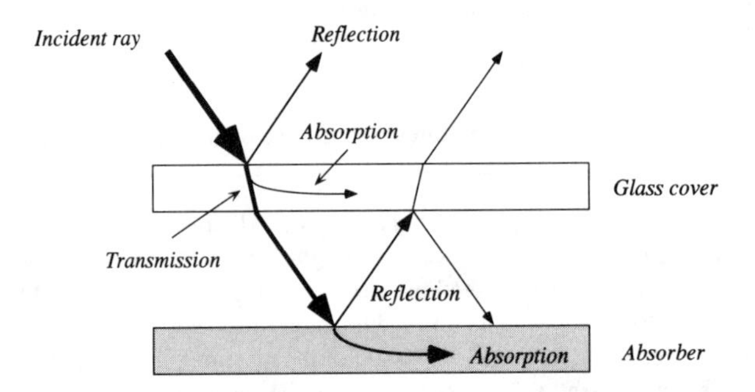

FIGURE 50. Light which is to be absorbed by a solar collector first has to pass through the cover system. Reflection and absorption by the glass cover reduce the amount of radiation incident upon the absorber.

The transmittance-absorptance product of a collector. Let us start with the final step in the chain of events which leads to absorption of light. We introduce the effective transmittance τ of the cover system which we are going to calculate later. If α is the absorptance of the absorber plate, a fraction $\tau\,\alpha$ of incident light is absorbed while the rest, which is $(1 - \alpha)\,\tau$, is reflected back up to the cover. Then, if ρ_d is the reflectance of the cover system for diffuse reflected light from the absorber, radiation with a relative intensity of $(1 - \alpha)\,\tau\rho_d$ is reflected back down again in the direction of the absorber; see Figure 50. This process continues infinitely many times, which leads to

$$(\tau\alpha) = \tau\alpha\sum_{i=0}^{\infty}\left[(1-\alpha)\rho_d\right]^i$$

for the effective transmittance-absorptance product. Note that $(\tau\alpha)$ should be considered as a new quantity and not as the product of τ and α. The result can be

40. Often, in engineering, the terms absorptance and emittance are used for absorptivity and emissivity, respectively, rather than for the rates of absorption and emission per unit area by a surface. See Section 3.2.5 for the original definitions.

41. Duffie and Beckman (1991), Chapter 5.7.

transformed to yield[42]

$$(\tau\alpha) = \frac{\tau\alpha}{1-(1-\alpha)\rho_d} \tag{222}$$

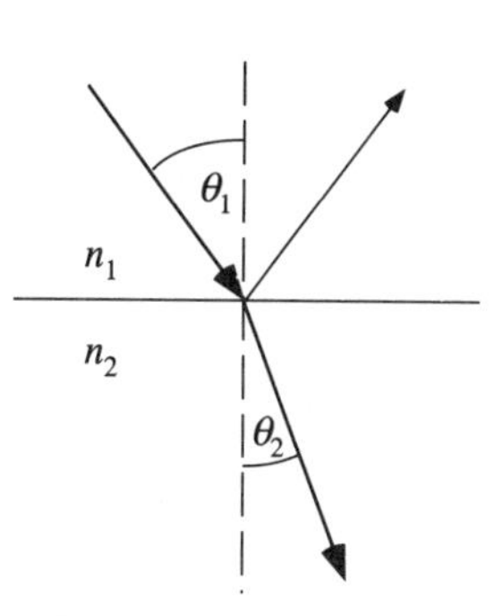

FIGURE 51. Refraction of light at an interface also leads to partial reflection.

Transmittance of the cover without absorption. Transmittance through one or more sheets of glass is not equal to unity, even if we neglect absorption, because of the reflection of light at the surfaces of the covers. When a ray of light is refracted at the surface of a body, part of it also is reflected (Figure 51). The reflection depends both on the nature of the refracting materials, i.e., upon the index of refraction *n*, and on the nature of light, i.e., its state of polarization. It is common to split a ray of light into two plane polarized components, one perpendicular to the plane as in Figure 51, the other parallel to that plane. The fraction of incident light of a particular component which is reflected is calculated according to[43]

$$r_{\parallel} = \frac{\tan^2(\theta_2-\theta_1)}{\tan^2(\theta_2+\theta_1)}$$
$$r_{\perp} = \frac{\sin^2(\theta_2-\theta_1)}{\sin^2(\theta_2+\theta_1)} \tag{223}$$

for the parallel and the perpendicular components, respectively. If the incident light is unpolarized, the total reflection is equal to the average of the values calculated using the previous result:

$$r = \frac{1}{2}(r_{\parallel}+r_{\perp}) \tag{224}$$

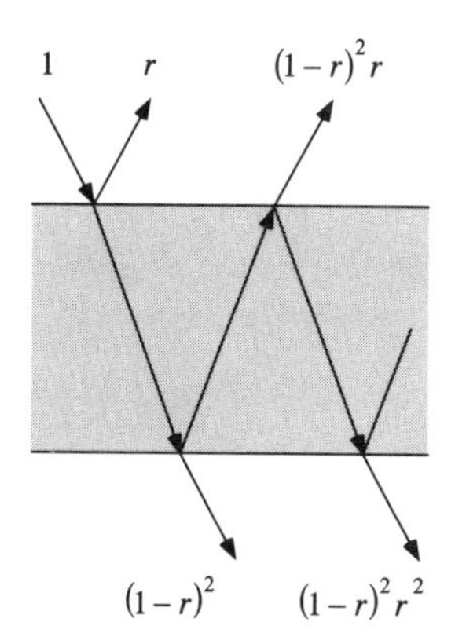

FIGURE 52. Multiple reflection and transmission of a ray of light in a sheet of glass; absorption is neglected. The total light transmitted is smaller than what passes through the first interface.

Now, the transmittance τ of unpolarized light is not simply $1 - r$, since part of a transmitted ray is reflected back up at the second interface of a cover sheet (as in Figure 52), and so on. This results in a total transmittance of

$$\tau_{\perp} = (1-r_{\perp})^2 \sum_{i=0}^{\infty} r_{\perp}^{2i} = \frac{(1-r_{\perp})^2}{1-r_{\perp}^2} = \frac{1-r_{\perp}}{1+r_{\perp}} \tag{225}$$

for, say, the normal component of polarization. The other component is treated in the same manner. Therefore, our final result for the transmittance of a single nonabsorbing sheet of glass will be

---

42. For $x < 1$, we have the series expansion

$$\frac{1}{1-x} = 1 + x + x^2 + x^3 + \ldots$$

43. For a derivation, see Siegel and Howell (1992), p. 102-108.

$$\tau_r = \frac{1}{2}\left(\frac{1-r_\perp}{1+r_\perp} + \frac{1-r_\parallel}{1+r_\parallel}\right) \tag{226}$$

for unpolarized incident radiation; the index r reminds us that this is the transmittance due to reflection. For N identical covers the formula becomes

$$\tau_r = \frac{1}{2}\left(\frac{1-r_\perp}{1+(2N-1)r_\perp} + \frac{1-r_\parallel}{1+(2N-1)r_\parallel}\right) \tag{227}$$

Transmittance with absorption in the cover. The absorption of radiation in a cover sheet made of glass can be calculated in the same manner as was done with absorption of radiation in the Earth's atmosphere; see Section 3.6.2. The length of the path of light in a glass cover is $L/\cos(\theta_2)$ where L is the thickness of the cover. If κ is the coefficient of absorption (the extinction coefficient) in the glass, the transmittance due only to absorption is

$$\tau_a = \exp\left(-\kappa\frac{L}{\cos(\theta_2)}\right) \tag{228}$$

For N covers, we simply have to use N times the thickness L. Now, with the effect of absorption present, the equations for transmittance, reflectance, and absorptance of the cover system are more complicated than what we would have without the effect. (See Problem 34.) A satisfactory approximation for solar collectors is the following:

$$\begin{aligned} \tau &\approx \tau_a\tau_r \\ \alpha &\approx 1-\tau_a \\ \rho &\approx \tau_a(1-\tau_r) \end{aligned} \tag{229}$$

This is a consequence of the fact that the transmittance due to absorption in the glass is nearly 1.

Transmission and absorption of diffuse radiation. Diffuse radiation, either scattered from the sky or reflected from the ground, cannot be handled that easily. To include these contributions to radiation, effective incidence angles should be introduced which can be used in the equations presented here. For diffuse light from the sky, an effective angle of 60° delivers satisfactory solutions for all slopes of collectors. (The slope is the angle between the plane of the collector and the horizontal.) For radiation reflected from the ground, however, the effective incidence angle used varies from 90° to 60° for collector slopes from 0° to 90°, respectively.[44]

---

44. Duffie and Beckman (1991), p. 227.

EXAMPLE 44. Normal transmittance-absorptance product of a solar collector.

A collector with a single sheet of glass as a cover is oriented directly toward the sun. The glass has a thickness of 3 mm and an extinction coefficient of 12 m^{-1}. Its refractive index is 1.53. The absorptance of the absorber for solar light is 0.92. a) Calculate the transmittance-absorptance product for direct light and for diffuse light from the sky. b) What is the rate of absorption of sunlight by this collector (per surface area) if the direct and the diffuse irradiances are 600 W/m^2 and 400 W/m^2, respectively? c) Consider direct radiation only and assume it to be blackbody radiation with a temperature of 5777 K; also assume that the radiation which is not absorbed by the collector is unaffected by transmission and reflection. Calculate the rate of production of entropy for the process of the interaction of the direct component of sunlight with the collector if the absorber is at a temperature of 70°C.

SOLUTION: a) To find the desired result we need the quantities in Equation (229) which means that we have to compute the transmittances of the cover due to reflection and due to absorption. The angles of incidence of direct and of diffuse light are 0° and 60°, respectively. Therefore, we have to calculate the appropriate quantities for these two angles:

$$\tau_a(0) = \exp(-\kappa L) = \exp(-12 \cdot 0.0030) = 0.965$$

$$\tau_a(60) = \exp\left(-\kappa L/\cos(60°)\right) = \exp\left(-12 \cdot 0.0030/0.5\right) = 0.931$$

We need Snell's law to calculate the reflection factors r at normal incidence and at 60°:

$$\frac{n_1}{n_2} = \frac{\sin(\theta_2)}{\sin(\theta_1)}$$

For normal incidence we combine this law with Equation (224). Let both angles θ_1 and θ_2 go to zero. In this limit, the parallel and the perpendicular components of r become

$$r_{\parallel} = \frac{\tan^2(\theta_2 - \theta_1)}{\tan^2(\theta_2 + \theta_1)} \quad \rightarrow \quad \frac{(\theta_2 - \theta_1)^2}{(\theta_2 + \theta_1)^2}$$

$$r_{\perp} = \frac{\sin^2(\theta_2 - \theta_1)}{\sin^2(\theta_2 + \theta_1)} \quad \rightarrow \quad \frac{(\theta_2 - \theta_1)^2}{(\theta_2 + \theta_1)^2}$$

Therefore we have

$$r(0) = \frac{(\theta_2 - \theta_1)^2}{(\theta_2 + \theta_1)^2} = \frac{(\theta_2/\theta_1 - 1)^2}{(\theta_2/\theta_1 + 1)^2} = \frac{(n_1/n_2 - 1)^2}{(n_1/n_2 + 1)^2} = 0.044$$

At 60° incidence, θ_2 is 34.5°. The reflection coefficients turn out to be

$$r_{\parallel}(60) = \frac{\tan^2(-25.5)}{\tan^2(94.5)} = 0.00140$$

$$r_{\perp}(60) = \frac{\sin^2(-25.5)}{\sin^2(94.5)} = 0.187$$

The transmission coefficients due to the process of reflection now are calculated as

$$\tau_r(0) = \frac{1 - r(0)}{1 + r(0)} = \frac{1 - 0.044}{1 + 0.044} = 0.916$$

and

$$\tau_r(60) = \frac{1}{2}\left(\frac{1 - r_\perp(60)}{1 + r_\perp(60)} + \frac{1 - r_\parallel(60)}{1 + r_\parallel(60)}\right) = \frac{1}{2}\left(\frac{1 - 0.187}{1 + 0.187} + \frac{1 - 0.00140}{1 + 0.00140}\right) = 0.841$$

Now we are ready to calculate the approximate optical properties of the cover systems. According to Equation (229) we obtain

$$\begin{aligned}
\tau(0) &\approx 0.965 \cdot 0.916 = 0.88 \\
\alpha(0) &\approx 1 - 0.965 = 0.035 \\
\rho(0) &\approx 0.965(1 - 0.916) = 0.081 \\
\tau(60) &\approx 0.931 \cdot 0.841 = 0.78 \\
\alpha(60) &\approx 1 - 0.931 = 0.069 \\
\rho(60) &\approx 0.931(1 - 0.841) = 0.15
\end{aligned}$$

Equation (222) allows us to compute the transmittance-absorptance products for the two incidence angles:

$$\begin{aligned}
(\tau\alpha)_n &= \frac{\tau(0) \cdot \alpha}{1 - (1 - \alpha)\rho_d(60)} = \frac{0.88 \cdot 0.92}{1 - (1 - 0.92)0.15} = 0.82 \\
(\tau\alpha)_{60} &= \frac{\tau(60) \cdot \alpha}{1 - (1 - \alpha)\rho_d(60)} = \frac{0.78 \cdot 0.92}{1 - (1 - 0.92)0.15} = 0.73
\end{aligned}$$

Note that the diffuse reflectance of the cover for light reflected from the absorber is equal to the value calculated for 60°.

b) The rate of absorption of energy of the collector is calculated simply as the product of the transmittance-absorptance product and the irradiance:

$$\begin{aligned}
\Sigma_E/A\big|_b &= (\tau\alpha)_n G_b = 0.82 \cdot 600\,\mathrm{W/m^2} = 492\,\mathrm{W/m^2} \\
\Sigma_E/A\big|_d &= (\tau\alpha)_{60} G_d = 0.73 \cdot 400\,\mathrm{W/m^2} = 292\,\mathrm{W/m^2}
\end{aligned}$$

c) If we wish to consider only the first part of the interaction of sunlight with the collector, we may simply assume heat loss and removal of heat not to be present. (The total rate of production of entropy due to all processes will certainly be larger than the number calculated here.) The law of balance of entropy for the absorber then takes the form

$$\dot{S} = -I_{s,in} - I_{s,reflected} + \Pi_s$$

where

$$\dot{S} = \frac{1}{T_c}\dot{E}$$

$$I_{s,in} = -\frac{4}{3}\frac{1}{T_s} I_{E,in}$$

$$I_{s,reflected} = \frac{4}{3}\frac{1}{T_s} I_{E,reflected}$$

T_c and T_s are the temperature of the absorber and of solar radiation, respectively. Therefore, the rate of production of entropy per unit surface are due to solar radiation is given by

$$\begin{aligned} \Pi_s / A &= \frac{1}{T_c}(\tau\alpha)_n \mathcal{G}_b - \frac{4}{3}\frac{1}{T_s}(\tau\alpha)_n \mathcal{G}_b \\ &= \frac{1}{273+70} 0.82 \cdot 600 \mathrm{W/(K \cdot m^2)} - \frac{4}{3}\frac{1}{5777} 0.82 \cdot 600 \mathrm{W/(K \cdot m^2)} \\ &= 1.32 \mathrm{W/(K \cdot m^2)} \end{aligned}$$

Because of the high temperature of solar radiation, incident sunlight delivers only a small part of the entropy which appears in the absorber as the result of absorption. The loss of available power because of this process alone reaches almost 400 W/m^2 (for 20°C ambient temperature).

Questions and Problems

1. The ends of a copper bar having a length of 0.5 m and a diameter of 2.0 cm are placed in evaporating water and in freezing water, respectively. Heat is conducted in the axial direction only. a) Estimate the rate of production of entropy in the bar. b) Determine the rate of loss of availability.

2. An immersion heater in a water kettle is hooked up to 220 V. Its electrical resistance is 160 Ω at a temperature of 20°C; the temperature coefficient of the resistance is $4 \cdot 10^{-3}$ K^{-1}. If the heat transfer coefficient between heater and water is 100 W/(K · m^2) and the surface area of the heater is 0.020 m^2, how large will the energy current from the heater to the water be? How does the situation change if a layer of mineral deposit builds up around the heater?

3. Determine the thermal resistance of a long cylindrical shell made out of concrete having inner and outer diameters of 50 cm and 1.0 m, respectively. (The thermal conductivity of concrete can be found in Table A.9.)

4. Show that the energy current transmitted through a cylindrical shell of length L having inner and outer radii r_1 and r_2 is

$$I_E = \pi L \left[\frac{1}{2r_1 h_1} + \frac{1}{2r_2 h_2} + \frac{1}{2k_E} \ln\left(\frac{r_2}{r_1}\right) \right]^{-1} (T_1 - T_2)$$

where h_1 and h_2 are the inner and the outer convective heat transfer coefficients. The temperatures of the fluids on the inside and the outside are T_1 and T_2.

5. A hemispherical igloo has an inner radius of 1.4 m. Two people present in the igloo emit a total energy current of 180 W. The outside temperature is – 20°C. a) How thick should the wall be made if the steady-state temperature inside the igloo is supposed to reach 12°C? For the conductivity with respect to energy take the value for ice, 2.2 W/(K · m); the inner and the outer convective heat transfer coefficients are 10 W/(K · m^2) and 20 W/(K · m^2), respectively. b) Determine the surface temperatures of the wall on the inside and the outside.

6. A cylindrical volume of rock below ground has been heated uniformly to 50°C while the rest of the rock has a temperature of 10°C. (This might be done in solar seasonal heat storage applications.) For properties of the rock use the average values for granite from the tables in the Appendix. a) For heat loss from the cylindrical area to the surroundings make the following model. While the temperatures of the storage area and the surroundings remain uniform, heat flows through a cylindrical mantle with inner and outer radii equal to half and to twice the radius of the storage cylinder, respectively. Estimate the energy current due to heat loss for a radius of 5.0 m and a length of the cylindrical space of 40 m. b) How large should the radius be made for heat loss over a period of half a year not to exceed one quarter of the energy stored in the cylinder?

7. The apparent brightness of stars is measured on a scale of magnitudes which is based on the fact that the eye has a logarithmic sensitivity. This means that the eye sees equal ratios of the irradiance of light as equal differences in brightness. Therefore, the difference of the magnitudes (which is the term for the apparent brightness of a star) of two stars is defined by

$$m_1 - m_2 = -2.5\log_{10}\left(\mathcal{G}_1/\mathcal{G}_2\right)$$

where $\mathcal{G}$ is the irradiance due to the light of a star. a) Show that a difference in brightness of 5 magnitudes corresponds to a ratio of irradiances of 100 (the higher the brightness the smaller the associated value of the magnitude; this definition roughly places the stars visible by the naked eye on a scale of magnitudes ranging from 0^m to 6^m). b) The absolute magnitude of a star is introduced as follows: it is the brightness the star would have if it were placed at a distance of 10 pc (parsec; 1 pc = 3.26 light years). The Sun's apparent magnitude at the Earth's surface is – 26.77^m. Derive the absolute magnitude of the Sun. c) Show that the absolute magnitude M is derived from the apparent magnitude m by

$$M = m - 5\log_{10}\left(r/10\right)$$

where r is the distance of the star in parsec. d) Show that the luminosity of a star (its radiant power) is given by

$$M = M_s - 2.5\log_{10}\left(L/L_s\right)$$

where M_s = + 4.79^m is the absolute magnitude of the Sun, and L_s is its luminosity (which is 3.8 · 10^{26} W).

8. With the definition of the absolute brightness of stars given in magnitudes (Problem 7), show that a star's radius R and its surface temperature T are related by

$$M = 42.3 - 5\log_{10}(R/R_s) - 10\log_{10}(T)$$

if we take it to radiate as a black body.

9. The star 40 Eridani B (the companion of 40 Eridani) has an absolute magnitude of 9.5^m. (See Problem 7 for a definition of magnitudes.) Its surface temperature, estimated from its spectrum, is 15,000 K. Assuming a mass equal to half of the mass of the Sun, estimate its density. (This star is a white dwarf, and you should get a very large value for its density.)

10. A spherical satellite orbits the Earth at a distance of twice the Earth's radius from the center of the planet. Estimate its temperature in the shadow of our planet if you assume the surface of the satellite not to be selective.

11. Estimate the energy flux which could be carried upward through the clear atmosphere by radiation alone. (The mean distance over which solar radiation is absorbed must be of the order of 10 km; take this value to hold for the Earth's radiation as well.) Would you say that radiation could carry all the heat that has to be reradiated by our planet? Could the known energy flux be transported conductively?

12. A sheet of metal with a selective surface of 2.0 m^2 is lying horizontally on the ground. The bottom side of the sheet is well insulated. In the visible part of the spectrum the emission coefficient of the metal is 0.90, while in the infrared it is 0.30. Take the ambient temperature to be 20°C. The Sun stands 50° above the horizon, and 70% of the radiation outside the atmosphere penetrates the air. (Assume all the radiation from the sky to be direct and not diffuse.) a) Neglecting convection, how large should the temperature of the metal sheet be in the light of the Sun? b) Now take into consideration convective heat transfer at the surface of the sheet. The convective heat transfer coefficient is assumed to be 14 W/(K · m^2). Calculate the temperature attained by the sheet under these conditions.

13. Consider two arbitrary gray surfaces having surface areas A_1 and A_2, and emissivities e_1 and e_2. a) Show that the energy flux due to radiation from one to the other of the surfaces must be given by

$$I_E = \frac{\sigma\left(T_2^4 - T_1^4\right)}{\dfrac{1-e_1}{e_1 A_1} + \dfrac{1-e_2}{e_2 A_2} + \dfrac{1}{A_1 F_{12}}}$$

where F_{12} is the radiation shape factor for radiation from surface 1 to surface 2. b) Derive the special results presented in Equations (44) and (46) from this more general one.

14. Compare the top heat loss coefficient of two flat-plate solar collectors, one without a cover, the other with a single cover. Take the convective heat transfer coefficient due to wind to be 10 W/(K · m^2). For the value of the transfer coefficient for convection between the absorber and the cover take 3.5 W/(K · m^2). The emittances of the absorber and the cover are taken to be 0.95 and 0.88, respectively. The temperature of the absorber is set at 70°C, while those of the ambient and of the sky are both assumed to be 15°C.

15. In solar energy applications, parabolic troughs are used to focus light upon absorbers of cylindrical shape. Calculate the heat loss coefficient of such an absorber. Consider it to be made of a metal pipe of diameter of 5.5 cm, surrounded by a thin glass cover

with an outer diameter of 8.5 cm. The annulus between the pipe and the cover is evacuated. Take the convective heat transfer coefficient at the surface of the cover to be 35 $W/(K \cdot m^2)$. The emissivities of glass and the metal pipe are 0.88 and 0.92, respectively. Present the result as a function of absorber temperature for an ambient temperature of 20°C.

16. A spherical thin-walled water tank has a volume of 1.0 m^3. The water inside is kept at a constant temperature of 60°C by heating it with an energy current equal to 1.0 kW. The ambient temperature is 15°C. How long will it take for the water to reach a temperature of 40°C after the heater has been turned off?

17. A body of water having a volume of 1.0 m^3 loses heat to the surroundings. The temperatures are 80°C and 20°C for the water (initially) and the environment, respectively. The product of total heat transfer coefficient and surface area is 60 W/K. a) How long does it take for the temperature difference between the water and the surroundings to decrease to half its initial value? b) How large is the rate of production of entropy right at the beginning? c) How much entropy is produced in total from the beginning until the water has cooled down completely? d) How much energy could have been released by an ideal Carnot engine operating between the water and the environment as the water cools to ambient temperature?

18. To maintain an inner temperature of 20°C in a building situated in a 0°C environment, the required heating load is 5 kW. Without heating, the house is found to cool down as follows: every day, its temperature decreases by 1/5 of the temperature difference to the environment. a) Determine the product of surface area and total heat transfer coefficient. b) Model the building as a single node system. Calculate its temperature coefficient of energy. c) Assume the temperature inside the building to be 12°C. Calculate the heating power necessary if you wish the temperature to rise by 1°C per hour.

19. When Russian cosmonauts had to repair their previously abandoned space station, it was freezing inside and the thermometers were not working. One of the cosmonauts was said to have spat at the metal wall. From measuring the time it took for the spit to start freezing the temperature of the walls was determined. Show how a very simple model of the process could lead to a solution of the problem of measuring the temperature.

20. A high, well insulated cylinder of radius 0.75 m contains 10,000 kg of water. The lower 3500 kg has a temperature of 20°C, while the temperature of the rest of the water is 80°C. Such stratification may be attained approximately while charging a hot water storage tank in solar applications. a) Estimate how long it will take for the difference of the temperatures of the two segments of water to decrease to 30°C. (*Hint:* Model the segments as uniform bodies; for the thermal resistance take a distance from the center of the hotter to the center of the cooler part.) b) Calculate the initial rate of production of entropy. b) How large is the initial loss of available power?

21. As in Problem 15, oil is heated in parabolic trough concentrators and then supplied to the steam generator of a power plant. Irreversible processes occurring on the way from the concentrators to the power plant are a consequence of both heat loss and friction. Why should you calculate the optimal value of the radius of the pipes carrying the hot fluid on the basis of minimizing entropy production rather than calculating the minimum of the combined energy losses (heat loss plus pumping)?

22. An initially empty tank is filled with an externally heated fluid. Consider the irreversible processes due to fluid flow and heat loss from the tank to the environment. Take fluid friction to obey the law of Hagen and Poiseuille. Heat loss from the tank should be proportional to how much hot fluid is in the tank. a) Give qualitative reasons to show that there should be an optimal rate of charging of the tank. b) Show that under these conditions, the optimal charging time should be proportional to the square root of the frictional resistance (as in Ohm's law, see the Prologue), and inversely proportional to the square root of both the total heat transfer coefficient and the difference of temperatures between the hot fluid and the environment. (Assume the temperature decrease of the fluid in the tank due to cooling to be small; i.e., take the temperature of the heated fluid to remain constant.)

23. Consider the following strongly simplified model of the charging and the discharging of a spherical tank with a well mixed fluid. Heat loss from the tank to the environment at temperature T_a is calculated using an overall heat transfer coefficient h. Heating is accomplished using solar collectors, which leads to a decrease of the energy current charging the tank as the fluid temperature increases:

$$I_{E,charge} = A - B\left(T - T_a\right)$$

A is the useful energy current of the collectors when they are at ambient temperature. The tank is to be charged starting at a temperature $T_o > T_a$. a) Calculate the temperature of the fluid in the tank reached after charging time t_{charge}. b) Show that the amount of entropy produced as a result of heat loss during charging is approximated by

$$S_{prod} = C\ln\left(T_f / T_o\right) + \left(At_{charge} - C\left(T - T_o\right)\right)\frac{1}{T_a}$$

C and T_f are the temperature coefficient of energy of the fluid in the tank and the final temperature of the fluid after charging, respectively. c) Find the radius of the storage tank for which the entropy produced is a minimum. d) Extend the analysis to include the period of discharging. Take a constant energy current delivering useful energy to an environment of temperature T_o. Calculate the additional entropy produced, the total entropy produced, and the useful energy (for discharging down to T_o). Calculate these values as a function of the radius of the tank, and compare the minimum of total entropy production with the maximum of useful energy delivered.
The following values yield useful numerical results. Take water as the fluid and set $h = 20$ W/(K · m^2). Use $T_a = 300$ K, and $T_o = 320$ K; set $A = 100$ kW and $B = 1$ kW/K and let the constant useful energy current during discharging be 10 kW. Let the charging time be $2 \cdot 10^5$ s.

24. According to Table A.9, the conductivity with respect to entropy of water depends less upon temperature than its counterpart, the conductivity with respect to energy. Taking the former quantity as constant for steady-state conduction through a slab of water a) should the temperature gradient be steeper at the hotter or at the cooler side? b) Show that the field equation for temperature should take the form

$$T\frac{d^2T}{dx^2} + \left(\frac{dT}{dx}\right)^2 = 0$$

25. Consider the conduction of heat through the Earth's crust, whose geometry can be taken as flat. Allow for sources of entropy in the material which are assumed to be distributed evenly, and let the conductivity with respect to energy be constant. a) Show that the temperature profile from the base of the crust to the surface is

$$T(x) = T_L + \frac{1}{2}\frac{\sigma_E}{k_E}\left(L^2 - x^2\right) + \frac{1}{k_E} j_E(0)(L - x)$$

for a given energy flux $j_E(0)$ at the base and surface temperature T_L. (The thickness of the crust is L.) b) Determine the dependence of the temperature gradient near the surface upon the conductivity, the energy flow at the base, and the source rate of energy in the material. c) Calculate the surface temperature gradient for a thickness of the crust of 50 km, a thermal conductivity of 2.5 W/(K · m), and a source rate of $1.25 \cdot 10^{-6}$ W/m^3.

26. Repeat the calculation of the steady-state temperature profile in a slab of matter such as the Earth's crust (Problem 25), but this time for a source rate of energy which decreases exponentially from the surface. Again, boundary conditions are given at the bottom (the energy flux is fixed) and at the top (the temperature is specified).

27. Show that for pure conduction, with sources of heat in the material, the field equation for temperature must take the form

$$\frac{\partial T}{\partial t} = \frac{k_E}{\rho c}\frac{1}{r^2}\frac{\partial}{\partial r}\left(r^2\frac{\partial T}{\partial r}\right) + \frac{\sigma_E}{\rho c}$$

in radial symmetry if nonsteady-state conditions are considered.

28. Consider the transport of heat with radiation in the interior of a star which we model as being in spherically symmetric hydrostatic equilibrium; changes of volume of the gas are assumed not to disturb this situation. Nuclear reactions release energy, with the source rate given by the specific rate $\sigma_{E,r}$ (i.e., the rate divided by the mass). The luminosity $L(r)$ is the total energy flux penetrating the spherical surface at radius r. a) Model stellar matter as a simple fluid and show that the rate of change of the specific entropy s (entropy per mass) must be given by

$$T\dot{s} = \sigma_{E,r} - \frac{\partial L}{\partial m}$$

Here, the independent variable has been changed to the mass $m(r)$ inside the sphere of radius r. b) Show that the gradient of the luminosity $\partial L/\partial m$ is given by

$$\frac{\partial L}{\partial m} = \sigma_{E,r} - \frac{3}{2}\rho^{2/3}\frac{d}{dt}\left(\frac{P}{\rho^{5/3}}\right)$$

for a monatomic ideal gas. c) Show that the gradient of luminosity inside a star is determined by the source rate due to reactions only if steady-state conditions prevail.

29. The precise definition of the mass absorption coefficient for diffusion of radiation through matter, i.e., the opacity κ_ρ (Example 34), actually is given by

$$l_{mean} = \frac{4}{3}\frac{1}{\kappa_\rho \rho}$$

where l_{mean} is the mean distance traveled by radiation before being absorbed. Show that the luminosity inside a star where all the energy is transported by radiation is given by

$$L(r) = -\frac{64}{3}\pi\sigma\frac{r^2T^3}{\kappa_\rho\rho}\frac{dT}{dr}$$

30. Calculate the normal spectral irradiance for solar radiation at the distance of the Earth for a blackbody spectrum of temperature 5777 K. (Normal means for a plane perpendicular to solar rays.)

31. Integration of the spectral entropy intensity of solar radiation according to the WRC spectrum (Figure 43) yields a value of 4620 W/(K · m^2 · sr), while the integral value of the energy intensity is 2.011 · 10^7 W/(m^2 · sr). Derive the equivalent blackbody temperature and calculate the entropy intensity for such radiation.

32. An experiment on the disinfection of water using solar radiation showed[45] that a dose of 2000 kJ/m^2 of radiation in the wavelength band between 350 nm and 450 nm was required to kill bacteria in a particular sample. Estimate how long the sample has to be exposed to sunlight at midlatitudes around noon on a clear day in summertime.

33. Calculate the steady-state temperature reached by a cylindrical absorber at the line focus of a parabolic trough concentrator assuming that the absorber is surrounded by a glass pipe and that convective losses from the surface of the pipe have to be taken into account. (See Example 43 and Problem 15.)

34. Show by use of ray-tracing techniques such as the one used in Section 3.6.4, that the transmittance, reflectance, and absorptance of the glass cover of a flat-plate solar collector are given by

$$\tau_\perp = \tau_a\frac{(1-r_\perp)^2}{1-(r_\perp\tau_a)^2}$$

$$\rho_\perp = r_\perp + \frac{(1-r_\perp)^2\tau_a^2 r_\perp}{1-(r_\perp\tau_a)^2}$$

$$\alpha_\perp = (1-\tau_a)\frac{1-r_\perp}{1-r_\perp\tau_a}$$

for the component of light which is polarized at right angles to the plane of incidence (Figure 51). Show that Equation (229) is a good approximation to these equations.

35. Consider the absorptance of cavities and rooms. Light falls from the outside on the opening of a cavity (which might be a room with a window for the opening). The surface area of the opening is A_a, while the area of the inner surfaces is A_i. The absorptance of the inner walls is assumed to be α_i (independent of the angle of incidence and the wavelength). a) Show that the total absorptance is given by

45. Wegelin et al. (1994).

$$\alpha = \alpha_i \left[\alpha_i + (1 - \alpha_i)\frac{A_a}{A_i} \right]^{-1}$$

if the opening is not covered. (*Hint:* Consider rays bouncing off the interior walls and assume that after each reflection there is a probability of A_a/A_i for the ray to escape through the hole.) b) Show that the result must be

$$\alpha = \tau \cdot \alpha_i \left[\alpha_i + (1 - \alpha_i)\tau_d \frac{A_a}{A_i} \right]^{-1}$$

if there is a window with a transmittance to direct light τ and a transmittance to diffuse reflected light from the interior of τ_d.

36. For the collector and the situation discussed in Example 44, estimate the rate of production of entropy due to heat loss to the ambient if the energy current due to this effect is 40% of the irradiance. Compare this value to the rate of production of entropy due to absorption by the collector.

CHAPTER 4

Heat and the Transformation and Transport of Substances

One of the principal objects of practical research... is to find the point of view from which the subject appears in its greatest simplicity.

J.W. Gibbs

In the previous chapters, we considered only phenomena which did not include any changes of the substances. Such changes can occur for two reasons. For one thing, the substances making up a body can undergo chemical reactions, leading to a new composition. Chemical reactions include changes of phase as well as the more obvious processes in which the molecular composition changes. Second, a body may change because substances are added or withdrawn. This is the case in transport processes, which are of two general types—the obvious bulk transport of matter, and the less visible phenomenon of diffusion.

There are two new physical quantities which facilitate the description of processes involving the transport and transformation of substances: they are the *amount of substance* as the measure of how much stuff there is, and the *chemical potential* as the expression of the tendency of substances to change or flow.[1] By showing how thermal and other processes relate to chemical ones, the subject of this chapter is tied in with thermodynamics.

Both chemical reactions and flow processes are used to motivate the new concepts. After explaining how amount of substance is a basic property of physical systems (Section 4.1), the chemical potential is used to quantify chemical reactions in Section 4.2. This discussion leads up to the subject of phase changes of pure substances and the phenomena related to mixtures of substances, all of which are equally important to chemistry, physics, and engineering (Section 4.3). In Section 4.4, the chemical potential is introduced again, this time as the tendency of substances to flow from place to place. This theme naturally leads to

1. The didactic concept behind the first four sections of this chapter owes much to the work done by G. Job (1972), and by G. Falk and F. Herrmann (and their colleagues, 1977–1982). To my knowledge, G. Job first suggested that an exposition of the subject of chemical change should start directly with the chemical potential as an easily grasped concept.

applications of vapor power cycles and convective heat transfer which are discussed in Sections 4.5 and 4.6. There, we will deal mostly with cases which are of interest in renewable energy engineering and in environmental studies.

A more formal discussion of open and reactive uniform fluids will be taken up in the Epilogue (Section E.1), serving as a stepping stone toward a theory of continuum thermodynamics.

4.1 The Concept of Amount of Substance

Physical systems "possess" certain fundamental properties: a body has mass (energy), entropy, momentum (if it moves), and electrical charge (if it is not electrically neutral), to name some of the most important. This section will describe another basic property, namely that of amount of substance. Amount of substance is by no means a replacement for mass, or just a convenient means of bookkeeping for chemists. Every physical system, be it matter or light, has this property, and there are physical phenomena associated with its existence.

4.1.1 Substances, Chemical Reactions, and Electrolysis

The numerous substances undergoing a myriad of processes of material transformation furnish the first indication of the existence of the quantity called amount of substance. Chemical reactions and electrolysis will be described briefly after we look at what constitutes a basic or an elementary substance.

Basic substances. The many material objects known to us usually are composed of different substances which we call *basic* in the sense that we can explain the material make-up of objects in terms of a mixture of these constituents. Now, there are several levels of *fundamental building blocks*, where the levels are defined by those investigating the objects. For a particular cook, it might suffice to know that a Thai green curry consists of green curry paste, chicken, coconut milk, and some vegetables; while another cook will only be satisfied by knowing how the green curry paste is composed of "basic" foods. A biologist will want to know how the Thai eggplant in the curry is made up of cells; a biochemist might stop only at the level of the pure chemical substances constituting a part of a cell or might even want to know which chemical elements these substances are made up of. If these various levels of what different people call *fundamental* are not enough, consider the viewpoint of physicists, who are concerned with the structure of atoms in terms of electrons, protons, and neutrons.

To describe the composition of an object from more fundamental constituents, different forms might be used. Common to all these probably is some way of telling "how much" of each of the building blocks occurs in the system under investigation. The composition of dough for bread might be described as

$$flour_3 water_2 yeast_{0.1} salt_{0.05}$$

where the subscripted numbers denote amounts of each substance measured in cups. Granite could be described by

$$A_{v1}B_{v2}C_{v3}\ldots$$

where A, B, and C denote the various minerals making up the rock, and the numbers $v1$, $v2$, and $v3$ measure the relative volume of each of the constituents. Chemists use the elements as their basic building blocks, which means that chemical substances are understood in terms of their composition written in the form

$$C_2H_5OH$$

(for ethyl alcohol) where the letters stand for the chemical elements, and the numbers give the amount of substance of each of the elements. The rest of this section is devoted to clarifying the latter concept. For physicists, finally, one unit of amount of substance of pure helium-4 could be written as

$$p_2n_2e_2$$

In this expression, p, n, and e stand for the substances called protons, neutrons, and electrons, respectively,[2] and the formula tells us that two units of amount of substance of each of the three building blocks make one unit of helium-4.

Chemical reactions and the amount of substance. We can approach the concept of amount of substance by studying chemical reactions. By going through a multitude of different reactions, we find that pure chemical substances in mixtures undergo complete reactions only if their amounts occur in special proportions. *Complete* means that the species are used up completely. Lacking the concept of amount of substance, we have to start with some other measure of the amount of a species, for which we take its mass.

As an example consider the possible reactions involving any two of the three elements hydrogen, oxygen, and chlorine, each time involving two of them. If we start with a unit amount of substance of hydrogen gas having a mass of 2 grams, we find that 16 g of oxygen gas will completely react with the hydrogen to form water. However, exactly twice that amount of oxygen gas, i.e., oxygen with a mass of 32 g, will react completely with 35.5 g of chlorine. Finally, twice as much chlorine, i.e., 71 g, will use up 2 g of hydrogen gas. Written in the language of chemical reaction equations, the reactions read

2. We do not have names for the substances whose "atoms" are the proton, neutron, and electron, respectively. The names are used for the particles rather than for the substances. Neutron stars, for example, are made up of the n-substance.

$$1(H_2)+\frac{1}{2}(O_2) \rightarrow 1(H_2O)$$

$$1(O_2)+\frac{1}{2}(Cl_2) \rightarrow 1(ClO_2)$$

$$1(Cl_2)+1(H_2) \rightarrow 2(ClH)$$

If we count 2 g of hydrogen gas as *one unit of amount of substance*, called 1 mole, the reactions tell us that 1 mole of hydrogen gas (H_2) has a mass of 2g, 1 mole of oxygen gas (O_2) has a mass of 32 g, while the mass of 1 mole of chlorine (Cl_2) is 71 g. The mass of one mole of amount of substance is called the *molar mass* M_o of the substance, and is defined by

$$M_o = \frac{m}{n} \qquad (1)$$

These observations carry over to all elements and pure chemical substances. There always are constant and multiple proportions of the amounts of the constituents involved in complete reactions. A complete reaction is our way of defining what we mean by equivalent amounts of substances in a chemical sense. Equivalent amounts are determined by the phenomena involving chemical transformation, and not, as we might believe from our everyday usage of the term, by the phenomena of gravity quantified in terms of (gravitational) mass.

Naturally, much more experience is required with chemical reactions than the three examples provided above, if we want to figure out how much of a substance is equal to one unit of amount of substance. The examples used display some complexity in that we would have to answer the question of why hydrogen gas is H_2 and not H, and why 1/2 mole of oxygen gas (and not 1/3, or 2, or 3 mole) reacts completely with one mole of hydrogen gas to form one mole (and not 1/2, or 2 mole) of water. Part of the experience which tells us more about equivalent amounts of substance actually comes from physical phenomena involving gases (see Section 4.1.2).

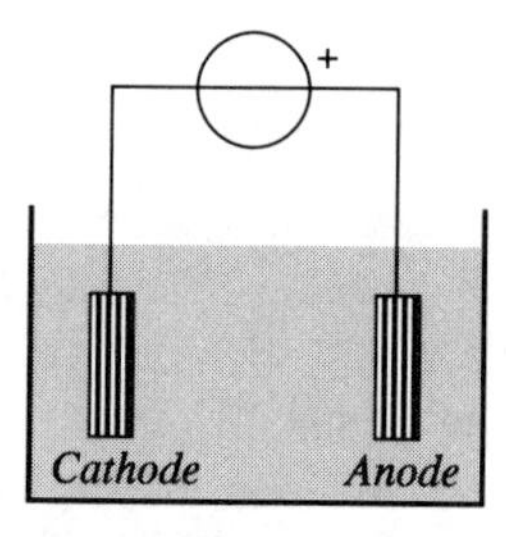

FIGURE 1. In electrolysis, chemical substances may be transformed by electrical currents flowing through them. In a melt of potassium chloride, for example, potassium is deposited at the cathode as long as the current is flowing.

Electrolysis. The combination of chemical reactions with electricity provides another strong indication of the special property measured in terms of amount of substance. In electrolytic reactions, electrical currents passing through conducting fluids are responsible for chemical transformations. Multiple proportions of electric charge are needed for multiple units of amount of substance to appear at the electrodes. Take the examples of the electrolysis of potassium chloride (KCl) and of copper sulphate ($CuSO_4$). It is found that a certain amount of charge has to be passed through the molten salt of KCl to deposit 1 mole of potassium at the cathode, while it takes exactly twice this charge to deposit 1 mole of copper. The reactions involved are

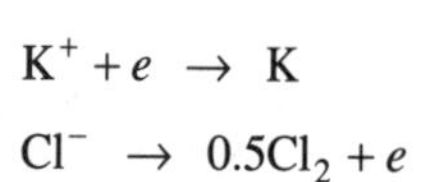

$$K^+ + e \rightarrow K$$

$$Cl^- \rightarrow 0.5Cl_2 + e$$

for potassium chloride, and

$$\begin{aligned} Cu^{++} + 2e &\rightarrow Cu \\ SO_4^{--} &\rightarrow SO_4 + 2e \end{aligned}$$

for the electrolysis of copper sulphate. Here, *e* again stands for the substance called electrons. These phenomena can be summarized in the relation

$$q = z \cdot \mathcal{F} \cdot n \tag{2}$$

An amount of substance n requires an amount of charge q for electrolysis which is determined by Faraday's constant $\mathcal{F} = 96487$ C/mole, and a (small) positive or negative integer number z.

What we should learn from this is what we have seen before: it is not the mass of a substance which scales simply with fixed amounts of electric charge involved; rather, it is the same quantity introduced above as a measure of equivalent amounts of a substance, namely the fundamental property called amount of substance. If we consider electrons to be a chemical substance, then electrolysis provides for a simple means of measuring amounts of substances equivalent to amounts of the substance made up of electrons.

EXAMPLE 1. Electrolysis of potassium copper sulphate.

A current of electric charge of 10 A is passed through melted copper sulphate for one hour. How much copper will be deposited at the cathode?

SOLUTION: The amount of charge passed through the melted substance is computed from the current:

$$q = I_q \Delta t = 10\text{A} \cdot 3600\text{s} = 3.6 \cdot 10^4\,\text{C}$$

With $z = 2$ for copper ions, Equation (2) yields

$$n = \frac{q}{z\mathcal{F}} = \frac{3.6 \cdot 10^4\,\text{C}}{2 \cdot 96487\text{C/mole}} = 0.187\text{mole}$$

which, with a molar mass of 0.0636 kg/mole for copper, is equivalent to $1.19 \cdot 10^{-2}$ kg of copper.

4.1.2 The Ideal Gas and Dilute Solutions

Substances can be brought into a state called gaseous in two rather distinct ways. First, we can increase the temperature or decrease the pressure to such an extent that the substance becomes a gas. Second, we can try to dissolve the body in a solvent; interestingly, the dissolved substance, called a *solute*, behaves just like a gaseous phase. In both cases, an interesting property emerges—the *amount of substance* of the gas. If its density is small, the substance obeys the ideal gas law, independent of whether it is a gas or a dilute dissolved substance.

The ideal gas law. A body changes its volume and pressure if its temperature is changed. While the relations between these quantities may be complex, it is easy to approximate them by linear relations for small ranges of the temperature. We have used such approximations in Section 1.1.2. They take the form

$$V(\theta) = V_o(1+\gamma\theta) \quad , \quad P = constant \tag{3}$$

$$P(\theta) = P_o(1+\beta\theta) \quad , \quad V = constant \tag{4}$$

for volume and pressure, respectively. The temperature θ is measured in degrees Celsius, and the reference state is taken at 0°C. The important point for us is this: if we make the density of the body smaller and smaller, all substances enter a phase for which both temperature coefficients of volume (γ) and of pressure (β) are equal and constant over a large range of temperatures. This value turns out to be

$$\gamma = \beta = 1/273.15°C \tag{5}$$

Based on this observation, the Kelvin scale of temperature was introduced. (See Section 1.1.2.) Also, in this phase, the product of pressure and volume is proportional to the Kelvin temperature T of the body:

$$PV \propto T \tag{6}$$

Naturally, this relation scales with the amount of gas present, where, for lack of a quantity measuring the amount of substance, we may take the mass of the body:

$$PV = R_m m T \tag{7}$$

Here, R_m is a constant representative of the type of gas present. Indeed, R_m, the *specific gas constant*, is different for every chemical substance. However, it is possible to make the gas law formulated in Equation (7) independent of the particular chemical nature of the gas. It is observed that 2 g of hydrogen gas, 32 g of oxygen gas, and 20 g of neon all have the same pressure for equal volumes and temperature. If we call these amounts of the gases equivalent in the sense of *equivalent amounts of substance*, we may write Equation (7) in the form

$$PV = RnT \tag{8}$$

This is the well-known equation of state of the ideal gas. R, the *universal gas constant*, has a value of R = 8.31441 J/(mole · K). Obviously, the quantity n measures an independent and fundamental property of the system; rather than taking an already known quantity such as the mass of a body, we have effectively introduced the coefficient of proportionality in Equation (6) as a new and independent measure of the amount of a substance. R only gives this new quantity its own physical unit.

As the particular numbers used in the example show, the quantity called amount of substance introduced on the basis of the behavior of dilute gases is the same as the quantity which makes amounts of different substances equivalent in the sense discussed above in the context of reactions. We now also know why 32 g of oxygen gas, not 16 g, is the same amount of substance as 2 g of hydrogen. (See Section 4.1.1.)

Dilute solutions. Substances can be made to behave according to the model of the ideal gas if they are dissolved in fluids. As long as the *concentration*

$$\bar{c} \equiv n/V \tag{9}$$

of the solute is small enough, it has a pressure which is related to the Kelvin temperature by the law

$$P = R\bar{c}T \tag{10}$$

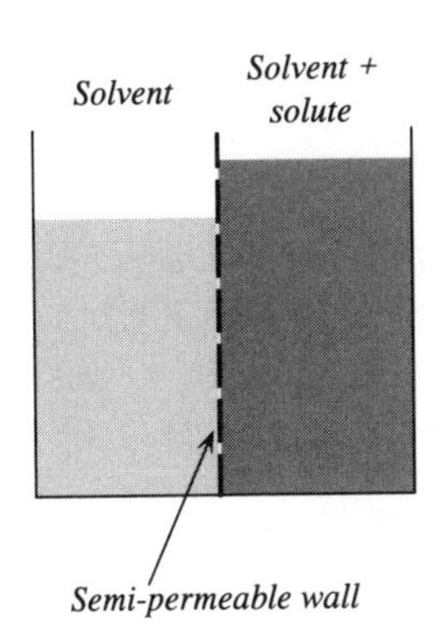

FIGURE 2. Osmosis occurs if two cells, one with only a solvent (water), the other with the solvent and a solute (sugar), communicate through a membrane which lets only the solvent pass. The solution draws more water which causes the fluid level to rise.

This relation is equivalent to the ideal gas law if the constant R in it takes the same value as the universal gas constant. Experiment tells us that this is the case. Most interestingly, it does not matter what type of stuff is dissolved; it may be a salt, or a substance composed of macromolecules, or even of macroscopic particles. Measurement of the pressure of the dissolved substance delivers its concentration, from which we can compute the amount of substance dissolved.

One way of observing the influence of a solute on the solvent is through the phenomenon of *osmosis*. If a cell containing a solution is separated from the pure solvent by a semipermeable membrane (permeable only to the solvent), more of the solvent is literally drawn into the cell to dilute the solution. As a result, the level of the fluid in a cell containing solvent and solute as in Figure 2 rises above the level in the other cell. The difference of the pressures of the fluids indicated by the respective levels is called the *osmotic pressure* of the solute.

While osmosis is easily observed, it does not lend itself that simply to precise measurements. Usually, the effect of the pressure of the solute is observed indirectly, in that it decreases the vapor pressure of the solvent. The change of vapor pressure, in turn, increases the temperature of vaporization and decreases the melting point. Measuring the latter effect is an important method of determining the molar mass of a dissolved substance.

The discussion also shows how we are to understand the concept of the pressure of the solute. The solvent naturally has a certain pressure of its own. Now, in the cell in Figure 2 containing the solution, the dissolved substance, i.e., the "gas" inside the solvent, adds to the pressure of the fluid. The total pressure of the fluid therefore is the sum of the pressures of the solvent and that of the solute:

$$P = P_f + P_s \tag{11}$$

The indices f and s stand for the solvent (fluid) and the solute, respectively. Since the total pressure is made up of different contributions, the terms on the right-hand side of Equation (11) are called *partial pressures*.

EXAMPLE 2. The osmotic pressure of salt in sea water.

Roughly 30 g of salt (NaCl) is dissolved in 1 liter of sea water. a) Calculate the osmotic pressure of the salt with respect to the pure solvent (water) at a temperature of 300 K. b) How high would the solution rise in an ideal cell separating it by a semipermeable membrane from the solvent?

SOLUTION: a) The concentration of the salt is determined using Equation (9). Since there are two ions per molecule of NaCl dissolved, we have:

$$\bar{c} = \frac{2n}{V} = \frac{2m/M_o}{V} = \frac{2 \cdot 0.030\,\text{kg}/0.0585\,\text{kg/mole}}{0.0010\,\text{m}^3} = 1026\frac{\text{mole}}{\text{m}^3}$$

The osmotic pressure follows from Equation (10):

$$P = \bar{c}RT = 1026\frac{\text{mole}}{\text{m}^3} \cdot 8.314\frac{\text{J}}{\text{K} \cdot \text{mole}} \cdot 300\,\text{K} = 2.56 \cdot 10^6\,\text{Pa} = 25.6\,\text{bar}$$

b) The osmotic pressure of the dissolved salt is rather large. It would be possible to let a column of water rise 260 m above the surroundings. The value also shows the pressure that would be exerted upon the walls of a living cell (having a membrane which lets water pass) were it filled with sea water and then placed in fresh water.

4.1.3 Mixtures of Ideal Gases

Many substances are mixtures of pure components, which were the focus of most of the previous chapters. In general, mixing pure substances leads to a variety of new effects. However, in this section, we will study only the simplest type of mixture, namely that of ideal gases. Such mixtures do not furnish a lot of new physics; still, they allow us to introduce concepts which are useful in applications.

The mole fraction. Our first task is to describe the composition of a mixture made up of different substances. Each of the N component furnishes a certain amount of substance n_i toward the total amount:

$$n = \sum_{i=1}^{N} n_i \qquad (12)$$

The relative amount

$$y_i = n_i / n \qquad (13)$$

is called the *mole fraction* of component i. Naturally, the mole fractions of all the parts together add up to 1. If the composition of a mixture is known in terms of its mole fractions, the average or apparent molar mass M_o may be calculated based on molar masses of the individual components M_{oi}. The molar mass of the mixture is equal to the ratio of its mass to its total amount of substance. Replacing the total mass by the sum of the masses of each component leads to

$$M_o = \sum_{i=1}^{N} y_i M_{oi} \tag{14}$$

In Section 2.3.1 we used this concept in a simple context for the case of air. Indeed, air at low densities furnishes the best everyday example of a mixture of ideal gases.

The partial pressure of a component of the mixture. Consider a mixture such as air. The interesting point about it is that it behaves just like a pure ideal gas. If we describe this "pure" substance by its apparent molar mass, we can treat it as if it were made up of pure helium or any other pure substance, for that matter. In other words, the mixture of ideal gases is an ideal gas itself. If we write the equation of state of the ideal gas for this substance expressed in terms of the amounts of substance of the components, we obtain

$$PV = nRT = \left(\sum_{i=1}^{N} n_i \right) RT \tag{15}$$

Using the mole fraction, the right-hand side of this equation can be written in the following form:

$$nRT = \left(\sum_{i=1}^{N} y_i \right) nRT = \left(\sum_{i=1}^{N} y_i \right) PV = \left(\sum_{i=1}^{N} y_i P \right) V \tag{16}$$

We now call the term

$$P_i = y_i P \tag{17}$$

the *partial pressure* of component i, and we see that the sum of the partial pressures is the total pressure of the mixture:

$$P = \sum_{i=1}^{N} P_i \tag{18}$$

There is a simple interpretation to these equations. In a mixture of ideal gases, each component exists as if it were completely independent of the others, filling the total volume V at the temperature T of the mixture. Each component then contributes its share P_i to the total pressure. Also, each part of the mixture obeys the equation of state of the ideal gas:

$$P_i V = n_i RT \tag{19}$$

Remember that these formulas hold only for mixtures of ideal gases. If components of a mixture interact, they cannot be considered independent of each other, which leads to effects not described by this model.

Molar entropy and energy. For thermodynamic purposes, we need to know such quantities as the entropy and the energy of a body. In the case of mixtures it might be necessary to compute these properties from those of the components.

Since, in our model, the components are independent of each other, each has its entropy and energy. Considering that these quantities are additive properties of a system, the total entropy or energy must be the sum of the parts:

$$S = \sum_{i=1}^{N} S_i \qquad E = \sum_{i=1}^{N} E_i \tag{20}$$

These equations can be written in terms of the molar quantities

$$n\bar{s} = \sum_{i=1}^{N} n_i \bar{s}_i \qquad n\bar{e} = \sum_{i=1}^{N} n_i \bar{e}_i \tag{21}$$

Divided by the total amount of substance, they take the form

$$\bar{s} = \sum_{i=1}^{N} y_i \bar{s}_i \qquad \bar{e} = \sum_{i=1}^{N} y_i \bar{e}_i \tag{22}$$

This means that the molar entropy and the molar energy of the mixture are given in terms of the mole fractions and the molar entropy and energy of the components. The latter quantities have to be evaluated at the conditions (P and T) of the mixture.

Spatially separated components. At first sight it appears as if mixtures of ideal gases could be described in an alternative but equivalent manner. The equation of state of the ideal gas suggests that we can consider the mixture to be made up of spatially separate components (each with its amount of substance n_i), all at the same pressure and temperature (Figure 3b). Each component would then occupy a partial volume V_i, and the sum of these would be the total volume V.

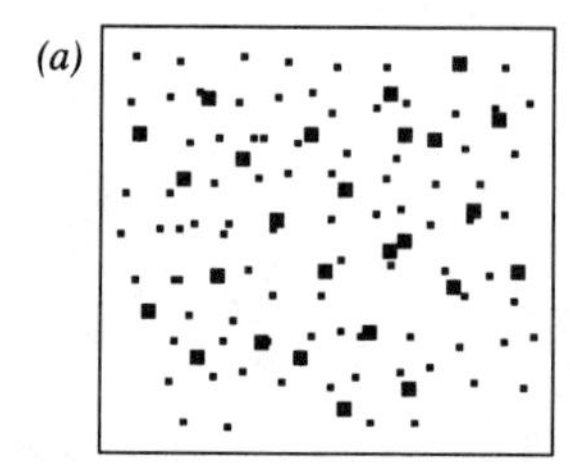

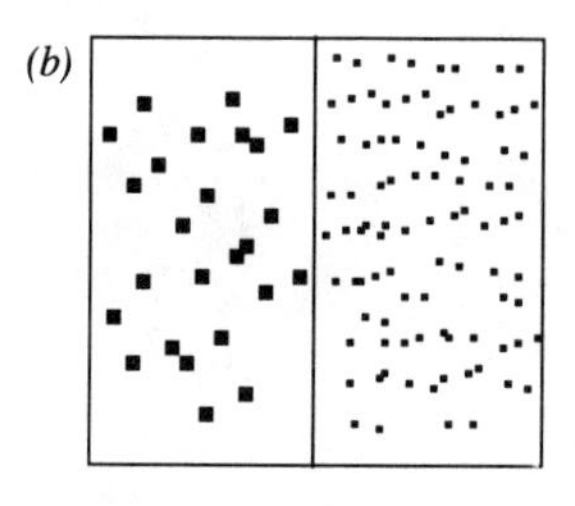

FIGURE 3. The cases of two different substances being spatially separate and mixed, are not equivalent. If the separating wall is removed, the gases diffuse and entropy is produced.

It is true that you get equivalent results for all quantities calculated above where we considered each component to occupy the total volume at some partial pressure, except for one: the entropy of the mixture is not equal to the sum of the entropies of the spatially separate components. Therefore, the two situations are by no means equivalent. We can see the difference easily in qualitative terms: if we were to remove the wall separating the two gases in Figure 3b, both would diffuse and finally occupy the entire volume as in Figure 3a. Since diffusion is dissipative, entropy must have been produced.

A quantitative analysis supports this view. The entropy of the system made up of the two separate ideal gases is calculated according to

$$\begin{aligned} S = S_A + S_B = S_{Ao} + n_A \bar{c}_{pA} \ln\left(\frac{T}{T_o}\right) - n_A R \ln\left(\frac{P}{P_o}\right) \\ + S_{Bo} + n_B \bar{c}_{pB} \ln\left(\frac{T}{T_o}\right) - n_B R \ln\left(\frac{P}{P_o}\right) \end{aligned} \tag{23}$$

The entropy of the mixture, on the other hand, is equal to

$$\begin{aligned} S_2 = S_A + S_B = S_{Ao} + n_A \bar{c}_{pA} \ln\left(\frac{T}{T_o}\right) - n_A R \ln\left(\frac{P_A}{P_o}\right) \\ + S_{Bo} + n_B \bar{c}_{pB} \ln\left(\frac{T}{T_o}\right) - n_B R \ln\left(\frac{P_B}{P_o}\right) \end{aligned} \tag{24}$$

where P_A and P_B are the partial pressures of the two components. The difference in entropy between the two cases is given by

$$S_2 - S_1 = -\left\{ n_A R \ln\left(\frac{P_A}{P}\right) + n_B R \ln\left(\frac{P_B}{P}\right) \right\} = -n\left\{ y_A R \ln(y_A) + y_B R \ln(y_B) \right\} \tag{25}$$

The y_i are the mole fractions of the parts of the mixture. Since these quantities are less than 1, the quantity calculated in Equation (25) is positive: entropy has been produced.

There seems to exist a paradoxical situation. If the two components in Figure 3 are identical, we know that nothing happens upon removal of the wall. On the other hand, Equation (25) suggests that entropy must be produced even in this case. Since this is obviously not so, we have to explain the difference between "identical" and "different" substances. The amount of entropy produced is independent of how "different" the two components are, so there cannot be a gradual change between the two situations. Either the gases are identical, or they are not, however slight their difference might be. The gases must notice the difference in the environment into which they are diffusing in the case of initially separate parts. If the substances are identical, there is no different environment, and there is no diffusion.

This phenomenon is called Gibbs' paradox. Obviously it calls for a quantum view of the nature of matter: there cannot be a gradual change of substances from "identical" to "different." The subject will be taken up again in Section 4.2.3, where we will discuss it from the point of view of the chemical potential.

EXAMPLE 3. The temperature coefficients of energy and enthalpy of a mixture.

Show that the molar temperature coefficients of energy and of enthalpy (the "specific heats") of a mixture can be calculated as the sum of the products of molar fractions and the coefficients for each component. Also, give the formulas for the entropy capacities of the mixture.

SOLUTION: The temperature coefficient of energy of the ideal gas is the derivative of the energy with respect to temperature. Applying this definition to Equation (22) (the second part) yields

$$\bar{c}_V = \frac{d\bar{e}}{dT} = \frac{d}{dT}\sum_{i=1}^{N} y_i \bar{e}_i = \sum_{i=1}^{N} y_i \frac{d}{dT}\bar{e}_i = \sum_{i=1}^{N} y_i \bar{c}_{Vi}$$

The molar temperature coefficient of enthalpy of the ideal gas is defined as the derivative of the enthalpy with respect to temperature. We first have to show that the molar enthalpy can be calculated similarly to Equation (22):

$$n\bar{h} = E + PV = \sum_{i=1}^{N} E_i + \left(\sum_{i=1}^{N} P_i\right)V = \sum_{i=1}^{N} \left(E_i + P_i V\right) = \sum_{i=1}^{N} n_i\left(\bar{e}_i + P_i \bar{v}_i\right) = \sum_{i=1}^{N} n_i \bar{h}_i$$

Therefore, the derivative of the molar enthalpy of the mixture is equal to

$$\bar{c}_p = \sum_{i=1}^{N} y_i \bar{c}_{pi}$$

The entropy capacities are related to the temperature coefficients through the temperature of the mixture. Since every component has the same temperature, the equations just derived apply to the entropy as well:

$$\bar{k}_p = \sum_{i=1}^{N} y_i \bar{k}_{pi}$$

The equation holds also for the entropy capacity at constant volume.

EXAMPLE 4. The molar mass of air and mass fractions of the components.

Air at standard conditions can be considered to be a mixture of ideal gases. The mole fractions of the major components of dry air are given in the following table.

TABLE 1. Composition of dry air

Component	Mole fraction	Molar mass / kg/mole
Nitrogen	0.7808	0.02802
Oxygen	0.2095	0.0320
Argon	0.0093	0.03994
Carbon dioxide	0.0003	0.04401
Others	0.0001	

a) Calculate the apparent molar mass of dry air. b) Calculate the mass fraction of each of the components.

SOLUTION: a) The average or apparent molar mass is computed according to Equation (14):

$$\begin{aligned} M_o &= \sum_{i=1}^{N} y_i M_{oi} \\ &\approx (0.781 \cdot 0.028 + 0.2095 \cdot 0.032 + 0.0093 \cdot 0.0399 + 0.0003 \cdot 0.044)\,\mathrm{kg/mole} \\ &= 0.0290\,\mathrm{kg/mole} \end{aligned}$$

b) The mass of each component is calculated using the amount of substance and the molar mass. Giving each component an amount of substance equal to the mole fraction makes the total amount of substance 1 mole. The results appear in Table 2.

TABLE 2. Mass fractions of the components of dry air

Component	n_i / mole	M_{oi} / kg/mole	m_i / kg	Mass fraction
N_2	0.7808	0.02802	0.0219	0.756
O_2	0.2095	0.0320	0.0067	0.231
Ar	0.0093	0.03994	0.00037	0.0128
CO_2	0.0003	0.04401	0.000013	0.00045
	1.0		0.02897	1.0

Note that the difference between the molar fraction and the mass fraction is small if the molar mass of the component is similar to the apparent molar mass of the mixture. This holds for nitrogen and for oxygen.

EXAMPLE 5. The temperature coefficients of energy and enthalpy of a mixture.

Take a sample of dry air at a temperature of 300 K and a pressure of 0.90 bar. a) Calculate the partial pressures of nitrogen, oxygen, and argon. b) Calculate the molar temperature coefficients of energy and of enthalpy of these components and of the air sample as a whole.

SOLUTION: a) With the total pressure known, the partial pressures of different components are calculated easily using the mole fractions according to Equation (17). The mole fractions of the three most abundant components are given in Table 1 of Example 4. With these values we obtain

$$P_{N_2} = y_{N_2} P = 0.781 \cdot 9 \cdot 10^4\,\text{Pa} = 7.03 \cdot 10^4\,\text{Pa}$$
$$P_{O_2} = y_{O_2} P = 0.2095 \cdot 9 \cdot 10^4\,\text{Pa} = 1.89 \cdot 10^4\,\text{Pa}$$
$$P_{Ar} = y_{Ar} P = 0.0093 \cdot 9 \cdot 10^4\,\text{Pa} = 837\,\text{Pa}$$

b) Table 1 displays our results. The molar temperature coefficients of energy and of enthalpy of the ideal gas are computed according to the results of Example 3. The values of the molar temperature coefficient of energy can be read from Figure 9 of Chapter 2. The temperature coefficients of enthalpy are obtained by adding the value of the universal gas constant.

TABLE 3. Temperature coefficients of energy and of enthalpy of dry air

Component	Mole fraction	c_V / J/(K · mole)	c_p / J/(K · mole)
nitrogen	0.7808	20.9	29.2
oxygen	0.2095	21.2	29.5
argon	0.0093	12.5	20.8

Neglecting other components, we get values of 20.88 J/(K · mole) and 29.17 J/(K · mole) for the temperature coefficients of energy and of enthalpy for the mixture. This is very nearly the value obtained for nitrogen alone.

EXAMPLE 6. Entropy produced in mixing two ideal gases.

Calculate the amount of entropy produced if two gases of 1 mole amount of substance each are allowed to mix at a temperature of 20°C. How much energy has been dissipated?

SOLUTION: We can use Equation (25) directly. The total amount of substance is 2 mole, and the molar fractions are both 0.5. Therefore:

$$S_2 - S_1 = -n\{y_A R\ln(y_A) + y_B R\ln(y_B)\}$$
$$= -2\,\text{mole} \cdot 8.31\,\text{J/(mole} \cdot \text{K)}\{0.5\ln(0.5) + 0.5\ln(0.5)\} = 11.5\,\text{J/K}$$

The energy dissipated is 11.5 J/K · 293 K = 3.38 kJ.

4.1.4 Particles and the Amount of Substance

Often, the forgoing discussion is cast in the form of a microscopic model of matter. Matter is said to be composed of particles (which, for the chemist, might be atoms or molecules), and the amount of substance is interpreted as the number of particles of the substance in question. The number of particles appearing in 1 mole of a substance is called Avogadro's number, and it can be calculated from different types of measurement. Let us mention only one measurement, which takes electrolysis as the starting point. Since we also have a microscopic model of charge, the "atom" of charge being the elementary charge ($e = 1.6022 \cdot 10^{-19}$ C), we may say that we need an integer number of elementary charges for each particle of the substance being electrolyzed and deposited at one of the electrodes. If we call τ the *atom of amount of substance*, i.e., the *elementary amount of substance*, and N is used for the number of particles as well as for the number of elementary charges, then Equation (2) can be transformed into

$$Ne = z\mathcal{F}N\tau \quad (26)$$

With $z = 1$ we obtain a value of

$$\tau = \frac{e}{\mathcal{F}} = \frac{1.6022 \cdot 10^{-19}\,\text{C}}{96485\,\text{C/mole}} = 1.661 \cdot 10^{-24}\,\text{mole} \quad (27)$$

for the smallest amount of substance possible. This translates into $6.022 \cdot 10^{23}$ particles per mole of a substance (Avogadro's number).

In general, we should be careful with the "particulate" interpretation of the structure of matter. The values of e and τ are understood in terms of quantum theory as the quanta of electric charge and amount of substance, respectively. Thinking of little particles roaming around in space might be handy, but can be misleading at times. Especially the microscopic picture of the ideal gas as a swarm of free particles does not help much in understanding the behavior of dilute solutions where the particles certainly are not free. And in the case of light it turns out that the quanta of amount of substance of light are not photons but rather combinations of photons.

4.2 Chemical Reactions and the Chemical Potential

Together with motion and heat, chemical transformations probably are the most noticeable and well-known phenomena in the world around us, whether in the biosphere, in stars and nebulae in outer space, or in technical systems. The chemical compositions of the bodies around us are perpetually changing. In this section we will introduce a concept and an image which explain why and to what extent substances transform into others. We will find that the chemical potential provides the basic information needed for answering many questions related to chemical change. This section will provide a first look at this new physical quantity. In Section 4.4, we will encounter a different approach to the same concept based on flow processes. Together, these two sections should prepare the ground for a more formal treatment of the thermodynamics of simple fluids in open and reactive systems (Section E.1).

4.2.1 The Tendency of Substances to Transform: the Chemical Potential

Chemical change is ubiquitous. Active ingredients of medication decay with time as do the radioactive elements in the earth's crust. Hydrogen inside the sun changes into helium, the humidity in the air condenses to form fog, and the chemicals inside a battery change when they drive an electric current through wires. We would like to cast this tendency of substances to change in the form of a physical principle. You certainly remember the questions of why water flows or what makes heat flow. In the same manner, we will ask what makes chemical species change. Why do some chemical reactions take place while others do not?

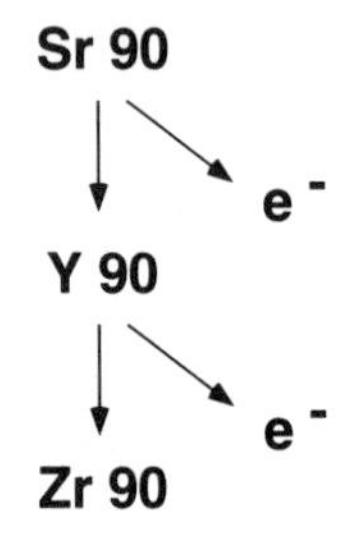

Reactions. Let us begin with the example of spontaneous transformation of one species into another, as in the decay of a molecule or the radioactive decay of an isotope. Here we most clearly see the innate tendency, or drive, to change. Change takes place until a form of matter is found which is stable. It is like water flowing downhill, over steps, until it reaches a place from which there is no possibility of falling further.

We now interpret the drive of a substance to change as a kind of level (Figure 4): the higher the level, the stronger the tendency to transform. Strontium-90 would be at a higher level than its decay product yttrium-90, which would be higher than zirconium-90. The answer to the question of why yttrium does not change into strontium would be simple: of the two possible reactions

$$^{90}\mathrm{Sr} \rightarrow {}^{90}\mathrm{Y} \;, \quad {}^{90}\mathrm{Y} \rightarrow {}^{90}\mathrm{Sr}$$

only the former would take place spontaneously.

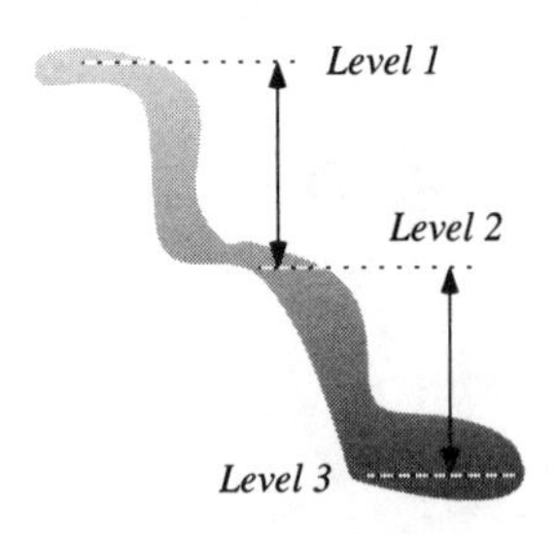

FIGURE 4. Decay of one substance into another is like balls rolling down a hill or water falling over rocks. In each case, we explain why the processes are taking place by calling upon the concept of a driving force.

The meaning of the strength of the tendency to transform becomes even clearer if we study a list of reactions of metals with sulphur. When we try to make magnesium, zinc, iron, copper, or gold react with sulphur, we observe a declining intensity of reaction. Magnesium and sulphur react explosively after initiation of the reaction, while gold hardly reacts at all. Interpreting the magnitude of the drive to react as a difference of levels, we would have to conclude that the first of the reactions corresponds to the largest "drop" in Figure 4. Assigning the initial elements the same level, with a value of zero (this is indeed what is done in chemistry), MgS should have the lowest level of the reaction products in the list (see Table 4).

The quantity introduced as a level or a driving force is called the *chemical potential* μ of a substance. For now we will give it its own unit, which will be called the *Gibbs* (G; J.W. Gibbs introduced the chemical potential in thermodynamics). We shall understand the chemical potential in this informal manner as the magnitude of the drive of each substance to undergo chemical transformation. Since each substance has its drive, i.e., its chemical potential, the question of whether a particular reaction takes place is answered by whether the potential of the product is lower than that of the original substance. For example, hydrogen (H_2) and oxygen (O_2) are both given a chemical potential of 0 kG. Water, on the other hand, is found to have a chemical potential of - 237 kG. Therefore, the sponta-

$$\underbrace{H_2 + 0.5O_2}_{0\,kG} \rightarrow \underbrace{H_2O}_{-237\,kG}$$

$$\underbrace{HgJ_{2,yellow}}_{-100.1\,kG} \rightarrow \underbrace{HgJ_{2,red}}_{-100.8\,kG}$$

FIGURE 5. The reaction which forms water out of the elements is spontaneous: it corresponds to a decrease (negative change) of the chemical potential. The reverse reaction does not occur spontaneously. Freshly prepared (yellow) mercuric iodide spontaneously changes from its yellow to its red modification.

neous reaction (first reaction in Figure 5) would have to be the one where water forms out of the elements; this is indeed what we observe. The (necessary) condition for a reaction to occur therefore is that the change of the chemical potential, i.e., the *driving force* of the reaction, has the proper sign: it must be negative; a spontaneous reaction goes "downhill."

TABLE 4. Chemical potential of some metal sulphides

Substance	MgS	ZnS	FeS	CuS	AuS
Chemical potential / kG	-347	-201	-100	-54	0

Phase changes. The condensation of water vapor, or the formation of ice out of water on a freezing day, are also examples of transformations. While here we do not have a change of one chemical species into another, we clearly are dealing with the transformation of one form of a substance into a distinctly different one. A similar case occurs when the yellow modification of mercuric iodide changes into its red modification (the second reaction in Figure 5). Returning to the phase changes of water, we would have to conclude that at room temperature water has a lower chemical potential than ice, since the latter spontaneously changes into the former (Figure 6). However, we know that the situation is reversed at lower temperatures: below the freezing point, liquid water spontaneously changes into solid water (ice).

$T = 298\mathrm{K}$:

$$\underbrace{H_2O_{solid}}_{-236.59\,kG} \rightarrow \underbrace{H_2O_{liquid}}_{-237.18\,kG}$$

$T = 198\mathrm{K}$:

$$\underbrace{H_2O_{solid}}_{-232\,kG} \leftarrow \underbrace{H_2O_{liquid}}_{-230\,kG}$$

FIGURE 6. H_2O changes its phase from one which has a higher chemical potential at a particular temperature to a phase having a lower chemical potential.

Dependence of the chemical potential on temperature and pressure. This brings up an important point about the chemical potential: its value must change with temperature, and probably with pressure as well. Therefore, if we give a value for the chemical potential of a substance, we have to state the particular conditions for which this value holds. Chemists have chosen a particular reference state, i.e., standard values of temperature and of pressure, for which the values of the chemical potential are reported. The *reference state* is at a temperature of 25°C (298 K) and a pressure of 1.0 atm (1.013 bar). All the values of chemical potentials given so far (Table 4, and Figures 5 and 6), with the exception of those for H_2O at 198 K, refer to this reference state.

How does the chemical potential depend upon temperature and pressure? We will spend some time in the following sections looking into this problem from a general perspective. Here, we will use a strongly simplified approach which will give us a first glimpse of the importance of this matter. We have dealt with temperature and pressure dependence before (see Section 4.1.2 and Chapter 2); it is always possible to approximate the relation between a quantity such as the chemical potential and temperature and pressure by using a linear relation of the form

$$\mu(T, P_{ref}) = \mu^o(T_{ref}, P_{ref}) + \alpha_\mu\left(T - T_{ref}\right) \tag{28}$$

and

$$\mu(T_{ref}, P) = \mu^o(T_{ref}, P_{ref}) + \beta_\mu\left(P - P_{ref}\right) \tag{29}$$

In Table A.13 you can find values of the temperature coefficients along with the chemical potential for some substances.

The zero point of the chemical potential. How is the zero point for the values of the chemical potential specified? Basically, this question has a simple and unambiguous answer: the chemical potential is an absolute potential just like temperature, and unlike the potentials of gravity, motion, or electricity. At conditions as they are found on earth, the chemical potential of a species is its mass (energy) divided by its amount of substance:

$$\mu \approx \frac{m \cdot c^2}{n} = M_o c^2 \tag{30}$$

where c is the speed of light. In Table 5, values for a few isotopes are presented. These values can be used for computing the energy released in nuclear reactions.

TABLE 5. Absolute chemical potentials of some isotopes

Isotope	M_o / kg/mole	μ / kG
^{1}H	0.001007825	$0.0905788 \cdot 10^{12}$
^{4}He	0.004002603	$0.359736 \cdot 10^{12}$
^{90}Sr	0.08990774	$8.080507 \cdot 10^{12}$
^{90}Y	0.08990716	$8.080455 \cdot 10^{12}$
^{90}Zr	0.08990470	$8.080233 \cdot 10^{12}$

Now, while this is simple and clear, it would be impossible to do calculations involving chemical reactions by using absolute values of the chemical potential; the differences occurring as the result of a reaction are so small that we would need figures with more than a dozen digits. Because of this, chemists have chosen a different approach: they arbitrarily set the chemical potential of the most stable form of the elements equal to zero. The values of the chemical potentials of compounds therefore represent the difference between their own absolute values and those of the elements they are formed of; for this reason, the chemical potential reported in tables such as Table A.13 are called the *chemical potential of formation* of a substance, and is denoted by μ_f^o, where f stands for formation, and the superscript o denotes standard conditions of temperature and pressure. This makes sense, because in a chemical reaction of compounds, the same elements occur on either side of the equation. The reaction

$$CaC_2 + 2H_2O \rightarrow Ca(OH)_2 + C_2H_2$$

may be understood as

$$\underbrace{CaC_2 + 2H_2O}_{(-68+2(-237))\,kG} \rightarrow \underbrace{Ca + 2C + 2H_2 + O_2}_{0\,kG}$$

$$\underbrace{Ca + 2C + 2H_2 + O_2}_{0\,kG} \rightarrow \underbrace{Ca(OH)_2 + C_2H_2}_{(-898+209)\,kG}$$

First, the reactants decay into their elements, then the products are formed out of these elements. Calculating the difference of the chemical potential for the reaction in this manner makes it unnecessary to use the absolute values of the potentials. The change is computed by adding up the chemical potentials of the compounds on either side of the reaction equation, noting that for a compound for which the amount of substance does not equal 1 mole, the potential has to be multiplied by the actual amount of substance:

$$\begin{gathered} n_1 A + n_2 B + \ldots \rightarrow n_3 C + n_4 D + \ldots \\ \Delta\mu_{reaction} = n_3\mu_C + n_4\mu_D + \ldots - \left(n_1\mu_A + n_2\mu_B + \ldots\right) \end{gathered} \qquad \textbf{(31)}$$

In summary, chemical potentials are evaluated in two steps. First, the potential of formation out of the elements is determined for the standard state of temperature and pressure; for the moment, such values are considered to be known and ready for us to be used in simple applications (Table A.13). Then, variations from the reference condition are taken into account. For the second step we have Equations (28) and (29), which provide an approximate solution to the problem.

EXAMPLE 7. Obtaining iron from iron oxide.

Iron metal is obtained from the reaction of iron oxide, Fe_2O_3, with carbon, which yields iron and carbon monoxide. a) Determine the reaction equation. b) Calculate the minimal temperature for which the reaction works spontaneously.

SOLUTION: a) To preserve the amount of substance of the elements involved, three moles of carbon must be combined with one mole of iron oxide:

$$Fe_2O_3 + 3C \rightarrow 2Fe + 3CO$$

b) At standard temperature and pressure, the reaction will not take place, as demonstrated by the values of the chemical potentials:

$$\underbrace{Fe_2O_3 + 3C}_{(-744+3\cdot 0)\,kG} \rightarrow \underbrace{2Fe + 3CO}_{(2\cdot 0+3(-137))\,kG}$$

If we want the reaction to proceed, the chemical potential on the left must be larger than that on the right-hand side of the equation. Since the chemical potential can be changed by changing the temperature, we have to look for that particular temperature for which the chemical potentials of the reactants and the products are equal:

$$\mu_{f,1}^{o} + \alpha_{\mu,1}\left(T - T_{ref}\right) + 3\left(\mu_{f,2}^{o} + \alpha_{\mu,2}\left(T - T_{ref}\right)\right) =$$
$$2\left(\mu_{f,3}^{o} + \alpha_{\mu,3}\left(T - T_{ref}\right)\right) + 3\left(\mu_{f,4}^{o} + \alpha_{\mu,4}\left(T - T_{ref}\right)\right)$$

or

$$\sum_{i} \nu_i\left(\mu_{f,i}^{o} + \alpha_{\mu,i}\Delta T\right) = \sum_{j} \nu_j\left(\mu_{f,j}^{o} + \alpha_{\mu,j}\Delta T\right)$$

for a general reaction. The numerical values are (from Table A.13)

$$-744 \cdot 10^3 - 87.4\Delta T + 3(0 - 5.69\Delta T) = 2(0 - 27.3\Delta T) + 3\left(-137 \cdot 10^3 - 197.5\Delta T\right)$$

from which we determine ΔT to be 614 K. In this linear approximation, we expect iron to be formed at temperatures above 640°C. Even though the temperature coefficients depend quite strongly upon temperature, the approximation is useful as a first guess for several hundreds of degrees above the reference value. This is so since changes of the coefficients are affected similarly by changes of temperature on both sides of the reaction equation.

EXAMPLE 8. The energy released in the decay of one nucleus of ^{90}Sr.

Calculate the energy released if one nucleus of ^{90}Sr spontaneously decays into ^{90}Y.

SOLUTION: The absolute value of the chemical potential of a species is approximated by its total energy per unit amount of substance, Equation (30). In this approximation, the energy released is related to the change of the chemical potential for the reaction:

$$\begin{aligned}\left|E_{rel}\right| &= \left|\Delta\mu_{reaction}\right| n = \left|\mu_Y - \mu_{Sr}\right| \tau \\ &= \left|8.080455 \cdot 10^{15}\,\text{G} - 8.080507 \cdot 10^{15}\,\text{G}\right| \cdot 1.661 \cdot 10^{-24}\,\text{mole} \\ &= 8.64 \cdot 10^{-14}\,\text{J} = 0.54\,\text{MeV}\end{aligned}$$

where τ is the elementary amount of substance (one particle; see Equation (27)), and MeV is the energy unit megaelectron volt. The values of the chemical potentials are given in Table 5. This closely corresponds to the observed value for energy released in the radioactive decay of ^{90}Sr into ^{90}Y.

4.2.2 Electrochemical Processes, the Chemical Potential, and Energy

The chemical potential of a species has been introduced as the tendency of that substance to change, and then has been compared to a level quantity such as the gravitational potential or the temperature. We know that potential differences are associated with the release or the use of quantities of energy; in other words, when a physical process takes place, the rate at which the substancelike quantities go through a potential difference is related to the power of the process. This concept was introduced for describing graphically what is happening in physical processes. We have studied the fall of water and the pumping of heat in terms of this image, and we will now do the same for processes involving chemical reactions.

The simplest phenomena allowing us to study the effects of chemical change are those involving a combination of chemistry and electricity. Technical applications range from batteries, accumulators, and fuel cells, to the production or the refining of metals in electrolysis.

An example of an electrochemical cell. Let us describe a particular electrochemical cell to see the details of reactions involved in such a device. We start with the observation that a rod of zinc will dissolve in a solution of copper sulfate. At the same time, metallic copper will deposit on the rod. Since we have Cu^{2+} ions in the solution, the reaction taking place can be described by the equation

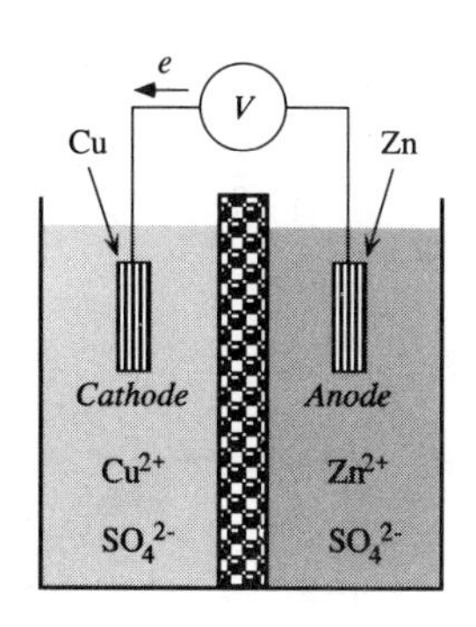

FIGURE 7. An electrochemical cell consisting of two compartments in which two reactions are taking place, each involving the exchange of electrons. The electrons are allowed to flow through an external wire.

$$\underbrace{\mathrm{Zn} + \mathrm{Cu}^{2+}}_{(0+65.52)\,\mathrm{kG}} \rightarrow \underbrace{\mathrm{Zn}^{2+} + \mathrm{Cu}}_{(-147.030)\,\mathrm{kG}}$$

The reaction is spontaneous in the given direction. Basically, it involves moving two electrons from a zinc atom to a copper atom, but we do not notice this because of how the process is allowed to proceed. The net result of the reaction is the dissolution of the rod while some copper is deposited and a lot of heat is produced.

If we wish to use the transfer of electrons directly in an electrical process; i.e., if the cell is to be turned into a battery, we somehow have to manage to separate the copper ions from the zinc rod and allow the electrons to pass through an external wire (Figure 7). This is accomplished in a cell with two compartments separated by a porous barrier. The barrier allows ions to pass but prevents mixing of the solutions. Since the positive copper ions do not diffuse toward the positive zinc electrode, they effectively are confined to their compartment with the copper electrode. Now, in each part of the cell a separate reaction takes place:

$$\underbrace{\mathrm{Zn}}_{0\,\mathrm{kG}} \rightarrow \underbrace{\mathrm{Zn}^{2+} + 2e}_{(-147.03+0)\,\mathrm{kG}}$$

$$\underbrace{\mathrm{Cu}^{2+} + 2e}_{(65.52+0)\,\mathrm{kG}} \rightarrow \underbrace{\mathrm{Cu}}_{0\,\mathrm{kG}}$$

The transfer of charge is accomplished as desired: the electrons involved in the reaction pass from one compartment to the other through the external wire.

Electricity pumps and chemical pumps. As with other phenomena, electrochemical processes can be divided into two groups, namely one in which chemical reactions drive the flow of electricity, and another in which the flow of electricity is used to drive a chemical reaction which would not run by itself in this direction. From the viewpoint of the driven process, devices applying these phenomena might be called electricity pumps (Figure 8) and chemical pumps

(Figure 9), respectively. In an electricity pump, electricity flows from a lower to a higher level, while in the chemical pump, substances are pumped from lower to higher chemical potentials.

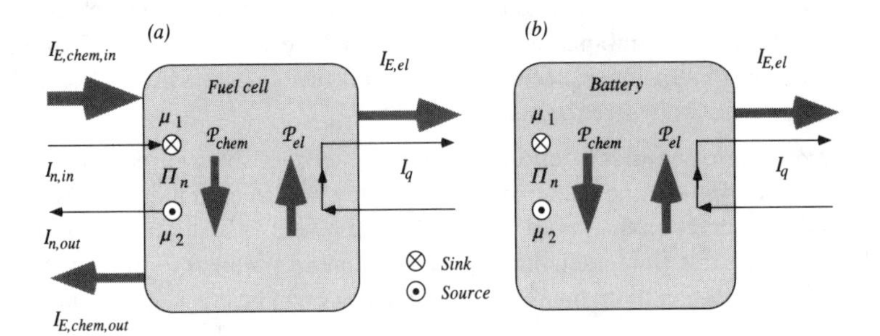

FIGURE 8. In chemical electricity pumps, chemical reactions drive the flow of electric charge. This requires energy which must be released in the chemical reactions taking place in the fuel cell (a) or the battery (b). The flow diagrams are those of ideal processes.

Examples of electricity pumps are fuel cells and the various forms of batteries and accumulators (storage batteries). There is an important practical difference between fuel cells and batteries. While in both the substances undergo chemical reactions, in the former the reactants are supplied continuously and the products are removed. In batteries, the chemicals are stored and are used up. This difference changes the form of equations of balance used to describe each species, but it does not affect the principle of operation of the device.

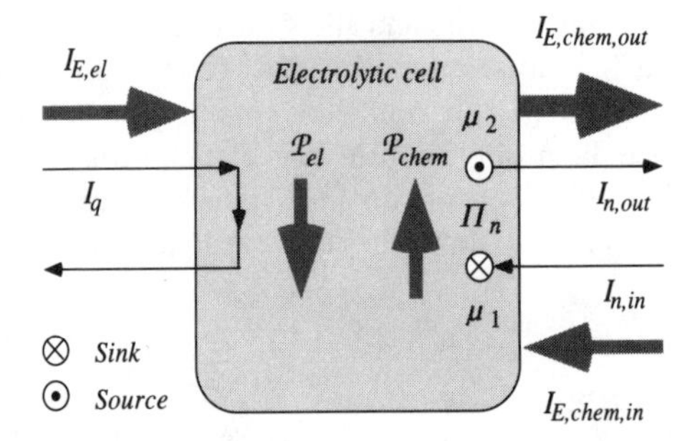

FIGURE 9. In an electrolytic cell, chemicals are changed by an electric current. Substances (such as water) enter the cell at a lower chemical potential and are destroyed, while new species (such as hydrogen and oxygen) are produced and leave the cell at a higher chemical potential.

In electrolysis and in the charging of storage batteries, we simply let the processes just described run in the reverse direction: electric currents are used to drive a chemical reaction (see Figures 1 and 9). The energy released in the fall of electric charge from higher to lower electric potential is used for the chemical transformation. Water can be split into hydrogen and oxygen; aluminum can be formed from its oxide; and copper can be refined. Reduction and oxidation reactions take place at the electrodes in the electrochemical cell. Often, electrolysis is the only means of getting the reaction to proceed at all: the necessary transfer of electrons is forced by the electrical process.

Energy in electrochemical processes. As a concrete example, consider a cell in which a reaction drives an electric current. We are accustomed to associating the energy used with a process which is driven, and we have introduced the notion of electrical power as the measure of what happens when an electric current flows through a potential difference:

$$\mathcal{P}_{el} = -\Delta\varphi_{el}\left|I_q\right| \tag{32}$$

The energy needed to drive the electric current must be supplied by the chemical reaction, which consists of the production of some species at different chemical potentials. Since the power associated with the process is proportional to the rates of the reaction, nature suggests a form for this quantity which we have seen many times before. For every species, there is a contribution

$$\mathcal{P}_{chem,i} = -\mu_i \Pi_{ni} \tag{33}$$

to the total power of the reaction:

$$\mathcal{P}_{chem} = -\sum_i \mu_i \Pi_{ni} \tag{34}$$

The quantities Π_{ni} denote the rates of production of the different species, destruction being counted as a negative rate of production. This means that the forgoing result can be written in the form

$$\mathcal{P}_{chem} = -\left[\sum_j \mu_j\left|\Pi_{nj}\right| - \sum_i \mu_i\left|\Pi_{ni}\right|\right] \tag{35}$$

where i and j count the reactants and the products, respectively. The simplest case is that which involves only one substance on either side of the reaction equation A → B; in this case, the rate of production of B is the negative of the corresponding rate of A. Therefore, the chemical power can be written as

$$\mathcal{P}_{chem} = -\left(\mu_A \Pi_{nA} + \mu_B \Pi_{nB}\right) = -\Delta\mu\left|\Pi_n\right| \tag{36}$$

This literally looks like the power of a current of water falling down over a series of rocks. From now on, we shall abbreviate the power of a chemical reaction by

$$\mathcal{P}_{chem} = -\left[\Delta\mu \cdot \Pi_n\right]_{reaction} \tag{37}$$

The term in brackets just has to be evaluated properly according to Equation (34). In general we may say that it is as if each disappearing substance released an amount of energy, while each substance appearing uses a certain amount. In the case of a spontaneous chemical reaction, the amount of energy released must be larger than the amount used. Literally, the net effect of a spontaneous reaction has to be the "running down" of chemical species.

All the electrochemical processes described so far have been assumed to work ideally, i.e., without production of entropy. Therefore, if the cells are operated in the steady state, the electrical and the chemical power add up to zero:

$$\left[\Delta\mu \cdot \Pi_n\right]_{reaction} + \Delta\varphi_{el}\left|I_q\right| = 0 \tag{38}$$

Remember that the amount of substance being transformed and the charge flowing through the cell are coupled as in Equation (2). If the production of one mole of a species involves z moles of electrons, the relation for each substance is

$$I_q = z_i \mathcal{F} \Pi_{ni} \tag{39}$$

and the balance of power requires that

$$\sum_i \left[\mu_i + z_i \mathcal{F} \Delta\varphi_{el}\right] \Pi_{ni} = 0 \tag{40}$$

for the complete reaction. The term in brackets is called the electrochemical potential of species i:

$$\eta_i = \mu_i + z_i \mathcal{F} \Delta\varphi_{el} \tag{41}$$

A reaction running in the steady state (or no reaction at all) may be interpreted as one which is in electrochemical equilibrium. These results demonstrate that the voltage of an ideal electrochemical cell can be used to measure the chemical potential of a reaction (see Example 9). Examples of values of chemical potentials and of voltages necessary for applications are given in Table A.14.

EXAMPLE 9. The chemical potential of the reaction of hydrogen and oxygen.

The voltage measured for an electrochemical cell (a fuel cell) converting hydrogen and oxygen to water has a maximum value of 1.23 V at standard conditions. Determine the chemical potential of water.

SOLUTION: We have to write down the reactions occurring at the electrodes of the fuel cell. H_2 is found to change into $2H^+$ and two electrons. At the other electrode, oxygen gas reacts with H^+ ions and electrons to form water:

$$H_2 \rightarrow 2H^+ + 2e$$

$$2H^+ + 2e + \frac{1}{2}O_2 \rightarrow H_2O$$

This means that 2 moles of electrons are involved if 1 mole of hydrogen gas reacts. According to Equation (38), the chemical potential of the reaction turns out to be

$$\left|\Delta\mu\right| = \left|z\mathcal{F}\, U_{el}\right| = 2 \cdot 96487 \cdot 1.23\,\text{J/mole} = 237\,\text{kG}$$

Since the chemical potentials of the hydrogen and oxygen gas are set equal to zero, this value also represents the chemical potential of water.

EXAMPLE 10. The voltage of a lead storage battery.

In a lead storage battery, an anode made out of lead and a cathode made out of lead dioxide are immersed in the same sulfuric acid solution. The reactions taking place at the electrodes are

$$\mathrm{Pb} + \mathrm{SO_4^{2-}} \rightarrow \mathrm{PbSO_4} + 2e$$

$$\mathrm{PbO_2} + 4H^+ + \mathrm{SO_4^{2-}} + 2e \rightarrow \mathrm{PbSO_4} + 2\mathrm{H_2O}$$

a) Calculate the voltage for each of the half reactions, and determine the voltage of the battery. b) Determine the equation of the overall reaction.

SOLUTION: a) The chemical potentials of the reactions can be taken from the previous tables:

$$\underbrace{\mathrm{Pb} + \mathrm{SO_4^{2-}}}_{(0+(-744.63))\,\mathrm{kG}} \rightarrow \underbrace{\mathrm{PbSO_4} + 2e}_{(-813.2+0)\,\mathrm{kG}}$$

$$\underbrace{\mathrm{PbO_2} + 4\mathrm{H^+} + \mathrm{SO_4^{2-}} + 2e}_{(-212.4+4\cdot 0+(-744.63)+2\cdot 0)\,\mathrm{kG}} \rightarrow \underbrace{\mathrm{PbSO_4} + 2\mathrm{H_2O}}_{(-813.2+2(-237.2))\,\mathrm{kG}}$$

The first reaction has a change of the chemical potential of - 68.6 kG, corresponding to a voltage of 0.36 V (there are two electrons involved). For the second reaction, the figures are - 330.6 kG and 1.71 V, respectively. The voltage of the battery therefore is 2.07 V.

b) The overall reaction is obtained by adding the equations of the partial reactions. This yields the following equation:

$$\mathrm{Pb} + \mathrm{PbO_2} + 2\mathrm{H_2SO_4} \rightarrow 2\mathrm{PbSO_4} + 2\mathrm{H_2O}$$

4.2.3 Thermochemical Processes: Entropy and the Chemical Potential

What will happen if a chemical reaction cannot drive an electrical process? This case certainly applies to fluids which do not allow electrical phenomena to occur, such as the simple fluids we have studied so far in thermodynamics. With no chance to do "useful work," chemical reactions dissipate the energy which is released.

Entropy production and the balance of entropy in chemical reactions. Dissipation means that entropy has been produced. The consequence of a chemical reaction in a simple fluid is just that: all the energy released is used to produce entropy (Figure 10).

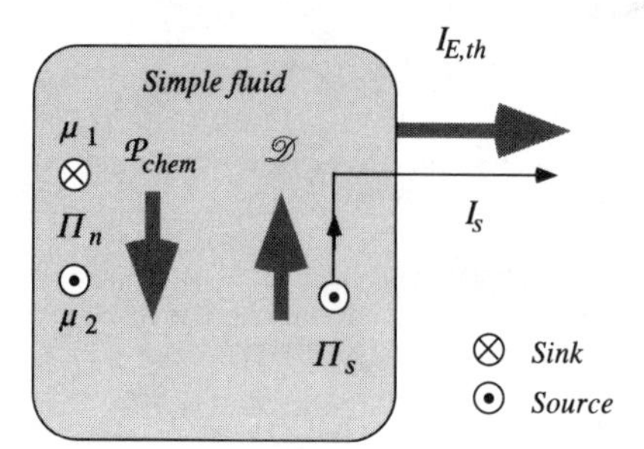

FIGURE 10. In a simple fluid, the consequence of a chemical reaction is the production of entropy. Since the reverse process cannot happen, chemical reactions in fluids run only in one direction.

According to what we have seen in Chapter 1, the rate of production of entropy is related to the power of dissipation by the temperature of the fluid. Since the rate of dissipation is the rate of release of energy by the reactions according to Equation (37):

$$\Pi_s = -\frac{1}{T}\left[\Delta\mu \cdot \Pi_n\right]_{reaction} \geq 0 \quad \textbf{(42)}$$

Since entropy can only be produced, we now understand the meaning of a spontaneous reaction and the sign of the change of the chemical potential. In a fluid which does not allow other processes, the reaction can proceed in only one direction.

Take a look at the balance of entropy and energy for a chemical reaction taking place in a fluid at constant temperature and pressure (Figure 11):

$$\begin{aligned} \dot{S} + I_s &= \Pi_s \\ \dot{E} + I_{E,th} + I_{E,mech} &= 0 \end{aligned} \quad \textbf{(43)}$$

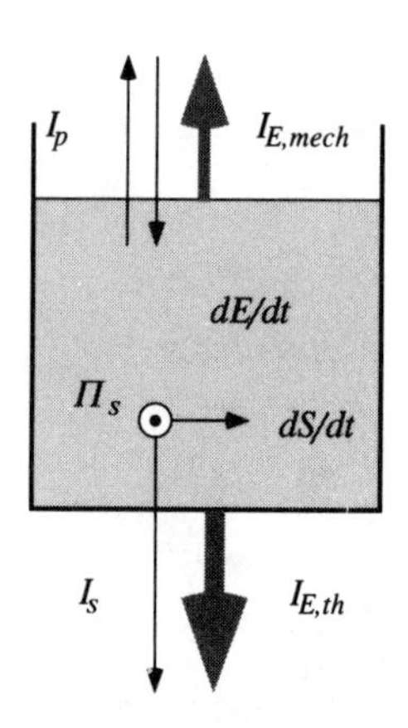

FIGURE 11. If a chemical reaction takes place in a simple fluid, all the energy released is used to produce entropy. The balance of entropy may require the reaction to be exothermic (as shown) or endothermic.

It should be clear from all we know about the thermodynamics of simple substances that the entropy content depends upon the chemical composition. Therefore, the entropy generated by the reaction goes toward the change of entropy of the chemical species which are involved; whatever is left is emitted. Naturally, it is also possible that the entropy of the resulting substances is smaller than that of the reactants, in which case even more entropy is emitted than has been produced. Either of these two cases represents what is called an *exothermic* reaction: the fluids emit heat. The opposite, an *endothermic reaction*, occurs if less entropy is produced than is required for the change of composition; in this case, some additional entropy is required to flow in from the environment.

Heat transfer and enthalpy of formation. One is often interested in the amount of energy emitted or absorbed together with the entropy exchanged. This quantity can be computed from the energy balance in Equation (43). If we also express the mechanical energy current of the simple fluid in terms of the rate of change of the volume, we obtain

$$I_{E,th} = -\left(\dot{E} + P\dot{V}\right) \tag{44}$$

It is customary to introduce the enthalpy of the fluid as a combination of the terms

$$H = E + PV \tag{45}$$

For processes at constant pressure, the rate of change of the enthalpy is the negative of the right-hand side of Equation (44). Therefore, the amount of energy exchanged in heating or cooling is equal to the change of the enthalpy of the fluid.

If we now introduce the expression of the balance of entropy in Equation (44), our basic assumption that the energy released by the reaction is dissipated will yield an equation involving properties of the fluid and the chemical potential:

$$\dot{H} = -TI_s = -T\left(\Pi_s - \dot{S}\right) = -T\left(-\frac{1}{T}\sum_i \mu_i \Pi_{ni} - \dot{S}\right) \tag{46}$$

where the summation is performed over all species involved in the reaction. Integrating the equation for a process at constant temperature and pressure yields

$$\Delta H = T\Delta S + \sum_i \mu_i \Delta n_i \tag{47}$$

The change of the entropy has to be computed for all the fluids in the mixture:

$$\Delta S = \sum_j S_j - \sum_i S_i \tag{48}$$

where i and j stand for the reactants and the products, respectively. This means that we can compute the amount of energy exchanged in cooling or heating of the fluid during the reaction if we know the chemical potentials and the entropies of the species involved.

Equation (47) can be used to obtain an expression of the chemical potential of a single fluid in terms of the molar enthalpy H/n and the molar entropy S/n:

$$\mu = \bar{h} - T\bar{s} \tag{49}$$

or of each component of a simple mixture. If the entropy and the enthalpy of the mixture can be written as the sum of the partial values of the components, then the result is the same for the parts:

$$\mu_i = \bar{h}_i - T\bar{s}_i \tag{50}$$

This equation will serve as the general result expressing the relation between the chemical potential and other fluid properties for simple fluids at rest. A more general result will be obtained in Section 4.4 where we will study flow processes.

Obviously, we need to know the entropy of a substance if we wish to calculate all the quantities involved in a reaction. The entropy can be obtained from measurements of entropy capacities down to a temperature of 0 K, and if the body is liquid or even gaseous, from measurements of the entropies of fusion and of vaporization. By setting the entropy of a pure substance equal to zero at 0 K, we obtain the absolute value of the required quantity. We will use values reported in the literature—Tables A.13 and A.15 are a small sampling. (Note that the temperature coefficient of the chemical potential is the negative value of the absolute entropy.)

It is customary to report values of the enthalpy of substances along with those of the chemical potential and the entropy. As in the case of the chemical potential, the value of the enthalpy is set equal to zero for the stable configurations of the elements at standard temperature and pressure. The result obtained for the enthalpy of a compound is then called the *enthalpy of formation* h_f:

$$\bar{h}_f^o = \bar{h}^o - \sum_{i=elements} \nu_i \bar{h}_i^o = \mu_f^o + T\bar{s}^o - \sum_{i=elements} \nu_i \left[\mu_i^o + T\bar{s}_i^o\right] \quad (51)$$

which is equivalent to

$$\bar{h}_f^o = \mu_f^o + T\bar{s}^o - T \sum_{i=elements} \nu_i \bar{s}_i^o \quad (52)$$

Here, ν_i is the factor multiplying the term for element i in the reaction equation

$$\nu_1 \mathrm{A} + \nu_2 \mathrm{B} + \ldots \rightarrow \textit{compound}$$

This equation should be written for one mole of the compound being formed. A, B, ... denote the elements which are used up as a result of the reaction.

EXAMPLE 11. Entropy produced and exchanged in the reaction forming water.

a) Consider the reaction which forms water out of hydrogen and oxygen. How much entropy is produced in forming one mole of water? b) Does the product contain more entropy than the reactants? c) Is the reaction exothermic or endothermic? How much entropy and energy are exchanged in heating or cooling? Remember that the reaction is assumed to proceed at constant temperature and pressure.

SOLUTION: a) The amount of entropy produced is equal to the energy dissipated divided by the temperature of the fluid. We have

$$\bar{s}_{prod} = -\frac{1}{T}\left[\mu_{H_2O} - \left(\mu_{H_2} + 0.5\mu_{O_2}\right)\right]$$
$$= -\frac{1}{298}\left[-237.2 \cdot 10^3 - (1 \cdot 0 + 0.5 \cdot 0)\right]\frac{J}{K \cdot mole} = 796\frac{J}{K \cdot mole}$$

b) The change of entropy of the substances calculated for the reaction is

$$\Delta\bar{s} = \bar{s}_{H_2O} - \left(\bar{s}_{H_2} + 0.5\bar{s}_{O_2}\right)$$
$$= 69.9\frac{J}{K \cdot mole} - (1 \cdot 130.6 + 0.5 \cdot 205)\frac{J}{K \cdot mole} = -163.2\frac{J}{K \cdot mole}$$

This means that water contains less entropy than the elements out of which it is formed.

c) Considering that entropy has been produced and that the entropy content has decreased, it is clear that the fluids must be cooled. The amount of entropy exchanged is given by

$$\bar{s}_e = \Delta\bar{s} - \bar{s}_{prod}$$
$$= -163.2\frac{J}{K \cdot mole} - 796\frac{J}{K \cdot mole} = -959\frac{J}{K \cdot mole}$$

This value, multiplied by the temperature of the fluid emitting the entropy, must be equal to the energy emitted:

$$\overline{Q} = T\bar{s}_e = 298\,\mathrm{K} \cdot \left(-959\frac{J}{K \cdot mole}\right) = -285.8 \cdot 10^3\frac{J}{mole}$$

Naturally, the same result could have been obtained by calculating the change of enthalpy of the fluids for the reaction. (See Example 12 and Table A.13.)

EXAMPLE 12. The enthalpy of formation of water.

Calculate the enthalpy of formation of water for the standard state from the values of chemical potentials and entropies found in Table A.13.

SOLUTION: Enthalpies of formation are calculated according to Equation (52). We need the chemical potentials (of formation) for water, hydrogen, and oxygen. (The latter two are equal to zero.) With the reaction

$$H_2 + 0.5O_2 \rightarrow H_2O$$

the enthalpy of formation is calculated according to

$$\bar{h}^o_{f,H_2O} = \left[\mu^o_f + T\bar{s}^o\right]_{H_2O} - T\left[\bar{s}^o_{i,H_2} + 0.5\bar{s}^o_{i,O_2}\right]$$
$$= -237.2 \cdot 10^3\,\mathrm{J} + 298 \cdot 70\,\mathrm{J} - 298[1 \cdot 130.6 + 0.5 \cdot 205]\,\mathrm{J} = -285.9 \cdot 10^3\,\mathrm{J}$$

This corresponds to the value found in Table A.13.

EXAMPLE 13. The heating value of a fuel.

The heating value of a fuel is defined as the energy exchanged as a consequence of cooling of the fluids. The higher heating value is calculated for liquid water in the products, while the lower heating value is defined for water as a vapor in the products.
Determine the higher and lower heating values of the combustion of methane, which combines with oxygen to form carbon dioxide and water.

SOLUTION: The amount of energy emitted in cooling of the fluids undergoing reaction is the change of enthalpy of the reactants and products. At standard conditions, this change can be expressed as

$$\Delta H = \left| \sum n_j \bar{h}^o_{fj} \right|_{products} - \left| \sum n_i \bar{h}^o_{fi} \right|_{reactants}$$

where the n_i and n_j can be read from the reaction equation, which for complete combustion takes the form

$$CH_4 + 2O_2 \rightarrow CO_2 + 2H_2O$$

The higher heating value is

$$\Delta H = |(-393520) + 2(-285900)|\,\text{kJ} - |(-74850) + 2 \cdot 0|\,\text{kJ} = 890\,\text{kJ}$$

for 1 mole of methane. The lower heating value turns out to be equal to

$$\Delta H = |(-393520) + 2(-241800)|\,\text{kJ} - |(-74850) + 2 \cdot 0|\,\text{kJ} = 802\,\text{kJ}$$

Again, the value is for 1 mole of methane. On a mass basis, the results are 55.6 MJ/kg and 50.1 MJ/kg, respectively.

EXAMPLE 14. Entropy produced in the diffusive mixing of two different ideal gases.

Take one mole of two different ideal gases, each at the same temperature and pressure. They are in the same container but separated by a wall (as in Figure 3). Show that you get the result derived in Section 4.1.3 for the production of entropy by considering the chemical potentials of the components with respect to each other, and the energy released in the diffusive "reaction."

SOLUTION: The amount of entropy produced for a chemical reaction taking place at constant temperature follows from Equation (42):

$$S_{prod} = -\frac{n}{T}\left[\mu_{Af} + \mu_{Bf} - (\mu_{Ai} + \mu_{Ai})\right]$$

where n is the amount of substance of each of the components, and the subscripts i and f refer to initial and final values, respectively. This can be expressed using the initial and the final (partial) pressures of the parts of the mixture (see Equation (58) below):

$$\begin{aligned} S_{prod} &= -\frac{n}{T}\left[(\mu_{Af} - \mu_{Ai}) + (\mu_{Bf} - \mu_{Bi})\right] \\ &= -\frac{n}{T}\left[RT\ln\left(\frac{P_{Af}}{P}\right) + RT\ln\left(\frac{P_{Bf}}{P}\right)\right] \\ &= -Rn\left[\ln(0.5) + \ln(0.5)\right] \end{aligned}$$

This means that we can understand the entropy produced as the result of a diffusive reaction. The reason why no entropy should be produced if the gases are identical is this: with identical gases, the chemical potentials, which also depend upon the environment of each of the species, would not change.

4.2.4 The Chemical Potential of Uniform Fluids

The values of chemical potential, entropies, and enthalpies reported in tables such as Table A.13, are calculated for the reference state. If we wish to obtain values for different temperatures, pressures, or concentrations, we have to know how they depend upon the state of the fluid. Basically, this information can be derived on the basis of the expression for the chemical potential of fluid given in Equation (49). If we write the enthalpy in terms of the quantities defining it, we have for the chemical potential

$$\mu = \bar{e} + P\bar{v} - T\bar{s} \tag{53}$$

where $\bar{v} = V/n$ is the molar volume of the fluid. This tells us that we must be able to compute energy, volume, and entropy as functions of temperature and pressure. Here we will present the calculations for the ideal gas and for incompressible fluids. We will again encounter the same results when we study the flow of substances in Section 4.4.

The chemical potential as a function of pressure. Let us begin with the dependence of the chemical potential on pressure for incompressible fluids. For such bodies, the energy and the entropy depend only upon temperature. (See Section 2.2.) Since we keep the temperature constant, the change of the chemical potential turns out to be

$$\mu(T,P) - \mu(T,P_o) = P\bar{v} - P_o\bar{v}_o \tag{54}$$

Now, the volume of an incompressible substance cannot be changed. For this reason, the molar volume is the same in both states, which means that the chemical potential is

$$\mu(T,P) = \mu(T,P_o) + \left(P - P_o\right)\bar{v} \tag{55}$$

For the ideal gas, the situation is somewhat different, since the entropy of the fluid also depends upon pressure. Again, as in the case of the incompressible fluid, the energy is a function of temperature only, which means that this term does not enter into our consideration. One more special point should be noticed: the product of pressure and of volume remains constant for constant temperature. In summary, only the entropy term in Equation (53) changes upon a change of pressure:

$$\mu(T,P) - \mu(T,P_o) = -T\bar{s} + T\bar{s}_o \tag{56}$$

We now need an expression which yields the entropy of the ideal gas as a function of pressure. According the results of Chapter 2, we have

$$\bar{s} - \bar{s}_o = -R\ln\left(\frac{P}{P_o}\right) \tag{57}$$

for the change of molar entropy. Combining the last two formulas, we obtain

$$\mu(T,P) = \mu(T,P_o) + RT\ln\left(\frac{P}{P_o}\right) \tag{58}$$

Remember that this result only holds for the ideal gas. Even though it only approximates the behavior of general gases, we will use it to discuss phase changes.

The temperature dependence of the chemical potential. We will derive the result for the chemical potential of a simple fluid as a function of temperature at constant pressure for the case of the ideal gas. You will notice, however, that the assumptions made also are valid for incompressible fluids under many circumstances.

We start again with the general expression for the chemical potential given in Equation (53):

$$\mu = \bar{u} + P\bar{v} - T\bar{s} = \bar{h} - T\bar{s} \tag{59}$$

For the difference of the chemical potential for values of temperature T and T_o we have the following relation:

$$\mu(T) - \mu(T_o) = \left[\bar{h} - T\bar{s}\right] - \left[\bar{h} - T\bar{s}\right]_o = \bar{h} - \bar{h}_o - T\bar{s} + T_o\bar{s}_o \tag{60}$$

Now, for the ideal gas, the change of the enthalpies at different temperatures is the product of the temperature coefficient of enthalpy and the temperature difference:

$$\bar{h} - \bar{h}_o = \bar{c}_p\left(T - T_o\right) \tag{61}$$

The entropy of the ideal gas as a function of temperature at constant pressure is

$$\bar{s}(T) = \bar{s}_o + \bar{c}_p\ln\left(\frac{T}{T_o}\right) \tag{62}$$

If we substitute these results into the expression for the difference of the chemical potential derived above we obtain

$$\mu(T,P) = \mu(T_o,P) + \bar{c}_p(T - T_o) - T\bar{c}_p \ln\left(\frac{T}{T_o}\right) - (T - T_o)\bar{s}_o \tag{63}$$

The main assumptions made concern the validity of Equations (61) and (62). These relations hold for constant temperature coefficients of enthalpy, a condition which may hold for incompressible fluids as well as for the ideal gas. Note that we have to know the absolute value of the entropy of the fluid if we wish to compute the temperature dependence of the chemical potential.

The temperature and pressure coefficients of the chemical potential. In Section 4.2.1, the coefficients used in the linear approximation of the chemical potential as a function of temperature and pressure were introduced. If you inspect Equations (28) and (29), you will see that they correspond to the partial derivatives of the chemical potential with respect to temperature or pressure. Here, we will derive them for the special cases treated in the paragraphs above. Starting with the pressure coefficient, the results for incompressible fluids and the ideal gas lead to the same general expression:

$$\beta_\mu \equiv \left.\frac{\partial \mu(T,P)}{\partial P}\right|_{T=T_o} = \bar{v}_o \tag{64}$$

The pressure coefficient corresponds to the molar volume of the fluid. The case of the temperature coefficient is a little bit more complicated. Evaluation of the derivative of Equation (63) with respect to temperature at $T = T_o$ shows that

$$\alpha_\mu \equiv \left.\frac{\partial \mu(T,P)}{\partial T}\right|_{T=T_o} = \left[\bar{c}_p - \bar{c}_p \ln\left(\frac{T}{T_o}\right) - T\bar{c}_p \frac{T_o}{T}\frac{1}{T_o} - \bar{s}_o\right]_{T=T_o} = -\bar{s}_o \tag{65}$$

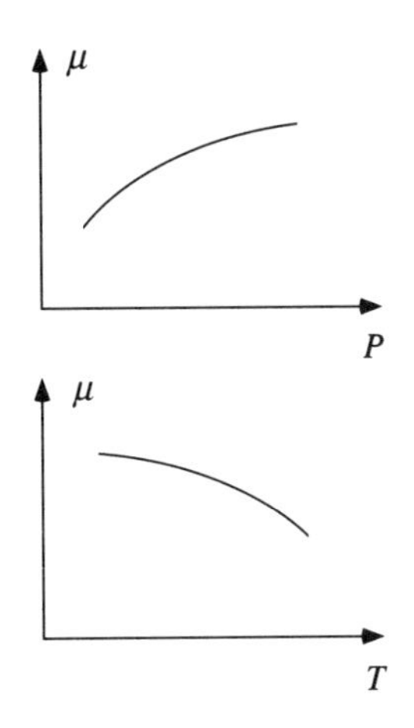

FIGURE 12. Chemical potential of a fluid as a function of pressure (top) and temperature (bottom).

The temperature coefficient is equal to the negative molar entropy of the reference state. This is what we already have observed in Table A.13. In Section E.1, it will be demonstrated that these results hold not only for the fluids investigated here, and not just for the reference state. For now they are enough to clarify some features of the tabulated values of chemical potentials we have been using in our examples. They show that the chemical potential increases with pressure at constant temperature, and it decreases with temperature at constant pressure (Figure 12). The latter result might come as a surprise, considering that chemical transformations tend to become more vigorous if temperature increases. However, you have to remember that the quantity responsible for a reaction is the difference of the chemical potentials for the reactants and the products; according to the equations, both sides of a reaction equation are influenced by a change of temperature.

EXAMPLE 15. Pressure coefficient and partial molar volume.

a) Determine the pressure coefficient of pure iron. The density of iron is 7874 kg/m^3, and its molar mass is 0.0559 kg/mole. b) The pressure coefficient of ammonia dissolved in water is $24.1 \cdot 10^{-6}$ G/Pa. Determine its partial molar volume and the equivalent density.

SOLUTION: a) The pressure coefficient of the chemical potential is the (partial) molar volume of the substance. The latter is related to the density by

$$\beta_\mu = \bar{\upsilon} = \frac{V}{n} = M_o \frac{V}{m} = M_o \frac{1}{\rho} = 0.0559 \frac{1}{7874} \frac{\text{m}^3}{\text{mole}} = 7.10 \cdot 10^{-6} \frac{\text{m}^3}{\text{mole}}$$

The unit is equivalent to G/Pa.

b) If the pressure coefficient of a substance is known, its partial molar volume can be determined:

$$\bar{\upsilon} = \beta_\mu = 24.1 \cdot 10^{-6}\,\text{G/Pa}$$

The molar volume is related to an equivalent density by

$$\rho = M_o \frac{1}{\bar{\upsilon}} = 0.017 \frac{1}{24.1 \cdot 10^{-6}}\,\text{kg/m}^3 = 705\,\text{kg/m}^3$$

Ammonia (NH_3) has a molar mass of 0.017 kg/mole. This means that ammonia dissolved in water without dissociating exists as if it had a density slightly smaller than that of water. Put differently, if 705 kg of ammonia are dissolved in a vast amount of water, the volume of the mixture will increase by 1 m^3.

4.2.5 Equilibrium in Chemical Reactions

What happens if a (spontaneous) reaction is allowed to run its course? You might think that, barring any resistance to the chemical transformation, all the reactants should be used up, leading to a complete change into the products. This is not quite the case. While it is necessarily true that the chemical potential of the reactants in a spontaneous reaction must be larger than the potential of the products, this condition will not remain valid. While the reactants are being used up, their concentration (or pressure) decreases leading to a change of the magnitude of the chemical potentials; the same is true for the products. Therefore, we must expect the reactions to reach a condition in which the difference of the chemical potentials becomes zero. This will happen before all the chemicals have been changed into others; in other words, a certain amount of the reactants will remain when the reaction has run its course. In the end, the drive of the reverse reaction will have become just as large, not allowing for any further transformations: this condition is called *chemical equilibrium*.

The condition of chemical equilibrium. Chemical equilibrium has been reached if the difference of the chemical potentials for a reaction has vanished. A reaction of the form

$$a\text{A} + b\text{B} + \ldots \leftrightarrow c\text{C} + d\text{D} + \ldots$$

where the lower case letters denote the stoichiometric coefficients, and the capital letters stand for the species, can also be written as a true equation:

$$0 = a\mathrm{A} + b\mathrm{B} + \ldots + c\mathrm{C} + d\mathrm{D} + \ldots \tag{66}$$

We give the first coefficients a, b, ... negative values, while the c, d, ... are taken to be positive:

$$a, b, \ldots < 0 \quad , \quad c, d, \ldots > 0$$

With these definitions in mind, the difference of the chemical potential for the reaction can be written conveniently as

$$[\Delta\mu]_{reaction} = \sum_{species} \nu_i \mu_i \tag{67}$$

Chemical equilibrium now means that this quantity must be zero:

$$[\Delta\mu]_{reaction} = 0 \tag{68}$$

The equilibrium concentration for ideal gases and dilute solutions. One is often interested in the outcome of a reaction, in the sense that one wishes to know the actual composition of the substances after equilibrium has been reached. Basically, this calls for the application of Equation (68) to each case. This leads to a relation between the partial pressures of the different substances; one value can be computed if the others are known.

To achieve this, we must be able to express the chemical potential of each species involved in terms of the actual temperature and pressure (or concentration). Let us assume that equilibrium has been achieved at standard conditions for the mixture. In this case, only the partial pressure of a species is different from P_o; the temperature has assumed the standard value of 298 K.

For the case of a mixture of ideal gases, the dependence of the chemical potentials upon pressure is expressed simply using Equation (58):

$$\mu_i = \mu_{fi}^o + RT\ln\left(\frac{P_i}{P_o}\right) \tag{69}$$

For dilute solutions, the partial pressure can be replaced by the concentration according to Equation (10):

$$P_i = RT\bar{c}_i \tag{70}$$

where you have to remember that the standard concentration for the calculation of the standard values of the chemical potential is 1 mole/liter, and not 1 mole/m^3. If we introduce Equation (69) in the condition of equilibrium, we obtain

$$
\begin{aligned}
0 &= \sum_{species} \nu_i \mu_i = \sum_{species} \nu_i \left[\mu_{fi}^o + RT \ln\left(\frac{P_i}{P_o}\right) \right] \\
&= \sum_{species} \nu_i \mu_{fi}^o + \sum_{species} \nu_i RT \ln\left(\frac{P_i}{P_o}\right)
\end{aligned}
\tag{71}
$$

The first term on the right is the change of the chemical potential for the reaction at standard conditions; if we also rewrite the second term, the condition reads

$$
0 = \left[\Delta\mu_f^o\right]_{reaction} + RT \ln\left(\frac{P_a^a}{P_o^a} \frac{P_b^b}{P_o^b} \cdots \frac{P_c^c}{P_o^c} \frac{P_d^d}{P_o^d} \cdots \right)
\tag{72}
$$

Remember to use the proper sign of the stoichiometric coefficients a, b,...as defined above. It is customary to call the quantity in parenthesis the equilibrium constant $\mathcal{K}_p$:

$$
\mathcal{K}_p \equiv \frac{P_a^a}{P_o^a} \frac{P_b^b}{P_o^b} \cdots \frac{P_c^c}{P_o^c} \frac{P_d^d}{P_o^d} \cdots
\tag{73}
$$

With this definition, the condition of chemical equilibrium in a mixture of ideal gases can be written in the following simple form:

$$
RT \ln\left(\mathcal{K}_p\right) = -\left[\Delta\mu_f^o\right]_{reaction}
\tag{74}
$$

The equilibrium constant has a value which is independent of pressure (since the right-hand side of the equation has been defined for the fixed standard pressure); i.e., it only depends upon the temperature. With the change of chemical potential for the reaction for standard conditions known, $\mathcal{K}_p$ can be calculated; this then yields the relation between the partial pressures of the species partaking in the reaction.

As we have seen before, changing the temperature at which the reaction takes place can influence the outcome, i.e., the equilibrium concentrations of the substances. The same is true of a change of the pressure. Even though the equilibrium constant is independent of pressure, the values of the partial pressures will change, Equation (73). Therefore, with changes both of temperature and of pressure, the outcome of a reaction can be shifted to favor one side.

Chemical equilibria do not occur only in outright chemical reactions. As we will see in the following section, the concept of equilibrium can be applied to such diverse situations as the vapor pressure of a liquid, the condensation of vapor, the solution of gases in water, and more.

EXAMPLE 16. Is there hydrogen gas in water vapor?

At standard conditions, water vapor is present in air at a partial pressure of 30 mb. How much hydrogen gas will be present relative to the amount of water?

SOLUTION: First, we express the reaction equation between hydrogen, oxygen, and water in the form of Equation (66):

$$0 = H_2O - H_2 - 0.5O_2$$

Then, the equilibrium constant can be computed from the known values of the standard chemical potentials:

$$\ln(\mathcal{K}_p) = -\frac{1}{RT}\left[\Delta\mu_f^o\right]_{reaction} = -\frac{1}{8.31\cdot 298}\left[-228.6\cdot 10^3 - 0 - 0.5\cdot 0\right] = 92.3$$

Remember that water is present in the gaseous state. According to the definition of the equilibrium constant, there exists the following relation between the pressures:

$$\mathcal{K}_p = \frac{P_{H_2O}^1}{P_o^1}\frac{P_{H_2}^{-1}}{P_o^{-1}}\frac{P_{O_2}^{-0.5}}{P_o^{-0.5}}$$

The ratio of the amounts of hydrogen and water vapor can be written using the respective partial pressures. Remembering that the amount of oxygen necessary for the reaction is half that of hydrogen, we have

$$\left(\frac{P_{H_2O}}{P_{H_2}}\right)^{3/2} = \frac{P_{H_2O}^{0.5}}{P_o^{0.5}}(0.5)^{0.5}\mathcal{K}_p$$

which yields a value of $1.3 \cdot 10^{26}$ for the ratio of water to hydrogen gas. Thus, there is about one molecule of H_2 for every 10^{26} molecules of water present, if the reaction can proceed freely.

EXAMPLE 17. The effect of pressure changes on the synthesis of ammonia.

Predict the effect of a change of the pressure of the mixture of nitrogen, hydrogen, and ammonia from P_o to P on the equilibrium composition. The ammonia synthesis proceeds according to

$$0 = 2NH_3 - 3H_2 + N_2$$

SOLUTION: The equilibrium constant takes the value

$$\ln(\mathcal{K}_p) = -\frac{1}{RT}\left[\Delta\mu_f^o\right]_{reaction} = -\frac{1}{8.31\cdot 298}\left[-2\cdot 16.4\cdot 10^3 - 3\cdot 0 - 0\right] = 13.2$$

which will not change. If the total pressure of the mixture is P_o, we have

$$\mathcal{K}_p = \left(\frac{y_{NH_3}P_o}{P_o}\right)^2\left(\frac{y_{H_2}P_o}{P_o}\right)^{-3}\left(\frac{y_{N_2}P_o}{P_o}\right)^{-1} = y_{NH_3}^2 y_{H_2}^{-3} y_{N_2}^{-1}$$

On the other hand, if the total pressure is P, the expression becomes

$$\mathcal{K}_p = \left(\frac{y_{NH_3}P}{P_o}\right)^2 \left(\frac{y_{H_2}P}{P_o}\right)^{-3} \left(\frac{y_{N_2}P}{P_o}\right)^{-1} = y_{NH_3}^2 y_{H_2}^{-3} y_{N_2}^{-1} \left(\frac{P}{P_o}\right)^{2-3-1}$$

Since the equilibrium constant does not change, the product of the proper mole fractions of the components will change according to the ratio of the total pressures (raised to a particular power dependent on the stoichiometric coefficients) in the two cases.

4.3 Phase Changes, Solutions, and Mixtures of Fluids

Except for a brief excursion in Section 1.3.6, where the purpose was to familiarize you with the notion of entropy as the quantity responsible for melting a body, we have studied the thermodynamics only of fluids which cannot undergo phase changes. However, no such theory would be complete without at least a brief look at this important phenomenon. In nature and in engineering applications, phase changes play an important role. In the following sections, we will discuss the entropy necessary for phase changes, the vapor pressure and Clapeyron's law, and property tables. Then we will describe the influence of dissolved substances upon melting points and temperatures of vaporization, and the concept of solubility, after which we will turn to mixtures of two phase fluids such as moist air.

4.3.1 A Description of Phase Changes

Here, we will describe in more concrete terms what happens to a fluid as a consequence of fusion, vaporization, or sublimation. So that we will not always have to mention all three processes, we usually will chose one of them as an example. We will see how much entropy and energy is involved in a phase change, and we will discover that there exist clear relations between the temperature and the pressure at the transition.

In Section 1.3.6, we described the melting or the vaporization of a body using the *TS* diagram of the process, a tool which we will use again. We will begin by describing the processes of vaporization and condensation.

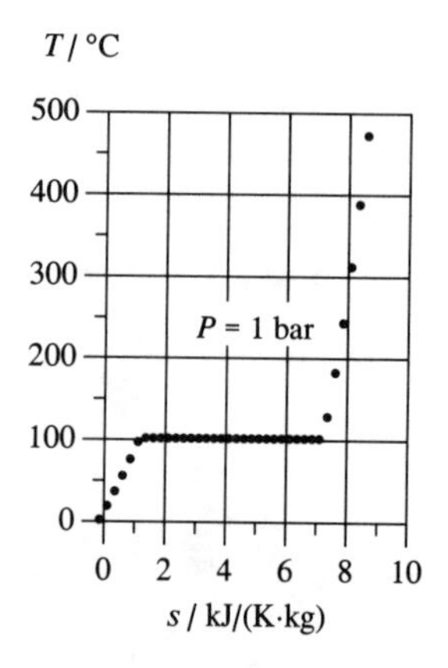

FIGURE 13. For 1 kg of water being heated at constant pressure of 1.0 bar, the temperature rises as a function of the entropy content. The horizontal section of the curve corresponds to the phase during which vaporization takes place: addition of entropy does not change the temperature of the fluid.

Vaporization of water. For the sake of argument, consider 1 kg of water being heated from 0°C at a constant pressure of 1 bar (Figure 13). For water being heated, entropy and temperature rise from values of zero along the section of the curve at the bottom left (for entropy, this is an arbitrary choice). During this phase, water is said to be a *compressed (subcooled) liquid*. Since water is nearly incompressible, the relation is approximated by the one known from Chapter 2:

$$s(T) = s(T_o) + c_p \ln\left(\frac{T}{T_o}\right) \tag{75}$$

This agrees well with what can be read from Figure 13. (Remember that c_p for water is about 4200 J/(kg · K).) At a temperature of 100°C, the water begins to boil. Just before the onset of boiling, the fluid is said to be a *saturated liquid.* While the entropy of the fluid increases, the temperature stays constant: the curve cuts horizontally through the *T-S* diagram at the temperature of vaporization. During this part of the process, the fluid is a *mixture of liquid and vapor.* Finally, when all the water is turned into steam, the temperature of the vapor begins to rise again as its entropy increases. Again, there is a name for the fluid just after all of it has turned into gas: it is called a *saturated vapor.* Pure vapor (without any liquid left) being heated is said to be a *superheated water vapor.* Equation (75) can be taken as an approximation for the actual relation between temperature and entropy, if we take steam as an ideal gas having constant temperature coefficient of enthalpy. (According to Figure 13, the average value of c_p between 100°C and 200°C is around 2000 J/(kg · K).)

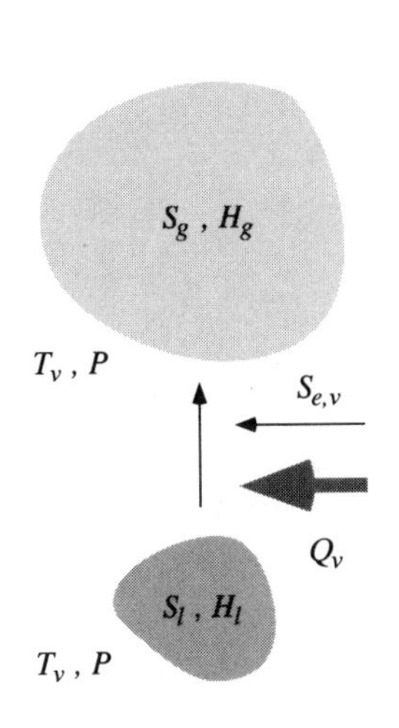

FIGURE 14. A certain amount of a liquid is vaporized. The entropy added remains in the body, while part of the energy added goes toward expansion. The energy added is equal to the change of enthalpy of the fluid body.

Balance of entropy and energy. Let us take a closer look at the transition and the balance of entropy and energy (Figure 14). Consider the pure liquid at the boiling point. Heat is added, and a certain amount of the liquid is transformed into vapor. Looked at from the viewpoint of a chemical reaction, we may say that only one substance is involved on either side of the reaction equation

$$\mathrm{A} \rightarrow \mathrm{B}$$

As a consequence, if the change of amount of substance of species A is Δn, then the corresponding change for B is - Δn:

$$\Delta n_l = -\Delta n_g \tag{76}$$

The subscripts l and g will indicate the liquid and vapor (gaseous) phases, respectively. If we take the change of the amount of substance of the vapor, the entropy necessary for vaporization is written in the form

$$S_e = \bar{l} \cdot \Delta n_g \tag{77}$$

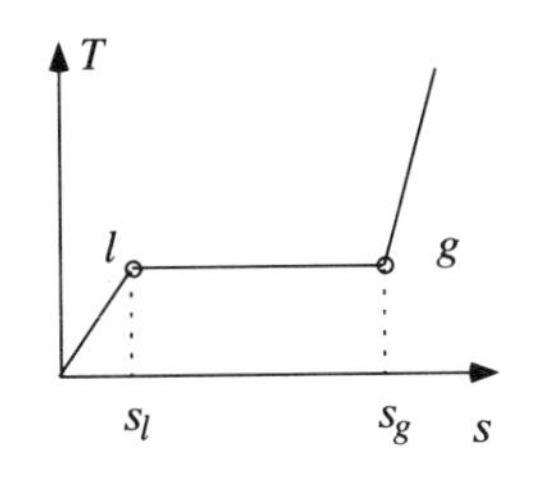

FIGURE 15. The entropy of vaporization can be read from the *T-S* diagram, where the value corresponds to the length of the horizontal section of the curve. Points l and g denote the states of saturated liquid and of saturated vapor (gas).

Observation tells us that vaporization at constant pressure and temperature proceeds reversibly, which means that the entropy added to the liquid phase will be present in the vapor phase (Figure 15). The *latent entropy* l of the phase change can therefore be expressed in terms of the entropies of the liquid and the gas:

$$\bar{l} = \bar{s}_g - \bar{s}_l \tag{78}$$

Normally, the phase change is also described in terms of the energy exchanged (see Figure 14). Since the process takes place at constant temperature, the energy added for vaporization is the product of T_v and $S_{e,v}$ as expressed by Equation (77). If we remember that the phase change also takes place at constant temperature, we can conclude that the energy added in heating is equal to the change of enthalpy of the fluid. (Part of the energy is used to raise the internal energy, while the rest is emitted because of the expansion of the fluid.) In summary, the change of entropy and the change of enthalpy are related by

$$\bar{h}_g - \bar{h}_l = T_v\left(\bar{s}_g - \bar{s}_l\right) \tag{79}$$

Naturally, these relations can also be expressed on the basis of mass rather than amount of substance. We simply introduce the specific entropy of vaporization l_v and relate it to the difference of the specific entropies of the fluid in the liquid and the vapor phases. Then, in Equation (79), we have the specific enthalpy and the specific entropy.

Vaporization at different pressures. From experience we know that water boils at different temperatures depending on the pressure of the fluid. If the pressure is lower than 1 bar, the boiling point is lower. Since the change in pressure hardly affects the properties of liquid water, we expect a *T-S* diagram of the process of heating to start off just as in the lower left of the curve in Figure 13. Then, however, the curve must break off at an earlier point and cut across the diagram horizontally. Measurements show that more entropy is needed to vaporize the same amount of water at lower pressure. Finally, when all the water has turned into steam, the temperature continues to climb again. For higher pressures, the changes with respect to the case of 1 bar are just the opposite. If the *T-S* curves for different values of the pressure are drawn in the same diagram, we get the result shown in Figure 16. How much energy is needed for vaporization depends not only upon the amount of entropy needed, but also on the temperature of vaporization. It turns out the amount of energy necessary for vaporization decreases with increasing pressure.

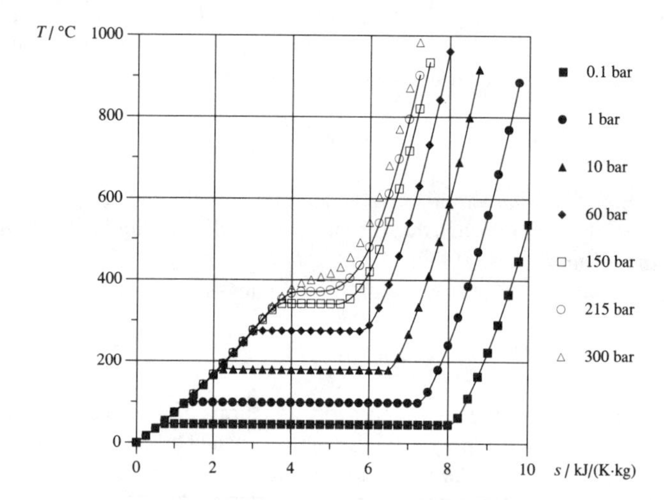

FIGURE 16. *T-S* diagram for 1 kg of water which is heated from 0°C at constant pressure. The curves are for different values of pressure, ranging from 0.1 bar to 300 bar. The last set of points shows that when the pressure is larger than the critical pressure of 220 bar, the vapor does not condense any more. The values were computed using the Steam-NBS function implemented in the program EES (Klein et.al, 1990).

There are some interesting points to note. The horizontal sections of the *T-S* curves, i.e., those parts which display the process of vaporization, form a bell-shaped area on the *T-S* diagram (Figure 17). The diagram can be divided into three sections, one on the left of the bell for the *compressed liquid phase*; one to the right, for the *superheated vapor*; and the bell area itself, where *mixtures of liquid and vapor* are present. The liquid phase basically occupies only a very thin strip along the saturated liquid line; this is so since the properties of liquid water do not depend strongly upon pressure. Only for very large pressures, of the order of hundreds or thousands of bar, do the properties of water differ considerably from the saturated state. The curve bounding the bell is divided into two parts; the left denotes the states of *saturated liquid* (pure liquid just at the verge of vaporization), and the right shows the states of *saturated vapor* (pure vapor on the

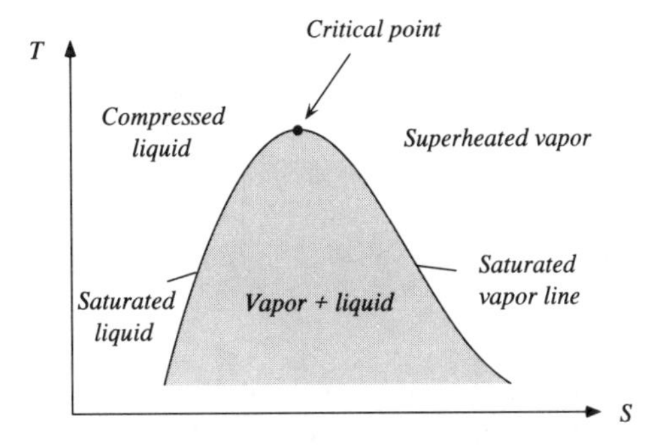

FIGURE 17. The line for saturated fluid (liquid and vapor) is shown in the *T-S* diagram. The dome shaped area below the line is occupied by the mixed phase of liquid and vapor. The line separates this area from those of the compressed liquid and the superheated steam.

verge of condensation). The two sections meet at the top of the bell, at a point called the *critical point*. If you have water vapor at pressures above the critical pressure of 220.9 bar, and you cool the vapor at constant pressure, it will no longer condense. Rather, as you can see in Figure 16, the temperature will pass above the critical temperature of 374.14°C for values of the specific entropy of 4.43 J/(K · kg).

The pressure-temperature relation for boiling. Obviously, there exists a relation between the boiling point, i.e., the temperature of vaporization, and the pressure at which the transition takes place. As Figure 16 demonstrates, the higher the pressure, the higher the temperature. Such a relation is rather different from what we know about liquids or gases. With liquids, taken as incompressible, the properties depend only upon the temperature, and the pressure does not come into play. For gases, on the other hand, two independent properties fix the temperature; one variable, such as the pressure, is not enough. For this reason, the relation between temperature and pressure at vaporization is remarkable. A typical pressure-temperature curve for a fluid appears in Figure 18. Note, that the line has a beginning and an end. The end we already know: it is the critical point of the fluid. The beginning of the line, however, will be understood shortly, when we transfer our description to the other two phase transitions as well. The beginning point is called the *triple point* of the substance.

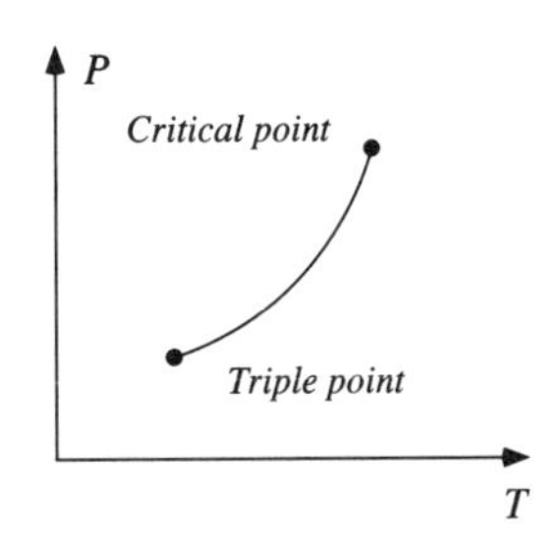

FIGURE 18. At vaporization, pressure and temperature of the fluid are directly related. *P* is called the vapor pressure at *T*.

Pressure-temperature relations for all three phase transitions. Most of the observations made about vaporization can be transferred to melting and sublimation (the direct transition between solid and vapor). Specific amounts of entropy are needed for these transformations (or are emitted for the reverse processes); the energy required is related to this quantity by the temperature of the phase transitions. The temperature stays constant while the process runs at constant pressure. And in each case, the pressure depends only upon the temperature for the change.

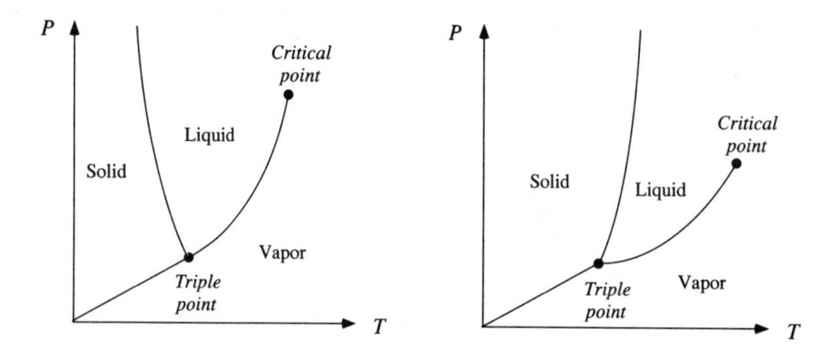

FIGURE 19. Phase diagrams for water (left) and other substances (right) show the pressure-temperature relations of the phase transitions solid-liquid, liquid-vapor, and solid-vapor. At the triple point, which has unique values of pressure and temperature, the three phases coexist. At temperatures (and pressures) above the critical point, liquid and vapor are indistinguishable.

Take the case of water freezing or ice melting. Even though it is harder to observe, changes of pressure affect the melting point; the reason for the difficulty lies in the fact that the pressure-temperature curve for melting is very steep (Figure 19). On the other hand, this effect leads to the phenomenon of freezing water being able to split rocks: water freezing at temperatures only slightly below 0°C develops a tremendous pressure. Having observed this, we have to conclude that the pressure-temperature relation for melting has a negative slope, unlike what we know of the vaporization of water, and, in fact, just about unlike anything else known in nature. Water expands upon freezing, while most other substances contract.

For the third phase transition, sublimation, an analogous pressure-temperature relation exists. The three lines meet in a single point, the so-called triple point, where all three phases coexist. Drawing the three $P(T)$ functions leads to the phase diagrams shown in the graphs above.

4.3.2 Phase Transformations and Chemical Equilibrium

Phase changes can be understood in terms of chemical reactions. If the chemical potential of a phase is higher at given pressure and temperature than that of another phase, the former will change into the latter. The conditions for the changes to occur can therefore be found by considering chemical equilibrium between the phases. This section will show how easily the chemical potential may be used to understand phenomena related to phase transitions.

Melting and vaporization as chemical reactions. Phase changes can be viewed as a particular kind of chemical transformation, subject to the same laws we have studied in Section 4.2. There we were able to calculate what happens as a consequence of a chemical reaction on the basis of the chemical potentials of the substances. Here, the substances are solid (*s*), liquid (*l*), and gaseous water (*g*). If you look at the following table, you see that at standard conditions, liquid water has the lowest chemical potential. This means that we should expect both ice and vapor to change into liquid water at a temperature of 25°C and a pressure of 1 atm.

TABLE 6. Properties of water at standard temperature and pressure

Phase	μ / kG	α_μ / G/K	β_μ / µG/Pa
Ice	- 236.59	- 44.8	19.7
Liquid water	- 237.18	- 69.9	18.1
Water vapor	- 228.60	- 189	24465

The values of the temperature coefficients of the chemical potential found in Table 6 allow us to compute approximate values for the temperatures of the melting point and the point of vaporization, if the pressure remains constant. Points of phase transitions obviously are those for which two phases coexist. If ice becomes liquid at a temperature of 25°C, because under this condition its chemical potential is larger than that of the liquid, we simply ask at which value of the temperature the tendency of water to change into ice has become equally large as the drive of ice to melt. In other words, we want to know the temperature for which the two chemical potentials have become equal:

$$\mu_{ice}\left(T_f, P_o\right) = \mu_{liquid}\left(T_f, P_o\right) \tag{80}$$

Since the potentials change with temperature according to Equation (28), the condition expressed in this equation becomes

$$\mu^o_{ice} + \alpha_{\mu,ice}\left(T - T_o\right) = \mu^o_{liquid} + \alpha_{\mu,liquid}\left(T - T_o\right) \tag{81}$$

Solving this simple equation, we obtain 274.7 K for the melting point of water. Considering that this is just a linear approximation, the result is quite acceptable.

Pressure dependence of the melting point. The problem of the change with pressure of the melting point and the temperature of vaporization is even more interesting, since it involves both changes in pressure and in temperature. Again, the chemical potentials of liquid and solid (or gaseous) water have to be equal at the actual melting point (or point of vaporization), and again the potentials change with temperature, and this time also with pressure. Therefore, the following condition must be satisfied:

$$\begin{aligned} &\mu^o_{ice} + \alpha_{\mu,ice}\left(T - T_o\right) + \beta_{\mu,ice}\left(P - P_o\right) \\ &\quad = \mu^o_{liquid} + \alpha_{\mu,liquid}\left(T - T_o\right) + \beta_{\mu,liquid}\left(P - P_o\right) \end{aligned} \tag{82}$$

For simplicity, take as the reference temperature the value of the melting point T_f at the reference pressure, in which case the standard potentials of ice and water are equal. With this in mind, we arrive at a relation between the change of pressure and temperature:

$$\frac{\Delta P}{\Delta T} = -\frac{\alpha_{\mu,liquid} - \alpha_{\mu,ice}}{\beta_{\mu,liquid} - \beta_{\mu,ice}} \tag{83}$$

Using the values for water from Table 6, we obtain the interesting result that the melting point of ice decreases with increasing pressure (with the corresponding values at 0°C we get the result of $\Delta P/\Delta T = -135$ bar/K). This is so since the pressure coefficient of ice is larger than that of water, a condition which holds only for water, and maybe one or two other substances. Generally, the temperature of the melting point increases with increasing pressure. The same is true for the temperature of vaporization, which increases if more pressure is applied. This holds also for water: water boils at higher temperature if the pressure is increased (see the discussion about vaporization at different pressures presented below).

In Section 4.2.4, we saw that at least for the bodies we have studied, the temperature coefficient of the chemical potential is the negative molar entropy of the substance, while the pressure coefficient is the molar volume. Applying this and writing l and s for liquid and solid, respectively, Equation (83) becomes

$$\frac{\Delta P}{\Delta T} = \frac{\bar{s}_l - \bar{s}_s}{\bar{v}_l - \bar{v}_s} \tag{84}$$

This result is of much more general importance than could be guessed from the derivation. It is called *Clapeyron's law,* and it shows that the change of pressure accompanying a change of temperature of phase transformation depends upon the entropy necessary for the phase change and the change of volume of the fluid. A derivation of this result based on a cyclic process will be presented below. Note that you can apply Equation (84) to vaporization as well. Since the change of volume is much larger in vaporization, the change of pressure with a change of boiling point should be much less dramatic than what we have observed for melting ice.

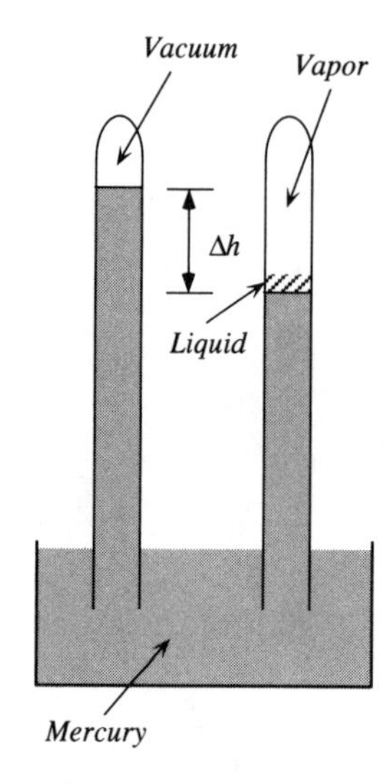

FIGURE 20. Detecting the pressure of the vapor which forms from the liquid at the top of a mercury gauge.

Vapor pressure. What is it that makes the states of saturated liquid or vapor special? What is the relation between the pressure of the fluid and the temperature at which it changes its phase? To answer these questions, consider some liquid put into an otherwise empty container. It is easy to see what happens if you place a drop of the liquid at the top of the mercury column in a pressure gauge (Figure 20). Even though the weight of the drop is so small as not to add to the pressure of the column of mercury, the top of the latter is observed to go down; obviously, there is some fluid in the previously empty space at the top of the gauge which has a noticeable pressure. We interpret this observation by assuming that a part of the drop has vaporized, and the pressure of the vapor is responsible for the change of height of the mercury column. Naturally, the pressure of the remaining liquid at the top of the gauge is the same as that of the vapor, and this pressure is called the *vapor pressure* of the fluid. It is further observed that the vapor pressure depends only upon the temperature of the fluids involved, as long as there always is at least a little bit of liquid left. How can we understand this relation between the temperature of the fluids and the vapor pressure?

For a given temperature, after the physical variables have assumed constant values, the situation of the fluid and its vapor is that of an insulated system inside a container (Figure 21). Since volume and energy are kept constant, the only possible changes inside the container are the transformation of vapor into liquid (or vice versa), and the production of entropy. According to Equation (42), the rate of production of entropy is given by

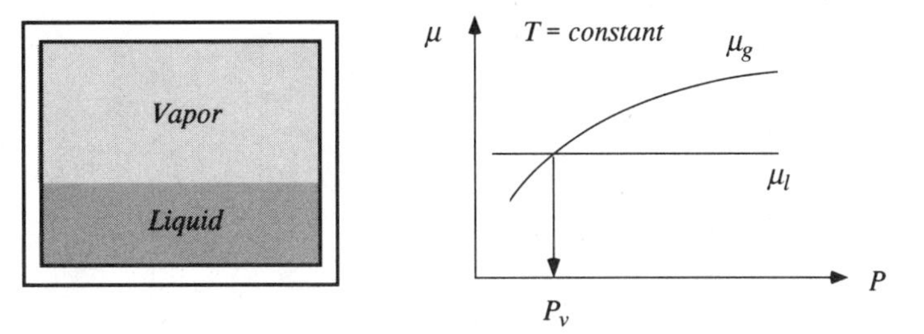

FIGURE 21. The chemical potential of the vapor inside the container changes much more strongly with pressure than the liquid. Therefore, it is easy to conceive of conditions for which μ_g is either smaller or larger than μ_l. In the latter case, some of the vapor will condense. When the chemical potentials are equal, the phases are in equilibrium.

$$\Pi_s = \frac{1}{T}\left|\mu_g - \mu_l\right|\left|\Delta n\right| \tag{85}$$

Assume for the moment that the chemical potential of the vapor is larger than that of the liquid. In this case, we expect some of the substance to go from the vapor into the liquid phase. Since this process will stop when equilibrium conditions have been reached, we have to conclude that the chemical potentials of the vapor and of the liquid phase have to be equal in equilibrium; it is equally likely for liquid to change into vapor as it is for vapor to condense:

$$\mu_g(P,T) = \mu_l(P,T) \tag{86}$$

We can estimate the value of the vapor pressure of water at a temperature of 25°C from the values found in Table 6. With the temperature given, the values of the chemical potential depend only upon pressure. Since, in Table 6, they are given for the standard pressure of 1 atm, we have to change the potentials to the condition of the still unknown vapor pressure. If we treat water vapor as an ideal gas, and water as an incompressible fluid, we have

$$\mu_g^o + RT\ln\left(\frac{P_v}{P_o}\right) = \mu_l^o + \left(P_v - P_o\right)\overline{v}_{lo} \tag{87}$$

as the condition of chemical equilibrium between liquid water and its vapor at 25°C. If you plug in some values, you will notice that the pressure term for the liquid is very small compared to the other terms, which means that the actual pressure of the liquid does not influence the result greatly. (For example, giving the liquid a larger pressure than the value P in the diagram of Figure 21 will shift

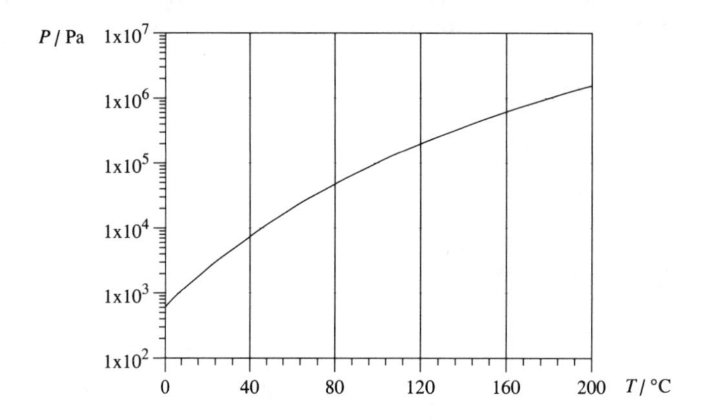

FIGURE 22. The vapor pressure of water rises rapidly with increasing temperature. The values must be obtained from measurements, since the fluids do not obey simple constitutive relations.

the horizontal line upward by a very small amount.) The important parameter is the vapor pressure. Solving the condition for the value of P_v yields

$$\ln\left(\frac{P_v}{P_o}\right) = -\frac{1}{RT}\left(\mu_g^o - \mu_l^o\right) \quad \textbf{(88)}$$

With the values taken from Table 6, the vapor pressure of water at 298 K is estimated to be 3150 Pa, which is close to the measured value of 3169 Pa (Table A.16).

Interpreting the condition of equilibrium in Equation (86), we conclude that there exists a particular relation between temperature and pressure as long as there is equilibrium between the phases. The pressure of the vapor cannot be changed independently of temperature as in the case of a single phase gas with constant amount of substance. The main difference is that here the gas is in equilibrium with its condensate. When the temperature is increased, more vapor is added to the vapor phase; when the temperature is decreased, some of the vapor can change into the liquid phase. This free exchange of amount of substance is responsible for the fact that the vapor pressure of a substance is a unique function of temperature. The answer to the question posed at the beginning of the section can now be given: the points of the bell-shaped curve in Figure 17 are those for which the chemical potentials are the same at a given temperature. They correspond to the conditions of chemical equilibrium between liquid and vapor.

Estimate of the temperature dependence of vapor pressure. We can get a first impression of the dependence of the vapor pressure on temperature, if we use the formulas for the temperature dependence of the ideal gas and incompressible fluids from Section 4.2.4. In other words, we treat the vapor as an ideal gas, and the liquid as incompressible. With Equation (63), the chemical potentials of the two fluids are

$$\mu_g = \mu_g^o + RT\ln\left(\frac{P_v}{P_o}\right) + \bar{c}_{pg}(T - T_o) - T\bar{c}_{pg}\ln\left(\frac{T}{T_o}\right) - (T - T_o)\bar{s}_{og}$$
$$\mu_l = \mu_l^o + (P_v - P_o)\bar{v}_{lo} + \bar{c}_{pl}(T - T_o) - T\bar{c}_{pl}\ln\left(\frac{T}{T_o}\right) - (T - T_o)\bar{s}_{ol} \qquad (89)$$

In addition to the pressure term in the chemical potential of the liquid, the terms with the temperature coefficients of enthalpy are relatively small compared to the entropy terms. Accepting this additional approximation, the condition of equilibrium leads to

$$\ln\left(\frac{P_v}{P_o}\right) = -\frac{1}{RT}\left\{\left(\mu_g^o - \mu_l^o\right) - (T - T_o)\left(\bar{s}_{og} - \bar{s}_{ol}\right)\right\} \qquad (90)$$

We can use this result to compute the temperature for which the vapor pressure is just the standard pressure of 1 atm. The values of Table 6 yield a temperature of 97°C, which is acceptably close to the actual value considering the approximations made.

The Clapeyron equation. Considering that the vapor pressure is a function of temperature only, its values could be found, if in addition to a starting value, the change of vapor pressure with temperature were known. As you will see shortly, the laws of thermodynamics provide for a relation between the derivative of the vapor pressure with respect to temperature and changes of properties of the fluid. The initial value is delivered by a single observation such as the one which tells us that at a temperature of 100°C the vapor pressure of water must be 1 atm.

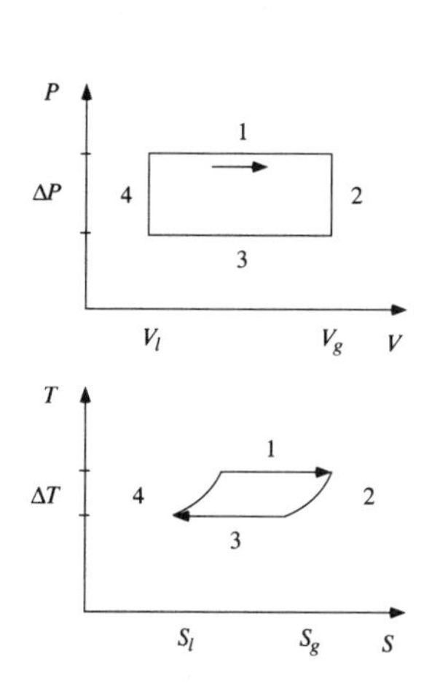

FIGURE 23. A pure simple fluid is allowed to go through a cycle including vaporization and condensation. The energy released in the fall of entropy is equal to the energy exchanged in the mechanical process. The curves are to be found in the bell-shaped area, where a mixture of liquid and vapor exists.

In Section E.1.2, a formal derivation of the following result will be given. Here, we will consider a cyclic process of a substance which undergoes evaporation and condensation. Take a four step cycle which starts with the evaporation of a fluid at constant temperature and pressure (Figure 23). The change of volume is from V_l to V_g, while the entropy changes from S_l to S_g (both changes are positive). Next, let the gas be cooled just a little bit at constant volume so that the temperature decreases by a very small amount ΔT. At this slightly lower temperature, allow the vapor to condense; this process will again take place at constant pressure. The volume returns to V_l, and very nearly the same amount of entropy is emitted as was absorbed during vaporization. This will be the case if the second step is made so small that it does not add considerably to the overall balances. Finally, the liquid is brought back to the initial state by a slight heating at constant volume.

The cycle described is that of a heat engine. We can evaluate the energy released in the fall of entropy, and the energy used for the mechanical process, and equate them (remember that steps 2 and 4 of the cycle do not contribute much to the balances):

$$\left(\bar{s}_g - \bar{s}_l\right)\Delta T = \left(\bar{\upsilon}_g - \bar{\upsilon}_l\right)\Delta P \qquad (91)$$

This is the relation between changes of temperature and changes of vapor pressure we have been looking for; it can be brought into the form known as *Clapeyron's equation*:

$$\frac{dP}{dT} = \frac{\bar{s}_g - \bar{s}_l}{\bar{\upsilon}_g - \bar{\upsilon}_l} \qquad (92)$$

If you wish, you can express the change of the entropy from the liquid to the vapor state using the enthalpy of the fluid, Equation (79).

EXAMPLE 18. Change of melting point of water with a change of pressure.

Determine the change of pressure accompanying a change of the melting point of ice.
a) Calculate the temperature and the pressure coefficients of the chemical potential of water and ice for 0°C, and then use Clapeyron's equation. The temperature coefficients of enthalpy of water and of ice are 4200 J/(K · kg) and 2100 J/(K · kg), respectively. b) Use the fact that the energy needed to melt ice is 334 kJ/kg.

SOLUTION: a) Modeling water and ice as bodies with constant temperature coefficients of enthalpy, we can calculate their change of entropy from standard temperature to 0°C:

$$\bar{s}\left(T_f\right) = \bar{s}_o + \bar{c}_p \ln\left(\frac{T_f}{T_o}\right)$$

Considering that the molar entropy is the negative temperature coefficient of the chemical potential given in Table 6, we obtain for it:

$$\bar{s}(0°\mathrm{C}) = 44.8 + 0.018 \cdot 2100 \cdot \ln\left(\frac{273}{298}\right) = 41.5 \frac{\mathrm{J}}{\mathrm{K \cdot mole}}$$

For liquid water, the figure is 63.3 J/(K · mole). Note that the pressure coefficients of the chemical potential, i.e., the molar volumes, of ice and of water will not change much from their values in Table 6. Now, with Equation (83), the result is

$$\frac{\Delta P}{\Delta T} = -\frac{\alpha_{\mu,liquid} - \alpha_{\mu,ice}}{\beta_{\mu,liquid} - \beta_{\mu,ice}} = -\frac{-63.3 - (-41.5)}{(18.1 - 19.7) \cdot 10^{-6}} \frac{\mathrm{Pa}}{\mathrm{K}} = -1.36 \cdot 10^{7} \frac{\mathrm{Pa}}{\mathrm{K}}$$

b) Starting with Clapeyron's equation in the form of Equation (84), we see that the right-hand side can be expressed in terms of the energy supplied to ice as it melts. The entropy and the energy necessary for melting are related by Equation (79), which means that Clapeyron's equation can be written as follows:

$$\frac{\Delta P}{\Delta T} = \frac{\bar{s}_l - \bar{s}_s}{\bar{\upsilon}_l - \bar{\upsilon}_s} = \frac{1}{T_f} \frac{\bar{h}_l - \bar{h}_s}{\bar{\upsilon}_l - \bar{\upsilon}_s} = \frac{1}{T_f} \frac{\Delta \bar{h}_{fusion}}{\bar{\upsilon}_l - \bar{\upsilon}_s}$$

With the value of the latent enthalpy of fusion, we obtain – 138 bar/K for $\Delta P/\Delta T$.

EXAMPLE 19. Changes of vapor pressure and enthalpy of vaporization.

a) Express the approximate result for vapor pressure as a function of temperature in terms of the change of the enthalpy of vaporization. Use as a reference the temperature of vaporization at 1 atm. b) Determine the vapor pressure and the standard chemical potential of mercury vapor at 25°C.

SOLUTION: a) First, with a pressure of 1 atm and its corresponding temperature of vaporization as the reference point, the standard chemical potentials of vapor and liquid are equal. Using the relation between the entropy and the enthalpy of vaporization leads to

$$\ln\left(\frac{P_v}{P_o}\right)=\frac{1}{R}\left(1-\frac{T_{bp}}{T}\right)\left(\bar{s}_{bp,g}-\bar{s}_{bp,l}\right)=\frac{1}{R}\left(\frac{1}{T_{bp}}-\frac{1}{T}\right)\Delta\bar{h}_{ev,bp}$$

where *bp* stands for *boiling point*. This means that a measurement of the enthalpy of vaporization yields fundamental information about a fluid (as an ideal gas).

b) We can apply the result just obtained with T_{bp} = 630 K, and 57 kJ/mole for the molar enthalpy of vaporization. This leads to

$$\ln\left(\frac{P_v}{P_o}\right)=\frac{1}{8.31}\left(\frac{1}{630}-\frac{1}{298}\right)57000=-12.1$$

or P_v = 0.54 Pa. This value, in turn, can be used to calculate the chemical potential of gaseous mercury at 25°C and 1 atm. We use the same derivation which led to Equation (88):

$$\ln\left(\frac{P_v}{P_o}\right)=-\frac{1}{8.31\cdot 298}\left(\mu_g^o-0\right)$$

Remember that the standard value of the chemical potential of liquid mercury is set equal to zero. This gives a value of + 30.0 kG (compared to the more accurate figure of 31.84 kG).

EXAMPLE 20. The Clausius-Clapeyron equation.

Derive the expression of the Clapeyron equation as it holds for the ideal gas. Show that this is equivalent to the approximation of the pressure-temperature relation for vapor pressure derived in Equation (90) or above in Example 19. Assume the enthalpy of vaporization to be constant.

SOLUTION: Written using the enthalpy of vaporization, Clapeyron's equation becomes

$$\frac{dP}{dT}=\frac{\bar{s}_g-\bar{s}_l}{\bar{v}_g-\bar{v}_l}=\frac{1}{T}\frac{\bar{h}_g-\bar{h}_l}{\bar{v}_g-\bar{v}_l}=\frac{1}{T}\frac{\Delta\bar{h}_{ev}}{\bar{v}_g-\bar{v}_l}$$

For the ideal gas, we use the equation of state, and we neglect the volume of the liquid phase compared to its value in the vapor phase. Therefore, the equation becomes

$$\frac{dP}{dT} = \frac{1}{T}\frac{\Delta \bar{h}_{ev}}{\bar{\upsilon}_g} = \frac{1}{T}\frac{\Delta \bar{h}_{ev}}{RT/P} = \frac{P}{RT^2}\Delta \bar{h}_{ev}$$

Integration of this equation yields

$$\ln\left(\frac{P}{P_o}\right) = \frac{1}{R}\left(\frac{1}{T_o} - \frac{1}{T}\right)\Delta \bar{h}_{ev}$$

which is equivalent to what was derived before. Clapeyron's equation specialized to the ideal gas is called the Clausius-Clapeyron equation.

4.3.3 The Chemical Potential of Dilute Solutions

In Section 4.1.2 we studied dilute solutions and the phenomenon of osmosis. Now we will take up the discussion once again; this time, however, we will be able to compute the chemical potentials involved. As a consequence, the effects of a solute on the vapor pressure, the melting and the boiling point, and the solubility of gases in water can be calculated.

The chemical potentials of solvent and solute. First, we should note how dissolving a little bit of a substance in a solvent decreases the latter's chemical potential. This effect is understood quite readily: since the total pressure of the solvent plus solute remains constant, the pressure of the solvent alone must have decreased by the amount by which the pressure of the dissolved substance has increased. Since we expect the chemical potential of a substance to decrease with decreasing pressure, we should obtain a lowering of the chemical potential of the solvent.

Formally, the derivation proceeds as follows. If we write P for the total pressure of the solution and use the index l (liquid) for the solvent, we have

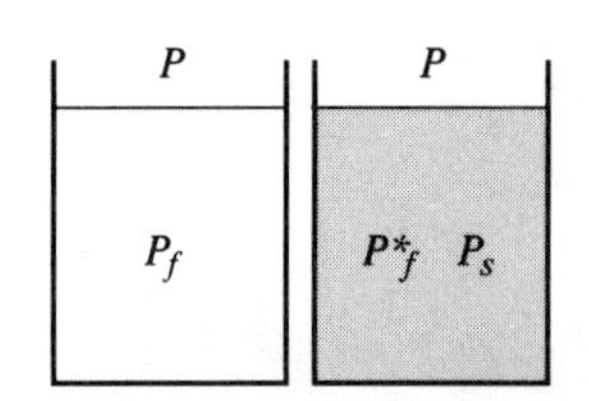

FIGURE 24. Two containers with the same solvent. On the left, we have only the pure solvent. Its pressure is equal to the ambient pressure. In the solution on right, the pressure is divided up among the two constituents.

$$P_l = P - P_s \tag{93}$$

where P_s is the pressure of the dissolved substance (the solute). Since the solvent is assumed to be an incompressible fluid, its chemical potential is expressed simply in terms of its pressure, Equation (55):

$$\mu_l(P) = \mu_l(P_o) + (P - P_o)\bar{\upsilon}_l \tag{94}$$

Applied to the solvent this becomes

$$\mu_l(T, P_l) = \mu_l(T, P_{lo}) + \left((P - P_s) - (P - 0)\right)\bar{\upsilon}_l \tag{95}$$

or

$$\mu_l(T, P_l) - \mu_l(T, P_{lo}) = -\bar{\upsilon}_l \cdot P_s \tag{96}$$

It is customary to introduce the mole fraction of the solute. Since the amount of the substance dissolved is taken to be small compared to the amount of substance of the solvent, this quantity is nearly equal to

$$y = \frac{n_s}{n_l + n_s} \approx \frac{n_s}{n_l} = \frac{\overline{v}_l}{\overline{v}_s} \tag{97}$$

If we accept that the solute obeys the equation of state of the ideal gas (as we discussed in Section 4.1.2), we arrive at the final result

$$\mu_l(T, P_l) = \mu_l(T, P_{lo}) - RTy \tag{98}$$

for the dependence of the chemical potential of the solvent on the amount of substance dissolved in it.

Next, we will compute the chemical potential of the solute as a function of its concentration at constant temperature. This is an easy task, since we know the equation of state of the dissolved substance: it is that of the ideal gas. The solute behaves like a gas all by itself with a vacuum as its "solvent"; therefore we can treat it just like an ideal gas in empty space, which means that its chemical potential must obey the relation

$$\mu_s(T, \overline{c}) = \mu_s(T, \overline{c}_o) + RT \ln\left(\frac{\overline{c}}{\overline{c}_o}\right) \tag{99}$$

Since concentration and pressure are proportional at constant temperature, this result is equivalent to what was derived in Equation (58).

The effect of the solute on the solvent. The dissolved substance has a few interesting and important effects upon the solvent: it changes its vapor pressure and its melting and boiling points. We can now understand how this happens.

Beginning with the vapor pressure of the solvent, we note that the chemical potentials of the liquid and its vapor must be equal. In the context of a pure liquid this condition led to Equation (87). The only difference as a result of the presence of the solute is the extra term calculated above in Equation (98). If we neglect the pressure term of the solvent, the relation reads

$$\mu_g^o + RT \ln\left(\frac{P_v}{P_o}\right) = \mu_l^o - RTy \tag{100}$$

Putting sugar into water so that there exists a solution with a mole fraction of 0.05 for the solute will decrease the vapor pressure from 3120 Pa to 2960 Pa. Remember that the standard temperature and pressure taken for this equation are 25°C and 1 atm.

Lowering the vapor pressure must have an influence on the melting and the boiling points of the liquid. In fact, the temperature of the melting point is decreased, while that of the boiling point is increased. Here we would only like to demonstrate the derivation for the melting point and apply the result to sea water. The condition of chemical equilibrium at the new melting point means that

$$\mu_s^o + \alpha_{\mu s}\Delta T = \mu_l^o + \alpha_{\mu l}\Delta T - RTy \tag{101}$$

On the left-hand side of this equation we have the chemical potential of the solid (s) substance, while on the right we have the values for the solution. If we take as the reference temperature the value for the melting point, then the standard values of the chemical potentials μ^o on the left and the right are equal. Solving for the change of temperature then leads to

$$\Delta T = -RT\frac{y}{\alpha_{\mu s} - \alpha_{\mu l}} = -RT\frac{y}{\bar{s}_l - \bar{s}_s} \tag{102}$$

Since the entropy of the liquid is larger than that of the solid, the change of temperature of the melting point is negative. If we apply this to sea water with a mole fraction of 1 mole of ions for 55.5 mole of water, the lowering of the melting point is 1.86 K. Sea water freezes at about - 2°C. Note that we need the values of the entropy of ice and water at 0°C which were computed in Example 18.

The solubility of gases in water. Another application of the idea of chemical equilibrium of two phases of a substance is the determination of the solubility of gases in a solvent such as water. This case is important in environmental sciences where we would like to be able to calculate the solubility of oxygen or of carbon dioxide in water, or in medicine where we are interested in the solubility of gases in blood.

Consider some water with air above it. There should be some oxygen dissolved in the water. The exact amount is determined by the relative tendencies of dissolved oxygen to go into the gas phase, and of oxygen in the air to dissolve. In other words, we are dealing with another case of chemical equilibrium.

Take standard values of temperature and pressure, and of concentration (1 mole/l). Then, the chemical potential of oxygen in air depends upon its partial pressure P_g, while that of oxygen dissolved in water changes as a function of the actual concentration c_{aq}. The condition of equilibrium states that

$$\mu_g^o + RT\ln\left(\frac{P_g}{P_o}\right) = \mu_{aq}^o + RT\ln\left(\frac{\bar{c}_{aq}}{\bar{c}_o}\right) \tag{103}$$

The index *aq* stands for *aqueous* solution. Knowledge of the standard potentials for gaseous and for dissolved oxygen let us calculate the solubility of the gas in water. (Some values can be found in Table A.13.) For oxygen in water we find a concentration of $2.7 \cdot 10^{-4}$ mole/l, about 3600 times less than the reference value.

EXAMPLE 21. The effect of temperature on the solubility of CO_2.

a) Calculate the standard value of the concentration of carbon dioxide in water. (The concentration of CO_2 in air is about 350 ppm.) b) Determine the effect of changes of temperature upon the solubility. Take an increase of the temperature of 5°C.

SOLUTION: a) We can directly apply Equation (103) to the first problem. The concentration of CO_2 is proportional to its partial pressure, which means that $P_g/P_o = 3.5 \cdot 10^{-4}$. Therefore,

$$\frac{\bar{c}_{aq}}{\bar{c}_o} = \exp\left[\frac{\mu_g^o - \mu_{aq}^o}{RT} + \ln\left(\frac{P_g}{P_o}\right)\right] = 1.18 \cdot 10^{-5}$$

b) The effect of temperature can be taken into account via the chemical potentials. Since the major contribution comes from the molar entropy, we can write

$$\mu_g^o + RT\ln\left(\frac{P_g}{P_o}\right) - \Delta T\bar{s}_g = \mu_{aq}^o + RT\ln\left(\frac{\bar{c}_{aq}}{\bar{c}_o}\right) - \Delta T\bar{s}_{aq}$$

Solving this condition for the concentration of the dissolved gas yields

$$\frac{\bar{c}_{aq}}{\bar{c}_o} = \exp\left[\frac{\mu_g^o - \mu_{aq}^o - (\bar{s}_g - \bar{s}_{aq})\Delta T}{RT} + \ln\left(\frac{P_g}{P_o}\right)\right]$$

$$= \exp\left[\frac{\mu_g^o - \mu_{aq}^o}{RT} + \ln\left(\frac{P_g}{P_o}\right)\right]\exp\left(-\frac{(\bar{s}_g - \bar{s}_{aq})\Delta T}{RT}\right) = 1.02 \cdot 10^{-5}$$

which corresponds to a decrease of 14% in solubility for an increase in temperature of 5°C.

EXAMPLE 22. Henry's law.

The problem of the solubility of gases in water often is expressed in terms of the vapor pressure of the dissolved gas. It is found that the vapor pressure is proportional to the mole fraction of the solute, which is Henry's law:

$$P_g = K_x x$$

Express the constant K_x using the chemical potentials.

SOLUTION: We begin by transforming Equation (103), which yields

$$\mu_g^o - \mu_{aq}^o = RT\ln\left(\frac{\bar{c}_{aq}}{\bar{c}_o}\frac{P_o}{P_g}\right)$$

Now, the expression of Henry's law is introduced, and the definition of the mole fraction is applied. For dilute solutions this leads to

$$\mu_g^o - \mu_{aq}^o = RT\ln\left(\frac{\bar{c}_{aq}}{\bar{c}_o}\frac{P_o}{K_x \cdot x}\right) = RT\ln\left(\frac{\bar{c}_{aq}}{\bar{c}_o}\frac{P_o \cdot n_l}{K_x \cdot n_s}\right)$$

$$= RT\ln\left(\frac{n_s}{V}\frac{P_o}{\bar{c}_o}\frac{1}{K_x}\frac{n_l}{n_s}\right) = RT\ln\left(\frac{P_o}{\bar{c}_o}\frac{1}{K_x}\frac{n_l}{V}\right)$$

where n_l / V is the concentration of the solvent. (For water, it is 55500 mole/m^3.) From this we obtain the formula for K_x:

$$\ln(K_x) = -\frac{\mu_g^o - \mu_{aq}^o}{RT} + \ln\left(\frac{P_o}{\bar{c}_o}\frac{n_l}{V}\right)$$

For carbon dioxide dissolved in water, K_x is $1.66 \cdot 10^8$ Pa, while for hydrogen gas, it turns out to be $7.10 \cdot 10^9$ Pa. The latter number means that in an atmosphere of hydrogen at 1 atm, the mole fraction of dissolved gas would be $1.4 \cdot 10^{-5}$.

4.3.4 Mixtures of Two-Phase Fluids: the Case of Moist Air

In this section we will discuss one more application of the ideas of chemical transformation, namely the thermodynamics of mixtures of two ideal gases when one can undergo phase changes. Moist air furnishes such an example—dry air and water vapor. While the air itself is a mixture, it can be treated as a single component, since in the range of temperature and pressure found in our atmosphere, its constituents all behave identically as ideal gases; in other words, we do not notice that dry air itself is composed of different parts. The case of the second component, however, is quite different. With liquid (or solid) water present as well, water vapor in the air can condense (or directly turn into frost or ice); or water can evaporate (and ice can sublimate). Thus, the conditions in the atmosphere are such that water can undergo phase transitions. We are all familiar with water condensing on window panes on a cold day, droplets forming on a pipe carrying cold water, dew accumulating on the grass in the morning, or frost forming on the ground.

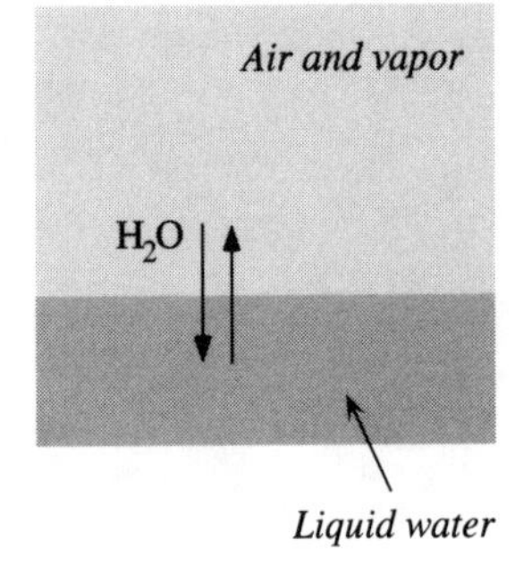

FIGURE 25. If a mixture of dry air and water vapor is present together with liquid water, the vapor component of the gas can undergo phase transitions. At chemical equilibrium, the air is said to be saturated with water vapor.

These and the reverse processes can again be understood as chemical phenomena. When vapor condenses out of moist air, this simply means that the chemical potential of the water vapor is larger under the given circumstances than that of the liquid phase. On the other hand, when water evaporates from the surface of a pond, the tendency of liquid to go into the vapor phase is stronger than the reverse drive.

We can use our knowledge of the concepts of vapor pressure and chemical equilibrium between phases to state these observations in more precise language. Consider evaporation of water into the air above a liquid surface (as in Figure 25). Obviously, if water diffuses from the liquid into the air, chemical equilibrium has not been attained. There is less water vapor in the air than there could be under the condition of equilibrium, or, put differently, the gas above the water surface is said to have a humidity which is smaller than the humidity at saturation. When the air is saturated with water vapor, the chemical potential of the vapor phase and of the liquid phase are equal, and there are equal rates of water evaporating and vapor condensing.

To give you an impression of the amount of water vapor present in the air, consider standard conditions of temperature and pressure. You know that at 25°C, the vapor pressure of water is about 3000 Pa or 0.030 bar. As we have seen before, it hardly matters that the pressure of the liquid is 1 atm because of the presence of the air: the chemical potential of liquid water only changes slightly as a function of pressure. Therefore, the value of the vapor pressure calculated under the assumption that only water is present can be used perfectly well in the present case. We can now say that at saturation the amount of water in the air is 0.03 mole per 1 mole of air (equivalent to about 20 g of H_2O in 1 m^3 of air). With such numbers, it is clear that the vapor component also behaves as an ideal gas. As far as the body of moist air is concerned, we can treat it as a mixture of two ideal gases and apply the rules found in Section 4.1.3.

Description of the state of moist air. It is customary to define two interrelated quantities with whose help the state of moist air is described. They are the *humidity ratio* (specific humidity or absolute humidity), and the *relative humidity.* First, it should be clear that the pressure of the mixture is P, while the partial pressures of dry air and of vapor are P_a and P_v, respectively, with

$$P = P_a + P_v \tag{104}$$

The humidity ratio ω is defined as the ratio of the mass of water vapor m_v and the mass of dry air m_a:

$$\omega \equiv m_v / m_a \tag{105}$$

Using the equation of state of the ideal gas this can be transformed into

$$\omega = \frac{M_{ov} P_v}{M_{oa} P_a} = 0.633 \frac{P_v}{P - P_v} \tag{106}$$

The numerical value 0.633 applies for dry air and water. The second quantity introduced, the relative humidity ϕ, is defined as the ratio of the amount of water vapor actually present and the amount present when the air is saturated with vapor:

$$\phi \equiv n_v / n_{v,sat} \tag{107}$$

The laws of ideal gases let us change this to

$$\phi = \frac{y_v}{y_{v,sat}} = \frac{P_v}{P_g} \tag{108}$$

where P_g is the saturation vapor pressure at the temperature under consideration. Note that the relative humidity and the humidity ratio are convenient definitions for expressing the case of moist air in terms of the laws of mixtures of ideal gases introduced in Section 4.1.3. As far as physics is concerned, nothing new has been formulated. What we can say, however, is that at a relative humidity smaller than 1, the chemical potential of the vapor μ_v is smaller than its value for the condition at saturation μ_g, the change stemming from the difference in actual vapor pressure from the saturation pressure. Therefore

$$\mu_v = \mu_g + RT \ln \phi \tag{109}$$

Since the temperature is the same for the actual condition and for the assumed condition of saturation, the enthalpy of the vapor is what it would be at saturation. This holds for an ideal gas; see Figure 48 for actual values:

$$h_v = h_g \tag{110}$$

Therefore, the entropy of the vapor deviates from its value at saturation by

$$\bar{s}_v = \bar{s}_g - R \ln \phi \tag{111}$$

The dew point. Assume a state of moist air with a relative humidity less than the saturation value of 1. This means that the chemical potential of the vapor at the given temperature is smaller than the potential of the liquid phase (Figure 26). Now what happens if the temperature of the mixture and the liquid water decreases? The chemical potentials of the vapor and the liquid phase increase, but not at the same rate. From Table 6, we see that the magnitude of the slope for the gaseous phase is larger, which means that the potentials become equal at a particular value of the temperature. At this point, water vapor in the air will begin to condense, and, for obvious reasons, this temperature is called the *dew point*. Using a table of saturation vapor pressure, dew points are computed easily (Example 23). Here, we will give a formal derivation assuming the validity of the ideal gas model.

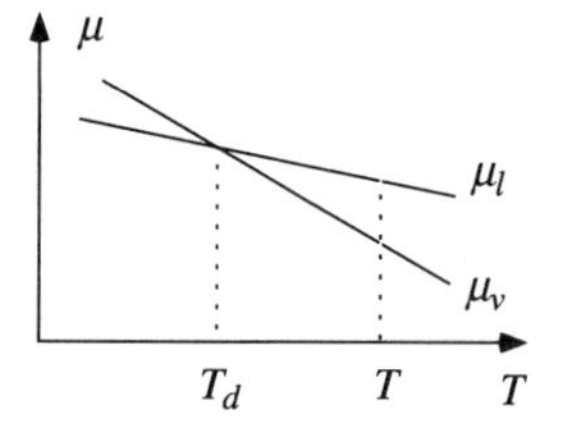

FIGURE 26. At a value of the relative humidity which is less than 1, the chemical potential of the vapor is smaller than that of liquid water. Decreasing the temperature leads to a condition for which the potentials are equal: condensation of water vapor sets in.

We have to find the temperature for which the chemical potentials of the vapor in the air and of the liquid water phase are equal, given a starting condition at temperature T_o and pressure P_o. The chemical potential of the liquid changes from the standard value only because of the change of temperature:

$$\mu_l(T_d) = \mu_l^o - (T_d - T_o)\bar{s}_{ol} \tag{112}$$

The potential of the vapor, on the other hand, is different from the standard value for two reasons: the pressure P_v at temperature T_o is different from P_o; and the temperature is now at the dew point:

$$\mu_v(T_d) = \mu_g^o + RT_d \ln\left(\frac{P_v}{P_o}\right) - (T_d - T_o)\bar{s}_{og} \tag{113}$$

Now, the actual vapor pressure P_v can be expressed in terms of the saturation value if we introduce the relative humidity:

$$\mu_v(T_d) = \mu_g^o + RT_d \ln\left(\frac{\phi P_{go}}{P_o}\right) - (T_d - T_o)\bar{s}_{og}$$
$$= \mu_g^o + RT_d \ln\left(\frac{P_{go}}{P_o}\right) + RT_d \ln\phi - (T_d - T_o)\bar{s}_{og} \tag{114}$$

Equating the chemical potentials, i.e., Equations (112) and (114), furnishes a nonlinear condition for the unknown temperature of the dew point. We need only the saturation pressure P_{go} at the original temperature T_o, which can be calculated from the approximation given in Equation (90). The formulas are rather accurate, and compare well with the measured values found in steam tables. Remember that, apart from the assumption of validity of the ideal gas model, we have neglected some terms in the temperature dependence of the chemical potentials.

Wet bulb temperature and adiabatic saturation. The condition of an ideal gas of constant amount of substance is fixed by two values such as the pressure and the temperature. A third value is obviously needed if a mixture such as moist air is described for which the amount of vapor can change. A suitable third value might be the humidity ratio which sets the amount of water in the air. However, humidity is not so readily measured, which is why one would like to be able to specify an additional temperature. Here, we will show how the temperature measured by a thermometer whose bulb is surrounded by a wet wick, can be used to specify the humidity of the air.

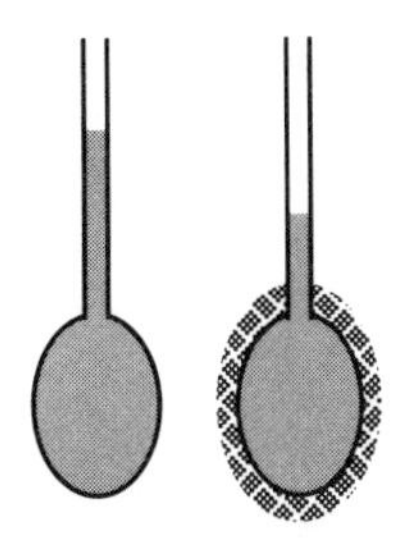

FIGURE 27. Dry and wet bulb thermometers for measuring the humidity of air. A wick saturated with water surrounds the second thermometer. If the air is not saturated, this thermometer will show a lower temperature than the dry bulb instrument.

The "normal" temperature measured by a simple thermometer is called the dry bulb temperature of the moist air. If you surround the bulb of the thermometer with a wick saturated with water, it will show a lower temperature for the same conditions of the air, the value being called the wet bulb temperature (Figure 27). The reason for the lower temperature is this. As the air, which is assumed to be not saturated, passes over the bulb with the wick, water will evaporate, increasing the humidity of the air. Indeed, we will assume the air to become saturated by the process. As a steady state is attained, the entropy necessary to vaporize additional water cannot come from the rest of the water (since its temperature remains constant). Rather, it has to come from the air itself, which means that the temperature of the stream of air leaving the wick is lower than the normal air temperature. This new temperature is the temperature taken by the water in the wick; therefore, the wet bulb thermometer shows the lower temperature of the saturated air.

Now, the process taking place here, which is a close approximation to what is called *adiabatic saturation*, is specified by the temperatures and humidities of the original moist air and of the saturated air. The value of the humidity ratio can then be calculated according to

$$\omega = \frac{h_a(T_{wb}) - h_a(T) + \omega'\left[h_g(T_{wb}) - h_l(T_{wb})\right]}{h_g(T) - h_l(T_{wb})} \tag{115}$$

where ω' is computed according to Equation (106) for the saturated air at the wet bulb temperature, i.e.,

$$\omega' = 0.633 \frac{P_g(T_{wb})}{P - P_g(T_{wb})} \tag{116}$$

The derivation of Equation (115) will be considered in Section 4.4.6 as an application of a simple flow process. There it will become clear that the condition of adiabatic saturation may be satisfied by the wet bulb thermometer only to a limited extent. The real wet bulb phenomenon is influenced by the rate of evaporation and the rate of diffusion through the wick. Still, the agreement is acceptable, and the wet bulb temperature is commonly used as the representation of the adiabatic saturation temperature.

EXAMPLE 23. Calculating a dew point using tabular data.

---

a) At a temperature of 30°C and standard pressure, the relative humidity of air is measured to be 75%. Determine the temperature at which water vapor would begin to condense. Use the data given in Table A.16. b) The temperature of the air drops to 20°C. What fraction of the vapor present in the air will condense?

SOLUTION: a) At 30°C, the saturation vapor pressure is 4246 Pa. With a relative humidity of 75%, the actual vapor pressure (the partial pressure of the water vapor in the air) is

$$P_v = \phi P_g = 0.75 \cdot 4246\,\text{Pa} = 3185\,\text{Pa}$$

The value of 3185 Pa almost precisely corresponds to the saturation vapor pressure at 25°C. This means that at 25°C the air would be saturated with the amount of water vapor it actually contains, and the vapor would begin to condense.

b) At 25°C, with a vapor pressure of 3185 Pa, the air could just retain the initial amount of water present at a relative humidity of 100%. At 20°C, however, the vapor pressure only is 2340 Pa. If we allow for 100% relative humidity at that state, the air could contain only the fraction 2340/3190 = 0.73 of the initial amount of water. An initial amount of

$$n_{vo}/V = \frac{P}{RT} = \frac{3184}{8.314 \cdot 298} \frac{\text{mole}}{\text{m}^3} = 1.29 \frac{\text{mole}}{\text{m}^3}$$

will be reduced to 0.94 mole/m^3. For every cubic meter of air, 6.3 g of water will condense.

EXAMPLE 24. Calculating the wet bulb temperature of moist air.

Take the same conditions for moist air as in Example 23 (75% relative humidity at 30°C and 1 atm). How large is the corresponding wet bulb temperature?

SOLUTION: In addition to the definition of relative humidity in terms of the vapor pressure and the saturation vapor pressure, we will need to simultaneously solve Equations (106), (190) and (120). For the stated problem, namely finding the wet bulb temperature, the equations are nonlinear. We will have to find the values of enthalpies and pressures for the quantities appearing in the equations, partly for the still unknown value T_{wb}. Below you will find the relations to be set up if the thermodynamic property data are given in the form of programmed functions. Compare the solution of these equations with a solution attempted using tabulated values.

Given data:
```
phi=0.75    T1=30        P1=101.3
```
Properties:
```
ha_T=Enthalpy(Air,T=T1)
ha_Twb=Enthalpy(Air,T=Twb')
hg_T=Enthalpy(Steam,T=T1,X=1),
hg_Twb=Enthalpy(Steam,T=Twb,X=1)
hl_Twb=Enthalpy(Steam,T=Twb,X=0)
Pg_Twb=Pressure(Steam,T=Twb,X=1)
Pg_T=Pressure(Steam,T=T1,X=1)
```
Relations:
```
w=((ha_Twb-ha_T)+w'*(hg_Twb-hl_Twb))/(hg_T-
hl_Twb)
w'=0.633*Pg_Twb/(P1-Pg_Twb)
w=0.633*Pv_T/(P1-Pv_T)
phi=Pv_T/Pg_T
```
Solution:
```
ha_T = 303.6 kJ/kgha_Twb = 299.9hg_T = 2556.8
hg_Twb = 2550.2hl_Twb = 110.2P1 = 101.3 kPa
Pg_T = 4.238Pg_Twb = 3.416phi = 0.750
Pv_T = 3.179T1 = 30.0°CTwb = 26.3
w = 0.0205 w' = 0.0221
```

The solution of the problem was performed completely within the program EES (Klein et al., 1991). The form of the property functions is pretty much self explanatory. *Air* stands for dry air, while *Steam* denotes water (liquid and vapor). The temperature and its wet bulb counterpart are denoted by `T` and `Twb`, respectively, while `w` is used for the humidity ratio. `X` is the quality (defined in Section 4.5.1; `X`=0 is for saturated liquid, while `X`=1 is for saturated vapor).

EXAMPLE 25. The height above ground of cloud formation

Make the following model of the vertical circulation of air. Moist air (not saturated) rises adiabatically due to convective instability. With a relative humidity of 0.50, and standard pressure and temperature at the ground, calculate the level above ground at which condensation of the water vapor should set in. Explain why this happens for a temperature which is lower than the dew point calculated for the ground.

TABLE 7. **Properties of moist air at constant entropy and humidity ratio**

P / kPa	T / K	υ / m³/kg	h / m	μ_l / kG	μ_v / kG
100	298.0	0.869	0	-0.0838	-1.7880
98	296.3	0.882	178	-0.0729	-1.5647
96	294.6	0.895	360	-0.0625	-1.3383
94	292.8	0.909	543	-0.0527	-1.1085
92	291.0	0.923	730	-0.0436	-0.8752
90	289.2	0.937	920	-0.0350	-0.6382
88	287.4	0.952	1110	-0.0271	-0.3976
86	285.5	0.968	1310	-0.0200	-0.1530
84	283.6	0.985	1510	-0.0135	-0.0135

SOLUTION: The quantities which are constant for the process are the entropy and the humidity ratio. (Before the onset of condensation, the amount of water vapor in the air remains constant.) One therefore should calculate these two quantities for the conditions at the ground. In functional form, this might look like

```
so=Entropy(AirH2O,T=298,P=100,R=0.5)
w=HumRat(AirH2O,T=298,P=100,R=0.5)
```

Tables (or computer programs) yield values of 5.788 kJ/(K · kg) and 0.00992, respectively. After this preparation, values of the temperature, and the specific volume can be taken from tables as a function of decreasing pressure. Again, using functions implemented in EES, we have

```
so=Entropy(AirH2O,T=T,P=P,w=w)
v=Volume(AirH2O,T=T,P=P,w=w)
```

Then, using the law of hydrostatic equilibrium, the volume can be integrated over the pressure to yield the height as a function of pressure:

$$\frac{dh}{dP} = -\frac{1}{g\rho} = -\frac{\upsilon}{g}$$

Finally, the chemical potential of liquid water and of water vapor have to be expressed:

```
mu_l=Mo*(Enthalpy(Water,T=T,X=0)-T*Entropy(Water, T=T,X=0))
mu_v=Mo*(Enthalpy(Water,T=T,P=Pv)-T*Entropy(Water, T=T,P=Pv))
w=0.622*Pv/(P-Pv)
```

Results are given in Table 7 and in the accompanying graph. The graph shows that, up to a certain height (about 1400 m, temperature of 284.5 K), the chemical potential of the vapor is smaller than that of liquid water (the chemical potential of liquid water increases because of the effect of decreasing temperature). At this point, the relative humidity becomes 1. We therefore expect cloud formation to begin at about this level above ground.

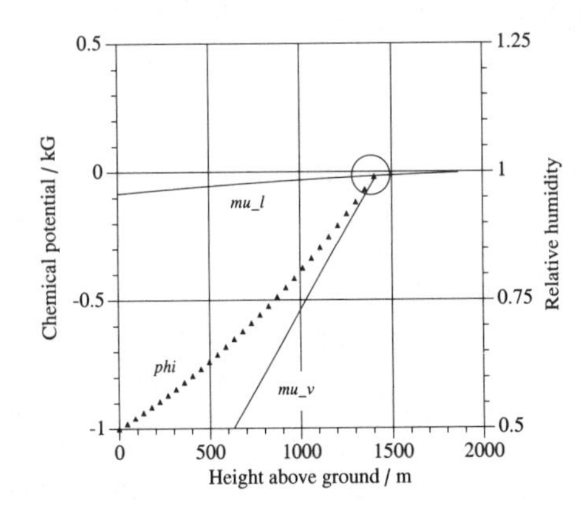

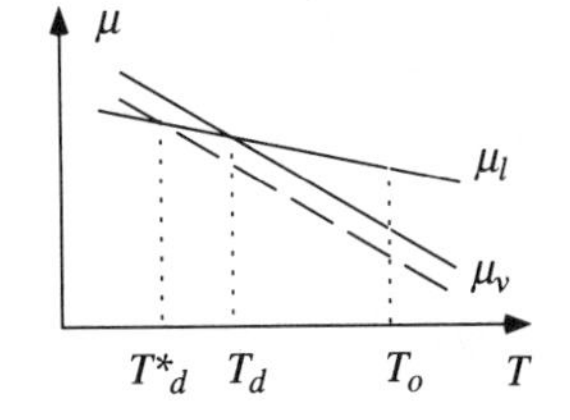

The dew point calculated for the values at the ground, on the other hand, is 286.9 K. A plot of the chemical potential of liquid water and water vapor as a function of temperature can demonstrate the influence of a change of pressure upon the dew point. Lowering the pressure leaves the potential of the liquid more or less unchanged, while that of the vapor decreases. This shifts the line representing the chemical potential of the vapor downward, leading to an intersection with the line for the liquid at a lower temperature.

4.4 Flow Processes and the Chemical Potential

Matter not only changes, it also has the tendency to flow, to spread out in space. Consider a simple example in which you put a piece of dry bread together with a piece of fresh bread into a sealed container. You will notice that the dried bread receives some moisture while the fresher piece dries up to some degree. Obviously, water has moved from a place where there is plenty to one where there is less. Similar phenomena abound in nature: salt or sugar dissolve in water and then slowly spread through the still fluid; a cell takes up chemicals through its membrane if their concentration outside is larger than on the inside. In all these cases we speak of the tendency of matter to flow as long as the conditions allow, and we hold a driving force responsible for the process to occur. As the example with the two pieces of bread demonstrates, the process stops and the driving force vanishes when the concentration of the diffusing substance has become the same at different locations (under otherwise equal conditions). It is quite clear that the driving force for these processes can be interpreted as the difference of the chemical potential of the species in different locations.

Now look at the bulk flow of matter. Usually we associate it with a mechanical process, and we say that a pressure difference is the driving force of the phenomenon. However, if we do not artificially distinguish between "normal" flow processes and the phenomena of diffusion and osmosis listed above, we should use the same driving force with bulk flow as with diffusion. In doing so, we will introduce the chemical potential from a new perspective which will provide additional insight into this quantity.

4.4.1 The Water Turbine and Chemical Energy Currents

In this section we will introduce the simplest engine which uses the flow of matter, a water turbine. By calling the transport of a fluid across a system boundary a chemical process, we can associate the flow of energy with a current of amount of substance and the chemical potential. If we calculate the energy delivered by the engine in a different manner, we will be able to express the as yet unknown chemical potential in terms of other variables. Let us begin with a model of an isothermal water turbine (Figure 28).

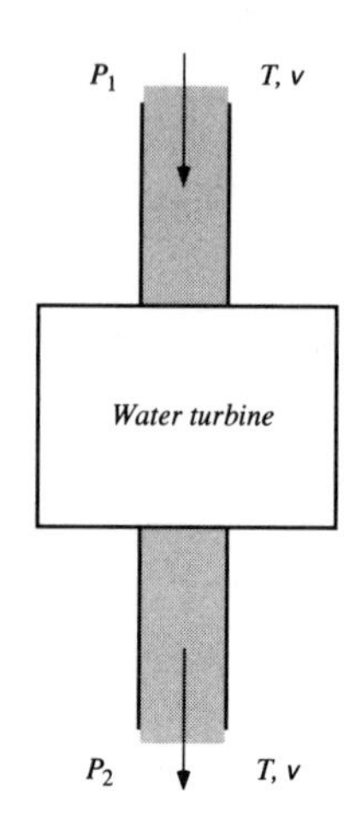

FIGURE 28. A water turbine takes up an incompressible fluid at higher pressure and releases it at lower pressure. The energy released in the flow process is used to drive an engine. The other variables specifying the state of the fluid, i.e., temperature and velocity, are held constant across the engine.

By considering engines like water turbines we will be able to derive the chemical potential of incompressible fluids at constant temperature. The waterfall diagram we have used so often will again clarify the relationships (Figure 29). Now we interpret the processes as follows. A fluid enters the system at a higher "level" and leaves at a lower one. The "level" associated with fluid flow will be interpreted as its chemical potential. Since we have related the chemical potential of substances to the energy released when they are used up in chemical reactions, we should use the equivalent relation in the case of the transport of substances. The rate at which energy is released (or used) as a result of the rate of destruction (or production) of a species in a chemical reaction has been expressed in Equation (33):

$$\mathcal{P}_{chem} = -\mu \Pi_n \tag{117}$$

Analogously, the rate of transfer of energy together with a species, i.e., the *chemical energy current*, should be expressed as

$$I_{E,chem} = \mu I_n \tag{118}$$

We take the current of amount of substance into or out of a system as the carrier of energy, with the chemical potential as the associated level.

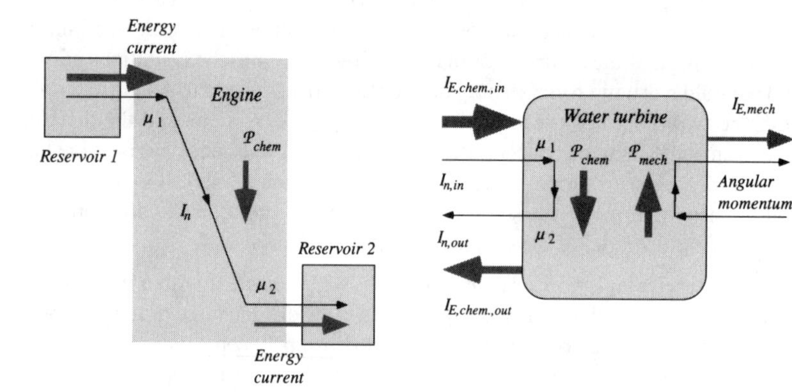

FIGURE 29. Waterfall diagram and energy flow diagram representing the processes of a water turbine. Water enters the system at higher pressure, i.e., at higher chemical potential, and it is ejected at a lower value of the potential. The power of the process is proportional to the difference of the chemical potentials and to the rate of flow expressed by the current of amount of substance.

Equation (118) introduces the meaning of the chemical potential in the context of the transport of substances, just as Equation (33) does for chemical reactions. We can express the rates of energy flow due to the transport of water both using the new relation and with what we know about "flow work" from the discussion in Section P.1; from this analysis we obtain:

$$(\mu_1 - \mu_2)I_n = (P_1 - P_2)I_V \tag{119}$$

(P is the pressure of the fluid.) Now we have only to express the volume flux in terms of the current of substance. Since in engineering we also introduce the current of mass as a measure of the flux of amount of substance, it will be convenient to write down their mutual relationships for later reference:

$$\begin{aligned} I_m &= \rho I_V \\ I_m &= M_o I_n \\ I_V &= \frac{M_o}{\rho} I_n \end{aligned} \tag{120}$$

If we use Equation (120c), we arrive at the result we have been looking for, i.e., the relation between the chemical potential of an incompressible (and isothermal) fluid and its pressure:

$$\mu(P) = \mu(P_o) + \frac{M_o}{\rho}(P - P_o) \tag{121}$$

or

$$\mu(P) = \mu(P_o) + (P - P_o)\overline{\upsilon} \tag{122}$$

where $\overline{\upsilon} = V/n$ is the molar volume of the fluid.

You should note that the expression of the energy current solely in terms of the current of substance depends on the fact that we were able to ignore both entropy and momentum which are also carried across the boundary of the system. Remember that we are dealing with open systems, where matter carries other quantities convectively across surfaces. We were able to neglect the other quantities because the temperature and velocity of the fluid have been left unchanged, and because the convective currents of both entropy and momentum are the same at the inlet and the outlet of the engine. A simple change of the engine to one which uses air instead of water already renders the conditions just stated invalid. If we wish to analyze compressed air engines (or compressors) and heat exchangers, we can no longer neglect the contribution of convective currents of entropy and momentum. We will see shortly how to include the convective currents of other fundamental quantities with the expression for the total energy current due to flow processes.

4.4.2 Convective Currents of Entropy and of Momentum

This section serves to prepare our discussion in Section 4.4.4. We need to express convective currents of entropy and of momentum in terms of the flow of substances, and to compute the total flux of energy accompanying the flow of a fluid across a surface (Section 4.4.3).

Convective entropy currents. Let us first discuss the case of entropy. A convective current of entropy is associated directly with the current of the fluid carrying the quantity. If substances flow, the entropy stored in them flows as well. For this reason we have to be able to tell how much entropy and energy are stored in a given fluid body, and how this body is flowing. Both factors determine the convective fluxes crossing the surface of a control volume. To express a convective current of entropy, either the density of entropy stored in the fluid or the amount of entropy per mass are introduced. So far we have encountered two types of densities, a volume density of the production of entropy, and a surface density of conductive transport of entropy (Chapter 3). Remember that densities are used to describe situations where quantities change spatially. The volume density of mass, for example, is defined in such a way that the integral of the density over the volume is the total amount of mass in the entire system. In much the same way we introduce the *density of entropy* ρ_s:

$$S = \int_{\mathcal{V}} \rho_s dV \tag{123}$$

and the entropy per mass s which is called *specific entropy*:

$$S = \int_{m(\mathcal{V})} s dm \tag{124}$$

Finally, we introduce the *molar entropy* $\bar{s}$ as

$$S = \int_{n(\mathcal{V})} \bar{s} dn \tag{125}$$

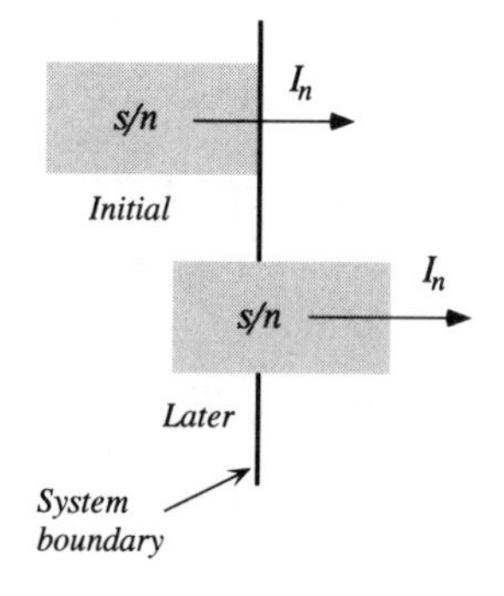

FIGURE 30. The density of entropy in a fluid determines how much of it is carried over a surface as a consequence of the flow of matter. Just as the flux of mass is given by the density of mass and the volume flux, so the convective flux of entropy is the product of the entropy density and the volume flux.

The integrations are performed over the volume of a system and over its mass or amount of substance, respectively. Often we shall be able to take the density, or the specific and molar quantities as constant for the system considered, in which case the entropy can be expressed simply by

$$\begin{aligned} S &= \rho_s V \\ S &= sm \\ S &= \bar{s}n \end{aligned} \tag{126}$$

This certainly applies to spatially uniform bodies. For now, these simple cases are all we need. The density of entropy and the entropy per mass are related by

$$\rho_s = \rho s \tag{127}$$

where ρ is the mass density. If we want to specify the density of entropy of a body, we shall need a constitutive theory such as that of incompressible fluids or the ideal gas (Chapter 2).

Now, the convective fluxes must be the product of the entropy per amount of substance and the flux of amount of substance:

$$I_{s,conv} = \bar{s} I_n \tag{128}$$

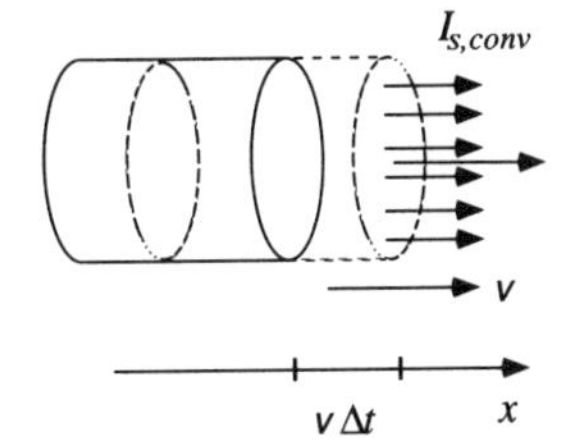

FIGURE 31. A fluid containing entropy is flowing in one dimension with speed v. In time Δt, the fluid contained in the cylinder of length $v\Delta t$ is swept across the surface A. The current of mass therefore is $A\rho v$, where ρ is the mass density of the fluid. The entropy contained in the fluid is equal to the heat per mass s, multiplied by the mass. As a consequence, the convective current of entropy is given by $s\rho A v$.

The other two possible expressions can be derived in a similar manner. In place of the molar entropy we use the specific entropy or the density of entropy and multiply by the flux of mass or of volume, respectively:

$$I_{s,conv} = s I_m \tag{129}$$

$$I_{s,conv} = \rho_s I_V \tag{130}$$

Finally, we may express the fluxes in terms of mass density, speed of flow, and surface area perpendicular to the flowing substance (Figure 31). Note that the last few expressions only hold if the densities or specific quantities are constant across the surface.

Convective momentum currents. In one spatial dimension, the case of convective currents of momentum is perfectly analogous to that of entropy. Instead of the specific entropy we have the momentum per unit mass of the fluid, which is equal to its speed of flow. Therefore, the convective flux of momentum is

$$I_{p,conv} = v I_m \tag{131}$$

if it is expressed using the current of mass. In analogy to the equations presented above, there are two more possibilities of writing the convective current.

Equations of balance. The equations of balance must include expressions for all the means by which a certain quantity is transported into and out of a system. For this reason, the laws of balance used so far have to be changed in that convective fluxes have to be integrated with all the other currents. We have seen how this is done in the case of single dimensional motion (Section P.2.5), and for entropy (Section 3.1.2). The balance of entropy, including convective currents, will accompany us throughout the rest of this chapter. Naturally, energy transport due to the flow of matter across system boundaries also has to be included. How this is done will be shown in the following section.

EXAMPLE 26. Convective currents with water.

Water at a temperature of 50°C is flowing through a pipe whose diameter is 5.0 cm. The mass flux is 10 kg/s. Calculate the magnitude of a) the convective current of entropy, and b) the convective current of momentum.

SOLUTION: a) If we choose to base the calculation on the flux of mass, we need the entropy per mass of the fluid, which is given by

$$s(T) = s\left(T_{ref}\right) + c_p \ln\left(\frac{T}{T_{ref}}\right) = 69.9/0.018 + 4200\ln\left(\frac{323}{298}\right) = 4220\frac{\text{J}}{\text{K}\cdot\text{kg}}$$

$s(T_{ref})$ can be found in the tables for the chemical potential (Table A.13). The convective flux of entropy follows from Equation (129):

$$\left|I_{s,conv}\right| = s(T)\left|I_m\right| = 4220\cdot 10\frac{\text{W}}{\text{K}} = 4.22\cdot 10^4\,\frac{\text{W}}{\text{K}}$$

b) To calculate the convective momentum current, we need to know the speed of flow, which follows from the flux of mass (or volume) and the cross section of the pipe:

$$\left|I_m\right| = \rho\left|I_V\right| = \rho A v \quad \Rightarrow \quad v = 5.1\,\text{m/s}$$

The current of momentum is the product of the momentum per mass (the speed) and the mass flux:

$$\left|I_{p,conv}\right| = v\left|I_m\right| = 51\,\text{N}$$

If the pressure of the water is 1 bar, then the value of the convective momentum flux approaches that of the conductive momentum current.

4.4.3 Energy Transport as a Consequence of Fluid Flow

Now we will turn to the problem of computing the energy transfer at those parts of a system boundary where currents of matter are flowing. Energy is transferred across boundaries by a fluid as a consequence of two distinct processes. First, as in the case of the convective current of entropy (or of momentum or charge, for that matter), there is a proper convective energy current due to the energy contained in the fluid. Since the fluid must be flowing, this part is again resolvable into two parts, namely the internal energy (the part noticed by an observer moving with the fluid) and the kinetic energy. To calculate the convective energy current we have to determine the *density of energy* ρ_E, the *molar energy* $\bar{e}$, or the energy per mass (*specific energy*) e. The latter is given by

$$e = E/m = \frac{U + mv^2/2}{m} = u + \frac{1}{2}v^2 \tag{132}$$

We will use the symbol u for the intrinsic part of the energy of a fluid, i.e., for the contribution which is independent of the motion of the fluid; this part is commonly called the *internal energy*. The first part of the energy flux due to the flow of matter therefore takes the form

$$eI_m = \left(u + \frac{1}{2}v^2\right)I_m \tag{133}$$

if it is expressed in terms of the current of mass. Again, these expressions hold only in the spatially homogeneous case.

We still have to calculate the second contribution to the energy current. This part is made up of the hydraulic energy current described before in the Prologue: the fluid is under pressure and thus sections of it are pushed across surfaces when it flows. (This contribution sometimes is called *flow work*; this is what has been calculated in Section 4.4.1 for the water turbine.) Interestingly, it is a consequence of a conductive current of momentum, which develops its energy carrying capacity because of the transfer of matter:

$$PI_V = \frac{P}{\rho} I_m. \tag{134}$$

Now we add the contributions calculated in Equations (133) and (134):

$$I_{E,flow} = \left[u + \frac{P}{\rho} + \frac{1}{2} v^2 \right] I_m \tag{135}$$

Let us introduce a quantity which abbreviates a term occurring very frequently in equations involving flow processes. The term consists of the first two parts within the brackets in Equation (135), and it is called the *specific enthalpy* of the fluid:

$$h = u + \frac{P}{\rho} = u + P\upsilon \tag{136}$$

where $\upsilon = 1/\rho$ is called the *specific volume* of the fluid. The specific enthalpy introduced here is the same quantity defined in the Interlude (where it was used to demonstrate that the so-called "heat capacity at constant pressure" is the temperature coefficient of enthalpy), and used in the sections above. In summary, we may say that the term $u + P\upsilon$ occurs in energy balances of both closed and open systems, together with heating and with flow.

EXAMPLE 27. Flow of an ideal incompressible fluid: Bernoulli's law.

Derive the expression for the dependence of pressure on the speed of flow for the case of an incompressible ideal fluid, without exchange of energy with the environment.

SOLUTION: The law of balance of energy for the fluid system between points 1 and 2 (see the accompanying figure) follows from Equation (135):

$$u + \frac{P_1}{\rho} + \frac{1}{2} v_1^2 = u + \frac{P_2}{\rho} + \frac{1}{2} v_2^2$$

This follows from our assumption that there is no exchange of energy with the surroundings; therefore, the only currents of energy with respect to the system considered are the fluxes associated with fluid flow at the inlet and the outlet of the system.

Since the flow is ideal, there will not be any production of entropy between points 1 and 2. Equal density of entropy of the incompressible fluid at the inlet and the outlet finally means that the temperature of the fluid must remain constant. This in turn implies constant internal energy. Therefore we find that:

$$P_1 + \frac{1}{2}\rho v_1{}^2 = P_2 + \frac{1}{2}\rho v_2{}^2$$

This relation is what physicists normally call Bernoulli's Law (for horizontal flow).

EXAMPLE 28. Measuring the temperature coefficient of enthalpy of air.

In Example 19 of Chapter 2, a way of measuring the "heat capacity at constant temperature", i.e., the temperature coefficient of enthalpy, was presented on the basis of a closed systems analysis. a) Derive the same result using the control volume approach (Section 4.4.5). b) Estimate the size of the velocity term in the expression for the flow energy current.

SOLUTION: a) We have three energy currents with respect to the vertical pipe in which heated air is rising. Therefore, the equation of balance takes the form

$$-\left|U_{el} I_q\right| + h_1 I_{m1} + h_2 I_{m2} = 0$$

Here, the velocity term has been neglected. If the temperature coefficient of enthalpy increases linearly with temperature, the coefficient can be estimated by

$$c_p = \frac{h_2 - h_1}{T_2 - T_1} = \frac{1}{T_2 - T_1}\frac{\left|U_{el} I_q\right|}{\left|I_m\right|}$$

With

$$I_m = \rho_2 A v_2 = \frac{M_o P}{R T_2} A v_2$$

b) The speed of flow will be of the order of a few meters per second, which gives the change of the velocity term in Equation (135) a magnitude of less than 100 m^2/s^2 between the bottom and the top of the pipe. The change of the enthalpy, on the other hand, has a magnitude of about 100 K · 1000 J/(K · kg). Therefore, we are justified in neglecting the kinetic energy of the fluid.

4.4.4 Flow Processes and the Chemical Potential of a Fluid

At this point we will generalize the concept of an energy current $I_{E,flow}$ associated with the flow of matter crossing the surface of an open system. The expression will then be compared to Equation (135), which will lead to a general form of the chemical potential of a simple flowing fluid. It is customary to include in this energy current the chemical energy current $I_{E,chem}$ along with the products of the convective currents of entropy and of momentum and their respective potentials. Let us start with the entropy of the fluid, excluding, for the moment, the analogous case of momentum:

$$I_{E,flow} = I_{E,chem} + TI_{s,conv} \tag{137}$$

Using Equation (118) for the chemical energy flux we arrive at

$$I_{E,flow} = \mu I_n + TI_{s,conv} \tag{138}$$

The definition of the chemical potential has been chosen in such a way as to explicitly exclude the contribution from the other basic convective currents. Note that the temperature is not the driving force of convective entropy currents; still it appears in Equation (138) in the same form as if it were the proper potential. Why this form was chosen will become clearer after the formal treatment of open and reacting systems in Section E.1. For now, we may say that the definition determines the form the chemical potential of a flowing substance will take, and this form should be equivalent to what we already know from chemical processes.

Convective currents have been calculated in Section 4.4.2. Introducing Equation (128) in Equation (138) leads to

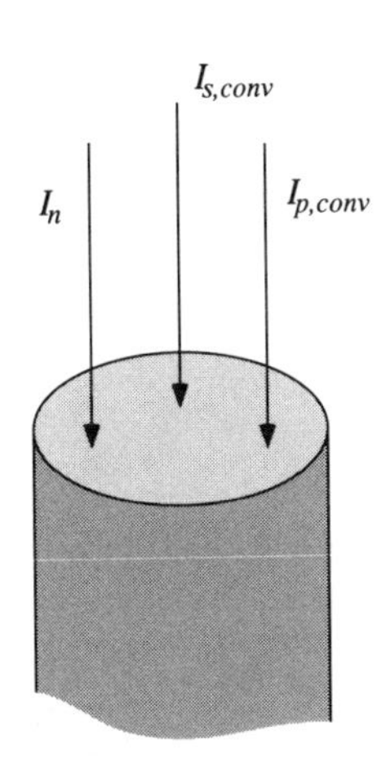

FIGURE 32. Matter carries a number of physical quantities across the boundary of a system.

$$I_{E,flow} = \mu I_n + T\bar{s}I_n = \left[\mu + T\bar{s}\right]I_n \tag{139}$$

This relation must be compared to the expression for the flow energy current presented in Equation (135); we have to exclude the kinetic energy term since the effect of changes of momentum have not yet been taken into consideration. The comparison yields:

$$\mu = \bar{u} + P\bar{v} - T\bar{s} \tag{140}$$

This is the first step toward the final expression for the chemical potential of a simple fluid. Since it does not yet contain the contribution resulting from the movement of the fluid, it is called the *intrinsic* part of the chemical potential.

Let us now turn to cases which involve nontrivial balances of momentum. In the case of fluids changing their speed, we also should include the convective part of the momentum flux. The current crossing the surface of the system is composed of amount of substance, entropy, and momentum (Figure 32). Therefore we write

$$I_{E,flow} = \mu I_n + T I_{s,conv} + v I_{p,conv} \tag{141}$$

for the flux of energy associated with the flow of matter. Since the convective current of momentum must be the product of the momentum per unit mass (the speed) and the flux of mass, we have

$$I_{E,flow} = \left[\mu + T\bar{s} + M_o v^2\right] I_n \tag{142}$$

If we set this equal to the expression known from Equation (135), we can calculate the dependence of the chemical potential of the fluid upon its speed:

$$\mu = \bar{u} + P\bar{\upsilon} - T\bar{s} - \frac{1}{2} M_o v^2 \tag{143}$$

This is the most general form of the chemical potential of a fluid we are going to use in the following sections. It shows that the chemical potential of a flowing substance decreases with increasing velocity.

EXAMPLE 29. Isothermal incompressible fluid flow with friction.

As in the case of Bernoulli's law (Example 27), consider the isothermal flow of an incompressible fluid through a pipe. Include the effect of friction and derive an expression for the rate of production of entropy in terms of the pressures and the speeds involved.

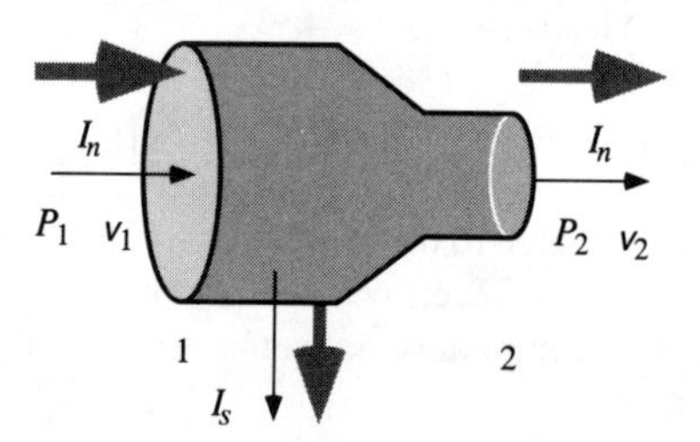

SOLUTION: A control volume analysis (Section 4.4.5) is performed on the system displayed in the figure of Example 27. In addition to the currents associated with the flow at points 1 and 2, there are fluxes of entropy and energy due to the cooling of the fluid. Since entropy is produced and the temperature of the incompressible fluid is assumed to be constant, entropy has to be emitted. The law of balance of entropy therefore reads

$$I_{s,conv,1} + I_{s,conv,2} + I_{s,cond} = \Pi_s$$

The contribution of the convective currents cancels since the specific entropy of the isothermal incompressible fluid remains constant. Therefore, the entropy current due to cooling is equal to the rate of production of entropy.

The equation of balance of energy includes the currents due to fluid flow and to cooling. They can be written as

$$\mu_1 I_{n1} + T I_{s,conv,1} + v I_{p,conv,1} + \mu_2 I_{n2} + T I_{s,conv,2} + v I_{p,conv,2} + T I_{s,cond} = 0$$

If we introduce the general expression for the chemical potential of the fluid (Equation (143)), we obtain the following result

$$M_o\left[\left(\frac{P_1}{\rho}+\frac{1}{2}v_1^2\right)-\left(\frac{P_2}{\rho}+\frac{1}{2}v_2^2\right)\right]I_{n1} + T I_{s,cond} = 0$$

which can be transformed into

$$\Pi_s = \left[\left(\frac{P_1}{\rho}+\frac{1}{2}v_1^2\right)-\left(\frac{P_2}{\rho}+\frac{1}{2}v_2^2\right)\right]\frac{|I_V|}{T}$$

In contrast to the case of ideal flow, the sum of the pressure term and the velocity term is not constant along the fluid stream.

4.4.5 Compressors and Heat Exchangers

The following examples of compressors and heat exchangers are presented as applications of the previous expressions for the chemical potential of a fluid. For their description we shall choose the widely used control volume approach (Figure 33) which singles out a certain part of space for analysis and considers all the fluxes with respect to the surface of the system. Since a control volume normally is an open system, its boundary will be crossed by currents of matter transporting physical quantities stored in a fluid. In other words, we have to include convective currents along with the fluxes known to exist in the case of closed systems.

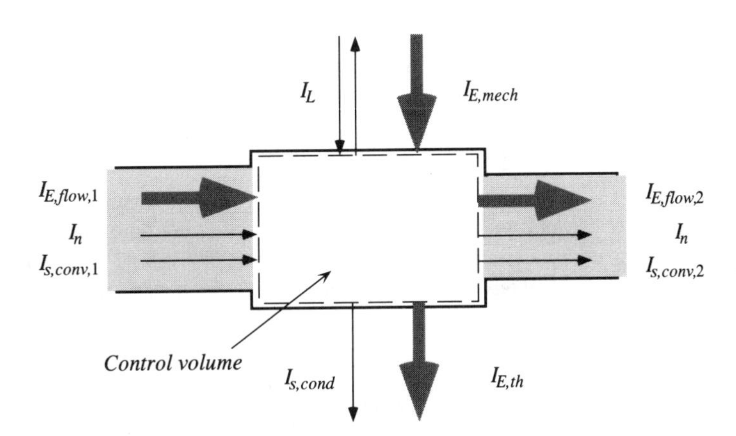

FIGURE 33. Control volume analysis of an (isothermal) compressor. It includes both conductive and convective currents of entropy. I_L denotes the flux of angular momentum.

The isothermal compressor. The principle of operation of an (isothermal) air compressor is shown in Figure 33, which displays the currents of amount of substance, entropy, and energy with respect to the chosen (stationary) control volume. The balance of amount of substance is trivial. For entropy and energy the equations are

$$I_{s,conv,1} + I_{s,conv,2} + I_{s,cond} = 0 \tag{144}$$

$$I_{E,flow,1} + I_{E,flow,2} + I_{E,mech} + I_{E,th} = 0 \tag{145}$$

Air, which is to be compressed isothermally, must be cooled. If we use Equation (138) for the energy current associated with fluid flow, and the well-known relation for the thermal energy current, we obtain

$$\mu_1 I_{n1} + T I_{s,conv,1} + \mu_2 I_{n2} + T I_{s,conv,2} + T I_{s,cond} = -I_{E,mech} \tag{146}$$

or

$$(\mu_2 - \mu_1)|I_n| + T\left[I_{s,conv,1} + I_{s,conv,2} + I_{s,cond}\right] = -I_{E,mech} \tag{147}$$

Since the net flux of entropy is zero, the mechanical power for an isothermal compressor can be calculated from the chemical energy currents alone; on the other hand, since we can compute the energy transferred in the mechanical interaction on the basis of independent information, we can get the chemical potential of the fluid for the conditions stated here. For the ideal gas the mechanical energy current is derived simply on the basis of what we know about this fluid. If we want to compress a certain amount of an ideal gas at constant temperature, the energy needed is calculated according to

$$W = nRT \ln\left(\frac{P_2}{P_1}\right) \tag{148}$$

For a constant stream of air, we get the energy current needed if we replace the amount of substance by the flux of this quantity:

$$\left|I_{E,mech}\right| = RT \ln\left(\frac{P_2}{P_1}\right)|I_n| \tag{149}$$

Equating this result and the difference of the chemical energy currents leaving and entering the compressor, we finally obtain the same expression for the *chemical potential of the ideal gas* for constant temperature which was derived in Section 4.2.4:

$$\mu(T,P) = \mu(T,P_o) + RT \ln\left(\frac{P}{P_o}\right) \tag{150}$$

We may take this derivation as an indication of the validity of the general form of the chemical potential introduced before in Equation (140). If you pursue an analysis based on a four step cycle undergone by an amount of ideal gas in a compressor, you get the same result (Problem 20).

Note that this is the second example other than an isothermal incompressible fluid where we see that the chemical potential increases with increasing pressure. (We have encountered this behavior before in the context of chemical transformations.) This result can be derived for general fluids using the relations presented in Section E.1.

Constant pressure heat exchangers. So far we have looked at examples of flow engines for which the temperature of the fluid was kept constant. If we wish to include the effect of changes of temperature while at the same time excluding the mechanical process studied above, we should investigate a simple model of a heat exchanger (Figure 34). In its simplest form, a heat exchanger is a pipe of constant diameter which transports a hot fluid. On its way, the fluid loses heat to the surroundings, leading to a decrease of temperature. Here, in contrast to the case of turbo machinery, we will keep the pressure constant. While it is true that the speed of flow will change if the fluid changes its density, we can neglect the influence of this effect under most circumstances.

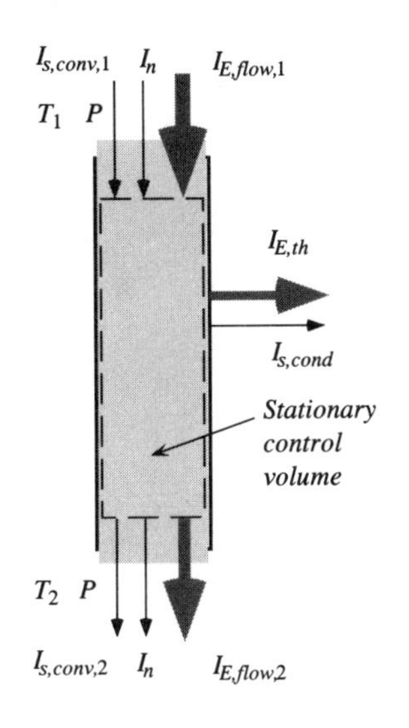

FIGURE 34. A heat exchanger in its simplest form lets a hot fluid flow through a pipe. The fluid loses heat to the surroundings. Note the convective current of entropy included in the analysis.

We should now formulate the equations of balance of energy and of entropy for the control volume for steady state processes (Figure 34):

$$I_{E,flow,1} + I_{E,flow,2} + I_{E,th} = 0 \tag{151}$$

$$I_{s,conv,1} + I_{s,conv,2} + I_{s,cond} = 0 \tag{152}$$

For the energy exchanged through the wall of the heat exchanger, we have the well-known relation between the entropy current and the temperature at the system boundary. We can first calculate the energy flux in terms of the chemical potentials, the temperatures, and the flux of amount of substance and the convective entropy currents:

$$\begin{aligned} I_{E,th} &= -I_{E,flow,1} - I_{E,flow,2} \\ &= -\left[\mu \cdot I_n + T I_{s,conv}\right]_1 - \left[\mu I_n + T I_{s,conv}\right]_2 \end{aligned} \tag{153}$$

From this we may calculate the difference of the chemical potential of the fluid (for constant pressure and as a function of the temperatures at the inlet and the outlet):

$$(\mu_2 - \mu_1) I_{n2} = I_{E,flow,net} - \left(T_2 \bar{s}_2 - T_1 \bar{s}_1\right) I_{n2} \tag{154}$$

Similarly to what we did in the case of the compressor, we can compute the chemical potential of a fluid if we manage to get independent information on the energy current associated with the fluid flowing through the heat exchanger. This information is provided by Equation (135):

$$I_{E,flow,net} = (h_2 - h_1)|I_m| \quad \textbf{(155)}$$

which is equivalent to

$$I_{E,flow,net} = c_p(T_2 - T_1)|I_m| \quad \textbf{(156)}$$

if the value of c_p is taken to be independent of temperature, and the pressure is constant. (See the Interlude for the relation between changes of enthalpy and the temperature coefficient of enthalpy c_p.) The net convective flux of entropy is important for applications of the balance of entropy. According to Equation (129) it is given by

$$I_{s,conv,net} = (s_2 - s_1)|I_m| \quad \textbf{(157)}$$

which can be written

$$I_{s,conv,net} = c_p \ln\left(\frac{T_2}{T_1}\right)|I_m| \quad \textbf{(158)}$$

Again, in the latter expression, both c_p and the pressure have to be held constant for the equation to be valid. This expression and Equation (156) can be applied frequently in many types of flow heaters (Example 30).

If you now combine Equation (154) with Equation (155), you get an expression for the chemical potential which could have been derived directly from our general form presented in Equation (140):

$$\mu_2 - \mu_1 = M_o\left[h_2 - h_1 - (T_2 s_2 - T_1 s_1)\right] \quad \textbf{(159)}$$

EXAMPLE 30. A solar hot water heater.

Solar radiation is used to heat water which is flowing through a collector. The collector surface is tilted at an angle of 30° to the direction of radiation (the surface area measures 7.0 m^2). The energy current density of solar radiation is 1360 W/m^2 outside the Earth's atmosphere. 30% of the incoming energy flux is reflected back to space by the atmosphere, and the collector loses 40% of the energy absorbed. (a) How much water can be pumped through the collector in steady-state operation if its temperature is to be raised from 20°C to 60°? (b) Do we have to consider the kinetic energy term in the convective energy flux? The water pipes have a diameter of 4.0 cm.

SOLUTION: (a) The rate at which energy is absorbed from solar radiation is determined according to Section 3.6.2, which yields the following expression:

$$\Sigma_E = 0.70 G_{sc} A \cos(30°) = 5.80\,\text{kW}$$

with G_{sc} = 1360 W/m^2. Only 60% of the energy incident on the body is retained by the water. In steady-state operation, the sum of all currents (and source rates) of energy with respect to the collector must be zero. We have the rate of absorption due to solar radiation, an incoming and an outgoing convective flux (or a net convective flux), and the flux due to losses:

$$I_{E,conv,net} + I_{E,loss} = \Sigma_E$$

We compute the convective fluxes according to Equation (156). If we neglect the kinetic energy term, we get:

$$c_p\left(T_{out} - T_{in}\right)I_m + 0.40\,\Sigma_E = \Sigma_E$$

We can solve for the value of the current of mass:

$$I_m = \frac{0.60\,\Sigma_E}{c_p\left(T_{out} - T_{in}\right)} = \frac{0.60 \cdot 5.80 \cdot 10^3}{4200 \cdot (333 - 293)}\frac{\text{kg}}{\text{s}} = 0.021\frac{\text{kg}}{\text{s}}$$

This corresponds to roughly 75 liters of warm water per hour.

(b) If the diameter of the pipe does not change, the speed of flow at the inlet and the outlet is the same. For this reason, the kinetic energy terms cancels in the equation of balance. Even if a change had to be taken into account, it would be rather small compared to those due to differences of the internal energy. The specific kinetic energy of the fluid flowing through the pipes turns out to be only 0.14 mJ/kg.

EXAMPLE 31. The balance of entropy for an electric flow heater with heat loss.

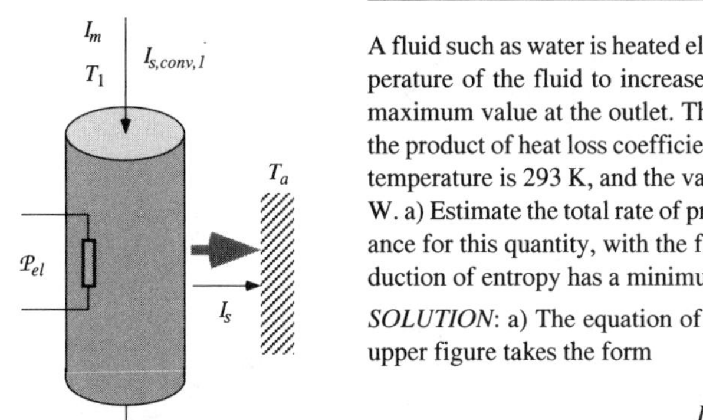

A fluid such as water is heated electrically while flowing through a pipe. Assume the temperature of the fluid to increase linearly along the pipe from ambient temperature to a maximum value at the outlet. There is heat loss to the environment. Take the average of the product of heat loss coefficient and surface area of the pipe to be 20 W/K; the ambient temperature is 293 K, and the value of c_p is 4200 J/(K · kg). The electrical power is 1000 W. a) Estimate the total rate of production of entropy by writing down the equation of balance for this quantity, with the flux of mass as a parameter. b) Show that the rate of production of entropy has a minimum for a particular value of the mass flux of the fluid.

SOLUTION: a) The equation of balance of entropy for the control volume shown in the upper figure takes the form

$$I_{s,conv,1} + I_{s,conv,2} + I_s = \Pi_s$$

If we take the boundary of the control volume to touch the surroundings at ambient temperature, the contribution to the production of entropy due to heat loss will be included. The sum of the convective entropy currents is

$$c_p\left|I_m\right|\ln\left(\frac{T_2}{T_1}\right)$$

See Equation (158). The current of entropy entering the environment at T_a can be determined by

$$\frac{1}{T_a}UA\left(\frac{1}{2}(T_1+T_2)-T_a\right)$$

UA is the product of heat transfer coefficient and surface area. The temperature of the fluid losing heat has been approximated by the arithmetic mean of inlet and outlet temperatures.

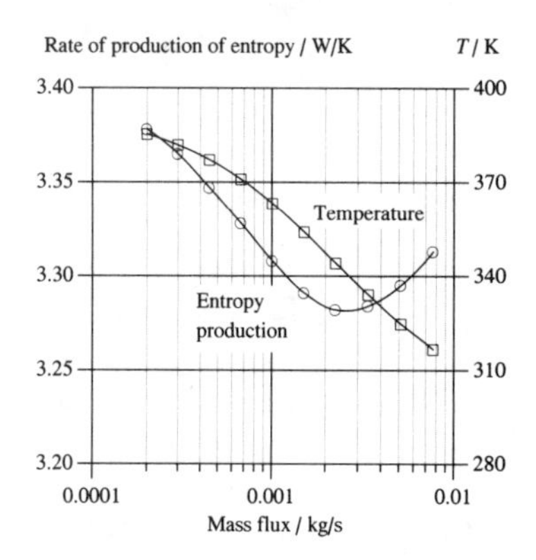

We still need to know the outlet temperature of the fluid. It can be calculated using the law of balance of energy:

$$c_p\left|I_m\right|(T_1-T_2)+UA\left(\frac{1}{2}(T_1+T_2)-T_a\right)=\left|\mathcal{P}_{el}\right|$$

b) Even though the equations do not look all that complicated, finding the minimum of the rate of production of entropy is best accomplished numerically. The graph shows this quantity as a function of the flux of mass through the heater.

EXAMPLE 32. Charging a hot water tank.

A tank containing 1000 liters of water at 20°C is charged with a stream of hot water at 60°C while the same amount of water is withdrawn from an outlet. The mass flux is set equal to 0.05 kg/s. The water supplied to the tank instantaneously mixes with the fluid already there. There is heat loss to the surroundings at ambient temperature of 20°C; the heat loss coefficient-surface area product is 10 W/K.

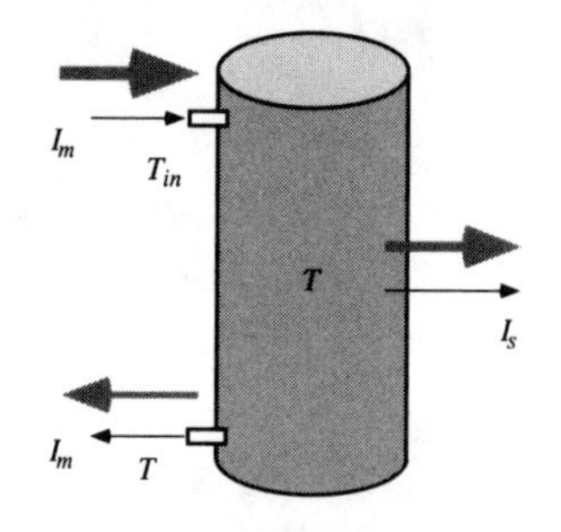

a) Calculate the temperature of the water in the tank as a function of time. b) Upon which parameters does the maximum temperature reached in the tank depend? c) Calculate the rate of production of entropy.

SOLUTION: a) The law of balance of energy yields the differential equation for the temperature of the water as a function of time:

$$\dot{E}=I_{E,flow,1}+I_{E,flow,2}+I_{E,loss}$$

The terms in this equation can be expressed using the constitutive relations derived so far:

$$C\dot{T} = c_p|I_m|(T_{in} - T) - UA(T - T_a)$$

C is the temperature coefficient of energy of the water in the tank. The differential equation can be integrated directly:

$$C\int_{T_{init}}^{T} \frac{dT}{a - bT} = t$$

$$a = c_p|I_m|T_{in} + UAT_a$$

$$b = c_p|I_m| + UA$$

which leads to

$$T(t) = \frac{a}{b} - \left(\frac{a}{b} - T_{init}\right)\exp\left(-\frac{b}{C}t\right)$$

b) The maximum temperature reached by the fluid inside the tank is equal to

$$T_{max} = \frac{a}{b} = \frac{c_p|I_m|T_{in} + UA \cdot T_a}{c_p|I_m| + UA}$$

Increasing the loss factor UA decreases the maximum possible temperature. If there is no heat loss, the maximum temperature is the temperature of the water supplied to the tank.

c) The equation of balance of entropy for the control volume coinciding with the water in the tank is

$$\dot{S} - c_p|I_m|\ln\left(\frac{T_{in}}{T}\right) + \frac{UA(T - T_a)}{T} = \Pi_s$$

Since the rate of change of the entropy of the incompressible body of water is related to the rate of change of its energy by the Gibbs fundamental form, we obtain

$$\Pi_s = \frac{1}{T}c_p|I_m|(T_{in} - T) - c_p|I_m|\ln\left(\frac{T_{in}}{T}\right)$$

Performing a balance of entropy on flow systems is as important as it is for other processes. Take the example of a hot water tank hooked up to solar collectors. If you manage to have the water stratified according to temperature (warmer water at the top), rather than having it mixed, you will improve upon the efficiency of the system. Naturally, with stratification, the amount of entropy produced, and therefore the magnitude of the irreversibility of the process, is smaller than with mixing.

4.4.6 Fluid Flow Without Heating or Cooling

The last section presented two examples of flow processes involving interaction with the surroundings. Here we will discuss four examples where the flow is thermally isolated, namely isentropic flow of the ideal gas, the throttling process, adiabatic saturation of air, and the flow through a reactor of a fuel undergoing combustion without losing heat to the surroundings. The first process might model a rocket engine, while the second occurs in refrigeration systems using vapor compression, and in processes for liquefying gases. The fourth example will lead to the notion of adiabatic flame temperature. We already have discussed one example of the kind treated here: if the fluid is incompressible and the flow isentropic, we have the case for which Bernoulli's law applies.

Steady state equations of balance. Independent of what happens inside the pipe, as long as there is no exchange of energy through the walls channelling the flow, the only energy currents are the result of fluid flow at positions 1 and 2 (Figure 35). The equation of balance of energy therefore takes the form

$$\left[\mu_1 + T_1\bar{s}_1 + M_o v_1^2\right] I_{n1} + \left[\mu_2 + T_2\bar{s}_2 + M_o v_2^2\right] I_{n2} = 0 \qquad \text{(160)}$$

Since the fluxes of amount of substance have the same magnitude at the inlet and the outlet, this is equivalent to

$$\mu_1 + T_1\bar{s}_1 + M_o v_1^2 = \mu_2 + T_2\bar{s}_2 + M_o v_2^2 \qquad \text{(161)}$$

If we use Equation (143) for the chemical potential of the fluid, we obtain

$$\bar{h}_2 - \bar{h}_1 = -\frac{1}{2} M_o\left[v_2^2 - v_1^2\right] \qquad \text{(162)}$$

In applications, the state of the fluid at the inlet of the pipe might be specified, and we might wish to compute the quantities at the outlet. The equation of balance of energy does not suffice in general for getting the required answer. We therefore have to be more specific about the details of the flow, and apply the other laws of balance as well. The balance of entropy simply states that

$$\bar{s}_1 I_{n1} + \bar{s}_2 I_{n2} = \Pi_s \qquad \text{(163)}$$

The balance of momentum must be applied if information about forces upon sections of the fluid (or the pipes) is desired; it will not be considered at this point.

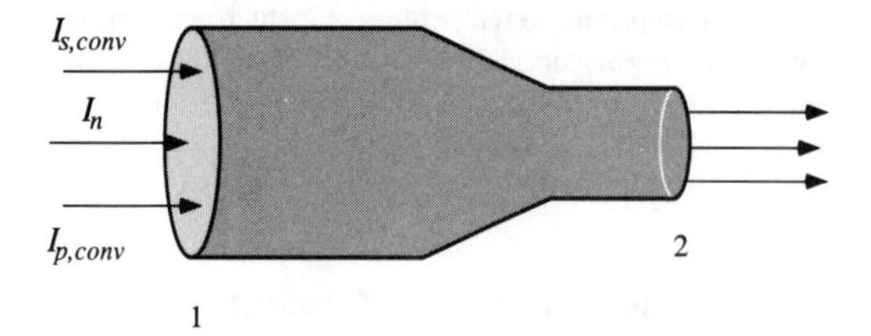

FIGURE 35. Assume the fluid flowing through the pipe not to exchange any heat with the walls. As a consequence, the only energy currents are due to fluid flow.

Isentropic flow of the ideal gas. In the case of the ideal gas we have the necessary constitutive information for computing what happens to the fluid. Equation (162) becomes

$$\bar{c}_p(T_1 - T_o) + \frac{1}{2}M_o v_1^2 = \bar{c}_p(T_2 - T_o) + \frac{1}{2}M_o v_2^2 \tag{164}$$

This shows that the temperature of the fluid decreases with increasing speed of flow. However, we are not able to calculate the state at the outlet from this information and the specification of the state at the inlet. We need the equations of balance of amount of substance (or mass) and of entropy:

$$|I_{n1}| = |I_{n2}| \quad \Rightarrow \quad \rho_1 A_1 v_1 = \rho_1 A_1 v_2 \quad \Rightarrow \quad \frac{P_1}{T_1} A_1 v_1 = \frac{P_2}{T_2} A_1 v_2 \tag{165}$$

$$|I_{s,conv,1}| = |I_{s,conv,2}| \quad \Rightarrow \quad \bar{s}_1 = \bar{s}_2 \quad \Rightarrow \quad -R\ln\left(\frac{P_2}{P_1}\right) + \bar{c}_p \ln\left(\frac{T_2}{T_1}\right) = 0 \tag{166}$$

Remember that the flow was assumed to be isentropic; in other words, it must be reversible. The last forms hold for the ideal gas with a constant temperature coefficient of enthalpy. Note that Equation (166) leads to the well-known formula for adiabatic changes of the ideal gas which was derived in Chapter 2.

The throttling process. In the throttling process, the fluid is allowed to flow through some sort of valve or porous plug, which reduces its pressure. If the valve is insulated, it will not lead to an exchange of energy and entropy with the surroundings, leaving our main assumption valid. If the diameter of the pipe is changed in such a way as to leave the speed of flow unchanged upon expansion of the fluid, the expression of the balance of energy reduces to

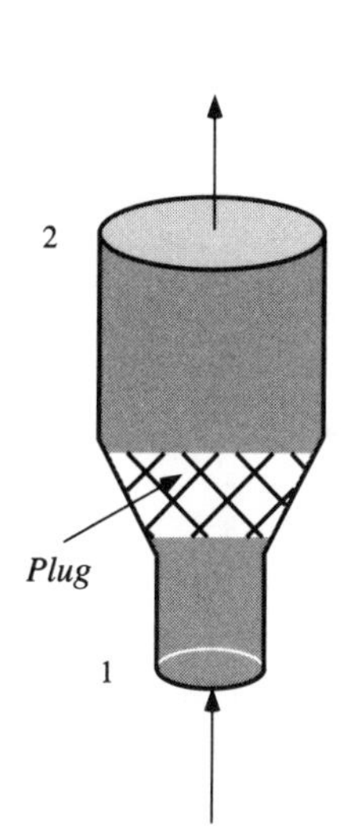

FIGURE 36. A gas is allowed to expand through a porous plug. If the fluid is near the point of condensation, the reduction in pressure can lead to a decrease of temperature.

$$\bar{h}_2 = \bar{h}_1 \tag{167}$$

The amount of substance and entropy have laws of balance which take the forms

$$\frac{A_1}{\bar{\upsilon}_1} = \frac{A_2}{\bar{\upsilon}_2} \tag{168}$$

and

$$\bar{s}_1 I_{n1} + \bar{s}_2 I_{n2} = \Pi_s \tag{169}$$

respectively. Clearly, the flow is irreversible, i.e., entropy is produced. Note that for the ideal gas the condition of constant enthalpy requires the temperature to remain constant: ideal gases do not change their temperature as a result of throttling, and they cannot be liquefied. For other gases, property data have to be known for the chemical potential, entropy, energy, enthalpy, and specific volume.

Adiabatic saturation. In Section 4.3.4, an equation was used which relates the humidity ratio of moist air to the wet bulb temperature. The process taking place around a wet bulb thermometer is similar to the model of adiabatic saturation which will be discussed below. Take a stream of moist air passing over the surface of some water as in Figure 37. The air picks up extra moisture until it is saturated. Since no conductive transfer of entropy is involved, the process is called *adiabatic saturation.*

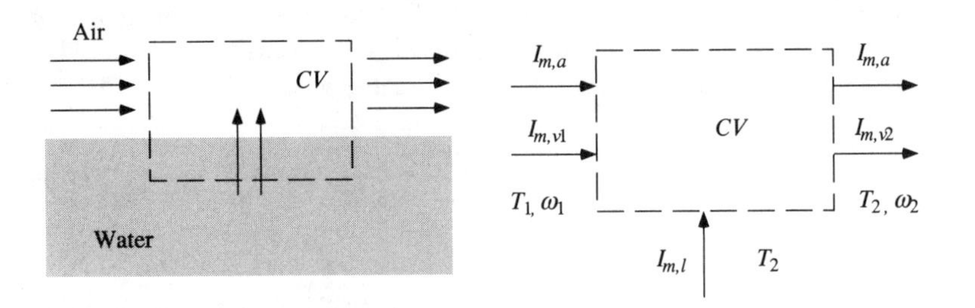

FIGURE 37. A stream of moist air passes over a water surface and picks up additional water until it is saturated. The air cools to T_2, and in the steady state, the water will have the same temperature (if the body of water is not too large).

Consider a control volume encompassing a section of the air stream and a small part of the water which delivers the moisture. In a steady state, the amount of water vapor flowing out with the saturated air will be the sum of the mass entering with the somewhat dryer air, and the mass of the water evaporating from the surface (this latter mass is replenished through the control surface):

$$\left|I_{m,a}\right|+\left|I_{m,v1}\right|+\left|I_{m,l}\right|=\left|I_{m,a}\right|+\left|I_{m,v2}\right| \tag{170}$$

Here, a, l, and v denote dry air, liquid water, and water vapor, respectively. For the balance of energy, we will assume the velocity terms to be negligible. Also, the moist air will be considered to be a simple mixture of dry air and water vapor at total pressure P. At the inlet, the temperature is T_1, and the humidity ratio is ω_1, while at the outlet we have T_2 and ω_2. Now, the energy currents can be written in terms of the enthalpy of the fluids:

$$h_a(T_1)\left|I_{m,a}\right|+h_v(T_1)\left|I_{m,v1}\right|+h_l(T_2)\left|I_{m,l}\right|=h_a(T_2)\left|I_{m,a}\right|+h_g(T_2)\left|I_{m,v2}\right| \tag{171}$$

The index g is used for the saturated vapor state. Using the balance of mass we can express the current of water from the reservoir. Dividing by the current of mass of dry air introduces the humidity ratio:

$$h_a(T_1)+\omega_1 h_g(T_1)+(\omega_2-\omega_1)h_l(T_2)=h_a(T_2)+\omega_2 h_g(T_2) \tag{172}$$

The only additional change has been introduced in approximating the enthalpy of the unsaturated vapor at T_1 by the corresponding value for the saturated state; see Equation (110). If you solve this equation for ω_1, you obtain the result used in Section 4.3.4. The humidity ratio at the exit of the control volume is calculated for saturated moist air at T_2, Equation (116).

Flow with combustion: the adiabatic flame temperature. How hot is a flame from burning fuel? The temperature is determined by three factors: first, by the amount of entropy produced by the reaction; second, by how much entropy is retained by the combustion products; and, finally, by the temperature dependence of the entropy of the substances involved in the reaction and in the flow through the reactor. If you let the reaction take place at constant pressure, the rise in temperature certainly cannot be the result of compression. Rather, the temperature of the combustion products directly depends upon how much entropy they contain.

Consider a simple flow reactor, with reactants entering at standard conditions, and products of the reaction leaving at the unknown temperature T. The value of T will be highest if there is no entropy lost to the surroundings: this is the condition to which the name *adiabatic flame temperature* applies. Our steady-state analysis will begin with the balance of entropy. The entropy leaving convectively with the products of the reaction must be the entropy produced during combustion, plus the entropy supplied by the reactants, i.e., by the fuel and the oxidizer:

$$\left|I_{s,conv,out}\right| = \Pi_s + \left|I_{s,conv,in}\right| \tag{173}$$

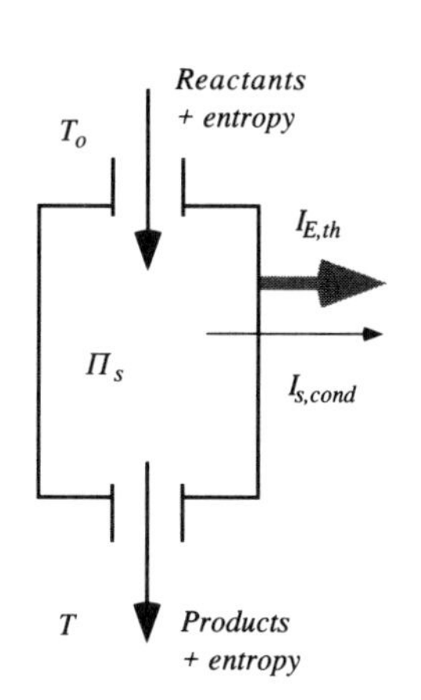

FIGURE 38. Fuel and oxidizer are burned in a flow reactor. If there is no loss of entropy to the surroundings, the process is adiabatic, and the temperature of the products is the adiabatic flame temperature.

During adiabatic operation, there will be no other entropy flux to or from the surroundings. The quantity which determines the entire outcome is the amount of entropy produced, which is governed by the magnitude of dissipation, which in turn depends upon whether or not other processes are involved. Since the substances burn at a different temperature from that of the entering fluid stream, there indeed is something else happening: the entropy entering the reactor with the reactants is lifted from temperature T_o to a final value of T. Therefore, the energy released by the combustion is divided among two processes. Equation (37) gives the rate at which energy is released; the expressions for the energy dissipated, and for the energy used in lifting an amount of energy, are well known. Therefore, the balance of power reads:

$$-\left[\Delta\mu \cdot \Pi_n\right]_{reaction} = T\Pi_s + \left(T - T_o\right)\left|I_{s,conv,in}\right| \tag{174}$$

All three terms contain the unknown temperature of the combustion products. If we manage to properly express all the quantities in the previous two equations, we can solve the expressions for T, which normally are nonlinear. First, the change of the chemical potential for the reaction is computed in the following manner. Write the reaction equation as follows:

$$fuel + \nu_2 oxidizer + \nu_3 \mathrm{A} + \ldots = \nu_4 \mathrm{C} + \nu_5 \mathrm{D} + \ldots \tag{175}$$

Note that the stoichiometric coefficient multiplying the fuel has been set equal to one. On the left, we have the reactants, which may include substances which do not necessarily take part in the combustion. This is the case, for example, if a fuel is mixed with air, which mostly contains nitrogen. Now, the change of chemical potential takes the form

$$[\Delta\mu \cdot \Pi_n]_{reaction} =$$

$$|\Pi_{n,fuel}| \left\{ \sum_{products} v_j\left[\bar{h}_j(T) - T\bar{s}_j(T)\right] - \sum_{reactants} v_i\left[\bar{h}_i(T_o) - T_o\bar{s}_i(T_o)\right] \right\} \quad (176)$$

$\Pi_{n,fuel}$ is the rate at which fuel is burned in the reactor. Next, the convective entropy currents have to be written down for the particular chemical constituents of the fluids, and for the particular physical conditions:

$$\begin{aligned} |I_{s,conv,in}| &= |\Pi_{n,fuel}| \sum_{reactants} v_i\bar{s}_i(T_o) \\ |I_{s,conv,out}| &= |\Pi_{n,fuel}| \sum_{products} v_j\bar{s}_j(T) \end{aligned} \quad (177)$$

Combining the different pieces, we arrive at the following expression for the rate of production of entropy:

$$\Pi_s = \frac{|\Pi_{n,fuel}|}{T} \left\{ \sum_{products} v_j\left[\bar{h}_j(T) - T\bar{s}_j(T)\right] - \sum_{reactants} v_i\left[\bar{h}_i(T_o) - T\bar{s}_i(T_o)\right] \right\} \quad (178)$$

Note that in the last term on the right the exit temperature appears (and not T_0). These equations tell us that we must know the entropy and the enthalpy of reactants and products at their respective temperatures (and at constant pressure), with the temperature of the products still unknown, if we want to do the calculations. One way of doing this is by approximating the dependence of enthalpy and entropy on temperature using average temperature coefficients of enthalpy c_p for the species. (Such values may be read off the graph in Figure 9 of Chapter 2). The nonlinear equations can be solved iteratively.

Apart from the fuel and the oxidizer, there might be several other substances involved in the flow through the reactor, as in the case of fuel mixed with air. Also, we might use too much of the oxidizer or of the air, or too little; in the latter case some of the fuel will not be burned. In either case, all these factors influence the outcome of the combustion process negatively compared to the case of a mixture of fuel and the right amount of the oxidizer (and nothing else). The amount of air which delivers the proper mass of oxygen for complete combustion of the fuel is called *theoretical amount of air*.

EXAMPLE 33. Temperature reduction in steam as a result of throttling.

Use the following steam table (Table 8) to estimate the reduction of the temperature resulting from the irreversible expansion of steam. The diameter of the pipe is increased in such a way that the pressure of the steam is reduced from 1.0 bar to 0.70 bar. Initially, the steam is at a temperature of 120°C. Also calculate the new specific volume and the necessary change of the diameter of the pipe; finally, compute the entropy produced per kilogram of steam passing through the porous plug in the throttling device.

SOLUTION: According to the balance of energy, the enthalpy of the fluid must remain constant as a result of the process. The specific enthalpy at the inlet is $h(P=1.0, T=120) =$ 2716.6 kJ/kg. Finding this same value in the table for a pressure of 0.70 bar yields a temperature of $T_2 = 118.3$°C by linear interpolation between the appropriate values.

TABLE 8. A small portion of the steam tables

P / bar	T / °C	υ / m^3/kg	u / kJ/kg	h / kJ/kg	s / kJ/(K · kg)
0.70	100	2.434	2509.7	2680.0	7.534
	120	2.571	2539.7	2719.6	7.638
	160	2.841	2599.4	2798.2	7.828
1.0	100	1.696	2506.7	2676.2	7.361
	120	1.793	2537.3	2716.6	7.467
	160	1.984	2597.8	2796.2	7.660

Using this new temperature, we interpolate to find the specific volume and the specific entropy of the fluid: $\upsilon_2 = 2.559$ m^3/kg and $s_2 = 7.629$ kJ/(K · kg).

With a value of $\upsilon_1 = 1.793$ m^3/kg at the inlet, we can infer that the cross section of the pipe has to increase by a factor of 1.427, which means that the radius at the outlet must be larger than at the inlet by a factor of 1.195.

The specific entropy at the inlet is 7.467 kJ/(K · kg). Therefore, the amount of entropy produced is 162 J/(K · kg).

EXAMPLE 34. Burning methane with oxygen and with different amounts of air.

a) Let methane burn completely with oxygen. Determine the adiabatic flame temperature for this reaction. b) Calculate the theoretical amount of air for the combustion of methane. Assume air to be composed of nitrogen and oxygen. c) Determine the adiabatic flame temperature for the theoretical amount of air. d) Vary the amount of air (both up and down), and compute both the flame temperature and the rate of production of entropy (per unit amount of methane). e) Demonstrate that Equation (178) and the balance of entropy lead to the expression

$$\sum_{reactants} \nu_i \bar{h}_i(T_o) = \sum_{products} \nu_j \bar{h}_j(T)$$

The enthalpy of the entering mixture is equal to the enthalpy of the fluid stream exiting the reactor.

SOLUTION: a) Methane and oxygen combine to form carbon dioxide and water according to

$$CH_4 + 2O_2 = CO_2 + 2H_2O$$

With only these substances present in their stoichiometric ratios, the rate of production of entropy turns out to be

$$\Pi_s = \frac{|\Pi_{n,fuel}|}{T}\left\{\left[\bar{h}_{CO_2} + 2\bar{h}_{H_2O} - \left(\bar{h}_{CH_4} + 2\bar{h}_{O_2}\right)\right]_T - T\left[\bar{s}_{CO_2} + 2\bar{s}_{H_2O} - \left(\bar{s}_{CH_4} + 2\bar{s}_{O_2}\right)\right]_{T_o}\right\}$$

while the convective entropy currents at the inlet and the outlet are equal to

$$|I_{s,conv,in}| = |\Pi_{n,fuel}|\left(\bar{s}_{CH_4} + 2\bar{s}_{O_2}\right)_{T_o}$$

$$|I_{s,conv,out}| = |\Pi_{n,fuel}|\left(\bar{s}_{CO_2} + 2\bar{s}_{H_2O}\right)_T$$

These last two relations provide for a second expression of the rate of production of entropy. Numerical data about the entropies and the enthalpies of the species for different temperatures are given in Table 9. Table 10 lists the convective entropy currents and the rate of entropy production calculated according to the two forms given above. Since the rates have to be equal at the condition of the adiabatic flame temperature, interpolation in the last two columns gives the desired result (5240 K).

b) In air, there are 3.76 mole of N_2 for every mole of O_2. Therefore, if we write the reaction equation with the theoretical amount of air, it takes the form

$$CH_4 + 2(O_2 + 3.76N_2) = CO_2 + 2H_2O + 2 \cdot 3.76N_2$$

c) Repetition of this procedure with nitrogen included, gives an adiabatic flame temperature of 2330 K. Obviously, having inert nitrogen present changes the result considerably. Even though nitrogen does not take part in the reaction, it has to be heated to the flame temperature, which reduces the effect.

TABLE 9. Molar enthalpy and entropy of substances involved in combustion[a]

T / K	*h*_CH4	*h*_O2	*h*_CO2	*h*_H2O	*s*_CH4	*s*_O2	*s*_CO2	*s*_H2O
298	-74870	0	-393520	-241820	186	205	214	189
1000	-36686	22705	-360161	-215795	248	243	269	233
2000	48678	59197	-302123	-169166	306	269	309	264
3000	145200	98084	-240730	-115419	345	284	334	286
4000	238396	139013	-177782	-58825	372	296	352	302
5000	320356	182045	-112279	-2589	390	306	367	315
6000	386171	227323	-41970	50808	402	314	379	325
7000	432540	275012	36083	99376	409	321	391	332

a. [*h*] = J/mole; [*s*] = J/(K · mole).

d) With a positive surplus of air over the theoretical amount, the reaction equation can be written in the following form:

$$CH_4 + (2+X)(O_2 + 3.76N_2) = CO_2 + 2H_2O + (2+X)3.76N_2 + XO_2$$

For negative X, the equation does not apply. (It has to be changed appropriately to take into consideration that not all methane is burned.) Repeating the calculation for various values of X gives the graph shown below.

TABLE 10. Entropy currents and entropy production rate[a]

T / K	I_{s_in}	I_{s_out}	Π_{s1}	Π_{s2}
298	596	591	-5	2687
1000	596	734	138	855
2000	596	838	242	525
3000	596	906	310	443
4000	596	957	361	416
5000	596	997	401	409
6000	596	1029	433	411
7000	596	1056	460	416

a. $[I_s]$ = W/K; $[\Pi_s]$ = W/K.

The curves for the rate of entropy production (calculated per unit amount of methane burned), and for the adiabatic flame temperature demonstrate that the lowest irreversibility leads to the largest value of the flame temperature.

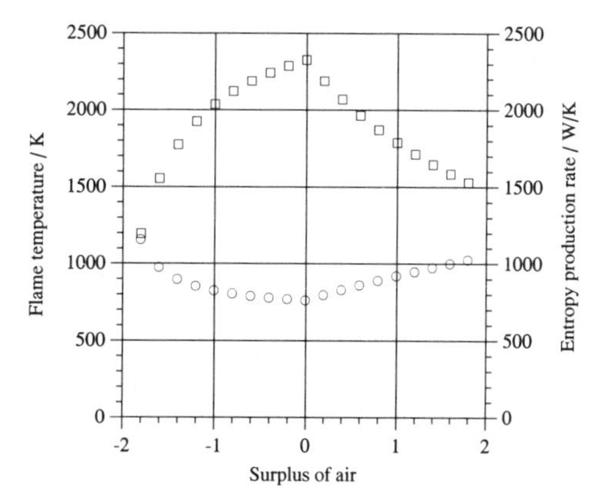

e) Look again at the equation for the rate of production of entropy for case (a). The second expression in brackets is itself equal to the rate of production of entropy (according to the balance of entropy). Therefore, the first term in brackets must be zero, proving the statement about the enthalpies of the two fluid streams. We could have started our analysis with this result; however, we wanted to base the derivation upon a consideration of entropy.

4.4.7 Transporting Matter in the Gravitational Field

At the surface of our planet, matter is transported up and down; this means that the gravitational field plays a role in the phenomena involving the transport of substances. Therefore we have to include in our laws of balance the transfer of energy as a consequence of the interaction of bodies and the gravitational field. If we consider a fluid flowing at the surface of the earth we have to add the source rates of energy associated with the transfer from field to body or vice versa.

A simple analysis of the influence of the gravitational field upon fluid flow can be performed using an energy balance involving a fluid entering a system at a particular gravitational level (potential) and leaving at another (Figure 39): with the fluid entering and leaving we have the two substancelike quantities mass and amount of substance flowing through the system. Therefore we write for the energy current at the inlet or outlet

$$I_{E,flow} = \mu I_n + \varphi_G I_m = \left[\mu + M_o \varphi_G\right] I_n \tag{179}$$

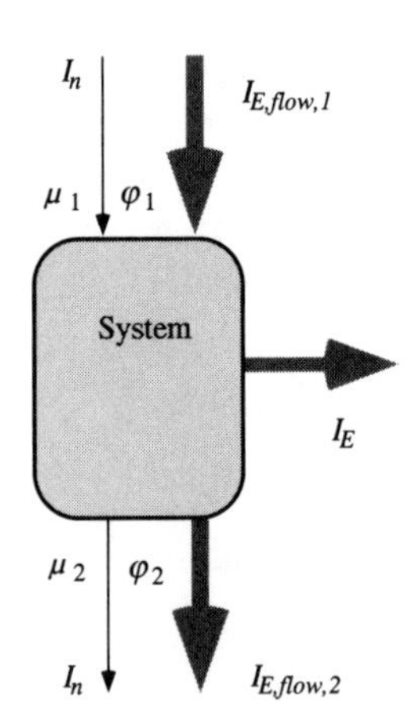

FIGURE 39. The effect of the interaction of fluids and the gravitational field may be included in the chemical potential of the fluid. We only have to write the energy current due to fluid flow as the sum of the chemical and the gravitational parts.

The quantity $\mu + M_o \varphi_G$, where M_o is the molar mass of the fluid, is often called the *gravito-chemical potential.* μ would be the normal chemical potential with no field; in other words, we can apply the results obtained so far if we wish to calculate the proper chemical potential. Equation (179) simply means that we can include the effect of the gravitational field in the chemical potential of the substance flowing in the field.

Now consider the balance of energy for the system depicted in Figure 39. If the fluid releases energy for a follow-up process, the balance in the steady state reads

$$\left[\mu + M_o \varphi_G\right]_1 I_{n1} + \left[\mu + M_o \varphi_G\right]_2 I_{n2} + I_E = 0 \tag{180}$$

If we have a stream of freely falling water or a fluid or air resting at the surface of the Earth, there is no other process involved, and we speak of gravito-chemical equilibrium, meaning that the potentials at the inlet and the outlet are equal:

$$\left[\mu + M_o \varphi_G\right]_1 = \left[\mu + M_o \varphi_G\right]_2 \tag{181}$$

Since in the case of a waterfall we have to include the effect of changes of velocity of the fluid, let us look only at a fluid at rest at the surface of our planet. Equation (181) tells us that the chemical potential, and therefore the pressure of the fluid, increases with decreasing height in the gravitational field. If we take the case of our atmosphere and model it as an ideal gas at constant temperature, we find that

$$P(h) = P_o \exp\left(-\frac{M_o g}{RT} h\right) \tag{182}$$

This is the well-known result for the dependence of pressure on height in an isothermal atmosphere (Chapter 2).

4.4.8 Osmosis and Sedimentation

Let us now turn to the discussion of diffusive processes. What we have learned so far can be used to prove that the osmotic pressure is very nearly equal to the partial pressure of the dissolved substance. Consider, as we have done in Section 4.1.2, two cells which are separated by a membrane which is permeable to the solvent only (Figure 40). In one cell we have the pure solvent, for example water, and in the other we have water with a dissolved substance which we may take to be sugar. Since the water may pass freely from one cell into the other its chemical potential in both, and therefore its pressure, must be the same:

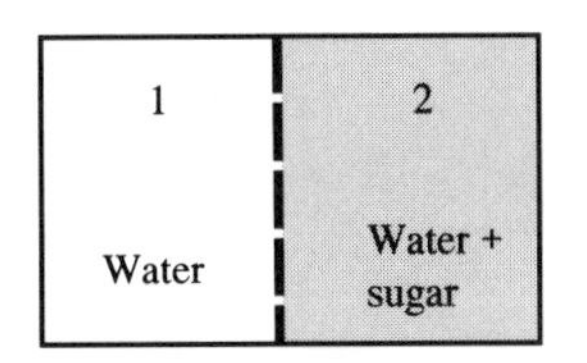

FIGURE 40. A solution (2) and its pure solvent (1) are separated by a semipermeable wall.

$$\mu_f(T, P_1) = \mu_f(T, P_2 - P_s) \tag{183}$$

Therefore, the total pressure in the second cell (the one with the solution) must be higher than the pressure of the pure water. Remember that in an isolated cell the solute decreases the pressure of the solvent, leaving the total pressure unchanged. If we apply the relation for the chemical potential of the incompressible solvent we get

$$M_o v_{f1}(P_1 - P_o) + \mu_f(T, P_o) = M_o v_{f2}(P_2 - P_s - P_o) + \mu_f(T, P_o) \tag{184}$$

The specific volume of the solvent is nearly the same in both cells, leading to

$$P_{osm} \equiv P_2 - P_1 = P_s \tag{185}$$

The pressure difference measured between the two cells is what we call the *osmotic pressure*. Therefore, Equation (185) yields the proof of the statement made above.

Let us consider another application of the concepts developed here, namely that of the chemical potential of a substance dissolved in an incompressible fluid, both being subject to a gravitational field (Figure 41). We can pretty much combine results derived before, leaving us with a fairly simple problem (with the exception of a small, but important, point). The phenomenon plays an important role if we create a suspension of fairly large particles, such as macromolecules, in a solvent. The resulting interplay between the solute, the solvent, and the gravitational field leads to *sedimentation*. It is possible to use the observed vertical concentration gradient to compute the molar mass of the suspended substance.

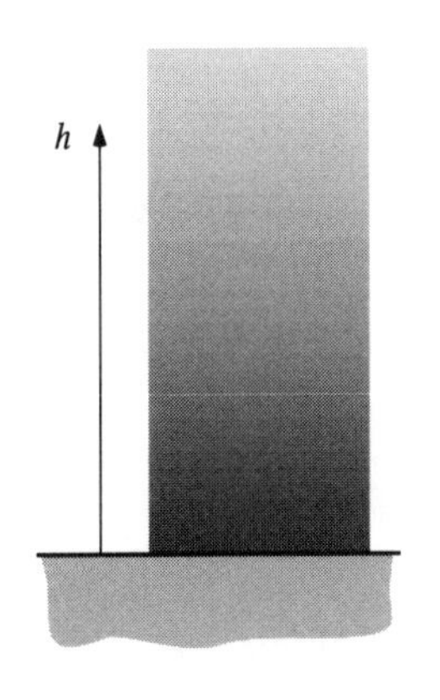

FIGURE 41. Solute sedimenting in a solvent. Both are subject to a gravitational field. Often, but not always, the concentration of the solute increases downward.

Take a container filled with an incompressible solvent and add a substance which will be dissolved. In the simplest case it is observed that more of the solute is concentrated further towards the bottom of the column of liquid. This is explained quite simply. The solute behaves like an ideal gas which tends to fill the space given to it: it spreads from points where the concentration, and therefore the chemical potential, are higher to places where they are lower. In a gravitational field, however, we have an effect which counters this tendency. In equilibrium, we should expect a certain density distribution as a function of height. Since the solute has the properties of an ideal gas, its pressure and its density will obey the relation we have seen several times before for an isothermal atmosphere in a gravitational field, Equation (182):

$$\rho_n(h) = \rho_{no} \exp\left(-\frac{M_{os}g}{RT}h\right) \tag{186}$$

M_{os} is the molar mass of the dissolved substance. This result may be used to derive the unknown molar mass of the solute.

There is a sometimes important effect which we have not taken into consideration in deriving the result of Equation (186). It is possible for the concentration of a solute to decrease downward in a solvent. (Such is the case for dissolved nitrogen in sea water; its solubility decreases with increasing depth.) Equation (186) does not predict such a result. We have neglected the buoyancy of the dissolved substance. This phenomenon may be included by noting that its effect is that of seemingly reducing the strength of the gravitational field. If we apply the standard analysis for the forces exerted upon a body in a fluid subject to a gravitational field, we find that the net force is

$$F_{net} = F_G - F_B = mg\left(1 - \frac{V}{m}\rho_f\right) \tag{187}$$

We therefore may introduce an apparent gravitational field g^*:

$$g^* = g\left(1 - \upsilon\rho_f\right) \tag{188}$$

where υ is the partial specific volume of the solute. Note that the volume V in the previous equation is not the volume of the entire solution but only its increase due to the presence of the solute. Including the effect of buoyancy leads to

$$RT \ln\left(\frac{\rho_n}{\rho_{no}}\right) = -M_{os}g\left(1 - \upsilon\rho_f\right)h \tag{189}$$

for the law of the sedimentation equilibrium.

4.4.9 The Transport of Substances as a Chemical Reaction

In this section, we will return to a point raised earlier: the description of transport processes can generally be put in the language developed for chemical reactions: the flow of a substance from point A to point B is like it disappeared at A and appeared at B, and the reason for the process can be found in a difference of the chemical potential. The substance will flow if

$$\mu(\mathrm{A}) > \mu(\mathrm{B}) \tag{190}$$

We have seen that it is certainly possible to use the language provided by the chemical potential for bulk flows and diffusion. We gain something by using an analogy, but we might not really have to do so. It would be nice to have examples of the flow of substances which are hard to cast in the usual language of pressure differences or of differences of concentration. Indeed, there are cases where water may flow against both pressure and concentration gradients.

Let us discuss a spectrum of cases of the flow of water through a pipe. At one end we have simple bulk flow through an empty pipe, while at the other end of the spectrum we will find a case for which the chemical potential provides the only explanation for the process.[3]

The first case is that of water, considered to be a viscous fluid, flowing through a pipe. If the flow remains laminar, there is a simple constitutive law for the current, namely the law of Hagen and Poiseuille (see the Prologue). Since the chemical potential is a function of the pressure it can be written in the following form:

$$j_n = -\sigma_n \frac{d\mu}{dx} \tag{191}$$

The law clearly resembles Ohm's law in electricity, and it is analogous to the law of diffusion, which is the second case to be discussed. Assume first that the water flowing is in the form of steam; then fill the pipe with sand or another porous medium. If we make the flow resistance larger and larger, we come to a point where we do not speak of the (bulk) flow of water any longer: we will now say that it is diffusing from A to B. In this case we introduce the constitutive law of diffusion:

$$j_n = -D\frac{d\rho_n}{dx} \tag{192}$$

where D is the diffusion constant. Since the chemical potential of the substances diffusing through the obstacle depends upon the concentration, we could just as well have written the law using the gradient of the potential instead of the gradient of concentration. If we accept this view, then bulk transport and diffusion do not look all that different.

Finally, let us add a hygroscopic substance to the stuff filling the pipe, and let the concentration of this substance increase from A in the direction of B. This apparently minor change brings about a situation which is hard to describe in "classical" terms of pressure and concentration of the flowing substance. If the effect of the hygroscopic substance can be made strong enough, water can flow from A to B even if there is a (slight) pressure difference in the opposite direction from what we had before, and even if the concentration of water increases from A to B. Water therefore flows both against the gradient of pressure and that of its own concentration. But it flows from A to B because its chemical potential at A is higher than at B due to the effect of the hygroscopic substance. The only reason we can give for the behavior of the water must be found in the values of its chemical potential. In summary, we may say that if other potential differences are not too large to counteract the chemical potential, a substance will always flow in the direction of decreasing chemical potential.

---

3. See G. Falk in Falk and Herrmann (1981), p. 38

4.5 Vapor Power and Refrigeration Cycles

This is the first of two sections dealing mostly with engineering applications of the thermodynamics of fluid systems. In the present section, we will deal with vapor power cycles and refrigeration systems,[4] while Section 4.6 will be devoted to convective heat transfer.

Fluid processes which include phase changes will allow us to apply what we have learned about pure ideal fluids in Section 4.3, and to combine the information with ideas discussed in the parts on fluid flow. In practical applications, fluids which can be treated as pure, run through cyclic processes while flowing through different parts of an engine, such as compressors or turbines, pumps or throttles, and so on. In contrast to what we discussed in earlier chapters, the fluids will be allowed to evaporate and to condense at various stages of the cycles.

4.5.1 Property Data and the Computation of Processes

If you want to calculate processes of simple fluids undergoing phase changes in addition to the changes discussed in previous chapters, you will need detailed information about the properties of the substances involved. These properties are reported in form of tables or graphs, of which the steam tables for water are the most famous. Even though you can also take advantage of computer programs providing the information in function format, it helps to look at the actual numbers when you are learning about the subject. Table A.16 provides values for saturated water (liquid or vapor; mixtures of liquid and vapor are calculated from the values of the saturated fluids). Also, if you wish to be able to calculate changes in the region of the superheated vapor, appropriate tables have to be given (Table A.17). Properties of subcooled or compressed liquid water complete the list of necessary material (Figure 44). An overview of the three main segments of fluid properties is shown in Figure 42.

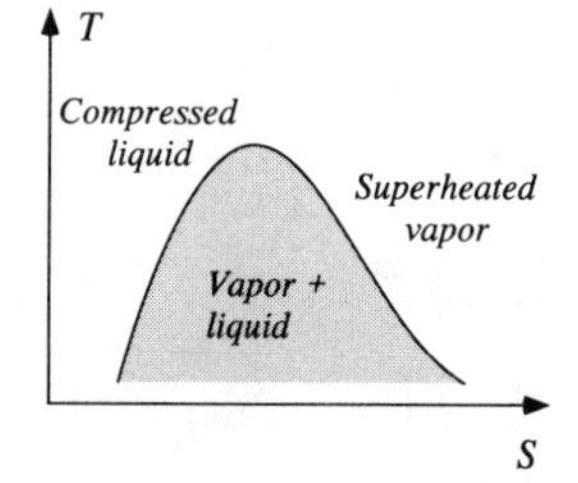

FIGURE 42. Compressed (subcooled) liquid, superheated steam, and mixtures of vapor and liquid are the main forms of a pure fluid.

Saturated fluids. To orient ourselves about the properties of fluids, it is convenient to start with the state of saturation, since the saturation line naturally divides the domain of independent variables into the different sections discussed above. The transition from the liquid to the gaseous state of a fluid is outlined in the *T-S* diagram. To catch this transition, one computes the values of thermodynamic properties of the saturated fluid, i.e., of saturated liquid and saturated vapor. We have seen what distinguishes the saturated fluid (and mixtures of the phases) from the rest: it is the nature of the chemical potential at those particular conditions. As we have learned during the description of phase changes, the chemical potential remains unchanged. Therefore, the potential is the same for both the liquid and the vapor at a given temperature.

4. For more information, see Moran and Shapiro (1992).

In Table A.16, the chemical potential of both phases is listed in the third column. As you can show by directly calculating the potential using energy, volume, and entropy, its values are indeed equal for the liquid and the vapor:

$$u_l + Pv_l - Ts_l = u_g + Pv_g - Ts_g \tag{193}$$

Continuing with Table A.16, the second column reports the vapor pressure of the fluid, while the last six columns list entropy, specific volume, and specific energy of both the liquid and the vapor. The values are given on a mass basis (specific values).

So far, we have used the *T-S* diagram only for reporting fluid properties, but other diagrams are used as well. In Figure 43, the distinctive saturation curve is presented for the refrigerant R123. The information is given in the *T-S* and the pressure-enthalpy diagrams. Depending on your purpose, you might use either diagram to discuss vapor processes.

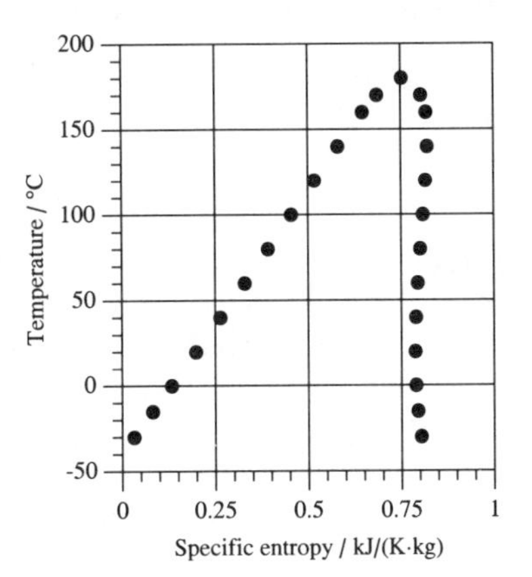

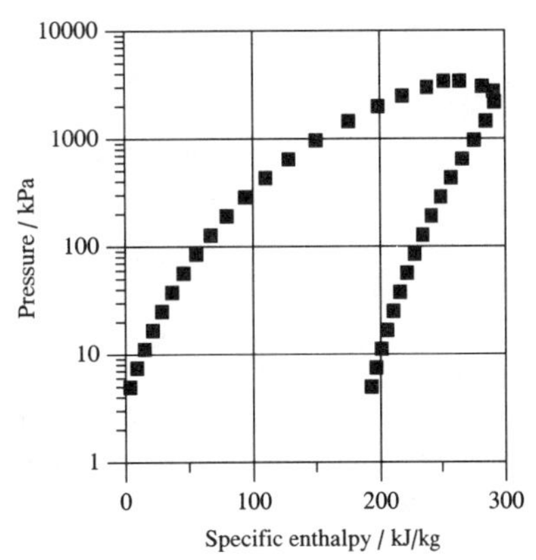

FIGURE 43. The saturation line for the refrigerant R123 looks quite different from that of water. While in the case of water, vapor will condense if it is compressed isentropically, this does not always happen with R123. (Consider, for example, a process starting with the saturated state at a temperature of 100°C, and leading to one at a lower temperature.) The curves have been computed using the program EES (Klein et at., 1991).

Subcooled (compressed) liquid. If a fluid exists in liquid form only, it may either satisfy the special condition of saturation (in which case its states would be found on the saturation line in Figure 42), or it may be found in the subcooled (compressed) region. Since liquids are hard to compress, their condition usually does not deviate much from that of the saturated liquid, if the pressure is not too high. Figure 44 shows a small region of the *T-S* diagram of water near the liquid saturation line. It demonstrates that for a pressure of 300 bar, the temperature is larger by only about 5 K at constant entropy. Put differently, for pressures in the range of up to a few hundred bar, the states lie in a narrow band around the saturation line. Entropy, energy, and volume of the liquid may be approximated by the saturated liquid data for the desired temperature, independently of pressure. Enthalpy, on the other hand, can be given by

$$h(T,P) = h_l(T) + (P - P_l)v(T) \tag{194}$$

where the index l refers to the state of liquid saturation.

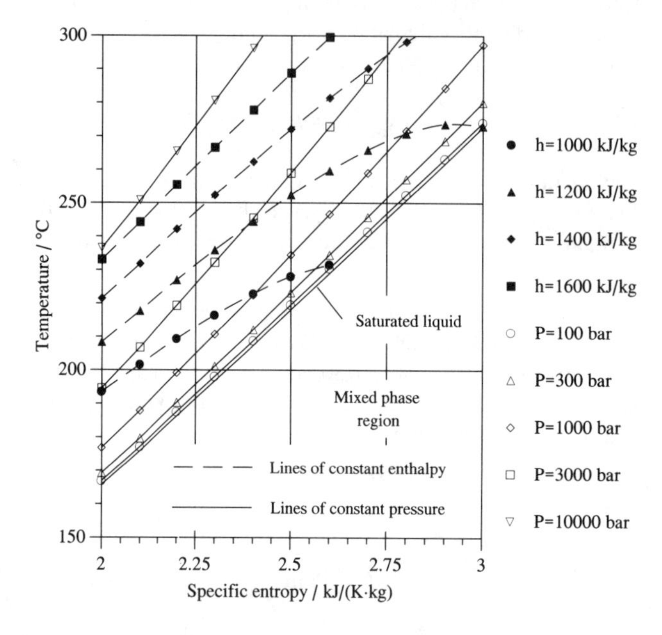

FIGURE 44. A small segment of the *T-S* diagram of subcooled water. For pressures that are not too large, the state of water is well approximated by that of the saturated liquid. At very high pressures, deviations become increasingly important. The diagram shows lines of constant pressure and enthalpy. Computations were done using the program EES (Klein et at., 1991).

Note that the lines of constant enthalpy are nearly horizontal under these conditions, meaning that this quantity is constant for constant temperature. (We can draw the same conclusion from the equation above for small specific volume and pressures that are not too large.) This is what we would expect of an incompressible substance. Take a closer look at the model of an *incompressible fluid.* Such a fluid is commonly defined as one which has constant specific volume, and whose energy depends only upon temperature. The enthalpy is then given by

$$h(T,P) = u(T) + P\upsilon \tag{195}$$

Under these circumstances, the temperature coefficient of energy (i.e., the specific heat at constant volume) is a function of temperature only, and the temperature coefficient of enthalpy (i.e., the specific heat at constant pressure) must be equal to the former:

$$\begin{aligned} c_V(T) &\equiv \frac{\partial u(T)}{\partial T} = \frac{du}{dT} \\ c_p(T) &\equiv \frac{\partial h(T,P)}{\partial T} = \frac{du}{dT} \end{aligned} \tag{196}$$

Changes of energy and of enthalpy are obtained by integrating the appropriate expressions. For constant coefficients, the results take the form

$$\begin{aligned} u(T) &= u(T_o) + c(T - T_o) \\ h(T) &= h(T_o) + c(T - T_o) + (P - P_o)\upsilon \end{aligned} \tag{197}$$

The same arguments applied to the entropy of the incompressible fluid yield the result, which again holds for constant c:

$$s(T) = s(T_o) + c\ln\left(\frac{T}{T_o}\right) \tag{198}$$

These are all relations we have used before in applications. In summary, to compute property data for the compressed liquid, we may use data of the saturated liquid for a desired temperature, and adjust the enthalpy for pressure as in Equation (194). Changes in temperature are taken into account according to the equations given above.

Mixtures of liquid and vapor: the quality. The conditions for which mixtures of a liquid and its vapor exist need special attention. It is customary to describe the properties in terms of the composition of the mixture which is quantified by stating how much vapor is present relative to the total fluid:

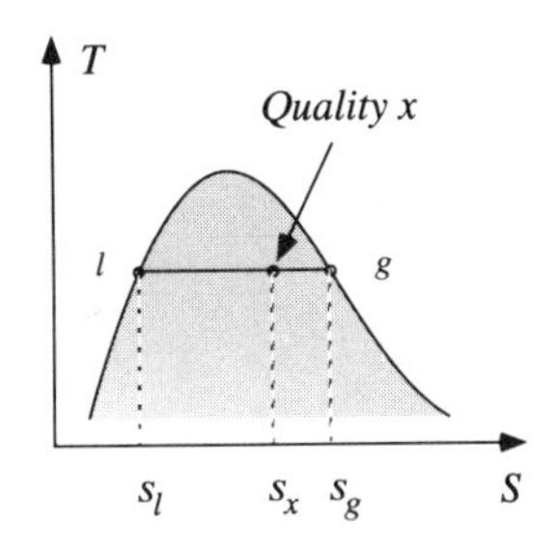

FIGURE 45. A state of quality x and the corresponding pure states (liquid and vapor).

$$x = \frac{n_g}{n_l + n_g} = \frac{m_g}{m_l + m_g} \tag{199}$$

This quantity is called the *quality* of the mixture (Figure 45). Note that it is equivalent to the notion of the mole fraction (of vapor) of a two-component mixture. All the intermediate states of the mixed fluid (liquid plus vapor) can be computed on the basis of the values of the saturated fluid, which means that we do not need additional data for this set of conditions. A fluid with a quality of 0 or 1 simply corresponds to pure liquid or pure vapor, respectively. States with a quality between these values are found on the horizontal line in the T-S diagram connecting the conditions of liquid and vapor. (See Figure 46.)

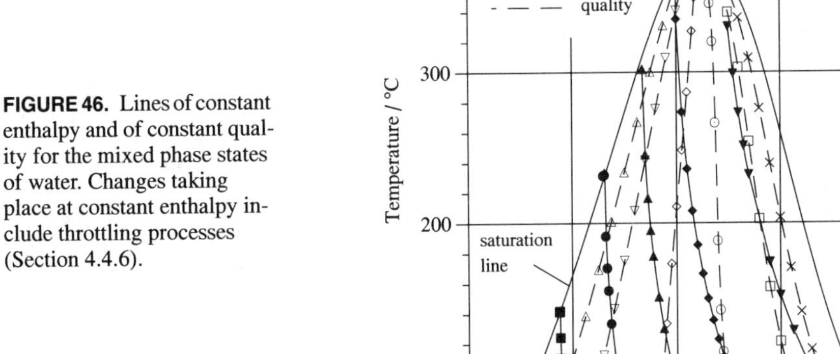

FIGURE 46. Lines of constant enthalpy and of constant quality for the mixed phase states of water. Changes taking place at constant enthalpy include throttling processes (Section 4.4.6).

The total volume, entropy, energy, and enthalpy of the fluid mixture are expressed as the sum of the corresponding quantities for the liquid and the vapor. To start with the volume, the total specific volume is defined as the total volume divided by the total mass

$$\upsilon = \frac{V}{m} = \frac{V_l + V_g}{m} = \frac{m_l \upsilon_l + m_g \upsilon_g}{m}$$
$$= \frac{m_l}{m}\upsilon_l + \frac{m_g}{m}\upsilon_g$$

which is equivalent to

$$\upsilon = (1-x)\upsilon_l + x\upsilon_g \tag{200}$$

Exactly the same forms can be shown to apply to the other properties:

$$\begin{aligned} s &= (1-x)s_l + xs_g \\ e &= (1-x)e_l + xe_g \\ h &= (1-x)h_l + xh_g \end{aligned} \tag{201}$$

Note that the values of s_l and s_g, for example, are those of the saturated states, which correspond to points on the saturation curve of Figure 45. With this information, the values corresponding to any mixture of liquid and vapor can be computed.

Superheated vapor (steam). When we finally have pure vapor, the conditions of the fluid can be changed quite drastically with relatively modest changes of pressure and temperature. This is demonstrated in Figure 47, which shows the

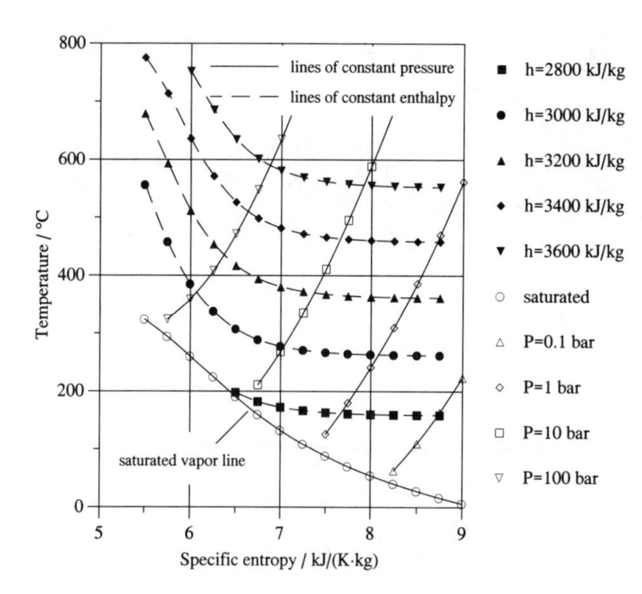

FIGURE 47. Superheated vapor region of the *T-S* diagram provides data about the vapor state of the fluid. Lines of constant pressure and enthalpy are given. Where the lines of constant enthalpy are horizontal, we can consider the vapor to be in the ideal gas state. Computations were done using the program EES (Klein et at., 1991).

part of the *T-S* diagram of water to the right of the saturation line. The most important feature of the data presented in the figure is this: vapor can be treated as an ideal gas only if the pressure is not too high. The conditions for which the ideal gas model applies are marked by the horizontal sections of the lines of constant enthalpy. As you remember, this model requires the enthalpy of the fluid to be a function of temperature only. If the conditions of the ideal gas are satisfied, we can apply all the simple relations derived in the previous sections and chapters. Figure 48 indicates that water vapor contained in the atmosphere can be treated in this simple manner. This simplifies applications in atmospheric physics and air conditioning to a certain degree. In the realm of the "real" gases, however, detailed property data, again in the form of tables, graphs, or computer programs, have to be provided (Table A.17). Calculations by hand are tedious but instructive, at least during the learning phase. Sketching processes in diagrams, however, always helps in visualizing the appropriate information.

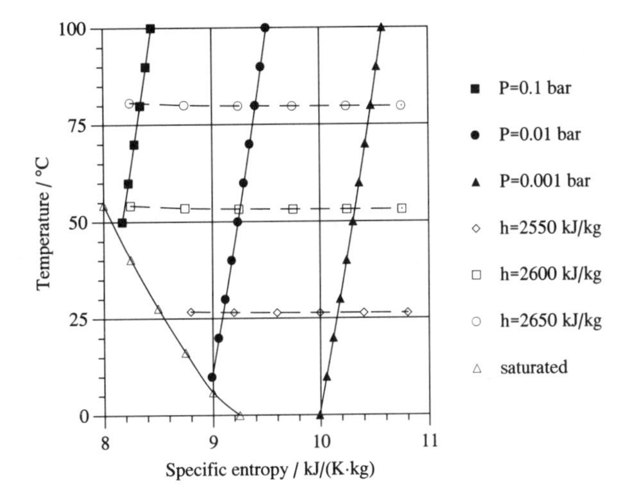

FIGURE 48. Superheated water vapor in the range interesting for moist air here on Earth. The ideal gas model may be applied to the computation of the properties of the water vapor contained in the air.

Calculating processes including phase transitions. Pure fluids which are allowed to go through phase changes play an important role in the sciences and in engineering. So far, we have a theory of uniform reversible processes, as discussed in the previous section. Even though real cases hardly ever conform to the conditions of this model, its results can still be applied. As long as we can provide information about the values of some variables at certain states, changes occurring between those states can be computed for the fluid even if irreversibilities are present. In the absence of more detailed information about actual processes, it is important to be able to approximate them by simple models. Experience shows that the results derived provide for a good basis from which to discuss of concrete cases in engineering and the sciences.

EXAMPLE 35. The entropy of a mixture of water and water vapor.

Water is found to vaporize at a temperature of 200°C. If there are 12 kg of liquid water and 18 kg of water vapor present, how large is the entropy of the mixture? Estimate the necessary values from Figure 46.

SOLUTION: First, we determine the quality of the mixture of steam and water. It is given by

$$x = \frac{m_{vapor}}{m_{liquid} + m_{vapor}} = \frac{18}{12+18} = 0.60$$

Now we need only the specific entropies of saturated water and of saturated steam at a transitional temperature of 200°C. According to Figure 46, the values are 2300 J/(K · kg) and 6400 J/(K · kg), respectively. Using Equation (201a), we calculate the specific entropy of the mixture to be equal to

$$s = (1-x)s_{liquid} + xs_{vapor} = (1-0.6)\cdot 2300 + 0.6\cdot 6400 = 4760\frac{\text{J}}{\text{K}\cdot\text{kg}}$$

With a total mass of 30 kg, the final result is 143 kJ/K.

EXAMPLE 36. Isentropic expansion of superheated water vapor.

Superheated water vapor at a pressure of 30 bar and a temperature of 300°C expands isentropically to a state with a temperature of 100°C. (Such a change might occur for adiabatic expansion of an ideal fluid in a turbine.) a) Calculate the specific entropy and enthalpy of the fluid. b) Determine the pressure and the temperature at which the fluid begins to condense. c) Calculate the pressure and the quality of the fluid mixture at the final state. Perform the calculations by interpolation of the graphs provided above.

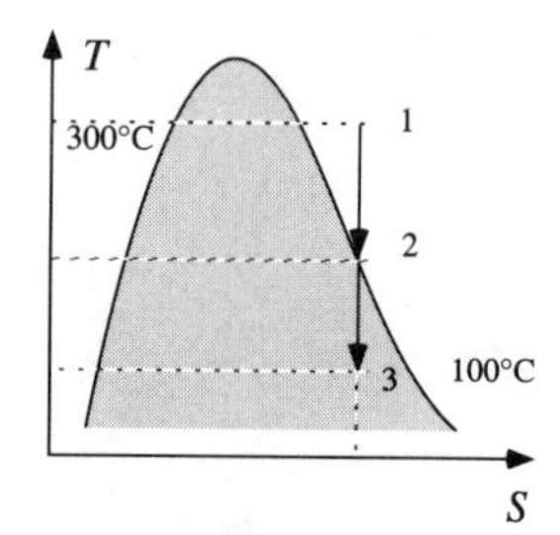

SOLUTION: a) In Figure 47, you can identify the initial state 1 (see the accompanying figure). For factors of 10, the lines of constant pressure are spaced at roughly equal distances. Therefore, the line for 30 bar is to be found between those for 10 bar and for 100 bar. At 300°C, the corresponding value of the entropy is 6.50 kJ/(K · kg). The enthalpy is read off the same graph; we get a value of 3000 kJ/kg. (Accurate values: s = 6.537 kJ/(K · kg), h = 2993 kJ/kg)

b) Following the vertical line in the T-S diagram leads to a point on the saturation curve at about 190°C. Both the vapor pressure curve in Figure 22 and the continuation of the 10 bar curve in Figure 47 indicate that the corresponding pressure must be slightly above 10 bar. (Accurate values: T = 186°C, P = 11.50 bar)

c) Water vapor condenses at a pressure of 1.0 bar if the temperature is 100°C. Interpolation in Figure 46 gives a rough value for the quality of 0.90 at the final state. The specific enthalpy turns out to be about 2500 kJ/kg. (Accurate value: x = 0.865)

EXAMPLE 37. Entropy and energy exchanged in a constant pressure process.

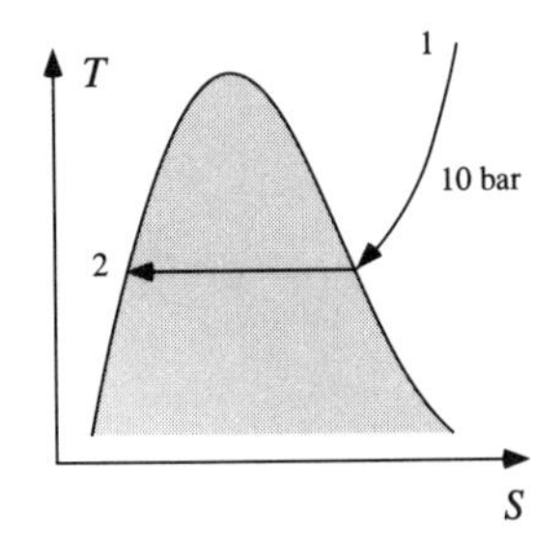

Superheated water vapor at a pressure of 10 bar and a temperature of 400°C is expanded in an isobaric process until the state of saturated liquid is reached. a) How much entropy must be emitted or absorbed by 10 kg of the fluid? b) How much energy is transferred in heating or in cooling? c) How much energy has been transferred as a consequence of the mechanical process? Use tables provided in the Appendix for calculations.

SOLUTION: a) The initial specific entropy is 7.465 kJ/(K · kg). Saturated liquid water at a pressure of 10 bar has a specific entropy of 2.139 kJ/(K · kg). Since the process is modeled as an ideal one, the change of entropy corresponds to the amount emitted: $S_e = -53.26$ kJ/K.

b) The process is isobaric. Therefore, the amount of energy emitted in cooling of the fluid is equal to the change of enthalpy: $Q = m\,(h_2 - h_1) = 10\,(763 - 3264)$ kJ = - 25.01 MJ. Enthalpies have been computed from the internal energy and the specific volume data in the tables.

c) At constant pressure, the energy exchanged because of expansion or contraction can be calculated from the change of the volume: $W = -P\Delta V = -10 \cdot 10^5 \cdot 10\,(0.00113 - 0.3066)$ J = 3.05 MJ.

This last result could also have been obtained from applying the law of balance of energy. The change of the energy of the system can be taken from the tables. Together with the energy exchanged in cooling, W is computed from

$$\Delta E = W + Q$$

We can use this to calculate the change of the energy of the vapor. (It is equal to - 22 MJ.)

4.5.2 Vapor Power Cycles

Detailed fluid properties play an important role in the design of power plants which use vapor power cycles such as the steam power plant of Figure 49. The thermal part of the plant consists of a boiler, turbine, condenser, and feedwater

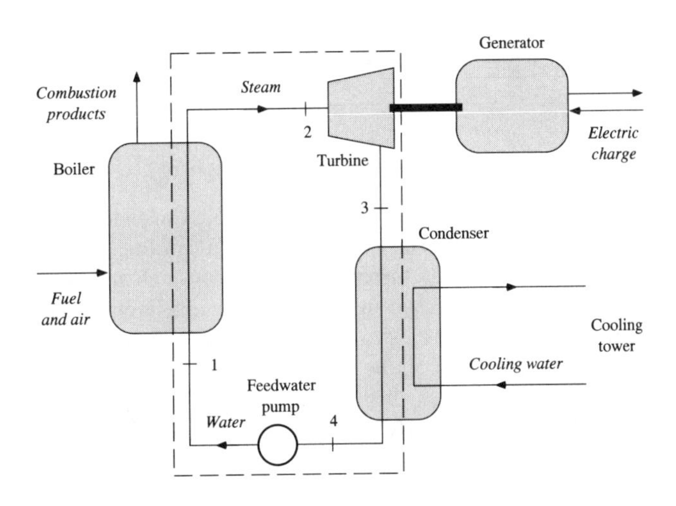

FIGURE 49. Steam power plant. Four devices operate on the water. The feedwater pump increases the pressure of the cold water. This water is then heated and evaporated in the boiler. Steam at high pressure drives the turbine and is thereby expanded adiabatically. Finally, the condenser turns the low pressure steam into water. The dashed line is the surface of the system which encloses the fluid operating in the power plant.

pump. A fluid such as water circulates through the boiler where steam is produced (points 1 to 2 in Figure 49). Then the steam drives the turbine (points 2 to 3) and enters the condenser, where it is turned into liquid water (points 3 to 4). The feedwater pump increases the pressure of the fluid to the value at the inlet to the turbine (points 4 to 1), completing the cycle.

Principle of operation of a steam power plant. We can model the cycle undergone by the fluid as a sequence of steps which we have studied before. First consider each step to be reversible. Starting with the process in the boiler, liquid water at high pressure (P_1) is first heated and then evaporated; assume the step to be finished when all the water is turned into saturated vapor (point 2 in Figure 50). We shall learn later about the consequences of superheating for the vapor power cycle. The step leading from point 1 to point 2 is supposed to take place at constant pressure ($P_2 = P_1$). A simplified model of what happens to the fluid expresses the energy current of heating in the boiler as follows:

$$\left|I_{E,12}\right| \approx \left(\frac{1}{2}(T_1 + T_2)(s_{2'} - s_1) + T_2 \cdot (s_2 - s_{2'})\right) I_m \tag{202}$$

I_m is the current of mass of the fluid flowing through the main loop of the power plant, and T_1 can be approximated by T_4. (The feedwater pump does not strongly increase the temperature of liquid water.) The upper and the lower temperatures of the cycle are determined by the value of the pressure associated with the appropriate point. Alternatively, the first term in Equation (202) can be computed using the enthalpy of the fluid.

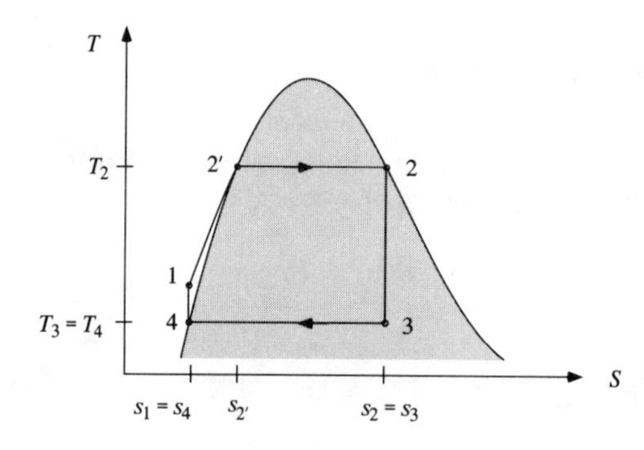

FIGURE 50. Vapor cycle undergone by an ideal fluid in a steam power plant. The sequence of steps is called the *ideal Rankine cycle*.

The second step of the cycle, from point 2 to point 3, is taken to be an isentropic (adiabatic and reversible) expansion of the fluid. According to Figure 50, the quality of the steam leaving the turbine is less than 1, which means that droplets of water form in the turbine. This effect can easily lead to problems with operating the engine. Therefore one tries to keep the quality as high as possible, which can be achieved by superheating the steam (see below). The energy released in this step is transferred to the electric generator of the power plant; it can be calculated either directly or indirectly by calculating the other rates of energy transfer in the cycle.

The condenser, which basically is a heat exchanger operating at constant pressure, turns the steam which has not already condensed, into liquid water. Here, all the entropy added to the cycle in heating is removed and transferred to the environment through a cooling tower or similar device. The rate of energy transfer is easily calculated to be equal to

$$\left|I_{E,24}\right| = T_3\left(s_3 - s_4\right)I_m \tag{203}$$

The final step from point 4 back to point 1 consists of raising the pressure of the liquid to its upper value. The pump is supposed to operated reversibly and adiabatically. Since the density of the liquid usually does not change appreciably, the energy current necessary for operating the pump is computed as follows:

$$\left|I_{E,41}\right| \approx v_4\left(P_1 - P_4\right)I_m \tag{204}$$

Overall, the current of energy delivered by the power plant is calculated from a steady state balance of energy for the system within the dashed line in Figure 49:

$$\begin{aligned}\left|I_{E,23}\right| &= \left|I_{E,12}\right| - \left|I_{E,24}\right| - \left|I_{E,41}\right| \\ &\approx \left(h_2 - h_1 + T_2\left(s_{2'} - s_1\right) - T_3\left(s_3 - s_4\right) - v_4\left(P_1 - P_4\right)\right)I_m\end{aligned} \tag{205}$$

Superheating in the power cycle. If heating of the fluid in the boiler is not stopped at the point where all the liquid has turned into saturated vapor, the cycle depicted in Figure 50 changes as demonstrated in Figure 51. The superheating is done in a heat exchanger separate from the boiler, called the superheater. The boiler and superheater are known as the steam generator. Superheating has two main effects: first, the average temperature of heating of the fluid is higher than without the additional process, leading to increased efficiency of the steam engine; second, the problem of low quality of the steam leaving the turbine is alleviated. You may even get pure vapor (quality equal to 1) at the exit of the turbine.

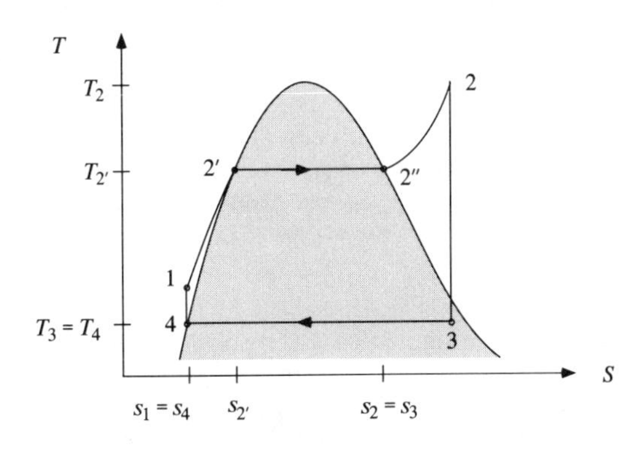

FIGURE 51. Vapor power cycle (ideal Rankine cycle) with superheating of the steam.

Irreversibilities in a vapor power plant. There are numerous sources of entropy production in a vapor power plant. If you look at Figure 49, you can identify different processes which lead to irreversibilities. First, there is the case of generation of heat, which normally is accomplished by combustion. As you follow the path of entropy through the plant, you next have to consider the effect of the heat exchangers in the boiler and the condenser. In the fluid undergoing the Rankine cycle, entropy production chiefly occurs in the turbine and the pump, with the turbine usually contributing much more strongly. This latter effect is depicted in Figure 52.

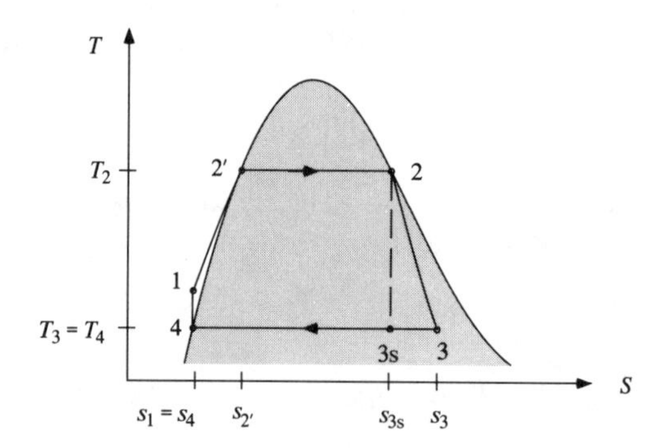

FIGURE 52. Entropy production in the vapor power cycle. The largest contribution to irreversibility in the fluid usually comes from the operation of the turbine.

Considering the rates of entropy production in the entire plant is important if we wish to quantify the losses occurring as a result of irreversibility (Example 40).

EXAMPLE 38. Energy currents and efficiency of an ideal Rankine cycle.

An ideal Rankine cycle is operated for water at a high pressure of 80 bar and a low pressure of 0.1 bar. Steam is not superheated. a) With a current of water of 100 kg/s, estimate the energy current with respect to the fluid in the boiler; use the diagrams showing fluid properties provided in this chapter. b) How large is the energy current leaving the power plant due to cooling? c) How much power is necessary to operate the feedwater pump? d) Calculate the thermal efficiency of the cycle. e) If the water used for cooling the plant enters at a temperature of 20°C and leaves at 35°C, how large does the current of mass of the cooling liquid have to be? f) Calculate the quality of the fluid after isentropic expansion. g) If the quality at the end of isotropic expansion is equal to 1, what is the efficiency of the cycle?

SOLUTION: Since you should solve the problem by reading property values from the graphs, you will get only approximate results. (Try solving this same problem with values read from tables or computed using appropriate programs.) First consider Figure 16, which has been redrawn in a slightly different manner in the diagram above. Both cycles, i.e., those without and with superheating, have been superimposed on the graph below. The former joins the points 1 - 2′ - 2″ - 3′ - 4 - 1, while the latter includes point 2 and 3.

The lines of constant pressure for 0.1 bar, 60 bar, and 150 bar have been included. For the upper pressure of 100 bar, we have to interpolate in the diagram. We can read the following values (Table 11) from this graph and from Figures 22 and 46:

TABLE 11. Steam properties at low and high pressure in the cycle

Pressure	$T_{evaporation}$	s_l	s_g
0.1 bar	50°C	0.7 kJ/(K · kg)	
80 bar	300°C	3.2 kJ/(K · kg)	5.7 kJ/(K · kg)

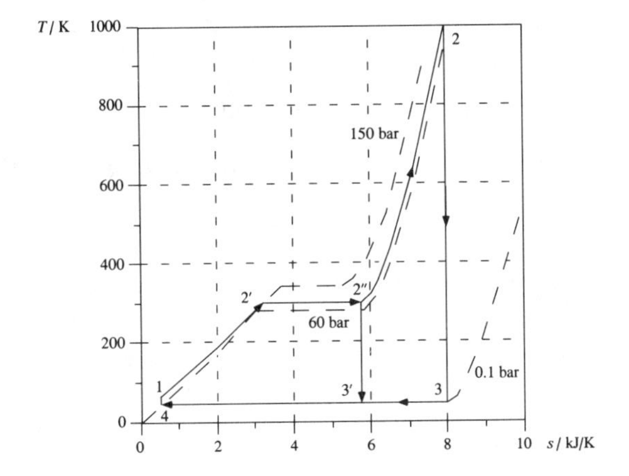

a) The energy current added to the steam in heating from point 1 to point 2″ can be estimated as in Equation (202):

$$I_{E,1-2''} = \left[T_{m,1-2'}\left(s_{l,2'} - s_{l,1}\right) + T_H\left(s_{g,2''} - s_{l,2'}\right)\right]I_m$$
$$= \left[450 \cdot (3.2 - 0.7) + 573 \cdot (5.7 - 3.2)\right]100\,\text{kW} = 255\,\text{MW}$$

b) Condensation takes place at constant temperature from point 3′ to point 4:

$$I_{E,3'-4} = T_l\left(s_{g,2''} - s_{l,1}\right)I_m = 323(5.7 - 0.7)100\,\text{kW} = 162\,\text{MW}$$

c) The energy current necessary for compressing the liquid from the state at point 4 to the one at point 1 turns out to be rather small:

$$I_{E,4-1} \approx \upsilon\left(P_1 - P_4\right)I_m \approx 1.2 \cdot 10^{-3} \cdot 8 \cdot 10^6 \cdot 100\,\text{W} = 1\text{MW}$$

d) The numbers computed so far let us calculate the thermal efficiency of the plant:

$$\eta = \frac{\left(I_{E,1-2''} + I_{E,4-1}\right) - I_{E,3'-4}}{I_{E,1-2''} + I_{E,4-1}} = \frac{256 - 162}{256} = 0.37$$

e) The entire energy current discharged in the condenser must be carried away by the cooling water. For this fluid, we can assume conditions of constant pressure and constant temperature coefficient of enthalpy. Therefore we have

$$c_p I_{m,cooling}\left(T_{c,H} - T_{c,L}\right) = I_{E,3'-4} \quad \Rightarrow \quad I_{m,cooling} = 2570\,\text{kg/s}$$

f) By extrapolating the lines of constant quality in Figure 46 down to a temperature of 50°C, we obtain a value of $x = 0.7$ for the quality of the vapor at point 3′.

g) To answer this question, we have to recalculate the quantities obtained above, this time including superheating of the steam from point 2″ to point 2. The important quantity to be read from the graph is the specific entropy of 8.0 kJ/(K · kg) at a temperature of about 1000°C at point 2. Estimates of the additional or new energy currents are

$$I_{E,2''-2} = T_{m,2''-2}\left(s_{g,2''} - s_{g,2}\right)I_m = 923(8.0 - 5.7)100\,\text{kW} = 210\,\text{MW}$$

$$I_{E,3-4} = T_l\left(s_{g,2'} - s_{l,1}\right)I_m = 323(8.0 - 0.7)100\,\text{kW} = 236\,\text{MW}$$

The new thermal efficiency turns out to be ((256+210)–236)/(256+210) = 0.49.

EXAMPLE 39. Entropy production and efficiency of a turbine.

The isentropic efficiency of a turbine is defined as the ratio of its actual power and the power it would have if it were operated isentropically. a) Draw a flow diagram for the operation of the turbine. b) Derive an expression for the rate of entropy production of the turbine in terms of the efficiency, the enthalpies of the fluid at points 2 and 3*s* in Figure 52, and the current of mass of the fluid.

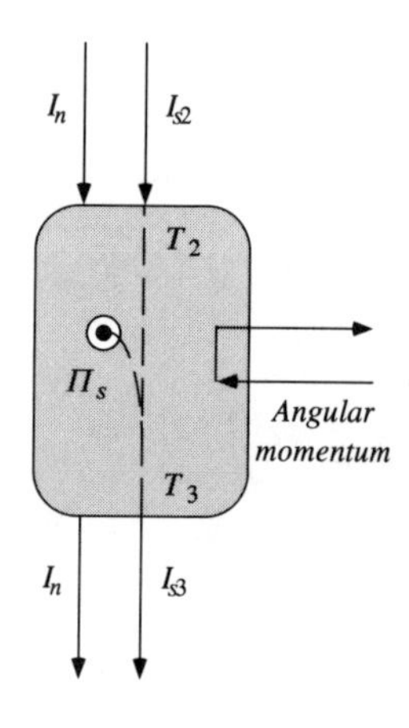

SOLUTION: a) According to the discussion of flow processes in Section 4.4, the flow diagram including the effect of irreversibility must look as shown in the figure. While the current of amount of substance is constant, the flux of entropy increases because of irreversibility.

b) The energy current associated with the mechanical process of the turbine can be calculated as the difference of the energy fluxes entering and leaving the system together with the steam. According to Section 4.4.5, the latter currents can be written in terms of the flux of amount of substance, the chemical potentials, and the molar entropies:

$$\left|I_{E,mech}\right| = \left|I_n\right|\left[\left(\mu_2 + T_2\bar{s}_2\right) - \left(\mu_3 + T_3\bar{s}_3\right)\right] = \left|I_n\right|\left[\left(\mu_2 + T_2\bar{s}_2\right) - \left(\mu_3 + T_3\left(\bar{s}_2 + \frac{1}{I_n}\Pi_s\right)\right)\right]$$

$$= \left|I_n\right|\left[\left(\mu_2 + T_2\bar{s}_2\right) - \left(\mu_3 + T_3\bar{s}_2\right)\right] - T_3\Pi_s$$

The term multiplying the current I_n is equal to the difference of the molar enthalpies of the fluid at points 2 and 3*s* in Figure 52. Therefore we have

$$\left|I_{E,mech}\right| = \left|I_n\right|\left[\bar{h}_2 - \bar{h}_{3s}\right] - T_3\Pi_s$$

$$= \left|I_m\right|\left[h_2 - h_{3s}\right] - T_3\Pi_s$$

While the quantity on the left-hand side is the real power of the turbine, the first term on the right is the isentropic power. With the definition

$$\eta_{turbine} = \frac{\left|I_{E,mech}\right|}{\left|I_{E,mech}\right|_s}$$

we can express the rate of production of entropy as follows:

$$\begin{aligned} T_3 \Pi_s &= \left|I_{E,mech}\right|_s - \left|I_{E,mech}\right| = \left|I_{E,mech}\right|_s - \eta_{turbine}\left|I_{E,mech}\right|_s \\ &= \left(1 - \eta_{turbine}\right)\left|I_m\right|\left[h_2 - h_{3s}\right] \end{aligned}$$

EXAMPLE 40. Contributions to irreversibility in a steam power plant.

Consider a power plant running a vapor cycle as described in Example 38 (without superheating). Allow for the adiabatic expansion in the turbine to be irreversible. Additional information is provided about the situation in the heat exchangers and in the burner. Calculate the relative importance of the different sources of irreversibility. a) Assume that methane is burned as fuel using the theoretical amount of air. The hot gases are cooled to 500 °C in the heat exchanger, where steam is produced, before they are emitted through the stack of the power plant. Calculate the flux of mass of methane and air necessary to operate the cycle. Calculate the rates of production of entropy for the combustion and for heat transfer in the heat exchanger. b) Let the turbine have an isentropic efficiency of 90%. Calculate the rate of entropy production in the turbine. c) What is the rate of entropy production in the heat exchanger where the steam is condensed? d) Express each contribution to the rate of production of entropy as a fraction of the total irreversibility.

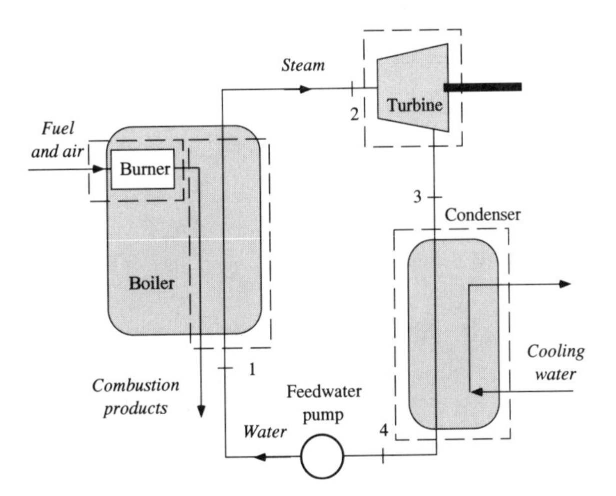

SOLUTION: There are four sources of irreversibility to be considered (the contribution to the production of entropy in the pump is neglected). The schematic of the plant shows the corresponding four control volumes.

a) Burning of methane with air was discussed in Example 34. In the accompanying diagram, the flame temperature and the rate of entropy production per mole of methane burned in unit time can be found.

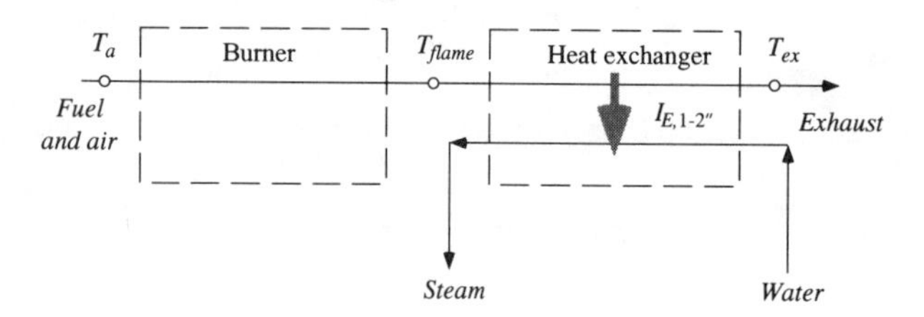

The energy current $I_{E,1\text{-}2''}$ was calculated in Example 38 as 255 MW. Now, two expressions for the balance of energy for both the burner and the heat exchanger, and for the heat exchanger alone, can be established:

$$I_{E,1-2''} = I_n\left[\left(\bar{h}_{CH4} + 2\bar{h}_{O2} + 3.76\cdot 2\bar{h}_{N2}\right)_{Ta} - \left(\bar{h}_{CO2} + 2\bar{h}_{H2O} + 3.76\cdot 2\bar{h}_{N2}\right)_{Tex}\right]$$

$$I_{E,1-2''} = I_n\left[\left(\bar{h}_{CO2} + 2\bar{h}_{H2O} + 3.76\cdot 2\bar{h}_{N2}\right)_{Tflame} - \left(\bar{h}_{CO2} + 2\bar{h}_{H2O} + 3.76\cdot 2\bar{h}_{N2}\right)_{Tex}\right]$$

I_n is the flux of amount of substance for methane, and T_{ex} denotes the temperature of the exhaust gases. T_a is set equal to 298 K. With T_{flame} = 2330 K and T_{ex} = 773 K, the flux of amount of substance of methane turns out to be 400 mole/s, which translates into a mass flux of 6.4 kg/s for methane and 110 kg/s for air. Now, with the results of Example 34, the rate of production of entropy in the burner must be equal to

$$\Pi_s = 763\frac{\text{W}}{\text{K}\cdot\left(\text{mole}\cdot\text{s}^{-1}\right)}\cdot 400\frac{\text{mole}}{\text{s}} = 3.05\cdot 10^5\,\text{W/K}$$

The balance of entropy for the heat exchanger, on the other hand, takes the form

$$\Pi_s = I_{m,steam}\left(s_{2''} - s_1\right)$$
$$- I_{n,CH4}\left[\left(\bar{s}_{CO2} + 2\bar{s}_{H2O} + 3.76\cdot 2\bar{s}_{N2}\right)_{Tflame} - \left(\bar{s}_{CO2} + 2\bar{s}_{H2O} + 3.76\cdot 2\bar{s}_{N2}\right)_{Tex}\right]$$

or

$$\Pi_s = 100\frac{\text{kg}}{\text{s}}(5700 - 700)\frac{\text{J}}{\text{K}\cdot\text{kg}} - 400\frac{\text{mole}}{\text{s}}\left(2.80\cdot 10^3 - 2.35\cdot 10^3\right)\frac{\text{J}}{\text{K}\cdot\text{mole}}$$
$$= 3.20\cdot 10^5\,\text{W/K}$$

b) The rate of entropy production in the turbine can be calculated according to the result of Example 39. The numbers computed in Example 38 tell us that the isentropic power of the turbine is 256MW - 162 MW = 94 MW. Therefore, the rate of entropy production is

$$\Pi_s = \frac{1}{T_L}\left(1 - \eta_{turbine}\right)I_{E,2''-3'} = \frac{1}{323}(1 - 0.90)94\cdot 10^6\,\text{W/K} = 0.29\cdot 10^5\,\text{W/K}$$

c) We have to recalculate the specific entropy of the steam leaving the turbine and entering the condenser. This is done simply by adding the entropy produced in the turbine to the value at point 2 in the power plant. With $s_3 = s_{3'} + 0.29$ kJ/(K · kg) = 6.0 J/(K · kg), the balance of entropy for the condenser is given by

$$\Pi_s = -I_{m,steam}(s_3 - s_1) + I_{m,cooling} c_{p,water} \ln\left(\frac{T_{c,out}}{T_{c,in}}\right)$$

$$= -100(6000 - 700)\text{W/K} + 2720 \cdot 4200 \ln\left(\frac{308}{293}\right)\text{W/K} = 0.40 \cdot 10^5 \text{ W/K}$$

The mass flux of the cooling water has to be recalculated as well; its new value is 2720 kg/s.

d) Relative contributions to irreversibility and fluxes of entropy with respect to the entire plant appear in Table 12.

TABLE 12. Irreversibilities in a power plant

Contribution	Fraction of total
Entropy production as a result of combustion	0.44
Entropy production in boiler heat exchanger	0.46
Entropy production in turbine	0.04
Entropy production in condenser heat exchanger	0.06
Entropy flux with fuel and exhaust	0.18
Entropy flux to environment through condenser	0.82

4.5.3 Vapor Refrigeration and Heat Pump Systems

Refrigeration and the application of heat pumps to heating systems are two areas where intensive research and development are taking place. The common refrigerants used up to date have to be replaced because they are responsible for reducing the ozone layer when released into the environment; and heat pumps have to compete with cheap fossil fuel for a place in heating systems. Finally, in the future, we may wish to replace these sources of energy by those provided directly or indirectly by the Sun, again forcing us to adapt the technical systems.

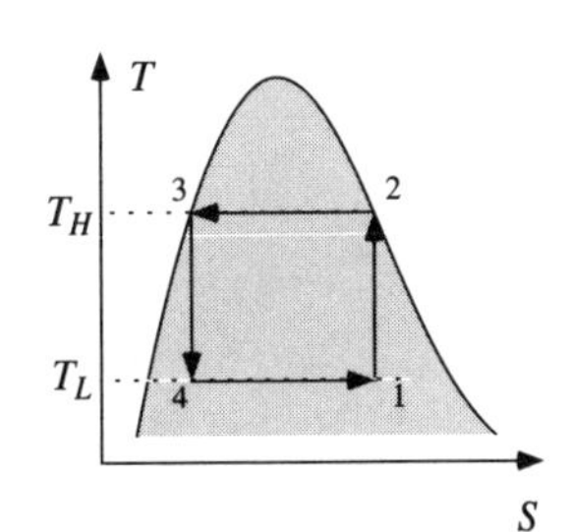

FIGURE 53. Carnot refrigeration cycle in the liquid-vapor region of a fluid. The cycle runs in the opposite direction from a Carnot power process.

A vapor Carnot refrigeration or heat pump cycle. We discussed the principle of operation of refrigerators or heat pumps in Chapter 1. There we saw that entropy is pumped from a lower temperature space, using supplied energy, to a higher temperature environment. A simple device for achieving this is a fluid running through a reverse Carnot cycle. If a substance is used which changes its phase in the range of temperatures and pressures encountered, the cycle may look like the one depicted in Figure 53. Starting at point 1, the fluid which has just absorbed the entropy removed from the cold environment, is compressed isentropically to a state corresponding to point 2. Its temperature has therefore changed from the lower value, T_L, to the higher one, T_H. The fluid is then condensed at T_H to form a saturated liquid at state 3, rejecting the entropy it received during the step from point 4 to point 1. To ready the working fluid for picking up entropy again, its temperature must be reduced back to T_L which is achieved by an isentropic expansion to state 4. Finally, the fluid is evaporated while entropy is transferred into it from the cold space. This completes the cycle leading through points 1–2–3–4–1.

A technical realization of the Carnot cycle described uses four elements as shown in Figure 54. We need a compressor to let the fluid undergo the step from point 1 to point 2. As in previous discussions, we first assume the process undergone by the fluid to be reversible; in other words, step 1–2 is a reversible adiabatic compression. Condensation of the high temperature and high entropy fluid takes place at constant pressure and temperature in the condenser, which is in contact with the high temperature environment. During step 2–3, the entropy picked up from the cold environment is rejected to the space at high temperature. The subsequent isentropic expansion requires a turbine which delivers useful energy for driving the compression. Finally, we need an evaporator in contact with the cold space.

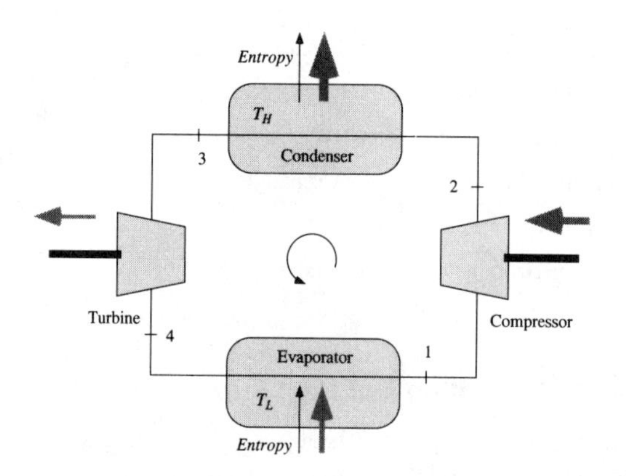

FIGURE 54. Reversible four-step vapor refrigeration cycle requires four components. Evaporator and condenser are in contact with the cold and the warm spaces, respectively. Commonly, the turbine is replaced by a throttling valve as in Figure 55.

A more realistic vapor refrigeration cycle. Even if we could achieve this reversible operation of the Carnot cycle, we still would not realize it in practice for two reasons. First, since the fluid in state 1 is a mixture of liquid and vapor, the presence of liquid droplets might damage the compressor. Therefore, step 1–2 in Figure 53 is replaced by one where we have only vapor (see Figure 55). The latter process is called dry compression, in contrast to the wet compression discussed above.

Second, the turbine normally is left out of the cycle. First of all, step 3–4 delivers only a small amount of energy compared to the energy needed for compression. Additionally, turbines operate rather poorly under the conditions called for in a refrigeration cycle. Therefore, the turbine is replaced by a simple throttling valve; the liquid at point 3 is allowed to expand freely while conserving its specific enthalpy:

$$h_3 = h_4 \tag{206}$$

Naturally, this process is irreversible as is step 1-2, under realistic conditions in the compressor (not shown in Figure 55).

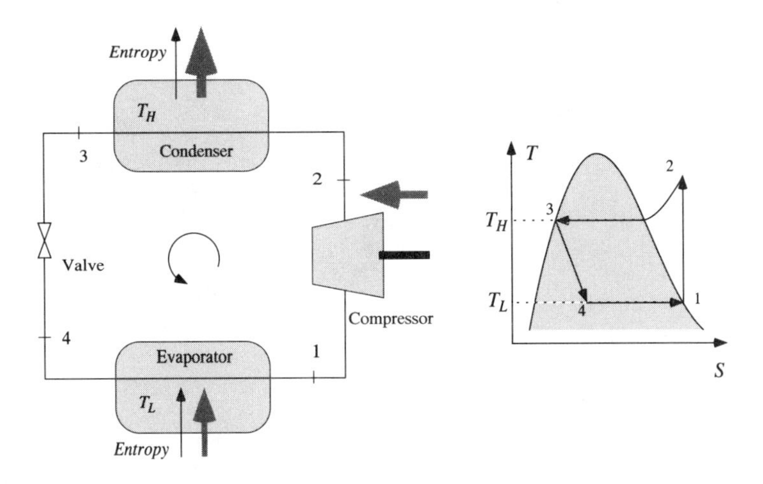

FIGURE 55. Realistic vapor power cycle includes a throttling valve rather than a turbine for the expansion step. Also, step 1-2 is performed with vapor only instead of with a mixture of vapor and liquid.

Absorption refrigeration. Energy is needed to raise entropy from a cold space to warmer surroundings. In the refrigeration cycles discussed, this energy is provided in the compression stage of the refrigerator or the heat pump. The compressor needs to be driven mechanically (or electrically) requiring other than thermal energy sources. There is a way, however, to use heat from a high temperature source to drive a refrigeration cycle.

Consider the setup as Figure 56 which provides the technical means for an absorption refrigeration cycle. At point 1 a refrigerant, such as ammonia, leaves the evaporator as vapor. Next, the vapor enters the first of three elements which replace the compressor of a standard refrigerator. This device is an absorber, in which ammonia is absorbed by liquid water to form a strong water-ammonia solution. This step is exothermic, meaning that entropy will be rejected to the environment, requiring a means of cooling the absorber. The liquid solution then enters a pump which increases the pressure of the fluid to the level needed subsequently in the condenser. Since the fluid is a liquid, compressing it requires much less energy than has to be supplied in the compression step of a normal refrigerator. The strong solution leaves the pump at high pressure and enters a generator where a high-temperature source of entropy drives the ammonia out of the solution.

Now the process splits into two paths. The weak solution (essentially water obtained after the ammonia has been driven out) returns to the absorber through a valve which allows for the pressure of the fluid to be reduced to its value in the lower portion of the cycle. Ammonia vapor at high pressure and temperature, on the other hand, enters the condenser, where the entropy picked up from the space to be cooled is rejected to the environment. The vapor finishes its cycle by passing through the throttling valve from point 3 to point 4, and by subsequently flowing through the evaporator where it again absorbs entropy.

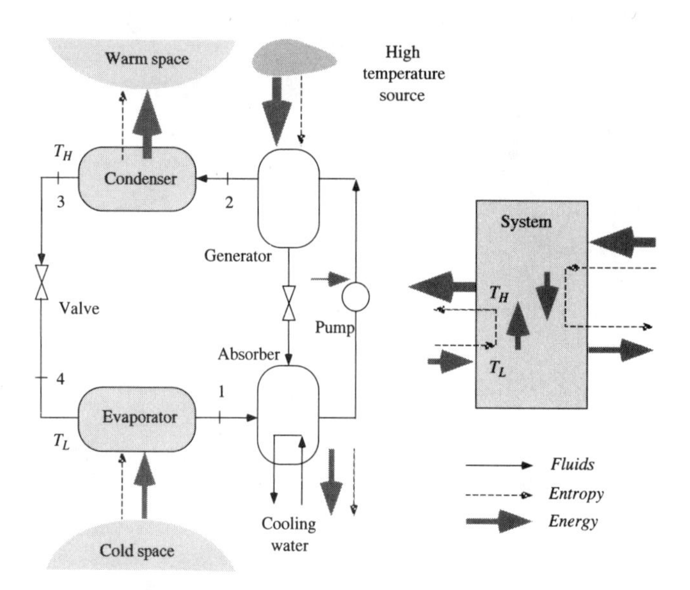

FIGURE 56. In an absorption refrigeration or heat pump system, the energy needed to pump entropy from a cold to a warm space is provided by a thermal process running between a generator and an absorber. In the absorber, the fluid of the refrigeration cycle (ammonia) is absorbed by a liquid (water). In the generator, heat from a high temperature source drives the refrigerant out of the strong solution. The former step is exothermic, while the latter is endothermic.

If you consider the thermodynamics of this entire process, you will notice that the energy needed to pump entropy from the cold space to the warmer surroundings is provided by lowering entropy from the high temperature source through the generator to the absorber and into the fluid used to cool the latter device. In effect, we have a "thermal transformer" in analogy to electrical or mechanical transformers (a gear box would provide an example for the latter). The flow diagram in Figure 56 provides an explanation of the process.

Having a means of pumping heat without the need for electrical power allows for refrigeration and heat pump processes to be directly driven by the Sun. Absorption refrigeration has been studied extensively in solar energy engineering. The solar collectors, absorber, pump, and generator, however, may increase the cost of such systems considerably, making them too expensive where electrical power is cheap.

4.6 Applications of Convective Heat Transfer

Convective transport of heat leads to many interesting natural phenomena and to important technical applications. From the point of theory the subject is very demanding. For this reason we have dealt with convection only briefly so far: in Chapter 3, we used a simple expression to calculate heat transfer at a solid-fluid interface. Knowing the heat transfer coefficient, we can establish a relation which serves as a boundary condition for a particular body.

Here we will take a second step by also considering the effect of convection at surfaces upon the fluid carrying away, or delivering, entropy. If we still assume the heat transfer coefficients at interfaces to be given, and if we treat simple fluid flow as in Section 4.4, we can calculate the performance of devices such as solar collectors and heat exchangers. In addition to these technical applications, convective heat transfer in an atmosphere or inside a star will be discussed. It should be obvious that the fundamental problem of how to calculate heat transfer coefficients still has not been solved. We shall attempt a first step in that direction in the Epilogue, which will deal with fluid flow in more depth.

4.6.1 Flat-Plate Solar Collectors

Basically, any object which absorbs the light of the Sun is a solar collector. Therefore, we should include in this list the leaves of trees, buildings, soil, the oceans, or photovoltaic cells, just to name a few. You can see the range of phenomena induced by the Sun's rays. The "collectors" may produce substances, they may lead to the flow of electric charge, or they may produce heat (i.e., entropy). Here we shall look only at the latter effect in simple technical devices.

A thermal solar collector is a device which absorbs a part of the solar radiation falling upon it, leading to entropy production in the absorber. If we let a fluid flow across the absorber, we may harness some of the entropy which has been created (Figure 57). In the simplest possible geometrical arrangement, we have a flat absorber plate possibly made out of some thin metal sheet. We may then let a fluid (liquid or air) flow through a rectangular channel behind the absorber, with the fluid wetting its entire surface area. Commonly, the collector is insulated at the bottom, and a transparent cover is placed above the absorber plate; both devices reduce the loss of heat to the environment. The ducts for fluid flow in the collector often are different from (and more complicated than) what we have assumed here, leading to more difficult geometrical arrangements for the transfer of heat between an absorber and fluid. This point will be discussed further below.

The balance of energy for a solar collector. Figure 57 shows the fluxes of energy with respect to the absorber. They will be used to express the balance of energy (and of entropy; see below) with respect to the absorber; in a second step, we will perform the balance of energy with respect to the entire collector. First, we have the flux associated with solar radiation falling upon the collector.[5] As you know, unless the surface is a black body, only part of the radiation will be absorbed, while the rest will be reflected back to the environment. The ratio of radiation absorbed to radiation falling upon the surface defines the optical properties of a collector which depend upon the absorber, the cover, and the type of radiation. Usually, for the purpose of an overall balance, their combined effect is described by a factor known as the *transmission-absorption product* $(\tau\alpha)$.

5. Naturally, radiation falling upon the collector includes the radiation of the atmosphere as a body at or near environmental temperature. This contribution is not included with the solar irradiance $\mathcal{G}$, but rather with the losses of the collector to the environment.

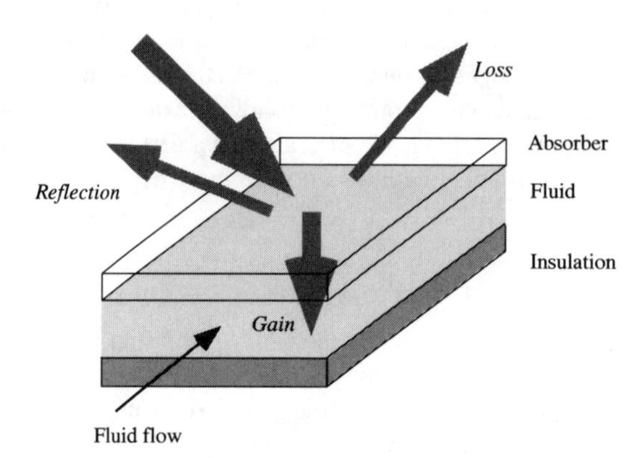

FIGURE 57. Simple flat-plate solar collector consists of a flat absorber plate (possibly including a cover for reducing the losses to the top). Absorption of solar radiation leads to the production of entropy in the plate. A fluid flowing across the bottom of the plate can carry away some of the entropy. The figure shows the simplest possible geometry for absorber and fluid flow. The fat arrows denote the energy fluxes with respect to the absorber.

The losses of the collector plate to the environment, on the other hand, are calculated in terms of a heat loss coefficient U_t (top loss coefficient). In the following analysis, we shall assume losses to occur only through the top of the collector. The last energy flux is due to convective transfer at the absorber-fluid interface. To describe its effect we introduce the heat transfer coefficient from the plate to the fluid (abbreviated by U_{pf}). All three coefficients characterizing a collector, i.e., $(\tau\alpha)$, U_t, and U_{pf}, may depend upon the conditions under which it is operated.

If we collect all the terms, a steady-state balance of energy for the absorber takes the form (Figure 58):

$$(\tau\alpha)\mathcal{G} - U_t\left(T_p - T_a\right) = U_{pf}\left(T_p - T_f\right) \tag{207}$$

The indices p and f refer to the absorber plate and the fluid, respectively. As usual, a stands for *ambient*. $\mathcal{G}$ is the total irradiance with respect to the surface of the collector. This equation holds for every point of the surface, with the temperatures of the fluid and the plate changing in the direction of fluid flow (but, for our simple geometry, not in the direction perpendicular to the flow). Naturally, we have a problem concerning the meaning of the temperatures T_f and T_p if we want to apply Equation (207) directly for the entire collector of surface are A. In this case, just think of the temperatures as some appropriate average value for the respective system.

The equation of balance of entropy will be written below for the collector as the system. First, let us express the overall balance of energy. In this case we have to include the convective currents of energy due to fluid flow into and out of the collector, while the flux from the plate to the fluid drops out:

$$c_p I_m\left(T_{f,out} - T_{f,in}\right) = A\left[(\tau\alpha)\mathcal{G} - U_t\left(T_{pm} - T_a\right)\right] \tag{208}$$

As just mentioned, in this equation T_{pm} represents the proper average value. Normally, we take both the temperature coefficient of enthalpy of the fluid and the pressure as constants, which permits us to use Equation (156) for the convective current.

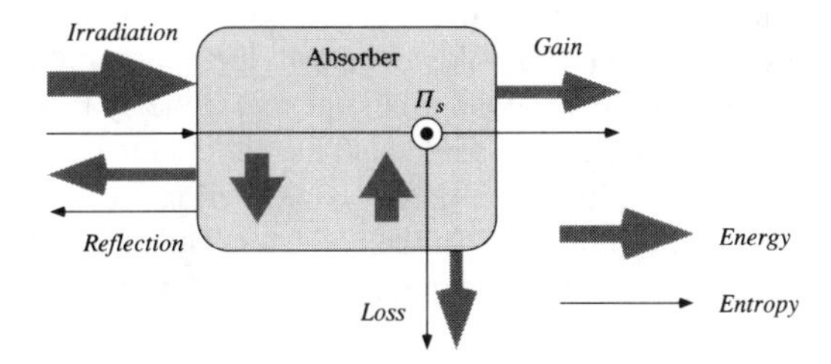

FIGURE 58. Flow diagram for the absorber showing fluxes of entropy and energy. Almost all the entropy transferred to the fluid and to the environment is produced in the system.

Since the temperature of the collector plate is not easily accessible, one replaces the temperature of the absorber in Equation (208) using Equation (207). As a result, the (average) temperature of the fluid appears in the law of balance:

$$I_{E,use} = A\frac{U_{pf}}{U_{pf} + U_t}\left[(\tau\alpha)\mathcal{G} - U_t\left(T_{fm} - T_a\right)\right]$$

$I_{E,use}$ is the useful energy current, i.e., the convective current with respect to the collector. The factor multiplying the term in parenthesis is called the *collector efficiency factor* and it is commonly abbreviated F'. Using this definition leads to the expression for the useful energy current in terms of the fluid temperature:

$$I_{E,use} = AF'\left[(\tau\alpha)\mathcal{G} - U_t\left(T_{fm} - T_a\right)\right]$$
$$F' = \frac{U_{pf}}{U_{pf} + U_t} \qquad \textbf{(209)}$$

If you rewrite the efficiency factor in a slightly different form, you can see that it corresponds to the ratio of the thermal resistances between the absorber and the environment on the one hand, and between the fluid and the environment on the other.

Different fluid duct geometries. Usually, and especially for collectors using liquids for heat transfer, the geometrical arrangement of the fluid ducts is different from the simple case assumed so far. Therefore, if U_{pf} is still used for the heat transfer coefficient from metal to fluid, the collector efficiency factor F' cannot be computed as in Equation (209b). Take the case of a liquid flowing through thin pipes attached to the absorber. Obviously, heat has to flow through the absorber sheet to the pipes before it can enter the fluid. The efficiency will be reduced both because of this process and because of the fact that the fluid surface may be smaller than that of the flat absorber. The efficiency factor will depend not only on the heat transfer coefficient but also on the distance between the pipes, the thickness and the conductivity of the absorber sheet, and the inner surface area of the pipes carrying the fluid. Therefore, the task of computing the efficiency factor can be quite complicated.[6] However, with the efficiency factor known, a reduced heat transfer coefficient can be calculated and used as in the equations presented above (Example 42).

The temperature distribution in the direction of fluid flow. As an overall balance of energy, Equation (209) holds only with the appropriate average of the temperature of the fluid. We may approximate this value by the arithmetic average of the fluid temperatures at the inlet and the outlet of the collector (Example 41). A better expression, however, requires considering the change of temperatures in the direction of fluid flow in the collector. This is achieved by treating the example as a continuous problem in one spatial dimension (Figure 59).

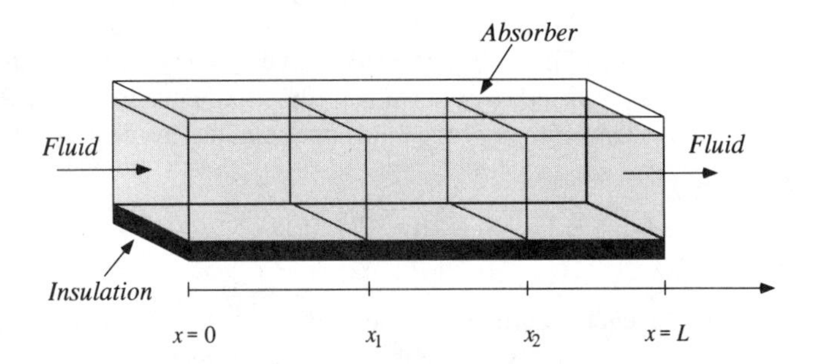

FIGURE 59. Fluid flowing through the collector heats up in the direction of flow. The law of balance of energy for the control volume between x_1 and x_2 leads to an expression for the temperature as a function of position.

The law of balance of energy for the control volume of length x_2-x_1 can be used to derive the expression for the temperature of the fluid as a function of position x. The fluid portion in the control volume is heated from above, and it flows into and out of the volume. If we use Equation (209) in the form

$$j_{E,use} = F'\left[(\tau\alpha)\mathcal{G} - U_t\left(T_f(x) - T_a\right)\right]$$

the law of balance for the control volume is

$$c_p I_m\left(T_f(x_2) - T_f(x_1)\right) = F'\int_{x_1}^{x_2} W\left[(\tau\alpha)\mathcal{G} - U_t\left(T_f(x) - T_a\right)\right]dx$$

Here, W is the width of the collector. The left-hand side represents the convective current due to the fluid entering and leaving, while the right-hand side is the integral of the energy current density over the top surface area of the control volume. The equation transforms into the differential equation

6. See for example Duffie and Beckman (1991), Chapter 6.

$$c_p I_m \frac{dT_f}{dx} = F'W\left[(\tau\alpha)G - U_t\left(T_f(x) - T_a\right)\right] \tag{210}$$

whose solution is

$$T_f(x) = \frac{(\tau\alpha)G}{U_t} + T_a - \left[\frac{(\tau\alpha)G}{U_t} + T_a - T_{f,in}\right]\exp\left(-\frac{F'U_t W}{c_p I_m}x\right) \tag{211}$$

If we take the value of T_f at the outlet, subtract it at the inlet, and multiply by the product of c_p and I_m, we obtain the useful energy current:

$$I_{E,use} = A\frac{c_p I_m}{AU_t}\left[1 - \exp\left(-\frac{F'U_t A}{c_p I_m}\right)\right]\left[(\tau\alpha)G - U_t\left(T_{f,in} - T_a\right)\right]$$

Basically, this result represents a transformation of the expression used previously in Equation (209) such that the mean fluid temperature is replaced by the temperature of the fluid at the inlet to the collector. As a consequence, the collector efficiency factor F' is replaced by the *heat removal factor* F_R:

$$I_{E,use} = AF_R\left[(\tau\alpha)G - U_t\left(T_{f,in} - T_a\right)\right] \tag{212}$$

where

$$F_R = \frac{c_p I_m}{AU_t}\left[1 - \exp\left(-\frac{F'U_t A}{c_p I_m}\right)\right] \tag{213}$$

Equation (212) allows us to calculate the useful energy current of a flat-plate solar collector in terms of environmental parameters, the flux of mass through the collector, and the inlet temperature of the fluid. The collector can be characterized by two parameters, $F_R(\tau\alpha)$ and $F_R U_t$; considering that F_R includes the efficiency factor, we may say that there are three values defining the device—$(\tau\alpha)$, U_t, and F'.

The balance of entropy for a solar collector. While it is convenient to obtain the results concerning the thermal performance of a collector in terms of the balance of energy, this law alone does not suffice for a complete description of its operation. If we wish to include the evaluation of irreversibilities in our analysis, we have to perform a balance of entropy for the collector.

An overall balance is written down quite easily. Again, we shall treat steady-state operation. Essentially, we have to take into consideration four currents of entropy with respect to the collector as a whole, and the rate of production of entropy. The four currents are due to radiation, heat loss, and the fluid entering and leaving the collector. Production is the result of the absorption of radiation, heat transfer to the fluid and heat flow into the environment. The overall balance then takes the form

$$\Pi_s = I_{s,rad} + I_{s,loss} + I_{fluid,in} + I_{fluid,out} \quad (214)$$

If we properly place the boundary of the control volume, all sources of irreversibility will be included. In particular, this means that for the incoming radiation we take the surface of the cover of the collector, while for the losses the boundary will coincide with the environment at ambient temperature. Now the terms in Equation (214) take the following form:

$$\Pi_s = -A(\tau\alpha)G\frac{4}{3T_s} + \frac{1}{T_a}\left|I_{E,loss}\right| + \left(s_{f,out} - s_{f,in}\right)\left|I_m\right| \quad (215)$$

This result allows us to calculate the rate of production of entropy since, in principle, all the other quantities are known. As before, T_S is the temperature of solar radiation. The energy current due to the loss to the environment can be computed using the balance of energy expressed in Equation (212), while the specific entropy of the fluid at the inlet and the outlet is obtained in terms of the values of temperature and pressure at the respective points. For constant pressure and constant value of the temperature coefficient of enthalpy, the net convective current of entropy can be calculated according to Equation (158).

Analysis of a solar air heater. Consider the following example of the balance of entropy for a collector using air as the fluid. Two points which were not discussed above tend to play an important role with air collectors. First, depending on the geometry of the air duct and the magnitude of the air current, the pressure of the fluid might drop considerably across the collector; second, the heat transfer coefficient from the absorber to the fluid strongly depends upon the fluid speed. Now take a simple rectangular duct behind the absorber as in Figure 57 and imagine a pump to be built into the collector which is operated in such a way that the air pressure will be the same at the inlet and the outlet. The balance of energy must become

$$c_p I_m\left(T_{f,out} - T_{f,in}\right) = A\left[(\tau\alpha)G - U_t\left(T_{pm} - T_a\right)\right] + \mathcal{P}_{pump}$$

while the heat transfer coefficient U_{pf} will be approximated by an increasing function of fluid speed.

If we wish to obtain air with a prescribed outlet temperature, we might be interested in the question of whether there is an optimal way of arranging a given field of collectors in series or in parallel. Alternatively, we may ask if, for fixed collector area and constant environmental parameters, there is an optimal length for the collector. Changing the length of the collector changes the air current necessary to reach the desired air temperature at the outlet. With it, other quantities such as the temperature of the absorber, losses, flow speed, and heat transfer coefficients change. Indeed, you should expect both extremes, i.e., very short and very long collectors, to be less favorable than one of intermediate length. At short lengths, the plate temperature must be high for the air to be heated appropriately, leading to large losses. If the collector is made very long, on the other hand, irreversibility will grow because of an increase in the power of the pump necessary to balance frictional losses. Now, the most interesting feature of this system is the fact that the net gain of energy (convective useful energy current minus the pumping power) keeps growing for increasing collector length (Figure 60). The rate of production of entropy, in contrast, displays the behavior discussed: there is a minimum of irreversibility for a particular length.

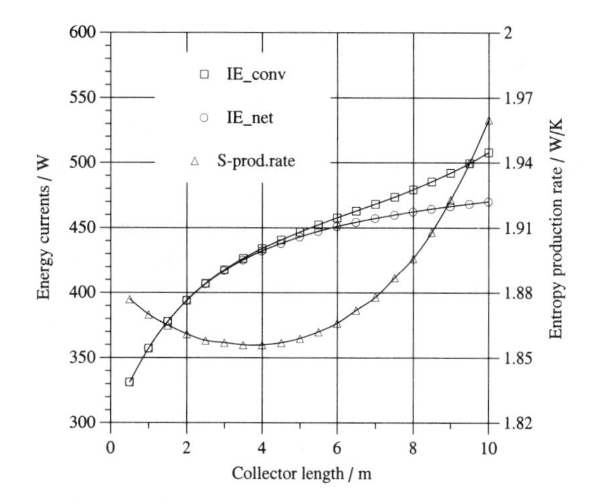

FIGURE 60. Balance of entropy performed on the model of a flat-plate solar air heater. The collector is operated in such a way that a constant value for the air temperature at the outlet is maintained while the length of the collector is changed at fixed surface area. Gross and net energy currents are shown with the rate of production of entropy as a function of the length of the collector. The collector area and height of the air duct are 1 m^2 and 1.2 cm, respectively. Environmental parameters: $\mathcal{G}$ = 800 W/m^2 and T_a = 290 K. Collector parameters: U = 4 W/(K · m^2) and $(\tau\alpha)$ = 0.75. The heat transfer coefficient is approximated by a linear function of speed of flow, while fluid friction is calculated as appropriate for the particular circumstances. The temperature of the air was fixed at 290 K and 340 K at the inlet and the outlet. The model is a simplified version of one presented by Oppliger (1993).

This behavior can be understood quite easily. The low value of the energy current at small lengths is due to high thermal losses; its continued increase with growing length is the result of the increase of the heat transfer coefficient between absorber and fluid. If the collector is made longer, the flux of air must be made larger to maintain a constant temperature at the outlet. Even though the thermal performance continues to improve with increasing length, long collectors (i.e., collectors placed in series) basically are heated by the pump instead of the Sun. The analysis of irreversibility tells us, that this strategy is not worth pursuing.

Dynamical model of a solar collector. A solar collector is a dynamical system, storing entropy and energy. A steady-state analysis therefore might not be appropriate. However, coupled with a heat storage element of much higher capacity, the simplification can be justified, although, in some cases, it might be necessary to consider the dynamical behavior of the collector.

We may start with the law of balance of energy. In analogy to what we have seen in Section 3.2.8, we can write

$$
\begin{aligned}
C_c \frac{dT_f}{dt} &= F'\left[(\tau\alpha)\mathcal{G} - U_t\left(T_f - T_a\right)\right] - c_p I_m\left(T_{f,out} - T_{f,in}\right) \\
T_f &= \left(T_{f,in} + T_{f,out}\right)/2
\end{aligned}
\tag{216}
$$

Here, C_c is the temperature coefficient of energy of the collector, or possibly of a collector element. (In the latter case, the collector can be thought of as being composed of several elements put in series which improves upon the approximation.) This value must include the absorber, the cover, the fluid, and the insulation. Equation (216) represents a particularly simple approximation of the terms in the equation of balance of energy of a uniform collector or collector element. The temperature of the element is approximated by the mean temperature of the fluid in it. Typical values of C_c are between 5000 J/K and 10000 J/K for one square meter of collector area for a liquid collector.

EXAMPLE 41. Measuring the characteristic parameters of a flat-plate collector.

Measurements of the thermal efficiency of a flat-plate solar collector are made for conditions where the collector directly faces the Sun. The efficiency is the plotted as a function of

$$
x = \frac{T_{f,am} - T_a}{\mathcal{G}}
$$

where $T_{f,am}$ is the arithmetic mean of the fluid inlet and outlet temperatures. Measurements were performed with a constant flux of mass of 40 liters per hour per square meter of collector area. The temperature coefficient of enthalpy of the fluid is 3800 J/(K · kg). a) Derive the expression for the efficiency of a solar collector as a function of x. b) Determine the values of $F_R U_t$ and $F_R(\tau\alpha)_n$ for the values shown in the accompanying graph. The index n with the transmission-absorption product stands for normal incidence of solar radiation. c) From detailed calculations it is known that the collector efficiency factor F' is 0.92. Calculate the actual optical efficiency $(\tau\alpha)_n$ and the actual heat loss factor U_t.

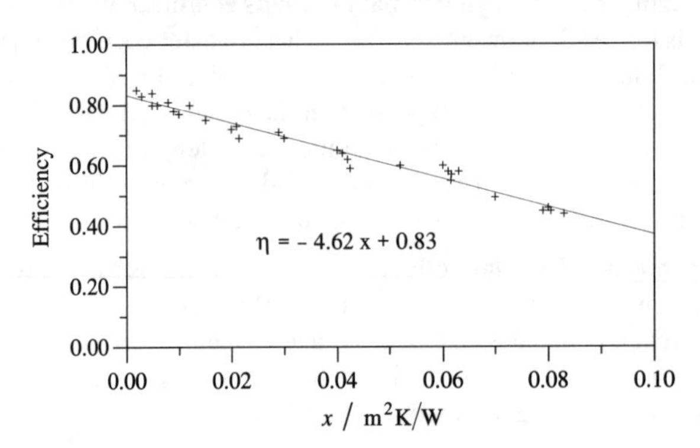

SOLUTION: a) The thermal efficiency of the collector is defined as the ratio of the energy delivered by the heated fluid and the energy incident upon the collector:

$$\eta = I_{E,use} / (A\mathcal{G})$$

The useful energy current should be expressed similarly to Equation (209), with F' replaced by F_{am} since T_{fm} has been replaced by $T_{f,am}$:

$$I_{E,use} = AF_{am}\left[(\tau\alpha)_n \mathcal{G} - U_t\left(T_{f,am} - T_a\right)\right]$$

We may expect the efficiency factor F_{am} to be nearly equal to F', considering that the arithmetic mean of the collector fluid temperature will not deviate all that much from the proper average value. Therefore, the efficiency turns out to be

$$\eta = F_{am}(\tau\alpha)_n - F_{am}U_t \frac{T_{f,am} - T_a}{\mathcal{G}}$$

which is close to

$$\eta = F'(\tau\alpha)_n - F'U_t \frac{T_{f,am} - T_a}{\mathcal{G}}$$

This shows that efficiency data plotted as a function of the coefficient x introduced above should yield a linear relation if the collector parameters are independent of the conditions under which the device is operated.

b) The data plotted in the graph yield approximate values of $F'(\tau\alpha)_n$ and $F'U_t$. Approximating the measured values by a least-squares linear fit, we obtain

$$F'(\tau\alpha)_n = 0.83$$

$$F'U_t = 4.62\,\mathrm{W/(K \cdot m^2)}$$

Since the measured values of the efficiency were obtained with the collector directly facing the Sun, we get the transmission-absorption product for normal incidence. To calculate the desired parameters of the collector, we have to transform the expressions. A simple way of doing this is the following. Since the heat removal factor F_R is used with the inlet temperature of the fluid in Equation (212) for the useful energy current, we should replace the arithmetic mean of the fluid temperature by the inlet value, which leads to

$$I_{E,use} = AF'\left[1 + \frac{1}{2}\frac{F'U_t A}{c_p I_m}\right]^{-1}\left[(\tau\alpha)\mathcal{G} - U_t\left(T_{f,in} - T_a\right)\right]$$

You can show that this is an approximation to the expression given in Equation (212) with the exact heat removal factor of Equation (213). A comparison of this result with the first equation for the useful energy current yields

$$F_R(\tau\alpha)_n = F'(\tau\alpha)_n\left[1+\frac{1}{2}\frac{F'U_tA}{c_pI_m}\right]^{-1}$$

$$F_RU_t = F'U_t\left[1+\frac{1}{2}\frac{F'U_tA}{c_pI_m}\right]^{-1}$$

The numerical results turn out to be

$$F_R(\tau\alpha)_n = 0.83\cdot\left[1+\frac{4.62}{2\cdot 3800\cdot 40/(3600)}\right]^{-1} = 0.79$$

$$F_RU_t = 4.62\cdot\left[1+\frac{4.62}{2\cdot 3800\cdot 40/(3600)}\right]^{-1}\mathrm{W/(K\cdot m^2)} = 4.38\,\mathrm{W/(K\cdot m^2)}$$

c) Knowing the collector efficiency factor, the actual collector parameters can be calculated:

$$F'(\tau\alpha)_n = 0.83 \quad\Rightarrow\quad (\tau\alpha)_n = 0.90$$

$$F'U_t = 4.62\,\mathrm{W/(K\cdot m^2)} \quad\Rightarrow\quad U_t = 5.02\,\mathrm{W/(K\cdot m^2)}$$

This is a collector with a fairly high optical efficiency, but with a mediocre heat loss factor.

EXAMPLE 42. A thermosyphon collector.

In a thermosyphon collector, the fluid circulates without a pump. Instead of forced convection, the process operates because of density gradients in the differentially heated fluid. (See the figure below.)

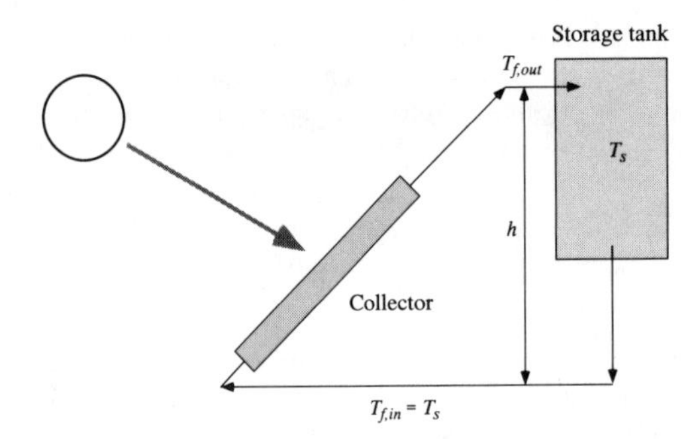

Assume the fluid in the tank to be well mixed, and approximate the temperature of the fluid in the rising part of the hydraulic circuit, including the collector, by the average temperature. Also assume the density of the fluid to be a linear function of temperature. Fluid resistance may be modeled by the law of Hagen and Poiseuille. Calculate the steady state flux of mass through the collector.

SOLUTION: a) The natural circulation is the result of different hydrostatic pressures in the rising and falling parts of the hydraulic circuit. In the steady state, the driving pressure difference is equal in magnitude to the one resulting from flow resistance:

$$\Delta P = \Delta\rho g h$$

$$I_V = \frac{1}{R_V}\Delta P$$

$\Delta\rho$ is the difference of densities of the fluid in the two parts of the circuit. It can be expressed as follows:

$$\Delta\rho = b\Delta T = b\left[\frac{1}{2}\left(T_{f,out} + T_{f,in}\right) - T_s\right]$$

where $T_{f,in}$ is the temperature T_s of the fluid in the storage tank. The parameter b is the temperature coefficient of the density of the fluid. The collector outlet temperature is calculated using the balance of energy expressed by Equation (209):

$$c_p \dot{I}_m\left(T_{f,out} - T_{f,in}\right) = AF'\left[(\tau\alpha)G - U_t\left(\frac{1}{2}\left(T_{f,out} + T_{f,in}\right) - T_a\right)\right] + \Delta P I_V$$

The last term is the rate of energy input to the system as a consequence of the natural circulation; the energy comes from the gravitational field. It turns out that this term is very small compared to the others in the equation of balance of energy; it will therefore be neglected.

Now, the temperatures can be eliminated from the last equation using the first three relations. This leaves us with a quadratic equation for the volume flux:

$$\frac{2c_p\rho R_v}{bgh}I_V^2 + \frac{AF'U_t R_v}{bgh}I_V - AF'\left[(\tau\alpha)G - U_t\left(T_s - T_a\right)\right] = 0$$

This demonstrates that the circulation depends upon the instantaneous temperature of the fluid in the storage tank. As the fluid temperature in the tank rises during charging, the circulation is found to decrease. Since the collector and hydraulic circuit probably react fast to changing conditions compared to the storage unit, the steady-state analysis appears to be appropriate.

EXAMPLE 43. Charging and discharging of seasonal ground heat storage.

A spatially uniform model of a cylindrical heat storage element in the ground hooked up to a solar collector field for charging during summer, and with a building for discharging during winter, can be used to approximate the much more complicated spatially continuous problem.[7]

7. Comparison with more detailed finite element approximations was done by F. Stucky (1994).

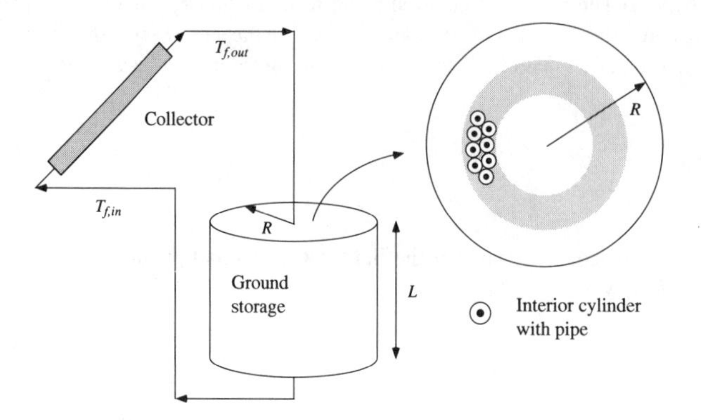

The system consists of a collector field of surface area A_c and described by the usual collector parameters. Pipes carry the heated fluid to the ground storage unit which is a cylinder of radius R and length L. (See the figure below.) There are N parallel pipes each heating an interior cylinder. (The interior cylinders uniformly cover the storage unit.) The ground storage cylinder loses heat to the surroundings. Discharging in winter occurs through the floor heating system of a building.

The model is described as follows. Charging is done during the six months from spring to fall. The irradiance is twice the average value for half the time (roughly $8 \cdot 10^6$ seconds), i.e., the usual amount of energy is delivered to the collectors. Collector parameters are assumed to be constant, as is the flux of mass of the collector fluid. The fluid, which flows through N parallel pipes, is cooled from $T_{f,out}$ to $T_{f,in}$ while flowing down through the storage unit (*in* and *out* refer to the collector).

The storage in the ground is supposed to have uniform values of density, temperature coefficient of energy, and thermal conductivity, and a single value of temperature at any time. Heat flow is supposed to be radial only (no heat loss through the top or bottom of the cylinder). Delivery of heat to each interior cylinder is described by an average value UA_{int} which is calculated from the fluid in the pipe through half of the radius of the cylinder. Heat loss from the entire storage unit to the surroundings is calculated with a value UA_{store}, representing a thick cylindrical shell with radii which are half and twice the radius R of the storage unit. The temperature of the ground surrounding the storage unit is supposed to be constant all the time.

During discharging in the winter, parameters for the storage space are the same. Heat is delivered to rooms in the building at a constant temperature. The value of UA_{floor} of the floor heating system is given.

a) Derive the differential equation of the temperature of the storage unit as a function of time for charging. Present its analytical solution. b) With an initial temperature of the ground storage unit at the beginning of winter, derive the temperature as a function of time. c) Give an expression for the total amount of entropy produced for an entire cycle of charging and discharging which carries the storage from an initial temperature back to the same value. d) If the only parameter varied in the entire system is the length L of the storage unit, do you expect a minimum of entropy production for a cycle for a particular value of L?

Neglect pipe losses and energy required for pumping the fluids during charging and discharging.

SOLUTION: a) The storage unit receives energy from the collector fluid and loses energy to the ground. The overall balance of energy for charging therefore takes the form

$$C_s \frac{dT_s}{dt} = \left|I_{E,fs}\right| - \left|I_{E,loss}\right|$$

with

$$C_s = \pi R^2 L \rho c_s$$
$$\left|I_{E,fs}\right| = UA_{fs}\left(T_{fm} - T_s\right)$$
$$\left|I_{E,loss}\right| = UA_{store}\left(T_s - T_{ground}\right)$$

For T_{fm} we take the arithmetic mean of the fluid inlet and outlet temperatures

$$T_{fm} = \frac{1}{2}\left(T_{f,in} + T_{f,out}\right)$$

which also depend upon the performance of the collectors:

$$\left|I_{E,fs}\right| = A_c F_R\left[(\tau\alpha)\mathcal{G} - U_t\left(T_{f,in} - T_a\right)\right]$$
$$\left|I_{E,fs}\right| = c_p I_m\left(T_{f,out} - T_{f,in}\right)$$

T_a is the average ambient temperature for the period of charging. The value of UA_{fs} for heat transfer from the fluid to an interior cylinder is given by

$$UA_{fs} = NUA_{int}$$

$$UA_{int} = 2\pi L\left[\frac{1}{r_{pipe}U_{pipe}} + \frac{1}{k}\ln\left(\frac{0.5R_{int}}{r_{pipe}}\right)\right]^{-1}$$

where R_{int} is the radius of an interior cylinder

$$R_{int} = \frac{1}{\sqrt{N}}R$$

and r_{pipe} and U_{pipe} are the radius and the heat transfer coefficient of the pipe; k is the thermal conductivity of the ground. The value of UA_{store} is calculated simply:

$$UA_{store} = 2\pi Lk\left[\ln\left(\frac{2R}{R/2}\right)\right]^{-1}$$

There are three equations for the energy current $I_{E,fs}$ and one for the arithmetic mean of the fluid temperature. We first can eliminate $T_{f,out}$ and obtain an expression for $T_{f,in}$:

$$T_{f,in} = T_s + \left(\frac{1}{AU_{fs}} - \frac{1}{2c_p I_m}\right)\left|I_{E,fs}\right|$$

If we plug this into the expression for the energy current delivered by the collector, we arrive at

$$|I_{E,fs}| = B\left[(\tau\alpha)G - U_t(T_s - T_a)\right]$$

$$B = \left(\frac{1}{A_c F_R} + \frac{U_t}{A U_{fs}} - \frac{U_t}{2c_p I_m}\right)^{-1}$$

This looks like a slightly different version of the equation for the energy delivered by a collector; note that the variable temperature occurring is the temperature T_S of the storage unit. Now, the differential equation for T_S is given by

$$\frac{dT_s}{dt} = a - bT_s$$

where

$$a = \frac{1}{C_s}\left(B\left[(\tau\alpha)G + U_t T_a\right] + UA_{store} T_{ground}\right)$$

$$b = \frac{1}{C_s}\left(BU_t + UA_{store}\right)$$

The analytical solution of this simple differential equation is

$$T_s(t) = \frac{a}{b} - \left(\frac{a}{b} - T_{so}\right)\exp(-bt)$$

Here, T_{SO} is the initial value of the temperature of the storage unit. Remember that the parameter t runs only for half the time span of charging, since we use twice the average insolation. (This models the fact that the Sun shines only for half the day.)

b) The process of discharging is modeled similarly using the equation of balance of energy for the uniform storage space:

$$C_s \frac{dT_s}{dt} = -|I_{E,fs}| - |I_{E,loss}|$$

The energy currents are calculated as follows:

$$|I_{E,fs}| = UA_{fs}\left(T_s - T_{fm}\right)$$

$$|I_{E,loss}| = UA_{store}\left(T_s - T_{ground}\right)$$

This time, however, the energy current $I_{E,fs}$ also obeys the expression

$$|I_{E,fs}| = UA_{floor}\left(T_{fm} - T_{building}\right)$$

Combining all the information yields the following differential equation

$$\frac{dT_s}{dt} = f - gT_s$$

with

$$f = \frac{1}{C_s}\left(DUA_{fs}T_{building} + UA_{store}T_{ground}\right)$$

$$g = \frac{1}{C_s}\left(DUA_{fs} + UA_{store}\right)$$

$$D = \left(1 + \frac{UA_{fs}}{UA_{floor}}\right)^{-1}$$

which has the simple analytic solution

$$T_s(t) = \frac{f}{g} - \left(\frac{f}{g} - T_{ss}\right)\exp(-gt)$$

where T_{SS} is the initial temperature for discharging, which is the temperature of the storage unit at the end of summer.

c) If you consider a system including the collectors, the storage unit, and the floor of the building, you will see that there are three currents of entropy and energy (each) leaving the system, while one is entering (associated with solar radiation). If we neglect the entropy delivered with the light of the Sun, the total amount of entropy produced in one cycle of the system can be written as follows:

$$S_{prod} = \frac{\left|Q_{coll.loss}\right|}{T_a} + \frac{\left|Q_{store.loss}\right|}{T_{ground}} + \frac{\left|Q_{building}\right|}{T_{building}}$$

The integrated form of the balance of energy for the period of charging is given by

$$A_c(\tau\alpha)\mathcal{G}t_{charge} = \left|Q_{coll.loss}\right| + \left|Q_{store.loss}\right|_{charge} + C_s\left(T_{s,final} - T_{so}\right)$$

$$\left|Q_{store.loss}\right|_{charge} = \int_{charge} UA_{store}\left(T_s - T_{ground}\right)dt$$

This allows for all relevant quantities to be computed for the process of charging. For discharging, we similarly obtain:

$$C_s\left(T_{s,final} - T_{so}\right) = \left|Q_{store.loss}\right|_{discharge} + \left|Q_{building}\right|$$

$$\left|Q_{store.loss}\right|_{discharge} = \int_{discharge} UA_{store}\left(T_s - T_{ground}\right)dt$$

d) Changing the length of the cylindrical storage unit (while leaving all other parameters untouched) will result in different storage temperatures. Increasing the length will decrease the maximum temperature reached. Now, having a high temperature after charging is going to increase losses (both from the collectors and from the storage to the ground). Very low storage temperatures, on the other hand, will make it difficult to discharge the unit. We should expect an optimal performance of the system for some value of L. Calculations show that this is indeed the case. Heating of the house will be optimal for minimal total entropy produced.

4.6.2 Heat Exchangers

Having to transfer heat from one system to another is a common task in engineering, and having to do so efficiently is becoming more and more important in applications such as energy engineering. Basically, any device which lets heat pass from one side to the other may be considered to be a heat exchanger. However, we normally speak of heat exchangers if heat is to be passed from a hot fluid stream to a colder one. In many cases, heat exchangers have to be used if different fluids are needed at different stages of operation in thermal equipment. There exist many basic types of heat exchanger designs which differ in their geometrical arrangements, in the fluids used, and in the processes undergone by the fluids when they emit or absorb heat. Well known examples of heat exchangers are radiators for heating rooms or cooling of car engines, or the steam generators used in power plants where steam is to be produced using a fluid heated by the Sun or by a nuclear reactor.

The principle of operation of heat exchangers. As an example of a particularly simple geometrical arrangement, consider a counter-flow heat exchanger with the two fluids passing through ducts in opposite directions (Figure 61) while heat flows from the hotter to the cooler body. Imagine both fluids to flow through thin but wide and long rectangular ducts separated by a wall which is responsible for the transfer of heat. Assume the temperature of each fluid to vary only in the x-direction.

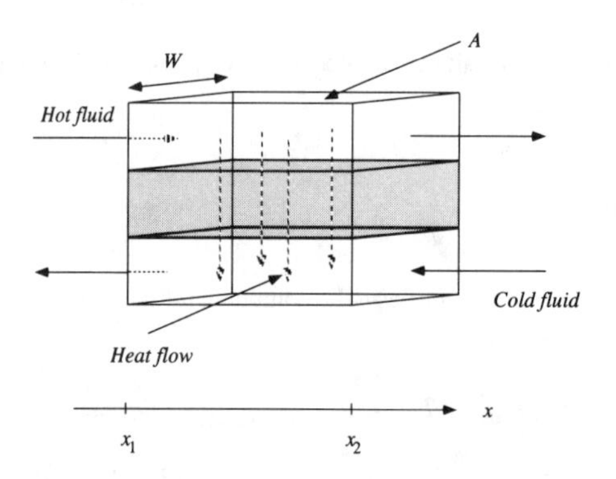

FIGURE 61. In the simplest possible geometrical arrangement, two fluids flowing through rectangular ducts exchange heat across the common flat surface. If the temperature of the fluids is allowed to vary only in the direction of flow, we have a one-dimensional problem. In a counter-flow heat exchanger, the fluids pass each other in opposite directions.

The temperature distribution in the fluids as a function of position is sketched in Figure 62. The temperature of the hotter fluid flowing in the upper duct decreases in the direction of x, while the hotness of the cooler fluid increases in the opposite direction. Naturally, the temperature of the upper fluid must always be larger than the corresponding value for the lower one.

In a parallel-flow heat exchanger, the fluids enter and leave the device on the same side. At the point where they enter, the fluids have the largest difference of temperatures (Figure 63). This difference then decreases in the direction of flow, a situation which is different from that of a counter-flow heat exchanger.

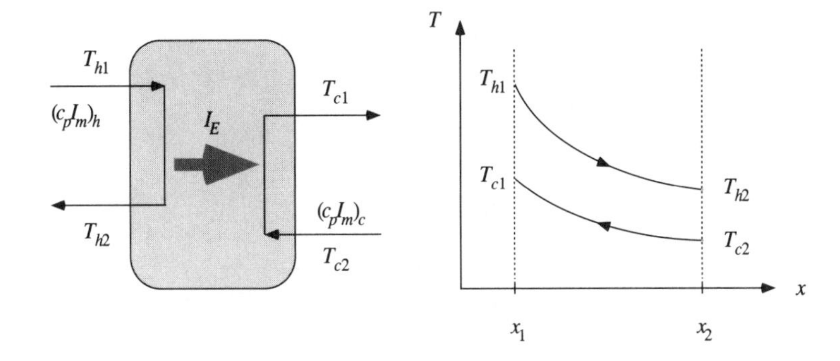

FIGURE 62. Flow diagram of counter-flow heat exchangers (left). The levels of the fluid currents symbolically represent the fact that heat is transferred from a higher to a lower thermal level. The temperature profiles of the two fluid streams have the forms shown (right). In counter-flow heat exchangers, the temperature difference between the fluids does not vary as strongly as in other arrangements.

The main factors which determine the quality of a heat exchanger are the surface area through which the two fluids exchange entropy and energy, and the heat transfer coefficients for the exchange of heat between the fluids and the wall separating them. (Assuming that the wall does not add much to thermal resistance, the two convective heat transfer coefficients determine the outcome.) At each point of the surface the energy current density is given by the total local heat transfer coefficient and the local difference of the temperatures of the fluids. Since at least the temperature difference is variable over the surface, the total energy flux from one side of the exchanger to the other is commonly expressed in terms of an average product of surface area and transfer coefficient Ah, and a mean temperature difference ΔT_m:

$$I_E = Ah\Delta T_m \tag{217}$$

The first of these factors depends upon the nature of heat transfer between the fluids and the wall separating them, and on the heat exchanger geometry, while the mean temperature difference[8] is a function of the inlet and outlet temperatures of the fluid and of the geometrical arrangement of the fluid currents. Naturally, the energy current transferred is equal in magnitude to the difference of the convective energy fluxes associated with each of the fluids entering and leaving the heat exchanger. Therefore we can write:

$$\begin{aligned} I_E &= \left(c_p I_m\right)_h \left(T_{h1} - T_{h2}\right) \\ I_E &= \left(c_p I_m\right)_c \left(T_{c1} - T_{c2}\right) \end{aligned} \tag{218}$$

The subscript h refers to the hotter of the two fluids, while c denotes the cooler one; one commonly calls the product of temperature coefficient of enthalpy and of the current of mass the "capacitance flow rate."

8. For reasons which will become clear shortly, this temperature difference often is called the *log mean temperature difference*.

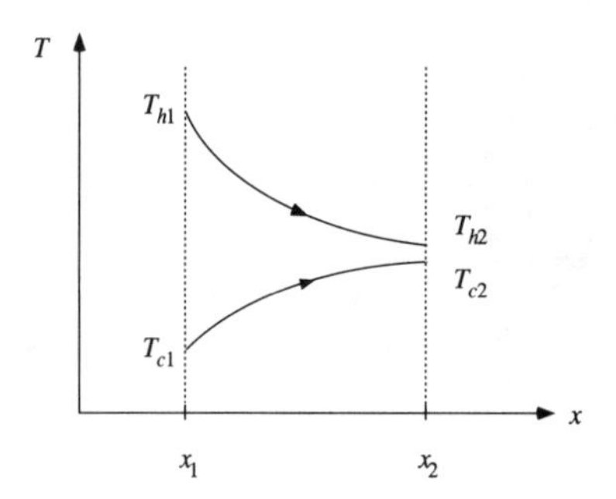

FIGURE 63. In a parallel-flow heat exchanger, the temperature difference between the fluids decreases in the direction of flow. The largest value of the difference occurs at the inlet to the exchanger. Usually, counter-flow heat exchangers are better than the parallel-flow type.

Heat exchanger effectiveness. The heat exchangers which we are going to study will all be "ideal" in the sense that they are not supposed to lose any heat to the surroundings; in other words, all the heat given up by the hotter fluid will be taken up by the second one. Still, the exchange of heat cannot be ideal since entropy is produced as a result of the transfer of entropy from higher to lower temperatures. The quality of a heat exchanger is therefore expressed in terms of a factor called the heat exchanger effectiveness. It is defined as the ratio of the actual transferred energy current and the current which could be passed between the fluids under ideal circumstances, i.e., if no temperature difference were needed:

$$\varepsilon = \frac{I_E}{\left(c_p I_m\right)_{min}\left(T_{h1} - T_{c2}\right)} \tag{219}$$

The latter of these quantities is determined by the largest temperature difference found and by the smaller of the two capacitance rates. It should be clear that the effectiveness of a heat exchanger depends upon the heat transfer factors and upon the capacitance rates of the fluids. At this point we will simply state the result for counter-flow and parallel-flow exchangers (a derivation for the balanced counter-flow type will be given below). In the case of *counter-flow heat exchangers*, the effectiveness is given by

$$\varepsilon = \frac{1 - \exp\left(-NTU(1 - C^*)\right)}{1 - C^* \exp\left(-NTU(1 - C^*)\right)} \tag{220}$$

where

$$C^* = \frac{\left(c_p I_m\right)_{min}}{\left(c_p I_m\right)_{max}} \qquad NTU = \frac{Ah}{\left(c_p I_m\right)_{min}} \tag{221}$$

The coefficient *NTU* is called the number of transfer units. This result does not apply if the two capacitance rates are equal. Under these special circumstances, the effectiveness of the counter-flow heat exchanger must be calculated according to

$$\varepsilon = \frac{NTU}{1 + NTU} \tag{222}$$

If the capacitance flow rates are equal, the heat exchanger is said to be *balanced.* The equation for the effectiveness of *parallel-flow heat exchangers* is the following:

$$\varepsilon = \frac{1 - \exp\left(-NTU(1 + C^*)\right)}{1 + C^*} \tag{223}$$

The definition of the factors in this result is the same as the one presented for counter-flow exchangers in Equation (221).

The balance of energy. The derivation of the governing differential equations is another example of the approach we used for flat-plate solar collectors. First, we are going to develop the appropriate equations of balance in integral form for an entire system from which the differential equations will be derived.

Consider one of the ducts of the counter-flow heat exchanger or a piece thereof as depicted in Figure 61. With respect to such a system we have three currents of entropy and the associated fluxes of energy, namely the convective currents entering and leaving a duct, and the conductive flow from the upper to lower duct through the wall separating the fluids. In the steady state, the balance of energy takes the following form:

$$c_p I_m \left(T_2 - T_1\right) + \int_{x_1}^{x_2} W j_E(x) dx = 0 \tag{224}$$

This law holds for both ducts if we apply the proper signs for the two cases.

The sum of the convective fluxes may be expressed in terms of the mass currents, the temperature coefficients of energy of the fluids, and the temperatures at the ends of the ducts. The conductive flux is the integral of the flux density over the surface of a duct. The flux density can be calculated using the transfer coefficient h and the temperature difference between the fluids of the upper and the lower ducts at position x:

$$j_E(x) = h(x)\left(T_h(x) - T_c(x)\right) \tag{225}$$

If we apply this relation to the fluids in the two ducts, we obtain

$$\begin{aligned} \left(c_p I_m\right)_h \left(T_{h2} - T_{h1}\right) + \int_{x_1}^{x_2} W h\left(T_h(x) - T_c(x)\right) dx = 0 \\ \left(c_p I_m\right)_c \left(T_{c1} - T_{c2}\right) - \int_{x_1}^{x_2} W h\left(T_h(x) - T_c(x)\right) dx = 0 \end{aligned} \tag{226}$$

Now we divide each of these equations by the difference of the positions of the end and the beginning of the ducts, i.e., by $x_2 - x_1$, and take the limit for very small Δx. In the first term of either equation we get the derivative of the temperature of the fluid with respect to position. From the second term we recover the integrand. The resulting differential equations therefore look like

$$\begin{aligned} \left(c_p I_m\right)_h \frac{dT_h}{dx} + Wh\left(T_h(x) - T_c(x)\right) = 0 \\ \left(c_p I_m\right)_c \frac{dT_c}{dx} + Wh\left(T_h(x) - T_c(x)\right) = 0 \end{aligned} \tag{227}$$

These are two coupled differential equations for T_h and T_c as unknown functions of position. The system of equations represents a two-point boundary value problem. Usually, the temperature of the fluids at their respective inlets would be given, and the variation of temperature with position is calculated.

Derivation of the results for balanced counter-flow heat exchangers. The derivation of the following equations can be carried over to more general examples. We begin by subtracting Equation (227b) from Equation (227a) to obtain a differential equation for the difference of the fluid temperatures. For balanced heat exchangers, the two capacitance flow rates are equal; i.e.,

$$\left(c_p I_m\right)_h = \left(c_p I_m\right)_c$$

which leads to

$$\frac{d\left(T_h - T_c\right)}{dx} = 0 \tag{228}$$

This tells us that the temperature difference between the two fluids is the same at every point in the direction of flow. We now apply the expressions for the currents of energy according to Equations (217) and (218). Noting that the temperature of the hot fluid at the outlet is equal to the value of the temperature of the cold fluid plus the constant temperature difference between the fluids, we get

$$Ah\Delta T = c_p I_m\left(T_{h1} - \left(T_{c2} + \Delta T\right)\right)$$

which can be transformed to yield

$$\Delta T = \frac{1}{1 + \dfrac{Ah}{c_p I_m}}\left(T_{h1} - T_{c2}\right) = \frac{1}{1 + NTU}\left(T_{h1} - T_{c2}\right) \tag{229}$$

Note that ΔT is the mean temperature difference ΔT_m in our case. Since the temperatures in this equation are those of the fluids at their respective inlets, the temperature difference can be calculated. If we apply the definition of the effectiveness according to Equation (219), we find that

$$\varepsilon = \frac{Ah\Delta T}{\left(Ah + c_p I_m\right)\Delta T} = \frac{NTU}{1 + NTU}$$

which is the result presented in Equation (222). Now, the temperature of the first fluid stream is easily calculated from the differential equation

$$c_p I_m \frac{dT_h}{dx} + Wh\Delta T = 0$$

since the temperature difference is constant. Using the result for this difference, we find that

$$T_h(x) = T_{h1} - \varepsilon \frac{T_{h1} - T_{c2}}{L} x \tag{230}$$

where L is the length of the heat exchanger duct. This shows that the fluid temperatures are linear functions of position for the example of the balanced counter-flow heat exchanger.

The rate of entropy production. Heat exchangers produce entropy as a matter of fact; the reversible exchanger does not exist. Therefore, it is imperative to quantify the magnitude of the irreversibility of heat transfer in such devices. Here, we will derive the expression for the rate of entropy production in a balanced counter-flow heat exchanger for negligible pressure drop of incompressible fluids in the ducts.

A look at the left side of Figure 62 shows that the only fluxes of entropy with respect to the heat exchanger are the four convective currents associated with fluid flow. In steady-state operation, the law of balance of entropy takes the form

$$\Pi_s = \left(s_{h2} - s_{h1}\right)\left|I_{mh}\right| + \left(s_{c1} - s_{c2}\right)\left|I_{mc}\right| \tag{231}$$

where s is the specific entropy of the fluid. (Remember that currents leading into the system are counted as negative quantities.) For our case, the specific entropy is

$$s(T) = s_o + c_p \ln\left(\frac{T}{T_o}\right)$$

which yields the expression for the rate of production of entropy

$$\Pi_s = c_p I_m \ln\left(\frac{T_{h2}T_{c1}}{T_{h1}T_{c2}}\right) = c_p I_m \ln\left(\frac{\left(T_{c2} + \Delta T\right)\left(T_{h1} - \Delta T\right)}{T_{h1}T_{c2}}\right)$$

Since the temperature difference is related to the effectiveness of the heat exchanger according to

$$\Delta T = (1 - \varepsilon)\left(T_{h1} - T_{c2}\right)$$

we can write

$$\begin{aligned} \Pi_s &= c_p I_m \ln\left[\left(1+\frac{\Delta T}{T_{c2}}\right)\left(1-\frac{\Delta T}{T_{h1}}\right)\right] \\ &= c_p I_m \ln\left[\left(1+(1-\varepsilon)\left(\frac{T_{h1}}{T_{c2}}-1\right)\right)\left(1-(1-\varepsilon)\left(1-\frac{T_{c2}}{T_{h1}}\right)\right)\right] \end{aligned} \tag{232}$$

You can tell very easily from this relation that the rate of production of entropy vanishes both for ideal heat exchangers ($\varepsilon = 1$) and for those with $\varepsilon = 0$; for intermediate values, there is a maximum of entropy production. While the first of these results is to be expected, the latter appears to be paradoxical. However, it can be understood of we realize that the condition $\varepsilon \rightarrow 0$ means that the exchanger lets less and less heat pass from the hotter to the cooler fluid. Therefore it represents a vanishing heat exchanger,[9] which certainly does not produce any entropy. This problem demonstrates that we cannot analyze heat exchangers without pressure drop by themselves if we wish to find optimal solutions for the design of a thermal system. The difficulty is resolved with ease, if the effect of the heat exchanger upon an entire system is considered. It is then found that the irreversibility of the total system decreases monotonically with increasing effectiveness of the heat exchanger. If we consider both heat transfer irreversibility and fluid flow irreversibility, however, there exist optimal designs for the exchangers.

EXAMPLE 44. Heat exchanger in a solar hot water system.

Consider a simple solar hot water heater with collector, balanced counter-flow heat exchanger, and hot water tank. On the collector side, the fluid is a mixture of water and glycol with a value for the temperature coefficient of enthalpy of 3800 J/(K · kg) and a mass flux of 0.10 kg/s. On the storage side, the fluid is water.

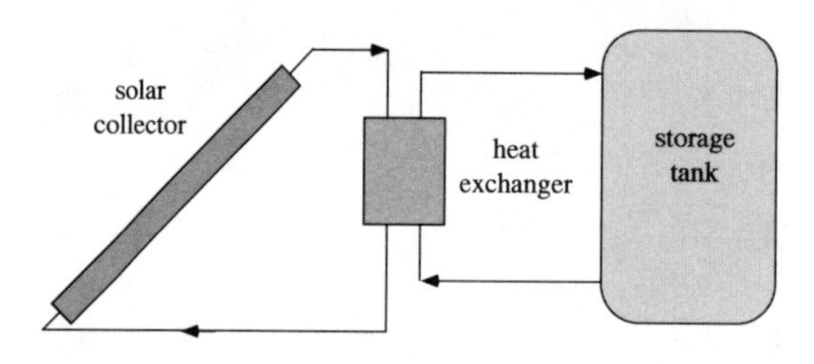

9. See Bejan (1988), p. 614–613, for a comprehensive discussion of heat exchanger irreversibilities (including those associated with friction).

a) Calculate the flux of mass necessary on the storage side if the heat exchanger is to operate in balanced mode. b) With an average heat transfer coefficient of 1000 W/(K · m²) from fluid to fluid, how large should the exchanger surface area be if we wish to have a heat exchanger effectiveness of 90%? c) It is found that the temperature of the water leaving the storage tank is 20°C, while the temperature of the water-glycol mixture leaving the collector is 80°C. Calculate the temperature of the water entering the storage tank, and the value of the energy current transferred. d) Could you just as well let the heat exchanger operate in parallel-flow mode?

SOLUTION: a) Since the heat exchanger is to be operated in balanced mode, the two capacitance flow rates are equal. Therefore we have

$$c_{p,coll} I_{m,coll} = c_{p,storage} I_{m,storage} \quad \Rightarrow$$

$$I_{m,storage} = \frac{c_{p,coll}}{c_{p,storage}} I_{m,coll} = \frac{3800}{4200} I_{m,coll} = 0.090\,\text{kg/s}$$

b) The effectiveness of the balanced counter-flow heat exchanger is given in Equation (222). With the definition of *NTU*, we obtain

$$NTU = \frac{Ah}{c_p I_m}$$

$$\varepsilon = \frac{NTU}{1+NTU} = \frac{Ah/c_p I_m}{1+Ah/c_p I_m} = \frac{Ah}{c_p I_m + Ah}$$

$$\Rightarrow \quad \left(c_p I_m + Ah\right)\varepsilon = Ah \quad \Rightarrow \quad Ah(1-\varepsilon) = \varepsilon c_p I_m$$

$$\Rightarrow \quad A = \frac{\varepsilon c_p I_m}{h(1-\varepsilon)} = \frac{0.9 \cdot 3800 \cdot 0.10}{1000(1-0.9)}\,\text{m}^2 = 3.42\,\text{m}^2$$

c) The current of energy transferred between the fluids can be calculated using the definition of the effectiveness in Equation (219):

$$I_E = \varepsilon c_p I_m \left(T_{h1} - T_{c2}\right)$$

Since the effectiveness and the inlet temperatures of the two fluid streams are known, we find that an energy current equal to

$$I_E = 0.9 \cdot 3800 \cdot 0.10(80-20)\,\text{W} = 20.5\,\text{kW}$$

is transferred from the collector loop to the storage loop. (This is a fairly large energy flux for a solar hot water system; these conditions can be achieved on a good day with some 40 - 50 m² of collector surface area. This result puts into perspective the surface area needed for the heat exchanger.) With this value known, we can compute the outlet temperatures of the two fluid streams:

$$I_E = \left(c_p I_m\right)_h \left(T_{h1} - T_{h2}\right) \quad \Rightarrow \quad T_{h2} = T_{h1} - \frac{I_E}{\left(c_p I_m\right)_h} = 80°\text{C} - 54°\text{C} = 26°\text{C}$$

$$I_E = \left(c_p I_m\right)_c \left(T_{c1} - T_{c2}\right) \quad \Rightarrow \quad T_{c1} = T_{c2} + \frac{I_E}{\left(c_p I_m\right)_c} = 20°\text{C} + 54°\text{C} = 74°\text{C}$$

d) With an effectiveness of 0.90 for the balanced counter-flow heat exchanger, the value of *NTU* is 9. Using this number also for the operation in parallel-flow mode, Equation (223) yields an effectiveness of 0.50, which is considerably less than the value of 0.90 in counter-flow.

EXAMPLE 45. Entropy production in an air-to air heat exchanger with friction.

Write the law of balance of entropy for steady-state operation of an air-to-air heat balanced counter-flow heat exchanger. Give the result for the limiting case of high effectiveness and low pressure drops.

SOLUTION: In this example, we need to know the specific entropy of air entering and leaving the heat exchanger. Since we wish to include the effects of changes of temperature and of pressure, we should give the specific entropy in terms of these two variables. If the air is modeled as an ideal gas, the specific entropy is given by

$$s(T,P) = s_o + c_p \ln\left(\frac{T}{T_o}\right) - \frac{R}{M_o}\ln\left(\frac{P}{P_o}\right)$$

See Chapter 2. In the steady state, the rate of production of entropy is equal to the sum of the four convective fluxes of entropy; this yields

$$\begin{aligned}\Pi_s &= (s_{h2} - s_{h1})|I_{mh}| + (s_{c1} - s_{c2})|I_{mc}| \\ &= c_p I_m \ln\left(\frac{T_{h2}T_{c1}}{T_{h1}T_{c2}}\right) - \frac{R}{M_o} I_m \ln\left(\frac{P_{h2}P_{c1}}{P_{h1}P_{c2}}\right)\end{aligned}$$

The first term was derived in Equation (232). This part of the expression therefore becomes

$$\begin{aligned}\Pi_{s,T} &= c_p I_m \ln\left[\left(1 + (1-\varepsilon)\left(\frac{T_{h1}}{T_{c2}} - 1\right)\right)\left(1 - (1-\varepsilon)\left(1 - \frac{T_{c2}}{T_{h1}}\right)\right)\right] \\ &\approx c_p I_m \ln\left[1 + (1-\varepsilon)\left(\frac{T_{h1}}{T_{c2}} - 1\right) - (1-\varepsilon)\left(1 - \frac{T_{c2}}{T_{h1}}\right)\right]\end{aligned}$$

if we set $1 - \varepsilon << 1$, which should hold for large values of the effectiveness. Further modification of the term leads to

$$\begin{aligned}\Pi_{s,T} &= c_p I_m \ln\left[1 - 2(1-\varepsilon) + (1-\varepsilon)\frac{T_{h1}^2 + T_{c2}^2}{T_{h1}T_{c2}}\right] \\ &= c_p I_m \ln\left[1 + (1-\varepsilon)\frac{(T_{h1} - T_{c2})^2}{T_{h1}T_{c2}}\right] \approx c_p I_m (1-\varepsilon)\frac{(T_{h1} - T_{c2})^2}{T_{h1}T_{c2}}\end{aligned}$$

The pressure term leads to the following contribution to the rate of production of entropy:

$$\Pi_{s,P} = -\frac{R}{M_o} I_m \ln\left(\frac{(P_{h1} - |\Delta P_h|)(P_{c2} - |\Delta P_c|)}{P_{h1} P_{c2}} \right) \approx \frac{R}{M_o} I_m \left[\frac{|\Delta P_h|}{P_{h1}} + \frac{|\Delta P_c|}{P_{c2}} \right]$$

where the last step is a consequence of the assumption that the pressure drops are small compared to the absolute values of the pressure of the fluids.

4.6.3 Stability in Free Convection

Sometimes it is not a pump which drives fluid motion in convective heat transfer. In the presence of a gravitational field, convection may take place as a result of buoyancy effects. This phenomenon is well known from boiling water, or from vertical motion of air in the Earth's atmosphere. The latter problem attracted a lot of attention more than 100 years ago when Kelvin tried to calculate the temperature gradient in the atmosphere. Later it was recognized that this type of convection is responsible, under certain circumstances, for heat transfer inside stars. Here we will discuss the question of how free convection takes place in an atmosphere of an ideal gas.

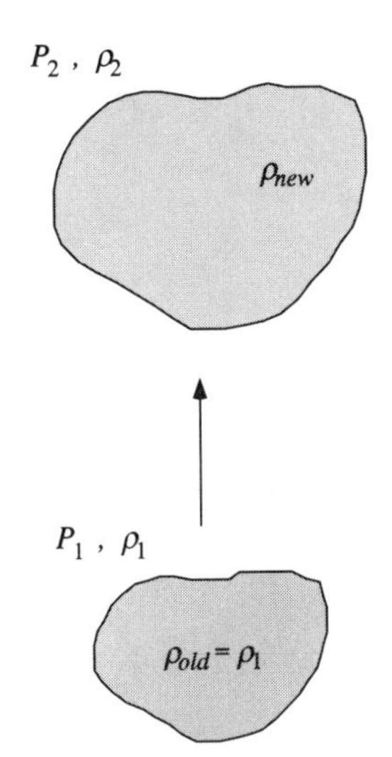

FIGURE 64. A blob of fluid rising in an atmosphere in a gravitational field. It is assumed to expand adiabatically over the distance it travels. If its density is greater than that of the surroundings at its new position, it will sink back down: convective motions will die out. The layer is said to be stable against convection.

The condition of convective stability. First we have to answer the question of when free convection occurs. The fluid may be at rest, in which case heat is transported conductively or radiatively. Therefore, there must exist special conditions which lead to the onset of convective motion, i.e., to *convective instability.* To find the condition of stability, consider the following model (Figure 64). A fluid parcel is somehow displaced upward from its normal environment. Before the displacement, the values of density, pressure, and temperature inside the parcel are the same as those of its surroundings. At its new position the fluid will expand to adjust its pressure to the new value. As a consequence, the density will decrease. If the new density inside the parcel is larger than that of the surroundings in the upper position, i.e., if

$$\rho_{new} > \rho_2 \tag{233}$$

the displaced mass will sink back to its initial position. In this case the fluid layer is stable against convection, and Equation (233) is said to be the *criterion of convective stability.* If Equation (233) is not satisfied, the layer is unstable and the parcel will keep moving since it is less dense than the environment; convection will start to take over.

We will be able to say more about the new density of the fluid parcel only if we know more about the conditions satisfied during the expansion. Assume convective motion to take place fast enough for *adiabatic conditions* to prevail. As a result, the blob of matter rising upward does not exchange heat with its environment. Only at the end, i.e., when the parcel has risen to the top of the convective layer, does it emit heat to the cooler surroundings. Since the expansion of the blob of ideal gas takes place adiabatically, we can compute the new density according to the Laplace-Poisson law of adiabatic change:

$$\rho_{new} = \rho_{old}\left(\frac{P_2}{P_1}\right)^{1/\gamma} \tag{234}$$

Here, P_1 and P_2 refer to the pressure in the lower and the upper positions, respectively. The values are the same inside the blob and the surroundings at any moment. The old density is the same inside the parcel and outside in the lower position. If we apply the law of adiabatic change to a small vertical displacement we can write the pressure and the density at the upper position in terms of the values at the lower point. With

$$P_2 = P_1 + \frac{dP}{dz}\Delta z$$

and

$$\rho_2 = \rho_1 + \frac{d\rho}{dz}\Delta z$$

we transform the stability criterion of Equation (233) to

$$\frac{1}{P}\frac{dP}{dz} > \gamma\frac{1}{\rho}\frac{d\rho}{dz} \tag{235}$$

It is convenient to express this condition in terms of the temperature gradient dT/dz. Since for the ideal gas

$$\frac{1}{\rho}\frac{d\rho}{dz} = \frac{1}{P}\frac{dP}{dz} - \frac{1}{T}\frac{dT}{dz}$$

the final form of the stability criterion is the following:

$$\frac{dT}{dz} > \left(1 - \frac{1}{\gamma}\right)\frac{T}{P}\frac{dP}{dz} \tag{236}$$

According to Example 15 of Chapter 2, the quantity on the right-hand side is the adiabatic temperature gradient. Since both the pressure and temperature gradients are negative quantities, the criterion expresses the following result: *If the magnitude of the actual temperature gradient dT/dz is smaller than the magnitude of the adiabatic temperature gradient, the fluid layer is stable against convective motion.*

Calculating the energy flux due to convection. Which value will the temperature gradient take if convection is responsible for the transport of heat? According to Equation (236), all we know is that its magnitude will be greater than the magnitude of the adiabatic gradient. The excess of the actual gradient over the adiabatic gradient is responsible for the magnitude of the net convective flux of heat or energy:

$$j_{E,conv} = \left(\left. \frac{dT}{dz} \right|_{ad} - \frac{dT}{dz} \right) \Delta z c_p \rho v = \Delta \nabla T \Delta z c_p \rho v \tag{237}$$

Here, Δz is the average distance travelled by the moving blobs, while c_p is the specific temperature coefficient of enthalpy of the gas; v is the average velocity of the parcels. The excess of the temperature gradient over the adiabatic one has been abbreviated by $\Delta \nabla T$. Note that we have neglected the kinetic energy term in Equation (237). In practice it may be very difficult to compute the actual temperature gradient. There are some cases, however, in which the excess is extremely small, and the gradient is very nearly the adiabatic one (Example 46).

In Equation (237), the average distance travelled by the blobs of air plays a central role. There exists a particular approach to calculating this quantity which is called the *mixing length approximation*. Assume the parcels of matter rising in free convection to be carried over an average distance which is called the mixing length l_{mix}. If we assume gradients to be constant over the mixing length, we can develop an approximation for the net energy flux carried over this distance. However, we still need an expression for the average velocity of the parcels in that equation. The velocity attained by a parcel is a consequence of the acceleration due to buoyancy. According to Equation (235), the excess of density $\Delta \rho_{ex}$ for a distance Δz is equal to

$$\Delta \rho_{ex} = \left(\frac{d\rho}{dz} - \frac{1}{\gamma} \frac{\rho}{P} \frac{dP}{dz} \right) \Delta z = \frac{\rho}{T} \Delta \nabla T \Delta z$$

where $\Delta \nabla T$ is the excess of temperature for the same distance which appears in Equation (237). Half of this density excess multiplied by the gravitational field strength g is the average net force per volume over the distance Δz. If we multiply this quantity by the distance Δz, we obtain the kinetic energy per volume gained by the parcel:

$$\frac{1}{2} \rho v_f^2 = \frac{1}{2} g \left(\frac{\rho}{T} \Delta \nabla T \Delta z \right) \Delta z$$

v_f is the final velocity of the blob, which we take to be twice the average value. This equation allows us to eliminate the flow speed from the expression for the net energy flux. Using the mixing length l_{mix} for the distance Δz, we obtain the following expression for the energy flux density:

$$j_{E,conv} = \frac{1}{2} c_p \rho \left(\frac{g}{T} \right)^{1/2} \Delta \nabla T^{3/2} l_{mix}^2 \tag{238}$$

EXAMPLE 46. Convection in the interior of a massive main sequence star.

Estimate the amount by which the actual temperature gradient inside a massive main sequence star surpasses the adiabatic gradient. In a star of about 5 solar masses (which makes it 10^{31} kg) heat is carried by convection in the innermost parts, and by radiation further out. The energy flux not too close to the center may be taken to be the total flux of $2.4 \cdot 10^{30}$ W (which is about three to four orders of magnitude larger than the luminosity of the Sun). The radius of such a star is $1.75 \cdot 10^9$ m, and its central temperature has a value of some 27 million K. (The values reported here are the result of numerical modeling.) Would you say that convection inside such a star follows the assumption that heat is not exchanged during upward (and downward) motion of the blobs?

SOLUTION: The expression for the energy flux in the mixing length approximation may be used to estimate the temperature excess

$$\Delta \nabla T = \left[\frac{2 j_E}{c_p \rho l_{mix}^2} \left(\frac{T}{g} \right)^{1/2} \right]^{2/3}$$

For our numerical estimate we need approximate values of the energy flux density, temperature, density, enthalpy capacity, gravitational field strength, and the mixing length at an intermediate point inside the star. The first of these turns out to be equal to

$$j_E = \frac{I_E}{4\pi (R/2)^2} = 2.5 \cdot 10^{11}\,\mathrm{W/m^2}$$

Let us take a temperature of 10 million K, and a density ten times the average density, which makes it 4500 kg/m^3. The temperature coefficient of enthalpy can be calculated according to the rule given in Section 2.4.3. The matter inside such a star, which consists of 70% hydrogen and 30% helium by mass, is completely ionized. This makes it a monatomic gas. Instead of one mole of hydrogen we have one mole of protons and one mole of electrons. For one mole of helium we end up with a total of three moles of amount of substance (one mole of nuclei and two moles of electrons). This leads to an average molar mass of stellar matter of 0.00062 kg/mole. With a molar enthalpy capacity of 20.8 J/(K · mole) we get a value of 33.6 kJ/(K · kg) for the specific value.

The gravitational field strength is estimated in the following manner. About half the mass of the star lies in the innermost part within a radius of half that of the entire sphere. g at the surface of this inner sphere is calculated according to Newton's law of gravitation, which yields a value of

$$g = G \frac{0.5 M}{(R/2)^2} = 440\,\mathrm{N/kg}$$

The mixing length, finally, cannot exceed the size of the convective zone near the center of the star. Let us use one tenth of the radius of the star for the missing parameter.

In summary we get an excess of the temperature gradient of $6.4 \cdot 10^{-8}$ K/m. We have to compare this to the average actual gradient of $1.1 \cdot 10^{-2}$ K/m. Obviously the excess is very small and can be neglected. In other words, convection takes place adiabatically which means that the temperature gradient in convective layers inside a star is given by

$$\frac{dT}{dr} = \left(1 - \frac{1}{\gamma}\right)\frac{T}{P}\frac{dP}{dr}$$

While the excess of the temperature gradient is crucial for the flux of energy carried convectively, it is negligible as far as the structure of the interior of a star is concerned. Even if we had grossly overestimated the magnitude of the mixing length, this result would still hold. However, near the surface of a star, the density and the mixing length become so small, that if there is convection, a large excess of the temperature gradient is required, and we may no longer calculate the temperature gradient according to the adiabatic gradient.

Questions and Problems

1. Imagine a power plant at the mouth of a river flowing into the ocean which uses the osmotic pressure difference between sea water and fresh water. If the river is carrying 1000 m^3/s of fresh water, how large could the power of an ideal plant be?
2. One mole of methane is burned with precisely the amount of air necessary for complete combustion. How much entropy and energy are given off by the combustion products if they are cooled from 800°C to 100°C?
3. Solely on the basis of energetics, would diamonds make a better fuel than coal (graphite)?
4. Calcite ($CaCO_3$) decomposes to lime (CaO) and CO_2 at a temperature of 1195 K. Show how to obtain an estimate of this value from chemical properties of the substances at 25°C listed in the tables of the Appendix.
5. Estimate the pressure necessary for transforming graphite into diamond.
6. Calculate the chemical potential of H_2O in its gaseous, liquid, and solid forms each at temperatures of 200 K, 300 K, and 400 K. From the results deduce which of the forms should be stable at the three temperatures listed.
7. Consider the optimum condition for helium burning in stars.[10] Carbon-12 nuclei can form from helium-4 nuclei inside stars in the following manner. First, beryllium-8 is formed from two helium nuclei, whereupon beryllium takes up another helium nucleus. However, beryllium spontaneously decays into its helium constituents, making the final reaction very difficult. Helium burning becomes possible if conditions inside a star are such as to provide for the highest possible beryllium concentration from which carbon can form. For a given concentration of helium-4 nuclei, the optimum temperature is found to be about $7 \cdot 10^8$ K. The absolute chemical potential of a nuclear species is given as a function of concentration by the results known for the ideal gas

$$\mu(T,\bar{c}) = \bar{e}_o + RT\ln\left(\frac{\bar{c}}{\bar{c}^*}\right)$$

---

10. Falk and Schmid in Falk and Herrmann (1977–82), Volume 4, p. 76.

if the concentration c is small compared to the concentration $\bar{c}^*$ at degeneracy; the concentration for a degenerate gas is proportional to $T^{-3/2}$. (For the following, you do not have to know more about degeneracy.) Here $\bar{e}_o$ is the molar rest energy of the nuclei under consideration. a) Show that the difference of the chemical potentials for the reaction 2 $p_2n_2 \rightarrow p_4n_4$ (for which $\Delta\bar{e}_o$ is - 95 keV per particle) is given by

$$\Delta\mu = \Delta\bar{e}_o + RT\ln\left(\frac{\bar{c}^2_{p_2n_2}}{\bar{c}^2_{p_2n_2}{}^*}\frac{\bar{c}_{p_4n_4}{}^*}{\bar{c}_{p_4n_4}}\right)$$

Here, p_2n_2 stands for helium 4 nuclei, while p_4n_4 represents the nuclei of beryllium. Note that on the basis of the term $\Delta\bar{e}_o$ alone the reaction proceeds from beryllium to helium. b) Show that the condition of chemical equilibrium for the reaction can be cast in the form

$$\frac{\bar{c}_{p_4n_4}}{\bar{c}^2_{p_2n_2}} = aT^{-3/2}\exp\left(-\frac{b}{T}\right) \quad , \quad b = 1.1\cdot 10^9\,\mathrm{K}$$

and demonstrate how the optimum temperature for helium burning can be obtained from this result.

8. Show how a measurement of the vapor pressure curve and the change of the volume of the fluid upon vaporization can be used to derive the changes of entropy, enthalpy, and energy of the fluid for the phase change.

9. Two grams of CO_2 are dissolved in a liter of bottled water. Assuming that there only is carbon dioxide in the space above the water, how large is the pressure of the gas?

10. Fast melting of snow (spring runoff) is thought not to be so much a result of warm rain but rather of fog flowing over snow covered slopes. Why should fog be much more effective in melting snow than warm rain?

11. Show that the relation between the mass of snow (modeled as ice) melting and the mass of water vapor condensing out of fog (see Problem 10) is given by

$$m_{snow} = 7.5 m_{vapor}$$

while the ratio of snow melting from condensation of fog and from cooling of rain from temperature T to T_{snow} is roughly $600/(T - T_{snow})$.[11]

12. Determine the vapor pressure of Refrigerant 123 (R123) as a function of temperature. It is known that at 0°C the vapor pressure is 32.7 kPa. Use data from Figure 43 and compare your result with Figure 65.

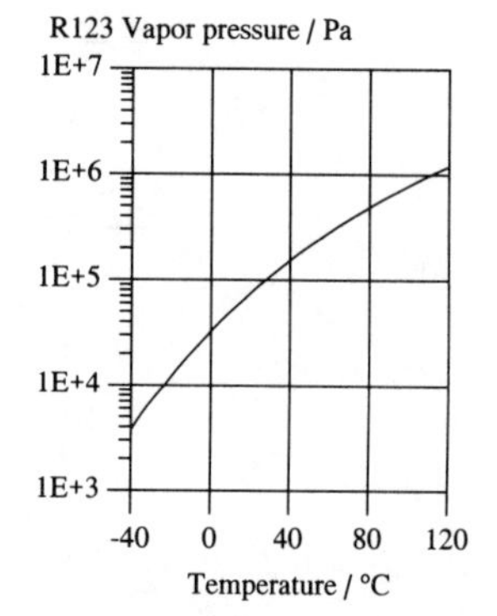

FIGURE 65. Problem 12.

13. Show that Equation (98) can be written in the form

$$\mu_l(T,p_n) = \mu_l(T,p_{no}) - M_o RT\upsilon_l(p_n - p_{no})$$

where the indices l and s refer to the solvent (fluid) and the solute, respectively.

14. The partial specific volume of nitrogen dissolved in sea water is 1.43 cm^3/g; for oxygen it is 0.97 cm^3/g. a) Determine the pressure coefficient of the chemical potential

11. See Bohren (1995), p. 79–81.

for the gases. b) With the density of sea water varying between 1.025 and 1.035 g/cm^3 (depending on its salinity), show that the solubility of nitrogen should decrease with increasing depth; the solubility of oxygen, on the other hand, might show both an increase and a decrease with depth, depending on the circumstances. Calculate the expected percent change of solubility for the two gases for a change of depth of 1000 m.

15. Model the water turbine discussed in Section 4.4.1 as a cylinder having a piston capable of taking in and ejecting water at the desired pressures. Create a four-step cycle and display the processes in the μ-n and the P-V diagrams.
16. Demonstrate that the simple result obtained for the chemical potential of an incompressible substance at constant temperature and velocity, Equation (126), follows from the general expression of Equation (143).
17. Show that in an incompressible fluid at rest at the surface of the Earth, the pressure is a linear function of depth.
18. Show that the chemical potential of a fluid takes the form

$$\mu = M_o\left[u + Pv - Ts\right]$$

if it is expressed using specific quantities of energy, volume, and entropy. How does this formula change if you include the flow velocity?

19. Explain whether (a) the conductive part of the momentum current and (b) the convective part are directly included with the chemical potential of a fluid.
20. Change the water turbine discussed in Section 4.4.1 to a compressed air engine. Model the engine as a cylinder having a piston, and an inlet and an outlet both with valves. Air is taken up at high chemical potential (pressure) and released again at lower potential. The piston moves in such a manner as to allow for an isothermal cycle of the gas. a) Show that you get the expression for the chemical potential of the ideal gas as a function of pressure (for constant temperature) derived above.
b) Why is it possible to get the result by equating the mechanical power of the engine to the net flow energy current expressed solely in terms of the chemical energy current, even though we must heat the gas for the cycle to be performed isothermally?
21. Apply a control volume analysis to the determination of the temperature coefficient of enthalpy of air proposed in Example 19 of Chapter 2.
22. Hydrogen is burned with oxygen in a rocket engine. Knowing that the exit speed of the gas is around 3000 m/s, estimate the temperature of the stream.
23. Derive Bernoulli's Law for a fluid flowing between points at different levels in a gravitational field. Neglect possible effects of friction.
24. Often, the terms resulting from the interaction of a fluid and the gravitational field and the flow speed are included with the definition of the specific enthalpy of a fluid:

$$h = u + \frac{P}{\rho} + \frac{1}{2}v^2 + \varphi_G$$

Show that you obtain this expression by defining the change of enthalpy as the difference of the generalized flow potential:

$$\varphi_{flow} = M_o^{-1}\mu + \varphi_G + v^2 + Ts$$

The difference of this potential may be interpreted as the generalized driving force of the combined convective flows of amount of substance, momentum, mass, and entropy.

25. Use the equation for the pressure as a function of height above ground in an isothermal atmosphere to derive the chemical potential of the ideal gas at constant temperature.

26. Show by direct evaluation of the terms that the partial derivative of the chemical potential with respect to the pressure is $M_o \upsilon$ for an ideal fluid. Then show that you obtain the differential equation of hydrostatic equilibrium of the fluid in the gravitational field from the condition of gravito-chemical equilibrium.

27. Derive the chemical potential for an incompressible fluid as a function of temperature at constant pressure.

28. Consider the following strongly simplified model of the accretion of a planet. Matter with a temperature T_o falls from far away onto the surface of a growing planet. (The planet is surrounded by a gas of temperature T_o.) Assume the surface of the planet to radiate as a gray body. a) If the effect of the rate of change of the temperature can be neglected, show that the surface temperature at an instantaneous value of the radius can be calculated using

$$|I_m| c_p (T - T_o) + e 4\pi r^2 \sigma (T^4 - T_o^4) = |I_m| G \frac{m(r)}{r}$$

where I_m is the flux of mass falling upon the planet, c_p is its specific temperature coefficient of enthalpy, and e is the emissivity of the surface. (*Hint:* Treat the surface as an open control volume and consider the law of balance of energy for this system; first derive the instationary model.) b) Show that this is equivalent to

$$\rho c_p (T - T_o) \frac{dr}{dt} + e\sigma (T^4 - T_o^4) = \rho G \frac{m(r)}{r} \frac{dr}{dt}$$

c) Model accretion such that the rate of change of the radius of the planet is given by[12]

$$\frac{dr}{dt} = k_1 t^2 \sin(k_2 t)$$

At $t = 0$ and at $t = t_a$ (total accretion time) this function is supposed to vanish, and the radius of the planet grows from 0 to R during this period. Show that this leads to the following expressions for the constants k_1 and k_2:

12. Anderson (1989), p. 3.

$$k_1 = \frac{R}{2t_a^3\left(1/\pi - 2/\pi^3\right)}$$

$$k_2 = \frac{\pi}{t_a}$$

d) Calculate $T(r)$ for the following values of the parameters. $T_o = 100$ K, $\rho = 5500$ kg/m^3, $c_p = 800$ J/(K · kg), $R = 6.4 \cdot 10^6$ m, $t_a = 5 \cdot 10^5$ years, $e = 1$. (You should get the largest temperature, roughly 1000 K, at a radius of 5000 km.)

29. Model the heating of the air in a room as follows. As entropy is added to the air it expands and diffuses through the walls such as to leave the pressure in the room at a constant value. Now take a constant value for the heating power. Show that in this case the temperature of the air rises according to

$$T(t) = T_o \exp\left(\frac{\mathcal{P}_{heating} t}{7/2\, PV}\right)$$

if a value of $7/2R$ is taken for the molar temperature coefficient of enthalpy of air. Note that it is assumed that the air remaining in the room does not lose any heat to the surroundings.

30. Express the efficiency of the Rankine cycle in terms of the average temperature of heating.

31. Discuss the effect of changing the upper and the lower operating pressures of the Rankine cycle. Why is a condenser used in vapor power plants if steam leaving the turbine could be discharged directly to the environment?

32. Calculate the Carnot efficiency of the cycles running between the upper and the lower operating temperatures occurring in the processes of Example 38 (without superheating). Does the difference between the Carnot efficiencies and the values calculated in that example result from dissipation?

33. Assume the furnace of the Carnot cycle proposed in Problem 32 to operate at 850°C and the condenser at 20°C, respectively. The cycle undergone by the working fluid is supposed to be the same as before. Calculate the rate of production of entropy and the rate of loss of availability.

34. a) Estimate the efficiency of a vapor power cycle without superheating designed for the fluid R123. The heat is supposed to be delivered by solar collectors such as vacuum tubes. Saturated liquid enters the evaporator at a pressure of 8.0 bar, while the condenser operates at a temperature of 30°C. (Use property data found in Figure 43 and Figure 65.) b) If the collectors deliver an energy current of 350 W per square meter of collector area, what is the minimum collector area needed per kW of power of the engine?[13]

35. Estimate the amount of entropy produced in the throttling process of a refrigerant if the following data are given: the initial and the final pressure, and the initial and the final specific volume of the fluid.

---

13. See Koch (1993) for a discussion of the merits of different types of fluids used in such a scheme with or without superheating.

36. If you include in the balance of energy the energy delivered by the pump to maintain the pressure of the air in an solar air collector, how does the form of the equation of balance of entropy in Equation (215) change?

37. The analysis of the solar air collector of Section 4.6.1 can be repeated for different heights of the air duct. Changing this parameter changes the location of the minimum of the rate of production of entropy. It is found that the minima for different heights approximately are found at collector lengths which leave the cross section of the duct constant. Can you explain this behavior?

38. Express the law of balance of entropy for a flat-plate solar collector in steady-state operation.

39. If 60°C water is to be produced with solar collectors, why should they be operated in such a way that they deliver water of exactly 60°C?

40. A fluid is flowing through a pipe which is being heated from outside. (This setup can be found in line focus concentrators for solar radiation.) Show that the differential equation for the temperature of the fluid as a function of position in the direction of flow is given by

$$c_p I_m \frac{dT}{dx} = F'\left[\sigma_E - 2\pi r_o U_L (T - T_a)\right]$$

where

$$F' = \frac{1/U_L}{1/U_{fa}}$$

can be called the efficiency factor of the tubular heater. U_L is the heat loss coefficient from the pipe to the surroundings (see Problem 15 of Chapter 3 for a numerical example) while U_{fa}, which represents the heat transfer coefficient from the fluid to the environment, was calculated in Problem 15 of Chapter 3. σ_E is the rate of absorption of energy per length of pipe, and T_a is the ambient temperature.

41. In Problem 40, assume F' and U_L to be constant. Show that the convective energy flux which is carried away by the fluid in the heated pipe can be expressed by

$$I_{E,useful} = F_R L\left[\sigma_E - 2\pi r_o U_L \left(T_{f,in} - T_a\right)\right]$$

where

$$F_R = \frac{c_p I_m}{2\pi r_o L U_L}\left[1 - \exp\left(-\frac{2\pi r_o L U_L F'}{c_p I_m}\right)\right]$$

is called the heat removal factor for the pipe of length L. $T_{f,in}$ is the inlet temperature of the fluid. Note that the result is similar in structure to the one obtained for flat-plate solar collectors (Section 4.6.1).

42. Consider a parabolic trough concentrator for solar light with a metal pipe running along the line focus. A fluid is pumped through the pipe. Take the concentration factor of the radiation to be C, and give the pipe a radius r and a length L. a) If you neglect convective losses and if the heat transfer coefficient from the surface of the pipe absorbing sunlight to the fluid is very large, show that the differential equation for the temperature of the fluid in the direction of flow is

$$c_p I_m \frac{dT}{dx} = 2r\left[C(\tau\alpha)\mathcal{G}_b - \pi e\sigma\left(T^4 - T_a^4\right)\right]$$

Here it has been assumed that all the light falling onto the mirrors (expressed in terms of direct solar irradiance $\mathcal{G}_b$, i.e., the irradiance excluding diffuse light) is concentrated onto the projection area of the pipe, and that a fraction $(\tau\alpha)$ of the incoming radiation is absorbed. The factor e is the emissivity of the pipe. b) Calculate the exit temperature of the fluid for given operating conditions and for inlet temperature $T_{f,in}$.

43. Compare balanced counter-flow and parallel-flow heat exchangers, both with an *NTU* of 5. Which of the exchangers has the higher effectiveness?

44. Compare balanced and unbalanced counter-flow heat exchangers. If they are built identically, do they have the same effectiveness? (For concreteness, take the smaller of the two capacitance flow rates to be equal to the one used in the balanced mode.)

45. Explain in qualitative terms why the effectiveness of a parallel-flow heat exchanger operated in balanced mode cannot be greater than 0.5. (*Hint:* see Figure 63.)

46. As we have found in examples in Chapter 3, the entropy generated in the Earth's interior cannot be transported by conduction because this mode of heat transfer is not effective enough under the given circumstances. It is therefore assumed that heat is transferred convectively, and the material of the mantle is modeled as an ideal fluid as discussed in the Interlude. Make a model of convective motion of the Earth's mantle in which a blob of matter rises from the interior to the surface adiabatically (as in the model for the Earth's atmosphere in Section 3.3.6, or as discussed in Section 4.6.3). a) Assume that the heating at a given radius is due to entropy production as the result of radioactive decay (which is taken to be distributed evenly in the entire mantle) at smaller radii, and friction which assumes all the energy released by the heat engine represented by the convective motion to be dissipated whereby more entropy is returned to the rising matter. Show that in this case the entropy current entering a thin shell at radius r and driving the heat engine is given by

$$I_s(r) = \frac{1}{T(r)} \frac{m(r) - m(r_c)}{m(R) - m(r_c)} I_E(R)$$

where r_c and R are the radius of the core (bordering on the mantle) and the radius of the Earth, respectively. $I_E(R)$ is the energy flux penetrating the Earth's surface from the interior of the planet. b) It is found that the gravitational field g is roughly constant in the entire mantle. Show that in this case

$$I_s(r) = \frac{1}{T(r)} \frac{r^2 - r_c^2}{R^2 - r_c^2} I_E(R)$$

c) Show that if you assume the temperature gradient through the mantle to be adiabatic, it can be expressed by

$$\mathcal{P} \approx g \cdot I_E(R) \int_{r_c}^{R} \frac{r^2 - r_c^2}{R^2 - r_c^2} \gamma^* \rho \cdot \kappa_s dr$$

where γ^* is the Grüneisen ratio defined in Problem 24 of the Interlude, and κ_S is the adiabatic compressibility of the material. (*Hint:* See Problem 24 of Chapter 1 and

Problem 27 in the Interlude.) d) The values of the different physical parameter of the Earth's interior have to be derived from seismic and other measurements.[14] Take an average density of 4500 kg/m^3, and average values of 10 N/kg, 0.8, and $3 \cdot 10^{-12}$ /Pa for the gravitational field, the Grüneisen ratio, and the adiabatic compressibility, respectively. The lower boundary of the mantle is at 3500 km from the center of the Earth, and the energy flux through the surface is $31 \cdot 10^{12}$ W. Estimate the efficiency of the convective motion interpreted as a heat engine.

14. See Stacey (1992) for numerical values for the Earth's interior.

EPILOGUE

Steps Toward Continuum Thermodynamics

The impression [that my dissertation (1879)] made upon the public of physics at that time was naught. ...Kirchhoff rejected its contents expressly with the remark that the concept of entropy ... could not legitimately be applied to irreversible processes.

M. Planck, 1948

In this closing chapter I hope to give you a final impression of the power of the concepts used in this book. The general law of balance of entropy has guided us through the development of introductory thermodynamics. What seemed to be so difficult for Planck's teachers to accept has served us well so far. This should encourage us to extend thermodynamics to the general case of continuous processes. The benefits will be immediate: you will get a first brief look at some of the more advanced examples of thermal processes to which we could only allude before.

We shall introduce the particular approach to thermodynamic theory[1] used in this chapter in the context of the dynamics of uniform fluids (Section E.1). You may take this section as a formal summary of what you have learned, and as a stepping stone into new territory. Section E.2 provides a derivation of the equations of continuity of entropy, momentum, and mass; and the law of balance of energy for continuous processes will be treated in Section E.3. Finally, two examples of continuum thermodynamics will be discussed. The first treats the case of a viscous heat-conducting fluid, leading to a theory which combines the action of two major sources of irreversibility in an otherwise simple material (Section E.4). The second will solve the problem of the infinite speed of propagation of thermal disturbances predicted by the classical theory of conductivity by extending the scope of irreversible thermodynamics (Section E.5).

---

1. Continuum thermodynamics has old roots, but consistent attempts toward clarifying the foundations and extending the applications date back only to about 1960. (For a summary see Truesdell, 1984.) Of the different routes leading toward continuum thermodynamics, we shall choose the one charted by I. Müller (Müller, 1985) since his approach is the most direct extension of the use of laws of balance as we have seen them. Note that laws of balance are only a small part of what is needed (we already have recognized the role played by constitutive laws), but they provide a perfectly simple starting point for an appreciation of physical processes.

E.1 Thermodynamics of Uniform Fluids

Now that we have learned about concepts, and have applied them to simple examples in the previous chapters, we will present a more formal treatment of the thermodynamics of uniform fluid systems. Speaking in general terms, we can divide the phenomena into the following three groups: first, there are processes involving pure fluids which may or may not undergo phase changes; second, we may consider fluid flow in nonreactive systems, including mixtures; and finally, we have to deal with cases which allow for chemical reactions and flow in open systems. In all cases, we will restrict the discussion to spatially uniform fluids.

We will use these examples for a first encounter with the particular approach to thermodynamic theory mentioned above; it was first developed in continuum thermodynamics, but it can be adjusted to apply to spatially uniform cases as well. It demonstrates the use of the laws of balance as the starting point of a description of nature. In contrast to the approach used in Chapters 1–4, it does not assume the form of the relationship between currents of entropy and of energy in heating and in cooling; therefore, the following sections can also be taken as an additional proof that the ideal gas temperature is the natural measure of the thermal potential. The method presented here will give you a firmer grounding in thermodynamics, and it will prepare you for the simple examples of continuum physics discussed further below.

E.1.1 A Single Viscous Fluid Without Phase Changes

We will introduce a generalized approach to thermodynamic theory by applying it to a simple fluid, namely a single viscous uniform body which cannot change its phase. With the exception of viscosity, this is the example of a material we have encountered in Chapter 2 and the Interlude where we have assumed constitutive relations which allow only for ideal processes. We have seen that a theory of thermodynamics of such ideal fluids leads to the same relations known from thermostatics, as you can verify by comparison with standard texts on this subject. In other words, the ideal fluids considered so far attain the same properties during dynamical processes as in equilibrium states. The materials for which the theory holds obey such simple constitutive laws as to deliver results which are independent of the speed and other details of processes. Even though the derivation is one of a theory of dynamics, time apparently drops out of the equations in the end.

If you change the conditions just a little bit, for example by introducing a viscous pressure term, time appears explicitly in the results. This is what we would like to demonstrate on the following pages. The previous results will then be obtained in the limit of vanishing viscosity.

Viscous pressure. First we should discuss how to include viscosity in the constitutive laws describing the behavior of the fluid. This will tell us something about the particular form the assumptions should take upon which we will base the following development.

Imagine a viscous fluid. Naturally, the effect of viscosity will be felt only as long as the fluid flows. In the case of a uniform body this means that its volume must be changing. As long as viscous effects are neglected, we would assume the pressure of the fluid to be expressed as a function of temperature and of volume; now, however, we will assume the pressure, which is equivalent to the momentum current density across a surface, also to depend upon the speed of the process:

$$P\left(V, T, \dot{V}\right) = P|_E(V, T) + a\dot{V} \tag{1}$$

$P|_E$ is the static pressure of the fluid, i.e., the value of the pressure attained when the volume does not change, or if viscosity is neglected. As you can see, the rate of change of the volume of the body therefore is included as one of the independent variables of the theory. This will have important consequences for the assumptions to be made regarding the behavior of the fluid.

The assumptions. We need to make a number of assumptions to develope a theory of thermodynamics of a particular type of material. Basically, the fundamental laws used so far, with the exception of the relation between currents of entropy and of energy in heating, and with Equation replacing a simpler form of the equation of state, will furnish the foundation. Naturally, leaving out one assumption calls for a replacement, unless it was an unnecessary one to begin with:

1. First, we have to agree on the independent variables of the theory. As before, they will be the volume and the temperature of the uniform fluid, which now will be joined by the rate of change of the volume. Therefore, entropy, energy, pressure, and other quantities will be functions of *V*, *T*, and *dV/dt:*

$$S = S\left(V, T, \dot{V}\right) \ , \quad E = E\left(V, T, \dot{V}\right) \ , \quad P = P\left(V, T, \dot{V}\right) \ , \quad \ldots$$

   This requirement carries over to quantities such as the entropy capacity and the latent entropy.

2. While we do not take the relation between currents of entropy, currents of energy, and temperature for granted (i.e., we do not assume the important law $I_{E,th} = T I_s$, with T the ideal gas temperature, to hold), we accept a constitutive law for the flux of energy in heating which makes it proportional to the flux of entropy.[2] If we heat twice as fast, i.e., if we double the current of entropy, we also double the current of energy associated with the heating:

2. In continuum physics, we have Fourier's law for the entropy flux density for fluids such as those described by the Navier-Stokes-Fourier equations:

$$I_{E,th} \propto I_s \tag{2}$$

We see that this is a particular constitutive assumption. The flux of energy associated with compression and expansion of the uniform fluid, on the other hand, is well known:

$$I_{E,mech} = P\dot{V} \tag{3}$$

3. The additional assumption which we need in place of knowledge of the full relation between the currents of entropy and of energy in heating can be furnished as follows. We introduce ideal walls separating different fluid systems.[3] Ideal walls are defined to be those which do not contribute to thermal processes; i.e., they do not produce entropy. This means that as part of the definition, the entropy flux across an ideal wall is continuous:

$$I_s(\mathrm{I}) = I_s(\mathrm{II}) \tag{4}$$

where the roman numerals refer to the fluids separated by the wall. As the actual assumption, we take for granted that two fluids separated by an ideal wall will have the same temperature at the wall:

$$T(\mathrm{I}) = T(\mathrm{II}) \tag{5}$$

Naturally, the flux of energy is continuous across this wall as well (remember the discussion in Section 1.6.5). Assuming the existence of such walls is necessary for the measurement of temperature to work the way we know it: one of the fluids separated by the wall would be the fluid of the thermometer, the other would be the body whose temperature we wish to measure. Only the fluids, not the wall separating them, may have an influence upon the temperatures (see Figure 1).

$$j_s = -\varphi(T,\rho)\frac{dT}{dx}$$

The same type of equation also applies to the thermal energy flux density:

$$j_{E,th} = -\beta(T,\rho)\frac{dT}{dx}$$

Therefore, the requirement of proportionality between these fluxes is satisfied. (See Section E.4 for a more detailed discussion.) We have assumed constitutive laws of heating of uniform fluids before for which this requirement is fulfilled as well.

3. For a discussion, see Müller (1985), p. 168–169. The assumption of the continuity of temperature at an ideal wall, Equation (5), replaces other assumptions which are made in different approaches to thermodynamics.

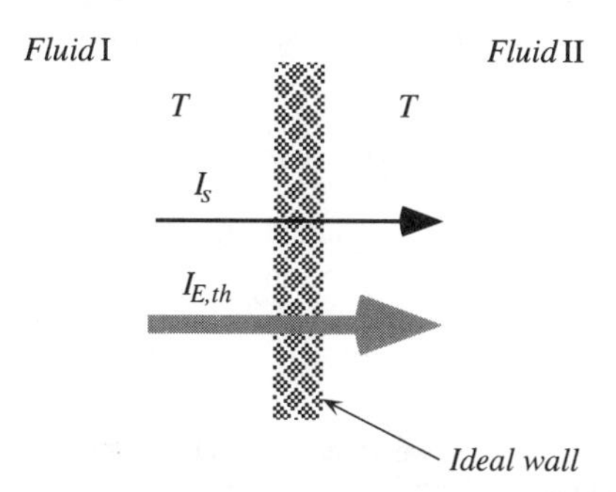

FIGURE 1. Two fluids of the same type are separated by an ideal wall. Across such a wall, entropy and energy flux are continuous. Also, it is assumed that the temperature is the same on both sides.

4. With these preliminaries, we can state the laws of balance which we take to be valid in the current case. They are the expressions of balance of entropy (for the thermal process), and of momentum (for the mechanical process). Therefore, we postulate that

$$\begin{aligned} \dot{S} + I_s &= \Pi_s \\ \dot{p} + I_p &= 0 \end{aligned} \tag{6}$$

Obviously, we should include the effect of irreversibility as a consequence of the viscosity of the material. Since the fluids are at rest, the balance of momentum is trivial and will not be required explicitly any more.

5. Processes also satisfy the requirement of the balance of energy, in addition to the laws of balance for the special processes taking place. Since the particular fluids under investigation allow for thermal and mechanical changes only, the equation of balance of energy takes the form

$$\dot{E} + I_{E,mech} + I_{E,th} = 0 \tag{7}$$

Remember that this law does not distinguish between different types of physical processes; this is the responsibility of the other laws of balance.

Consequences of the energy principle. The crucial point about processes undergone by physical systems is that in addition to the proper laws of balance, they always satisfy the energy principle, i.e., the law of balance of energy, as well. This requirement may be expressed as follows:

Physical processes obey the energy principle. If these processes are required to be thermal, they also satisfy the balance of entropy; this law is then said to be a restriction upon the processes. If the phenomena also are supposed to be mechanical, a second restriction applies to them, namely the law of balance of momentum. Each particular law of balance serves as a restriction upon the form the laws governing a process may take.

Now, such restrictions can be taken into account using Lagrange multipliers;[4] both Equations (6a) and (7) are satisfied simultaneously if and only if [5]

$$\dot{E} + I_{E,mech} + I_{E,th} - \lambda^s\left[\dot{S} + I_s - \Pi_s\right] = 0 \tag{8}$$

(Remember that we do not have to explicitly take the law of balance of momentum into consideration here.) This is the starting point of one of the approaches

---

4. Lagrange multipliers are known from extremal problems. Assume that a function $f(x_1, x_2, x_3)$ is to be maximized under the additional assumption that the two relations $g_i(x_1, x_2, x_3) = 0$, $i = 1{,}2$, have to be satisfied. These requirements mean that the three total derivatives have to be zero:

$$\frac{\partial f}{\partial x_1}\dot{x}_1 + \frac{\partial f}{\partial x_2}\dot{x}_2 + \frac{\partial f}{\partial x_3}\dot{x}_3 = 0$$

$$\frac{\partial g_1}{\partial x_1}\dot{x}_1 + \frac{\partial g_1}{\partial x_2}\dot{x}_2 + \frac{\partial g_1}{\partial x_3}\dot{x}_3 = 0$$

$$\frac{\partial g_2}{\partial x_1}\dot{x}_1 + \frac{\partial g_2}{\partial x_2}\dot{x}_2 + \frac{\partial g_2}{\partial x_3}\dot{x}_3 = 0$$

The dot denotes a derivative with respect to some parameter. The restrictions are taken into consideration using Lagrange multipliers λ^1 and λ^2. The last two equations are each multiplied by one of the factors and then added to the first. This leads to

$$\left[\frac{\partial f}{\partial x_1} + \lambda^1\frac{\partial g_1}{\partial x_1} + \lambda^2\frac{\partial g_2}{\partial x_1}\right]\dot{x}_1 + \left[\frac{\partial f}{\partial x_2} + \lambda^1\frac{\partial g_1}{\partial x_2} + \lambda^2\frac{\partial g_2}{\partial x_2}\right]\dot{x}_2 + \left[\frac{\partial f}{\partial x_3} + \lambda^1\frac{\partial g_1}{\partial x_3} + \lambda^2\frac{\partial g_2}{\partial x_3}\right]\dot{x}_3 = 0$$

The restrictions introduced mean that two of the unknowns x_i could be expressed in terms of the third. One may therefore chose the Lagrange multipliers in such a way that the first two terms in brackets are equal to zero. (These are equations for the two unknowns.) As a result, the third term in brackets also has to equal zero. We therefore obtain three new relations

$$\frac{\partial f}{\partial x_1} + \lambda^1\frac{\partial g_1}{\partial x_1} + \lambda^2\frac{\partial g_2}{\partial x_1} = 0$$

$$\frac{\partial f}{\partial x_2} + \lambda^1\frac{\partial g_1}{\partial x_2} + \lambda^2\frac{\partial g_2}{\partial x_2} = 0$$

$$\frac{\partial f}{\partial x_3} + \lambda^1\frac{\partial g_1}{\partial x_3} + \lambda^2\frac{\partial g_2}{\partial x_3} = 0$$

which have to be satisfied with the two restrictions. Together, we have five equations for the unknowns x_i, and for the Lagrange multipliers λ^i.

5. This proof, and the proof of the equivalence of these expressions to the requirement presented in Equation (11), has been given by Liu (1972) for the general case. See also Müller (1985, p. 170).

developed in continuum thermodynamics, which for lack of a better name, is called *thermodynamics with Lagrange multipliers.* Equation (8) is obtained by subtracting the entropy principle (multiplied by a Lagrange multiplier) from the law of balance of energy. Naturally, λ^s must have the dimension of temperature for the equation to be dimensionally correct.

The next steps consist of expressing the mechanical current of energy, which is a result of changes of volume, and determining the time derivatives of the energy and entropy functions. Since these quantities are functions of the volume, temperature, and rate of change of volume, we can write Equation (8) as follows:

$$\begin{aligned}&\frac{\partial E}{\partial V}\dot{V}+\frac{\partial E}{\partial T}\dot{T}+\frac{\partial E}{d\dot{V}}\ddot{V}+\left(P|_E+a\dot{V}\right)\dot{V}+I_{E,th}\\&\qquad-\lambda^s\left[\frac{\partial S}{\partial V}\dot{V}+\frac{\partial S}{\partial T}\dot{T}+\frac{\partial S}{d\dot{V}}\ddot{V}+I_s-\Pi_s\right]=0\end{aligned}\qquad(9)$$

To obtain this result, the derivatives of $E(V,T,dV/dt)$ and $S(V,T,dV/dt)$ have been written in terms of the partial derivatives, and the law for the mechanical energy flux for uniform viscous fluids has been applied. If we combine terms which are explicitly linear in the time derivatives of the independent variables, it becomes

$$\begin{aligned}&\left[\frac{\partial E}{\partial V}-\lambda^s\frac{\partial S}{\partial V}+P|_E\right]\dot{V}+\left[\frac{\partial E}{\partial T}-\lambda^s\frac{\partial S}{\partial T}\right]\dot{T}\\&\qquad+\left[\frac{\partial E}{\partial \dot{V}}-\lambda^s\frac{\partial S}{\partial \dot{V}}\right]\ddot{V}+I_{E,th}-\lambda^s I_s+a\cdot\dot{V}^2+\lambda^s\Pi_s=0\end{aligned}\qquad(10)$$

This equation must hold for all imaginable processes; i.e., it must be satisfied for all values of the derivatives of the independent variables. Assume this to be the case for a set of such values. Now change one of them just a little bit: in general, Equation (9) will not be satisfied anymore unless the factor multiplying this derivative is set equal to zero. Since this reasoning applies to all terms explicitly involving the derivatives, the following three equations must hold:

$$\begin{aligned}&\frac{\partial E}{\partial V}-\lambda^s\frac{\partial S}{\partial V}+P|_E=0\\&\frac{\partial E}{\partial T}-\lambda^s\frac{\partial S}{\partial T}=0\\&\frac{\partial E}{\partial \dot{V}}-\lambda^s\frac{\partial S}{\partial \dot{V}}=0\end{aligned}\qquad(11)$$

This leaves us with the residual equation

$$I_{E,th}-\lambda^s I_s+a\cdot\dot{V}^2+\lambda^s\Pi_s=0$$

Now, we can say a number of things about this equation. For one, the assumption that the fluxes of entropy and of energy in heating are proportional requires that

$$I_{E,th} = \lambda^s I_s \tag{12}$$

which at the same time means that

$$\Pi_s = -\frac{a}{\lambda^s}\dot{V}^2 \tag{13}$$

Obviously, the production of entropy is a direct consequence of the viscosity of the fluid.

Determination of the Lagrange multiplier. Equations (11) through (13) are the preliminary results of our theory. If we manage to determine the multiplier λ^s in terms of physical quantities, we will have the basis for deriving all desired results about the particular material investigated.

There still are a couple of assumptions we have not used so far. First, we shall make use of the idea regarding ideal walls. Imagine two uniform ideal fluids (I and II) to be separated by an ideal wall which only lets entropy and energy pass (Figure 1). For both fluids, Equation (12) holds, which means that

$$I_{E,th}(\mathrm{I}) - \lambda^s\left(V_\mathrm{I}, T_\mathrm{I}, \dot{V}_\mathrm{I}\right)I_s(\mathrm{I}) = I_{E,th}(\mathrm{II}) - \lambda^s\left(V_\mathrm{II}, T_\mathrm{II}, \dot{V}_\mathrm{II}\right)I_s(\mathrm{II}) \tag{14}$$

Since both fluxes and the temperature are continuous across the wall, this condition reduces to

$$\lambda^s\left(V_\mathrm{I}, T, \dot{V}_\mathrm{I}\right) = \lambda^s\left(V_\mathrm{II}, T, \dot{V}_\mathrm{II}\right) \tag{15}$$

The volume and the rate of change of the volume of the two fluids are independent and can be given any values; this means that the Lagrange multiplier may be a function only of temperature, and not of the other two independent variables. We have

$$\lambda^s = \lambda^s(T) \tag{16}$$

The multiplier is a universal function of temperature, the same for all fluids of the type considered here.

We have to perform one more step before we can apply the results to the ideal gas and find the Lagrange multiplier. Since λ^s only depends upon the temperature of the fluid, Equation (11c) reduces to

$$\frac{\partial}{\partial \dot{V}}\left(E - \lambda^s S\right) = 0$$

which means that $E - \lambda^s S$ is independent of dV/dt. Next, Equation (11b) is transformed into

$$\frac{\partial}{\partial T}\left(E - \lambda^s S\right) = -S\frac{d\lambda^s}{dT}$$

This means that both the entropy and the energy of the fluid must be independent of the rate of change of the volume, and Equation (11c) can be dropped from the list of results. These are strong restrictions upon the behavior of the material. We can say that

$$\begin{aligned} S &= S(V,T) \\ E &= E(V,T) \end{aligned} \tag{17}$$

Now we are ready to determine the Lagrange multiplier. For the moment, we will set the parameter a, which describes the effects of viscosity in Equation , equal to zero. Since we have additional constitutive information about the ideal gas, let us apply the results to this body. We solve the first two parts in Equation (11) for the derivatives of the energy function, and take the derivative with respect to the other independent variable:

$$\frac{\partial}{\partial T}\left(\frac{\partial E}{\partial V}\right) = \frac{\partial}{\partial T}\left(\lambda^s \frac{\partial S}{\partial V} - P|_E\right) = \lambda^s \frac{\partial^2 S}{\partial V \partial T} + \frac{d\lambda^s}{dT}\frac{\partial S}{\partial V} - \frac{\partial P|_E}{\partial T}$$

$$\frac{\partial}{\partial V}\left(\frac{\partial E}{\partial T}\right) = \frac{\partial}{\partial V}\left(\lambda^s \frac{\partial S}{\partial T}\right) = \lambda^s \frac{\partial^2 S}{\partial T \partial V}$$

If the functions are assumed to be sufficiently smooth, the mixed derivatives must be equal. Therefore, we obtain the condition

$$\frac{d\lambda^s}{dT}\frac{\partial S}{\partial V} = \frac{\partial P|_E}{\partial T}$$

or

$$\frac{d\lambda^s}{dT}\left(\frac{\partial E}{\partial V} + P|_E\right)\frac{1}{\lambda^s} = \frac{\partial P|_E}{\partial T}$$

This finally leads to

$$\frac{1}{\lambda^s}\frac{d\lambda^s}{dT} = \frac{\partial P|_E / \partial T}{\partial E/\partial V + P|_E} \tag{18}$$

This still holds for all fluids. In the case of the ideal gas, however, the special properties show that the right-hand side of Equation (18) is equal to the inverse of the ideal gas temperature. Remember that the energy of the ideal gas only depends upon temperature:

$$\frac{1}{\lambda^s}\frac{d\lambda^s}{dT} = \frac{1}{T} \tag{19}$$

Integration of this result shows that the unknown Lagrange multiplier is the ideal gas temperature:

$$\lambda^s = T \tag{20}$$

In other words, the ideal gas temperature takes the role of the thermal potential. This concludes the proof of the relation between currents of entropy and of energy in heating for the fluids under investigation. Considering that we allow for irreversibility, the result is even more interesting than the equivalent statement which we derived in the Interlude for ideal fluids.

Results for uniform viscous fluids. Now that the main unknown factor of the theory has been determined, we can collect the results and derive some more important expressions. First, the fluxes of entropy and of energy in heating are related by

$$I_{E,th} = T I_s \tag{21}$$

Second, the rate of production of entropy as a consequence of viscous pressure is

$$\Pi_s = -\frac{a}{T}\dot{V}^2 \tag{22}$$

Since Π_s cannot be negative we conclude that the parameter a in the law of viscous pressure cannot take positive values:

$$\Pi_s \geq 0 \quad \Rightarrow \quad a \leq 0$$

Furthermore, the properties of the fluids are such that the partial derivatives of energy and of entropy are related by

$$\begin{aligned} \frac{\partial E(V,T)}{\partial V} &= T\frac{\partial S(V,T)}{\partial V} - P|_E(V,T) \\ \frac{\partial E(V,T)}{\partial T} &= T\frac{\partial S(V,T)}{\partial T} \end{aligned} \tag{23}$$

This is a consequence of Equation (11). This result can be used to derive the Gibbs fundamental form of the fluid. Since the time derivative of the energy is composed of partial derivatives, according to

$$\dot{E} = \frac{\partial E(V,T)}{\partial V}\dot{V} + \frac{\partial E(V,T)}{\partial T}\dot{T}$$

this quantity can be expressed using Equation (23) to yield

$$\dot{E} = T\dot{S} - P|_E \dot{V} \tag{24}$$

Finally, if we introduce the entropy capacity and the latent entropy; i.e., if we write

$$K_V = \frac{\partial S(V,T)}{\partial T}$$
$$\Lambda_V = \frac{\partial S(V,T)}{\partial V} \tag{25}$$

we see that the flux of entropy in heating can be expressed as follows:

$$I_s = -\Lambda_V \dot{V} - K_V \dot{T} - \frac{a}{T}\dot{V}^2 \tag{26}$$

Obviously, in the case of viscous fluids heating depends upon the rate of the process; in other words, it is no longer reversible. If you now calculate processes undergone by the fluids described in this section, time will appear explicitly in the equations.

The results we have come across before in Chapter 2 and the Interlude can be derived as the limit of the present theory for vanishing viscosity: if you set $a = 0$, you will obtain the previous results. Aside from the inclusion of irreversibility, the new approach taken to thermodynamics is the most important aspect of this section; in particular, it demonstrates how we can derive the forms of potentials, such as the thermal potential which relates fluxes of entropy and of energy. Remember that historically, the route via the properties of heat engines was chosen. (See the Interlude for an example of this approach.) Today, we trust the law of balance of entropy as a general expression of the second law of thermodynamics; therefore, the development presented here seems to flow naturally from what we have learned about thermal processes.

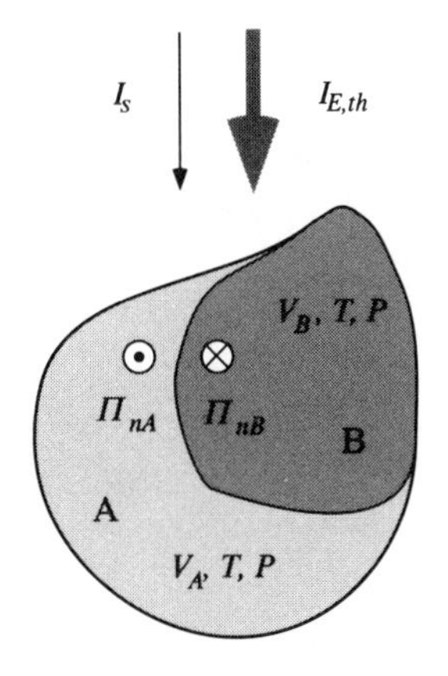

FIGURE 2. A system composed of two phases of the same fluid is heated. In addition to changes of volume and temperature, this case includes changes of amount of substance of the parts A and B. (When A loses an amount of substance, B gains just that much.) Temperature and pressure of the parts are assumed to be the same.

E.1.2 The Case of Pure Ideal Fluids With Phase Transformations

Consider a fluid composed of two phases of the same substance (Figure 2). Both phases are assumed to have the same temperature and pressure. Basically, this sounds like the prescription for two unrelated, noninteracting fluids of the kind described in the previous section (if you disregard viscosity). But here, the parts of the fluid are supposed to interact through a change of phase: one component can disappear while the amount of the other phase increases.

Assumptions. The assumptions which will be made in the course of the following derivation are mostly the same as those introduced above in Section E.1.1. Especially, we are dealing again with ideal fluids and processes which do not produce entropy. Experience tells us that fluxes of entropy are reversed upon reversal of the changes undergone by the two phase fluids in chemical equilibrium shown in Figure 2. Therefore, the equation of balance of entropy looks like the one used above:

$$\dot{S} + I_s = 0 \tag{27}$$

where the entropy of the system is now the sum of the entropies of the parts A and B:

$$S(V,T,n) = S_A(V_A,T,n_A) + S_B(V_B,T,n_B) \tag{28}$$

Note that in the current case, the independent variables include the amounts of substance of the components.

Currents of entropy and of energy in heating are again taken to be proportional, as in Equation (2), and we take the existence of ideal walls for granted.

We now have to add equations of balance for the two species A and B. There is no flow of any substance across system boundaries, but both A and B can undergo phase changes, leading to the production or destruction of the species. Therefore, the equations of balance read

$$\begin{aligned} \dot{n}_A - \Pi_{nA} &= 0 \\ \dot{n}_B - \Pi_{nB} &= 0 \end{aligned} \tag{29}$$

Since the chemical reactions taking place are of a very special nature, the rates of production and destruction are related by

$$\Pi_{nA} + \Pi_{nB} = 0 \tag{30}$$

Except for the balance of momentum, which again is trivial and does not have to be considered separately, these are the laws of balance of the processes taking place in the two phase fluids discussed here. As always, the processes also must satisfy the energy principle.

Consequences of the balance of energy. Even though we have extended the nature of the fluids under investigation, the law of balance of energy looks just like it did in the case of the single phase fluid. Since, with the exception of energy, the only physical quantities exchanged across the boundary of the system are entropy and momentum, the only energy currents are the thermal and the mechanical ones:

$$\dot{E}_A + \dot{E}_B + I_{E,mech} + I_{E,th} = 0 \tag{31}$$

The expression for the exchange of energy in the mechanical process is again that of the power of nonviscous uniform fluids. This time, however, we can write it with the volumes of both components in mind:

$$I_{E,mech} = P(\dot{V}_A + \dot{V}_B) \tag{32}$$

The method of taking the balance of energy into account simultaneously with the balance of all other relevant quantities, once more requires the introduction of Lagrange multipliers. There is one such multiplier for every basic law of balance that is added to the list, including the energy principle. In our case, this means that we have to introduce three such factors, which we label λ^s, λ^{nA}, and λ^{nB}, respectively. Then, the relation to be satisfied takes the form:

$$\begin{aligned}&\dot{E}_A + \dot{E}_B + P\left(\dot{V}_A + \dot{V}_B\right) + I_{E,th} \\ &\quad - \lambda^s\left[\dot{S}_A + \dot{S}_B + I_s\right] - \lambda^{nA}\left[\dot{n}_A - \Pi_{nA}\right] - \lambda^{nB}\left[\dot{n}_B - \Pi_{nB}\right] = 0\end{aligned} \quad (33)$$

The following steps in the evaluation of the consequences of the energy principle are very similar to those in the case of the single phase fluid. However, they involve some more algebra. First, we introduce the derivatives of the energy and the entropy with respect to the independent variables. After collecting the terms which are explicitly linear in the time derivatives of these independent variables, we obtain

$$\begin{aligned}&\left[\frac{\partial E_A}{\partial V_A} - \lambda^s \frac{\partial S_A}{\partial V_A} + P\right]\dot{V}_A + \left[\frac{\partial E_B}{\partial V_B} - \lambda^s \frac{\partial S_B}{\partial V_B} + P\right]\dot{V}_B + \left[\frac{\partial E_A}{\partial T} - \lambda^s \frac{\partial S_A}{\partial T}\right]\dot{T} \\ &\left[\frac{\partial E_B}{\partial T} - \lambda^s \frac{\partial S_B}{\partial T}\right]\dot{T} + \left[\frac{\partial E_A}{\partial n_A} - \lambda^s \frac{\partial S_A}{\partial n_A} - \lambda^{nA}\right]\dot{n}_A + \left[\frac{\partial E_B}{\partial n_B} - \lambda^s \frac{\partial S_B}{\partial n_B} - \lambda^{nB}\right]\dot{n}_B \\ &+\left[I_{E,th} - \lambda^s I_s\right] - \left[\lambda^{nA}\Pi_{nA} + \lambda^{nB}\Pi_{nB}\right] = 0\end{aligned}$$

The same arguments that led to Equations (11) and (12) also apply here. Starting with the second one, the fluxes of entropy and of energy will be proportional if

$$I_{E,th} = \lambda^s I_s \quad (34)$$

For the remainder of the equation, we can say that it has to be satisfied for all possible values of the rates of change of the independent variables. Since this requirement can easily be violated unless all factors are identically zero, we end up with the following relations:

$$\begin{aligned}&\frac{\partial E_A}{\partial V_A} - \lambda^s \frac{\partial S_A}{\partial V_A} + P = 0 \\ &\frac{\partial E_B}{\partial V_B} - \lambda^s \frac{\partial S_B}{\partial V_B} + P = 0 \\ &\frac{\partial E_A}{\partial T} - \lambda^s \frac{\partial S_A}{\partial T} = 0 \\ &\frac{\partial E_B}{\partial T} - \lambda^s \frac{\partial S_B}{\partial T} = 0 \\ &\frac{\partial E_A}{\partial n_A} - \lambda^s \frac{\partial S_A}{\partial n_A} - \lambda^{nA} = 0 \\ &\frac{\partial E_B}{\partial n_B} - \lambda^s \frac{\partial S_B}{\partial n_B} - \lambda^{nB} = 0\end{aligned} \quad (35)$$

and, for the production rates of the species:

$$\lambda^{nA}\Pi_{nA} + \lambda^{nB}\Pi_{nB} = 0 \tag{36}$$

This last relation will prove to be important when we interpret the meaning of the Lagrange multipliers that go with the species A and B. For now, we will use the reasoning presented when the Lagrange multiplier λ^s was determined. We start with a single-phase fluid for which the result of the previous section applies: λ^s is equal to the temperature. Then, such a fluid (II) is brought in contact with the two-phase fluid (I) via an ideal wall. Again, the argument which led to Equation (14) is used; here it means that

$$\lambda^s(\mathrm{II}) = \lambda^s(\mathrm{I}) = T \tag{37}$$

All that remains to be done is to determine the multipliers λ^{nA} and λ^{nB}.

The chemical potential of pure fluids. Take a closer look at Equation (36). If we combine it with the relation between the production rates, i.e., with Equation (30), we see that the Lagrange multipliers of the two phases of the fluid must be equal:

$$\left(\lambda^{nA} - \lambda^{nB}\right)\Pi_{nA} = 0 \quad \Rightarrow \quad \lambda^{nA} = \lambda^{nB} \tag{38}$$

Since the phases are in chemical equilibrium, it makes sense to interpret the as yet unknown factors as the chemical potentials of the fluid:

$$\lambda^{nA} = \mu_A \quad , \quad \lambda^{nB} = \mu_B \tag{39}$$

The rest of the equations derived above provide a means of relating the chemical potentials to other variables of the fluids. By introducing the Lagrange multipliers into the intermediate results, we obtain the laws that govern the behavior of pure fluids which may undergo phase transformations. First, the fluxes of entropy and of energy in heating must be related by the ideal gas temperature:

$$I_{E,th} = T I_s \tag{40}$$

Then, the partial derivatives of the energy and the entropy functions satisfy the relations presented in Equation (35). For each of the phases, we may now write

$$\begin{aligned}
\frac{\partial E(V,T,n)}{\partial V} &= T\frac{\partial S(V,T,n)}{\partial V} - P(V,T,n) \\
\frac{\partial E(V,T,n)}{\partial T} &= T\frac{\partial S(V,T,n)}{\partial T} \\
\frac{\partial E(V,T,n)}{\partial n} &= T\frac{\partial S(V,T,n)}{\partial n} + \mu(V,T,n)
\end{aligned} \tag{41}$$

The Gibbs fundamental form and the chemical potential. Since the energy is a function of volume, temperature, and amount of substance, its time derivative is

$$\dot{E} = \frac{\partial E(V,T,n)}{\partial V}\dot{V} + \frac{\partial E(V,T,n)}{\partial T}\dot{T} + \frac{\partial E(V,T,n)}{\partial n}\dot{n}$$

Introducing the laws derived in Equation (41) and collecting terms leads to the *Gibbs fundamental form* for fluids whose amount of substance may change:

$$\dot{E} = T\dot{S} - P\dot{V} + \mu\dot{n} \qquad \textbf{(42)}$$

This is an extension of the form used for single-phase fluids. Again, there is one such relation for each of the phases. Now, we are ready to derive the relationship between the chemical potential and other variables of the fluid. We write each of the extensive quantities in Equation (42) as the product of its density and the volume:

$$\frac{d}{dt}(\rho_E V) - T\frac{d}{dt}(\rho_s V) + P\frac{d}{dt}V - \mu\frac{d}{dt}(\rho_n V) = 0$$

Taking the derivatives leads to

$$[\rho_E - T\rho_s + P - \mu\rho_n]\dot{V} + [\dot{\rho}_E - T\dot{\rho}_s - \mu\dot{\rho}_n]V = 0$$

Since this relation must be satisfied for all possible (independent) values of the volume and its time derivative, both factors in brackets must be identically zero. Otherwise, the equation could be violated. The result of these considerations are the Gibbs fundamental form written in terms of the densities, and the relations between energy, entropy, density, and chemical potential:

$$\dot{\rho}_E = T\dot{\rho}_s + \mu\dot{\rho}_n \qquad \textbf{(43)}$$

and

$$\bar{e} = T\bar{s} - P\bar{v} + \mu \qquad \textbf{(44)}$$

The last formula has been rewritten so as to agree in form with what we derived in Section 4.2.4.

The temperature and pressure dependence of the chemical potential. We have made extensive use of the dependence of the chemical potential upon temperature and pressure in the course of the applications presented in the previous sections. There the proof was performed for special cases, but now we are in a position to deliver a general derivation.

In the development of the theory we started with temperature, volume, and amount of substance as the independent variables. With the Gibbs fundamental form, on the other hand, energy, temperature, pressure, and the chemical potential are variables of entropy, volume, and amount of substance. Other forms of Equation (42) are based on other sets of independent variables. We need the set including temperature and pressure in addition to amount of substance. This transformation is facilitated using the definition of the *Gibbs free energy* function G:

$$G = E + PV - TS \tag{45}$$

The time derivative of this quantity is equal to

$$\dot{G} = \dot{E} + P\dot{V} + \dot{P}V - T\dot{S} - \dot{T}S$$

Using Equation (42), this can be written

$$\dot{G} = V\dot{P} - S\dot{T} + \mu\dot{n} \tag{46}$$

which is an alternative fundamental form having independent variables P, T, and n. The Gibbs free energy is a computational aid in this transformation, just as the enthalpy is for other purposes. These additional quantities, which have the dimension of energy, are convenient abbreviations for terms which appear frequently in calculations. You should not try to associate any meaning with them.[6]

According to Equation (46), the partial derivatives of $G(P,T,n)$ are equal to the volume, the entropy, and the chemical potential, respectively:

$$\frac{\partial G(P,T,n)}{\partial P} = V(P,T,n) = n\bar{v}(P,T,n) \tag{47}$$

$$\frac{\partial G(P,T,n)}{\partial T} = -S(P,T,n) = -n\bar{s}(P,T,n) \tag{48}$$

$$\frac{\partial G(P,T,n)}{\partial n} = \mu(P,T,n) \tag{49}$$

Taking the derivative of Equation (47) with respect to amount of substance, the derivative of Equation (49) with respect to pressure, and equating the mixed derivatives

$$\frac{\partial}{\partial n}\frac{\partial G(P,T,n)}{\partial P} = \bar{v}(P,T)$$

$$\frac{\partial}{\partial P}\frac{\partial G(P,T,n)}{\partial n} = \frac{\partial \mu(P,T,n)}{\partial P}$$

6. Enthalpy and Gibbs free energy (together with still another function, which is called the *free energy* $F=E-TS$) are called thermodynamic potentials. (See the Interlude for a brief discussion). The word *potential* has an altogether different meaning from that used in this text; these quantities do not have anything to do with potentials such as temperature and the chemical potential.

we obtain the first of the desired results:

$$\frac{\partial \mu(P,T,n)}{\partial P} = \overline{v}(P,T) \tag{50}$$

The second one is derived in just the same way:

$$\frac{\partial \mu(P,T,n)}{\partial T} = -\overline{s}(P,T) \tag{51}$$

These are the relations we used above. They are an example of the restrictions put upon the multitude of quantities used for the description of phenomena; theory limits the number of independent functions which might have to be measured in the lab if property data are required (see Section 4.5.1).

The Clapeyron equation. In Section 4.3, we made extensive use of the fact that for phase changes, pressure and temperature are related directly. This is a consequence of the equality of the chemical potentials of both phases at the transition:

$$\mu_{\text{I}}(P,T) = \mu_{\text{II}}(P,T) \tag{52}$$

Since this holds for all values of the temperature, the derivative of the difference of the potentials with respect to T must be zero:

$$\frac{d(\mu_{\text{II}} - \mu_{\text{I}})}{dT} = 0 \tag{53}$$

Performing the differentiation leads to

$$\frac{\partial \mu_{\text{II}}}{\partial T} + \frac{\partial \mu_{\text{II}}}{\partial P}\frac{dP}{dT} - \left(\frac{\partial \mu_{\text{I}}}{\partial T} + \frac{\partial \mu_{\text{I}}}{\partial P}\frac{dP}{dT}\right) = 0$$

We can introduce the temperature and pressure dependence of the chemical potentials according to Equations (50) and (51), and obtain the Clapeyron equation:

$$-\overline{s}_{\text{II}} + \overline{s}_{\text{I}} + \left[\overline{v}_{\text{II}} - \overline{v}_{\text{I}}\right]\frac{dP}{dT} = 0 \tag{54}$$

This can be transformed into Equation (92), derived in Section 4.3.2.

E.1.3 Reaction in Uniform Mixtures of Fluids

Let us now treat the special case of reactions occurring in a closed system (no flow). A fluid is composed of several species which may react; it may exchange heat, and it is assumed to be deformable. Still, we will treat the system as a uniform, nonviscous chunk of matter. The effects of motion will be neglected.

Again, we have to consider the proper expressions of balance which hold for entropy and amount of substance. Since we have N different species, we arrive at

$$\begin{aligned} &\dot{S} + I_s = \Pi_s \\ &\dot{n}_1 = \Pi_{n1} \\ &\vdots \\ &\dot{n}_N = \Pi_{nN} \end{aligned} \tag{55}$$

While we do not have convective currents, we no longer can exclude the possibility of entropy being generated; indeed, we know that this is precisely what happens in reacting fluids which cannot undergo other processes. There is an equation of balance for the amount of substance of each of the species present.

Energy may be exchanged with the surroundings as a consequence of heating and of deformation. Therefore, we have an equation of balance of energy with two terms describing exchange, namely:

$$\dot{E} + I_{E,mech} + I_{E,th} = 0 \tag{56}$$

For the nonviscous, uniform fluid, the flux of energy due to deformation is

$$I_{E,mech} = P\dot{V} \tag{57}$$

where P is the total pressure of the fluid. Again, we will follow the derivation introduced above, which leads to

$$\dot{E} + P\dot{V} + I_{E,th} - T\left[\dot{S} + I_s - \Pi_s\right] - \sum_{i=1}^{N} \mu_i\left[\dot{n}_i - \Pi_{ni}\right] = 0 \tag{58}$$

as a consequence of the fact that all equations of balance should be satisfied simultaneously. To simplify the work, we have assumed the Lagrange multipliers to be known. This equation is easily rearranged as follows:

$$\underbrace{\dot{E} + P\dot{V} - T\dot{S} - \sum_{i=1}^{N} \mu_i \dot{n}_i}_{=0} + \underbrace{I_{E,th} - TI_s}_{=0} + \underbrace{T\Pi_s + \sum_{i=1}^{N} \mu_i \Pi_{ni}}_{=0} = 0 \tag{59}$$

For the same reasons as given above, the first and the second terms must be equal to zero, which leads to the main result of the present case:

$$\dot{E} + P\dot{V} - T\dot{S} - \sum_{i=1}^{N} \mu_i \dot{n}_i = 0 \tag{60}$$

$$T\Pi_s + \sum_{i=1}^{N} \mu_i \Pi_{ni} = 0 \tag{61}$$

The first of these expressions is the Gibbs fundamental form of a fluid composed of different substances, while the second demonstrates how much entropy is produced as a consequence of the irreversible reactions between species.

Consequences of the theory are derived in a manner very similar to what we have seen in Section E.1.2. We have taken the energy and the entropy of the fluid to be functions of volume, temperature, and N species. If we perform partial derivatives on the Gibbs fundamental form, Equation (60), and collect the terms, we get

$$\left[\frac{\partial E}{\partial V}-T\frac{\partial S}{\partial V}+P\right]\dot{V}+\left[\frac{\partial E}{\partial T}-T\frac{\partial S}{\partial T}\right]\dot{T}+\sum_{i=1}^{N}\left[\frac{\partial E}{\partial n_i}-T\frac{\partial S}{\partial n_i}-\mu_i\right]\dot{n}_i=0$$

Since every one of the independent variables can be changed independently of the others, each of the terms in brackets must be zero:

$$\begin{aligned}\frac{\partial E(V,T,n_1\ldots n_N)}{\partial V}&=T\frac{\partial S(V,T,n_1\ldots n_N)}{\partial V}-P(V,T,n_1\ldots n_N)\\\frac{\partial E(V,T,n_1\ldots n_N)}{\partial T}&=T\frac{\partial S(V,T,n_1\ldots n_N)}{\partial T}\end{aligned}\tag{62}$$

and

$$\frac{\partial E(V,T,n_1\ldots n_N)}{\partial n_i}=T\frac{\partial S(V,T,n_1\ldots n_N)}{\partial n_i}+\mu_i(V,T,n_1\ldots n_N)\tag{63}$$

for $i = 1\ldots N$. These relations are similar, but not equal, to what we have derived in Equation (41). There, the relations held for each individual phase, which was composed of only one species. In our current case, on the other hand, E and S are the energy and the entropy of the entire mixture, and we have N equations of the form given in Equation (63), where the derivatives of these functions are taken with respect to each of the species of the mixture.

In the following section, we will have a brief look at the effects of flow processes in open systems. As a result, you should gain an understanding of why the particular form of the energy current due to flow was chosen in Chapter 4.

E.1.4 Uniform Flow Systems and the Energy Current Associated With Flow

In Section 4.4.4, an important point was left open: why should the energy current associated with matter flowing across the surface of a system be written in the form of Equation (137)? Shouldn't we include the effects of convective transport of entropy directly with what was termed the *chemical* energy current? The same question can be raised with regard to momentum, Equation (141); however, for simplicity, we shall restrict the following considerations to the case of the exchange of entropy.

The case of a species flowing across a system boundary and carrying with it some entropy will be treated in a manner which is an abbreviated version of what we witnessed above. Take as a control volume the interior of a cylinder bounded by a rigid wall on the right, and by an imaginary surface on the left (Figure 3). Assume the volume to be filled with a fluid. More of this fluid is assumed to flow into the control volume; its speed as it flows past the imaginary surface is equal to *v*, and its temperature is the same as that of the fluid already present. The system enclosed by the cylinder and piston satisfies a number of laws of balance, namely those for entropy, amount of substance, and energy.[7] The first two laws take the form

$$\begin{aligned} \dot{S} + I_{s,cond} + I_{s,conv} &= 0 \\ \dot{n} + I_n &= 0 \end{aligned} \tag{64}$$

where the derivatives denote the rates of change of the quantities inside the control volume. The processes will be reversible since we continue to deal with uniform systems, assuming nondissipative constitutive laws of fluid flow and of heating; this makes the rate of production of entropy equal to zero. The convective current of entropy is due to the flow of matter across the imaginary surface. Now, the balance of energy must be expressed by

$$\dot{E} + I_{E,th} + I_{E,flow} = 0 \tag{65}$$

The only possibilities of exchanging energy are those at the rigid surface (heating or cooling) and at the imaginary boundary (flow).

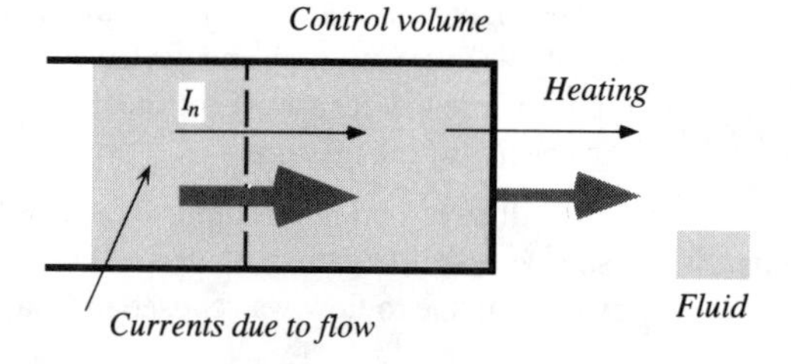

FIGURE 3. Open, stationary control volume. A current of matter I_n is crossing the imaginary surface on the left. The only fluxes of energy present are associated with heating and with the flow of matter.

Following the idea of combining all the equations of balance into one statement, we can multiply each of the expressions in Equation (64) by a factor and subtract both from Equation (65). If we already assume the temperature and the chemical potential of the fluid to be the proper factors,[8] we arrive at

---

7. As before, we take the effect of the balance of momentum to be small and thus negligible.

8. They certainly must have the units of temperature and chemical potential, respectively, for Equation (66) to be correct.

$$\dot{E} + I_{E,th} + I_{E,flow} - T\left[\dot{S} + I_{s,cond} + I_{s,conv}\right] - \mu\left[\dot{n} + I_n\right] = 0 \tag{66}$$

For a homogeneous system, we know the relation between the currents of entropy and of energy in heating:

$$I_{E,th} = T I_{s,cond} \tag{67}$$

Applying this, we can transform Equation (66) into

$$\dot{E} - T\dot{S} - \mu\dot{n} + I_{E,flow} - T I_{s,conv} - \mu I_n = 0 \tag{68}$$

If we also assume the Gibbs fundamental form for the fluid considered here to hold,[9]

$$\dot{E} - T\dot{S} - \mu\dot{n} = 0 \tag{69}$$

then we arrive at the form of the energy current associated with the flow of matter which we had assumed to be correct before:

$$I_{E,flow} = T I_{s,conv} + \mu I_n \tag{70}$$

(See Section 4.4.4.) The result is a consequence of the form which we assumed to hold for the balance of energy, i.e., Equation (64).

Treating the exchange of momentum associated with flow processes in the manner of spatially uniform systems does not make that much sense. While it is relatively easy to associate single values of temperature and pressure with a fluid, the same does not work with values of velocity. Therefore, it is better to consider the case of momentum in the context of continuous systems. (See the following sections of this chapter for a brief discussion.)

E.2 Equations of Balance for Continuous Processes

So far we have used laws of balance mostly in their integral forms; in other words we have written and applied the appropriate equations of balance for an entire body. Only in the context of the conduction of heat did we use the laws as they apply to the continuous case (Chapter 3). Now we will motivate and derive the proper equations of balance of mass, entropy, and momentum, for continuous bodies. This will prepare the ground for theories of thermodynamics of continuous processes.

---

9. We can take this to be a special case of the Gibbs fundamental form in Equation (59) if we only have one species to be considered.

We should emphasize that the following development will not be rigorous in either the mathematical or the physical senses. Mathematical derivations will be performed for the simplest geometrical cases of purely one-dimensional flow only, and the physical scope of the laws will be limited to relatively simple phenomena.[10] Still, we will use ideas in their most fundamental form to motivate the laws of balance of continuum physics, and the general results will be presented as extensions of the simpler ones. The result should be an uncluttered presentation of a subject often deemed difficult for beginners.

E.2.1 Densities and Current Densities

Equations of balance for a body relate the rate of change of system content, currents across the surface of the system, and possibly, source rates and rates of production. They tell us how quantities such as mass, entropy, and momentum of a body change as a result of flow processes, and where appropriate, as a consequence of absorption and production. While in integral form these quantities refer to an entire body, in continuous processes they have to be transformed to reflect their distribution in space. Entering these distributions in the well-known forms of the equations of balance will yield the desired result, namely equations of balance for continuous processes.

FIGURE 4. The density of a quantity tells us about its distribution over a volume. Where the density is higher, more "stuff" is contained in a part of space. If the density changes in space, we have to integrate the distribution over volume to obtain the amount of the quantity contained in a system.

The density of substancelike quantities. We have encountered densities of quantities such as mass, momentum, and entropy before in Section 3.4.6. The density of a substancelike quantity is our means of telling how the quantity is distributed over a given volume. The concept is easily grasped: if constant, the density multiplied by the volume tells us how much of the quantity is contained in the volume; if variable, we simply have to integrate the density over the volume to obtain the desired result (Figure 4). A density is commonly introduced for the mass of a body. Integrating the mass density ρ_m over volume yields the mass of the body contained in the volume:

$$m = \int_{\mathcal{V}} \rho_m dV \tag{71}$$

10. Neither restriction will fundamentally limit the applicability of the ideas. The general mathematical forms for more complicated three-dimensional flow processes have the same basic appearance as the simple one-dimensional ones. We will simply suggest that the derivations presented will carry over to the more advanced cases. When terms representing particular physical phenomena are left out of an equation for simplicity, it is assumed that they could be included in the manner of other terms for which the derivation is presented. Those of you who wish to see a more rigorous treatment of the subject should turn to books on fluid dynamics and continuum mechanics. (See, for example, Malvern, 1969; Mase, 1970; Landau and Lifshitz, 1959; Lai, Rubin, and Krempl, 1978; and Whitaker, 1968.)

This carries over to all quantities of a similar type, including entropy, momentum, and energy. If we write ρ_Q for the density of a particular quantity, the amount Q of this quantity contained in a given volume is calculated according to

$$Q = \int_{\mathcal{V}} \rho_Q \, dV \tag{72}$$

For our current purpose, this definition is applied to entropy, mass, and momentum of a system. For mass it has been presented in Equation (71); as is customary in this special case, we will not use the subscript m to denote mass: the symbol ρ is used for the mass density of a body. Often, the density of a particular quantity is written in terms of the specific value $q = Q/m$ of the quantity and the mass density of the material containing the quantity:

$$\rho_Q = \rho q \tag{73}$$

Using this form, Equation (72) becomes

$$Q = \int_{\mathcal{V}} \rho q dV \tag{74}$$

The case of entropy was treated in Section 3.4.6. There we obtained the following expression for the entropy of a system contained in volume $\mathcal{V}$:

$$S = \int_{\mathcal{V}} \rho_s \, dV = \int_{\mathcal{V}} \rho s dV \tag{75}$$

The second form is a consequence of Equation (73). Now, for momentum, we obtain a perfectly analogous equation, at least in the one-dimensional case:

$$p = \int_{\mathcal{V}} \rho_p \, dV = \int_{\mathcal{V}} \rho v dV \tag{76}$$

The velocity v of matter represents its specific momentum, i.e., its momentum per mass. Since momentum is a vector, we either have to write Equation (76) as one vector equation or, equivalently, as three component equations (in a Cartesian coordinate system):

$$\mathbf{p} = \int_{\mathcal{V}} \rho \mathbf{v} dV \tag{77}$$

or

$$\begin{aligned} p_x &= \int_{\mathcal{V}} \rho v_x dV \\ p_y &= \int_{\mathcal{V}} \rho v_y dV \\ p_z &= \int_{\mathcal{V}} \rho v_z dV \end{aligned} \tag{78}$$

If we work with the components of momentum in a Cartesian coordinate system,[11] we can deal with definitions and laws as if momentum consisted of three independent scalar quantities with properties similar to those of mass or entropy.

Source rate densities and production rate densities. Some of the substancelike quantities change as a consequence of source or production processes. Sources are used to describe the interaction of bodies and fields where quantities are transported into and out of bodies without crossing their surfaces. Quantities transferred in this way either originate or end up in another system; they are not produced in this manner. Since a source rate tells us the rate at which a substancelike quantity appears or disappears inside a body, and since this rate may be different in different parts of a body, we again introduce the concept of a distribution function, this time of the source rate over a volume. In other words, we use the spatial density σ_Q of the source rate Σ_Q to quantify its distribution. The relation between the source rate and its density is known from the density of a quantity:

$$\Sigma_Q = \int_{\mathcal{V}} \sigma_Q \, dV = \int_{\mathcal{V}} \rho f_Q \, dV \tag{79}$$

where f_Q is the specific source rate (source rate per mass). If you wish to use this expression for a particular physical quantity such as entropy or momentum,[12] you have to substitute that quantity for Q. Remember, that in the case of momentum you have to write three independent equations for each of the Cartesian components of the momentum vector.

The same idea applies to the phenomenon of the production of a quantity. Again, the rate of production may vary over the volume of a system, in which case we should introduce the spatial density π_Q of the rate of production Π_Q:

$$\Pi_Q = \int_{\mathcal{V}} \pi_Q \, dV \tag{80}$$

Of the quantities we are dealing with in this chapter—mass, entropy, momentum, and energy—only entropy admits a production term. As you know, the rate of production of entropy may not be negative.

Current densities. The current density (or flux density) of a substancelike quantity is used as the measure of the distribution of a current over a surface, just as

11. For reasons of simplicity, whenever multidimensional forms of equations of balance will be written in component form, a rectangular Cartesian coordinate system will be chosen. More general coordinate systems are treated in books on continuum mechanics (for example, Malvern, 1969).

12. As you know, there is no source term of mass if we use mass as a measure of amount of substance (as we currently do). Substances cannot be transferred radiatively. If we actually had to deal with mass as in gravitational processes, the equivalence of mass and energy tells us that there could be radiative sources of mass in a system.

the density is used to describe the distribution of a substancelike quantity in space (Figure 5).

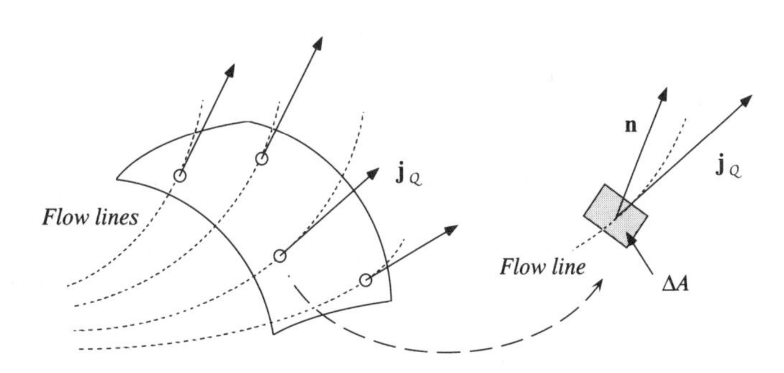

FIGURE 5. Flow lines cutting through a surface present an image of the distribution of a current. The distribution is measured in terms of the current density over the surface. Note that the flux density of a scalar quantity is a vector with a direction tangential to the flow line. The orientation of this vector with respect to the surface normal vector **n** is used when calculating the flux of the quantity transported through space.

Let us start with the simplest possible case of a current density and its associated current or flux, namely that of a flow perpendicular to a flat surface having a constant value of the current density on the surface. Naturally, in this case the flux is calculated simply as the product of the absolute value of the current density and the surface area. With a given orientation of the surface, the value of the flux is taken to be positive if the current flows in the direction of the normal vector representing the orientation. This means that, in this simple case, we can write the flux in the following form:

$$I_Q = A\mathbf{j}_Q \cdot \mathbf{n} \tag{81}$$

The dot denotes the scalar product of the vectors $\mathbf{j}_Q$ and **n**. As before, for a surface of a body, we will take the orientation positive for outward direction (Figure 6). We now can relax the condition that the flow must be perpendicular to the flat surface. The same form as in Equation (81) still applies; and it does so as well for a small part of a curved surface cut by arbitrary flow lines depicted in Figure 5. In general, then, the flux of an arbitrary flow field cutting through a curved surface must be given by the surface integral

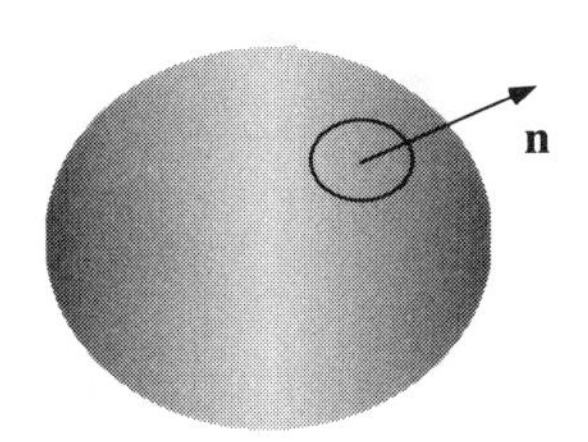

FIGURE 6. The orientation of the surface of a body is taken to be positive for outward direction. The normal vector **n** points away from the body.

$$I_Q = \int_{\mathcal{A}} \mathbf{j}_Q \cdot \mathbf{n} dA \tag{82}$$

where $\mathcal{A}$ is the surface under consideration. This surface may be the closed surface of a body or a part thereof.

Remember that we have to distinguish between two different types of currents if we are dealing with material systems,[13] namely those associated with convective and conductive transports. If a substancelike quantity flows through matter, the

13. There is a third type of current or flux associated with the transport of a substancelike quantity through a field.

flux is said to be conductive; if it flows with matter, we call the transport convective. Conductive transports are quantified in terms of a conductive current density $\mathbf{j}_Q^{(c)}$. The convective current density, on the other hand, can be expressed in terms of the density of the quantity which is transported by matter and the flux of mass; see Equation (129). The convective flux is equal to

$$I_{Q,conv} = qI_m \tag{83}$$

All we must do to obtain the continuum form of this relation is to calculate the flux of mass for the spatially variable case. You know that the flux of mass is equal to the product of the density of the material and its flux of volume. The latter quantity must be given by the integral of the volume flux density over the surface:

$$I_V = \int_{\mathcal{A}} \mathbf{j}_V \cdot \mathbf{n} dA = \int_{\mathcal{A}} \mathbf{v} \cdot \mathbf{n} dA \tag{84}$$

As you can see, the volume flux density is equal to the speed of flow:

$$\mathbf{j}_V = \mathbf{v} \tag{85}$$

Therefore, the flux density of mass is the product of the density and the speed of flow:

$$\mathbf{j}_m = \rho \mathbf{v} \tag{86}$$

This tells us that the convective current density of a quantity Q is given by the product of the specific quantity and the mass flux density:

$$\mathbf{j}_{Q,conv} = q\rho \mathbf{v} \tag{87}$$

Now, we have the means of expressing the total flux density of a particular quantity, namely by adding up the conductive and the convective parts:

$$\mathbf{j}_Q = q\rho \mathbf{v} + \mathbf{j}_Q^{(c)} \tag{88}$$

and the flux turns out to be

$$I_Q = \int_{\mathcal{A}} \left(q\rho \mathbf{v} + \mathbf{j}_Q^{(c)} \right) \cdot \mathbf{n} dA \tag{89}$$

Current densities of mass, entropy, and momentum. To obtain the expressions pertaining to the particular quantities with which we are dealing, we simply have to replace the general quantity Q by either mass, entropy, and momentum. Since mass (amount of substance) does not have a conductive flux, the total flux density is given by the expression in Equation (86):

$$\mathbf{j}_m = \rho \mathbf{v} \tag{90}$$

Entropy is a fairly simple case since this quantity is a scalar just like mass or electric charge. The specific entropy is abbreviated by s, while the conductive current density of entropy is written as $\mathbf{j}_s^{(c)}$:

$$\mathbf{j}_s = s\rho\mathbf{v} + \mathbf{j}_s^{(c)} \tag{91}$$

Momentum, on the other hand, presents us with a more complicated case since we have to deal with a vectorial quantity. The problem is simplified if we treat each of the components of the vector independently. As you recall from the brief presentation in the Prologue, a component of momentum can be thought of as flowing through matter much like entropy or charge do. The flow of each of the components of momentum results in a flow field like those shown in the Prologue and below in Figure 7.

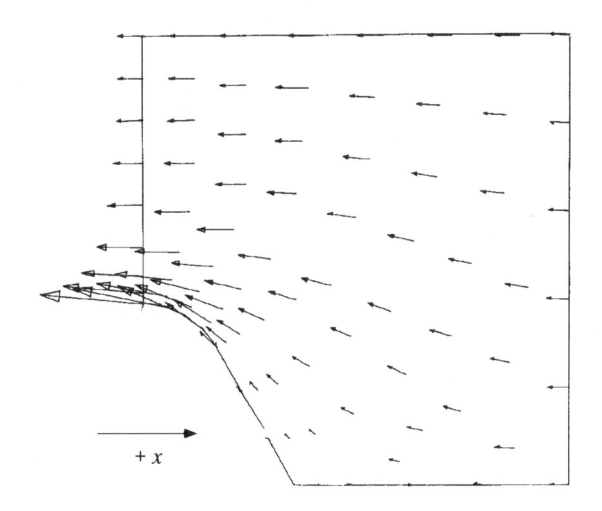

FIGURE 7. Flow pattern of one component of momentum resulting from tension in a flat strip having a notch. The component of momentum whose flow is depicted here is the one identified with the direction of tension.

Let us present the equations for a single component of momentum. You can build the complete result for all three components by combining the parts. If we choose the x-component, the specific value of x-momentum p_x is the x-component of the velocity. Therefore we have

$$\mathbf{j}_{px} = \rho v_x \mathbf{v} + \mathbf{j}_{px}^{(c)} \tag{92}$$

In this case it might be instructive to present all three components of the current density:

$$\begin{aligned} j_{pxx} &= \rho v_x v_x + j_{pxx}^{(c)} \\ j_{pxy} &= \rho v_x v_y + j_{pxy}^{(c)} \\ j_{pxz} &= \rho v_x v_z + j_{pxz}^{(c)} \end{aligned} \tag{93}$$

These quantities have a simple graphical representation; j_{pxx}, for example, represents the current density of x-momentum flowing in x-direction, while j_{pxy} is the current density of x-momentum flowing in y-direction. (See Figure 7, and Figure 7 of the Prologue.) Since there are three components of current density vectors belonging to the three components of momentum, a total of nine components[14] form the momentum current density tensor.

The transformation of a surface integral (divergence theorem). In Section 3.4.6 the transformation of the simplest possible case of a surface integral into an integral over the volume bounded by the surface. There, we treated the example of purely one-dimensional flow of heat. Since entropy is a scalar quantity, its current density is a vector describing the three possible directions of flow of this substancelike quantity. If the entropy current density vector has only one component, then entropy flows in only one direction. In this case, the entropy flux is

$$I_s = \int_{\mathcal{A}} j_{sx} \, dA$$

This integral can be transformed into a volume integral according to

$$I_s = \int_{\mathcal{A}} j_{sx} \, dA = \int_{\mathcal{V}} \frac{\partial}{\partial x} j_{sx} \, dV$$

14. This quantity cannot be represented as a vector anymore; rather, it is a tensor which may be written in matrix form

$$\mathcal{J}_p = \begin{pmatrix} \rho v_x v_x + j_{pxx}{}^{(c)} & \rho v_x v_y + j_{pxy}{}^{(c)} & \rho v_x v_z + j_{pxz}{}^{(c)} \\ \rho v_y v_x + j_{pyx}{}^{(c)} & \rho v_y v_y + j_{pyy}{}^{(c)} & \rho v_y v_z + j_{pyz}{}^{(c)} \\ \rho v_z v_x + j_{pzx}{}^{(c)} & \rho v_z v_y + j_{pzy}{}^{(c)} & \rho v_z v_z + j_{pzz}{}^{(c)} \end{pmatrix}$$

The negative conductive part of this quantity is commonly called the *stress tensor*

$$\mathcal{T} = \begin{pmatrix} t_{xx} & t_{xy} & t_{xz} \\ t_{yx} & t_{yy} & t_{yz} \\ t_{zx} & t_{zy} & t_{zz} \end{pmatrix} = -\begin{pmatrix} j_{pxx}{}^{(c)} & j_{pxy}{}^{(c)} & j_{pxz}{}^{(c)} \\ j_{pyx}{}^{(c)} & j_{pyy}{}^{(c)} & j_{pyz}{}^{(c)} \\ j_{pzx}{}^{(c)} & j_{pzy}{}^{(c)} & j_{pzz}{}^{(c)} \end{pmatrix}$$

while the complete quantity would be called the *momentum current tensor.* The surface integral of a row of the tensor (for one of the components of the coordinate systems) is called the component of the surface force

$$F_x = \int_{\mathcal{A}} \mathbf{T}_x \cdot \mathbf{n} dA$$

($\mathbf{T}_x$ is the first row of the stress tensor), while the surface integral for the stress tensor is the surface force vector

$$\mathbf{F} = \int_{\mathcal{A}} \mathcal{T} \cdot \mathbf{n} dA$$

We used this relation to derive the local form of the equation of balance of entropy for conduction in Chapter 3. In this form, the transformation is the simplest example of what is called the *divergence theorem* or *Gauss's theorem.* Let us briefly write down this relation without giving a proof.[15] If we have a current density vector $\mathbf{j}_Q$ defined on the closed surface of a body, the surface integral can be transformed into an integral over the volume enclosed by the surface:

$$\int_{\mathcal{A}} \mathbf{j}_Q \cdot \mathbf{n} dA = \int_{\mathcal{V}} \nabla \cdot \mathbf{j}_Q \, dV \tag{94}$$

where $\nabla \cdot \mathbf{j}_Q$ is called the *divergence* of $\mathbf{j}_Q$. In rectangular Cartesian coordinates

$$\nabla \cdot \mathbf{j}_Q = \frac{\partial}{\partial x} j_{Qx} + \frac{\partial}{\partial y} j_{Qy} + \frac{\partial}{\partial z} j_{Qz} \tag{95}$$

Often, the divergence of a vector written in component form is abbreviated as follows:

$$\frac{\partial}{\partial x} j_{Qx} + \frac{\partial}{\partial y} j_{Qy} + \frac{\partial}{\partial z} j_{Qz} \equiv \frac{\partial}{\partial x_i} j_{Qi} \tag{96}$$

In this notation it is assumed that a summation is carried out over all indices which appear twice in the same term; x_i, $i = 1,2,3$ stands for the three components (x,y,z) of the coordinate system. In this form, the divergence looks like the expression used in Chapter 3. In fact, the simplest examples usually suggest the proper form of more complicated cases.

E.2.2 The Balance of Mass

Let us start with the first of the three substancelike quantities for which we have to obtain laws of balance, namely the amount of substance. The balance of amount of substance is a necessary prerequisite for formulating theories applicable to fluid or otherwise deformable media. If we wish to quantify convective currents associated with processes in open systems, we have to be able to write down the currents of amount of substance. For practical reasons, however, engineers commonly use mass as a substitute for amount of substance, and as long as there are no chemical reactions taking place inside the material, there is no problem in doing so. Therefore we will use a formulation based on mass.

In the previous section we introduced the concepts and tools needed to formulate the continuum forms of the laws of balance of substancelike quantities. Starting with the integral form of the balance of mass

$$\dot{m} + I_m = 0 \tag{97}$$

15. For a derivation of the divergence theorem see Marsden and Weinstein (1985), Vol. III, p. 927.

we can easily show how to obtain the appropriate local equation applicable to the continuous case. Let us apply this law to a stationary control volume of simple shape (Figure 8), and assume the flow field to be one-dimensional. In this equa-

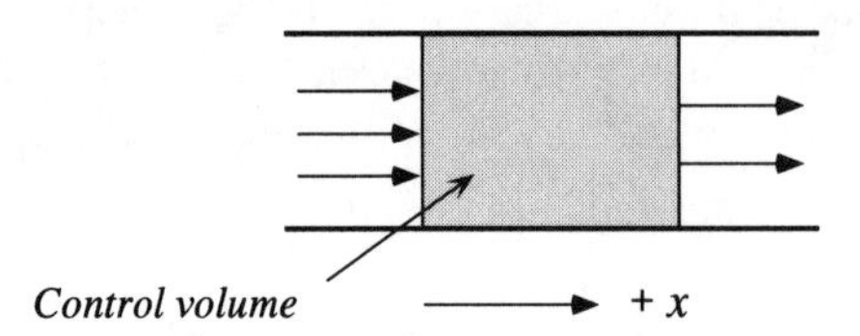

FIGURE 8. Simple one-dimensional flow with respect to an open control volume. Imagine a fluid flowing in the x-direction only.

tion, m is the mass inside the control volume, while I_m is the net current of mass across the surface of the control volume. We shall replace the mass by the volume integral of the mass density, and the flux by the surface integral of the flux density, as in Equations (71) and (90). This leads to

$$\frac{d}{dt}\int_{\mathcal{V}} \rho dV + \int_{\mathcal{A}} \rho v dA = 0 \tag{98}$$

If we use the divergence theorem for the surface integral and apply the time derivative to the integrand of the first integral, we obtain

$$\int_{\mathcal{V}} \frac{\partial \rho}{\partial t} dV + \int_{\mathcal{A}} \frac{\partial}{\partial x}(\rho v) dV = 0$$

or

$$\int_{\mathcal{V}} \left[\frac{\partial \rho}{\partial t} + \frac{\partial}{\partial x}(\rho v)\right] dV = 0$$

Since the integral must be zero for arbitrary volumes V, the last expression can only be satisfied if the terms in brackets are equal to zero:

$$\frac{\partial \rho}{\partial t} + \frac{\partial}{\partial x}(\rho v) = 0 \tag{99}$$

You can easily apply the transformations to the more general three-dimensional case

$$\frac{\partial \rho}{\partial t} + \frac{\partial}{\partial x_i}(\rho v_i) = 0 \tag{100}$$

or

$$\frac{\partial \rho}{\partial t} + \nabla \cdot (\rho \mathbf{v}) = 0 \tag{101}$$

This looks very similar to the simpler expression. In contrast to Equation (98) which is the integral form of the law of balance of mass, Equation (99) and its counterpart Equations (100) or (101) represent the local or differential form of this law. The balance of mass often is called the equation of continuity.

E.2.3 The Balance of Entropy

The phenomenon of one-dimensional conduction of heat served as the first example of a theory of continuous processes (Section 3.4). You may wish to review the derivations performed there. In the current section, we should try to add the effect of convection to the law of balance.

Consider as we did in Figure 8 the flow of a fluid in the x-direction only. As far as entropy is concerned, we will include in the derivation conductive and convective transports, and production of entropy in irreversible processes. Sources of entropy, however, will be excluded. The integral form of the equation of balance of entropy for the control volume in Figure 8 therefore looks like

$$\dot{S} + I_{s,conv} + I_{s,cond} = \Pi_s \qquad \textbf{(102)}$$

If we introduce densities and current densities as in Section E.2.1, the law becomes

$$\frac{d}{dt}\int_{\mathcal{V}} (\rho s) dV + \int_{\mathcal{A}} \left(s\rho v + j_s^{(c)} \right) dA = \int_{\mathcal{V}} \pi_s dV \qquad \textbf{(103)}$$

Remember that we are dealing with a purely one-dimensional case. If we now apply the transformation of the surface integral, we obtain

$$\int_{\mathcal{V}} \left[\frac{\partial}{\partial t}(\rho s) + \frac{\partial}{\partial x}\left(s\rho v + j_s^{(c)} \right) - \pi_s \right] dV = 0$$

The expression in brackets must be zero, which yields the local form of the law of balance:

$$\frac{\partial}{\partial t}(\rho s) + \frac{\partial}{\partial x}\left(s\rho v + j_s^{(c)} \right) = \pi_s \qquad \textbf{(104)}$$

The general three-dimensional case can be written in a form which looks just like the one derived for purely one-dimensional transports. Applying the divergence theorem to the generalized form of Equation (103) yields

$$\frac{\partial}{\partial t}(\rho s) + \nabla \cdot \left(s\rho \mathbf{v} + \mathbf{j}_s^{(c)} \right) = \pi_s \qquad \textbf{(105)}$$

or

$$\frac{\partial}{\partial t}(\rho s) + \frac{\partial}{\partial x_i}\left(s\rho v_i + j_{si}^{(c)} \right) = \pi_s \qquad \textbf{(106)}$$

Extending this result to include the effects of sources is pretty simple. How this is done will be demonstrated below for the case of momentum.

E.2.4 The Balance of Momentum

Basically, the law of balance of momentum is derived in analogy to what you have seen so far. While the fundamental ideas do not change, the current case can be rather complex if we try to deal with it in the most general form. Therefore it is all the more important to discuss the simplest possible nontrivial case. Fortunately, purely one-dimensional flow of momentum is meaningful in physical terms.

One-dimensional convective transport of momentum is a simple concept: if a fluid flows in one direction only, it just carries a single component of momentum. The case of one-dimensional conductive transport is just as well known. Let the direction of fluid flow define the spatial component we are talking about. Having the same component of momentum flowing through the fluid simply means that the material is under compression or tension in the same direction. A frictionless fluid flowing through a straight pipe demonstrates what we mean: the conductive momentum current density of the component parallel to the pipe's axis is the pressure of the fluid.

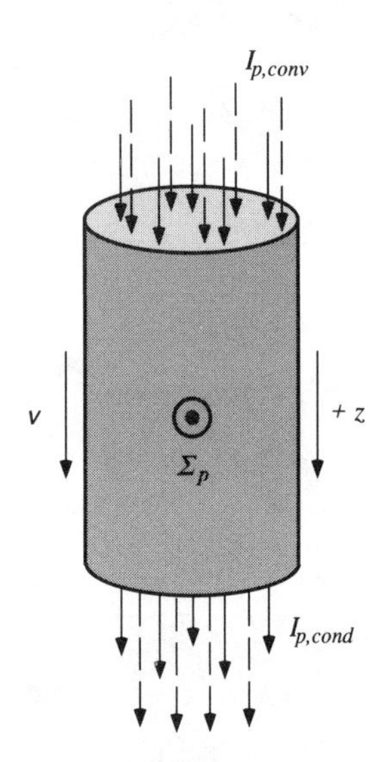

FIGURE 9. Flow lines depicting the convective and conductive transports of momentum are shown together with a source of momentum due to the interaction of the fluid with the gravitational field. The fluid is flowing downward leading to the convective downward flow of momentum together with the fluid (dashed lines). Since the material is under compression, momentum flows conductively into the positive direction (downward; solid lines). With the positive direction as chosen, the gravitational field supplies momentum to the fluid.

In addition to conductive and convective modes of transport, we will consider sources of momentum due to the interaction of the fluid with a field. If you imagine the fluid flowing through a vertical pipe (Figure 9), the action of the gravitational field leads to the flow of momentum of the same (vertical) component directly into or out of the body.

If we collect the different terms, the integral equation of balance of momentum for the z-direction looks like

$$\dot{p} + I_{p,conv} + I_{p,cond} = \Sigma_p \tag{107}$$

where p stands for the z-component of momentum. As before, we introduce the proper densities and flux densities and obtain

$$\frac{d}{dt}\int_{\mathcal{V}} (\rho v) dV + \int_{\mathcal{A}} \left(\rho v v + j_p^{(c)} \right) dA = \int_{\mathcal{V}} \sigma_p dV \tag{108}$$

Now, we can apply the divergence theorem to this equation:

$$\int_{\mathcal{V}} \left[\frac{\partial}{\partial t}(\rho v) + \frac{\partial}{\partial x}\left(\rho v v + j_p^{(c)} \right) - \sigma_p \right] dV = 0$$

leading to the desired differential form of balance of momentum for the purely one-dimensional example discussed so far:

$$\frac{\partial}{\partial t}(\rho v) + \frac{\partial}{\partial x}\left(\rho v v + j_p^{(c)} \right) = \sigma_p \tag{109}$$

The general case requires two additional steps. First, for a component of momentum, we have to treat the momentum flux as a vector; this yields an equation similar to Equation (105) for the component. Second, we have to write analogous equations for the other two components of momentum:

$$\frac{\partial}{\partial t}(\rho v_x) + \nabla \cdot \left(\rho v_x \mathbf{v} + \mathbf{j}_{px}{}^{(c)}\right) = \sigma_{px}$$
$$\frac{\partial}{\partial t}(\rho v_y) + \nabla \cdot \left(\rho v_y \mathbf{v} + \mathbf{j}_{py}{}^{(c)}\right) = \sigma_{py} \qquad (110)$$
$$\frac{\partial}{\partial t}(\rho v_z) + \nabla \cdot \left(\rho v_z \mathbf{v} + \mathbf{j}_{pz}{}^{(c)}\right) = \sigma_{pz}$$

These three equations can be written as a single one using the momentum current tensor introduced in Footnote 14. While the result can be presented in a compact form, the actual equations are rather lengthy and difficult to read in their component form.

Let us briefly return to the meaning of the conductive momentum flux and the source term in Equation (109). For a simple fluid the purely one-dimensional case of conductive momentum transport describes the state of compression of the fluid. In other words, the current density is the pressure of the fluid:

$$j_p{}^{(c)} = P \qquad (111)$$

The source term, on the other hand, arises as the result of the gravitational interaction. As described in the Prologue, the gravitational field supplies momentum to the body, leading to a source rate, which in integrated form, must be equal to the weight of the body. The source density therefore must equal

$$\sigma_p = \rho g \qquad (112)$$

which shows that the specific source rate f_p introduced in Equation (79) is the gravitational field strength g. The influence of fields other than the gravitational field can be included in the same manner. Forces of this nature are called body forces.

E.2.5 The Material Derivative and Laws of Balance

The local form of the equations of balance derived above has a couple of features which need explaining. The derivatives occurring in relations such as Equations (104) or (109) are partial derivatives, which means that, for example, the time derivative must be taken at a fixed position. This form is often called the Eulerian form of the partial differential equations describing the balance of substancelike quantities. The second feature is related to the first: the equations explicitly contain convective currents.

It is possible to transform the laws of balance using the balance of mass and to obtain alternative forms which no longer contain the convective currents. As such, the resulting equations look as if they were written for closed systems for which convection does not play a role. Instead, there appears a combination of derivatives of the density of the quantity investigated, a combination which can be interpreted as a *material derivative*. Let us see how this happens.

Take as an example the simple law of balance of entropy in Equation (104). If you take the derivative of some of the products occurring in the equation, you get

$$\rho\frac{\partial s}{\partial t}+s\frac{\partial \rho}{\partial t}+s\frac{\partial(\rho v)}{\partial x}+\rho v\frac{\partial s}{\partial x}+\frac{\partial j_s^{(c)}}{\partial x}=\pi_s$$

or

$$\rho\left[\frac{\partial s}{\partial t}+v\frac{\partial s}{\partial x}\right]+s\left[\frac{\partial \rho}{\partial t}+\frac{\partial(\rho v)}{\partial x}\right]+\frac{\partial j_s^{(c)}}{\partial x}=\pi_s$$

According to the law of balance of mass presented in Equation (99), the term in the second pair of brackets must equal zero. Therefore, the transformed law of balance looks like

$$\rho\left[\frac{\partial s}{\partial t}+v\frac{\partial s}{\partial x}\right]+\frac{\partial j_s^{(c)}}{\partial x}=\pi_s \tag{113}$$

As you can see, the convective current density has disappeared. Introducing the operator

$$\frac{D}{Dt}=\frac{\partial}{\partial t}+v\frac{\partial}{\partial x} \tag{114}$$

which commonly is called the *material time derivative* or the *substantial derivative*, Equation (113) can be written as follows:

$$\rho\frac{Ds}{Dt}+\frac{\partial j_s^{(c)}}{\partial x}=\pi_s \tag{115}$$

The time rate of change of the quantity s expressed by the material derivative is the one measured by someone carried along with the fluid rather than by someone at a fixed location.

E.3 The Energy Principle

The law of balance of energy should be considered separately. The reason for this is not that the energy principle is more important than any of the other fundamental laws discussed in Section E.2. Rather, we should stress again that energy is special in that it appears equally in all physical phenomena, and not just in one special subject area. The formal expression of this idea was used in Section E.1, where the energy principle was taken to deliver restrictions upon the processes investigated. It will be used again in Section E.4, where the theory of thermodynamics of viscous heat-conducting fluids will be presented.

Since energy is exchanged in all types of physical processes, it is almost impossible, and certainly not realistic, to try to present a totally general form of the law of balance. The following discussion is in line with the area of applications considered here, namely the thermomechanics of nonreacting fluid systems.

E.3.1 Energy Density and Energy Current Densities

Consider again a fluid flowing through a control volume such as in Figure 8, and let all transports be purely one-dimensional. In addition to mass, entropy, and momentum, energy is transported across the faces of the control volume, and just like the other quantities, it can be stored in the system as well. Since energy cannot be produced, we will have to consider the density of energy in the fluid, energy density currents across the surface, and sources of energy due to radiative transports.

The energy density of a fluid. A fluid contains an amount of energy that depends on its state, which in the current case, can be described by temperature, pressure, and velocity. It is common to associate a part of the energy content with the state of motion of the material (i.e., the kinetic energy) while the rest is lumped together as the internal energy of the fluid. This distinction was made in Chapter 4 where we discussed simple applications of flow processes. The density of energy of the fluid is therefore written as the sum of two parts, the density of internal energy, and the density of kinetic energy. If we use instead the specific values of internal energy and of kinetic energy, the expression takes the following form:

$$\rho e = \rho\left(u + \frac{1}{2}v^2\right) \tag{116}$$

Here, e is the specific energy, while u stands for the specific internal energy. Naturally, the second term on the right-hand side represents the kinetic energy per mass. The energy of a fluid system can then be written as follows:

$$E = \int_{\mathcal{V}} \rho\left(u + \frac{1}{2}v^2\right)dV \tag{117}$$

Energy current densities. Energy can be carried into or out of a control volume across the surface of the system. Such processes are described in terms of currents. Since the exchange of energy may be the result of numerous different interactions, we should be specific and discuss only those phenomena which are important to our current theme. Other cases can be treated in analogy to what we do here. Since energy can be transferred convectively and conductively, we can say that

$$j_E = j_{E,conv} + j_{E,cond} \tag{118}$$

In general, these current densities are vectors. The convective transport is the result of the fluid carrying stored energy across the system boundary. To quantify this process, we need the convective energy flux density; it can be expressed in terms of the specific energy and the flux of mass:

$$j_{E,conv} = \rho v e = \rho v\left(u + \frac{1}{2}v^2\right) \tag{119}$$

The conductive transport of energy presents a slightly more complicated problem, since we might have to consider all sorts of transfer processes. Limiting our attention to thermal and mechanical phenomena, however, leaves us with only two possibilities: energy may either be added or withdrawn together with entropy in heating and cooling, or it may flow across the system boundary as the result of the conductive exchange of momentum:

$$j_{E,cond} = j_{E,th} + j_{E,mech} \tag{120}$$

The former possibility gets short treatment; since we shall try to establish the relationship between transports of entropy and of energy, we cannot say more about this case and we simply introduce a thermal energy current density. The second term in this equation, on the other hand, is well known from mechanics. The exchange of energy is directly tied to the conductive flow of momentum, and to the speed of the fluid at the surface of the system:

$$j_{E,mech} = v j_p^{(c)} \tag{121}$$

In three dimensions, for each component of momentum, a component of the energy flux density vector is obtained by calculating the scalar product of the momentum current vector and the velocity. Addition of all the terms introduced so far yields the desired expression for the energy current density:

$$j_E = \rho v\left(u + \frac{1}{2}v^2\right) + v j_p^{(c)} + j_{E,th} \tag{122}$$

As mentioned before, this equation has to be changed or extended if other processes are taken into consideration.

The total energy flux with respect to the control volume in Figure 8 is obtained by integrating the expression in Equation (122) over the surface elements perpendicular to the direction of flow:

$$I_E = \int_{\mathcal{A}}\left[\rho v\left(u + \frac{1}{2}v^2\right) + v j_p^{(c)} + j_{E,th}\right]dA \tag{123}$$

Sources of energy. Energy can be directly supplied to the interior of a body or a control volume as a consequence of its interaction with fields. Three cases of interest to us are the supply of energy together with momentum if we include the action of a gravitational field; sources of energy due to the absorption of electromagnetic radiation; and sources of energy such as those related to nuclear decay or chemical reactions inside the system. Describing these effects calls for the introduction of an energy supply density such that

$$\Sigma_E = \int_{\mathcal{V}} \sigma_E dV \tag{124}$$

Gravitational interaction and absorption of radiation lead to a specific expression for the source density, namely

$$\sigma_E = \rho\left(f_p v + r\right) \tag{125}$$

where $f_p = g$ is the specific source rate of momentum (the specific body force), and r denotes the specific rate of absorption of energy together with radiation.

E.3.2 The Balance of Energy

Writing down the law of balance of energy as it applies to a fairly general case of the thermomechanics of fluid systems, is as simple as the examples we have seen before. The integral statement

$$\dot{E} + I_E = \Sigma_E \tag{126}$$

can be written using the densities and the current densities:

$$\frac{d}{dt}\int_{\mathcal{V}} \rho\left(u + \frac{1}{2}v^2\right)dV + \int_{\mathcal{A}}\left[\rho \cdot v\left(u + \frac{1}{2}v^2\right) + v j_p{}^{(c)} + j_{E,th}\right]dA = \int_{\mathcal{V}} \rho\left(f_p v + r\right)dV \tag{127}$$

Applying the integral transformation to the surface integral, collecting all the terms on one side, and noting that the resulting volume integral must vanish identically, yields the local form of the law of balance of energy:

$$\frac{\partial}{\partial t}\rho\left(u + \frac{1}{2}v^2\right) + \frac{\partial}{\partial x}\left[\rho v\left(u + \frac{1}{2}v^2\right) + v j_p{}^{(c)} + j_{E,th}\right] = \rho\left(f_p v + r\right) \tag{128}$$

Together with the laws of balance derived above for mass, entropy, and momentum, this equation will furnish the starting point for the development of our example of continuum thermodynamics in Section E.4.

E.3.3 The Balance of Energy in the Material Form

As mentioned in Section E.2.5, the local equations of balance often are written using the material derivative, Equation (114), in which case they look like laws of balance written for a particular body, i.e., for a closed system. Let us first motivate this particular form for supply-free processes before attempting a derivation.

Imagine following a fluid in its motion. For observers flowing with a particular body, the momentum of the material is zero all the time, and as a corollary, the kinetic energy vanishes; they see only the internal energy of the fluid. The exchange of energy with the environment takes two forms. The momentum flowing through the body does so with or against a gradient of the potential of motion, i.e., with or against the gradient of the velocity at the location of the body. The rate per volume at which energy is released (see the Prologue for a discussion) is equal to the product of the current density of momentum and the gradient of the velocity. A similar result holds for the thermal interaction; only here we do not know the relationship between the flux of energy and the flux of entropy. The rate of release of energy per volume in the body is equal to the gradient of the energy flux density. Adding up the terms should yield

$$\rho \frac{Du}{Dt} + j_p^{(c)} \frac{\partial v}{\partial x} + \frac{\partial j_{E,th}}{\partial x} = 0 \tag{129}$$

The second term in this equation is called the *stress power*, while the third might be called *thermal power.*

This result is a consequence of Equation (128), and the laws of balance of mass and of momentum Equations (99) and (109); remember that we do not include any of the source terms. In a first step, the law of balance of energy becomes

$$\rho \frac{\partial u}{\partial t} + u \frac{\partial \rho}{\partial t} + \frac{\partial}{\partial t}\left(\frac{1}{2}\rho v^2\right) + \rho v \frac{\partial u}{\partial x} + u \frac{\partial (\rho \cdot v)}{\partial x}$$
$$+ \frac{\partial}{\partial x}\left[\rho v \left(\frac{1}{2} v^2\right)\right] + \frac{\partial}{\partial x}\left(v j_p^{(c)}\right) + \frac{\partial j_{E,th}}{\partial x} = 0$$

Together, the second and the fifth terms vanish because of the law of balance of mass, while the first and the fourth terms yield the product of the density and the material derivative of the internal energy. Therefore we have

$$\rho \frac{Du}{Dt} + j_p^{(c)} \frac{\partial v}{\partial x} + \frac{\partial j_{E,th}}{\partial x} + \left\{ \frac{\partial}{\partial t}\left(\frac{1}{2}\rho v^2\right) + \frac{\partial}{\partial x}\left[\rho v \left(\frac{1}{2} v^2\right)\right] + v \frac{\partial j_p^{(c)}}{\partial x} \right\} = 0$$

If we can show that the expression inside the braces is zero, then the proof of Equation (129) is complete. Now, the spatial derivative of the momentum current density can be replaced using the law of balance of momentum:

$$\frac{\partial}{\partial t}\left(\frac{1}{2}\rho v^2\right) + \frac{\partial}{\partial x}\left[\rho v \left(\frac{1}{2} v^2\right)\right] + v \frac{\partial j_p^{(c)}}{\partial x}$$
$$= \frac{\partial}{\partial t}\left(\frac{1}{2}\rho v^2\right) + \frac{\partial}{\partial x}\left[\rho v \left(\frac{1}{2} v^2\right)\right] + v\left[-\frac{\partial}{\partial t}(\rho v) - \frac{\partial}{\partial x}(\rho v v)\right]$$

After some lengthy algebra we find that this expression is equal to

$$-\frac{1}{2}v^2\left(\frac{\partial\rho}{\partial t}+v\frac{\partial\rho}{\partial x}+\rho\frac{\partial v}{\partial x}\right)$$

This quantity vanishes because of the law of balance of mass. Naturally, for the three-dimensional case, the derivation is more involved, but the result still holds in a form similar to the one stated above.

E.4 Thermodynamics of Viscous Heat-Conducting Fluids

The bulk of engineering work in the fields of fluid dynamics and heat transfer is built on the basis of the theory of heat-conducting viscous fluids, the subject we will discuss in this section. Except for constructing the constitutive laws applicable to the field, we already have done most of the preparations necessary for the following development. We only must assemble the parts and derive the consequences of the assumptions, much like we did for the uniform viscous fluid in Section E.1.1. If the following lines look forbidding at first, keep in mind that it is the amount of algebra needed which creates this impression. The results can be understood intuitively in terms of what we have learned in the previous chapters.

The assumptions upon which the theory is based can be divided into three groups. The first, the laws of balance of mass, entropy, momentum, and energy, has been discussed above; we only must assemble them in the form which applies to our case. The second is made up of the preliminary expressions for the constitutive laws which will be introduced in Section E.4.2. The third group, finally, deals with the assumption of ideal walls which we will not go into again.

E.4.1 The Laws of Balance

The laws of balance which are used in the field of nonreactive heat-conducting viscous fluids are those of mass, entropy, momentum, and energy. For the following derivation, we will stay with the simple one-dimensional case discussed in the previous sections. For source-free processes, the appropriate forms of the first three laws are[16]

---

16. In component notation for three dimensions, the four equations of balance are as follows:

$$\frac{\partial \rho}{\partial t}+\frac{\partial}{\partial x}(\rho v)=0$$
$$\frac{\partial}{\partial t}(\rho s)+\frac{\partial}{\partial x}\left(s\rho v+j_s{}^{(c)}\right)=\pi_s \tag{130}$$
$$\frac{\partial}{\partial t}(\rho v)+\frac{\partial}{\partial x}\left(\rho v v+j_p{}^{(c)}\right)=0$$

while the balance of energy looks like

$$\frac{\partial}{\partial t}\left[\rho\left(u+\frac{1}{2}v^2\right)\right]+\frac{\partial}{\partial x}\left[\rho v\left(u+\frac{1}{2}v^2\right)+v j_p{}^{(c)}+j_{E,th}\right]=0 \tag{131}$$

As you know from all the examples in the previous chapters, equations of balance by themselves do not solve a problem. Moreover, they contain as their most important quantities those which often are not accessible to measurement. If we wish to specify the condition of a fluid in terms of density, velocity, and temperature at different points in space for different times, i.e., if we wish to obtain the functions $\rho(\mathbf{x},t)$, $\mathbf{v}(\mathbf{x},t)$, and $T(\mathbf{x},t)$, we obviously will have to eliminate quantities such as entropy and energy from the equations. This goal is achieved if we know the proper constitutive laws with whose help the equations of balance can be transformed into field equations for the measurable quantities.

E.4.2 Constitutive Relations for a Navier-Stokes-Fourier Fluid

In this section we will present the constitutive laws which govern the behavior of viscous heat-conducting fluids in a strongly simplified form. It is possible to include all the interesting physical effects with the purely one-dimensional case, which considerably reduces the complexity of the calculations. We will need preliminary forms of the constitutive laws for the specific entropy and energy,

---

$$\frac{\partial \rho}{\partial t}+\frac{\partial}{\partial x_j}(\rho v_j)=0$$
$$\frac{\partial}{\partial t}(\rho s)+\frac{\partial}{\partial x_j}\left(s\rho v_j+j_{sj}{}^{(c)}\right)=\pi_s$$
$$\frac{\partial}{\partial t}(\rho v_i)+\frac{\partial}{\partial x_j}\left(\rho v_i v_j+j_{pij}{}^{(c)}\right)=0$$
$$\frac{\partial}{\partial t}\left[\rho\left(u+\frac{1}{2}v^2\right)\right]+\frac{\partial}{\partial x_j}\left[\rho v_j\left(u+\frac{1}{2}v^2\right)+v_i j_{pij}{}^{(c)}+j_{E,th,j}\right]=0$$

Remember, that we have to sum over the indices which appear twice in an expression.

and for the flux densities of entropy, momentum, and energy (for the case of heating). A suitable thermal equation of state has to be added to these.[17]

In many instances, gradients of temperature and velocity in fluids can be assumed to be small, which leads to particularly simple forms of the constitutive laws with only linear dependence on the gradients. We shall assume that

$$s = s(\rho, T) \tag{132}$$

and

$$u = u(\rho, T) \tag{133}$$

for the specific quantities, and

$$\begin{aligned} j_p^{(c)} &= P(\rho, T) - \mu'(\rho, T)\frac{\partial v}{\partial x} \\ j_s^{(c)} &= -k_s(\rho, T)\frac{\partial T}{\partial x} \\ j_{E,th} &= -\beta(\rho, T)\frac{\partial T}{\partial x} \end{aligned} \tag{134}$$

for the fluxes of momentum, entropy, and energy (for heating).[18] Here, μ' is a viscosity which may include the bulk viscosity with the normal viscosity μ of the fluid, while k_s is the thermal conductivity with respect to entropy (i.e., the entro-

---

17. For more details on the proper constitutive laws, you should turn to books such as those by Müller (1985), or Malvern (1969).

18. Actually, the preliminary forms are slightly more complicated. First, one exploits the principle of material frame indifference and representation theorems for isotropic functions to obtain rather general expressions for the constitutive laws (Müller, p. 4–7). These expressions are linearized, which leads to

$$\begin{aligned} s &= s(\rho, T) + a(\rho, T)d_{ii} \\ u &= u(\rho, T) + b(\rho, T)d_{ii} \\ j_{pij}^{(c)} &= P(\rho, T)\delta_{ij} - \lambda(\rho, T)d_{kk}\delta_{ij} - 2\mu(\rho, T)d_{ij} \\ j_s^{(c)} &= -k_s(\rho, T)\frac{\partial T}{\partial x} \\ j_{E,th} &= -\beta(\rho, T)\frac{\partial T}{\partial x} \end{aligned}$$

for the Navier-Stokes-Fourier fluid for the three-dimensional case. When the equations of balance are exploited, the functions a and b turn out to be equal to zero which means that the entropy and the density take the same form in equilibrium as in non-steady-state processes. To simplify the derivation, we have omitted the terms relating to the gradient of velocity from Equation (132). Equation (134a) is the purely one-dimensional version of the expression for the stress tensor for nonvanishing bulk viscosity, where py conductivity);[19] β is a function which later will be related to k_S. We have used constitutive expressions of these forms before for the simple materials treated in previous chapters. Equations (134b,c) are the Fourier law of heat conduction, while Equation (134a) reminds us of the special pressure law used in Section E.1.1, where we dealt with a uniform viscous fluid. Velocity gradients in one dimension mean that the fluid is either being compressed or expanded, leading to viscous effects.

E.4.3 Evaluation of the Energy Principle

We wish to be able to derive more detailed results about the constitutive laws and quantities of a Navier-Stokes-Fourier fluid. As you remember, the laws of balance serve as restrictions upon these relations. In particular, all processes must satisfy the energy principle in addition to the laws of balance of amount of substance (mass), entropy, and momentum. In Section E.1.1, the approach using Lagrange multipliers was introduced; we will now use it for the present case. The laws of balance are satisfied simultaneously if and only if the following relation holds:

$$d_{ij} = \frac{1}{2}\left(\frac{\partial v_i}{\partial x_j} + \frac{\partial v_j}{\partial x_i}\right)$$

is the symmetric part of the velocity gradient tensor, and λ and μ are two independent parameters characterizing the influence of viscosity of the fluid.

19. The bulk viscosity is defined as

$$\kappa = \lambda + \frac{2}{3}\mu$$

where λ and μ have been introduced in the constitutive relations of the Navier-Stokes fluid in Footnote 18. In terms of the deviators

$$j_{pij}' = j_{pij} - P^*\delta_{ij} \quad , \quad d_{ij}' = d_{ij} - \frac{1}{3}d_{kk}\delta_{ij}$$

the (conductive) momentum current can be written as follows:

$$j_{pij}' = -2\mu d_{ij}' \quad , \quad P^* = P - \kappa d_{kk}$$

(see Malvern (1967), p. 299). The pressure P^* includes a bulk viscosity term in addition to the pressure P in equilibrium. For fluids with vanishing bulk viscosity, the pressure term to be used is the normal pressure, and the viscosity of a "one-dimensional" gas only depends upon the normal viscosity μ.

$$
\begin{aligned}
&\frac{\partial}{\partial t}\left[\rho\left(u+\frac{1}{2}v^2\right)\right]+\frac{\partial}{\partial x}\left[\rho v\left(u+\frac{1}{2}v^2\right)+v j_p^{(c)}+j_{E,th}\right] \\
&-\lambda^{\rho}\left[\frac{\partial \rho}{\partial t}+\frac{\partial}{\partial x}(\rho v)\right]-\lambda^{s}\left[\frac{\partial}{\partial t}(\rho s)+\frac{\partial}{\partial x}\left(s\rho v+j_s^{(c)}\right)-\pi_s\right] \\
&-\lambda^{v}\left[\frac{\partial}{\partial t}(\rho v)+\frac{\partial}{\partial x}\left(\rho v v+j_p^{(c)}\right)\right]=0
\end{aligned}
\tag{135}
$$

This expression is obtained by multiplying each of the equations of balance of mass, entropy, and momentum by its own Lagrange multiplier, which here are called λ^{ρ}, λ^{s}, and λ^{v}, respectively, and then subtracting them from the energy principle Equation (131). For the following calculations, note that the constitutive quantities in the laws of balance are functions of ρ, T, dT/dx, and dv/dx. You can verify this by looking at Equations (132) and (134). We should now calculate the derivatives with respect to time and to position in Equation (135), keeping in mind the dependencies just mentioned. This yields a long expression[20] with terms containing the derivatives

$$
\frac{\partial v}{\partial t},\ \frac{\partial^2 v}{\partial x^2},\ \frac{\partial \rho}{\partial t},\ \frac{\partial \rho}{\partial x},\ \frac{\partial T}{\partial t},\ \frac{\partial^2 T}{\partial x^2},\ \text{and}\ \frac{\partial T}{\partial x},\ \frac{\partial v}{\partial x}
$$

20. There essentially are four parts associated with the four laws of balance included in the combined expression of Equation (135). Together they read:

$$
\begin{aligned}
&\rho\frac{\partial u}{\partial \rho}\frac{\partial \rho}{\partial t}+\rho\frac{\partial u}{\partial T}\frac{\partial T}{\partial t}+u\frac{\partial \rho}{\partial t}+\rho v\frac{\partial v}{\partial t}+\frac{1}{2}v^2\frac{\partial \rho}{\partial t}+\rho v\frac{\partial u}{\partial \rho}\frac{\partial \rho}{\partial x}+\rho v\frac{\partial u}{\partial T}\frac{\partial T}{\partial x}+uv\frac{\partial \rho}{\partial x} \\
&+\rho v^2\frac{\partial v}{\partial x}+v\frac{1}{2}v^2\frac{\partial \rho}{\partial x}+\rho\left(u+\frac{1}{2}v^2\right)\frac{\partial v}{\partial x}+j_p^{(c)}\frac{\partial v}{\partial x}+v\frac{\partial j_p^{(c)}}{\partial \rho}\frac{\partial \rho}{\partial x}+v\frac{\partial j_p^{(c)}}{\partial T}\frac{\partial T}{\partial x} \\
&+v\frac{\partial j_p^{(c)}}{\partial v_{,x}}\frac{\partial^2 v}{\partial x^2}+\frac{\partial j_E}{\partial \rho}\frac{\partial \rho}{\partial x}+\frac{\partial j_E}{\partial T}\frac{\partial T}{\partial x}+\frac{\partial j_E}{\partial T_{,x}}\frac{\partial^2 T}{\partial x^2} \\
&-\lambda^{\rho}\left[\frac{\partial \rho}{\partial t}+\rho\frac{\partial v}{\partial x}+v\frac{\partial \rho}{\partial x}\right] \\
&-\lambda^{s}\left[s\frac{\partial \rho}{\partial t}+\rho\frac{\partial s}{\partial \rho}\frac{\partial \rho}{\partial t}+\rho\frac{\partial s}{\partial T}\frac{\partial T}{\partial t}+sv\frac{\partial \rho}{\partial x}+s\rho\frac{\partial v}{\partial x}+\rho v\frac{\partial s}{\partial \rho}\frac{\partial \rho}{\partial x}+\rho v\frac{\partial s}{\partial T}\frac{\partial T}{\partial x}\right. \\
&\qquad\left.+\frac{\partial j_s^{(c)}}{\partial \rho}\frac{\partial \rho}{\partial x}+\frac{\partial j_s^{(c)}}{\partial T}\frac{\partial T}{\partial x}+\frac{\partial j_s^{(c)}}{\partial T_{,x}}\frac{\partial^2 T}{\partial x^2}-\pi_s\right] \\
&-\lambda^{v}\left[\rho\frac{\partial v}{\partial t}+v\frac{\partial \rho}{\partial t}+2\rho v\frac{\partial v}{\partial x}+v^2\frac{\partial \rho}{\partial x}+\frac{\partial j_p^{(c)}}{\partial \rho}\frac{\partial \rho}{\partial x}+\frac{\partial j_p^{(c)}}{\partial T}\frac{\partial T}{\partial x}+\frac{\partial j_p^{(c)}}{\partial v_{,x}}\frac{\partial^2 v}{\partial x^2}\right]
\end{aligned}
$$

Here, $T_{,x}$ and $v_{,x}$ are the gradients of temperature and velocity, respectively.

(If we had included the terms depending on the gradient of velocity in the constitutive expressions for specific entropy and energy (Footnote 18), there also would be the mixed derivative of the velocity with respect to time and position to be taken into account.) While terms containing the first six of these are expressly linearly in them (these derivatives do not occur again in the terms multiplying them), this is not true for the last two derivatives, since these are among the list of independent variables of the constitutive quantities. We separately assemble all the terms dependent upon the derivatives listed above.[21] Noting that the expressions multiplying the first six of the derivatives must vanish identically (otherwise the expression in Equation (135) may be violated), we obtain six conditions which must be satisfied by the fluid, plus one residual equation containing the rate of production of entropy; the second of these is identical to the first, so we have only five conditions:

---

21. After collecting the different terms, the expression in Footnote 20 becomes

$$\frac{\partial v}{\partial t}\left\{v\rho - \lambda^v \rho\right\} + \frac{\partial^2 v}{\partial x^2}\left\{v\frac{\partial j_p^{(c)}}{\partial v_{,x}} - \lambda^v \frac{\partial j_p^{(c)}}{\partial v_{,x}}\right\} +$$

$$\frac{\partial \rho}{\partial t}\left\{\rho\frac{\partial u}{\partial \rho} + u + \frac{1}{2}v^2 - \lambda^\rho - \lambda^s s - \lambda^s \rho\frac{\partial s}{\partial \rho} - \lambda^v v\right\} +$$

$$\frac{\partial \rho}{\partial x}\left\{\rho v\frac{\partial u}{\partial \rho} + uv + v\frac{1}{2}v^2 + v\frac{\partial j_p^{(c)}}{\partial \rho} + \frac{\partial j_E}{\partial \rho} - \lambda^\rho v\right.$$

$$\left. -\lambda^s\left[sv + \rho v\frac{\partial s}{\partial \rho} + \frac{\partial j_s^{(c)}}{\partial \rho}\right] - \lambda^v\left[v^2 + \frac{\partial j_p^{(c)}}{\partial \rho}\right]\right\} +$$

$$\frac{\partial T}{\partial t}\left\{\rho\frac{\partial u}{\partial T} - \lambda^s \rho\frac{\partial s}{\partial T}\right\} + \frac{\partial^2 T}{\partial x^2}\left\{\frac{\partial j_E}{\partial T_{,x}} - \lambda^s \frac{\partial j_s^{(c)}}{\partial T_{,x}}\right\} +$$

$$\frac{\partial T}{\partial x}\left\{\rho v\frac{\partial u}{\partial T} + v\frac{\partial j_p^{(c)}}{\partial T} + \frac{\partial j_E}{\partial T} - \lambda^s\left[\rho v\frac{\partial s}{\partial T} + \frac{\partial j_s^{(c)}}{\partial T}\right] - \lambda^v \frac{\partial j_p^{(c)}}{\partial T}\right\} +$$

$$\frac{\partial v}{\partial x}\left\{\rho v^2 + \rho\left(u + \frac{1}{2}v^2\right) + j_p^{(c)} - \lambda^\rho \rho - \lambda^s s\rho - \lambda^v 2\rho v\right\} + \lambda^s \pi_s = 0$$

Each of the first six expressions in braces must vanish; otherwise Equation (135) may be violated. This leaves the residual equation

$$\frac{\partial T}{\partial x}\left\{\rho v\frac{\partial u}{\partial T} + v\frac{\partial j_p^{(c)}}{\partial T} + \frac{\partial j_E}{\partial T} - \lambda^s\left[\rho v\frac{\partial s}{\partial T} + \frac{\partial j_s^{(c)}}{\partial T}\right] - \lambda^v \frac{\partial j_p^{(c)}}{\partial T}\right\} +$$

$$\frac{\partial v}{\partial x}\left\{\rho v^2 + \rho\left(u + \frac{1}{2}v^2\right) + j_p^{(c)} - \lambda^\rho \rho - \lambda^s s\rho - \lambda^v 2\rho v\right\} + \lambda^s \pi_s = 0$$

which will let us calculate the rate of production of entropy as a consequence of conduction of heat and viscosity.

$$
\begin{aligned}
& v = \lambda^{v} \\
& \rho \frac{\partial u}{\partial \rho} + u - \frac{1}{2} v^2 - \lambda^{\rho} - \lambda^{s} s - \lambda^{s} \rho \frac{\partial s}{\partial \rho} = 0 \\
& \frac{\partial j_E}{\partial \rho} - \lambda^{s} \frac{\partial j_s^{(c)}}{\partial \rho} = 0 \\
& \frac{\partial u}{\partial T} - \lambda^{s} \frac{\partial s}{\partial T} = 0 \\
& \frac{\partial j_E}{\partial T_{,x}} - \lambda^{s} \frac{\partial j_s^{(c)}}{\partial T_{,x}} = 0
\end{aligned}
\tag{136}
$$

and the following residual equation involving the gradients of temperature and velocity:

$$
\begin{aligned}
& \frac{\partial T}{\partial x} \left\{ \frac{\partial j_E}{\partial T} - \lambda^{s} \frac{\partial j_s^{(c)}}{\partial T} \right\} \\
& \qquad + \frac{\partial v}{\partial x} \left\{ \rho \left(u - \frac{1}{2} v^2 - \lambda^{\rho} - \lambda^{s} s \right) + j_p^{(c)} \right\} + \lambda^{s} \pi_s = 0
\end{aligned}
\tag{137}
$$

This relation will be used to determine the rate of production of entropy as a consequence of heat conduction and of viscosity.

Determination of the Lagrange multipliers. A first look at Equation (136a) tells us that the Lagrange multiplier associated with the equation of balance of momentum is equal to the velocity, a result which should not surprise us at all: the velocity is the potential of motion.

Next we are going to introduce the constitutive expressions for the fluxes of entropy and of energy in heating Equations (134b,c) into Equation (136e). This yields a relation between the conductivity and the factor β:

$$\lambda^{s} k_s = \beta$$

which tells us that the Lagrange multiplier associated with entropy may depend only upon density and temperature. Also, it yields an equivalent relation between the fluxes themselves. You can introduce this latter relation in Equation (136c) which leads to the following result:

$$\frac{\partial j_E}{\partial \rho} - \lambda^{s} \frac{\partial}{\partial \rho} \left(\frac{1}{\lambda^{s}} j_E \right) = 0 \quad \Rightarrow \quad \frac{\partial \lambda^{s}}{\partial \rho} = 0$$

This proves that the Lagrange multiplier may be a function only of temperature, and we have

$$j_E = \lambda^{s}(T) j_s^{(c)} \tag{138}$$

The residual equation now takes the following form:

$$\lambda^s \pi_s = -j_s^{(c)} \frac{d\lambda^s}{dT} \frac{\partial T}{\partial x} - \frac{\partial v}{\partial x} \left\{ \rho \left(u - \frac{1}{2} v^2 - \lambda^\rho - \lambda^s s \right) + j_p^{(c)} \right\} \tag{139}$$

This result has important consequences. Introducing the constitutive law for the flux density of entropy and for the current density of momentum yields

$$\lambda^s \pi_s = k_s \frac{d\lambda^s}{dT} \frac{\partial T}{\partial x} \frac{\partial T}{\partial x} - \rho \left(u - \frac{1}{2} v^2 + \frac{P}{\rho} - \lambda^\rho - \lambda^s s \right) \frac{\partial v}{\partial x} + \mu' \frac{\partial v}{\partial x} \frac{\partial v}{\partial x}$$

In equilibrium, where the gradients of temperature and of velocity are zero, the rate of production of entropy will vanish. Moreover, for the rate of production of entropy to be minimal in equilibrium, its derivatives with respect to the gradients must vanish:

$$\left. \frac{\partial \pi_s}{\partial T_{,x}} \right|_E = 0$$

and

$$\left. \frac{\partial \pi_s}{\partial v_{,x}} \right|_E = 0$$

The first condition is satisfied identically, while the latter delivers an interesting relation for the Lagrange multiplier associated with mass:

$$\lambda^\rho = u - \frac{1}{2} v^2 + \frac{P}{\rho} - \lambda^s s \tag{140}$$

As you may note, the Lagrange multiplier belonging to mass (amount of substance) already looks very similar to the chemical potential of the simple fluids discussed in Chapter 4.

Determination of the Lagrange multiplier for entropy. If we manage to determine the multiplier associated with entropy, the essential results of this section will have been derived. We will know the relation (138) between flux densities of entropy and of energy in heating; the expression (139) for the rate of production of entropy; and the Lagrange multiplier for mass, Equation (140).

The derivation roughly goes as follows. We now use the condition of existence of ideal walls. In analogy to what we did in Section E.1.1, we derive the result that for two fluids separated by an ideal wall the Lagrange multiplier must be equal:

$$\lambda_{\mathrm{I}}^s(T) = \lambda_{\mathrm{II}}^s(T)$$

This means that the Lagrange multiplier is a universal function; i.e., it must be the same for different Navier-Stokes-Fourier fluids.

We can derive an expression for the rate of change of the specific entropy by combining Equations (136b) and (136d). Taking into consideration the result derived in Equation (140) we arrive at

$$\dot{s} = \frac{1}{\lambda^s}\left(\frac{\partial u}{\partial \rho} - \frac{P}{\rho^2}\right)\dot{\rho} + \frac{1}{\lambda^s}\frac{\partial u}{\partial T}\dot{T}$$

Since we assume the functions to be sufficiently smooth, this implies an integrability condition of the form

$$\frac{\partial}{\partial T}\left\{\frac{1}{\lambda^s}\left(\frac{\partial u}{\partial \rho} - \frac{P}{\rho^2}\right)\right\} = \frac{\partial}{\partial \rho}\left\{\frac{1}{\lambda^s}\frac{\partial u}{\partial T}\right\}$$

Keep in mind that the Lagrange multiplier depends only upon temperature; you can now derive the following differential equation for the multiplier:

$$\frac{1}{\lambda^s}\frac{d\lambda^s}{dT} = \frac{\dfrac{\partial}{\partial T}\left(\dfrac{P}{\rho^2}\right)}{\dfrac{P}{\rho^2} - \dfrac{\partial u}{\partial \rho}} \tag{141}$$

Since this relation must hold for all types of fluids, it should also hold for the ideal gas. For this material, the right-hand side turns out to be equal to the inverse of the ideal gas temperature. Therefore, we finally have the important result that

$$\lambda^s = T \tag{142}$$

As in previous examples of materials, the thermal potential is the ideal gas temperature for the Navier-Stokes-Fourier fluid as well.

The results for Navier-Stokes-Fourier fluids. We can now assemble the results of the theory. Expressions for the relation between fluxes of entropy and energy, for the rate of production of entropy, and for the Lagrange multiplier for mass (i.e., the chemical potential) can be written in their final form:

$$j_{E,th} = T j_s^{(c)} \tag{143}$$

$$\pi_s = \frac{1}{T}k_s\frac{\partial T}{\partial x}\frac{\partial T}{\partial x} + \frac{1}{T}\mu'\frac{\partial v}{\partial x}\frac{\partial v}{\partial x} \tag{144}$$

$$\mu_m = u - \frac{1}{2}v^2 + \frac{P}{\rho} - Ts \tag{145}$$

Equation (144) also shows that the viscosity may not be a negative quantity (the rate of production of entropy should be positive also if no conduction of heat is present), and Equation (143) lets us calculate the factor β and the energy current due to conduction:

$$\beta = k_E = Tk_S \quad \Rightarrow \quad j_{E,th} = -k_E \frac{\partial T}{\partial x} \tag{146}$$

The index m in Equation (145) reminds us that the chemical potential of the fluid has been written for mass instead of for amount of substance. Multiplication of the expression by the molar mass delivers the result which we know from the treatment of uniform fluids in Chapter 4. Also, we can derive the Gibbs fundamental form for the heat-conducting viscous fluids investigated here. Interestingly, it is the same as the one known from the theory of uniform ideal fluids:

$$\dot{u} = T\dot{s} + \frac{P}{\rho^2}\dot{\rho} \tag{147}$$

Here, it is written using the density of the fluid instead of the volume of a particular body. The fact that some results look the same for reversible and for irreversible fluids often is interpreted as meaning that Navier-Stokes-Fourier fluids do not deviate much from equilibrium (a condition which is called *local thermodynamic equilibrium*). We normally get such results for materials having linear constitutive relations like Fourier's law or Newton's law of viscosity. A simple case of where all of this does not hold is the example of heat conduction with inductive behavior.

The thermal energy equation for the one-dimensional NSF fluid. It is common to present the balance of energy in the material form with the constitutive quantities introduced. The resulting relation is called the thermal energy equation. If you start with Equation (129), i.e., with

$$\rho \frac{Du}{Dt} + j_p{}^{(c)} \frac{\partial v}{\partial x} + \frac{\partial j_{E,th}}{\partial x} = 0$$

and use the constitutive relations in the form

$$u = u(\rho, T)$$

$$j_p{}^{(c)} = P(\rho, T) - \mu'(\rho, T)\frac{\partial v}{\partial x}$$

$$j_{E,th} = -k_E(\rho, T)\frac{\partial T}{\partial x}$$

then the balance of energy becomes

$$\rho \frac{Du(\rho, T)}{Dt} + \left(P(\rho, T) - \mu'(\rho, T)\frac{\partial v}{\partial x} \right)\frac{\partial v}{\partial x} + \frac{\partial}{\partial x}\left(-k_E(\rho, T)\frac{\partial T}{\partial x} \right) = 0$$

or

$$\rho\left[\frac{\partial u}{\partial \rho}\frac{D\rho}{Dt}+\frac{\partial u}{\partial T}\frac{DT}{Dt}\right]+P\frac{\partial v}{\partial x}-\mu'\left(\frac{\partial v}{\partial x}\right)^2-\frac{\partial}{\partial x}\left(k_E\frac{\partial T}{\partial x}\right)=0 \quad (148)$$

It is interesting to specialize this result to an incompressible fluid. Because of this new condition, the first term with the rate of change of the density disappears, and so do the terms involving pressure and viscosity. (In one dimension, the speed of the incompressible fluid may not change with position.) If we assume constant fluid properties, we finally get

$$\rho c\left(\frac{\partial T}{\partial t}+v\frac{\partial T}{\partial x}\right)-k_E\frac{\partial^2 T}{\partial x^2}=0 \quad (149)$$

With the exception of the second term, which is the result of fluid flow, i.e., of convection, this equation is identical to the one derived for time-dependent conduction of heat through a stationary body, Equation (131) of Chapter 3. An incompressible fluid with all quantities (mass, entropy, and momentum) flowing only in a single direction, displays the effects of the transport of entropy both due to conduction and to convection. Since there are no velocity gradients and no shear forces (which require sideways flow of momentum), there are no viscous effects, and the momentum flux density (i.e., the pressure) must remain constant in the direction of flow. As a result, there is no reference to the mechanical processes to be found anymore in the resulting energy equation. This will not be the case for two-dimensional incompressible flow which we will study in the following section.

E.4.4 Application to Convective Heat Transfer

While we have derived more or less general results for the Navier-Stokes-Fourier fluid in the previous sections, the present section will demonstrate how the equations can be specialized to apply to a simple case of boundary layer flow. Convective heat transfer at a solid-fluid interface leads to such a type of flow; by looking at the proper laws of balance and constitutive relations, we will get a first encounter with how heat transfer coefficients can be calculated.

The application to be treated here will first require a generalization of the equations to the two-dimensional case. Second, we will apply a number of special assumptions which hold for steady laminar and incompressible boundary layer flow, which will lead to a considerable reduction of complexity of the equations. Finally, we will show how the relations can be applied to the problem of forced flow over a flat plate.

The boundary layer. We first encountered the problem of heat transfer at a solid-fluid interface in Section 3.2.3, where the boundary layer developing at the surface of a solid body was briefly described. We would like to add some remarks to the previous treatment.

The boundary layer is the region near a surface in which the conditions of the fluid change from those at the surface to those in the free stream. All the interesting action takes place in this normally very thin layer. Naturally, there is no sharp line dividing the boundary layer from the undisturbed fluid. Therefore, it is common to specify that, say, 99% of the change of velocity and temperature perpendicular to the surface will occur between the surface and the line depicting the "edge" of the boundary layer in Figure 10. Note, that the change of temperature and of velocity do not have to take place over the same normal distance, which means that the thermal boundary layer and the velocity boundary layer do not have to have the same thickness (these thicknesses are denoted by δ_t and δ, respectively). It is clear that to solve a convective heat transfer problem, we will have to determine the temperature and velocity profiles over the range of the surface.

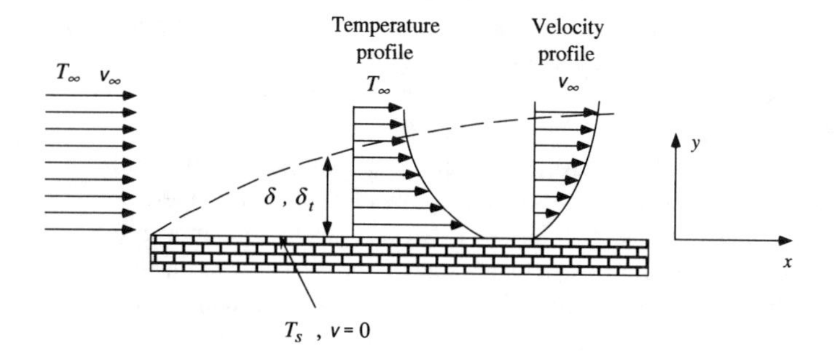

FIGURE 10. Boundary layer flow was discussed briefly in Section 3.2.3. Temperature and speed vary in a thin layer from their values at the interface to the free stream values of the fluid. If the geometrical arrangement extends far in the direction perpendicular to the drawing, we basically can treat the flow as two-dimensional.

For the determination of the heat transfer coefficient, this statement follows from the discussion in Chapter 3. There, we found that this quantity can be calculated as follows:

$$h = -\frac{k_{Ef}}{T_s - T_\infty} \left. \frac{\partial T}{\partial y} \right|_{y=0} \qquad (150)$$

(remember, that h is used to calculate the energy flux associated with the conductive flux of entropy at the interface). Obviously, in addition to fluid properties, we need to know the temperature gradient at the surface, which in turn depends upon the conditions in the boundary layer.

Since the velocity and the thermal boundary layers are related, we need to know the mechanical conditions as well. Indeed, we may say that the entire problem starts with the effect of friction at the surface over which the fluid is flowing. To specify this phenomenon, a dimensionless friction coefficient is introduced which is defined as the ratio of the shear stress at the surface and the density of kinetic energy of the fluid in the free stream:

$$C_f = \frac{-j_{pxy}(y=0)}{\rho v_\infty / 2} \qquad (151)$$

Introduction of Newton's law of friction yields

$$C_f = \frac{\mu}{\rho v_\infty/2} \frac{\partial v_x}{\partial y}\bigg|_{y=0} \tag{152}$$

Therefore, the conditions in the velocity boundary layer determine the friction coefficient, or vice versa. Again, knowing the conditions of the fluid at the interface is of central importance in calculating the outcome of convective heat transfer.

The general equations for two dimensional flow. The laws of balance and the constitutive relations were listed for the general case of three-dimensional flow in Footnotes 16 and 18. The only difference between the two- and the three-dimensional cases is to be found in the indices i and j, which for the present case, extend only from1 to 2. Taking into consideration the results of the thermodynamic constitutive theory presented above, the laws of balance are

$$\begin{aligned}
&\frac{D\rho}{Dt} + \rho \frac{\partial v_j}{\partial x_j} = 0 \\
&\rho \frac{Dv_i}{Dt} + \frac{\partial j_{pij}{}^{(c)}}{\partial x_j} = 0 \\
&\rho \frac{Du}{Dt} + j_{pij}{}^{(c)} d_{ij} + \frac{\partial j_{E,th,j}}{\partial x_j} = 0
\end{aligned} \tag{153}$$

if they are written in the material form.[22] It is customary to leave out the law of balance of entropy, and include the results of the constitutive theory of the thermal processes with the energy equation. The material derivative and the symmetric part of the velocity gradient tensor are given by

$$\frac{D}{Dt} = \frac{\partial}{\partial t} + v_j \frac{\partial}{\partial x_j} \tag{154}$$

$$d_{ij} = \frac{1}{2}\left(\frac{\partial v_i}{\partial x_j} + \frac{\partial v_j}{\partial x_i}\right) \tag{155}$$

For us, i,j = 1,2.. The constitutive laws pertaining to the Navier-Stokes-Fourier fluid are

22. See Malvern (1969), p. 206–230, for a derivation of the laws of balance in the form given here. They are motivated by the one-dimensional case treated in the previous sections. As before, it is assumed that a summation is performed over indices appearing twice in a term.

$$u = u(\rho, T)$$
$$j_{pij}{}^{(c)} = P\delta_{ij} - \nu d_{kk}\delta_{ij} - 2\mu d_{ij} \qquad (156)$$
$$j_{E,th} = -k_E \frac{\partial T}{\partial x}$$

The laws relating to entropy are included, for example, in the expression for the thermal energy current. (You can start with Fourier's law for conduction of entropy, use the relation between fluxes of entropy and of energy derived for the NSF fluid, and arrive at Equation (156c).)

The Navier-Stokes equations for steady laminar and incompressible flow in two dimensions. As suggested by Figure 10, we can treat our example as a case of two-dimensional flow. If we assume the flow to be developed, we no longer take changes in time into account. Also, with the exception of high speed flow of gases, fluids can normally be considered to be nearly incompressible. Now, we will write out the equations of balance for two dimensions, introduce the constitutive laws, and apply the restrictions. This leads to

$$\frac{\partial v_x}{\partial x} + \frac{\partial v_y}{\partial y} = 0$$
$$\rho v_x \frac{\partial v_x}{\partial x} + \rho v_y \frac{\partial v_x}{\partial y} + \frac{\partial P}{\partial x} - \mu\left(\frac{\partial^2 v_x}{\partial x^2} + \frac{\partial^2 v_x}{\partial y^2}\right) = 0 \qquad (157)$$
$$\rho v_x \frac{\partial v_y}{\partial x} + \rho v_y \frac{\partial v_y}{\partial y} + \frac{\partial P}{\partial y} - \mu\left(\frac{\partial^2 v_y}{\partial x^2} + \frac{\partial^2 v_y}{\partial y^2}\right) = 0$$

for mass and for the x- and the y-components of momentum.[23] In general, we can say that the divergence of the velocity vector is zero for incompressible fluids.

---

23. The derivation for the case of the balance of mass is quite simple. Since the density is taken to be constant (both in space and in time), Equation (153a) takes the form

$$\frac{\partial \rho}{\partial t} + v_x \frac{\partial \rho}{\partial x} + v_y \frac{\partial \rho}{\partial y} + \rho\left(\frac{\partial v_x}{\partial x} + \frac{\partial v_y}{\partial y}\right) = 0 \;\Rightarrow\; \frac{\partial v_x}{\partial x} + \frac{\partial v_y}{\partial y} = 0$$

There are two momentum balance equations. We shall perform the derivation for the x-component only. First, because of steady-state conditions, we have

$$\rho\left(v_x \frac{\partial v_x}{\partial x} + v_y \frac{\partial v_x}{\partial y}\right) + \frac{\partial j_{pxx}{}^{(c)}}{\partial x} + \frac{\partial j_{pxy}{}^{(c)}}{\partial y} = 0$$

The momentum equations, on the other hand, contain terms referring to convection, to stress power due to the spatial change of pressure, and to viscous friction. The energy equation (with the internal energy replaced by the enthalpy) turns out to be[24]

$$\rho v_x \frac{\partial h}{\partial x} + \rho v_y \frac{\partial h}{\partial y} = v_x \frac{\partial P}{\partial x} + v_y \frac{\partial P}{\partial y}$$
$$+ 2\mu\left[\left(\frac{\partial v_x}{\partial x}\right)^2 + \left(\frac{\partial v_y}{\partial y}\right)^2\right] + \mu\left(\frac{\partial v_x}{\partial y} + \frac{\partial v_y}{\partial x}\right)^2 - \frac{\partial j_{E,th,x}}{\partial x} - \frac{\partial j_{E,th,y}}{\partial y} \tag{158}$$

The constitutive law for the momentum current density is given by

$$j_{pij}{}^{(c)} = P\delta_{ij} - \lambda d_{kk}\delta_{ij} - 2\mu d_{ij}$$

Because of the particular form of the balance of mass for the present case, the second term on the right-hand side is zero:

$$d_{kk} = \frac{\partial v_x}{\partial x} + \frac{\partial v_y}{\partial y} = 0$$

which means that

$$j_{pxx}{}^{(c)} = P - 2\mu\frac{\partial v_x}{\partial x}$$

$$j_{pxy}{}^{(c)} = -\mu\left(\frac{\partial v_x}{\partial y} + \frac{\partial v_y}{\partial x}\right)$$

The sum of the two spatial derivatives of the momentum current density vector is equal to

$$\frac{\partial j_{pxx}{}^{(c)}}{\partial x} + \frac{\partial j_{pxy}{}^{(c)}}{\partial y} = \frac{\partial P}{\partial x} - \left\{\frac{\partial}{\partial x}\left[2\mu\frac{\partial v_x}{\partial x}\right] + \frac{\partial}{\partial y}\left[\mu\left(\frac{\partial v_x}{\partial y} + \frac{\partial v_y}{\partial x}\right)\right]\right\}$$
$$= \frac{\partial P}{\partial x} - \left\{\mu\left(\frac{\partial^2 v_x}{\partial x^2} + \frac{\partial^2 v_x}{\partial y^2}\right) + \mu\left(\frac{\partial}{\partial x}\frac{\partial v_x}{\partial x} + \frac{\partial}{\partial y}\frac{\partial v_y}{\partial x}\right)\right\}$$
$$= \frac{\partial P}{\partial x} - \left\{\mu\left(\frac{\partial^2 v_x}{\partial x^2} + \frac{\partial^2 v_x}{\partial y^2}\right) + \mu\frac{\partial}{\partial x}\left(\frac{\partial v_x}{\partial x} + \frac{\partial v_y}{\partial y}\right)\right\} = \frac{\partial P}{\partial x} - \mu\left(\frac{\partial^2 v_x}{\partial x^2} + \frac{\partial^2 v_x}{\partial y^2}\right)$$

Addition of the parts yields the desired result. The derivation proceeds analogously for the second component of momentum.

24. The derivation of the law of balance of energy can be obtained as follows. In two dimensions, Equation (153c) looks like

The boundary layer approximations. Several approximations apply to the case of boundary layer flow. The consequences of steady-state incompressible flow have already been worked out. First, the law of balance of mass applies as derived above in Equation (157a), namely:

$$\frac{\partial v_x}{\partial x} + \frac{\partial v_y}{\partial y} = 0 \tag{159}$$

$$v_x \frac{\partial u}{\partial x} + v_y \frac{\partial u}{\partial y} + j_{pxx}{}^{(c)} \frac{\partial v_x}{\partial x} + j_{pxy}{}^{(c)} \frac{1}{2}\left(\frac{\partial v_x}{\partial y} + \frac{\partial v_y}{\partial x} \right)$$
$$+ j_{pyx}{}^{(c)} \frac{1}{2}\left(\frac{\partial v_y}{\partial x} + \frac{\partial v_x}{\partial y} \right) + j_{pxy}{}^{(c)} \frac{\partial v_y}{\partial y} + \frac{\partial j_{E,th,x}}{\partial x} + \frac{\partial j_{E,th,y}}{\partial y} = 0$$

The four components of the momentum current density tensor (the conductive part) are the following

$$j_{pxx}{}^{(c)} = P - 2\mu \frac{\partial v_x}{\partial x}$$
$$j_{pxy}{}^{(c)} = -\mu\left(\frac{\partial v_x}{\partial y} + \frac{\partial v_y}{\partial x} \right)$$
$$j_{pyx}{}^{(c)} = -\mu\left(\frac{\partial v_y}{\partial x} + \frac{\partial v_x}{\partial y} \right)$$
$$j_{pyy}{}^{(c)} = P - 2\mu \frac{\partial v_y}{\partial y}$$

Inserting these into the energy equation yields

$$v_x \frac{\partial u}{\partial x} + v_y \frac{\partial u}{\partial y} + \left(P - 2\mu \frac{\partial v_x}{\partial x} \right) \frac{\partial v_x}{\partial x} - \mu\left(\frac{\partial v_x}{\partial y} + \frac{\partial v_y}{\partial x} \right) \frac{1}{2}\left(\frac{\partial v_x}{\partial y} + \frac{\partial v_y}{\partial x} \right)$$
$$- \mu\left(\frac{\partial v_y}{\partial x} + \frac{\partial v_x}{\partial y} \right) \frac{1}{2}\left(\frac{\partial v_y}{\partial x} + \frac{\partial v_x}{\partial y} \right) + \left(P - 2\mu \frac{\partial v_y}{\partial y} \right) \frac{\partial v_y}{\partial y}$$
$$+ \frac{\partial j_{E,th,x}}{\partial x} + \frac{\partial j_{E,th,y}}{\partial y} = 0$$

The terms containing the pressure add up to zero because of the equation of balance of mass. The rest yields the result given in Equation (158); as a last step, you only must replace the specific internal energy u by the specific enthalpy $h = u + P/\rho$.

Now, we take into account (1) that the speed of flow parallel to the flat surface (as in Figure 10) should be considerably larger than that normal to it, and (2) that the gradients of velocity and of temperature in the y-direction are much larger than those in the x-direction. Moreover, we assume (3) the variation or pressure in the direction of flow to be small and (4) fluid properties to be constant. Then, as the first consequence of assumption (1), all terms in the second momentum equation (157c) involving the velocity in the normal direction are very small, leaving us with

$$\frac{\partial P}{\partial y} = 0$$

which means that the pressure should be approximately constant in the y-direction. In the first momentum equation, however, only the pressure gradient and the friction term involving the x-gradient of the parallel flow speed are small compared to the other terms—assumptions (1–3). This leads to

$$\rho v_x \frac{\partial v_x}{\partial x} + \rho v_y \frac{\partial v_x}{\partial y} = \mu \frac{\partial^2 v_x}{\partial y^2}$$

or

$$v_x \frac{\partial v_x}{\partial x} + v_y \frac{\partial v_x}{\partial y} = \nu \frac{\partial^2 v_x}{\partial y^2} \quad , \quad \nu = \mu / \rho \qquad \textbf{(160)}$$

where the ratio of the dynamic viscosity and the density of the fluid is called the *kinematic viscosity*. The energy equation, finally, attains a structure quite similar to the law of balance of momentum in the x-direction. In addition to the previous assumptions we also assume (5) the effect of viscous production of entropy to be small, and (6) conductive heat transfer to essentially take place in the direction normal to the flow. This yields

$$\rho v_x \frac{\partial h}{\partial x} + \rho v_y \frac{\partial h}{\partial y} = k_E \frac{\partial^2 T}{\partial y^2} \quad \Rightarrow$$

$$\rho v_x \frac{\partial h}{\partial T}\frac{\partial T}{\partial x} + \rho v_y \frac{\partial h}{\partial T}\frac{\partial T}{\partial y} = k_E \frac{\partial^2 T}{\partial y^2}$$

If we take the internal energy of the incompressible fluid to be a linear function of temperature; i.e., if

$$h = h_o + c_p (T - T_o)$$

we obtain the following relation for the energy equation

$$v_x \frac{\partial T}{\partial x} + v_y \frac{\partial T}{\partial y} = \alpha \frac{\partial^2 T}{\partial y^2} \quad , \quad \alpha = \frac{k_E}{\rho c_p} \qquad \textbf{(161)}$$

The quantity α is called the *thermal diffusivity* (which is proportional to the entropy diffusivity). If we compare this last result with Equation (160), we can see that the kinematic viscosity plays the role of momentum diffusivity. Even though we use the energy equation to express this process, heat transfer is the transfer of entropy. On the basis of the fundamental quantities entropy and momentum, the thermal and mechanical phenomena in a boundary layer are structurally analogous. It is interesting to note that the diffusion of a chemical species at an interface can be treated in the same manner, leading to another deep analogy.[25] Obviously, diffusive transport occurs with entropy, momentum, and charge as much as it does with chemical substances.

Nondimensional form of the boundary layer equations and similarity parameters. Solving the differential equations for boundary layer flow, even in their reduced form of Equations (159) - (161), is not a small feat. Having to do this for different geometries and fluids having different properties under varying conditions would be even more daunting. Therefore, it is imperative to be able to reduce the amount of work by introducing dimensionless forms of the equations and dimensionless groups of quantities and fluid properties. It is found that the results depend for a given geometry upon these dimensionless groups only.

First, we have to introduce the following dimensionless independent and dependent variables:

$$x^* = x/L$$
$$y^* = y/L$$
$$v_x{}^* = v_x/v_\infty$$
$$v_y{}^* = v_y/v_\infty$$
$$T^* = \frac{T - T_s}{T_\infty - T_s}$$

Here, L is some characteristic length associated with the geometry of the body over which the fluid is flowing. In the case of a flat plate, we may take L to represent its length. With these quantities, the simplified boundary layer equations become

$$\frac{\partial v_x{}^*}{\partial x^*} + \frac{\partial v_y{}^*}{\partial y^*} = 0$$

$$v_x{}^* \frac{\partial v_x{}^*}{\partial x^*} + v_y{}^* \frac{\partial v_x{}^*}{\partial y^*} = \frac{1}{Re_L} \frac{\partial^2 v_x{}^*}{\partial y^{*2}} \tag{162}$$

$$v_x{}^* \frac{\partial T^*}{\partial x^*} + v_y{}^* \frac{\partial T^*}{\partial y^*} = \frac{1}{Re_L Pr} \frac{\partial^2 T^*}{\partial y^{*2}}$$

25. See, for example, Incropera and DeWitt, 1981.

where the dimensionless Reynolds and Prandtl numbers have been introduced:

$$\text{Reynolds number:} \quad Re_L = \frac{v_\infty L}{\nu}$$
$$\text{Prandtl number:} \quad Pr = \frac{\nu}{\alpha} \tag{163}$$

The Reynolds number approximately represents the ratio of inertial to viscous effects in the fluid, while the Prandtl number is the ratio of the momentum and the thermal diffusivities. The Reynolds number can also be used to express the friction coefficient:

$$C_f = \frac{2}{Re_L} \frac{\partial v_x{}^*}{\partial y^*}\bigg|_{y^*=0} \tag{164}$$

If the dimensionless variables are introduced in the relation for the heat transfer coefficient in Equation (150), we obtain

$$h = \frac{k_{Ef}}{L} \frac{\partial T^*}{\partial y^*}\bigg|_{y^*=0}$$

A third dimensionless group is introduced, which relates the convective heat transfer coefficient to the conductive one:

$$\text{Nusselt number:} \quad Nu = \frac{hL}{k_{Ef}} \tag{165}$$

As you can tell from the definition, this factor is the dimensionless temperature gradient at the surface of the solid body.

Results of the determination of convective heat transfer coefficients are now cast in the form of relations between the dimensionless groups which hold for a given geometry independently of fluid properties and flow conditions as long as the basic restrictions apply (i.e., those restrictions which led to the reduced form of the differential equations). The dimensionless equations suggest that the x-velocity is some universal function of the dimensionless coordinates and the Reynolds number, the friction factor depends only upon x^* and the Reynolds number; and the dimensionless temperature may be obtained from a universal function of the coordinates, the Reynolds number and the Prandtl number. Therefore, for prescribed geometry, the Nusselt number is a universal function of the position along the surface, the Reynolds number, and the Prandtl number:

$$Nu = f(x^*, Re_L, Pr) \tag{166}$$

To give an example, for laminar flow over a flat plate the equations can be integrated to yield

$$C_{f,x} = 0.664\,Re_x^{-1/2} \quad , \quad Nu_x = 0.332\,Re_x^{1/2}\,Pr^{1/3} \tag{167}$$

as long as the Prandtl number is larger than 0.6.[26] The subscript x refers to the position in x-direction along the flat plate. The local Nusselt number Nu_x has to be integrated if we wish to obtain the average value up to position x. For this case, it turns out that the average value up to position x is twice the local value at x.

Naturally, other geometries and flow conditions lead to different and commonly much more complicated problems. Turbulent flow, for example, cannot be treated analytically, at least not without the introduction of additional strong assumptions. Turbulent mixing in the boundary layer leads to greatly increased thermal and momentum diffusivities[27] for which we do not have simple expressions, since they depend upon the state of motion and not just upon fluid properties. In many heat transfer applications of practical interest, experimental determination of the heat transfer coefficient is required. Fortunately, the dimensionless groups help to reduce the complexity of the problem just as they did in the theoretical example. In other words, as in the case of an analytical calculation, we should try to measure the Nusselt number in terms of the Reynolds and the Prandtl numbers to obtain empirical relations analogous to what we have seen in Equation (167).

E.5 Inductive Thermal Behavior: Extended Irreversible Thermodynamics

So far, irrespective of the material undergoing thermal processes, all the examples discussed have led to the same basic result concerning the functional dependence of entropy and energy. We have found that even in irreversible processes the local behavior of materials is the same as that encountered in the simplest of all cases, namely, in static conditions or in the dynamics of uniform ideal fluids. To be precise, the entropy and the energy of a simple thermomechanical material have always been found to depend only on temperature and density. We have discovered the same type of Gibbs fundamental relation in each example. Because of this similarity, we speak of local thermodynamic equilibrium prevailing even in the case of irreversible processes such as those treated for heat-conducting viscous fluids.

Despite the success of the models of materials treated so far, in at least one case it has been known for a long time that something is amiss: the simple theory of conduction of heat based upon Fourier's law predicts infinite speed of thermal disturbances. (See Section 3.4.6.) We do not have to advance theories of more

26. See, for example, Incropera and DeWitt (1981), p. 313–318.

27. Incropera and DeWitt (1981), p. 293–296.

complicated materials to find the need to extend the usual results of thermodynamics. We will see in this section that pure conduction of heat calls for a treatment where the energy and the entropy of a body are functions of nonequilibrium variables in addition to local temperature and density, if we wish to solve the problem of infinite speed of propagation of thermal pulses. To give an impression of what is involved, we will first describe a simple example which should clarify the difference between pure diffusion on the one hand, and wave propagation on the other. We will see that we need to introduce an inductive term in the constitutive law for the transport of heat. Then, a simple derivation of the consequences for thermodynamics of the additional assumption will be given for the case of conduction in a rigid conductor of constant density.[28]

E.5.1 Diffusion Versus Wave Propagation

Take a simple example of the conduction of heat through a long rigid bar where, initially, steady-state conditions prevail. Think of a case with the temperature decreasing linearly from left to right as in the bar of Figure 11. For simplicity, consider the body to be divided into a few uniform finite elements, each with its own temperature. Now let the temperature at the left end suddenly decrease from its initial value to a much lower one. You can easily imagine the consequences of this jump in the boundary value. The temperature of the first element will drop relatively fast (if the heat transfer coefficient at the left wall is not too small compared to the conduction transfer coefficient between the elements), with the current of entropy between elements 1 and 2 immediately responding to the changing conditions. This is a consequence of our assumption as to how heat transfer in a bar comes about: Fourier's law predicts that the current of entropy (or of energy transported with entropy) directly depends upon the instantaneous value of the temperature difference between adjacent elements:

$$j_s = -k_s \frac{\Delta T}{\Delta x} \tag{168}$$

or

$$j_{E,th} = -k_E \frac{\Delta T}{\Delta x} \tag{169}$$

This also holds for all the fluxes between the other elements, which demonstrates that the change occurring at the left is felt immediately at the far right.[29] Now,

28. This example was discussed in a paper by Jou and Casas-Vázquez (1988).

29. If you think of what happens in the course of several consecutive small time steps, you get the impression that it will take a finite time for the disturbance to propagate from left to right; it takes one time step for a change taking place in an element to be felt in the neighboring element as well. Now, make the time step as small as you wish, and you see that the last element feels the change occurring in the first as quickly as you wish.

as soon as the temperature of the first element has become smaller than that of the second one, the current reverses its sign and flows from the second to the first element. This is what is observed with high accuracy in normal heat conduction.

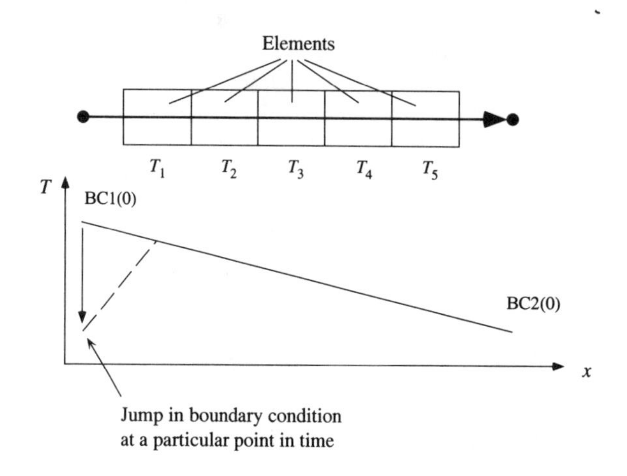

FIGURE 11. Flow of heat through a long bar. The figure suggests a finite element representation of the process. Initially, we have steady-state conditions. At a particular time the outside temperature at the left edge is lowered suddenly. Depending on whether we include "inertia" with a current of entropy, the behavior of heat conduction will be very different.

What would happen in a chain of interconnected water containers? Compare the transport of entropy along the bar divided into elements to water flowing between containers joined by pipes (Figure 12). Again take an initial steady state to prevail. At some time, the pressure at the left end is decreased, which leads to water flowing out of the first container to the left. (Assume this to happen immediately.) Depending on circumstance, it may happen that the water level in the

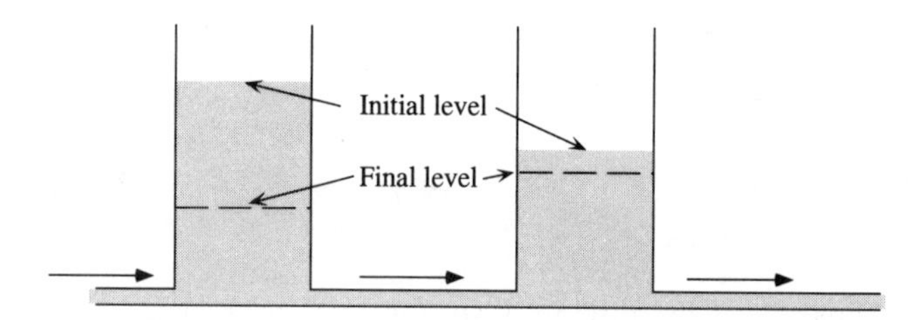

FIGURE 12. Water flowing through a series of containers joined by pipes. Even in cases where the normal pressure gradient along a pipe would indicate flow in one direction, water still may flow in the other direction due to the effect of its inertia.

first container drops below that of the second. Will the water in the pipe joining containers 1 and 2 reverse its direction immediately? You know that the answer is *no*, since water has a certain inertia which will let it continue to flow for some time from left to right even after the pressure at the bottom of container 1 has become smaller than that at the bottom of container 2. However, if we had used a law such as that of Hagen and Poiseuille to calculate the current of water, we would have found this quantity to react immediately to the changing values of pressure at the bottom of the containers, in accordance to what we expect of the flow of heat, but contrary to what we know about the behavior of water. In the

Prologue we solved this problem by extending the constitutive law for the current of water to include an inductive term representing the phenomenon of the inertia of a flowing substance:

$$I_V = -\frac{1}{R_V}\Delta P - \frac{L_V}{R_V}\frac{dI_V}{dt}$$

This constitutive relation not only predicts the real behavior of the water flowing in a situation such as the one in Figure 12, it also leads to a wave equation for currents or for pressure if combined with the law of balance of volume:

$$\frac{\partial^2 P}{\partial t^2} = \frac{1}{L_V * K_V *}\frac{\partial^2 P}{\partial x^2} - \frac{R_V *}{L_V *}\frac{\partial P}{\partial t}$$

where the constant values R^*, K^*, and L^* are the hydraulic resistance, capacitance, and inductance per unit length, respectively. If we assumed a current of heat to have some type of "inertia" associated with it, we should simply include a thermal inductance with the law of conduction of heat, i.e., with Fourier's law, which then would become

$$\tau\frac{\partial j_{E,th}}{\partial t} + j_{E,th} = -k_E\frac{\partial T}{\partial x} \quad (170)$$

where τ, the relaxation time of the current of energy in heating, is equal to the ratio of thermal inductance to thermal resistance; i.e., it is equivalent to the inductive time constant known from electricity or from hydraulics. Equation (170) is known as the law of Maxwell-Cattaneo for the conduction of heat. Combined with the law of balance of energy in pure heat conduction, this leads to

$$\tau\frac{\partial^2 T}{\partial t^2} + \frac{\partial T}{\partial t} = \frac{k_E}{\rho c}\frac{\partial^2 T}{\partial x^2} \quad (171)$$

which is a wave equation (telegrapher's equation) for heat conduction. As we have seen before, such a field equation means that thermal disturbances propagate with finite speed, which in this case is given by

$$v = \sqrt{\frac{k_E}{\tau\rho c}} \quad (172)$$

In the last two equations, c and ρ are the specific temperature coefficient of energy and the density of the material, respectively. Zero relaxation time for currents of heat, i.e., immediate reaction to changes of temperature gradients, therefore leads to infinite speed of propagation as predicted by Fourier's law of conduction of heat. Aside from the question of whether we are satisfied with the normal theory of the conduction of heat, phenomena which predict finite speed disturbances in thermal processes are known to exist. This is the case, for example, in high-frequency phenomena where frequencies become comparable to the inverse of the inductive time constant (ultrasonic propagation in gases, neutron scattering in liquids), or in solids at very low temperatures (second sound).

A qualitative explanation of inductive phenomena usually involves the notion of the energy associated with the processes. In electrodynamics, the energy of the magnetic field is increased or decreased in accordance with the change of the magnitude of the currents of charge. In hydraulics, on the other hand, the energy change of the flowing fluid, i.e., the kinetic energy, is associated with changes of currents. At first, we might look in vain for the energy related to the "inertia" of currents of entropy. However, the energy flowing together with entropy represents a flux of mass: energy has inertia. Changing the currents of entropy leads to a change of the associated flux of energy (mass), which should provide the effect we have been looking for.

E.5.2 Conduction Including Inductive Effects

Normal conduction of heat through a rigid conductor is described by the laws of balance of entropy and energy:

$$\rho\dot{s} + \frac{\partial}{\partial x} j_s = \pi_s$$
$$\rho\dot{u} + \frac{\partial}{\partial x} j_{E,th} = 0 \tag{173}$$

(where we have assumed constant density for the conductor), by Fourier's law of conduction

$$j_{E,th} = -k_E \frac{\partial T}{\partial x} \tag{174}$$

and, as a consequence, by the expression for the rate of production of entropy

$$\pi_s = -\frac{1}{T^2} j_{E,th} \frac{\partial T}{\partial x} \tag{175}$$

and by the simple Gibbs fundamental relation

$$\dot{u} = T\dot{s} \tag{176}$$

This means that the entropy and the energy of the body are functions of temperature only. It will be shown now that we should include the flux of energy as one of the independent variables for these quantities if we wish to account for phenomena which predict finite speed of propagation of heat. Having equations like those presented in the previous section calls for evolution equations not only for the entropy (and the energy), but for their currents as well. Therefore, let us assume a Gibbs fundamental relation of the form

$$\dot{u} = T\dot{s} + a j_{E,th} \frac{dj_{E,th}}{dt} \tag{177}$$

which implies

$$s = s(T, j_s) \tag{178}$$

and

$$u = u(T, j_s) \tag{179}$$

The equations of balance of entropy and of energy remain unchanged. If we plug the Gibbs relation into the law of balance energy, we obtain

$$\rho\left(T\dot{s} + a j_{E,th} \frac{dj_{E,th}}{dt}\right) + \frac{\partial}{\partial x} j_{E,th} = 0$$

which is equivalent to

$$\rho T\dot{s} + \rho a j_{E,th} \frac{dj_{E,th}}{dt} + T\frac{\partial}{\partial x}\left(\frac{j_{E,th}}{T}\right) + \frac{1}{T} j_{E,th} \frac{\partial T}{\partial x} = 0$$

Now compare this result to the equation of balance of entropy. We obtain the following expressions for the flux of energy in terms of the flux of entropy, and for the rate of production of entropy:

$$\begin{aligned} j_s &= \frac{j_{E,th}}{T} \\ \pi_s &= \frac{1}{T}\left(-\rho a \frac{dj_{E,th}}{dt} - \frac{1}{T}\frac{\partial T}{\partial x}\right) j_{E,th} \end{aligned} \tag{180}$$

Note that we again have the well-known relation between currents of entropy and energy in conduction. Now we exploit the fact that the entropy production may not be negative. The simplest possibility for this to be the case is if

$$-\rho a \frac{dj_{E,th}}{dt} - \frac{1}{T}\frac{\partial T}{\partial x} = b j_{E,th} \quad \text{with} \quad b > 0 \tag{181}$$

Comparison with the constitutive law assumed to hold for the current of energy, i.e., with the law of Maxwell-Cattaneo (170) allows us to identify the coefficients a and b:

$$k_E = \frac{1}{Tb}$$

$$\frac{\rho a}{b} = \tau$$

Therefore, the rate of production of entropy turns out to be

$$\pi_s = -\frac{1}{T}\left(\frac{\tau}{k_E T}\frac{dj_{E,th}}{dt} + \frac{1}{T}\frac{\partial T}{\partial x}\right)j_{E,th} \tag{182}$$

while the Gibbs fundamental relation is given by

$$\dot{u} = T\dot{s} + \frac{\tau}{\rho k_E T} j_{E,th} \frac{dj_{E,th}}{dt} \tag{183}$$

Naturally, for vanishing relaxation time τ (or for vanishing time rates of change of fluxes) the results are equivalent to the usual case of heat conduction according to Fourier's law. It is obvious that the requirement of finite speed of propagation of thermal disturbances requires the fluxes to be included in the list of the independent variables of the theory. The resulting entropy and energy of a body are called *nonequilibrium entropy* and *energy*; in addition to terms which depend only upon temperature and density, i.e., upon the equilibrium conditions, they include a term involving the square of the entropy flux.[30] This requirement distinguishes the extended version of irreversible thermodynamics from the usual one.

E.6 The Lessons of Continuum Thermodynamics

Even though we only have had a very brief encounter with continuum thermodynamics up to this point, we can draw some important conclusions.

In view of our normal tendency to divide the world into small and neat compartments, maybe the most important of the insights gained from continuum thermodynamics is the fact that the classical field of thermodynamics and the subject of heat transfer form a natural unity. Despite all the claims in books on thermodynamics that heat transfer cannot be included, and of books on heat transfer that thermodynamics is an altogether different thing, we do not have to artificially separate the two. Thermodynamics of uniform bodies can be presented as a theory of the dynamics of heat using the proper laws of balance of entropy, momentum, energy, etc., and associated constitutive relations; these same equations of balance are then used as the starting point for investigations of the flow of heat.

Problems in engineering and in the sciences are not solved if we put up high walls around every subject. Rather, if we accept the images inherent in continuum physics for the description of simpler examples as well, these walls are brought down. Introductory physics and thermodynamics will profit considerably from a careful look at what continuum processes have to tell us. Students who have grown up with this kind of mental picture will more easily venture into the exciting combined field of thermodynamics and of heat transfer.

---

30. See Jou and Casas-Vázquez (1988), p.332.

Continuum processes are clearly irreversible. While the quest for a description of reversible processes is understandable, and while theories of the dynamics of such processes can be built and applied successfully, the belief that reversibility requires equilibrium in the sense of static conditions has led classical thermodynamics into a tight corner out of which it can escape only if the lessons of continuum physics are accepted. If you have followed the development of ideas in this book, you will have seen that everyday phenomena suggest an image of heat as the fundamental extensive thermal quantity which directly leads to the equation of balance of entropy. Applying this law to continuum processes therefore is natural and straightforward, and its success there tells us that self-imposed limitations in classical thermodynamics are unjustified and unnecessary.

The development of continuum physics in this century should give us the courage to strengthen and renew our teaching of classical physics, to expose students of the sciences and engineering to the foundations of the classical fields, so that these foundations may be applied to the solution of current and future problems.

Questions and Problems

1. Consider a uniform viscous fluid such as the one introduced in Section E.1.1. Let the fluid have a pressure, and entropy capacity, and a latent entropy like the ideal gas, but let friction be present. a) Show that the energy added in heating of the fluid at constant volume is

$$Q = E(V, T_f) - E(V, T_i)$$

where the indices i and f refer to the initial and the final states, respectively.

b) Consider heating at constant pressure. Show that in this case the energy added in heating can be expressed as

$$Q = H(P, T_f) - H(P, T_i) + \int_{t_i}^{t_f} a\dot{V}^2 dt$$

Show that this quantity is always less than the difference of the enthalpies at the end and the beginning.

c) Assume the friction factor a, the rate of change of the volume, and the energy current in heating to be given. Take the rate of change of the volume to be constant. Show that the initial value problem of the fluid takes the form

$$\frac{C_V}{T}\dot{T} - \frac{1}{T}I_E(t) + \frac{a}{T}\dot{V}^2 = -\frac{nR}{V_o + \dot{V}t}\dot{V}$$

$$T(t = 0) = T_o$$

where V_o is the volume at $t = 0$.

2. Consider the uniform viscous fluid described in Problem 1. Show that for isothermal heating at a temperature T, with prescribed entropy current $I_S(t)$, the differential equation for the volume becomes

$$\frac{a}{T}\dot{V}^2 + \frac{nR}{V}\dot{V} = -I_s(t)$$

3. Show that the component form of the general equation of balance of momentum should be written as follows:

$$\frac{\partial}{\partial t}(\rho v_i) + \frac{\partial}{\partial x_j}\left(\rho v_i v_j + j_{pij}^{(c)}\right) = \sigma_{pi}$$

Write all of the equations in fully expanded form.

4. Show that the law of balance of entropy in the form of Equation (115) looks as if it represented the case of pure conductive transport of entropy. Does the law expressed by Equation (115) exclude the phenomenon of convection? Can you explain the name *material derivative* given to the operator in Equation (114)?

5. Write the law of balance of mass using the material derivative. Do the same for the simple case of pure one-dimensional transport of momentum.

6. Derive the general three-dimensional form of the substantial derivative. Write the result both in coordinate-independent form and using Cartesian coordinates.

7. Consider a flat car filled with water travelling horizontally. Water flows out of the bottom through a hole. Determine the flux densities and the fluxes of momentum with respect to a stationary control volume.

8. Perform a direct derivation of the balance of energy for purely one-dimensional convective and conductive flow of entropy as in Equation (149). (*Hint:* Use a stationary control volume and compute convective and conductive currents of energy associated with entropy transfer only.)

9. Derive the complete expression for the rate of production of entropy for general three-dimensional flow of a Navier-Stokes-Fourier fluid.

10. Consider laminar flow over a flat plate. Show that the heat transfer coefficient increases as the square root of the velocity of the free stream. Also show that the average heat transfer coefficient over a length L of the plate is equal to twice its local value at L.

11. Consider a pebble bed through which air is pumped. (Pebble bed heat storage is one of the means of storing energy from the sun for heating purposes.) Show that in the purely one-dimensional case the differential equations for the temperature of the air and of the pebbles as a function of time and of axial position are given by

$$\left(\rho \cdot c_p\right)_a e \frac{\partial T_a}{\partial t} = -\frac{1}{A}\left(c_p I_m\right)_a \frac{\partial T_a}{\partial x} - h^*\left(T_a - T_p\right)$$

$$\left(\rho \cdot c_p\right)_a (1-e) \frac{\partial T_p}{\partial t} = h^*\left(T_a - T_p\right)$$

Here, A is the cross section of the pebble bed, while e denotes the bed void fraction. h^* is the heat transfer coefficient between air and pebbles multiplied by the pebble surface per unit bed volume. The following additional assumptions have been made: no heat loss to the environment and no temperature gradient within the pebbles.

APPENDIX

Tables, Symbols, Glossary, and References

A.1 Tables of Thermodynamic Properties

Table 1	Density and compressibility of water
Table 2	Temperature coefficient of expansion
Table 3	Resistivity and temperature coefficient of resistance
Table 4	Latent heats of fusion
Table 5	Latent heats of vaporization
Table 6	Heat capacities of rigid substances
Table 7	Heat capacity of water
Table 8	Curie constants
Table 9	Thermal conductivities
Table 10	Heat transfer coefficients
Table 11	Emissivities and solar absorptivities
Table 12	Solar radiation spectrum
Table 13	Chemical potentials
Table 14	Chemical potentials of some ions in aqueous solution
Table 15	Chemical potential, entropy, and enthalpy of some substances
Table 16	Properties of saturated water (liquid-vapor)
Table 17	Properties of superheated water vapor

TABLE A.1. **Density and compressibility of water at 101.3 kPa**[a]

$\theta/°C$	$\rho/kg \cdot m^{-3}$	κ_T / GPa^{-1}	$\theta/°C$	$\rho/kg \cdot m^{-3}$	κ_T / GPa^{-1}	$\theta/°C$	$\rho/kg \cdot m^{-3}$	κ_T / GPa^{-1}
0	999.83	0.51	30	995.65	0.448	65	980.56	0.449
3.98	1000	0.495	35	994.03	0.445	70	977.78	0.452
5	999.99	0.492	38	992.21	0.443	75	974.86	0.457
10	999.74	0.478	40	990.21	0.442	80	971.82	0.462
15	999.14	0.467	45	988.03	0.442	85	968.64	0.468
18	998.23	0.458	50	985.69	0.444	90	965.35	0.475
20	997.06	0.452	55	983.2	0.446	95	961.93	0.483
25	999.83	0.51	60	995.65	0.448	100	958.4	0.487

a. Values computed according to the Steam-NBS function implemented in the program EES (Klein et.al, 1990).

TABLE A.2. **Temperature coefficient of expansion (for room temperature, in 10^{-6} K^{-1})**

Solids (linear coefficient α_l)		**Liquids (volume coefficient α_V)**	
Copper	16.8	Alcohol	1100
Glass (pyrex)	3.2	Gasoline	1060
Glass (quartz)	0.45	Glycerine	500
Granite	3....8	Mercury	181
Ice (0°C)	0.502	Sulfuric acid	570
Iron	12.1	Water	207
Lead	29		
Marble	≈ 11		
Sandstone	7....12		
Sodium	71		
Steel	10....16		
Teflon	60....100		
Titanium	9		

TABLE A.3. Resistivity and temperature coefficients of resistance (at 20°C)

Substances	Resistivity	Temperature coefficient	Temperature coefficient
	ρ / 10^{-8} Ωm	α / 10^{-3} K^{-1}	β / 10^{-6} K^{-2}
Aluminum	2.8	3.9	0.6
Carbon	3500	- 0.5	
Copper	1.7	3.9	0.6
Iron	9.8	56	6.0
Platinum	10.5	3.0	0.6
Silver	1.6	3.8	0.7
Tungsten	5.6	4.1	0.96

TABLE A.4. Latent enthalpy and entropy of fusion

Substance	Molar mass	Temperature of fusion	Enthalpy of fusion	Entropy of fusion
	M_o / kg · $mole^{-1}$	T_f / K	$\bar{q}$ / J · $mole^{-1}$	$\bar{l}_f$ / J · K^{-1} · $mole^{-1}$
Aluminum	0.027	887	10720	12.1
Copper	0.064	1356	13120	9.68
Gallium	0.070	303	5656	18.7
Gold	0.197	1336	12943	9.69
Iron	0.056	1808	15512	8.58
Lead	0.207	600	4761	7.94
Mercury	0.201	234	2372	10.1
Platinum	0.195	2042	21645	10.6
Silicon	0.028	1693	4592	2.17
Sodium	0.023	371	2599	7.01
Tungsten	0.184	3653	35300	9.67
Water	0.018	273	6010	22.0

TABLE A.5. Latent enthalpy and entropy of vaporization

Substance	Molar mass	Temperature of vaporization	Enthalpy of vaporization	Entropy of vaporization
	M_o / kg · mole^{-1}	T_v / K	$\bar{r}$ / J · mole^{-1}	$\bar{l}_v$ / J · K^{-1}mole^{-1}
Aluminum	0.027	2720	294000	108
Copper	0.064	2860	306000	107
Gallium	0.070	2500	255000	102
Gold	0.197	2950	325000	110
Iron	0.056	3000	354000	118
Lead	0.207	2020	1780000	880
Mercury	0.201	630	57300	91
Platinum	0.195	4270	448000	105
Silicon	0.028	2900	394000	136
Tungsten	0.184	5770	802000	139
Argon	0.040	87	6520	75
Carbon dioxide	0.044	195	6020	30.9
Chlorine	0.035	239	10200	42.5
Helium	0.004	4	82	20.5
Hydrogen	0.002	20	908	45.4
Krypton	0.084	120	9070	75.6
Neon	0.020	27	1820	67.6
Nitrogen	0.028	77	5540	72.0
Oxygen	0.032	90	6820	75.7
Water	0.018	373	40700	109

TABLE A.6. Entropy capacity and temperature coefficient of energy of some substances at 20°C

Substances	Specific entropy capacity k / J · K^{-2}kg^{-1}	Molar entropy capacity $\bar{k}$ / J · K^{-2}mole^{-1}	Specific temperature coefficient of energy c / J · K^{-1}kg^{-1}
Aluminum	3.06	0.0826	896
Concrete	2.87		840
Copper	1.31	0.0837	383
Glass (quartz)	0.58		170
Granite	2.56		750
Ice (0°C)	7.69	0.123	2100
Iron	1.54	0.0864	452
Lead	0.44	0.0911	129
Lithium	1.16	0.0080	339
Sand (dry)	2.87		840
Silicon	2.40	0.0672	703
Sodium	4.16	0.0958	1220
Steel (average)	1.57		460
Wood (average)	8.53		2500
Mercury	0.47	0.0954	139
Petroleum	7.30		2140
Water	14.26	0.257	4180

TABLE A.7. Entropy capacity and temperature coefficient of enthalpy of water

θ / °C	k_p / J · K^{-2}kg^{-1}	c_p / J · K^{-1}kg^{-1}	θ / °C	k_p / J · K^{-2}kg^{-1}	c_p / J · K^{-1}kg^{-1}
0	15.44	4217.7	60	12.56	4184.4
10	14.81	4192.2	70	12.21	4189.6
20	14.27	4181.9	80	11.88	4196.4
30	13.78	4178.5	90	11.58	4205.1
40	13.34	4178.6	100	11.30	4216.0
50	12.94	4180.7			

TABLE A.8. Curie constant [a]

PARAMAGNETIC SALT	M_o / kg [b]	C^* / 10^{-5} m^{3}K · mole^{-1}
$Cr_2(SO_4)_3 \cdot K_2SO_4 \cdot 24H_2O$	0.499	2.31
$Fe_2(SO_4)_3 \cdot (NH_4)2SO_4 \cdot 24H_2O$	0.482	5.52
$Gd_2(SO_4)_3 \cdot 8H_2O$	0.373	9.80
$2Ce(NO_3)_3 \cdot 3Mg(NO_3)_2 \cdot 24H_2O$	0.765	0 [c]
		0.389[d]

a. Values taken from Zemansky and Dittman (1981), p. 473.
b. Mass of crystal containing NA magnetic ions (NA is Avogadro's constant).
c. Parallel component.
d. Perpendicular component.

TABLE A.9. Thermal conductivity, enthalpy capacity, and density of some materials

Substance	Conditions	Conductivity (entropy) k_s / W · K^{-2}m^{-1}	Conductivity (energy) k_E / W · K^{-1}m^{-1}	Temperature coefficient of enthalpy c_p / J · K^{-1}kg^{-1}	Density ρ / kg · m^{-3}
Gases at atmospheric pressure					
Air	200 K	$9.05 \cdot 10^{-5}$	0.0181	1006.1	1.7684
	300 K	$8.73 \cdot 10^{-5}$	0.0262	1005.7	1.1774
	400 K	$8.41 \cdot 10^{-5}$	0.0337	1014.0	0.8826
Helium	200 K	$5.89 \cdot 10^{-4}$	0.1177	5200	0.2435
CO_2	250 K	$5.16 \cdot 10^{-5}$	0.0129	804	2.1657
	300 K	$5.53 \cdot 10^{-5}$	0.0166	871	1.7973
H_2O vapor	400 K	$6.52 \cdot 10^{-5}$	0.0261	2014	0.5542
	500 K	$6.76 \cdot 10^{-5}$	0.0339	1985	0.4405
	600 K	$7.03 \cdot 10^{-5}$	0.0422	2026	0.3652
Saturated liquids					
mercury	293 K	$2.97 \cdot 10^{-2}$	8.69	139.4	13579

TABLE A.9. Thermal conductivity, enthalpy capacity, and density of some materials

Substance	Conditions	Conductivity (entropy) k_s / W · K^{-2}m^{-1}	Conductivity (energy) k_E / W · K^{-1}m^{-1}	Temperature coefficient of enthalpy c_p / J · K^{-1}kg^{-1}	Density ρ / kg · m^{-3}
Water	273 K	$2.02 \cdot 10^{-3}$	0.552	4217.8	1002.3
	293 K	$2.04 \cdot 10^{-3}$	0.597	4181.8	1000.5
	313 K	$2.01 \cdot 10^{-3}$	0.628	4178.4	994.6
	333 K	$1.96 \cdot 10^{-3}$	0.651	4184.3	985.5
	353 K	$1.89 \cdot 10^{-3}$	0.668	4196.4	974.1
	373 K	$1.82 \cdot 10^{-3}$	0.680	4216.1	960.6
Liquid metals					
Sodium	366 K	0.232	84.96	1384	931.6
Solids at 20°C					
Aluminum		0.70	204	896	2707
Brick (building brick)		0.0024	0.7	840	1600
Bronze (75% Cu, 25% Sn)		0.089	26	343	8666
Clay	30°C	0.0043	1.3	880	1460
Concrete (cinder)		0.0026	0.76		
Copper		1.32	386	383	8954
Fibre (insulating board)		0.00016	0.048		240
Glass (window)		0.0027	0.8	840	2700
Glass wool		0.00013	0.038	700	24
Granite		0.0058-0.014	1.7-4.0	820	2640
Iron		0.25	73	452	7897
Lead		0.12	35	130	11373
Limestone	100-300°C		1.3	900	2500
Paper	30°C	$3.6 \cdot 10^{-5}$	0.011	1340	930
Pyrex	30°C	0.0046	1.4	835	2225
Rubber (vulcanized, hard)	30°C	$4.3 \cdot 10^{-5}$	0.013	1190	
Sand	30°C	$8.9 \cdot 10^{-5}$	0.027	800	1515
Sandstone	40°C	0.0058	1.83	710	2200
Steel	1.0% C	0.15	43	473	7801
	20% Ni	0.065	19	460	7933

TABLE A.9. Thermal conductivity, enthalpy capacity, and density of some materials

Substance	Conditions	Conductivity (entropy) k_s / W · K^{-2}m^{-1}	Conductivity (energy) k_E / W · K^{-1}m^{-1}	Temperature coefficient of enthalpy c_p / J · K^{-1}kg^{-1}	Density ρ / kg · m^{-3}
Tissue	30°C				
	Fat layer	0.00066	0.20		
	Muscle	0.0014	0.41		
Tungsten		0.556	163	134.4	19350
Wood	Balsa	0.00019	0.055		140
	Fir	0.00038	0.11	2720	420
	Oak	0.00057	0.166	2400	540
	White pine	0.00038	0.112		430

TABLE A.10. Heat transfer coefficients with respect to energy [a]

Substance	Transport mode	h / W · K^{-1}m^{-2}
Air	Free convection	6–30
Air in rooms	Inside wall	8
	Window	8
	Floors and ceilings	6–8
Superheated steam or air	Forced convection	30–300
Oil	Forced convection	60–1800
Water	Forced convection	300–6000
	Boiling	3000–60000
Steam	Condensing	6000–120000

a. Order of magnitude.

TABLE A.11. Emissivities and solar absorptivities

Substance		Surface temperature T / K	Normal emissivity	Normal absorptivity for solar radiation, surface at 295 K
Aluminum	Highly polished	480–870	0.04–0.06	0.10
	Polished	373	0.095	0.20
	Heavily oxidized	370–810	0.20–0.33	
Brass	Polished	370	0.09	
	Dull	320–620	0.22	
	Oxidized	480–810	0.60	
Brick	White refractory	1370	0.29	
	Red	310	0.93	0.75
Carbon	Lampsoot	310	0.95	
Clay			0.39	
Concrete	Rough	310	0.94	
	Roofing tile, uncolored			0.73
	Roofing tile, black			0.91
Copper	Highly polished	310	0.02	0.18
	Tarnished			0.64
	Oxidized	480–810	0.78	0.70
Earth	Plowed field			0.75
Gold	Polished	400	0.018	
	Bright foil			0.29
Graphite			0.88	
Grass			0.8	
Gravel			0.29	
Ice	Smooth	273	0.966	
Lead	Polished	310–530	0.06–0.08	
	Oxidized	310	0.63	
Paint	Aluminum	373	0.92–0.98	0.55
	Oil, zinc white			0.30
	Oil, light green			0.50
	Oil, light gray			0.75
	Oil, black on galvanized Iron			0.90
Paper	White	310	0.95	0.28

TABLE A.11. Emissivities and solar absorptivities

Substance		Surface temperature T / K	Normal emissivity	Normal absorptivity for solar radiation, surface at 295 K
Plaster		310	0.91	
Sandstone		310–530	0.83–0.90	
Silver	Polished	310–810	0.01–0.03	0.13
	Commercial sheet			0.30
Snow	Clean	270	0.82	0.2–0.35
Soot, coal			0.95	
Stainless steel	Type 301, polished	297	0.16	0.37
Steel	Polished sheet	273–420	0.08–0.14	
	Sheet, oxidized	295	0.81	
Tungsten	Filament	300	0.032	
	Filament	3590	0.39	
Water	Deep	273–373	0.96	
Wood	Sawdust	310	0.75	
	Oak, planed	295	0.90	

TABLE A.12. Solar radiation WRC spectrum and molecular absorption [a]

λ μm	G_λ [b] W/m²μm	$k_{O\lambda}$ [c]	$k_{g\lambda}$ [d]	$k_{wa\lambda}$ [e]	λ μm	G_λ [b] W/m²μm	$k_{O\lambda}$ [c]	$k_{g\lambda}$ [d]	$k_{wa\lambda}$ [e]
0.250	64.56	38.00	0.00E+00	0.00E+00	0.400	1702.49	0.000	0.00E+00	0.00E+00
0.255	91.25	38.00	0.00E+00	0.00E+00	0.405	1643.75	0.000	0.00E+00	0.00E+00
0.260	122.50	38.00	0.00E+00	0.00E+00	0.410	1710.00	0.000	0.00E+00	0.00E+00
0.265	253.75	38.00	0.00E+00	0.00E+00	0.415	1747.50	0.000	0.00E+00	0.00E+00
0.270	275.00	38.00	0.00E+00	0.00E+00	0.420	1747.50	0.000	0.00E+00	0.00E+00
0.275	212.50	38.00	0.00E+00	0.00E+00	0.425	1629.51	0.000	0.00E+00	0.00E+00
0.280	162.50	38.00	0.00E+00	0.00E+00	0.430	1492.50	0.000	0.00E+00	0.00E+00
0.285	286.25	38.00	0.00E+00	0.00E+00	0.435	1761.25	0.000	0.00E+00	0.00E+00
0.290	535.00	38.00	0.00E+00	0.00E+00	0.440	1755.02	0.000	0.00E+00	0.00E+00
0.295	560.00	20.00	0.00E+00	0.00E+00	0.445	1922.49	0.003	0.00E+00	0.00E+00
0.300	527.50	10.00	0.00E+00	0.00E+00	0.450	2099.99	0.003	0.00E+00	0.00E+00
0.305	557.50	4.800	0.00E+00	0.00E+00	0.455	2017.51	0.004	0.00E+00	0.00E+00
0.310	602.51	2.700	0.00E+00	0.00E+00	0.460	2032.49	0.006	0.00E+00	0.00E+00
0.315	705.00	1.350	0.00E+00	0.00E+00	0.465	2000.00	0.008	0.00E+00	0.00E+00
0.320	747.50	0.800	0.00E+00	0.00E+00	0.470	1979.99	0.009	0.00E+00	0.00E+00
0.325	782.50	0.380	0.00E+00	0.00E+00	0.475	2016.25	0.012	0.00E+00	0.00E+00
0.330	997.50	0.160	0.00E+00	0.00E+00	0.480	2055.00	0.014	0.00E+00	0.00E+00
0.335	906.25	0.075	0.00E+00	0.00E+00	0.485	1901.26	0.017	0.00E+00	0.00E+00
0.340	960.00	0.040	0.00E+00	0.00E+00	0.490	1920.00	0.021	0.00E+00	0.00E+00
0.345	877.50	0.019	0.00E+00	0.00E+00	0.495	1965.00	0.025	0.00E+00	0.00E+00
0.350	955.00	0.007	0.00E+00	0.00E+00	0.500	1862.52	0.030	0.00E+00	0.00E+00
0.355	1044.99	0.000	0.00E+00	0.00E+00	0.505	1943.75	0.035	0.00E+00	0.00E+00
0.360	940.00	0.000	0.00E+00	0.00E+00	0.510	1952.50	0.040	0.00E+00	0.00E+00
0.365	1125.01	0.000	0.00E+00	0.00E+00	0.515	1835.01	0.045	0.00E+00	0.00E+00
0.370	1165.00	0.000	0.00E+00	0.00E+00	0.520	1802.49	0.048	0.00E+00	0.00E+00
0.375	1081.25	0.000	0.00E+00	0.00E+00	0.525	1894.99	0.057	0.00E+00	0.00E+00
0.380	1210.00	0.000	0.00E+00	0.00E+00	0.530	1947.49	0.063	0.00E+00	0.00E+00
0.385	931.25	0.000	0.00E+00	0.00E+00	0.535	1926.24	0.070	0.00E+00	0.00E+00
0.390	1200.00	0.000	0.00E+00	0.00E+00	0.540	1857.50	0.075	0.00E+00	0.00E+00
0.395	1033.74	0.000	0.00E+00	0.00E+00	0.545	1895.01	0.080	0.00E+00	0.00E+00

TABLE A.12. Solar radiation WRC spectrum and molecular absorption [a]

λ μm	G_λ [b] W/m^{2}μm	$k_{O\lambda}$ [c]	$k_{g\lambda}$ [d]	$k_{wa\lambda}$ [e]	λ μm	G_λ [b] W/m^{2}μm	$k_{O\lambda}$ [c]	$k_{g\lambda}$ [d]	$k_{wa\lambda}$ [e]
0.550	1902.50	0.085	0.00E+00	0.00E+00	0.790	1142.50	0.000	0.00E+00	1.75E-02
0.555	1885.00	0.095	0.00E+00	0.00E+00	0.800	1144.70	0.000	0.00E+00	3.60E-02
0.560	1840.02	0.103	0.00E+00	0.00E+00	0.810	1113.00	0.000	0.00E+00	3.30E-01
0.565	1850.00	0.110	0.00E+00	0.00E+00	0.820	1070.00	0.000	0.00E+00	1.53E+00
0.570	1817.50	0.120	0.00E+00	0.00E+00	0.830	1041.00	0.000	0.00E+00	6.60E-01
0.575	1848.76	0.122	0.00E+00	0.00E+00	0.840	1019.99	0.000	0.00E+00	1.55E-01
0.580	1840.00	0.120	0.00E+00	0.00E+00	0.850	994.00	0.000	0.00E+00	3.00E-03
0.585	1817.50	0.118	0.00E+00	0.00E+00	0.860	1002.00	0.000	0.00E+00	1.00E-05
0.590	1742.49	0.115	0.00E+00	0.00E+00	0.870	972.00	0.000	0.00E+00	1.00E-05
0.595	1785.00	0.120	0.00E+00	0.00E+00	0.880	966.00	0.000	0.00E+00	2.60E-03
0.600	1720.00	0.125	0.00E+00	0.00E+00	0.890	945.00	0.000	0.00E+00	6.30E-02
0.605	1751.25	0.130	0.00E+00	0.00E+00	0.900	913.00	0.000	0.00E+00	2.10E+00
0.610	1715.00	0.120	0.00E+00	0.00E+00	0.910	876.00	0.000	0.00E+00	1.60E+00
0.620	1715.00	0.105	0.00E+00	0.00E+00	0.920	841.00	0.000	0.00E+00	1.25E+00
0.630	1637.50	0.090	0.00E+00	0.00E+00	0.930	830.00	0.000	0.00E+00	2.70E+01
0.640	1622.50	0.079	0.00E+00	0.00E+00	0.940	801.00	0.000	0.00E+00	3.80E+01
0.650	1597.50	0.067	0.00E+00	0.00E+00	0.950	778.00	0.000	0.00E+00	4.10E+01
0.660	1555.00	0.057	0.00E+00	0.00E+00	0.960	771.00	0.000	0.00E+00	2.60E+01
0.670	1505.00	0.048	0.00E+00	0.00E+00	0.970	764.00	0.000	0.00E+00	3.10E+00
0.680	1472.50	0.036	0.00E+00	0.00E+00	0.980	769.00	0.000	0.00E+00	1.48E+00
0.690	1415.02	0.028	0.00E+00	1.60E-02	0.990	762.00	0.000	0.00E+00	1.25E-01
0.700	1427.50	0.023	0.00E+00	2.40E-02	1.000	743.99	0.000	0.00E+00	2.50E-03
0.710	1402.50	0.018	0.00E+00	1.25E-02	1.050	665.98	0.000	0.00E+00	1.00E-05
0.720	1355.00	0.014	0.00E+00	1.00E+00	1.100	606.04	0.000	0.00E+00	3.20E+00
0.730	1355.00	0.011	0.00E+00	8.70E-01	1.150	551.04	0.000	0.00E+00	2.30E+01
0.740	1300.00	0.010	0.00E+00	6.10E-02	1.200	497.99	0.000	0.00E+00	1.60E-02
0.750	1272.52	0.009	0.00E+00	1.00E-03	1.250	469.99	0.000	7.30E-03	1.80E-04
0.760	1222.50	0.007	3.00E+00	1.00E-05	1.300	436.99	0.000	4.00E-04	2.90E+00
0.770	1187.50	0.004	2.10E-01	1.00E-05	1.350	389.03	0.000	1.10E-04	2.00E+02
0.780	1195.00	0.000	0.00E+00	6.00E-04	1.400	354.03	0.000	1.00E-05	1.10E+03

TABLE A.12. Solar radiation WRC spectrum and molecular absorption [a]

λ μm	G_λ [b] $W/m^2\mu m$	$k_{O\lambda}$ [c]	$k_{g\lambda}$ [d]	$k_{wa\lambda}$ [e]	λ μm	G_λ [b] $W/m^2\mu m$	$k_{O\lambda}$ [c]	$k_{g\lambda}$ [d]	$k_{wa\lambda}$ [e]
1.450	318.99	0.000	6.40E−02	1.50E+02	3.300	17.25	0.000	1.90E+00	1.20E+02
1.500	296.99	0.000	6.30E−04	1.50E+01	3.400	15.75	0.000	1.30E+00	1.95E+01
1.550	273.99	0.000	1.00E−02	1.70E−03	3.500	14.00	0.000	7.50E−02	3.60E+00
1.600	247.02	0.000	6.40E−02	1.00E−05	3.600	12.75	0.000	1.00E−02	3.10E+00
1.650	234.02	0.000	1.45E−03	1.00E−02	3.700	11.50	0.000	1.95E−03	2.50E+00
1.700	215.00	0.000	1.00E−05	5.10E−01	3.800	10.50	0.000	4.00E−03	1.40E+00
1.750	187.00	0.000	1.00E−05	4.00E+00	3.900	9.50	0.000	2.90E−01	1.70E−01
1.800	170.00	0.000	1.00E−05	1.30E+02	4.000	8.50	0.000	2.50E−02	4.50E−03
1.850	149.01	0.000	1.45E−04	2.20E+03	4.100	7.75	0.000	0.00E+00	0.00E+00
1.900	136.01	0.000	7.10E−03	1.40E+03	4.200	7.00	0.000	0.00E+00	0.00E+00
1.950	126.00	0.000	2.00E+00	1.60E+02	4.300	6.50	0.000	0.00E+00	0.00E+00
2.000	118.50	0.000	3.00E+00	2.90E+00	4.400	6.00	0.000	0.00E+00	0.00E+00
2.100	93.00	0.000	2.40E−01	2.20E−01	4.500	5.50	0.000	0.00E+00	0.00E+00
2.200	74.75	0.000	3.80E−04	3.30E−01	4.600	5.00	0.000	0.00E+00	0.00E+00
2.300	63.25	0.000	1.10E−03	5.90E−01	4.700	4.50	0.000	0.00E+00	0.00E+00
2.400	56.50	0.000	1.70E−04	2.03E+01	4.800	4.00	0.000	0.00E+00	0.00E+00
2.500	48.25	0.000	1.40E−04	3.10E+02	4.900	3.75	0.000	0.00E+00	0.00E+00
2.600	42.00	0.000	6.60E−04	1.50E+04	5.000	3.47	0.000	0.00E+00	0.00E+00
2.700	36.50	0.000	1.00E+02	2.20E+04	6.000	1.75	0.000	0.00E+00	0.00E+00
2.800	32.00	0.000	1.50E+02	8.00E+03	7.000	0.95	0.000	0.00E+00	0.00E+00
2.900	28.00	0.000	1.30E−01	6.50E+02	8.000	0.55	0.000	0.00E+00	0.00E+00
3.000	24.75	0.000	9.50E−03	2.40E+02	9.000	0.35	0.000	0.00E+00	0.00E+00
3.100	21.75	0.000	1.00E−03	2.30E+02	10.00	0.20	0.000	0.00E+00	0.00E+00
3.200	19.75	0.000	8.00E−01	1.00E+02	20.00	0.12	0.000	0.00E+00	0.00E+00

a. Values have been taken from tables in Iqbal (1983).

b. G, normal solar irradiance at distance of Earth.

c. k_O, absorption coefficient due to ozone.

d. k_g, absorption coefficient due to uniformly mixed gases.

e. k_{wa}, absorption coefficient due to water vapor.

TABLE A.13. **Some values of chemical potential and temperature and pressure coefficients**[a]

Substance		State[b]	μ / kG	α_μ / G/K	β_μ / µG/Pa
C	Carbon (graphite)	s	0	- 5.69	5.4
	Diamond	s	2.90	- 2.38	3.42
C_2H_2	Acetylene (ethyne)	g	209.17	- 200.85	24465
CH_4	Methane	g	- 50.81	- 186.10	24465
CO	Carbon monoxide	g	- 137.16	- 197.5	24465
CO_2	Carbon dioxide	g	- 394.40	- 213.68	24465
		aq	- 386.00	- 113.00	
CaC_2	Calcium carbide	s	- 67.78	- 70.29	28.9
$CaCO_3$	Calcium carbonate (calcite)	s	- 1128.78	- 92.88	36.92
CaO	Calcium oxide (lime)	s	- 604.17	- 39.75	16.5
$Ca(OH)_2$	Calcium hydroxide	s	- 896.76	- 76.15	33.2
ClO_2		g	122.31	- 257.12	24465
Fe	Iron	s	0	- 27.3	7.1
Fe_2O_3	Iron oxide	s	- 743.58	- 87.4	30.4
H_2	Hydrogen	g	0	-131	24465
H_2O	Water	g	- 228.60	- 188.72	24465
		l	- 237.18	- 69.91	18.07
		s	- 236.59	- 44.77	19.73
N_2	Nitrogen	g	0	- 191.50	24465
		aq	18.19		
NH_3	Ammonia	g	- 16.38	- 192.59	24465
		aq	- 26.57	- 111.29	24.1
NO_2	Nitrogen dioxide	g	51.24	- 239.91	24465
NaCl	Sodium chloride ("salt")	s	- 384.04	- 72.13	27.02
O_2	Oxygen	g	0	- 205.02	24465
		aq	16.44		
O3	Ozone	g	163.16	- 238.02	24465
PbO_2	Lead dioxide	s	- 212.42	- 76.57	25.5
PbS	Lead sulphide	s	- 98.74	- 91.21	31.9
$PbSO_4$	Lead sulfate	s	- 813.20	- 148.57	48.2
SiO_2	Quartz	s	- 856.48	- 41.46	22.6
ZnS	Zinc sulphide	s	- 201.29	- 57.74	23.89

a. Source: Job in Falk and Herrmann (1981), p. 100–107.

b. g, gaseous; l, liquid; s, solid; aq, dissolved in water.

TABLE A.14. Chemical potentials of some ions in aqueous solution[a]

Ion	μ / kG	U_{el} / V
Ag^+	77.12	0.80
CO_3^{2-}	- 527.90	- 2.74
Ca^{2+}	- 553.04	- 2.87
Cl^-	- 131.26	- 1.36
Cu^{2+}	65.52	0.34
Fe^{2+}	- 78.87	- 0.41
H^+	0	0
Hg^{2+}	164.43	0.85
K^+	- 283.26	- 2.94
Li^+	- 293.80	- 3.05
Mn^{2+}	- 228.03	- 1.18
Na^+	- 261.89	- 2.71
OH^-	- 157.29	- 1.63
Pb^{2+}	- 24.39	- 0.13
SO_4^{2-}	- 744.63	- 3.86
Zn^{2+}	- 147.03	- 0.76

a. At a concentration of 1 mole/l, and standard temperature and pressure. Values have been taken from Job in Falk and Herrmann (1981), p. 100–107.

TABLE A.15. Chemical potential, entropy, and enthalpy of some substances[a]

Substance		State[b]	μ / kG [c]	s / J/K	h / kJ
C	Carbon (graphite)	s	0	5.69	0
CO	Carbon monoxide	g	- 137.15	197.5	- 110.53
CO_2	Carbon dioxide	g	- 394.38	213.7	- 393.52
CH_4	Methane	g	- 50.79	186.2	- 74.85
C_3H_8	Propane	g	- 23.49	269.9	- 103.85
C_4H_{10}	Butane	g	- 15.71	310.0	- 126.15
C_2H_5OH	Ethyl alcohol	l	- 174.89	160.7	- 277.69
Fe	Iron	s	0	27.3	0
Fe_2O_3	Iron oxide	s	- 743.58	87.4	- 825.5
H_2	Hydrogen	g	0	130.6	0
H_2O	Water	g	- 228.60	188.7	- 241.8
		l	- 237.18	69.9	- 285.9
N_2	Nitrogen	s	0	191.5	0
NO_2	Nitrogen dioxide	g	51.24	239.91	33.1
O_2	Oxygen	g	0	205.0	0

a. Values have been taken from Job in Falk and Herrmann (1981), p. 100–107.

b. g, gaseous; l, liquid; s: solid.

c. At standard condition of 298 K and 101.3 kPa.

TABLE A.16. Properties of saturated water (liquid–vapor)[a]

T °C	P_v bar	$\mu_l = \mu_g$ kG	s_l kJ/K · kg	s_g kJ/K · kg	v_l*1000 m^3/kg	v_g m^3/kg	e_l kJ/kg	e_g kJ/kg
0.01	0.00612	0.000	0.0000	9.1541	1.0002	205.987	0.00	2374.5
5	0.00873	-0.003	0.0763	9.0236	1.0001	147.0239	21.02	2381.4
10	0.01228	-0.014	0.1510	8.8986	1.0003	106.3229	41.99	2388.3
15	0.01706	-0.031	0.2242	8.7792	1.0009	77.8971	62.92	2395.2
20	0.02339	-0.054	0.2962	8.6651	1.0018	57.7777	83.83	2402.0
25	0.03169	-0.084	0.3670	8.5558	1.0030	43.3566	104.75	2408.9
30	0.04246	-0.120	0.4365	8.4513	1.0044	32.8955	125.67	2415.7
35	0.05627	-0.162	0.5050	8.3511	1.0060	25.2204	146.58	2422.5
40	0.07381	-0.211	0.5723	8.2550	1.0079	19.5283	167.50	2429.2
45	0.09590	-0.265	0.6385	8.1629	1.0099	15.2634	188.41	2435.9
50	0.12344	-0.326	0.7037	8.0745	1.0122	12.0367	209.31	2442.6
60	0.19932	-0.464	0.8312	7.9080	1.0171	7.6743	251.13	2455.8
80	0.47373	-0.807	1.0753	7.6112	1.0290	3.4088	334.88	2481.6
100	1.01322	-1.235	1.3069	7.3545	1.0434	1.6736	418.96	2506.1
120	1.98483	-1.743	1.5278	7.1297	1.0603	0.8922	503.57	2529.1
140	3.61195	-2.329	1.7394	6.9302	1.0797	0.5090	588.85	2550.0
160	6.17663	-2.987	1.9429	6.7503	1.1019	0.3071	674.97	2568.3
180	10.01927	-3.714	2.1397	6.5853	1.1273	0.1940	762.12	2583.4
200	15.53650	-4.508	2.3308	6.4312	1.1564	0.1273	850.58	2594.7
220	23.17846	-5.364	2.5175	6.2847	1.1900	0.0862	940.75	2601.6
240	33.44673	-6.281	2.7013	6.1423	1.2292	0.0597	1033.12	2603.1
260	46.89449	-7.256	2.8838	6.0009	1.2758	0.0422	1128.4	2598.4
280	64.13154	-8.287	3.0669	5.8565	1.3324	0.0302	1227.53	2585.7
300	85.83784	-9.371	3.2534	5.7042	1.4037	0.0217	1332.00	2562.8
320	112.79318	-10.506	3.4476	5.5356	1.4984	0.0155	1444.35	2525.2
340	145.94085	-11.691	3.6587	5.3345	1.6373	0.0108	1569.93	2463.9
360	186.55306	-12.924	3.9153	5.0542	1.8936	0.0070	1725.64	2352.2
370	210.29877	-13.558	4.1094	4.8098	2.2068	0.0050	1843.33	2235.2
373	217.98862	-13.751	4.2259	4.6537	2.4852	0.0041	1912.45	2153.2

a. Values computed according to the Steam-NBS function implemented in the program EES (Klein et. al, 1990).

TABLE A.17. Properties of superheated water vapor[a]

	0.30 bar			1.0 bar			3.0 bar		
T / °C	s	v	e	s	v	e	s	v	e
80	7.8282	5.4015	2483.8						
120	8.0358	6.0273	2542.4	7.4665	1.7931	2537.0			
160	8.2229	6.6485	2601.0	7.6591	1.9838	2597.5	7.1274	0.6506	2586.9
200	8.3944	7.2676	2660.1	7.8335	2.1723	2657.6	7.3108	0.7163	2650.2
240	8.5535	7.8855	2719.9	7.9942	2.3594	2718.1	7.4765	0.7804	2712.6
280	8.7021	8.5026	2780.7	8.1438	2.5458	2779.2	7.6292	0.8438	2775.0
320	8.842	9.1193	2842.3	8.2844	2.7317	2841.1	7.7716	0.9067	2837.8
360	8.9743	9.7356	2904.9	8.4171	2.9173	2904.0	7.9057	0.9692	2901.2
400	9.1001	10.3518	2968.5	8.5432	3.1027	2967.7	8.0327	1.0315	2965.4
440	9.2201	10.9677	3033.2	8.6634	3.2879	3032.5	8.1536	1.0937	3030.5
480	9.3349	11.5836	3098.9	8.7785	3.4730	3098.3	8.2692	1.1557	3096.6
520	9.4452	12.1994	3165.7	8.8889	3.6581	3165.2	8.3800	1.2177	3163.7

a. Units as in Table 16.

	10 bar			30 bar			100 bar		
T / °C	s	v	e	s	v	e	s	v	e
200	6.6932	0.2059	2621.5						
240	6.8805	0.2274	2692.2	6.2251	0.0682	2618.9			
280	7.0454	0.2479	2759.6	6.4445	0.0771	2709.0			
320	7.1954	0.2678	2825.6	6.6232	0.0850	2787.6	5.7093	0.0192	2588.2
360	7.3344	0.2873	2891.3	6.7794	0.0923	2861.3	6.0043	0.0233	2728.0
400	7.4648	0.3066	2957.2	6.9210	0.0994	2932.7	6.2114	0.0264	2832.0
440	7.5882	0.3257	3023.6	7.0521	0.1062	3003.0	6.3807	0.0291	2922.3
480	7.7055	0.3447	3090.6	7.1750	0.1129	3073.0	6.5287	0.0316	3005.8
520	7.8177	0.3635	3158.5	7.2913	0.1195	3143.2	6.6625	0.0339	3085.9
560	7.9254	0.3823	3227.2	7.4022	0.1260	3213.8	6.7862	0.0362	3164.0
600	8.0292	0.4011	3297.0	7.5084	0.1324	3285.0	6.9022	0.0384	3241.1
640	8.1293	0.4198	3367.7	7.6105	0.1388	3357.0	7.0119	0.0405	3317.9

A.2 Symbols used in the text

The following tables list the most important symbols used in the text, along with their meanings and their units. The last column of each table identifies the chapter in which the symbol was introduced.

Table S.1 Symbols using latin letters
Table S.2 Symbols using greek letters
Table S.3 Subscripts and superscripts
Table S.4 Operators

TABLE S.1: Symbols using latin letters

Symbol	Meaning	SI-units	Chapter [a]
a	Steffan-Boltzmann constant	$J \cdot m^{-3}K^{-4}$	2
A	Surface area, cross section	m^2	Prol
$\mathcal{A}$	Rate of absorption of radiant energy per unit area	$W \cdot m^{-2}$	3
a	Absorptivity		3
B	Magnetic flux density	T	Prol
c	Speed of sound	$m \cdot s^{-1}$	2
c	Specific temperature coefficient of energy	$J \cdot K^{-1}kg^{-1}$	2
$\bar{c}_P$	Molar temperature coefficient of enthalpy	$J \cdot K^{-1}mole^{-1}$	2
$\bar{c}_V$	Molar temperature coefficient of energy	$J \cdot K^{-1}mole^{-1}$	2
$\bar{c}$	Concentration	$mole \cdot m^{-3}$	4
C	Coulomb (unit of electrical charge)		Prol
C	Capacity (electrical)	F	Prol
C	Temperature coefficient of energy	$J \cdot K^{-1}$	2
C	Concentration ratio		3
COP	Coefficient of performance		1
$\mathscr{C}$	Cycle (closed path)		Inter
C_f	Friction coefficient		Ep
C_V	Temperature coefficient of energy	$J \cdot K^{-1}$	2
C_p	Temperature coefficient of enthalpy	$J \cdot K^{-1}$	2
C^*	Momentum capacitance per length	$kg \cdot m^{-1}$	Prol
C^*	Ratio of capacitance rates		4
Ct	Carnot (unit of heat—entropy)		1
°C	Degrees Celsius		1

TABLE S.1: Symbols using latin letters

Symbol	Meaning	SI-units	Chapter [a]
d_{ij}	Components of the velocity gradient tensor (symmetric part)	s^{-1}	Ep
D	Diffusion constant	m^2s^{-1}	4
$\mathcal{D}$	Rate of dissipation of energy	W	1
e	Specific energy (energy of system divided by mass)	$J \cdot kg^{-1}$	4
$\bar{e}$	Molar energy	$J \cdot mole^{-1}$	4
E	Energy of system (energy content)	J	1
E	Young's modulus	$N \cdot m^{-2}$	1
$\mathcal{E}$	Energy function $E(S, V)$	J	Inter
$\mathbf{E}$	Electric flux density	$V \cdot m^{-1}$	Prol
E_{kin}	Kinetic energy of a system	J	Prol
$\mathcal{E}$	Emissive power (rate of radiant energy emitted per unit area)	$W \cdot m^{-2}$	3
$\mathcal{E}_b$	Emissive power of black body	$W \cdot m^{-2}$	3
f	Degrees of freedom		2
f_Q	Specific source rate of quantity Q		Ep
F	Carnot's function (special axiom)		Inter
F	Free energy	J	Inter
$\mathbf{F}$	Force (flux of momentum)	N	Prol
F_{12}	Radiation shape factor		3
F'	Solar collector efficiency factor		4
F_R	Solar collector heat removal factor		4
$\mathcal{F}$	Faraday's constant	$C \cdot mole^{-1}$	4
g	Gravitational field strength	$N \cdot kg^{-1}$	Prol
G	Gravitational constant	$N \cdot m^2kg^{-2}$	2
G	Carnot's function (general axiom)		Inter
G	Gibbs free energy	J	Ep
$\mathcal{G}$	Irradiance (rate of incident radiant energy per unit area)	$W \cdot m^{-2}$	3
$\mathcal{G}_{sc}$	Solar constant	$W \cdot m^{-2}$	3
h	Height	m	Prol
h	Specific enthalpy	$J \cdot kg$	4
h	Planck's constant	$J \cdot s$	3

TABLE S.1: Symbols using latin letters

Symbol	Meaning	SI-units	Chapter [a]
h	Overall heat transfer coefficient with respect to energy	$W \cdot K^{-1}m^{-2}$	1
h	Convective heat transfer coefficient with respect to energy	$W \cdot K^{-1}m^{-2}$	3
h_a	Average convective heat transfer coefficient	$W \cdot K^{-1}m^{-2}$	3
$\bar{h}$	Molar enthalpy	$J \cdot mole^{-1}$	4
H	Enthalpy	J	2
$\mathcal{H}$	Enthalpy function $H(S,P)$	J	Inter
$\mathbf{H}$	Magnetic field strength	$A \cdot m^{-1}$	Prol
i_s	Entropy intensity of radiation	$W \cdot K^{-1}m^{-2}sr^{-1}$	3
$i_{s\nu}$	Spectral entropy intensity (with respect to frequency)	$W \cdot K^{-1}m^{-2}s \cdot sr^{-1}$	3
$i_{s\lambda}$	Spectral entropy intensity (with respect to wavelength)	$W \cdot K^{-1}m^{-2}m^{-1}sr^{-1}$	3
i_E	Energy intensity of radiation	$W \cdot m^{-2}sr^{-1}$	3
$i_{E\nu}$	Spectral energy intensity (with respect to frequency)	$W \cdot m^{-2}s \cdot sr^{-1}$	3
$i_{E\lambda}$	Spectral energy intensity (with respect to wavelength)	$W \cdot m^{-2}m^{-1}sr^{-1}$	3
I	Current; flux		Prol
I_E	Flux of energy	W	Prol
I_L	Flux of angular momentum	$kg \cdot m^2s^{-2}$	Prol
I_m	Flux of gravitational mass	$kg \cdot s^{-1}s$	Prol
I_{mag}	Hertz magnetic current	A	Prol
I_n	Flux of amount of substance	$mole \cdot s^{-1}$	Prol
I_p	Flux of momentum	N	Prol
I_q	Flux of electrical charge	A	Prol
I_s	Flux of entropy	$W \cdot K^{-1}$	1
I_V	Volume flux	m^3s^{-1}	Prol
J	Joule (unit of energy)		Prol
j	Flux density		3
$\mathbf{j}$	Flux density vector		3
j_E	Energy flux density	$W \cdot m^{-2}$	3
j_p	Momentum flux density	$N \cdot m^{-2}$	Prol
j_s	Entropy flux density	$W \cdot K^{-1}m^{-2}$	3
$\mathcal{J}_p$	Momentum current density tensor	$N \cdot m^{-2}$	Ep
k	Specific entropy capacity	$J \cdot K^{-2}kg^{-1}$	2
k	Boltzmann's constant	$J \cdot K^{-1}$	3

TABLE S.1: Symbols using latin letters

Symbol	Meaning	SI-units	Chapter [a]
k_E	Thermal conductivity with respect to energy	$W \cdot K^{-1}m^{-1}$	3
$\bar{k}$	Molar entropy capacity	$J \cdot K^{-2}mole^{-1}$	2
k_S	Thermal conductivity with respect to entropy	$W \cdot K^{-2}m^{-1}$	3
K	Kelvin (unit of temperature)		1
K	Hydraulic capacity	m^3Pa^{-1}	Prol
K	Entropy capacity	$J \cdot K^{-2}$	2
K_V	Entropy capacity at constant volume	$J \cdot K^{-2}$	2
K_p	Entropy capacity at constant pressure	$J \cdot K^{-2}$	2
$\mathcal{K}_p$	Chemical equilibrium constant		4
l	Length	m	1
l_f, l_v	Specific entropy of fusion (vaporization)	$J \cdot K^{-1}kg^{-1}$	1
$\bar{l}_f, \bar{l}_v$	Molar latent entropy of fusion (vaporization)	$J \cdot K^{-1}mole^{-1}$	1
L	Electrical inductance	H	Prol
L	Luminosity of star	W	3
L	Thickness of cover	m	3
L_V	Hydraulic inductance	$Pa \cdot s^2m^{-3}$	Prol
L_V	Product of latent entropy and absolute temperature	$J \cdot m^{-3}$	2
L^*	Momentum inductance per length	N^{-1}	Prol
$\mathcal{L}$	Loss of available power	W	1
m	Meter (unit of length)		Prol
m	Mass of a body	kg	Prol
m_a	Air mass		3
mole	Unit of amount of substance		Prol
M	Mass	kg	4
M_o	Molar mass	$kg \cdot mole^{-1}$	Prol
n	Amount of substance	mole	Prol
$\mathbf{n}$	Unit normal vector on surface (directed outward)		3
N_A	Avogadro's constant		4
N	Newton (unit of momentum flux—force)		Prol
Nu	Nusselt number		Ep
NTU	Number of transfer units		4
$\mathbf{p}$	Momentum	$N \cdot s$	Prol

TABLE S.1: **Symbols using latin letters**

Symbol	Meaning	SI-units	Chapter [a]
p	Pressure function $P(V,T)$	Pa	Inter
p^*	Pressure function $P(\rho,T)$	Pa	Inter
P	Pressure	Pa	Prol
$\mathcal{P}$	Power	W	Prol
$\mathcal{P}_{av}$	Available power	W	1
$\mathscr{P}$	Path		Inter
Pa	Pascal (unit of pressure)		2
Pr	Prandtl number		Ep
q	Electrical charge	C	Prol
q	Specific enthalpy of fusion	$J \cdot kg^{-1}$	1
q_n	Molar enthalpy of fusion	$J \cdot mole^{-1}$	1
Q	Energy exchanged in heating or cooling	J	1
$\mathcal{Q}$	Substancelike quantity		Ep
q	Specific quantity		Ep
r	Radial variable	m	3
r	Specific enthalpy of vaporization	$J \cdot kg^{-1}$	1
r	Specific rate of absorption of energy	$W \cdot kg^{-1}$	Ep
r_n	Molar enthalpy of vaporization	$J \cdot mole^{-1}$	1
R	Universal gas constant, 8.31 J/(K · mole)	$J \cdot K^{-1} mole^{-1}$	2
R	Electrical resistance	Ω	Prol
R	Radius	m	3
Re	Reynolds number		Ep
R_E	Thermal resistance with respect to energy	$K \cdot W^{-1}$	3
R_s	Thermal resistance	W^{-1}	3
R_m	Specific gas constant	$J \cdot K^{-1} kg^{-1}$	4
R_V	Hydraulic resistance	$Pa \cdot s \cdot m^{-3}$	Prol
s	Second (unit of time)		Prol
s	Entropy per mass (specific entropy)	$J \cdot K^{-1} kg^{-1}$	3
$\bar{s}$	Molar entropy	$J \cdot K^{-1} mole^{-1}$	4
S	Entropy; entropy content of a body	$J \cdot K^{-1}$	1
$\mathscr{S}$	Entropy function $S(V,T)$	$J \cdot K^{-1}$	Inter
$\mathscr{S}^*$	Entropy function $S(P,T)$	$J \cdot K^{-1}$	Inter

TABLE S.1: Symbols using latin letters

Symbol	Meaning	SI-units	Chapter [a]
$\mathcal{S}^{\#}$	Entropy function $S(P,\rho)$	$J \cdot K^{-1}$	Inter
S_e	Entropy exchanged in a process	$J \cdot K^{-1}$	1
S_{gen}	Amount of entropy produced in a system	$J \cdot K^{-1}$	1
t	Time	s	Prol
T	Ideal gas temperature, absolute temperature	K	1
$\mathcal{T}$	Conductive part of momentum current density tensor (stress tensor)	$N \cdot m^{-2}$	Ep
u	Specific internal energy	$J \cdot kg^{-1}$	4
U	Voltage	V	Prol
U	Internal energy	J	3
U	Overall heat transfer coefficient	$W \cdot K^{-1}m^{-2}$	4
U_{mag}	Magnetic tension	V	Prol
u	Specific energy, internal energy per mass	$J \cdot kg^{-1}$	3
v	Velocity	$m \cdot s^{-1}$	Prol
ν	Stoichiometric coefficient		4
V	Volume of a body	m^3	Prol
$\mathcal{V}$	Volume function $V(P,T)$	m^3	Inter
$\mathcal{V}^*$	Volume function $V(P,S)$	m^3	Inter
w	Amount of precipitable water		3
W	Watt (unit of energy flux or power)		Prol
W	Energy exchanged in mechanical process	J	1
W	Width	m	4
W_{el}	Energy exchanged in electrical process	J	1
x	Position variable	m	Prol
x	Quality		4
y	Mole fraction		4
z	Vertical distance	m	2
z	Ionization number of atom		4

a. Prol, Prologue; Inter, Interlude; Ep, Epilogue

TABLE S.2: **Symbols using Greek letters**

Symbol	Meaning	SI-units	Chapter [a]
α	Absorptivity (absorptance)	m^2s^{-1}	3
α	Thermal diffusivity		Ep
α_l	Linear temperature coefficient of expansion	K^{-1}	1
α_R	Linear temperature coefficient of electrical resistance	K^{-1}	1
α_V	Temperature coefficient of expansion of volume	K^{-1}	1
α_μ	Temperature coefficient of chemical potential	mole · K^{-1}	4
β	Temperature coefficient of pressure	K^{-1}	1
β_R	Quadratic temperature coefficient of electrical resistance	K^{-2}	1
β_E	Scattering coefficient	m^{-1}	3
β_μ	Pressure coefficient of chemical potential	mole · Pa^{-1}	4
γ	Adiabatic exponent, ratio of entropy capacities		2
γ'	Polytropic exponent		2
γ	Volume coefficient of thermal expansion	K^{-1}	4
δ	Kronecker symbol		Ep
ε	Energy function $E(V,T)$	J	Inter
ε	Heat exchanger effectiveness		4
η	Efficiency		1
η_c	Carnot efficiency		1
η_1	Thermal efficiency, first law efficiency		1
η_2	Second law efficiency		1
θ	Celsius temperature	°C	1
κ	Bulk viscosity	Pa · s	Ep
κ_E	Absorption coefficient	m^{-1}	3
κ_s	Adiabatic compressibility	Pa^{-1}	Inter
κ_T	Isothermal compressibility	Pa^{-1}	Inter
λ	Wavelength	m	3
λ	Second viscosity coefficient	Pa · s	Ep
λ^x	Lagrange multiplier for quantity x		Ep
Λ_V	Latent entropy with respect to volume	J · $K^{-1}m^{-3}$	2
Λ_p	Latent entropy with respect to pressure	J · $K^{-1}Pa^{-3}$	2
μ	Viscosity	Pa · s	Prol
μ	Chemical potential	G = J · $mole^{-1}$	4

TABLE S.2: Symbols using Greek letters

Symbol	Meaning	SI-units	Chapter [a]
μ	Attenuation coefficient	m^{-1}	3
μ_o	Permeability constant	$H \cdot m^{-1}$	Prol
ν	Frequency	s^{-1}	3
ν	Kinematic viscosity	$m^2 s^{-1}$	Ep
π	Volume density of rate of production		3
π_s	Volume density of rate of production of entropy	$W \cdot K^{-1} m^{-3}$	3
Π	Rate of production		1
Π_s	Rate of production of entropy	$W \cdot K^{-1}$	1
ρ	Density (general)		3
ρ	Mass density of a body	$kg \cdot m^{-3}$	1
ρ	Reflectivity (reflectance)		
ρ_E	Energy density	$J \cdot m^{-3}$	3
ρ_s	Density of entropy of body	$J \cdot K^{-1} m^{-3}$	3
σ	Steffan-Boltzmann constant	$W \cdot m^{-2} K^{-4}$	3
σ	Volume density of source rate		3
σ_E	Volume density of source rate of energy	$W \cdot m^{-3}$	3
σ_n	Diffusivity of amount of substance	$mole^2 J^{-1} s^{-1} m^{-1}$	3
σ_s	Volume density of source rate of entropy	$W \cdot K^{-1} m^{-3}$	3
Σ	Source rate		Prol
Σ_E	Source rate of energy	W	3
Σ_s	Source rate of entropy	$W \cdot K^{-1}$	3
τ	Time constant	s	Prol
τ	Transmittance		3
τ	Quantum of amount of substance	mole	4
τ	Relaxation time	s	Ep
$(\tau\alpha)$	Transmission-absorption product		3
υ	Specific volume (inverse density)	$m^3 kg^{-1}$	4
$\overline{\upsilon}$	Molar volume	$m^3 mole^{-1}$	4
φ	Potential		Prol
ϕ	Relative humidity		4
Ω	Ohm (unit of electrical resistance)		Prol

TABLE S.2: **Symbols using Greek letters**

Symbol	Meaning	SI-units	Chapter [a]
ω	Angular velocity	s^{-1}	Prol
ω	Humidity ratio		4

a. Prol, Prologue; Inter, Interlude; Ep, Epilogue

TABLE S.3: **Subscripts and superscripts**

Symbol	Meaning
a	Air
a	Ambient
a	average
ad	Adiabatic
av	Available
b	Body
b	Beam
b	Blackbody
c	Carnot
(c)	Conductive part of a flux
chem	Chemical
cond	Conductive
conv	Convective
d, diff	Diffuse
e	Exchanged
el	Electrical
E	Energy, with respect to energy
E	Equilibrium
f	Final
f	Fluid
f	Fusion
f	Formation
g	Gas, gaseous, vapor

TABLE S.3: Subscripts and superscripts

Symbol	Meaning
gen	Generated
grav	Gravitation
h	Horizontal
hp	Heat pump
hydro	Hydraulic
i	Initial
in	In, flowing inward
kin	Kinetic
l	Liquid
l	Linear
L	Angular momentum
m	Mass
m	Mean
mag	Magnetic
max	Maximum
mech	Mechanical
min	Minimum
net	Sum, total (net current)
o	environment, ambient
o	Reference point
out	Out, flowing outward
p	Momentum
p	Absorber plate
p	Pressure, with respect to pressure, at constant pressure
q	Charge
Q	With respect to quantity *Q*
r	Radiation
rad	Radiative
refr	Refrigerator
R	Electrical resistance
s	Entropy, with respect to entropy
s	Surface

TABLE S.3: Subscripts and superscripts

Symbol	Meaning
s	Sun, solar
s	Solid
s	Solute
t	Top
th	Thermal
v	Vaporization, vapor
V	Volume, with respect to volume, at constant volume
wb	Wet bulb
x,y,z	Spatial coordinates, with respect to spatial coordinate
λ	With respect to wavelength
ν	With respect to frequency
$+$	Upper, absorbed
$-$	Lower, emitted
$*$	Per length

TABLE S.4: Operators

Symbol	Meaning
$\Delta(\ldots)$	Difference
$\frac{D}{Dt}$	Material derivative
$\int_{\mathscr{P}} \ldots d \ldots$	Path integral
$\int_{\mathscr{C}} \ldots d \ldots$	Integral along closed path (cycle)
$\int_{\mathcal{A}} \ldots dA$	Integral over surface
∇	Del operator
$,x$	Partial derivative with respect to x
$\int_{\mathcal{V}} \ldots dV$	Integral over volume
$\int_{2\pi} \ldots d\Omega$	Integral over hemisphere
$\int_{4\pi} \ldots d\Omega$	Integral over sphere

A.3 Glossary

The following short glossary is provided since the generalized version of thermodynamics presented in this book requires a generalization of and sometimes a change from usual terminology. Only the most important terms are included. Expressions in italics can be found elsewhere in the glossary.

Amount of substance Formal measure of an amount of substance as used in the sense of chemistry (the "number of moles", the "number of particles").

Availability The amount of energy which can be released (see *release of energy*) in the fall of *entropy* from points of high to points of low *temperature*.

Binding of energy Binding energy to the current of a substancelike quantity which thereby is lifted from a lower to a higher *potential*. Opposite of *release of energy*.

Caloric Used as an alternative term for *heat*. The caloric theory of heat can be rendered formal and correct in a modern sense if it is accepted that caloric is not conserved (that it can be produced). In this case it turns out to be equivalent to the *entropy* of a body.

Chemical driving force The difference of the *chemical potential*.

Chemical potential The *potential* associated with processes which have to do with the change or the flow of *amount of substance*.

Continuous processes Processes which are spatially continuous, i.e. processes in which the variables change from point to point inside a body or a system.

Constitutive relations The laws which are not generic (i.e. the *laws of balance*) but differentiate between bodies and circumstances.

Current Informal term for the phenomenon of the transport of a *substancelike quantity*. Also used colloquially for the formal measure which is called *flux*.

Current density Formal measure of the local condition of a current. The *flux* is the surface integral of the current density. For a scalar *substancelike quantity*, the current density is a vector.

Density Spatial density of a *substancelike quantity*. The integral of the density of such a quantity over the volume of a system delivers the amount of the substancelike quantity stored in the system.

Dissipation rate Rate at which energy is bound (see *binding of energy*) as the result of the *production* of *entropy*.

Dissipative process A process during which *entropy* has been produced, i.e. an *irreversible process*.

Driving force Informal term for the difference of a *potential*. The thermal driving force is the difference of the thermal potentials at two points in space, i.e. the difference of *temperatures*.

Dynamics A theory of dynamics requires the formulation of the *laws of balance* and the *constitutive relations* appropriate for a particular case. Models of dynamical processes rely upon the clear distinction between laws of balance and the constitutive relations.

Energy current The amount of energy crossing the surface of a system in unit time as the result of a transport process. It must be distinguished from *power.*

Entropy Formal for a quantity of *heat* or *caloric*. Entropy is the *substancelike quantity* of thermal processes and thus obeys a *law of balance*. It can be stored (see *heat function*), it can flow (*entropy current*), and it can be created (see *production*).

Entropy current Measure of the transfer of *entropy* across the surface a a system.

Entropy production The process of the *production* of *entropy* as the result of an *irreversible process.*

Exchanged quantity The amount of a *substancelike quantity* which has crossed the surface of a system together with a *current* in a certain interval of time. Formally equal to the integral of the *flux* over time.

Extensive quantities Quantities which scale with the size of a system are said to be extensive. The *substancelike quantities* are a subset of the extensive quantities. An example of a nonsubstancelike extensive quantity is provided by the volume.

First law of thermodynamics The *law of balance* of energy. It includes only rates of change of the energy content, energy *currents*, and energy *source rates.*

Flux Formal measure of the amount of a *substancelike quantity* crossing the surface of a system in unit time (informally, the same quantity is called a *current*). The flux is counted as positive for a current flowing out of the system.

Flux density The surface density of a *flux*. The surface integral of a flux density delivers the flux. Equivalent to *current density.*

Heat Informal term for *entropy*. Equivalent to *caloric*. (Commonly the energy *exchanged* in *heating* is called heat; this usage is not followed in this text.)

Heating The process of the transfer of *heat* (*entropy*) across the surface of a body excluding convective transports. The opposite process is cooling.

Heat capacity Used in the sense of *entropy* capacity, i.e. as the derivative of the entropy function with respect to temperature. The usual "heat capacities" are called the *temperature coefficients of energy* and *of enthalpy.*

Heat function The formal expression of the assumption that a body contains a certain amount of *heat*, where the heat stored is a function of the independent variables describing the properties of the body. This heat function turns out to be equivalent to the *entropy* of the body.

Hotness The hotness manifold is the primitive concept for describing the ordering of bodies according to the sensation of how hot they are. The numerical measure of the hotness is the *temperature*.

Intensive quantities The quantities which remain the same if a body is divided into parts. A subset of the intensive quantities are the *potentials*.

Irreversible process A process which leads to the *production* of *entropy*.

Irreversibility Opposite of *reversibility*. The condition of irreversibility means that *entropy* is produced during a process.

Laws of balance The formal relation which holds for the rate of change of the *substancelike quantity* of a body and its currents (and possibly its *source rates* and *production rates*).

Level Informal term for *potential*. Levels are the conjugate quantities (conjugate with respect to energy) of the *substancelike quantities*.

Minimization of entropy production Minimizing *irreversibility* is achieved by minimizing the rate of *production* of *entropy*.

Potential Formal term for the quantities which take the role of physical *levels*, otherwise known as the *intensive quantities*. There is a potential associated with each of the *substancelike quantities*. The classical potentials are velocity (for momentum), angular velocity (for angular momentum), temperature (for *entropy*), the electrical potential (for charge), the *chemical potential* (for *amount of substance*), and the gravitational potential (for gravitational mass).

Power The rate of *release of energy* or the rate of *binding of energy*. Power is associated with an internal process as opposed to an external process (i.e. a transport process which is quantified by *energy currents*).

Power of heat Colloquial for the *power* associated with the fall of *entropy* from points of higher to points of lower temperature. This is Carnot's *puissance du feu*. Integrating the power of heat over time delivers the *availability*.

Production Informal term for the phenomenon of production of a *substancelike quantity*. A quantity which is produced can accumulate inside a system even without being transported into the system. Production (or destruction) is associated with nonconservation of a quantity.

Production rate Formal measure of the production of a *substancelike quantity*. It describes the amount of the quantity produced inside a system per unit time. A negative production rate means the quantity is destroyed.

Production (rate) density The spacial density of the *production rate*. Its volume integral delivers the production rate.

Release of energy Release of energy when the current of a *substancelike quantity* goes from higher to lower *potential*. Opposite of *binding of energy*.

Reversibility The condition of reversibility means that there is no *entropy production* during a process.

Second law of thermodynamics The *law of balance* of *entropy.*

Source Informal for processes as by which a *substancelike quantity* is transferred into a system without having to cross the surface of the systems. This happens as the result of the interaction of bodies and fields.

Source rate Formal measure of a *source* of a *substancelike quantity.* It determines the amount of the quantity delivered to the system per unit time.

Source (rate) density The spatial density of the *source rate*. Its volume integral delivers the source rate.

Substancelike quantities Physical quantities which possess a *density* and a *current density* (and possibly *source densities* and *production densities*) are called substancelike. *Laws of balance* can be written for them. They form a subset of the *extensive quantities*. The classical substancelike quantities are momentum, angular momentum, *entropy*, charge, amount of substance, and (gravitational) mass.

Superconducting process A transport process of a *substancelike quantity* which does not require a *driving force*.

Temperature Measure of the *hotness* of a body. *Temperature* is like the coordinate on the hotness manifold. Temperature serves the role of the thermal *potential.*

Temperature coefficient of energy The derivative of the energy with respect to temperature at constant volume. Normally called *heat capacity at constant volume*.

Temperature coefficient of enthalpy The derivative of the enthalpy with respect to temperature at constant pressure. Normally called *heat capacity at constant pressure*.

Thermostatics Theories of thermostatics try to derive the conditions pertaining only to static thermal situations. Usually, these conditions are derived by maximizing or minimizing functions such as the *entropy* or the energy of a system.

Uniform processes Spatially uniform processes, i.e. processes in which variables of a system have the same value at every point at a given moment.

Using energy Same as *binding of energy.*

A.4 References

D.L. Anderson (1989): *Theory of the Earth.* Blackwell Scientific Publications, Boston MA.

B. Andresen, P. Salamon, and R.S. Berry (1984): Thermodynamics in finite time. *Phys. Today* **37**, September 1984, pp. 62–70.

A. Bejan (1988): *Advanced Engineering Thermodynamics.* John Wiley & Sons, New York.

Bennet and Myers (1965): *Momentum, Heat, and Mass Transfer.* Chemical Engineering Series, McGraw-Hill, New York.

R.B. Bird, W.E. Stewart, and E.N. Lightfoot (1960): *Tansport Phenomena.* John Wiley & Sons, New York.

C.F. Bohren (1995): Rain, Snow, and Spring Runoff Revisited. *The Physics Teacher* **33**, 79–81.

R.S. Brodkey, and H.C. Hershey (1988): *Transport Phenomena.* Chemical Engineering Series, McGraw-Hill, New York.

H. Burkhardt (1987): Systems physics: A uniform approach to the branches of classical physics. *Am.J.Phys.* **55**, 344–350.

H.L. Callendar (1911): The caloric theory of heat and Carnot's principle. *Proc. Phys. Soc.* (London) **23**, 153–189.

S. Carnot (1824): *Reflections on the Motive Power of Fire.* Translated by R.H. Thurston, Dover Publications, New York 1960.

S. Chandrasekhar (1960): *Radiative Transfer.* Dover Publications, New York.

(1967): *An Introduction to the Study of Stellar Structure.* Dover Publications, New York.

D.D. Clayton (1968): *Principles of Stellar Evolution and Nucleosynthesis.* McGraw-Hill, New York.

F.L. Curzon, and B. Ahlborn (1975): Efficiency of a carnot engine at maximum power output. *Am. J. Phys.* **43**, 22–24.

H. Davy (1839): The Collected Works of Sir Humphry Davy. Edited by J. Davy, London, 1839.

S.R. de Groot, and P. Mazur: *Non-Equilibrium Thermodynamics.* Interscience, New York, 1962.

A. DeVos (1985): Efficiency of some heat engines at maximum-power conditions. *Am. J. Phys.* **53**, 570–573.

J.A. Duffie, and W.A. Beckman: *Solar Engineering of Thermal Processes.* 2nd ed. Wiley, New York, 1991.

G.Falk, and F. Herrmann eds. (1977–1982): *Konzepte eines zeitgemässen Physikunterrichts*, Heft 1-5. Schroedel, Hannover, Germany.

G. Falk, F. Herrmann, and G.B. Schmid (1983): Energy forms or energy carriers? *Am. J. Phys.* **51**, 1074–1077.

G. Falk, and W. Ruppel (1973): *Mechanik, Relativität, Gravitation*. Springer-Verlag, Berlin.

(1976): *Energie und Entropie*. Springer-Verlag, Berlin.

H.D. Försterling, and H. Kuhn (1983): *Moleküle und Molekülanhäufungen*. Springer-Verlag, Berlin.

R. Fox (1971): *The Caloric Theory of Gases from Lavoisier to Regnault*. Clarendon Press, Oxford.

H.U. Fuchs (1986): A surrealistic tale of electricity. *Am. J. Phys.* **54**, 907–909.

(1987a): Entropy in the teaching of introductory thermodynamics. *Am. J. Phys.* **55**, 215–219.

(1987b): Do we feel forces? In J.D. Novak ed.: *Proceedings of the Second International Seminar on Misconceptions in Science and Mathematics*, Vol.III, p. 152–159. Cornell University, Ithaca, New York.

(1987c): Thermodynamics - A "misconceived" theory. In J.D. Novak ed.: *Proceedings of the Second International Seminar on Misconceptions in Science and Mathematics*, Vol.III, p. 160–167. Cornell University, Ithaca, New York.

(1987d): Introductory mechanics on the basis of the continuum point of view. *Berichte der Gruppe Physik*, No.3a. Technikum Winterthur.

(1987e): Classical continuum mechanics and fields: The local view. *Berichte der Gruppe Physik*, No.13. Technikum Winterthur.

(1989): The transport and the balance of momentum. *Berichte der Gruppe Physik*, Nr.21. Technikum Winterthur.

P.M. Gerhart and R.J. Gross (1985): *Fundamentals of Fluid Mechanics*. Addison-Wesley, Reading MA.

P. Germain (1973): The role of thermodynamics in continuum mechanics. In J.J. Delgado Domingos, M.N.R. Nina, and J.H. Whitelaw, eds.: *Foundations of Continuum Mechanics*. The Macmillan Press, London.

J.W. Gibbs (1892): Graphical methods in the thermodynamics of fluids. Transactions of the Connecticut Academy of Sciences **2**, 309–342.

(1928) The Collected Works of J.Willard Gibbs. Longmans, Green & Co., New York.

J.M. Gordon (1990): Observations on efficiency of heat engines operating at maximum power. *Am. J. Phys.* **58**(4), 370–375.

J.M. Gordon, and Y. Zarmi (1989): Wind energy as a solar-driven heat engine: A thermodynamic approach. *Am. J. Phys.* **57**, 995–998.

S. Gysel (1994): *Auslegung eines luftgekühlten Sonnenkollektors.* Diploma Thesis, Technikum Winterthur, Winterthur, Switzerland.

G. Heiduck, F. Herrmann, and G.B. Schmid (1987): Momentum flow in the gravitational field. *Eur. J. Phys.* **8**, 41–43.

F. Herrmann, and G.B. Schmid (1985): Momentum currents in the electromagnetic field. *Am. J. Phys.* **53**, 415–418.

(1986): The Poynting vector field and the energy flow within a transformer. *Am. J. Phys.* **54**, 528–531.

I. Iben (1967): Stellar Evolution within and off the Main Sequence. *Annual Review of Astronomy and Astrophysics*, Vol. 5, p. 599.

F.P. Incropera and D.P. DeWitt (1985): *Fundamental of Heat and Mass Transfer.* John Wiley & Sons, New York.

M. Iqbal (1983): *An Introduction to Solar Radiation.* Academic Press, Toronto, 1983.

J. Ivory (1827): Investigation of the heat extricated from air when it undergoes a given condensation, *Phil. Mag.* (n.s.) **1**, 89–94.

G. Job (1972): *Neudarstellung der Wärmelehre.* Akademische Verlagsgesellschaft, Frankfurt a.M., Germany

D. Jou and J. Casas-Vázqzez (1988): Extended irreversible thermodynamics of heat conduction. *Eur. J. Phys.* **9**, 329–333.

S. Klein, et al (1991): EES. *Engineering Equation Solver.* F-Chart Software. Madison WI.

T. Koch (1993): *Elektrische Stromerzeugung mit Sonnenkollektoren.* Diploma Thesis, Technikum Winterthur, Winterthur, Switzerland.

F. Kreith, and M.S. Bohn (1986): *Principles of Heat Transfer*, 4th ed. Harper & Row Publishers, New York.

W.M. Lai, D. Rubin, and E. Krempel (1978): *Introduction to Continuum Mechanics.* Pergamon Press, Oxford UK.

L.D. Landau, and E.M. Lifshitz (1959): *Fluid Mechanics.* Pergamon Press, Oxford UK.

K.R. Lang (1980): *Astrophysical Formulae.* 2nd ed. Springer-Verlag, New York.

R.S. Lindzen (1990): *Dynamics in Atmospheric Physics.* Cambridge University Press, UK.

E. Mach (1923): *Die Prinzipien der Wärmelehre, historisch-kritisch entwickelt.* 4. Auflage, Barth, Leipzig.

L.E. Malvern (1969): *Introduction to the Mechanics of a Continuous Medium.* Prentice-Hall, Englewood Cliffs.

J. Marsden and A. Weinstein (1985): *Calculus III*, 2nd ed. Springer-Verlag, New York.

G.E. Mase (1970): *Continuum Mechanics*, Schaum's Outline Series. McGraw-Hill, New York.

W. Maurer (1989): Spannungszustände im Festkörper - einmal anders. *Techinfo* **3**, No.2, 20–22. Technikum Winterthur.

(1990a): Neues von den Impulsströmen. *Techinfo* **4**, No.3, 18–19. Technikum Winterthur.

(1990b): Ingenieurphysik auf neuen Wegen. *Technische Rundschau* **82** (29/30), 12–16.

(1990c): Drehimpuls. *Techinfo* **4**, No.4, 30–32. Technikum Winterthur.

M.J. Moran and H.N. Shapiro (1992): *Fundamentals of Engineering Thermodynamics*. John Wiley and Sons, New York.

I. Müller (1985): *Thermodynamics*. Pitman, Boston.

J. Oppliger (1993): *Berechnungsgrundlagen für Luft-Wasser Solaranlagen*. Diploma Thesis, Technikum Winterthur, Winterthur, Switzerland.

C. Payne-Gaposhkin (1979): *Stars and Clusters*. Harvard University Press, Cambridge MA.

P.J.E. Peebles (1971): *Physical Cosmology*. Princeton University Press, Princeton, New Jersey.

M. Pitteri (1982): Classical thermodynamics of homogeneous systems based upon Carnot's general axiom. *Arch. Rat. Mech. Anal.* **80**, 333–385.

M. Planck (1906): *Theorie der Wärmestrahlung*. Original and translation in The History of Modern Physics, vol. 11. American Institute of Physics, 1988.

(1921): *Theorie der Wärmestrahlung*, 4th ed. Johann Amprosius Barth, Leipzig.

(1926): *Treatise on Thermodynamics*, 3rd ed. Translated by A.Ogg, Dover Publications 1945.

A. Rabl (1985): *Active Solar Collectors and Their Applications*. Oxford University Press, New York.

P. Ridgely (1987): Life in the heart of the computer: A bit of thermal physics. *The Physics Teacher* **25**, 276–279.

D. Roller (1950): *The Early Development of the Concepts of Temperature and Heat*. J.B. Conant, ed.: Harvard Case Histories in Experimental Science, Case 3. Harvard University Press, Cambridge.

P. Salamon, A. Nitzan, B. Andresen, and R.S. Berry (1980): Minimum entropy production and the optimization of heat engines. *Phys. Rev. A* **21**, 2115–2129.

G.B. Schmid (1982): Energy and its carriers. *Phys. Educ.* **17**, 212–218.

(1984): An up-to-date approach to Physics. *Am. J. Phys.* **52**, 794-799.

M. Schwarzschild (1958): *Structure and Evolution of the Stars.* Princeton University Press. (Also Dover Publications, New York, 1965)

I.S. Shklovskii (1978): *Stars. Their Birht, Life, and Death.* W.H. Freeman & Co., San Francisco.

F.H. Shu (1982): *The Physical Universe.* Oxford University Press, Oxford UK.

J. Silk (1980): *The Big Bang.* W.H. Freeman & Co., San Francisco.

F.D. Stacey (1992): *Physics of the Earth*, 3rd ed. Brookfield Press, Brisbane, Australia.

F. Stucky (1994): *Modell zur Grobauslegung diffusiver Erdwärmespeicher.* Diploma Thesis, Technikum Winterthur, Winterthur, Switzerland.

J.S. Thomsen, and T.J. Hartka (1962): Strange Carnot cycles. *Am. J. Phys.* **30**, 26–33,388–389.

J.E. Trevor (1928): *Sibley Journal of Engineering* **42**, 274–278

C. Truesdell (1979): Absolute temperatures as a consequence of Carnot's general axiom. *Archive for History of Exact Sciences* **20**, Number 3/4, 357–380.

(1980): *The Tragicomical History of Thermodynamics*, Springer-Verlag, New York.

(1984): *Rational Thermodynamics.* 2nd ed. Springer-Verlag, New York.

C. Truesdell, and S. Bharatha (1977): *The Concepts and Logic of Classical Thermodynamics as a Theory of Heat Engines.* Springer-Verlag, New York.

C. Truesdell, and W. Noll (1965): The Non-Linear Field Theories of Mechanics, in *Encyclopedia of Physics*, v. III/3, S.Flügge ed. Berlin, Springer-Verlag.

C. Truesdell, and R.A. Toupin (1960): The Classical Field Theories, in *Encyclopedia of Physics*, v. III/1, S.Flügge ed. Berlin, Springer-Verlag.

L.A. Turner (1962): *Am. J. Phys.* **30**, 804–806

G. Walker (1973a): The Stirling Engine. *Sci. Am.* **229**, No.2 (August), 80–87.

(1973b): *Stirling Cycle Machines.* Oxford University Press.

J. Walker (1985): Experiments with the external-combustion fluidyne engine. *Sci. Am.* **252**, No.4 (April), 108–112.

A.J. Watson, J.E. Lovelock (1983): Biological Homeostasis of the Global Environment: The Parable of the "Daisy" World. *Tellus* **35**b, 284–289.

M. Wegelin et al. (1994): Solar water disinfection: Scope of the process and analysis of radiation experiments. *J. Water SRT - Aqua* **43** (3), 154–169.

L.P. Wheeler (1952): *Josiah Willard Gibbs*, 2nd ed. Yale University Press, New Haven.

S. Whitaker (1968): *Introduction to Fluid Dynamics*. Prentice Hall, Englewood Cliffs, NJ.

H.J. White, and S. Tauber (1969): *Systems Analysis*. Saunders, Toronto.

C.J. Winter et al., eds. (1991): *Solar Power Plants*. Springer-Verlag, Berlin.

J. Wu (1990): Are sound waves isothermal or adiabatic? *Am. J. Phys.* **58**, 694-696.

Y.B. Zel'dovich, and Y.P. Raizer (1966): *Physics of Shock Waves and High-Temperature Hydrodynamic Phenomena*. Vol. I and II. Academic Press, New York.

M.W. Zemansky, and R.H. Dittman (1981): *Heat and Thermodynamics*, 6th ed. McGraw-Hill, Singapore.

Subject Index

A

absorptance, see *radiation*
absorption refrigeration 553
absorptivity, see *radiation*
accounting, see *laws of balance*
adiabat
  fluid with neutral points 260
  T-S diagram 75
  T-V diagram 260
adiabatic compressibility, see *compressibility*
adiabatic demagnetization 216
adiabatic flame temperature 526–528
adiabatic process 74
  black body radiation 214
  differential equation 183, 260
  fluid flow 523–528
  ideal fluid 259–261
  ideal gas 181–184
  irreversible 116
  saturation 525–526, see *mixtures, moist air*
  temperature gradient, see *atmosphere*
  work 200
air
  mass fractions 458
  moist air 500
  molar mass 458
  temperature coefficient of enthalpy 514
  temperature coefficients of energy and enthalpy 459
  theoretical amount 528
air mass, see *solar radiation*
air thermometer 71
ammonia synthesis 483
amount 4
amount of substance 448–454
  balance, see *balance of amount of substance*
  chemical reactions 19, 449–450
  electrolysis 450–451
  equation of state of ideal gas 171
  measuring amounts of stuff 2, 19
  nonconservation 19
  particles 460
  unit 450
amount of water stored 3
analogies
  diffusion 646
  electrical and chemical pumps 467
  entropy and momentum 646
  hydraulics and electricity 2–11
  momentum and charge 25
atmosphere
  adiabatic temperature gradient 187, 300, 580
  cloud formation 505
  isothermal 173
  winds 353–357
attenuation, see *radiation*
availability
  available power 95, 103, 124–131
  Carnot's power of heat 125
  exergy 125
  loss of available power 103, 125–129, 134
  maximum of 136
  of a body of water 167

Avogadro's number 461

B
balance of amount of substance 19, 602, 608
balance of energy 33
 and entropy transfer 311
 combined with balance of entropy 226-227
 conduction 315-316, 357-362
 consequences 602
 continuous processes 627-629
 first law of thermodynamics 220, 223
 general form for uniform processes 220-223
 heat exchangers 573
 ideal fluids 286-287
 ideal gas 197
 integral form 223
 material form 627
 purely thermal processes 166
 sources 368
 thermal energy equation in fluid flow 638
 thermomechanical processes 221
balance of entropy 109-119
 chemical reactions 472
 combined with balance of energy 226-227
 conduction 303, 357-362
 convection 304
 equation of 111
 general form 310
 heat exchangers 520, 575
 ideal gas 178
 local form 621
 phase change 485
 production of entropy 114
 radiation 307, 308-309
 reversible processes 110
 second law 111, 601
 solar collectors 559
 sources 309, 368
 supply and conduction 366-369
 time dependent conduction 372-376
 time dependent heat transfer 339
 transport processes 302-310
 uniform reversible process 159
balance of heat
 derived from simple assumptions 65
 ideal fluids 254-255
 phase change 80
 see *balance of entropy*
balance of mass 18
 continuous processes 619
 equation of continuity 620
balance of momentum 13
 as restriction upon processes 595
 for control volume 18
 local form 621
 Newton's law 13
 uniform fluid 595
balance of volume 4
battery 471
Bernoulli's law 513
black body, see *radiation*
body
 as distinct from control volume 305
body force, see *force*
Boltzmann's constant 405
boundary layer 319, 639
 approximations 644
 thermal boundary layer 319, 640
 velocity boundary layer 319, 640
Boyle and Mariotte, law of 171

C
caloric
 absorption and emission 253
 and Carnot's axiom 276
 and entropy 65
 and heat 63
 and propagation of sound 264
 capacity 287
 Carnot cycle 269
 current 252, 272
 currents of caloric and energy 283
 energy exchanged with caloric 271
 equation of balance 255
 exchanged 254, 273
 fall of caloric 275
 heat function 243
 in adiabatic processes 260
 motive power of heat engine 282
 production 252
 re-establishment of equilibrium 275
caloric equation of state
 paramagnetic substance 216
caloric theory 251, 289-296
 Carnot's Axiom 277
 heat capacities 294-296
 heat function 251
calorimetry
 problem of 107
 theory of 253, 255, 291-292
capacitance 6
 charge 16
 hydraulic 7
 momentum 16
capacitance flow rate, see *heat exchangers*
capacitor 3
Carnot cycle, see *cycles*

Carnot engine, see *heat engines*
Carnot process 124
  dissipative 228-229
Carnot's Axiom 274-279
  and caloric theory of heat 277
Carnot, see *entropy, units*
Carnot's function 281
Carnot-Clapeyron theorem 280, 283
Celsius scale, see *temperature*
changes of state 230-232
charge
  and entropy 66
  and momentum 25-26
  change of charge 6
  electrolysis 451
  exchanged 6
chemical potential
  and chemical equilibrium 481
  as a function of pressure 477
  as a function of temperature 478
  dependence upon temperature and pressure 463-464, 605
  dilute solutions 496-498
  driving force for chemical change 463
  driving force for transports 535
  electrochemical reactions 466-470
  entropy production 472
  flow processes 507-509, 515-516
  gravito-chemical 532
  ideal gas 518
  intrinsic part 515
  isotopes 464
  melting and vaporization 488
  mixtures of fluids 474
  moist air 502, 503
  Navier-Stokes-Fourier fluids 638
  of formation 464
  osmosis 533
  phase change 463
  pressure coefficient 479
  pure fluids 604
  reactions 461-474
  solvent and solute 496-497
  temperature coefficient 479
  transformation of substances 462
  uniform fluids 474, 477-479
  unit 462
  vapor and liquid 491
  zero point 464
chemical pumps 467-468
chemical reactions 3
  amount of substance 449-450
  battery 471
  chemical potential 461-474
  combustion 526
  electrochemical 466-470
  electrochemical cell 467
  endothermic 472
  enthalpy of formation 472-474
  equilibrium 480-482
  exothermic 472
  heat transfer 472-474
  heating value 476
  melting and vaporization 488
  power 469
  transport of substances 534-535
Clapeyron equation 607
Clausius-Clapeyron equation, see *phase change*
collectors, see *solar collectors*
combustion, see *fluid flow*
compressed liquid 484
compressibility
  adiabatic 261
  ideal fluid 299
  isothermal 250, 261
compressor 518-519
concentration 453
conduction 7
  balance of energy 315-316, 361
  balance of entropy 303, 359-361
  charge 25
  continuous processes 357-376
  current density 358-359
  currents of entropy and energy 314
  energy current density 626
  entropy 617
  entropy production 361-362
  entropy transfer 302-303
  field equation for temperature 363-364, 369
  Fourier's law 312-314
  infinite speed 375-376, 648
  influence of induction 651
  Maxwell-Cattaneo equation 651
  momentum 7, 12-15, 25, 617-618
  production of entropy 105
  supply (sources) 366-368
  upper layers of soil 376-378
  wave equation 651
conductivity
  diffusion of radiation 395
  electrical 25
  entropy 313
  momentum conductivity 25
conservation
  and laws of balance 19
  balance of power 85
  charge 6
  energy 33
  momentum 13
  volume 5

conservation of energy, see *balance of energy*
constitutive domain 246
constitutive laws 6, 20-28
  and dynamics 20
  black body radiation 210-211
  capacitance 6
  dissipative transports 7
  flow of charge 6
  flow of water 6
  heat and thermal processes 62
  inductance 6
  inequalities for uniform fluids 247, 248
  melting 79
  Navier-Stokes-Fourier fluid 630, 632, 638
  Newton's law of gravitation 21
  resistance 6
constitutive quantities of the ideal gas 283-284
constitutive theories
  determining currents of entropy 119
  ideal gas 179
continuity, equation of 620
continuity, see *balance of mass*
continuous processes
  laws of balance 611-629
  Navier-Stokes-Fourier fluid 629-638
continuum physics 38
continuum thermodynamics 591
control volume 18, 305
convective currents, see *convection* and *currents*
convection
  adiabatic of moist air 194
  boundary layer 319
  boundary layer equations 639
  convective current density 616
  convective stability 579
  energy current density 625
  energy flux in free convection 580
  entropy transfer 303-306
  forced 304
  free convection 579
  free or natural 304
  heat transfer applications 554-581
  heat transfer coefficient 320
  in Earth's mantle 299
  in the interior of a star 582
  of entropy 510-511, 617
  of momentum 17, 511
convective heat transfer coefficient, see *heat transfer coefficient*
cooling 54
cosmic background radiation 213
creation of heat 59
critical point 487
Curie constant 216
current and flux 110
current density 614-616
  conduction 358-359
  energy 625
  entropy 616
  mass 616
  momentum 14, 617-618
cycles
  calculation 270-274
  caloric in Carnot cycle 269
  Carnot cycle 84, 179, 268-270
    caloric absorbed and emitted 269
    most efficient 284
  heat exchanged in a Carnot cycle 269
  heat pump 551-553
  motive power of Carnot cycle 274
  Otto cycle 204-206
  Rankine cycle 544
  refrigeration 551-554
  Stirling cycle 203
  strange Carnot cycles 269
  superheating in the power cycle 545
  T-S diagram of Carnot cycle 85
  vapor power 543-546

D

daisy world 412
Debye temperature 165
density 360, 374, 612-614
  current density 360
  energy 625
  entropy 373
  momentum 614
  production rate density 360, 614
  source rate density 368, 614
dew point, see *mixtures, moist air*
diffusion 7, 646
diffusivity
  entropy 646
  momentum 646
  thermal 376, 646
dimensionless groups 646
  Nusselt number 647
  Prandtl number 647
  Reynolds number 647
discharging
  capacitors 8
  containers 8
displacement law, see *radiation, Wien's displacement law*
dissipation 119-124
  and irreversibility 114
  as internal process 117

combined balance of entropy and of energy 226
electrical 227
entropy production 120–121, 226
flow of momentum in viscous medium 106
in conduction 105
loss of available power 124
loss of power 103
pumping entropy from absolute zero 123
rate of dissipation of energy 121
see *entropy production*
dissipative processes
compared 106
conduction 105
free expansion 106
friction 106
distribution function, see *radiation*
divergence 619
divergence theorem 618–619
doctrine of latent and specific heats 253
Doppler effect, see *radiation*
driving force 4
chemical 463, 535
electric 25
for convection 304
hydraulic 6
thermal 57
thermal (in conduction) 302
dynamics
compared to statics 20
laws of motion 20
role of constitutive laws 20

E

Earth
as a selective absorber 414
conduction through mantle 317
conduction with sources 371
convection in mantle 299, 589
surface temperature 327, 333
winds in atmosphere 353–357
effectiveness of heat exchangers, see *heat exchangers*
efficiency
energy efficiency 93
first law efficiency 94
ideal Carnot engine 92–94
second law efficiency 92, 94, 126
of ideal Carnot engine 92
thermal efficiency 93
electrical potential, see *potential*
electricity 3
current of charge 3
electrical level 4
electrical pump compared to chemical pump 467
electrochemical processes 469–470
electrolytic cell 468
energy 10
flow of charge 2
induction 8
Ohm's law 7
potential 3
electrochemistry, see *chemical reactions*
electrolysis
amount of substance 450–451
Faraday's constant 451
of water 224
emissive power, see *radiation*
emissivity, see *radiation*
emittance, see *radiation*
endoreversible engine, see *heat engines*
energy
and entropy in heating 87–89
balance of power 85
balance of, see *balance of energy*
conservation, see *balance of energy*
coupling of processes 28
currents, see *energy current*
electrochemical processes 469–470
energy content of water 167
energy function as a "fundamental" relation 288
equivalence to mass 36
exchanged in heating 89, 222
external process 29
flow diagram 10–11, 32–33
electrochemical 468
electrolytic cell 468
heat engine 86
turbine 508
general physical processes 28–34
in electricity 10
in hydraulics 10
internal 220, 285–288, 512
internal processes 28
molar (in gas mixture) 456
molar energy 512
power 11
releasing 10, 628
rigid bodies 166
source rate density in radiation 389
sources 311, 626
specific energy 512
storage 11
transfer 11
and substancelike quantities 29
fluid flow 512–513
ideal gas 200
through radiation field 378–382

transport processes 29
using 11
voluntary and involuntary processes 29
energy carrier 32
energy current
and energy carrier 32
and potential 32
chemical 508
compression of fluid 35
in fluid flow 512–513, 609
in heating 88
load factor 32
magnetic processes 35
mechanical 31
thermal, in Navier-Stokes-Fourier fluids 637
energy current density 625
conductive 626
convective 625
mechanical 626
thermal 626
energy density 512, 625
energy liberated 11
by a fall of entropy 85–87, 92, 97
energy principle 2
consequences for Navier-Stokes-Fourier fluids 632
consequences for viscous uniform fluid 595
in continuum physics 624–629
energy transfer, see *energy*
engines 11
Otto engine 203
Stirling engine 203
vapor power 543–546
enthalpy 199
ideal fluids 288
of formation 474
phase change 486
specific enthalpy 513
thermodynamic potential 606
entropy 53
and chemical potential 471–474
as heat 65
balance of, see *balance of entropy*
black body radiation 211
caloric 65
content 110–112, 159, 164–165
content of body with constant temperature coefficient of energy 167
density in conduction 373
density in fluid 510
density in radiation 402, 403
diffusivity 646
entropy and energy in heating 87–89
entropy content of the ideal gas 190
entropy content of water 167
exchanged 111
falling and releasing energy 85–87, 92, 97
generation of, see *entropy production*
latent, see *latent entropy*
maximum entropy postulate 234–236
minimum rate of production 139
mixing of gases 456
moist air 502
molar (in gas mixture) 456
molar entropy 510
of fusion 91
of radiation, see *radiation*
of solids 165
reservoir 134
source rate density in radiation 389
source rate of entropy and energy 368
sources 309, 366–368
specific entropy 510
spectral density of radiation 397
spectral intensity of radiation 397
thermal charge 64
transport processes 302–310
units 64
entropy capacity 157–158
as constitutive quantity 158
at constant magnetization 217
at constant pressure 178
at constant volume 157, 176
black body radiation 209
ideal gas
determination 188
measurement 160
molar 162
ratio of entropy capacities of fluids 184
ratio, for air 184
specific 162
water 161
entropy conductivity 313–314, 631
entropy current 110
absorbed 89
and energy current in conduction 105
and energy current in heating 88
as distinct from flux 110
at ideal wall 118
at solid-fluid interface 321
available power of 95, 127
conduction 312–314
conductive 303
continuity at ideal wall 118
convective 510–511
cooling heat engine 93
flow across surfaces 303

Fourier's law 312-314
in heating 88
increase in conduction 104
increase through irreversible engine 101
power of 92
pumped by heat pump 100
through Carnot engine 86
to be determined by constitutive theory 119
entropy flux, see *entropy current*
entropy production 101-109
and balance of energy 226-227
avoiding of 134
balance of entropy 114
chemical reactions 471-472, 609
combustion 531
conduction 361-362
dissipation 119-144, 226
extended theory of conduction 653
flow heater with loss 521
fluid flow 345
free expansion of gas 233
from constitutive theory 136
in calorimetry 107
in conduction 105
in heat engine at maximum power 140-141
in heat exchanger with pressure drop 578
in heat exchangers 132, 575-576
in heat transfer 343-350
in radiation 349
in scattering 427
irreversibility 59-61
minimal in equilibrium 636
minimization for heating purposes 135
minimization of 134-138, 345-348, 351-357
mixing of ideal gases 476
Navier-Stokes-Fourier fluids 637
solar air heater 561
uniform heating at constant volume 159
vapor power plant 546, 549-551
entropy transfer
conductive transport 302-303
convection 303-306
overview 302-310
radiation 307-309
through radiation field 378-382
equation of state
black body radiation 209
gas with radiation 212
ideal gas 171-172
paramagnetic substance 216
thermal 171, 244-250
transformation of variables 248
equations of balance, see *laws of balance*
equilibrium
chemical 488-494
final temperature reached in thermal contact 232
hydrostatic 173, 196
in chemical reactions 480-482
local thermodynamic equilibrium 638, 648
re-establishment of 134, 275
thermal 230
evaporation, see *phase change*
exergy, see *availability*
extended irreversible thermodynamics 652
extensive quantities 3, 4, 36
and energy transfer 33
and processes 29, 30
magnetic 36
external process 29
extinction coefficient, see *radiation, attenuation*

F

Faraday's constant 451
field equation for temperature 363, 369
fields 307-310
interaction with bodies 308, 388-394
finite-time thermodynamics 143
flame temperature 134
see *adiabatic flame temperature*
flow diagram 10
flow processes
chemical potential 507-509
uniform fluids 609
flow work 509
fluid flow
adiabatic 523-528
combustion 526-528
energy current 512-513, 609
similarity parameters, see *dimensionless groups*
thermal energy equation 638
throttling process 525
turbulence 648
fluids
Bernoulli's law 513
boundary layer flow 639
chemical potential 509, 515-516
entropy density 510
ideal
balance of energy 286-288
incompressible 538
energy equation 639
mixing length approximation 581

- molar entropy 510
- molar volume 509
- Navier-Stokes equations 642
- Navier-Stokes-Fourier 594
  - constitutive laws 630, 631
  - GFF 638
  - laws of balance 629
- Newton's law for viscous fluids 24
- Newtonian 24
- property data 536-541
- saturated 536-537
- specific entropy 510
- uniform
  - general thermodynamic theory 592-611
  - independent variables 245
  - pressure as an independent variable 257-259
  - theory of heat 250-259
- uniform viscous 592-601
- with phase transformation 601-607

flux 3
- electric and magnetic 5
- mass 17
- net flux 5
- positive and negative sign 5
- rate of change 9

flux and current 110
flux density 5, 360, 614
flux of momentum, see *momentum current*
force
- body force 16
- momentum current 13
- surface force 13, 618

Fourier's law 312-314, 594, 632, 648
free energy 606
free energy, see *thermodynamic potentials*
fuel
- adiabatic flame temperature 527
- burning of 134
- oxidizer 527
- theoretical amount of air 528

fuel cells 134, 468
fundamental relation 288

G

gas constant 171
Gauss's theorem, see *divergence theorem*
Gay-Lussac, law of 71, 171
generic laws, see *laws of balance*
GFF, see *Gibbs fundamental relation*
Gibbs (unit of chemical potential) 462
Gibbs free energy 606
- see *thermodynamic potentials*

Gibbs fundamental form, see *Gibbs fundamental relation*
Gibbs fundamental relation
- and chemical potential 605
- black body radiation 209
- extended theory of conduction 654
- heating at constant volume 167
- ideal fluids 287
- ideal gas 197
- Navier-Stokes-Fourier fluids 638
- paramagnetism 218
- similarity in different cases 648
- uniform reactive fluid 609

Gibbs' paradox 458
Gouy- Stodola rule 126
gradient
- temperature 631
- velocity 631

gravitation
- gravito-chemical potential 532
- source of momentum 15
- transporting matter 532

gravitational power 85
gray surface, see *radiation*
Green's transformation, see *path integrals*
Grüneisen ratio 299

H

Hagen and Poiseuille 6, 24
heat
- absorbed 254
- and energy 52
- and entropy 53, 65
- and hotness 56-58
- energy exchanged in heating 89
- exchanged 253
- extensive thermal quantity 56
- heat function 251, 255
- in Carnot cycles 269
- magnetic process 215
- measuring amounts of entropy 77
- motion of the least parts 296
- motive power 282
- nonconservation 56, 58
- power of heat 61
- production and irreversibility 59
- production of 58
- production of, see *entropy production*
- storage 55, 251, 347-348
- substancelike quantity 56, 63
- the power of heat (*la puissance du feu*) 95
- thermal charge 64
- what is heat? 290

heat capacity
- and Carnot's function 281
- and heat function 255

at constant volume, see *temperature coefficient of energy*
ideal gas 266
uniform fluid 252-253
heat content 57
and specific heat 295
as distinct from hotness 77
change in phase changes 80
change of 74
heat as extensive quantity 57
heat as substancelike quantity 63
heat function 265
heating and production of heat 79
heating at constant volume 75
in adiabatic change 75, 260
isothermal process 76
rate of change 255
reversible change 255
see *entropy, content*
heat engines 83-85
at maximum power 135-142
atmosphere 354
Carnot engine 83
Carnot's comparison to hydraulic engine 62
Curzon-Ahlborn 139-142
Curzon-Ahlborn efficiency 142
dissipative 226
efficiency of Carnot engine 92-94
endoreversible 140-143
flow diagram 86, 87
ideal Carnot engine 139
irreversible Carnot engine 101
motive power of heat 83
need for cooling 108
solar thermal 351-353
heat exchangers 131, 519-520, 570-576
balance of energy 519, 573
balance of entropy 519-520
balanced 573
capacitance flow rate 571
counter-flow 570, 574
effectiveness 572
counter-flow 572
parallel-flow 573
entropy production 575
flow diagram 132
mean temperature difference 132, 571
number of transfer units 572
parallel-flow 570
principle of operation 570
product of surface area and transfer coefficient 571
simple model 131
with heat engines 131
heat function
and heat capacity and latent heat 265
heat pumps 96-100
coefficient of performance 98
dissipative 129, 226
flow diagram 97
vapor cycles 551
heat storage
charging and discharging 347-348
heat transfer
and thermodynamics 654
at a solid-fluid boundary 318-322
between uniform bodies 169
conduction, see *conduction*
convective, see *convection*
entropy production 343-350
heating 54
in chemical reactions 472
in heat engines 139
incompressible one-dimensional flow 639
see *conduction*
see *convection*
see *entropy transfer*
see *radiation*
through cylindrical shells 324
through several layers 322
time dependent, discrete 338-340
heat transfer coefficient 320-321
for heat exchanger 132
for laminar flow over flat plate 647
for loss from solar collectors 337
local 320
radiative 328, 337
total 321-322
heating 54
absorption and emission of radiation 334-335
and cooling 54
at constant temperature 175-176
at constant volume 75, 157-165
equation of balance of entropy 159
ideal gas 176-177
reversibility 159
constitutive law for uniform heating at constant volume 158-159
difference to charging or moving 52
examples 73-82
flow diagram 88
heat transfer 54
ideal gas 175-179
isothermal process 76
of simple fluids 74
relation between entropy and energy 88
Reversal Theorem 254

room at constant pressure 587
viscous fluids 601
with a heat pump 163
heating value 476
helium 418
Helmholtz free energy, see *thermodynamic potentials*
Henry's law 499
Hertzsprung-Russell diagram, see *stars*
Hook's law 69
hot air balloon 173
hotness 53
absolute zero 54
and heat 56-58
and heat content 77
and temperature 53
hydraulic power 10
hydraulic resistance 7
hydraulic work 10
hydraulics
amount of water 2
capacitance 7
change of amount of water 4
current of volume 3, 4
current of water 2
energy 10
fluxes of volume 4
inductance 9
induction 8
law of balance of volume 4
law of Hagen and Poiseuille 6
LC-circuits 9
rate of change of volume 4
rate of change of volume flux 9
resistance 6
time constant 8
volume flux 3
hydrogen
burning to water 470
hydrostatic equilibrium, see *equilibrium*

I

ice calorimeter 81
ideal gas 171, 599
adiabatic process 201
constitutive law of heating 178
constitutive laws 179-189
energy 198-199
enthalpy 198-199
entropy capacities 188
entropy content 190
free expansion 198, 233
gas constant
specific 452
universal 452
heating 175-179
isentropic flow 524
mixtures 454-458
mole fraction 454
temperature 71
thermal equation of state 171-172, 452-453
ideal gas temperature, see *temperature*
ideal walls 117-119, 594, 598
incompressible, see *fluids*
inductance 6
hydraulic 9
mechanical 26, 27
induction 8
thermal 28, 651
integral transformation
divergence theorem 618
intensity 4
intensity, see *radiation*
intensive quantities 4
level 4
pressure 7
thermal 57
internal energy, see *energy*
internal processes 28
irradiance, see *radiation*
irreversibility 59-61
and production of heat 59
and time 206
isochoric heating, see *heating at constant volume*
isothermal process 76
isotopes
chemical potential 464

K

Kelvin scale, see *temperature*
Kirchhoff's law, see *radiation*

L

Lagrange multipliers 596
Navier-Stokes-Fourier fluids 633, 635
thermodynamics with 596
uniform fluids with phase changes 602
uniform viscous fluids 598
latent entropy 175
black body radiation 209
fusion 91
ideal gas 176
determined 181
phase change 485
with respect to magnetization 217
with respect to pressure 178
with respect to volume 175
latent heat 76, 175, 252-253, 255
sign 278
latent heat (entropy) of fusion 79, 91

latent heat (entropy) of vaporization 80
latent heat, see *latent entropy*
laws of balance 2
accounting 4
charge 6
conservation 19
continuous processes 611-624
energy, see *balance of energy*
entropy, see *balance of entropy*
heat, see *balance of heat or entropy*
integral form 619
local form 620
Navier-Stokes-Fourier fluid 629
open systems 511
see *balance of...*
transport processes 12-20
volume 6
LC-circuits 9
level
difference of levels 4
electrical level 4
hydraulic level 4
level, see *potential*
liquids
subcooled 537-539
local thermodynamic equilibrium 638
log mean temperature difference, see *heat exchangers*
loss
dissipation 124
of available power in conduction 129
of power 103
luminosity, see *stars*

M

magnetic current 35
magnetic tension 35
magnetism and heat 215-219
mass 2
equivalence to energy 36
gravitational 3
inertial (momentum capacitance) 16
see *balance of mass*
material derivative 305, 623, 641
material laws, see *constitutive laws*
mean free path, see *radiation*
mean temperature difference 132
mechanical theory of heat 292-293
melting, see *phase change*
misconceptions
learning and teaching of physics 41
mixing length approximation 581
mixtures
moist air 501-502, 541
adiabatic saturation 503
dew point 502-503
humidity ratio 501
relative humidity 501
wet bulb temperature 503-504
quality 539
T-S diagram for superheated water vapor 541
two phase fluids 500-504
mixtures of gases, see *ideal gas*
moist air, see *mixtures*
molar mass 450
air 458
equation of state of ideal gas 172
matter inside stars 212
mole 19
mole fraction 454
momentum
diffusivity 646
sources 15
momentum capacitance (mass) 16
momentum current 12
force 13
momentum current tensor 618
momentum transport
comparison with charge 25
conductive 7, 12-15, 617
convection 17, 617
current density 14
flow pattern 14
graphical 14
Hagen-Poiseuille 7
light 37
radiative 15
stress 12-15
two-dimensional 13
viscosity 23
wave equation 26
motive power of a heat engine 279-283
motive power of Carnot cycle 274
motive power of heat 282

N

Navier-Stokes equations, see *fluids*
Newton's law of motion, see *balance of momentum*
Newton's law of motion
and convective momentum currents 18
NTU (number of transfer units), see *heat exchangers*
nuclear reactions 462, 466
number of transfer units, see *heat exchangers*
Nusselt number 647

O

Ohm's law 7, 25
opacity, see *radiation*

open system 610
optimization
 and minimization of entropy production 136
 of solar hot water heater 136
oscillations
 electromagnetic 9
 hydraulic 9
 thermal 649-652
osmosis 453, 533-534
osmotic pressure 453, 533

P

paramagnetic substance 216
partial pressure, see *pressure*
path integrals 271
 Green's transformation 272
paths 245-246
phase change 484-488
 as chemical reaction 488
 balance of energy 485
 balance of entropy 485
 change of enthalpy 486
 chemical equilibrium 488-494
 chemical potential 463, 488
 Clapeyron's law 490, 493-494
 Clausius-Clapeyron equation 495
 constitutive laws 79
 critical point 487
 latent entropy 485
 melting 77
 melting point
  change with solute 498
  dependence upon pressure 489-490
 moist air 501
 pressure-temperature relation 487-488
 sublimation 488
 temperature dependence of vapor pressure 492
 T-S diagram 77
 T-S phase diagram 487
 vapor pressure 490-493
 vaporization 77
 vaporization at different pressures 486
 vaporization of freon 82
 vaporization of water 225, 484-488
photon gas 208
Planck's formula, see *radiation*
Planck's constant 405
Poisson and Laplace
 law of 184
polarization 406-407
 degree of polarization 407
 energy intensity 407
 entropy intensity 407
 plane polarized 406
 principal values 406
 solar radiation 426
polytropic
 exponent 193
 gas spheres 195
 process 192-193
  adiabatic convection of moist air 194
  compression of ideal gas 193
  ideal fluid 262-263
potential 7, 29-31
 and energy 29-31
 chemical, see *chemical potential*
 electrical
 electrochemical 470
 induced potential difference 9
 thermal 57, 88, 601
 thermodynamic, see *thermodynamic potential*
power 10
 available power, see *availability*
 balance of 85
 chemical reactions 469
 electrical 10, 11
 exergetic power 95
 fluid flow 508
 gravitational 85
 heat
  the "power of heat" 95
 hydraulic 10
 loss of available power 103
 of an internal process 30
 of heat 61
 thermal
Prandtl number 647
pressure 2
 dissolved substance 496
 of ideal fluids 247-250
 osmotic 453
 partial 454, 455-456
pressure coefficient of chemical potential 479
primitive 2
production of entropy, see *entropy production*
production of heat
 irreversibility 59
 see *entropy production*
 see *heat*
production rate density 614
properties of bodies, see *constitutive laws*
property data 536-541

Q

quality, see *mixtures*

R

radiation 308-310, 324-336, 378-415
- absorptance 329
- absorption 334-335, 388-391
- absorption coefficient 390
- absorptivity 329
- attenuation 420
- balance of entropy 307, 308-309
- black body radiation 207, 324-326, 396
  - adiabatic process 214
  - energy 209
  - energy density 209, 211
  - entropy 211
  - entropy capacity 209
  - Gibbs fundamental relation 402
  - in cavity 207
  - intensity 380
  - isothermal process 214
  - Kirchhoff's law 399
  - latent entropy 209
  - pressure 209
  - spectral distribution 404-405
  - thermodynamics 402
- configuration factor, see *shape factor*
- cosmic background 213
- diffusion of radiation 395
- distribution functions 378-379
- Doppler effect 399, 400
- emission 334-335, 388-391
  - spontaneous 393
  - stimulated 393
- emission coefficient 390
- emissivity 330
- emittance 329
- entropy 402-404
- entropy density 402
- entropy intensity of solar radiation 418
- entropy production 349
- entropy transfer 307-310
- extended parallel plates 331
- extinction 393-394, 420
- extinction coefficient 436
- flux density 380
- from opaque surfaces 324-332
- from the surface of stars 417
- gray surfaces 329-331, 396
- heat transfer coefficient 328
- hemispherical emission 325
- hemispherical emissive power 325
- hemispherical flux density 381-382
- inside stars 211
- intensity 379
  - black body radiation 380
- interaction of bodies and fields 308-309
- interaction with matter 388
- irradiance 329
- Kirchhoff's law 329, 391-393, 398
- mean free path 394
- monochromatic radiation 396, 404
- networks 386-388
- normal spectrum (black body spectrum) 396
- opacity 395
- Planck's formula 404
- polarization of, see *polarization*
- radiosity 331
- reflectivity 329
- see *solar radiation*
- selective absorbers 410-413
- shape factor 384-385
- Snell's law 437
- source rate 390
- sources 309, 389-391
- spectral
  - density 397
  - distribution 396-406
  - distribution functions 397
  - entropy density 403, 405
- spectral intensity 397
  - transformation from frequency to wavelength 398
- Stefan-Boltzmann constant 211
- Stefan-Boltzmann radiation constant 381
- temperature 404
- thermal 207-208
- view factor 384
- wavelength 398
- Wien's displacement law 396, 399-402, 408

radiative transfer
- through a medium with a temperature gradient 394-395

radiosity, see *radiation*

Rankine cycle, see *cycles*

rate of change 4

rate of production
- amount of substance 19
- see *laws of balance*

reactions
- uniform mixtures of fluids 607-609

reflectivity, see *radiation*

refrigerant R123 537

refrigeration
- absorption cycle 553
- see *cycles*

refrigerator 96-100
- coefficient of performance 98

regenerator
- Stirling engine 204

resistance 6
  electric 8
  hydraulic 6
  thermal 155, 316
resistor
  electric 3
  thermal 139
reversibility 60, 254
Reynolds number 647

S

saturated liquid 485
saturated vapor 487
scattering
  entropy production 427
  Mie scattering 423
  Rayleigh scattering 423
  see *solar radiation*
second law efficiency, see *efficiency*
second law, see *balance of entropy*
second sound 28, 652
sedimentation 533-534
selective absorbers, see *radiation*
similarity groups, see *dimensionless groups*
sink, see *source*
sky temperature 338
Snell's law 437
solar collectors
  air heater 560-561
  balance of energy 555-557
  balance of entropy 559
  duct geometries 557
  dynamical model 561
  efficiency 562
  efficiency factor 557
  flat plate 555-562
  heat loss coefficient 336-338
  heat removal factor 559
  temperature distribution 558-559
  thermosyphon collector 564
  top loss coefficient 556
  transmission-absorption 433-436, 555
solar constant, see *solar radiation*
solar radiation 415-436
  absorption 419-422
  air mass 420
  at surface of the Sun 418
  attenuation coefficient 420
  cloudless atmosphere 420
  concentration 429-432
  diffuse radiation 424
  entropy 424
  entropy intensity 418
  extraterrestrial spectrum 418
  global radiation 424
  intensity 382
  maximum power of heat engine 351-353
  monochromatic temperature 419
  origin 415-418
  polarization 426
  scattering 422, 423
  solar constant 414, 420
  temperature 424-425
  temperature of diffuse light 425
  temperature of direct beam 425
  transmittance through atmosphere 421
  transmittance-absorptance of collector for diffuse radiation 436
  transmittance-absorptance of collector for direct radiation 435
  wind in Earth's atmosphere 353-357
solid angle 352, 379-382
  of the Sun 382, 425
solubility, see *solutions*
solutions
  chemical potential 496-498
  concentration 453
  dilute 453-454
  effect of solute on solvent 497
  Henry's law 499
  solubility of gases in water 498
  solute 496
  solvent 496
sound
  Laplace's theory 264
  propagation in fluids 264-265
  ratio of specific heats 265
  speed of sound 264
  speed of sound in ideal gas 185
source rate
  momentum 15
  see *laws of balance*
source rate density 389, 614
sources 614
  energy 626
  energy and entropy 368
  field equation with sources 369
  momentum 15
specific heat, see *temperature coefficient of energy*
specific source rate 614
  momentum 627
speed of sound 28, 185
stars
  brightness 440
  composition 416
  convection 416
  evolution 417
  Hertzsprung-Russell diagram 417
  hydrogen burning 416
  luminosity 417

magnitude 440
main sequence 415, 416
main sequence stars 417
mass 416
molar mass 212
polytropic gas spheres 195
radiation in the interior 211
structure 416–417
surface temperature (Sirius) 328
statics
versus dynamics 20
steam engine, see *heat engine*
steam power plant 544–545
Stefan and Boltzmann law 211
Stirling engine 203–204
stoichiometric coefficient 481, 482, 527
Stokes' law 22
storage 7
hot water tank 522
of heat 55
seasonal ground heat storage 565–569
stress 12–15, 618
and momentum flow 12–15
in tides 17
pure tension or compression 12
thermal 69
stress power 628
stress tensor, see *momentum current tensor*
subcooled liquid 484
sublimation 488
substancelike quantities 2
and energy transfer 31
and exchange of energy 29
and release of energy 30
as energy carrier 32
comparison 40
conservation and nonconservation 20
fall from high to low potential 25
transport processes 12
substances
amount of substance 448–451
basic building blocks 448–449
transport as chemical reaction 534–536
substantial derivative, see *material derivative*
summation 619
Sun
central temperature 395
composition 416
interior 416
mass 416
opacity 395
photosphere 418
radius 416
spectral type 415
surface temperature 326, 417
superconductivity
electrical 25
mechanical 25
thermal 154
superheated vapor 487
superheating in power cycle 545
surface force 618
surface integral 615
system
closed 306
open 306
system dynamics 39

T

telegrapher's equation 28
temperature
absolute 284
absolute scale 72, 88
absorbers on Earth 413
adiabatic flame temperature 526
Celsius scale 67
change of
length 68
resistance 69
volume 69
dry bulb 503
final temperature reached in thermal contact 232
ideal gas temperature 71–73
Kelvin scale 72
measure of hotness 53
monochromatic, of solar radiation 419
of radiation 405
of sky 338
reaching low temperatures 219
scales 54, 67
thermal potential 88
thermometry 66–73
universe 213
wet bulb 503
temperature coefficient
of chemical potential 479
of expansion 68, 69, 249
of pressure 71, 249
of resistance 69
temperature coefficient of energy 161, 188
magnetization 217
mixtures 458
name explained 166
solids 162
specific and molar values 162
values for gases 189

temperature coefficient of enthalpy 188
  for air 191
  mixtures 458
  name explained 199
  relation to temperature coefficient of energy 190
temperature gradient
  adiabatic 187, 300
  in atmosphere 187
  in conduction 312
temperature-entropy diagram, see *T-S diagram*
temperature-heat diagram, see *T-S diagram*
temperature-volume diagram, see *T-V diagram*
theory of calorimetry, see *calorimetry*
thermal charge 64
thermal conductivities 632
thermal driving force, see *driving force*
thermal power 628, see *power*
thermal resistor 139
thermal work 222
thermodynamic potentials 288-289, 606
  enthalpy 288
  Gibbs free energy 289
  Helmholtz free energy 289
thermodynamics
  and heat transfer 654
  extended irreversible 652-654
  finite-time 143
  heat-conducting fluids 629-638
  historical experiments 295
  uniform fluids 592-611
  viscous fluids 592-601
thermostatics 229-235
  maximum entropy postulate 234
third law of thermodynamics 162, 165
throttling process 525
tides 16
time and irreversibility 206
time constant 8
  inductive 9
transmittance, see *solar radiation*
transport processes
  conductive 12, 302-303
  convective 17, 303-306
  diffusive transport 646
  energy 29
  energy transfer 31
  fields 15
  heat transfer 55
  laws of balance 12-20
  momentum transport 12
  of substancelike quantities 12
  radiative 15, 307-309
T-S diagram 75
  adiabatic process 75, 182
  Carnot cycle 85
  heating at constant volume 75, 158
  isothermal process 76
  J.W. Gibbs 66
  liquid-vapor mixture of water 539
  Otto cycle 205
  phase change 77
  subcooled water 538
  superheated water vapor 540
  vaporization of water 486
turbulence, see *fluid flow*
T-V diagram
  general importance 246
  heating at constant temperature 176
  heating at constant volume 158
  Otto cycle 205
  Stirling cycle 204

U

uniform processes
  dynamics 154
  equilibrium 154
  model of 153-157
  thermal superconductors 154, 156
universal gas constant 171

V

vapor 485
  saturated 487
  superheated 487, 540-541
vapor pressure, see *phase change*
vaporization, see *phase change*
velocity gradient tensor 641
viscosity 7, 22
  bulk viscosity 631
  dynamic viscosity 645
  kinematic viscosity 645
  momentum conductivity 25
  momentum transport 23
  Newton's law for viscous fluids 24
viscous pressure 593
voltage 3
volume 2
  change of volume 5
  exchanged 6
  law of balance of volume 4, 6
  molar 477, 479
  partial molar volume 480
  specific volume 513

W

walls
  ideal, see *ideal walls*
  system walls 117

water
  adiabat around 2°C 261
  anomaly 68, 245
  chemical potential 489
  compressed liquid 484
  density 68
  liquid-vapor mixture 539
  mixture of liquid and vapor 485
  osmotic pressure of salt 454
  saturated liquid 485
  subcooled (*T-S* diagram) 538
  subcooled liquid 484
  superheated vapor (*T-S* diagram) 540
  vapor pressure 492
  vaporization 484–487
  vaporization at different pressures 486
  viscosity 7
waterfall 4
  image of internal processes 32
  waterfall diagram of heat engine 85
wave equation 26–28
  for heat conduction 651
wave guide 28
Wien's displacement law, see *radiation*
wind, see *atmosphere*
work 10
  hydraulic 10
  thermal 61

Y
Young's modulus 69